Two-Dimensional Self and Product Polynomial Systems

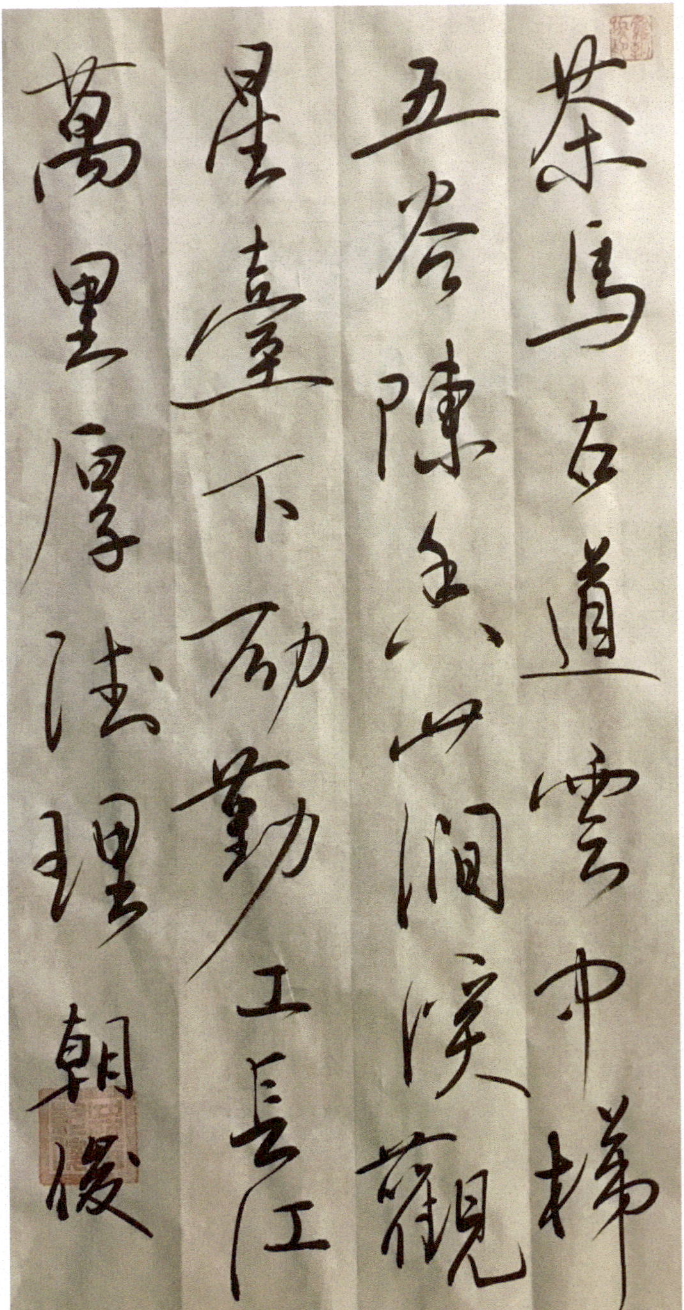

茶马古道云中梯，五谷陈香山洞溪，观星台下励勤工，长江万里厚德理。

—朝俊甲辰秋【四川轻化工大学六十年寄语】

Albert C. J. Luo

Two-Dimensional Self and Product Polynomial Systems

高等教育出版社
HIGHER EDUCATION PRESS

Springer

Albert C. J. Luo
Department of Mechanical
and Mechatronics Engineering
Southern Illinois University Edwardsville
Edwardsville, IL, USA

ISBN 978-981-96-5482-6 ISBN 978-981-96-5483-3 (eBook)
https://doi.org/10.1007/978-981-96-5483-3

Jointly published with Higher Education Press Limited Company
The print edition is not for sale in the mainland of China. Customers from the mainland of China please order the print book from: Higher Education Press Limited Company.

This Springer imprint is published by the registered company Springer Nature Singapore Pte Ltd.
The registered company address is: 152 Beach Road, #21-01/04 Gateway East, Singapore 189721, Singapore

Preface

This book is a monograph about hybrid networks of singular and non-singular, one-dimensional flows and equilibriums in self and product polynomial systems. The higher-order singular one-dimensional flows and singular equilibriums are for the appearing bifurcations of lower-order singular and non-singular one-dimensional flows and equilibriums. The singular equilibriums are singular sinks, sources, and saddles, saddle-sinks, saddle-sources, and double-saddles. The singular one-dimensional flows are singular hyperbolic flows, hyperbolic-to-hyperbolic-secant flows, inflection-source and sink flows, inflection-saddle flows. The infinite-equilibriums are the switching bifurcations for two associated networks of singular and non-singular, one-dimensional flows and equilibriums. The infinite-equilibriums include:

- increasing-to-decreasing inflection-sources (sinks), hyperbolic-sources (sinks), hyperbolic-upper (lower) saddles, increasing-to-decreasing inflection-upper (lower)-saddles;
- increasing (deceasing) inflection-sources (sinks), hyperbolic-to-hyperbolic-secant sources (sinks), hyperbolic-to-hyperbolic-secant upper (lower)-saddles, increasing (decreasing) inflection-upper (lower)-saddles;
- increasing-to-deceasing concave-sources (sinks), asymptote-sinks (sources), asymptote-upper (lower)-saddles, increasing-to-decreasing concave-upper (lower)-saddles;
- increasing (decreasing)-concave-sources (sinks), down (up)-asymptote sources (sinks), down (up)-asymptote upper (lower)-saddles, increasing (decreasing)-concave-upper (lower)-saddles.

The corresponding mathematical conditions are presented, and the theory for nonlinear dynamics of self and product polynomial systems is presented through a theorem. The mathematical proof is completed through the local analysis and the first integral manifolds. The illustrations of singular one-dimensional flows and equilibriums are completed, and the sampled networks of non-singular one-dimensional flows and equilibriums are presented in this book.

In this book, three chapters are included. Chapter 1 presents the nonlinear dynamics of singular flows and equilibriums with the corresponding infinite-equilibriums in self and product polynomial systems through a theorem. The appearing and switching bifurcations based on the singular flows and equilibriums, and infinite-equilibriums are presented. Chapter 2 proves the mathematical condition in the theorem through local analysis and the first integral manifolds. In Chap. 3, the singular one-dimensional flows and singular equilibriums with infinite-equilibriums are presented for a better understanding of singular dynamics in self and product polynomial systems. In addition, hybrid networks of non-singular one-dimensional flows and equilibriums are presented, and such hybrid networks appear, and two associated networks are switched through the singular one-dimensional flows and equilibriums with the infinite-equilibriums.

Finally, the author hopes the materials presented herein can provide a better understanding of the extended Hilbert sixteen problems and the corresponding dynamics.

Edwardsville, IL, USA Albert C. J. Luo

Contents

Chapter 1
Self and Product Polynomial Systems

In this chapter, the dynamics of self and product polynomial systems is presented through a theorem. The singular equilibriums and 1-diemsnional flows with infinite-equilibriums in the self and product polynomial systems are presented in the theorem. The singular equilibriums are singular sinks, saddles, sources, saddle-sources and saddle-sinks, double-saddles, and the singular flows are singular hyperbolic and hyperbolic-secant flows, hyperbolic-to-hyperbolic-secant flows, inflection-sources and inflection-sinks flows, inflection-saddle flows. The corresponding integral manifolds are obtained. The hybrid networks of singular equilibriums and 1-diemsnional flows in the self and product polynomial systems are also presented in the theorem, and the corresponding infinite-equilibriums are given for switching bifurcations. The hybrid network of simple equilibriums and 1-diemsnional flows with infinite-equilibriums in the self and product polynomial systems is stated in the theorem. The appearing and switching bifurcations of simple and singular 1-dimensional flows and equilibriums are stated in the theorem. Finally, the switching bifurcations of hybrid networks of simple and singular equilibriums and flows is also stated. The corresponding dynamical behaviors are presented in the theorem.

1.1 Theorem 1.1

(i) Consider a two-dimensional, nonlinear dynamical system as

$$\begin{aligned}\dot{x}_{j_2} &= a_{j_2j_20}(x_{j_2} - a_{j_2j_21})^{n_2},\\ \dot{x}_{j_1} &= a_{j_1j_10}(x_{j_1} - a_{j_1j_11})^{m_1}(x_{j_2} - a_{j_1j_21})^{n_1}.\end{aligned} \tag{1.1}$$

(i_1) For $a_{j_2j_21} \neq a_{j_1j_21}$ the first integral manifold is

A. C. J. Luo, *Two-Dimensional Self and Product Polynomial Systems*,
https://doi.org/10.1007/978-981-96-5483-3_1

$$
\begin{aligned}
&-\frac{1}{m_1-1}\left[\frac{1}{(x_{j_1}-a_{j_1j_11})^{m_1-1}}-\frac{1}{(x_{j_10}-a_{j_1j_11})^{m_1-1}}\right]\\
&=\frac{a_{j_1j_10}}{a_{j_2j_20}}\sum_{\alpha,\beta=0}^{\alpha+\beta=n_1}\frac{n_1!}{\alpha!\beta!}(a_{j_2j_21}-a_{j_1j_21})^{\alpha}\\
&\times\frac{1}{\beta-n_2+1}\left[(x_{j_2}-a_{j_2j_21})^{\beta-n_2+1}-(x_{j_20}-a_{j_2j_21})^{\beta-n_2+1}\right],\\
&\text{for } m_1\neq 1,\ \beta-n_2+1\neq 0;\\
&-\frac{1}{m_1-1}\left[\frac{1}{(x_{j_1}-a_{j_1j_11})^{m_1-1}}-\frac{1}{(x_{j_10}-a_{j_1j_11})^{m_1-1}}\right]=\ln\frac{|x_{j_1}-a_{j_1j_11}|}{|x_{j_10}-a_{j_1j_11}|},\\
&\text{for } m_1=1;\\
&\frac{1}{\beta-n_2+1}\left[(x_{j_2}-a_{j_2j_21})^{\beta-n_2+1}-(x_{j_20}-a_{j_2j_21})^{\beta-n_2+1}\right]=\ln\frac{|x_{j_2}-a_{j_2j_21}|}{|x_{j_20}-a_{j_2j_21}|},\\
&\text{for } \beta-n_2+1=0.
\end{aligned}
\tag{1.2}
$$

(i_{1a}) There is an equilibrium for 1-diemsnional flow at

$$
(x_{j_1}^*,x_{j_2}^*)=(a_{j_1j_11},a_{j_1j_21}) \tag{1.3}
$$

(i_{1a1}) For $m_1=2m_{11}+1$ and $n_1=2n_{11}+1$, the equilibrium of $(x_{j_1}^*,x_{j_2}^*)=(a_{j_1j_11},a_{j_1j_21})$ in the x_{j_1}-direction has the following properties.

- *For $a_{j_1j_10}>0$ and $a_{j_2j_20}(a_{j_1j_21}-a_{j_2j_21})^{n_2}>0$,*

$$
(a_{j_1j_11},a_{j_1j_21})=\underbrace{([(2m_{11}+1),(2n_{11}+1)]\text{DP:UP, pF})}_{[(2m_{11}+1),(2n_{11}+1)]\text{-hyperbolic flow }(+)}. \tag{1.4}
$$

The flow at $(x_{j_1}^*,x_{j_2}^*)=(a_{j_1j_11},a_{j_1j_21})$ is a ([(2m_{11}+1), (2n_{11}+1)]DP:UP, pF)*-positive hyperbolic flow.*

- *For $a_{j_1j_10}<0$ and $a_{j_2j_20}(a_{j_1j_21}-a_{j_2j_21})^{n_2}>0$,*

$$
(a_{j_1j_11},a_{j_1j_21})=\underbrace{([(2m_{11}+1),(2n_{11}+1)]\text{UP:DP, pF})}_{[(2m_{11}+1),(2n_{11}+1)]\text{-hyperbolic-secant flow }(+)}. \tag{1.5}
$$

The flow at $(x_{j_1}^*,x_{j_2}^*)=(a_{j_1j_11},a_{j_1j_21})$ is a ([(2m_{11}+1), (2n_{11} + 1)]UP:DP, pF)*-positive hyperbolic-secant flow.*

- *For $a_{j_1j_10}>0$ and $a_{j_2j_20}(a_{j_1j_21}-a_{j_2j_21})^{n_2}<0$,*

$$
(a_{j_1j_11},a_{j_1j_21})=\underbrace{([(2m_{11}+1),(2n_{11}+1)]\text{UP:DP, nF})}_{[(2m_{11}+1),(2n_{11}+1)]\text{-hyperbolic-secant flow }(-)}. \tag{1.6}
$$

The flow at $(x_{j_1}^*,x_{j_2}^*)=(a_{j_1j_11},a_{j_1j_21})$ is a ([(2m_{11}+1), (2n_{11} + 1)]UP:DP, nF)*-negative hyperbolic-secant flow.*

- *For* $a_{j_1j_10} < 0$ *and* $a_{j_2j_20}(a_{j_1j_21} - a_{j_2j_21})^{n_2} < 0$,

$$(a_{j_1j_11}, a_{j_1j_21}) = \underbrace{([(2m_{11}{+}1), (2n_{11}+1)]\text{DP:UP, nF})}_{[(2m_{11}+1),(2n_{11}+1)]\text{-hyperbolic flow }(-)}. \tag{1.7}$$

The flow at $(x^*_{j_1}, x^*_{j_2}) = (a_{j_1j_11}, a_{j_1j_21})$ *is a* ([(2m_{11}+1), (2n_{11} + 1)]DP:UP, pF)-*negative hyperbolic flow.*

(i_{1a2}) For $m_1 = 2m_{11}$ *and* $n_1 = 2n_{11}+1$, *the equilibrium of* $(x^*_{j_1}, x^*_{j_2}) = (a_{j_1j_11}, a_{j_1j_21})$ *for the* x_{j_1}*-direction has the following properties.*

- *For* $a_{j_1j_10} > 0$ *and* $a_{j_2j_20}(a_{j_1j_21} - a_{j_2j_21})^{n_2} > 0$,

$$(a_{j_1j_11}, a_{j_1j_21}) = \underbrace{([(2m_{11}), (2n_{11}+1)]\text{UP:UP, pF})}_{[(2m_{11}),(2n_{11}+1)]\text{-hyperbolic-secant-to-hyperbolic flow }(+)}. \tag{1.8}$$

The flow at $(x^*_{j_1}, x^*_{j_2}) = (a_{j_1j_11}, a_{j_1j_21})$ *is a* ([(2m_{11}), (2n_{11} + 1)]UP:UP, pF)-*positive hyperbolic-secant to hyperbolic flow.*

- *For* $a_{j_1j_10} < 0$ *and* $a_{j_2j_20}(a_{j_1j_21} - a_{j_2j_21})^{n_2} > 0$,

$$(a_{j_1j_11}, a_{j_1j_21}) = \underbrace{([(2m_{11}), (2n_{11}+1)]\text{DP:DP, pF})}_{[(2m_{11}),(2n_{11}+1)]\text{-hyperbolic-to-hyperbolic-secant flow }(+)}. \tag{1.9}$$

The flow at $(x^*_{j_1}, x^*_{j_2}) = (a_{j_1j_11}, a_{j_1j_21})$ *is a* ([(2m_{11}), (2n_{11} + 1)]DP:DP, pF)-*positive hyperbolic to hyperbolic-secant flow.*

- *For* $a_{j_1j_10} > 0$ *and* $a_{j_2j_20}(a_{j_1j_21} - a_{j_2j_21})^{n_2} < 0$,

$$(a_{j_1j_11}, a_{j_1j_21}) = \underbrace{([(2m_{11}), (2n_{11}+1)]\text{DP:DP, nF})}_{[(2m_{11}),(2n_{11}+1)]\text{-hyperbolic-to-hyperbolic-secant flow }(-)}. \tag{1.10}$$

The flow at $(x^*_{j_1}, x^*_{j_2}) = (a_{j_1j_11}, a_{j_1j_21})$ *is a* ([(2m_{11}), (2n_{11} + 1)]DP:DP, nF)-*negative hyperbolic-to-hyperbolic-secant flow.*

- *For* $a_{j_1j_10} < 0$ *and* $a_{j_2j_20}(a_{j_1j_21} - a_{j_2j_21})^{n_2} < 0$,

$$(a_{j_1j_11}, a_{j_1j_21}) = \underbrace{([(2m_{11}), (2n_{11}+1)]\text{UP:UP, nF})}_{[(2m_{11}),(2n_{11}+1)]\text{-hyperbolic-secant-to-hyperbolic flow }(-)}. \tag{1.11}$$

The flow at $(x^*_{j_1}, x^*_{j_2}) = (a_{j_1j_11}, a_{j_1j_21})$ *is a* ([(2m_{11}), (2n_{11} + 1)]UP:UP, nF)-*negative hyperbolic-secant-to-hyperbolic flow.*

(i_{1a3}) For $m_1 = 2m_{11}+1$ *and* $n_1 = 2n_{11}$, *the equilibrium of* $(x^*_{j_1}, x^*_{j_2}) = (a_{j_1j_11}, a_{j_1j_21})$ *for the* x_{j_1}*-direction has the following properties.*

- *For* $a_{j_1j_10} > 0$ *and* $a_{j_2j_20}(a_{j_1j_21} - a_{j_2j_21})^{n_2} > 0$,

$$(a_{j_1j_11}, a_{j_1j_21}) = \underbrace{([(2m_{11}+1), (2n_{11})]_{\text{DI:II}}\text{SO, pF})}_{[(2m_{11}+1),(2n_{11})]\text{-decreasing-to-increasing inflection source flow }(+)}. \quad (1.12)$$

The flow at $(x^*_{j_1}, x^*_{j_2}) = (a_{j_1j_11}, a_{j_1j_21})$ *is a* $([(2m_{11}+1), (2n_{11})]_{\text{DI:II}}\text{SO, pF})$-*positive decreasing-to-increasing inflection source flow.*

- *For* $a_{j_1j_10} < 0$ *and* $a_{j_2j_20}(a_{j_1j_21} - a_{j_2j_21})^{n_2} > 0$,

$$(a_{j_1j_11}, a_{j_1j_21}) = \underbrace{([(2m_{11}+1), (2n_{11})]_{\text{II:DI}}\text{SI, pF})}_{[(2m_{11}+1),(2n_{11})]\text{-increasing-to-decreasing inflection sink flow }(+)}. \quad (1.13)$$

The flow at $(x^*_{j_1}, x^*_{j_2}) = (a_{j_1j_11}, a_{j_1j_21})$ *is a* $([(2m_{11}+1), (2n_{11})]_{\text{II:DI}}\text{SI, pF})$-*positive increasing-to-decreasing inflection source flow.*

- *For* $a_{j_1j_10} > 0$ *and* $a_{j_2j_20}(a_{j_1j_21} - a_{j_2j_21})^{n_2} < 0$,

$$(a_{j_1j_11}, a_{j_1j_21}) = \underbrace{([(2m_{11}+1), (2n_{11})]_{\text{II:DI}}\text{SO, nF})}_{[(2m_{11}+1),(2n_{11})]\text{-increasing-to-decreasing inflection source flow }(-)}. \quad (1.14)$$

The flow at $(x^*_{j_1}, x^*_{j_2}) = (a_{j_1j_11}, a_{j_1j_21})$ *is a* $([(2m_{11}+1), (2n_{11})]_{\text{II:DI}}\text{SO, nF})$-*negative increasing-to-decreasing inflection source flow.*

- *For* $a_{j_1j_10} < 0$ *and* $a_{j_2j_20}(a_{j_1j_21} - a_{j_2j_21})^{n_2} < 0$,

$$(a_{j_1j_11}, a_{j_1j_21}) = \underbrace{([(2m_{11}+1), (2n_{11})]_{\text{DI:II}}\text{SI, nF})}_{[(2m_{11}+1),(2n_{11})]\text{-decreasing-to-increasing inflection sink flow }(-)}. \quad (1.15)$$

The flow at $(x^*_{j_1}, x^*_{j_2}) = (a_{j_1j_11}, a_{j_1j_21})$ *is a* $([(2m_{11}+1), (2n_{11})]_{\text{DI:II}}\text{SI, nF})$-*negative decreasing-to-increasing inflection sink flow.*

(i_{1a4}) For $m_1 = 2m_{11}$ *and* $n_1 = 2n_{11}$, *the equilibrium of* $(x^*_{j_1}, x^*_{j_2}) = (a_{j_1j_11}, a_{j_1j_21})$ *for the* x_{j_1}*-direction has the following properties.*

- *For* $a_{j_1j_10} > 0$ *and* $a_{j_2j_20}(a_{j_1j_21} - a_{j_2j_21})^{n_2} > 0$,

$$(a_{j_1j_11}, a_{j_1j_21}) = \underbrace{([(2m_{11}), (2n_{11})]_{\text{II:II}}\text{US, pF})}_{[(2m_{11}),(2n_{11})]\text{-increasing-inflection upper-saddle flow }(+)}. \quad (1.16)$$

The flow at $(x^*_{j_1}, x^*_{j_2}) = (a_{j_1j_11}, a_{j_1j_21})$ *is a* $([(2m_{11}), (2n_{11})]_{\text{II:II}}\text{US, pF})$-*positive increasing-inflection upper-saddle flow.*

- *For* $a_{j_1j_10} < 0$ *and* $a_{j_2j_20}(a_{j_1j_21} - a_{j_2j_21})^{n_2} > 0$,

$$(a_{j_1j_11}, a_{j_1j_21}) = \underbrace{([(2m_{11}), (2n_{11})]_{\text{DI:DI}}\text{LS, pF})}_{[(2m_{11}),(2n_{11})]\text{-decreasing-inflection lower-saddle flow }(+)}. \quad (1.17)$$

The flow at $(x_{j_1}^*, x_{j_2}^*) = (a_{j_1j_11}, a_{j_1j_21})$ *is a* ([(2m_1), (2n_1)]$_{\text{DI:DI}}$LS, pF)*-positive decreasing-inflection lower-saddle flow.*

- *For* $a_{j_1j_10} > 0$ *and* $a_{j_2j_20}(a_{j_1j_21} - a_{j_2j_21})^{n_2} < 0$,

$$(a_{j_1j_11}, a_{j_1j_21}) = \underbrace{([(2m_{11}), (2n_{11})]_{\text{DI:DI}}\text{US, nF})}_{[(2m_{11}),(2n_{11})]\text{-decreasing-inflection upper-saddle flow }(-)}. \tag{1.18}$$

The flow at $(x_{j_1}^*, x_{j_2}^*) = (a_{j_1j_11}, a_{j_1j_21})$ *is a* ([(2m_{11}), (2n_{11})]$_{\text{DI:DI}}$US, nF)*-negative decreasing-inflection upper-saddle flow.*

- *For* $a_{j_1j_10} < 0$ *and* $a_{j_2j_20}(a_{j_1j_21} - a_{j_2j_21})^{n_2} < 0$,

$$(a_{j_1j_11}, a_{j_1j_21}) = \underbrace{([(2m_{11}), (2n_{11})]_{\text{II:II}}\text{LS, nF})}_{[(2m_{11}),(2n_{11})]\text{-increasing-inflection lower-saddle flow }(-)}. \tag{1.19}$$

The flow at $(x_{j_1}^*, x_{j_2}^*) = (a_{j_1j_11}, a_{j_1j_21})$ *is a* ([(2m_{11}), (2n_{11})]$_{\text{II:II}}$LS, nF)*-negative increasing-inflection lower-saddle flow.*

(i_{Ib}) *There is an equilibrium for self-univariate equilibrium as*

$$(x_{j_1}^*, x_{j_2}^*) = (a_{j_1j_11}, a_{j_2j_21}) \tag{1.20}$$

(i_{Ib1}) *For* $m_1 = 2m_{11} + 1$ *and* $n_2 = 2n_{21} + 1$, *the equilibrium of* $(x_{j_1}^*, x_{j_2}^*) = (a_{j_1j_11}, a_{j_2j_21})$ *has the following properties.*

- *For* $a_{j_1j_10}(a_{j_2j_21} - a_{j_1j_21})^{n_1} > 0$ *and* $a_{j_2j_20} > 0$,

$$(a_{j_1j_11}, a_{j_2j_21}) = \underbrace{((2m_{11}+1)^{\text{th}}\,\text{SO}, (2n_{21}+1)^{\text{th}}\,\text{SO})}_{((2m_{11}+1),(2n_{21}+1))\text{-source}}. \tag{1.21}$$

The flow at $(x_{j_1}^*, x_{j_2}^*) = (a_{j_1j_11}, a_{j_2j_21})$ *is a* ((2m_{11}+1)th SO, (2n_{21}+1)th SO)*-source.*

- *For* $a_{j_1j_10}(a_{j_2j_21} - a_{j_1j_21})^{n_1} < 0$ *and* $a_{j_2j_20} > 0$,

$$(a_{j_1j_11}, a_{j_2j_21}) = \underbrace{((2m_{11}+1)^{\text{th}}\,\text{SI}, (2n_{21}+1)^{\text{th}}\,\text{SO})}_{((2m_{11}+1),(2n_{21}+1))\text{-saddle}}. \tag{1.22}$$

The flow at $(x_{j_1}^*, x_{j_2}^*) = (a_{j_1j_11}, a_{j_2j_21})$ *is a* ((2m_{11}+1)th SI, (2n_{21}+1)th SO)*-saddle.*

- *For* $a_{j_1j_10}(a_{j_2j_21} - a_{j_1j_21})^{n_1} > 0$ *and* $a_{j_2j_20} < 0$,

$$(a_{j_1j_11}, a_{j_2j_21}) = \underbrace{((2m_{11}+1)^{\text{th}}\,\text{SO}, (2n_{21}+1)^{\text{th}}\,\text{SI})}_{((2m_{11}+1),(2n_{21}+1))\text{-saddle}}. \tag{1.23}$$

The flow at $(x_{j_1}^*, x_{j_2}^*) = (a_{j_1j_11}, a_{j_2j_21})$ *is a* ((2m_{11}+1)th SO, (2n_{21}+1)th SI)*-saddle.*

- *For* $a_{j_1j_10}(a_{j_2j_21} - a_{j_1j_21})^{n_1} < 0$ *and* $a_{j_2j_20} < 0$,

$$(a_{j_1j_11}, a_{j_2j_21}) = \underbrace{((2m_{11}+1)^{\text{th}}\,\text{SI}, (2n_{21}+1)^{\text{th}}\,\text{SI})}_{((2m_{11}+1),(2n_{21}+1))\text{-sink}}. \tag{1.24}$$

The flow at $(x^*_{j_1}, x^*_{j_2}) = (a_{j_1j_11}, a_{j_2j_21})$ *is a* $((2m_{11}+1)^{\text{th}}\,\text{SI}, (2n_{21}+1)^{\text{th}}\,\text{SI})$*-sink.*

(i_{Ib2}) For $m_1 = 2m_{11}$ *and* $n_2 = 2n_{21} + 1$, *the equilibrium of* $(x^*_{j_1}, x^*_{j_2}) = (a_{j_1j_11}, a_{j_2j_21})$ *has the following properties.*

- *For* $a_{j_1j_10}(a_{j_2j_21} - a_{j_1j_21})^{n_1} > 0$ *and* $a_{j_2j_20} > 0$,

$$(a_{j_1j_11}, a_{j_2j_21}) = \underbrace{((2m_{11})^{\text{th}}\,\text{US}, (2n_{21}+1)^{\text{th}}\,\text{SO})}_{((2m_{11}),(2n_{21}+1))\text{-upper-saddle source}}. \tag{1.25}$$

The flow at $(x^*_{j_1}, x^*_{j_2}) = (a_{j_1j_11}, a_{j_2j_21})$ *is a* $((2m_{11})^{\text{th}}\,\text{US}, (2n_{21}+1)^{\text{th}}\,\text{SO})$*-upper-saddle source.*

- *For* $a_{j_1j_10}(a_{j_2j_21} - a_{j_1j_21})^{n_1} < 0$ *and* $a_{j_2j_20} > 0$,

$$(a_{j_1j_11}, a_{j_2j_21}) = \underbrace{((2m_{11})^{\text{th}}\,\text{LS}, (2n_{21}+1)^{\text{th}}\,\text{SO})}_{((2m_{11}),(2n_{21}+1))\text{-lower-saddle source}}. \tag{1.26}$$

The flow at $(x^*_{j_1}, x^*_{j_2}) = (a_{j_1j_11}, a_{j_2j_21})$ *is a* $((2m_{11})^{\text{th}}\,\text{LS}, (2n_{21}+1)^{\text{th}}\,\text{SO})$*-lower-saddle source.*

- *For* $a_{j_1j_10}(a_{j_2j_21} - a_{j_1j_21})^{n_1} > 0$ *and* $a_{j_2j_20} < 0$,

$$(a_{j_1j_11}, a_{j_2j_21}) = \underbrace{((2m_{11})^{\text{th}}\,\text{US}, (2n_{21}+1)^{\text{th}}\,\text{SI})}_{((2m_{11}),(2n_{21}+1))\text{-upper-saddle sink}}. \tag{1.27}$$

The flow at $(x^*_{j_1}, x^*_{j_2}) = (a_{j_1j_11}, a_{j_2j_21})$ *is a* $((2m_{11})^{\text{th}}\,\text{US}, (2n_{21}+1)^{\text{th}}\,\text{SI})$*-upper-saddle sink.*

- *For* $a_{j_1j_10}(a_{j_2j_21} - a_{j_1j_21})^{n_1} < 0$ *and* $a_{j_2j_20} < 0$,

$$(a_{j_1j_11}, a_{j_2j_21}) = \underbrace{((2m_{11})^{\text{th}}\,\text{LS}, (2n_{21}+1)^{\text{th}}\,\text{SI})}_{((2m_{11}),(2n_{21}+1))\text{-lower-saddle sink}}. \tag{1.28}$$

The flow at $(x^*_{j_1}, x^*_{j_2}) = (a_{j_1j_11}, a_{j_2j_21})$ *is a* $((2m_{11})^{\text{th}}\,\text{LS}, (2n_{21}+1)^{\text{th}}\,\text{SI})$*-lower-saddle sink.*

(i_{Ib3}) For $m_1 = 2m_{11} + 1$ *and* $n_2 = 2n_{21}$, *the equilibrium of* $(x^*_{j_1}, x^*_{j_2}) = (a_{j_1j_11}, a_{j_2j_21})$ *has the following properties.*

- *For* $a_{j_1j_10}(a_{j_2j_21} - a_{j_1j_21})^{n_1} > 0$ *and* $a_{j_2j_20} > 0$,

$$(a_{j_1j_11}, a_{j_2j_21}) = \underbrace{((2m_{11}+1)^{\text{th}}\,\text{SO}, (2n_{21})^{\text{th}}\,\text{US})}_{((2m_{11}+1),(2n_{21}))\text{-upper-saddle source}}. \tag{1.29}$$

The flow at $(x^*_{j_1}, x^*_{j_2}) = (a_{j_1j_11}, a_{j_2j_21})$ *is a* $((2m_{11}+1)^{\text{th}}$ SO, $(2n_{21})^{\text{th}}$ US)-upper-saddle *source.*

- *For* $a_{j_1j_10}(a_{j_2j_21} - a_{j_1j_21})^{n_1} < 0$ *and* $a_{j_2j_20} > 0$,

$$(a_{j_1j_11}, a_{j_2j_21}) = \underbrace{((2m_{11}+1)^{\text{th}}\,\text{SI}, (2n_{21})^{\text{th}}\,\text{US})}_{((2m_{11}+1),(2n_{21}))\text{-upper-saddle sink}}. \tag{1.30}$$

The flow at $(x^*_{j_1}, x^*_{j_2}) = (a_{j_1j_11}, a_{j_2j_21})$ *is a* $((2m_{11}+1)^{\text{th}}$ SI, $(2n_{21})^{\text{th}}$ US)*-upper-saddle sink.*

- *For* $a_{j_1j_10}(a_{j_2j_21} - a_{j_1j_21})^{n_1} > 0$ *and* $a_{j_2j_20} < 0$,

$$(a_{j_1j_11}, a_{j_2j_21}) = \underbrace{((2m_{11}+1)^{\text{th}}\,\text{SO}, (2n_{21})^{\text{th}}\,\text{LS})}_{((2m_{11}+1),(2n_{21}))\text{-lower-saddle source}}. \tag{1.31}$$

The flow at $(x^*_{j_1}, x^*_{j_2}) = (a_{j_1j_11}, a_{j_2j_21})$ *is a* $((2m_{11}+1)^{\text{th}}$ SO, $(2n_{21})^{\text{th}}$ LS)*-lower-saddle source.*

- *For* $a_{j_1j_10}(a_{j_2j_21} - a_{j_1j_21})^{n_1} < 0$ *and* $a_{j_2j_20} < 0$,

$$(a_{j_1j_11}, a_{j_2j_21}) = \underbrace{((2m_{11}+1)^{\text{th}}\,\text{SI}, (2n_{21})^{\text{th}}\,\text{LS})}_{((2m_{11}+1),(2n_{21}))\text{-lower-saddle sink}}. \tag{1.32}$$

The flow at $(x^*_{j_1}, x^*_{j_2}) = (a_{j_1j_11}, a_{j_2j_21})$ *is a* $((2m_{11}+1)^{\text{th}}$ SI, $(2n_{21})^{\text{th}}$ LS)-lower-saddle *sink.*

(i_{1b4}) For $m_1 = 2m_{11}$ *and* $n_2 = 2n_{21}$, *the equilibrium of* $(x^*_{j_1}, x^*_{j_2}) = (a_{j_1j_11}, a_{j_2j_21})$ *has the following properties.*

- *For* $a_{j_1j_10}(a_{j_2j_21} - a_{j_1j_21})^{n_1} > 0$ *and* $a_{j_2j_20} > 0$,

$$(a_{j_1j_11}, a_{j_2j_21}) = \underbrace{((2m_{11})^{\text{th}}\,\text{US}, (2n_{21})^{\text{th}}\,\text{US})}_{((2m_{11}),(2n_{21}))\text{-double-saddle}}. \tag{1.33}$$

The flow at $(x^*_{j_1}, x^*_{j_2}) = (a_{j_1j_11}, a_{j_2j_21})$ *is a* $((2m_{11})^{\text{th}}$ US, $(2n_{21})^{\text{th}}$ US)*-double-saddle.*

- *For* $a_{j_1j_10}(a_{j_2j_21}-a_{j_1j_21})^{n_1}<0$ *and* $a_{j_2j_20}>0$,

$$(a_{j_1j_11},a_{j_2j_21})=\underbrace{((2m_{11})^{\text{th}}\,\text{LS},(2n_{21})^{\text{th}}\,\text{US})}_{((2m_{11}),(2n_{21}))\text{-double-saddle}}. \tag{1.34}$$

The flow at $(x^*_{j_1},x^*_{j_2})=(a_{j_1j_11},a_{j_2j_21})$ *is a* $((2m_{11})^{\text{th}}\,\text{LS},(2n_{21})^{\text{th}}\,\text{US})$*-double-saddle.*

- *For* $a_{j_1j_10}(a_{j_2j_21}-a_{j_1j_21})^{n_1}>0$ *and* $a_{j_2j_20}<0$,

$$(a_{j_1j_11},a_{j_2j_21})=\underbrace{((2m_{11})^{\text{th}}\,\text{US},(2n_{21})^{\text{th}}\,\text{LS})}_{((2m_{11}),(2n_{21}))\text{-double-saddle}}. \tag{1.35}$$

The flow at $(x^*_{j_1},x^*_{j_2})=(a_{j_1j_11},a_{j_2j_21})$ *is a* $((2m_{11})^{\text{th}}\,\text{US},(2n_{21})^{\text{th}}\,\text{LS})$*-double-saddle.*

- *For* $a_{j_1j_10}(a_{j_2j_21}-a_{j_1j_21})^{n_1}<0$ *and* $a_{j_2j_20}<0$,

$$(a_{j_1j_11},a_{j_2j_21})=\underbrace{((2m_{11})^{\text{th}}\,\text{LS},(2n_{21})^{\text{th}}\,\text{LS})}_{((2m_{11}),(2n_{21}))\text{-double-saddle}}. \tag{1.36}$$

The flow at $(x^*_{j_1},x^*_{j_2})=(a_{j_1j_11},a_{j_2j_21})$ *is a* $((2m_{11})^{\text{th}}\,\text{LS},(2n_{21})^{\text{th}}\,\text{LS})$*-double-saddle.*

(i_2) For $a_{j_2j_21}=a_{j_1j_21}$ *the first integral manifold is*

$$\begin{aligned}
&-\frac{1}{m_1-1}\left[\frac{1}{(x_{j_1}-a_{j_1j_11})^{m_1-1}}-\frac{1}{(x_{j_10}-a_{j_1j_11})^{m_1-1}}\right]\\
&=\frac{a_{j_1j_10}}{a_{j_2j_20}}\frac{1}{n_1-n_2+1}\left[(x_{j_2}-a_{j_2j_21})^{n_1-n_2+1}-(x_{j_20}-a_{j_2j_21})^{n_1-n_2+1}\right],\\
&\text{for } m_1\neq 1,\ n_2-n_1\neq 1;\\
&-\frac{1}{m_1-1}\left[\frac{1}{(x_{j_1}-a_{j_1j_11})^{m_1-1}}-\frac{1}{(x_{j_10}-a_{j_1j_11})^{m_1-1}}\right]=\ln\frac{|x_{j_1}-a_{j_1j_11}|}{|x_{j_10}-a_{j_1j_11}|},\\
&\text{for } m_1=1;\\
&\frac{1}{n_1-n_2+1}\left[(x_{j_2}-a_{j_2j_21})^{n_1-n_2+1}-(x_{j_20}-a_{j_2j_21})^{n_1-n_2+1}\right]=\ln\frac{|x_{j_2}-a_{j_2j_21}|}{|x_{j_20}-a_{j_2j_21}|},\\
&\text{for } n_1-n_2+1=0.
\end{aligned} \tag{1.37}$$

(i_{2a}) For $n_1\geq n_2$*, the first integral manifold is*

$$-\frac{1}{m_1-1}\left[\frac{1}{(x_{j_1}-a_{j_1j_11})^{m_1-1}}-\frac{1}{(x_{j_10}-a_{j_1j_11})^{m_1-1}}\right]$$
$$=\frac{a_{j_1j_10}}{a_{j_2j_20}}\frac{1}{n_1-n_2+1}\left[(x_{j_2}-a_{j_1j_21})^{n_1-n_2+1}-(x_{j_20}-a_{j_1j_21})^{n_1-n_2+1}\right],$$
for $m_1 \neq 1$;
$$-\frac{1}{m_1-1}\left[\frac{1}{(x_{j_1}-a_{j_1j_11})^{m_1-1}}-\frac{1}{(x_{j_10}-a_{j_1j_11})^{m_1-1}}\right]=\ln\frac{|x_{j_1}-a_{j_1j_11}|}{|x_{j_10}-a_{j_1j_11}|},$$
for $m_1 = 1$. (1.38)

(i_{2a1}) There are 16 types of infinite-equilibriums at $x^*_{j_2}=a_{j_2j_21}=a_{j_1j_21}$.

(i_{2a1a}) For $n_1=2n_{11}+1$ and $n_2=2n_{21}+1$, the infinite-equilibrium of $x^*_{j_2}=a_{j_2j_21}=a_{j_1j_21}$ with $\overline{x}_{j_1}\neq a_{j_1j_11}$ has the following properties.

- *For $a_{j_1j_10}(\overline{x}_{j_1}-a_{j_1j_11})^{m_1}>0$ and $a_{j_2j_20}>0$,*

$$(\overline{x}_{j_1},a_{j_2j_21})=\underbrace{([2(n_{11}-n_{21})]^{\text{th}}\,\text{II},\,(2n_{21}+1)^{\text{th}}\,\text{SO})}_{([2(n_{11}-n_{21})],(2n_{21}+1))\text{-increasing inflection source}}. \tag{1.39}$$

  The infinite-equilibrium is a ([2($n_{11}-n_{21}$)]th II, ($2n_{21}+1$)th SO)-increasing-inflection source at $x^*_{j_2}=a_{j_2j_21}=a_{j_1j_21}$.
- *For $a_{j_1j_10}(\overline{x}_{j_1}-a_{j_1j_11})^{m_1}<0$ and $a_{j_2j_20}>0$,*

$$(\overline{x}_{j_1},a_{j_2j_21})=\underbrace{([2(n_{11}-n_{21})]^{\text{th}}\,\text{DI},\,(2n_{21}+1)^{\text{th}}\,\text{SO})}_{([2(n_{11}-n_{21})],(2n_{21}+1))\text{-decreasing inflection source}}. \tag{1.40}$$

  The infinite-equilibrium is a ([2($n_{11}-n_{21}$)]th DI, ($2n_{21}+1$)th SO)-decreasing-inflection source at $x^*_{j_2}=a_{j_2j_21}=a_{j_1j_21}$.
- *For $a_{j_1j_10}(\overline{x}_{j_1}-a_{j_1j_11})^{m_1}>0$ and $a_{j_2j_20}<0$,*

$$(\overline{x}_{j_1},a_{j_2j_21})=\underbrace{([2(n_{11}-n_{21})]^{\text{th}}\,\text{DI},\,(2n_{21}+1)^{\text{th}}\,\text{SI})}_{([2(n_{11}-n_{21})],(2n_{21}+1))\text{-decreasing inflection sink}}. \tag{1.41}$$

  The infinite-equilibrium is a ([2($n_{11}-n_{21}$)]th DI, ($2n_{21}+1$)th SI)-decreasing-inflection sink at $x^*_{j_2}=a_{j_2j_21}=a_{j_1j_21}$.
- *For $a_{j_1j_10}(\overline{x}_{j_1}-a_{j_1j_11})^{m_1}<0$ and $a_{j_2j_20}<0$,*

$$(\overline{x}_{j_1},a_{j_2j_21})=\underbrace{([2(n_{11}-n_{21})]^{\text{th}}\,\text{II},\,(2n_{21}+1)^{\text{th}}\,\text{SI})}_{([2(n_{11}-n_{21})],(2n_{21}+1))\text{-increasing inflection sink}}. \tag{1.42}$$

  The infinite-equilibrium is a ([2($n_{11}-n_{21}$)]th II, ($2n_{21}+1$)th SI)-increasing-inflection sink at $x^*_{j_2}=a_{j_2j_21}=a_{j_1j_21}$.

(i_{2a1b}) *For* $n_1 = 2n_{11}$ *and* $n_2 = 2n_{21} + 1$, *the infinite-equilibrium of* $x_{j_2}^* = a_{j_2j_21} = a_{j_1j_21}$ *with* $\overline{x}_{j_1} \neq a_{j_1j_11}$ *has the following properties.*

- *For* $a_{j_1j_10}(\overline{x}_{j_1} - a_{j_1j_11})^{m_1} > 0$ *and* $a_{j_2j_20} > 0$,

$$(\overline{x}_{j_1}, a_{j_2j_21}) = \underbrace{([2(n_{11} - n_{21}) - 1]^{\text{th}}\,\text{UP}, (2n_{21} + 1)^{\text{th}}\,\text{SO})}_{([2(n_{11}-n_{21})-1],(2n_{21}+1))\text{-up-parabola source}}. \tag{1.43}$$

  The infinite-equilibrium is a $([2(n_{11} - n_{21}) - 1]^{\text{th}}$ UP, $(2n_{21} + 1)^{\text{th}}$ SO)*-up-parabola source at* $x_{j_2}^* = a_{j_2j_21} = a_{j_1j_21}$.

- *For* $a_{j_1j_10}(\overline{x}_{j_1} - a_{j_1j_11})^{m_1} < 0$ *and* $a_{j_2j_20} > 0$,

$$(\overline{x}_{j_1}, a_{j_2j_21}) = \underbrace{([2(n_{11} - n_{21}) - 1]^{\text{th}}\,\text{DP}, (2n_{21} + 1)^{\text{th}}\,\text{SO})}_{([2(n_{11}-n_{21})-1],(2n_{21}+1))\text{-down-parabola source}}. \tag{1.44}$$

  The infnite-equilibrium is a $([2(n_{11} - n_{21}) - 1]^{\text{th}}$ DP, $(2n_{21} + 1)^{\text{th}}$ SO)*-down-parabola source at* $x_{j_2}^* = a_{j_2j_21} = a_{j_1j_21}$.

- *For* $a_{j_1j_10}(\overline{x}_{j_1} - a_{j_1j_11})^{m_1} > 0$ *and* $a_{j_2j_20} < 0$,

$$(\overline{x}_{j_1}, a_{j_2j_21}) = \underbrace{([2(n_{11} - n_{21}) - 1]^{\text{th}}\,\text{DP}, (2n_{21} + 1)^{\text{th}}\,\text{SI})}_{([2(n_{11}-n_{21})-1],(2n_{21}+1))\text{-down-parabola sink}}. \tag{1.45}$$

  The infinite-equilibrium is a $([2(n_{11} - n_{21}) - 1]^{\text{th}}$ DP, $(2n_{21} + 1)^{\text{th}}$ SI)-down-parabola sink at $x_{j_2}^* = a_{j_2j_21} = a_{j_1j_21}$.

- *For* $a_{j_1j_10}(\overline{x}_{j_1} - a_{j_1j_11})^{m_1} < 0$ *and* $a_{j_2j_20} < 0$,

$$(\overline{x}_{j_1}, a_{j_2j_21}) = \underbrace{([2(n_{11} - n_{21}) - 1]^{\text{th}}\,\text{UP}, (2n_{21} + 1)^{\text{th}}\,\text{SI})}_{([2(n_{11}-n_{21})-1],(2n_{21}+1))\text{-up-parabola sink}}. \tag{1.46}$$

  The infinite-equilibrium is a $([2(n_{11} - n_{21}) - 1]^{\text{th}}$ UP, $(2n_{21} + 1)^{\text{th}}$ SI)-up-parabola sink at $x_{j_2}^* = a_{j_2j_21} = a_{j_1j_21}$.

(i_{2a1c}) *For* $n_1 = 2n_{11} + 1$ *and* $n_2 = 2n_{21}$, *the infinite-equilibrium of* $x_{j_2}^* = a_{j_2j_21} = a_{j_1j_21}$ *with* $\overline{x}_{j_1} \neq a_{j_1j_11}$ *has the following properties.*

- *For* $a_{j_1j_10}(\overline{x}_{j_1} - a_{j_1j_11})^{m_1} > 0$ *and* $a_{j_2j_20} > 0$,

$$(\overline{x}_{j_1}, a_{j_2j_21}) = \underbrace{([2(n_{11} - n_{21}) + 1]^{\text{th}}\,\text{UP}, (2n_{21})^{\text{th}}\,\text{US})}_{([2(n_{11}-n_{21})+1],(2n_{21}))\text{-up-parabola upper-saddle}}. \tag{1.47}$$

  The infinite-equilibrium is a $([2(n_{11} - n_{21}) + 1]^{\text{th}}$ UP, $(2n_{21})^{\text{th}}$ US)*-up-parabola upper-saddle at* $x_{j_2}^* = a_{j_2j_21} = a_{j_1j_21}$.

- *For* $a_{j_1j_10}(\overline{x}_{j_1} - a_{j_1j_11})^{m_1} < 0$ *and* $a_{j_2j_20} > 0$,

$$(\overline{x}_{j_1}, a_{j_2j_21}) = \underbrace{([2(n_{11} - n_{21}) + 1]^{\text{th}}\,\text{DP}, (2n_{21})^{\text{th}}\,\text{US})}_{([2(n_{11}-n_{21})+1],(2n_{21}))\text{-down-parabola upper-saddle}}. \tag{1.48}$$

The infinite-equilibrium is a ($[2(n_{11} - n_{21}) + 1]^{\text{th}}$ DP, $(2n_{21})^{\text{th}}$ US)*-down-parabola upper-saddle at* $x^*_{j_2} = a_{j_2j_21} = a_{j_1j_21}$.
- *For* $a_{j_1j_10}(\overline{x}_{j_1} - a_{j_1j_11})^{m_1} > 0$ *and* $a_{j_2j_20} < 0$,

$$(\overline{x}_{j_1}, a_{j_2j_21}) = \underbrace{([2(n_{11} - n_{21}) + 1]^{\text{th}}\,\text{DP}, (2n_{21})^{\text{th}}\,\text{LS})}_{([2(n_{11}-n_{21})+1],(2n_{21}))\text{-down-parabola lower-saddle}}. \tag{1.49}$$

The infinite-equilibrium is a ($[2(n_{11} - n_{21}) + 1]^{\text{th}}$ DP, $(2n_{21} + 1)^{\text{th}}$ LS)*-down-parabola lower-saddle at* $x^*_{j_2} = a_{j_2j_21} = a_{j_1j_21}$.
- *For* $a_{j_1j_10}(\overline{x}_{j_1} - a_{j_1j_11})^{m_1} < 0$ *and* $a_{j_2j_20} < 0$,

$$(\overline{x}_{j_1}, a_{j_2j_21}) = \underbrace{([2(n_{11} - n_{21}) + 1]^{\text{th}}\,\text{UP}, (2n_{21})^{\text{th}}\,\text{LS})}_{([2(n_{11}-n_{21})+1],(2n_{21}))\text{-up-parabola lower-saddle}}. \tag{1.50}$$

The infinite-equilibrium is a ($[2(n_{11} - n_{21}) + 1]^{\text{th}}$ UP, $(2n_{21})^{\text{th}}$ LS)*-up-parabola lower-saddle at* $x^*_{j_2} = a_{j_2j_21} = a_{j_1j_21}$.

(i_{2a1d}) *For* $n_1 = 2n_{11}$ *and* $n_2 = 2n_{21}$, *the infinite-equilibrium of* $x^*_{j_2} = a_{j_2j_21} = a_{j_1j_21}$ *with* $\overline{x}_{j_1} \neq a_{j_1j_11}$ *has the following properties.*

- *For* $a_{j_1j_10}(\overline{x}_{j_1} - a_{j_1j_11})^{m_1} > 0$ *and* $a_{j_2j_20} > 0$,

$$(\overline{x}_{j_1}, a_{j_2j_21}) = \underbrace{([2(n_{11} - n_{21})]^{\text{th}}\,\text{II}, (2n_{21})^{\text{th}}\,\text{US})}_{([2(n_{11}-n_{21})],(2n_{21}))\text{-increasing-inflection upper-saddle}}. \tag{1.51}$$

The infinite-equilibrium is a ($[2(n_{11} - n_{21})]^{\text{th}}$ II, $(2n_{21})^{\text{th}}$ US)*-increasing-inflection upper-saddle at* $x^*_{j_2} = a_{j_2j_21} = a_{j_1j_21}$.
- *For* $a_{j_1j_10}(\overline{x}_{j_1} - a_{j_1j_11})^{m_1} < 0$ *and* $a_{j_2j_20} > 0$,

$$(\overline{x}_{j_1}, a_{j_2j_21}) = \underbrace{([2(n_{11} - n_{21})]^{\text{th}}\,\text{DI}, (2n_{21})^{\text{th}}\,\text{US})}_{([2(n_{11}-n_{21})],(2n_{21}))\text{-decreasing inflection upper-saddle}}. \tag{1.52}$$

The infinite-equilibrium is a ($[2(n_{11} - n_{21})]^{\text{th}}$ DI, $(2n_{21})^{\text{th}}$ US)*-decreasing-inflection upper-saddle at* $x^*_{j_2} = a_{j_2j_21} = a_{j_1j_21}$.
- *For* $a_{j_1j_10}(\overline{x}_{j_1} - a_{j_1j_11})^{m_1} > 0$ *and* $a_{j_2j_20} < 0$,

$$(\overline{x}_{j_1}, a_{j_2j_21}) = \underbrace{([2(n_{11} - n_{21})]^{\text{th}}\,\text{DI}, (2n_{21})^{\text{th}}\,\text{LS})}_{([2(n_{11}-n_{21})],(2n_{21}))\text{-decreasing inflection lower-saddle}}. \tag{1.53}$$

The infinite-equilibrium is a $([2(n_{11}-n_{21})]^{\text{th}}\,\text{DI}, (2n_{21})^{\text{th}}\,\text{LS})$*-decreasing-inflection lower-saddle at* $x^*_{j_2}=a_{j_2j_21}=a_{j_1j_21}$.

- *For* $a_{j_1j_10}(\overline{x}_{j_1}-a_{j_1j_11})^{m_1}<0$ *and* $a_{j_2j_20}<0$,

$$(\overline{x}_{j_1},a_{j_2j_21})=\underbrace{([2(n_{11}-n_{21})]^{\text{th}}\,\text{II}, (2n_{21})^{\text{th}}\,\text{LS})}_{([2(n_{11}-n_{21})],(2n_{21}))\text{-increasing inflection lower-saddle}}. \tag{1.54}$$

The infinite-equilibrium is a $([2(n_{11}-n_{21})]^{\text{th}}\,\text{II}, (2n_{21})^{\text{th}}\,\text{LS})$*-increasing-inflection lower-saddle at* $x^*_{j_2}=a_{j_2j_21}=a_{j_1j_21}$.

(i_{2a2}) There are 32 types of equilibrium $(x^*_{j_1},x^*_{j_2})=(a_{j_1j_11},a_{j_2j_21})$ *on the infinite-equilibrium of* $x^*_{j_2}=a_{j_2j_21}=a_{j_1j_21}$.

(i_{2a2a}) For $m_1=2m_{11}+1$, $n_1=2n_{11}+1$, *and* $n_2=2n_{21}+1$, *the equilibrium of* $(x^*_{j_1},x^*_{j_2})=(a_{j_1j_11},a_{j_2j_21})$ *on the infinite-equilibrium of* $x^*_{j_2}=a_{j_2j_21}=a_{j_1j_21}$ *has the following properties.*

- *For* $a_{j_1j_10}>0$ *and* $a_{j_2j_20}>0$,

$$(a_{j_1j_11},a_{j_2j_21})=\underbrace{([2(n_{11}-n_{21})]^{\text{th}}\,\text{DI:II}, [(2m_{11}+1),(2n_{21}+1)]^{\text{th}}\,\text{SO})}_{([2(n_{11}-n_{21})],[(2m_{11}+1),(2n_{21}+1)])\text{-decreasing-to-increasing inflection source}}. \tag{1.55}$$

The equilibrium is a $([2(n_{11}-n_{21})]^{\text{th}}\,\text{DI:II}, [(2m_{11}+1),(2n_{21}+1)]^{\text{th}}\,\text{SO})$*-decreasing-to-increasing-inflection source at* $(x^*_{j_1},x^*_{j_2})=(a_{j_1j_11},a_{j_2j_21})$.

- *For* $a_{j_1j_10}<0$ *and* $a_{j_2j_20}>0$,

$$(a_{j_1j_11},a_{j_2j_21})=\underbrace{([2(n_{11}-n_{21})]^{\text{th}}\,\text{II:DI}, [(2m_{11}+1),(2n_{21}+1)]^{\text{th}}\,\text{SO})}_{([2(n_{11}-n_{21})],[(2m_{11}+1),(2n_{21}+1)])\text{-increasing-to-decreasing inflection source}}. \tag{1.56}$$

The equilibrium is a $([2(n_{11}-n_{21})]^{\text{th}}\,\text{II:DI}, [(2m_{11}+1),(2n_{21}+1)]^{\text{th}}\,\text{SO})$*-increasing-to-decreasing-inflection source at* $(x^*_{j_1},x^*_{j_2})=(a_{j_1j_11},a_{j_2j_21})$.

- *For* $a_{j_1j_10}>0$ *and* $a_{j_2j_20}<0$,

$$(a_{j_1j_11},a_{j_2j_21})=\underbrace{([2(n_{11}-n_{21})]^{\text{th}}\,\text{II:DI}, [(2m_{11}+1),(2n_{21}+1)]^{\text{th}}\,\text{SI})}_{([2(n_{11}-n_{21})],[(2m_{11}+1),(2n_{21}+1)])\text{-increasing-to-decreasing inflection sink}}. \tag{1.57}$$

The equilibrium is a $([2(n_{11}-n_{21})]^{\text{th}}\,\text{II:DI}, [(2m_{11}+1),(2n_{21}+1)]^{\text{th}}\,\text{SI})$*-increasing-to-decreasing-inflection sink at* $(x^*_{j_1},x^*_{j_2})=(a_{j_1j_11},a_{j_2j_21})$.

- *For* $a_{j_1j_10}<0$ *and* $a_{j_2j_20}<0$,

$$(a_{j_1j_11},a_{j_2j_21})=\underbrace{([2(n_{11}-n_{21})]^{\text{th}}\,\text{DI:II}, [(2m_{11}+1),(2n_{21}+1)]^{\text{th}}\,\text{SI})}_{([2(n_{11}-n_{21})],[(2m_{11}+1),(2n_{21}+1)])\text{-decreasing-to-increasing inflection sink}}. \tag{1.58}$$

The equilibrium is a ([2(n_{11} − n_{21})]th DI:II, [(2m_{11}+1), (2n_{21} + 1)]th SI)-*decreasing-to-increasing-inflection sink at* $(x^*_{j_1}, x^*_{j_2}) = (a_{j_1j_1 1}, a_{j_2j_2 1})$.

(i_{2a2b}) *For* $m_1 = 2m_{11} + 1$, $n_1 = 2n_{11}$ *and* $n_2 = 2n_{21} + 1$, *the equilibrium of* $(x^*_{j_1}, x^*_{j_2}) = (a_{j_1j_1 1}, a_{j_2j_2 1})$ *on the infinite-equilibrium of* $x^*_{j_2} = a_{j_2j_2 1} = a_{j_1j_2 1}$ *has the following properties.*

- *For* $a_{j_1j_1 0} > 0$ *and* $a_{j_2j_2 0} > 0$,

$$(a_{j_1j_1 1}, a_{j_2j_2 1}) = \underbrace{([2(n_{11}-n_{21})-1]^{\text{th}}\ \text{DP:UP},\ [(2m_{11}+1), (2n_{21}+1)]^{\text{th}}\ \text{SO})}_{([2(n_{11}-n_{21})-1],[(2m_{11}+1),(2n_{21}+1)])\text{-hyperbolic source}}. \tag{1.59}$$

The equilibrium is a ([2(n_{11}−n_{21})−1]th DP:UP, [(2m_{11}+1), (2n_{21}+1)]th SO)-*hyperbolic source at* $(x^*_{j_1}, x^*_{j_2}) = (a_{j_1j_1 1}, a_{j_2j_2 1})$.

- *For* $a_{j_1j_1 0} < 0$ *and* $a_{j_2j_2 0} > 0$,

$$(a_{j_1j_1 1}, a_{j_2j_2 1}) = \underbrace{([2(n_{11}-n_{21})-1]^{\text{th}}\ \text{UP:DP},\ [(2m_{11}+1), (2n_{21}+1)]^{\text{th}}\ \text{SO})}_{([2(n_{11}-n_{21})-1],[(2m_{11}+1),(2n_{21}+1)])\text{-hyperbolic-secant source}}. \tag{1.60}$$

The equilibrium is a ([2(n_{11}−n_{21})−1]th UP:DP, [(2m_{11}+1), (2n_{21}+1)]th SO)-*hyperbolic-secant source at* $(x^*_{j_1}, x^*_{j_2}) = (a_{j_1j_1 1}, a_{j_2j_2 1})$.

- *For* $a_{j_1j_1 0} > 0$ *and* $a_{j_2j_2 0} < 0$,

$$(a_{j_1j_1 1}, a_{j_2j_2 1}) = \underbrace{([2(n_{11}-n_{21})-1]^{\text{th}}\ \text{UP:DP},\ [(2m_{11}+1), (2n_{21}+1)]^{\text{th}}\ \text{SI})}_{([2(n_{11}-n_{21})-1],[(2m_{11}+1),(2n_{21}+1)])\text{-hyperbolic-secant sink}}. \tag{1.61}$$

The equilibrium is a ([2(n_{11}−n_{21})−1]th UP:DP, [(2m_{11}+1), (2n_{21}+1)]th SI)-*hyperbolic-secant sink at* $(x^*_{j_1}, x^*_{j_2}) = (a_{j_1j_1 1}, a_{j_2j_2 1})$.

- *For* $a_{j_1j_1 0} < 0$ *and* $a_{j_2j_2 0} < 0$,

$$(a_{j_1j_1 1}, a_{j_2j_2 1}) = \underbrace{([2(n_{11}-n_{21})-1]^{\text{th}}\ \text{DP:UP},\ [(2m_{11}+1), (2n_{21}+1)]^{\text{th}}\ \text{SI})}_{([2(n_{11}-n_{21})-1],[(2m_{11}+1),(2n_{21}+1)])\text{-hyperbolic sink}}. \tag{1.62}$$

The equilibrium is a ([2(n_{11}−n_{21})−1]th DP:UP, [(2m_{11}+1), (2n_{21}+1)]th SI)-*hyperbolic sink at* $(x^*_{j_1}, x^*_{j_2}) = (a_{j_1j_1 1}, a_{j_2j_2 1})$.

(i_{2a2c}) *For* $m_1 = 2m_{11}+1, n_1 = 2n_{11}+1$, *and* $n_2 = 2n_{21}$, *the equilibrium* of $(x^*_{j_1}, x^*_{j_2}) = (a_{j_1j_1 1}, a_{j_2j_2 1})$ *on the infinite-equilibrium of* $x^*_{j_2} = a_{j_2j_2 1} = a_{j_1j_2 1}$ *has the following properties.*

- *For* $a_{j_1j_10} > 0$ *and* $a_{j_2j_20} > 0$,

$$(a_{j_1j_11}, a_{j_2j_21}) = \underbrace{([2(n_{11} - n_{21}) + 1]^{\text{th}}\,\text{DP:UP}, [(2m_{11}+1), (2n_{21})]^{\text{th}}\,\text{US}])}_{([2(n_{11}-n_{21})+1],[(2m_{11}+1),(2n_{21})])\text{-hyperbolic upper-saddle}}. \tag{1.63}$$

The equilibrium is a $([2(n_{11} - n_{21}) + 1]^{\text{th}}$ DP:UP, $[(2m_{11} + 1), (2n_{21})]^{\text{th}}$ US)-*hyperbolic upper-saddle at* $(x^*_{j_1}, x^*_{j_2}) = (a_{j_1j_11}, a_{j_2j_21})$.

- *For* $a_{j_1j_10} < 0$ *and* $a_{j_2j_20} > 0$,

$$(a_{j_1j_11}, a_{j_2j_21}) = \underbrace{([2(n_{11} - n_{21}) + 1]^{\text{th}}\,\text{UP:DP}, [(2m_{11}+1), (2n_{21})]^{\text{th}}\,\text{US})}_{([2(n_{11}-n_{21})+1],[(2m_{11}+1),(2n_{21})])\text{-hyperbolic-secant upper-saddle}}. \tag{1.64}$$

The equilibrium is a $([2(n_{11} - n_{21}) + 1]^{\text{th}}$ UP:DP, $[(2m_{11} + 1), (2n_{21})]^{\text{th}}$ US)-*hyperbolic-secant upper-saddle at* $(x^*_{j_1}, x^*_{j_2}) = (a_{j_1j_11}, a_{j_2j_21})$.

- *For* $a_{j_1j_10} > 0$ *and* $a_{j_2j_20} < 0$,

$$(a_{j_1j_11}, a_{j_2j_21}) = \underbrace{([2(n_{11} - n_{21}) + 1]^{\text{th}}\,\text{UP:DP}, [(2m_{11}+1), (2n_{21})]^{\text{th}}\,\text{LS})}_{([2(n_{11}-n_{21})+1],[(2m_{11}+1),(2n_{21})])\text{-hyperbolic-secant lower-saddle}}. \tag{1.65}$$

The equilibrium is a $([2(n_{11} - n_{21}) + 1]^{\text{th}}$ UP:DP, $[(2m_{11} + 1), (2n_{21})]^{\text{th}}$ LS)-*hyperbolic-secant lower-saddle at* $(x^*_{j_1}, x^*_{j_2}) = (a_{j_1j_11}, a_{j_2j_21})$.

- *For* $a_{j_1j_10} < 0$ *and* $a_{j_2j_20} < 0$,

$$(a_{j_1j_11}, a_{j_2j_21}) = \underbrace{([2(n_{11} - n_{21}) + 1]^{\text{th}}\,\text{DP:UP}, [(2m_{11}+1), (2n_{21})]^{\text{th}}\,\text{LS})}_{([2(n_{11}-n_{21})+1],[(2m_{11}+1),(2n_{21})])\text{-hyperbolic lower-saddle}}. \tag{1.66}$$

The equilibrium is a $([2(n_{11} - n_{21}) + 1]^{\text{th}}$ DP:UP, $[(2m_{11} + 1), (2n_{21})]^{\text{th}}$ LS)-*hyperbolic lower-saddle at* $(x^*_{j_1}, x^*_{j_2}) = (a_{j_1j_11}, a_{j_2j_21})$.

(i_{2a2d}) *For* $m_1 = 2m_{11} + 1$, $n_1 = 2n_{11}$, *and* $n_2 = 2n_{21}$, *the equilibrium* of $(x^*_{j_2}, x^*_{j_2}) = (a_{j_1j_11}, a_{j_2j_21})$ *on the infinite-equilibrium of* $x^*_{j_2} = a_{j_2j_21} = a_{j_1j_21}$ *has the following properties.*

- *For* $a_{j_1j_10} > 0$ *and* $a_{j_2j_20} > 0$,

$$(a_{j_1j_11}, a_{j_2j_21}) = \underbrace{([2(n_{11} - n_{21})]^{\text{th}}\,\text{DI:II}, [(2m_{11} + 1), (2n_{21})]^{\text{th}}\,\text{US})}_{([2(n_{11}-n_{21})],[(2m_{11}+1),(2n_{21})])\text{-decreasing-to-increasing inflection upper-saddle}}. \tag{1.67}$$

The equilibrium is a $([2(n_{11} - n_{21})]^{\text{th}}$ DI:II, $[(2m_{11} + 1), (2n_{21})]^{\text{th}}$ US)-*decreasing-to-increasing-inflection upper-saddle at* $(x^*_{j_1}, x^*_{j_2}) = (a_{j_1j_11}, a_{j_2j_21})$.

- *For* $a_{j_1j_10} < 0$ *and* $a_{j_2j_20} > 0$,

$$(a_{j_1j_11}, a_{j_2j_21}) = \underbrace{([2(n_{11}-n_{21})]^{\text{th}}\ \text{II:DI}, [(2m_{11}+1), (2n_{21})]^{\text{th}}\ \text{US})}_{([2(n_{11}-n_{21})],[(2m_{11}+1),(2n_{21})])\text{-increasing-to-decreasing inflection upper-saddle}}. \tag{1.68}$$

The equilibrium is a $([2(n_{11}-n_{21})]^{\text{th}}$ II:DI, $[(2m_{11}+1), (2n_{21})]^{\text{th}}$ US)-*increasing-to-decreasing-inflection upper-saddle at* $(x^*_{j_1}, x^*_{j_2}) = (a_{j_1j_11}, a_{j_2j_21})$.

- *For* $a_{j_1j_10} > 0$ *and* $a_{j_2j_20} < 0$,

$$(a_{j_1j_11}, a_{j_2j_21}) = \underbrace{([2(n_{11}-n_{21})]^{\text{th}}\ \text{II:DI}, [(2m_{11}+1), (2n_{21})]^{\text{th}}\ \text{LS})}_{([2(n_{11}-n_{21})],[(2m_{11}+1),(2n_{21})])\text{-increasing-to-decreasing inflection lower-saddle}}. \tag{1.69}$$

The equilibrium is a $([2(n_{11}-n_{21})]^{\text{th}}$ II:DI, $[(2m_{11}+1), (2n_{21})]^{\text{th}}$ LS)-*increasing-to-decreasing-inflection lower-saddle at* $(x^*_{j_1}, x^*_{j_2}) = (a_{j_1j_11}, a_{j_2j_21})$.

- *For* $a_{j_1j_10} < 0$ *and* $a_{j_2j_20} < 0$,

$$(a_{j_1j_11}, a_{j_2j_21}) = \underbrace{([2(n_{11}-n_{21})]^{\text{th}}\ \text{DI:II}, [(2m_{11}+1), (2n_{21})]^{\text{th}}\ \text{LS})}_{([2(n_{11}-n_{21})],[(2m_{11}+1),(2n_{21})])\text{-decreasing-to-increasing inflection lower-saddle}}. \tag{1.70}$$

The equilibrium is a $([2(n_{11}-n_{21})]^{\text{th}}$ DI:II, $[(2m_{11}+1), (2n_{21})]^{\text{th}}$ LS)-*decreasing-to-increasing-inflection lower-saddle at* $(x^*_{j_1}, x^*_{j_2}) = (a_{j_1j_11}, a_{j_2j_21})$.

(i_{2a2e}) For $m_1 = 2m_{11}$, $n_1 = 2n_{11}+1$, *and* $n_2 = 2n_{21}+1$, *the equilibrium of* $(x^*_{j_1}, x^*_{j_2}) = (a_{j_1j_11}, a_{j_2j_21})$ *on the infinite-equilibrium of* $x^*_{j_2} = a_{j_2j_21} = a_{j_1j_21}$ *has the following properties.*

- *For* $a_{j_1j_10} > 0$ *and* $a_{j_2j_20} > 0$,

$$(a_{j_1j_11}, a_{j_2j_21}) = \underbrace{([2(n_{11}-n_{21})]^{\text{th}}\ \text{II:II}, [(2m_{11}), (2n_{21}+1)]^{\text{th}}\ \text{SO})}_{([2(n_{11}-n_{21})],[(2m_{11}),(2n_{21}+1)])\text{-increasing inflection source}}. \tag{1.71}$$

The equilibrium is a $([2(n_{11}-n_{21})]^{\text{th}}$ II:II, $[(2m_{11}), (2n_{21}+1)]^{\text{th}}$ SO)-*increasing-inflection source at* $(x^*_{j_1}, x^*_{j_2}) = (a_{j_1j_11}, a_{j_2j_21})$.

- *For* $a_{j_1j_10} < 0$ *and* $a_{j_2j_20} > 0$,

$$(a_{j_1j_11}, a_{j_2j_21}) = \underbrace{([2(n_{11}-n_{21})]^{\text{th}}\ \text{DI:DI}, [(2m_{11}), (2n_{21}+1)]^{\text{th}}\ \text{SO})}_{([2(n_{11}-n_{21})],[(2m_{11}),(2n_{21}+1)])\text{-decreasing inflection source}}. \tag{1.72}$$

The equilibrium is a $([2(n_{11}-n_{21})]^{\text{th}}$ DI:DI, $[(2m_{11}), (2n_{21}+1)]^{\text{th}}$ SO)-*decreasing-inflection source at* $(x^*_{j_1}, x^*_{j_2}) = (a_{j_1j_11}, a_{j_2j_21})$.

- For $a_{j_1j_10} > 0$ and $a_{j_2j_20} < 0$,

$$(a_{j_1j_11}, a_{j_2j_21}) = \underbrace{([2(n_{11} - n_{21})]^{\text{th}}\,\text{DI:DI}, [(2m_{11}), (2n_{21} + 1)]^{\text{th}}\,\text{SI})}_{([2(n_{11}-n_{21})],[(2m_{11}),(2n_{21}+1)])\text{-decreasing inflection sink}}. \quad (1.73)$$

The equilibrium is a $([2(n_{11}-n_{21})]^{\text{th}}$ DI:DI,$[(2m_{11}),(2n_{21}+1)]^{\text{th}}$ SI)-*decreasing-inflection sink at* $(x^*_{j_1}, x^*_{j_2}) = (a_{j_1j_11}, a_{j_2j_21})$.

- For $a_{j_1j_10} < 0$ and $a_{j_2j_20} < 0$,

$$(a_{j_1j_11}, a_{j_2j_21}) = \underbrace{([2(n_{11} - n_{21})]^{\text{th}}\,\text{II:II}, [(2m_{11}), (2n_{21} + 1)]^{\text{th}}\,\text{SI})}_{([2(n_{11}-n_{21})],[(2m_{11}),(2n_{21}+1)])\text{-increasing inflection sink}}. \quad (1.74)$$

The equilibrium is a $([2(n_{11}-n_{21})]^{\text{th}}$ II:II, $[(2m_{11}), (2n_{21}+1)]^{\text{th}}$ SI)-*increasing-inflection source at* $(x^*_{j_1}, x^*_{j_2}) = (a_{j_1j_11}, a_{j_2j_21})$.

(i_{2a2f}) *For* $m_1 = 2m_{11}$, $n_1 = 2n_{11}$, *and* $n_2 = 2n_{21} + 1$, *the equilibrium of* $(x^*_{j_1}, x^*_{j_2}) = (a_{j_1j_11}, a_{j_2j_21})$ *on the infinite-equilibrium of* $x^*_{j_2} = a_{j_2j_21} = a_{j_1j_21}$ *has the following properties.*

- For $a_{j_1j_10} > 0$ and $a_{j_2j_20} > 0$,

$$(a_{j_1j_11}, a_{j_2j_21}) = \underbrace{([2(n_{11} - n_{21}) - 1]^{\text{th}}\,\text{UP:UP}, [(2m_{11}), (2n_{21} + 1)]^{\text{th}}\,\text{SO})}_{([2(n_{11}-n_{21})-1],[(2m_{11}),(2n_{21}+1)])\text{-double up-parabola source}}. \quad (1.75)$$

The equilibrium is a $([2(n_{11} - n_{21}) - 1]^{\text{th}}$ UP:UP, $[(2m_{11}), (2n_{21} + 1)]^{\text{th}}$ SO) *-double up-parabola source at* $(x^*_{j_1}, x^*_{j_2}) = (a_{j_1j_11}, a_{j_2j_21})$.

- For $a_{j_1j_10} < 0$ and $a_{j_2j_20} > 0$,

$$(a_{j_1j_11}, a_{j_2j_21}) = \underbrace{([2(n_{11} - n_{21}) - 1]^{\text{th}}\,\text{DP:DP}, [(2m_{11}), (2n_{21} + 1)]^{\text{th}}\,\text{SO})}_{([2(n_{11}-n_{21})-1],[(2m_{11}),(2n_{21}+1)])\text{-double down-parabola source}}. \quad (1.76)$$

The equilibrium is a $([2(n_{11} - n_{21}) - 1]^{\text{th}}$ DP:DP, $[(2m_{11}), (2n_{21} + 1)]^{\text{th}}$ SO) *-double down-parabola source at* $(x^*_{j_1}, x^*_{j_2}) = (a_{j_1j_11}, a_{j_2j_21})$.

- For $a_{j_1j_10} > 0$ and $a_{j_2j_20} < 0$,

$$(a_{j_1j_11}, a_{j_2j_21}) = \underbrace{([2(n_{11} - n_{21}) - 1]^{\text{th}}\,\text{DP:DP}, [(2m_{11}), (2n_{21} + 1)]^{\text{th}}\,\text{SI})}_{([2(n_{11}-n_{21})-1],[(2m_{11}),(2n_{21}+1)])\text{-double down-parabola sink}}. \quad (1.77)$$

The equilibrium is a $([2(n_{11} - n_{21}) - 1]^{\text{th}}$ DP:DP, $[(2m_{11}), (2n_{21} + 1)]^{\text{th}}$ SI) *-double down-parabola sink at* $(x^*_{j_1}, x^*_{j_2}) = (a_{j_1j_11}, a_{j_2j_21})$.

- *For* $a_{j_1j_10} < 0$ *and* $a_{j_2j_20} < 0$,

$$(a_{j_1j_11}, a_{j_2j_21}) = \underbrace{([2(n_{11} - n_{21}) - 1]^{\text{th}}\ \text{UP:UP}, [(2m_{11}), (2n_{21} + 1)]^{\text{th}}\ \text{SI})}_{([2(n_{11}-n_{21})-1],[(2m_{11}),(2n_{21}+1)])\text{-double up-parabola sink}}. \tag{1.78}$$

The equilibrium is a $([2(n_{11} - n_{21}) - 1]^{\text{th}}$ UP:UP, $[(2m_{11}), (2n_{21} + 1)]^{\text{th}}$ SI) *-double up-parabola sink at* $(x^*_{j_1}, x^*_{j_2}) = (a_{j_1j_11}, a_{j_2j_21})$.

(i_{2a2g}) *For* $m_1 = 2m_{11}, n_1 = 2n_{11} + 1$, *and* $n_2 = 2n_{21}$, *the equilibrium* of $(x^*_{j_1}, x^*_{j_2}) = (a_{j_1j_11}, a_{j_2j_21})$ *on the infinite-equilibrium of* $x^*_{j_2} = a_{j_2j_21} = a_{j_1j_21}$ *has the following properties.*

- *For* $a_{j_1j_10} > 0$ *and* $a_{j_2j_20} > 0$,

$$(a_{j_1j_11}, a_{j_2j_21}) = \underbrace{([2(n_{11} - n_{21}) + 1]^{\text{th}}\ \text{UP:UP}, [(2m_{11}), (2n_{21})]^{\text{th}}\ \text{US]})}_{([2(n_{11}-n_{21})+1],[(2m_{11}),(2n_{21}+1)])\text{-up-parabola upper-saddle}}. \tag{1.79}$$

The equilibrium is a $([2(n_{11} - n_{21}) + 1]^{\text{th}}$ UP:UP, $[(2m_{11}), (2n_{21})]^{\text{th}}$ US)*-up-parabola upper-saddle at* $(x^*_{j_2}, x^*_{j_2}) = (a_{j_1j_11}, a_{j_2j_21})$.

- *For* $a_{j_1j_10} < 0$ *and* $a_{j_2j_20} > 0$,

$$(a_{j_1j_11}, a_{j_2j_21}) = \underbrace{([2(n_{11} - n_{21}) + 1]^{\text{th}}\ \text{DP:DP}, [(2m_{11}), (2n_{21})]^{\text{th}}\ \text{US]})}_{([2(n_{11}-n_{21})+1],[(2m_{11}),(2n_{21}+1)])\text{-down-parabola upper-saddle}}. \tag{1.80}$$

The equilibrium is a $([2(n_{11} - n_{21}) + 1]^{\text{th}}$ DP:DP, $[(2m_{11}), (2n_{21})]^{\text{th}}$ US)*-down-parabola upper-saddle at* $(x^*_{j_1}, x^*_{j_2}) = (a_{j_1j_11}, a_{j_2j_21})$.

- *For* $a_{j_1j_10} > 0$ *and* $a_{j_2j_20} < 0$,

$$(a_{j_1j_11}, a_{j_2j_21}) = \underbrace{([2(n_{11} - n_{21}) + 1]^{\text{th}}\ \text{DP:DP}, [(2m_{11}), (2n_{21})]^{\text{th}}\ \text{LS]})}_{([2(n_{11}-n_{21})+1],[(2m_{11}),(2n_{21})])\text{-down-parabola lower-saddle}}. \tag{1.81}$$

The equilibrium is a $([2(n_{11} - n_{21}) + 1]^{\text{th}}$ DP:DP, $[(2m_{11}), (2n_{21})]^{\text{th}}$ LS) *-down-parabola lower-saddle at* $(x^*_{j_1}, x^*_{j_2}) = (a_{j_1j_11}, a_{j_2j_21})$.

- *For* $a_{j_1j_10} < 0$ *and* $a_{j_2j_20} < 0$,

$$(a_{j_1j_11}, a_{j_2j_21}) = \underbrace{([2(n_{11} - n_{21}) + 1]^{\text{th}}\ \text{UP:UP}, [(2m_{11}), (2n_{21})]^{\text{th}}\ \text{LS]})}_{([2(n_{11}-n_{21})+1],[(2m_{11}),(2n_{21})])\text{-up-parabola lower-saddle}}. \tag{1.82}$$

The equilibrium is a $([2(n_{11} - n_{21}) + 1]^{\text{th}}$ UP:UP, $[(2m_{11}), (2n_{21})]^{\text{th}}$ LS)*-up-parabola lower-saddle at* $(x^*_{j_1}, x^*_{j_2}) = (a_{j_1j_11}, a_{j_2j_21})$.

(i_{2a2h}) *For* $m_1 = 2m_{11}, n_1 = 2n_{11}$, *and* $n_2 = 2n_{21}$, *the equilibrium of* $(x^*_{j_1}, x^*_{j_2}) = (a_{j_1j_11}, a_{j_2j_21})$ *on the infinite-equilibrium of* $x^*_{j_2} = a_{j_2j_21} = a_{j_1j_21}$ *has the following properties.*

- *For* $a_{j_1j_10} > 0$ *and* $a_{j_2j_20} > 0$,

$$(a_{j_1j_11}, a_{j_2j_21}) = \underbrace{([2(n_{11} - n_{21})]^{\text{th}}\ \text{II:II}, [(2m_{11}), (2n_{21})]^{\text{th}}\ \text{US})}_{([2(n_{11}-n_{21})],[(2m_{11}),(2n_{21})])\text{-increasing inflection upper-saddle}}. \tag{1.83}$$

The equilibrium is a ($[2(n_{11} - n_{21})]^{\text{th}}$ II:II, $[(2m_{11}), (2n_{21})]^{\text{th}}$ US)*-increasing-inflection upper-saddle at* $(x^*_{j_1}, x^*_{j_2}) = (a_{j_1j_11}, a_{j_2j_21})$.

- *For* $a_{j_1j_10} < 0$ *and* $a_{j_2j_20} > 0$,

$$(a_{j_1j_11}, a_{j_2j_21}) = \underbrace{([2(n_{11} - n_{21})]^{\text{th}}\ \text{DI:DI}, [(2m_{11}), (2n_{21})]^{\text{th}}\ \text{US})}_{([2(n_{11}-n_{21})],[(2m_{11}),(2n_{21})])\text{-decreasing-inflection upper-saddle}}. \tag{1.84}$$

The equilibrium is a ($[2(n_{11}-n_{21})]^{\text{th}}$ DI:DI, $[(2m_{11}), (2n_{21})]^{\text{th}}$ US)*-decreasing-inflection upper-saddle at* $(x^*_{j_1}, x^*_{j_2}) = (a_{j_1j_11}, a_{j_2j_21})$.

- *For* $a_{j_1j_10} > 0$ *and* $a_{j_2j_20} < 0$,

$$(a_{j_1j_11}, a_{j_2j_21}) = \underbrace{([2(n_{11} - n_{21})]^{\text{th}}\ \text{DI:DI}, [(2m_{11}), (2n_{21})]^{\text{th}}\ \text{LS})}_{([2(n_{11}-n_{21})],[(2m_{11}+1),(2n_{21})])\text{-decreasing inflection lower-saddle}}. \tag{1.85}$$

The equilibrium is a ($[2(n_{11}-n_{21})]^{\text{th}}$ DI:DI, $[(2m_{11}), (2n_{21})]^{\text{th}}$ LS)*-decreasing-inflection lower-saddle at* $(x^*_{j_1}, x^*_{j_2}) = (a_{j_1j_11}, a_{j_2j_21})$.

- *For* $a_{j_1j_10} < 0$ *and* $a_{j_2j_20} < 0$,

$$(a_{j_1j_11}, a_{j_2j_21}) = \underbrace{([2(n_{11} - n_{21})]^{\text{th}}\ \text{II:II}, [(2m_{11}), (2n_{21})]^{\text{th}}\ \text{LS})}_{([2(n_{11}-n_{21})],[(2m_{11}),(2n_{21})])\text{-increasing inflection lower-saddle}}. \tag{1.86}$$

The equilibrium is a ($[2(n_{11} - n_{21})]^{\text{th}}$ DI:II, $[(2m_{11}), (2n_{21})]^{\text{th}}$ LS)*-increasing-inflection lower-saddle at* $(x^*_{j_1}, x^*_{j_2}) = (a_{j_1j_11}, a_{j_2j_21})$.

(i_{2b}) *For* $n_1 < n_2$, *the first integral manifold is*

$$\begin{aligned}
&\frac{1}{m_1 - 1}\left[\frac{1}{(x_{j_1} - a_{j_1j_11})^{m_1-1}} - \frac{1}{(x_{j_10} - a_{j_1j_11})^{m_1-1}}\right]\\
&= \frac{a_{j_1j_10}}{a_{j_2j_20}}\frac{1}{n_2 - n_1 - 1}\left[\frac{1}{(x_{j_2} - a_{j_2j_21})^{n_2-n_1-1}} - \frac{1}{(x_{j_20} - a_{j_2j_21})^{n_2-n_1-1}}\right],\\
&\text{for } m_1 \neq 1, n_2 - n_1 \neq 1;\\
&\frac{1}{m_1 - 1}\left[\frac{1}{(x_{j_1} - a_{j_1j_11})^{m_1-1}} - \frac{1}{(x_{j_10} - a_{j_1j_11})^{m_1-1}}\right] = -\ln\frac{|x_{j_1} - a_{j_1j_11}|}{|x_{j_10} - a_{j_1j_11}|},\\
&\text{for } m_1 = 1;\\
&\frac{1}{n_2 - n_1 - 1}\left[\frac{1}{(x_{j_2} - a_{j_2j_21})^{n_2-n_1-1}} - \frac{1}{(x_{j_20} - a_{j_2j_21})^{n_2-n_1-1}}\right] = -\ln\frac{|x_{j_2} - a_{j_2j_21}|}{|x_{j_20} - a_{j_2j_21}|},\\
&\text{for } n_2 - n_1 - 1 = 0.
\end{aligned} \tag{1.87}$$

(i_{2b1}) There are 16 types of infinite-equilibriums at $x^*_{j_2} = a_{j_2j_21} = a_{j_1j_21}$.

(i_{2b1a}) For $n_1 = 2n_{11} + 1$ *and* $n_2 = 2n_{21} + 1$, *the infinite-equilibrium of* $x^*_{j_2} = a_{j_2j_21}$ $= a_{j_1j_21}$ *and* $\bar{x}_{j_1} \neq a_{j_1j_11}$ *has the following properties.*

- *For* $a_{j_1j_10}(\bar{x}_{j_1} - a_{j_1j_11})^{m_1} > 0$ *and* $a_{j_2j_20} > 0$,

$$(\bar{x}_{j_1}, a_{j_2j_21}) = \underbrace{([2(n_{21} - n_{11})]^{\text{th}}\,\text{IC}, (2n_{21} + 1)^{\text{th}}\,\text{SO})}_{([2(n_{21}-n_{11})],(2n_{21}+1))\text{-increasing concave source}}. \tag{1.88}$$

The infinite-equilibrium is a $([2(n_{21} - n_{11})]^{\text{th}}$ IC, $(2n_{21} + 1)^{\text{th}}$ SO)*-increasing-concave source at* $x^*_{j_2} = a_{j_2j_21} = a_{j_1j_21}$.

- *For* $a_{j_1j_10}(\bar{x}_{j_1} - a_{j_1j_11})^{m_1} < 0$ *and* $a_{j_2j_20} > 0$,

$$\left(\bar{x}_{j1}, a_{j2j21}\right) = \underbrace{\left([2(n_{21} - n_{11})]^{\text{th}}\text{DC}, (2n_{21} + t)^{\text{th}}\text{SO}\right).}_{([2(n_{21}-n_{11})],(2n_{21}+1))\text{-decreasing concave source}} \tag{1.89}$$

The infinite-equilibrium is a $([2(n_{21} - n_{11})]^{\text{th}}$ DC, $(2n_{21} + 1)^{\text{th}}$ SO)*-decreasing-concave source at* $x^*_{j_2} = a_{j_2j_21} = a_{j_1j_21}$.

- *For* $a_{j_1j_10}(\bar{x}_{j_1} - a_{j_1j_11})^{m_1} > 0$ *and* $a_{j_2j_20} < 0$,

$$(\bar{x}_{j_1}, a_{j_2j_21}) = \underbrace{([2(n_{21} - n_{11})]^{\text{th}}\,\text{DC}, (2n_{21} + 1)^{\text{th}}\,\text{SI})}_{([2(n_{21}-n_{11})],(2n_{21}+1))\text{-decreasing concave sink}}. \tag{1.90}$$

The infinite-equilibrium is a $([2(n_{21} - n_{11})]^{\text{th}}$ DC, $(2n_{21} + 1)^{\text{th}}$ SI)*-decreasing-concave sink at* $x^*_{j_2} = a_{j_2j_21} = a_{j_1j_21}$.

- *For* $a_{j_1j_10}(\bar{x}_{j_1} - a_{j_1j_11})^{m_1} < 0$ *and* $a_{j_2j_20} < 0$,

$$(\bar{x}_{j_1}, a_{j_2j_21}) = \underbrace{([2(n_{21} - n_{11})]^{\text{th}}\,\text{IC}, (2n_{21} + 1)^{\text{th}}\,\text{SI})}_{([2(n_{21}-n_{11})],(2n_{21}+1))\text{-increasing concave sink}}. \tag{1.91}$$

The infinite-equilibrium is a $([2(n_{21} - n_{11})]^{\text{th}}$ IC, $(2n_{21} + 1)^{\text{th}}$ SI)*-increasing-concave sink at* $x^*_{j_2} = a_{j_2j_21} = a_{j_1j_21}$.

(i_{2b1b}) For $n_1 = 2n_{11}$ *and* $n_2 = 2n_{21} + 1$, *the infinite-equilibrium of* $x^*_{j_2} = a_{j_2j_21} =$ $a_{j_1j_21}$ *with* $\bar{x}_{j_1} \neq a_{j_1j_11}$ *has the following properties.*

- *For* $a_{j_1j_10}(\bar{x}_{j_1} - a_{j_1j_11})^{m_1} > 0$ *and* $a_{j_2j_20} > 0$,

$$(\bar{x}_{j_1}, a_{j_2j_21}) = \underbrace{([2(n_{21} - n_{11}) + 1]^{\text{th}}\,\text{DA}, (2n_{21} + 1)^{\text{th}}\,\text{SO})}_{([2(n_{21}-n_{11})+1],(2n_{21}+1))\text{-down-asymptote source}}. \tag{1.92}$$

The infinite-equilibrium is a $([2(n_{21} - n_{11}) + 1]^{\text{th}}$ DA, $(2n_{21} + 1)^{\text{th}}$ SO)*-down-asymptote source at* $x^*_{j_2} = a_{j_2j_21} = a_{j_1j_21}$.

- *For* $a_{j_1j_10}(\overline{x}_{j_1} - a_{j_1j_11})^{m_1} < 0$ *and* $a_{j_2j_20} > 0$,

$$(\overline{x}_{j_1}, a_{j_2j_21}) = \underbrace{([2(n_{21} - n_{11}) + 1]^{\text{th}}\,\text{UA}, (2n_{21} + 1)^{\text{th}}\,\text{SO})}_{([2(n_{21}-n_{11})+1],(2n_{21}+1))\text{-up-asymptote source}}. \tag{1.93}$$

The infinite-equilibrium is a $([2(n_{21} - n_{11}) + 1]^{\text{th}}\,\text{UA}, (2n_{21} + 1)^{\text{th}}\,\text{SO})$*-up-asymptote source at* $x^*_{j_2} = a_{j_2j_21} = a_{j_1j_21}$.

- *For* $a_{j_1j_10}(\overline{x}_{j_1} - a_{j_1j_11})^{m_1} > 0$ *and* $a_{j_2j_20} < 0$,

$$(\overline{x}_{j_1}, a_{j_2j_21}) = \underbrace{([2(n_{21} - n_{11}) + 1]^{\text{th}}\,\text{UA}, (2n_{21} + 1)^{\text{th}}\,\text{SI})}_{([2(n_{21}-n_{11})+1],(2n_{21}+1))\text{-up-asymptote sink}}. \tag{1.94}$$

The infinite-equilibrium is a $([2(n_{21} - n_{11}) + 1]^{\text{th}}\,\text{UA}, (2n_{21} + 1)^{\text{th}}\,\text{SI})$*-up-asymptote sink at* $x^*_{j_2} = a_{j_2j_21} = a_{j_1j_21}$.

- *For* $a_{j_1j_10}(\overline{x}_{j_1} - a_{j_1j_11})^{m_1} < 0$ *and* $a_{j_2j_20} < 0$,

$$(\overline{x}_{j_1}, a_{j_2j_21}) = \underbrace{([2(n_{21} - n_{11}) + 1]^{\text{th}}\,\text{DA}, (2n_{21} + 1)^{\text{th}}\,\text{SI})}_{([2(n_{21}-n_{11})+1],(2n_{21}+1))\text{-down-asymptote sink}}. \tag{1.95}$$

The infinite-equilibrium is a $([2(n_{21} - n_{11}) + 1]^{\text{th}}\,\text{DA}, (2n_{21} + 1)^{\text{th}}\,\text{SI})$*-down-asymptote sink at* $x^*_{j_2} = a_{j_2j_21} = a_{j_1j_21}$.

(i_{2b1c}) *For* $n_1 = 2n_{11} + 1$ *and* $n_2 = 2n_{21}$, *the infinite-equilibrium of* $x^*_{j_2} = a_{j_2j_21} = a_{j_1j_21}$ *and* $\overline{x}_{j_1} \neq a_{j_1j_11}$ *has the following properties.*

- *For* $a_{j_1j_10}(\overline{x}_{j_1} - a_{j_1j_11})^{m_1} > 0$ *and* $a_{j_2j_20} > 0$,

$$(\overline{x}_{j_1}, a_{j_2j_21}) = \underbrace{([2(n_{21} - n_{11}) - 1]^{\text{th}}\,\text{DA}, (2n_{21})^{\text{th}}\,\text{US})}_{([2(n_{21}-n_{11})-1],(2n_{21}))\text{-down-asymptote upper-saddle}}. \tag{1.96}$$

The infinite-equilibrium is a $([2(n_{21} - n_{11}) - 1]^{\text{th}}\,\text{DA}, (2n_{21})^{\text{th}}\,\text{US})$*-down-asymptote upper-saddle* at $x^*_{j_2} = a_{j_2j_21} = a_{j_1j_21}$.

- *For* $a_{j_1j_10}(\overline{x}_{j_1} - a_{j_1j_11})^{m_1} < 0$ *and* $a_{j_2j_20} > 0$,

$$(\overline{x}_{j_1}, a_{j_2j_21}) = \underbrace{([2(n_{21} - n_{11}) - 1]^{\text{th}}\,\text{UA}, (2n_{21})^{\text{th}}\,\text{US})}_{([2(n_{21}-n_{11})-1],(2n_{21}))\text{-up-asymptote upper-saddle}}. \tag{1.97}$$

The infinite-equilibrium is a $([2(n_{21} - n_{11}) - 1]^{\text{th}}\,\text{UA}, (2n_{21})^{\text{th}}\,\text{US})$*-up-asymptote upper-saddle at* $x^*_{j_2} = a_{j_2j_21} = a_{j_1j_21}$.

- *For* $a_{j_1j_10}(\overline{x}_{j_1} - a_{j_1j_11})^{m_1} > 0$ *and* $a_{j_2j_20} < 0$,

$$(\overline{x}_{j_1}, a_{j_2j_21}) = \underbrace{([2(n_{21} - n_{11}) - 1]^{\text{th}}\,\text{UA}, (2n_{21})^{\text{th}}\,\text{LS})}_{([2(n_{21}-n_{11})-1],(2n_{21}))\text{-up-asymptote lower-saddle}}. \tag{1.98}$$

The infinite-equilibrium is a $([2(n_{21}-n_{11})-1]^{\text{th}}\,\text{UA}, (2n_{21})^{\text{th}}\,\text{LS})$*-up-asymptote lower-saddle at* $x^*_{j_2} = a_{j_2j_21} = a_{j_1j_21}$.

- *For* $a_{j_1j_10}(\overline{x}_{j_1} - a_{j_1j_11})^{m_1} < 0$ *and* $a_{j_2j_20} < 0$,

$$(\overline{x}_{j_1}, a_{j_2j_21}) = \underbrace{([2(n_{21}-n_{11})-1]^{\text{th}}\,\text{DA}, (2n_{21})^{\text{th}}\,\text{LS})}_{([2(n_{21}-n_{11})-1],(2n_{21}))\text{-down-asymptote lower-saddle}}. \tag{1.99}$$

The infinite-equilibrium is a $([2(n_{21}-n_{11})-1]^{\text{th}}\,\text{DA}, (2n_{21})^{\text{th}}\,\text{LS})$*-down-asymptote lower-saddle at* $x^*_{j_2} = a_{j_2j_21} = a_{j_1j_21}$.

(i_{2b1d}) For $n_1 = 2n_{11}$ *and* $n_2 = 2n_{21}$*, the infinite-equilibrium of* $x^*_{j_2} = a_{j_2j_21} = a_{j_1j_21}$ *with* $\overline{x}_{j_1} \neq a_{j_1j_11}$ *has the following properties.*

- *For* $a_{j_1j_10}(\overline{x}_{j_1} - a_{j_1j_11})^{m_1} > 0$ *and* $a_{j_2j_20} > 0$,

$$(\overline{x}_{j_1}, a_{j_2j_21}) = \underbrace{([2(n_{21}-n_{11})]^{\text{th}}\,\text{IC}, (2n_{21})^{\text{th}}\,\text{US})}_{([2(n_{21}-n_{11})],(2n_{21}))\text{-increasing-concave upper-saddle}}. \tag{1.100}$$

The equilibrium is a $([2(n_{21}-n_{11})]^{\text{th}}\,\text{IC}, (2n_{21})^{\text{th}}\,\text{US})$*-increasing-concave upper-saddle* at $x^*_{j_2} = a_{j_2j_21} = a_{j_1j_21}$.

- *For* $a_{j_1j_10}(\overline{x}_{j_1} - a_{j_1j_11})^{m_1} < 0$ *and* $a_{j_2j_20} > 0$,

$$(\overline{x}_{j_1}, a_{j_2j_21}) = \underbrace{([2(n_{21}-n_{11})]^{\text{th}}\,\text{DC}, (2n_{21})^{\text{th}}\,\text{US})}_{([2(n_{21}-n_{11})],(2n_{21}))\text{-decreasing concave upper-saddle}}. \tag{1.101}$$

The equilibrium is a $([2(n_{21}-n_{11})]^{\text{th}}\,\text{DC}, (2n_{21})^{\text{th}}\,\text{US})$*-decreasing-concave upper-saddle at* $x^*_{j_2} = a_{j_2j_21} = a_{j_1j_21}$.

- *For* $a_{j_1j_10}(\overline{x}_{j_1} - a_{j_1j_11})^{m_1} > 0$ and $a_{j_2j_20} < 0$,

$$(\overline{x}_{j_1}, a_{j_2j_21}) = \underbrace{([2(n_{21}-n_{11})]^{\text{th}}\,\text{DC}, (2n_{21})^{\text{th}}\,\text{LS})}_{([2(n_{21}-n_{11})],(2n_{21}))\text{-decreasing concave lower-saddle}}. \tag{1.102}$$

The equilibrium is a $([2(n_{21}-n_{11})]^{\text{th}}\,\text{DC}, (2n_{21})^{\text{th}}\,\text{LS})$*-decreasing-concave lower-saddle at* $x^*_{j_2} = a_{j_2j_21} = a_{j_1j_21}$.

- *For* $a_{j_1j_10}(\overline{x}_{j_1} - a_{j_1j_11})^{m_1} < 0$ *and* $a_{j_2j_20} < 0$,

$$(\overline{x}_{j_1}, a_{j_2j_21}) = \underbrace{([2(n_{21}-n_{11})]^{\text{th}}\,\text{II}, (2n_{21})^{\text{th}}\,\text{LS})}_{([2(n_{21}-n_{11})],(2n_{21}))\text{-increasing concave lower-saddle}}. \tag{1.103}$$

The equilibrium is a $([2(n_{21}-n_{11})]^{\text{th}}\,\text{II}, (2n_{21})^{\text{th}}\,\text{LS})$*-increasing-concave lower-saddle at* $x^*_{j_2} = a_{j_2j_21} = a_{j_1j_21}$.

(i_{2b2}) There are 32 types of equilibrium $(x^*_{j_1}, x^*_{j_2}) = (a_{j_1j_11}, a_{j_2j_21})$ *on the infinite-equilibrium of* $x^*_{j_2} = a_{j_2j_21} = a_{j_1j_21}$.

(i_{2b2a}) For $m_1 = 2m_{11} + 1, n_1 = 2n_{11} + 1$, *and* $n_2 = 2n_{21} + 1$, *the equilibrium of* $(x^*_{j_1}, x^*_{j_2}) = (a_{j_1j_11}, a_{j_2j_21})$ *on the equilibrium of* $x^*_{j_2} = a_{j_2j_21} = a_{j_1j_21}$ *has the following properties.*

- *For* $a_{j_1j_10} > 0$ *and* $a_{j_2j_20} > 0$,

$$(a_{j_1j_11}, a_{j_2j_21}) = \underbrace{([2(n_{21} - n_{11})]^{\text{th}}\,\text{DC:IC}, [(2m_{11} + 1), (2n_{21} + 1)]^{\text{th}}\,\text{SO})}_{([2(n_{21}-n_{11})],[(2m_{11}+1),(2n_{21}+1)])\text{-decreasing-to-increasing concave source}}. \tag{1.104}$$

The equilibrium is a $([2(n_{21} - n_{11})]^{\text{th}}$ DC: IC, $[(2m_{11} + 1), (2n_{21} + 1)]^{\text{th}}$ SO)-*decreasing-to-increasing-concave source at* $(x^*_{j_1}, x^*_{j_2}) = (a_{j_1j_11}, a_{j_2j_21})$.

- *For* $a_{j_1j_10} < 0$ *and* $a_{j_2j_20} > 0$,

$$(a_{j_1j_11}, a_{j_2j_21}) = \underbrace{([2(n_{21} - n_{11})]^{\text{th}}\,\text{IC:DC}, [(2m_{11} + 1), (2n_{21} + 1)]^{\text{th}}\,\text{SO})}_{([2(n_{21}-n_{11})],[(2m_{11}+1),(2n_{21}+1)])\text{-increasing-to-decreasing concave source}}. \tag{1.105}$$

The equilibrium is a $([2(n_{21} - n_{11})]^{\text{th}}$ IC:DC, $[(2m_{11} + 1), (2n_{21} + 1)]^{\text{th}}$ SO)-*increasing-to-decreasing-concave source at* $(x^*_{j_1}, x^*_{j_2}) = (a_{j_1j_11}, a_{j_2j_21})$.

- *For* $a_{j_1j_10} > 0$ *and* $a_{j_2j_20} < 0$,

$$(a_{j_1j_11}, a_{j_2j_21}) = \underbrace{([2(n_{21} - n_{11})]^{\text{th}}\,\text{IC:DC}, [(2m_{11} + 1), (2n_{21} + 1)]^{\text{th}}\,\text{SI})}_{([2(n_{21}-n_{11})],[(2m_{11}+1),(2n_{21}+1)])\text{-increasing-to-decreasing concave sink}}. \tag{1.106}$$

The equilibrium is a $([2(n_{21} - n_{11})]^{\text{th}}$ IC:DC, $[(2m_{11} + 1), (2n_{21} + 1)]^{\text{th}}$ SI)-*increasing-to-decreasing concave sink at* $(x^*_{j_1}, x^*_{j_2}) = (a_{j_1j_11}, a_{j_2j_21})$.

- *For* $a_{j_1j_10} < 0$ *and* $a_{j_2j_20} < 0$,

$$(a_{j_1j_11}, a_{j_2j_21}) = \underbrace{([2(n_{21} - n_{11})]^{\text{th}}\,\text{DC:IC}, [(2m_{11} + 1), (2n_{21} + 1)]^{\text{th}}\,\text{SI})}_{([2(n_{21}-n_{11})],[(2m_{11}+1),(2n_{21}+1)])\text{-decreasing-to-increasing concave sink}}. \tag{1.107}$$

The equilibrium is a $([2(n_{21} - n_{11})]^{\text{th}}$ DC:IC, $[(2m_{11} + 1), (2n_{21} + 1)]^{\text{th}}$ SI)-*decreasing-to-increasing-concave sink at* $(x^*_{j_1}, x^*_{j_2}) = (a_{j_1j_11}, a_{j_2j_21})$.

(i_{2b2b}) For $m_1 = 2m_{11}+1, n_1 = 2n_{11}$, *and* $n_2 = 2n_{21}+1$, *the equilibrium of* $(x^*_{j_1}, x^*_{j_2}) = (a_{j_1j_11}, a_{j_2j_21})$ *on the infinite-equilibrium of* $x^*_{j_2} = a_{j_2j_21} = a_{j_1j_21}$ *has the following properties*

- *For* $a_{j_1j_10} > 0$ *and* $a_{j_2j_20} > 0$,

$$(a_{j_1j_11}, a_{j_2j_21}) = \underbrace{([2(n_{21} - n_{11}) + 1]^{\text{th}}\ \text{UA:DA}, [(2m_{11} + 1), (2n_{21} + 1)]^{\text{th}}\ \text{SO})}_{([2(n_{21}-n_{11})+1],[(2m_{11}+1),(2n_{21}+1)])\text{-up-to-down asymptote source}}. \tag{1.108}$$

The equilibrium is a $([2(n_{21}-n_{11})+1]^{\text{th}}$ UA:DA, $[(2m_{11}+1), (2n_{21}+1)]^{\text{th}}$ SO)-*up-to-down asymptote source* at $(x^*_{j_1}, x^*_{j_2}) = (a_{j_1j_11}, a_{j_2j_21})$.

- *For* $a_{j_1j_10} < 0$ *and* $a_{j_2j_20} > 0$,

$$(a_{j_1j_11}, a_{j_2j_21}) = \underbrace{([2(n_{21} - n_{11}) + 1]^{\text{th}}\ \text{DA:UA}, [(2m_{11} + 1), (2n_{21} + 1)]^{\text{th}}\ \text{SO})}_{([2(n_{21}-n_{11})+1],[(2m_{11}+1),(2n_{21}+1)])\text{-down-to-up asymptote source}}. \tag{1.109}$$

The equilibrium is a $([2(n_{21}-n_{11})+1]^{\text{th}}$ DA:UA, $[(2m_{11}+1), (2n_{21}+1)]^{\text{th}}$ SO)-*down-to-up asymptote source at* $(x^*_{j_1}, x^*_{j_2}) = (a_{j_1j_11}, a_{j_2j_21})$.

- *For* $a_{j_1j_10} > 0$ *and* $a_{j_2j_20} < 0$,

$$(a_{j_1j_11}, a_{j_2j_21}) = \underbrace{([2(n_{21} - n_{11}) + 1]^{\text{th}}\ \text{DA:UA}, [(2m_{11} + 1), (2n_{21} + 1)]^{\text{th}}\ \text{SI})}_{([2(n_{21}-n_{11})+1],[(2m_{11}+1),(2n_{21}+1)])\text{-down-to-up asymptote sink}}. \tag{1.110}$$

The equilibrium is a $([2(n_{21}-n_{11})+1]^{\text{th}}$ DA:UA, $[(2m_{11}+1), (2n_{21}+1)]^{\text{th}}$ SI)-*down-to-up asymptote sink at* $(x^*_{j_1}, x^*_{j_2}) = (a_{j_1j_11}, a_{j_2j_21})$.

- *For* $a_{j_1j_10} < 0$ *and* $a_{j_2j_20} < 0$,

$$(a_{j_1j_11}, a_{j_2j_21}) = \underbrace{([2(n_{21} - n_{11}) + 1]^{\text{th}}\ \text{UA:DA}, [(2m_{11} + 1), (2n_{21} + 1)]^{\text{th}}\ \text{SI})}_{([2(n_{21}-n_{11})+1],[(2m_{11}+1),(2n_{21}+1)])\text{-up-to-down asymptote sink}}. \tag{1.111}$$

The equilibrium is a $([2(n_{21}-n_{11})+1]^{\text{th}}$ UA:DA, $[(2m_{11}+1), (2n_{21}+1)]^{\text{th}}$ SI)-*up-to-down asymptote sink at* $(x^*_{j_1}, x^*_{j_2}) = (a_{j_1j_11}, a_{j_2j_21})$.

(i_{2b2c}) *For* $m_1 = 2m_{11}+1$, $n_1 = 2n_{11}+1$, *and* $n_2 = 2n_{21}$, *the equilibrium of* $(x^*_{j_1}, x^*_{j_2}) = (a_{j_1j_11}, a_{j_2j_21})$ *on the infinite-equilibrium of* $x^*_{j_2} = a_{j_2j_21} = a_{j_1j_21}$ *has the following properties.*

- *For* $a_{j_1j_10} > 0$ *and* $a_{j_2j_20} > 0$,

$$(a_{j_1j_11}, a_{j_2j_21}) = \underbrace{([2(n_{21} - n_{11}) - 1]^{\text{th}}\ \text{UA:DA}, [(2m_{11}+1), (2n_{21})]^{\text{th}}\ \text{US})}_{([2(n_{21}-n_{11})-1],[(2m_{11}+1),(2n_{21})])\text{-up-to-down asymptote upper-saddle}}. \tag{1.112}$$

The equilibrium is a $([2(n_{21} - n_{11}) - 1]^{\text{th}}$ UA:DA, $[(2m_{11} + 1), (2n_{21})]^{\text{th}}$ US)-*up-to-down asymptote upper-saddle at* $(x^*_{j_1}, x^*_{j_2}) = (a_{j_1j_11}, a_{j_2j_21})$.

- *For* $a_{j_1j_10} < 0$ *and* $a_{j_2j_20} > 0$,

$$(a_{j_1j_11}, a_{j_2j_21}) = \underbrace{([2(n_{21} - n_{11}) - 1]^{\text{th}}\ \text{DA:UA}, [(2m_{11}+1), (2n_{21})]^{\text{th}}\ \text{US})}_{([2(n_{21}-n_{11})-1],[(2m_{11}+1),(2n_{21})])\text{-down-to-up asymptote upper-saddle}}. \tag{1.113}$$

The equilibrium is a $([2(n_{21} - n_{11}) - 1]^{\text{th}}$ DA:UA, $[(2m_{11} + 1), (2n_{21})]^{\text{th}}$ US) *-down-to-up asymptote upper-saddle* at $(x^*_{j_1}, x^*_{j_2}) = (a_{j_1j_11}, a_{j_2j_21})$.

- *For* $a_{j_1j_10} > 0$ *and* $a_{j_2j_20} < 0$,

$$(a_{j_1j_11}, a_{j_2j_21}) = \underbrace{([2(n_{21} - n_{11}) - 1]^{\text{th}}\ \text{DA:UA}, [(2m_{11}+1), (2n_{21})]^{\text{th}}\ \text{LS})}_{([2(n_{21}-n_{11})-1],[(2m_{11}+1),(2n_{21})])\text{-down-to-up asymptote lower-saddle}}. \tag{1.114}$$

The equilibrium is a $([2(n_{21} - n_{11}) - 1]^{\text{th}}$ DA:UA, $[(2m_{11} + 1), (2n_{21})]^{\text{th}}$ LS) *-down-to-up asymptote lower-saddle at* $(x^*_{j_1}, x^*_{j_2}) = (a_{j_1j_11}, a_{j_2j_21})$.

- *For* $a_{j_1j_10} < 0$ *and* $a_{j_2j_20} < 0$,

$$(a_{j_1j_11}, a_{j_2j_21}) = \underbrace{([2(n_{21} - n_{11}) - 1]^{\text{th}}\ \text{UA:DA}, [(2m_{11}+1), (2n_{21})]^{\text{th}}\ \text{LS})}_{([2(n_{21}-n_{11})-1],[(2m_{11}+1),(2n_{21})])\text{-up-to-down asymptote lower-saddle}}. \tag{1.115}$$

The equilibrium is a $([2(n_{21} - n_{11}) - 1]^{\text{th}}$ UA:DA, $[(2m_{11} + 1), (2n_{21})]^{\text{th}}$ LS)*-up-to-down asymptote lower-saddle at* $(x^*_{j_1}, x^*_{j_2}) = (a_{j_1j_11}, a_{j_2j_21})$.

(i_{2b2d}) For $m_1 = 2m_{11} + 1$, $n_1 = 2n_{11}$, *and* $n_2 = 2n_{21}$, *the equilibrium of* $(x^*_{j_1}, x^*_{j_2}) = (a_{j_1j_11}, a_{j_2j_21})$ *on the infinite-equilibrium of* $x^*_{j_2} = a_{j_2j_21} = a_{j_1j_21}$ *has the following properties.*

- *For* $a_{j_1j_10} > 0$ *and* $a_{j_2j_20} > 0$,

$$(a_{j_1j_11}, a_{j_2j_21}) = \underbrace{([2(n_{21} - n_{11})]^{\text{th}}\ \text{DC:IC}, [(2m_{11} + 1), (2n_{21})]^{\text{th}}\ \text{US})}_{([2(n_{21}-n_{11})],[(2m_{11}+1),(2n_{21})])\text{-deceasing-to-increasing concave upper-saddle}}. \tag{1.116}$$

The equilibrium is a $([2(n_{21} - n_{11})]^{\text{th}}$ DC:IC, $[(2m_{11} + 1), (2n_{21})]^{\text{th}}$ US)*-decreasing-to-increasing-concave upper-saddle at* $(x^*_{j_1}, x^*_{j_2}) = (a_{j_1j_11}, a_{j_2j_21})$.

- *For* $a_{j_1j_10} < 0$ *and* $a_{j_2j_20} > 0$,

$$(a_{j_1j_11}, a_{j_2j_21}) = \underbrace{([2(n_{21} - n_{11})]^{\text{th}}\ \text{IC:DC}, [(2m_{11} + 1), (2n_{21})]^{\text{th}}\ \text{US})}_{([2(n_{21}-n_{11})],[(2m_{11}+1),(2n_{21})])\text{-inceasing-to-decreasing concave upper-saddle}}. \tag{1.117}$$

The equilibrium is a $([2(n_{21} - n_{11})]^{\text{th}}$ IC:DC, $[(2m_{11} + 1), (2n_{21})]^{\text{th}}$ US)*-increasing-to-decreasing-concave upper-saddle at* $(x^*_{j_1}, x^*_{j_2}) = (a_{j_1j_11}, a_{j_2j_21})$.

- *For* $a_{j_1j_10} > 0$ *and* $a_{j_2j_20} < 0$,

$$(a_{j_1j_11}, a_{j_2j_21}) = \underbrace{([2(n_{21} - n_{11})]^{\text{th}}\,\text{IC:DC}, [(2m_{11} + 1), (2n_{21})]^{\text{th}}\,\text{LS})}_{([2(n_{21}-n_{11})],[(2m_{11}+1),(2n_{21})])\text{-inceasing-to-dccreasing concave lower-saddle}}. \quad (1.118)$$

The equilibrium is a $([2(n_{21} - n_{11})]^{\text{th}}\,\text{IC:DC}, [(2m_{11} + 1), (2n_{21})]^{\text{th}}\,\text{LS})$*-increasing-to-decreasing-concave lower-saddle at* $(x^*_{j_1}, x^*_{j_2}) = (a_{j_1j_11}, a_{j_2j_21})$.

- *For* $a_{j_1j_10} < 0$ *and* $a_{j_2j_20} < 0$,

$$(a_{j_1j_11}, a_{j_2j_21}) = \underbrace{([2(n_{21} - n_{11})]^{\text{th}}\,\text{DC:IC}, [(2m_{11} + 1), (2n_{21})]^{\text{th}}\,\text{LS})}_{([2(n_{21}-n_{11})],[(2m_{11}+1),(2n_{21})])\text{-deceasing-to-increasing concave lower-saddle}}. \quad (1.119)$$

The equilibrium is a $([2(n_{21} - n_{11})]^{\text{th}}\,\text{DC:IC}, [(2m_{11} + 1), (2n_{21})]^{\text{th}}\,\text{LS})$*-decreasing-to-increasing-concave lower-saddle* at $(x^*_{j_1}, x^*_{j_2}) = (a_{j_1j_11}, a_{j_2j_21})$.

(i_{2b2e}) For $m_1 = 2m_{11}, n_1 = 2n_{11}+1$, *and* $n_2 = 2n_{21}+1$, *the equilibrium of* $(x^*_{j_1}, x^*_{j_2}) = (a_{j_1j_11}, a_{j_2j_21})$ *on the infinite-equilibrium of* $x^*_{j_2} = a_{j_2j_21} = a_{j_1j_21}$ *has the following properties.*

- *For* $a_{j_1j_10} > 0$ *and* $a_{j_2j_20} > 0$,

$$(a_{j_1j_11}, a_{j_2j_21}) = \underbrace{([2(n_{21} - n_{11})]^{\text{th}}\,\text{IC:IC}, [(2m_{11}), (2n_{21} + 1)]^{\text{th}}\,\text{SO})}_{([2(n_{21}-n_{11})],[(2m_{11}),(2n_{21}+1)])\text{-increasing concave source}}. \quad (1.120)$$

The equilibrium is a $([2(n_{21} - n_{11})]^{\text{th}}\,\text{IC:IC}, [(2m_{11}), (2n_{21} + 1)]^{\text{th}}\,\text{SO})$*-increasing-concave source at* $(x^*_{j_1}, x^*_{j_2}) = (a_{j_1j_11}, a_{j_2j_21})$.

- *For* $a_{j_1j_10} < 0$ *and* $a_{j_2j_20} > 0$,

$$(a_{j_1j_11}, a_{j_2j_21}) = \underbrace{([2(n_{21} - n_{11})]^{\text{th}}\,\text{DC:DC}, [(2m_{11}), (2n_{21} + 1)]^{\text{th}}\,\text{SO})}_{([2(n_{21}-n_{11})],[(2m_{11}),(2n_{21}+1)])\text{-decreasing concave source}}. \quad (1.121)$$

The equilibrium is a $([2(n_{21} - n_{11})]^{\text{th}}\,\text{DC:DC}, [(2m_{11}), (2n_{21} + 1)]^{\text{th}}\,\text{SO})$*-decreasing-concave source at* $(x^*_{j_2}, x^*_{j_2}) = (a_{j_1j_11}, a_{j_2j_21})$.

- *For* $a_{j_1j_10} > 0$ *and* $a_{j_2j_20} < 0$,

$$(a_{j_1j_11}, a_{j_2j_21}) = \underbrace{([2(n_{21} - n_{11})]^{\text{th}}\,\text{DC:DC}, [(2m_{11}), (2n_{21} + 1)]^{\text{th}}\,\text{SI})}_{([2(n_{21}-n_{11})],[(2m_{11}),(2n_{21}+1)])\text{-decreasing concave sink}}. \quad (1.122)$$

The equilibrium is a $([2(n_{21} - n_{11})]^{\text{th}}\,\text{DC:DC}, [(2m_{11}), (2n_{21} + 1)]^{\text{th}}\,\text{SI})$*-decreasing concave sink at* $(x^*_{j_1}, x^*_{j_2}) = (a_{j_1j_11}, a_{j_2j_21})$.

- *For* $a_{j_1j_10} < 0$ *and* $a_{j_2j_20} < 0$,

$$(a_{j_1j_11}, a_{j_2j_21}) = \underbrace{([2(n_{21} - n_{11})]^{\text{th}}\,\text{IC:IC}, [(2m_{11}), (2n_{21} + 1)]^{\text{th}}\,\text{SI})}_{([2(n_{21}-n_{11})],[(2m_{11}),(2n_{21}+1)])\text{-increasing concave sink}}. \quad (1.123)$$

The equilibrium is a $([2(n_{21} - n_{11})]^{\text{th}}\,\text{IC:IC}, [(2m_{11}), (2n_{21} + 1)]^{\text{th}}\,\text{SI})$-*increasing-concave sink at* $(x^*_{j_1}, x^*_{j_2}) = (a_{j_1j_11}, a_{j_2j_21})$.

(i_{2b2f}) *For* $m_1 = 2m_{11}$, $n_1 = 2n_{11}$, *and* $n_2 = 2n_{21} + 1$, *the equilibrium of* $(x^*_{j_2}, x^*_{j_2}) = (a_{j_1j_11}, a_{j_2j_21})$ *on the infinite-equilibrium of* $x^*_{j_2} = a_{j_2j_21} = a_{j_1j_21}$ *has the following properties.*

- *For* $a_{j_1j_10} > 0$ *and* $a_{j_2j_20} > 0$,

$$(a_{j_1j_11}, a_{j_2j_21}) = \underbrace{([2(n_{21} - n_{11}) + 1]^{\text{th}}\,\text{DA:DA}, [(2m_{11}), (2n_{21} + 1)]^{\text{th}}\,\text{SO})}_{([2(n_{21}-n_{11})+1],[(2m_{11}),(2n_{21}+1)])\text{-down-asymptote source}}. \quad (1.124)$$

The equilibrium is a $([2(n_{21} - n_{11}) + 1]^{\text{th}}\,\text{DA:DA}, [(2m_{11}), (2n_{21} + 1)]^{\text{th}}\,\text{SO})$-*down asymptote source at* $(x^*_{j_1}, x^*_{j_2}) = (a_{j_1j_11}, a_{j_2j_21})$.

- *For* $a_{j_1j_10} < 0$ *and* $a_{j_2j_20} > 0$,

$$(a_{j_1j_11}, a_{j_2j_21}) = \underbrace{([2(n_{21} - n_{11}) + 1]^{\text{th}}\,\text{UA:UA}, [(2m_{11}), (2n_{21} + 1)]^{\text{th}}\,\text{SO})}_{([2(n_{21}-n_{11})+1],[(2m_{11}),(2n_{21}+1)])\text{-up asymptote source}}. \quad (1.125)$$

The equilibrium is a $([2(n_{21} - n_{11}) + 1]^{\text{th}}\,\text{UA:UA}, [(2m_{11}), (2n_{21} + 1)]^{\text{th}}\,\text{SO})$ *-up-asymptote source at* $(x^*_{j_1}, x^*_{j_2}) = (a_{j_1j_11}, a_{j_2j_21})$.

- *For* $a_{j_1j_10} > 0$ *and* $a_{j_2j_20} < 0$,

$$(a_{j_1j_11}, a_{j_2j_21}) = \underbrace{([2(n_{21} - n_{11}) + 1]^{\text{th}}\,\text{UA:UA}, [(2m_{11}), (2n_{21} + 1)]^{\text{th}}\,\text{SI})}_{([2(n_{21}-n_{11})+1],[(2m_{11}),(2n_{21}+1)])\text{-up asymptote sink}}. \quad (1.126)$$

The equilibrium is a $([2(n_{21} - n_{11}) + 1]^{\text{th}}\,\text{UA:UA}, [(2m_{11}), (2n_{21} + 1)]^{\text{th}}\,\text{SI})$-*up asymptote sink at* $(x^*_{j_1}, x^*_{j_2}) = (a_{j_1j_11}, a_{j_2j_21})$.

- *For* $a_{j_1j_10} < 0$ *and* $a_{j_2j_20} < 0$,

$$(a_{j_1j_11}, a_{j_2j_21}) = \underbrace{([2(n_{21} - n_{11}) + 1]^{\text{th}}\,\text{DA:DA}, [(2m_{11}), (2n_{21} + 1)]^{\text{th}}\,\text{SI})}_{([2(n_{21}-n_{11})+1],[(2m_{11}),(2n_{21}+1)])\text{-down asymptote sink}}. \quad (1.127)$$

The equilibrium is a $([2(n_{21}-n_{11})+1]^{\text{th}}$ DA:DA, $[(2m_{11}),(2n_{21}+1)]^{\text{th}}$ SI) *-down-asymptote sink at* $(x_{j_1}^*, x_{j_2}^*) = (a_{j_1j_11}, a_{j_2j_21})$.

(i_{2b2g}) For $m_1 = 2m_{11}$, $n_1 = 2n_{11}+1$, *and* $n_2 = 2n_{21}$, *the equilibrium of* $(x_{j_1}^*, x_{j_2}^*) = (a_{j_1j_11}, a_{j_2j_21})$ *on the infinite-equilibrium of* $x_{j_2}^* = a_{j_2j_21} = a_{j_1j_21}$ *has the following properties.*

- *For* $a_{j_1j_10} > 0$ *and* $a_{j_2j_20} > 0$,

$$(a_{j_1j_11}, a_{j_2j_21}) = \underbrace{([2(n_{21}-n_{11})-1]^{\text{th}}\text{ DA:DA}, [(2m_{11}),(2n_{21})]^{\text{th}}\text{ US})}_{([2(n_{21}-n_{11})-1],[(2m_{11}),(2n_{21})])\text{-down asymptote upper-saddle}}. \quad (1.128)$$

The equilibrium is a $([2(n_{21}-n_{11})-1]^{\text{th}}$ DA:DA, $[(2m_{11}),(2n_{21})]^{\text{th}}$ US)*-down asymptote upper-saddle at* $(x_{j_1}^*, x_{j_2}^*) = (a_{j_1j_11}, a_{j_2j_21})$.

- *For* $a_{j_1j_10} < 0$ *and* $a_{j_2j_20} > 0$,

$$(a_{j_1j_11}, a_{j_2j_21}) = \underbrace{([2(n_{21}-n_{11})-1]^{\text{th}}\text{ UA:UA}, [(2m_{11}),(2n_{21})]^{\text{th}}\text{ US})}_{([2(n_{21}-n_{11})-1],[(2m_{11}),(2n_{21})])\text{-up asymptote upper-saddle}}. \quad (1.129)$$

The equilibrium is a $([2(n_{21}-n_{11})-1]^{\text{th}}$ UA:UA, $[(2m_{11}),(2n_{21})]^{\text{th}}$ US)*-up asymptote upper-saddle at* $(x_{j_1}^*, x_{j_2}^*) = (a_{j_1j_11}, a_{j_2j_21})$.

- *For* $a_{j_1j_10} > 0$ *and* $a_{j_2j_20} < 0$,

$$(a_{j_1j_11}, a_{j_2j_21}) = \underbrace{([2(n_{21}-n_{11})-1]^{\text{th}}\text{ UA:UA}, [(2m_{11}),(2n_{21})]^{\text{th}}\text{ LS})}_{([2(n_{21}-n_{11})-1],[(2m_{11}),(2n_{21})])\text{-up-asymptote lower-saddle}}. \quad (1.130)$$

The equilibrium is a $([2(n_{21}-n_{11})-1]^{\text{th}}$ UA:UA, $[(2m_{11}),(2n_{21})]^{\text{th}}$ LS)*-up asymptote lower-saddle at* $(x_{j_1}^*, x_{j_2}^*) = (a_{j_1j_11}, a_{j_2j_21})$.

- *For* $a_{j_1j_10} < 0$ *and* $a_{j_2j_20} < 0$,

$$(a_{j_1j_11}, a_{j_2j_21}) = \underbrace{([2(n_{21}-n_{11})-1]^{\text{th}}\text{ DA:DA}, [(2m_{11}),(2n_{21})]^{\text{th}}\text{ LS})}_{([2(n_{21}-n_{11})-1],[(2m_{11}),(2n_{21})])\text{-down asymptote lower-saddle}}. \quad (1.131)$$

The equilibrium is a $([2(n_{21}-n_{11})-1]^{\text{th}}$ DA:DA, $[(2m_{11}),(2n_{21})]^{\text{th}}$ LS)*-down-asymptote lower-saddle at* $(x_{j_1}^*, x_{j_2}^*) = (a_{j_1j_11}, a_{j_2j_21})$.

(i_{2b2h}) For $m_1 = 2m_{11}$, $n_1 = 2n_{11}$, *and* $n_2 = 2n_{21}$, *the equilibrium of* $(x_{j_1}^*, x_{j_2}^*) = (a_{j_1j_11}, a_{j_2j_21})$ *on the infinite-equilibrium of* $x_{j_2}^* = a_{j_2j_21} = a_{j_1j_21}$ *has the following properties.*

- *For* $a_{j_1j_10} > 0$ *and* $a_{j_2j_20} > 0$,

$$(a_{j_1j_11}, a_{j_2j_21}) = \underbrace{([2(n_{21} - n_{11})]^{\text{th}}\,\text{IC:IC}, [(2m_{11}), (2n_{21})]^{\text{th}}\,\text{US})}_{([2(n_{21}-n_{11})],[(2m_{11}),(2n_{21})])\text{-increasing concave upper-saddle}}. \tag{1.132}$$

The equilibrium is a ($[2(n_{21} - n_{11})]^{\text{th}}$ IC:IC, $[(2m_{11}), (2n_{21})]^{\text{th}}$ US)*-increasing-concave upper-saddle at* $(x^*_{j_1}, x^*_{j_2}) = (a_{j_1j_11}, a_{j_2j_21})$.

- *For* $a_{j_1j_10} < 0$ *and* $a_{j_2j_20} > 0$,

$$(a_{j_1j_11}, a_{j_2j_21}) = \underbrace{([2(n_{21} - n_{11})]^{\text{th}}\,\text{DC:DC}, [(2m_{11}), (2n_{21})]^{\text{th}}\,\text{US})}_{([2(n_{21}-n_{11})],[(2m_{11}),(2n_{21})])\text{-decreasing concave upper-saddle}}. \tag{1.133}$$

The equilibrium is a ($[2(n_{21} - n_{11})]^{\text{th}}$ DC:DC, $[(2m_{11}), (2n_{21})]^{\text{th}}$ US)*-decreasing-concave upper-saddle at* $(x^*_{j_1}, x^*_{j_2}) = (a_{j_1j_11}, a_{j_2j_21})$.

- *For* $a_{j_1j_10} > 0$ *and* $a_{j_2j_20} < 0$,

$$(a_{j_1j_11}, a_{j_2j_21}) = \underbrace{([2(n_{21} - n_{11})]^{\text{th}}\,\text{DC:DC}, [(2m_{11}), (2n_{21})]^{\text{th}}\,\text{LS})}_{([2(n_{21}-n_{11})],[(2m_{11}),(2n_{21})])\text{-dccreasing concave lower-saddle}}. \tag{1.134}$$

The equilibrium is a ($[2(n_{21} - n_{11})]^{\text{th}}$ DC:DC, $[(2m_{11}), (2n_{21})]^{\text{th}}$ LS)*-decreasing-concave lower-saddle at* $(x^*_{j_1}, x^*_{j_2}) = (a_{j_1j_11}, a_{j_2j_21})$.

- *For* $a_{j_1j_10} < 0$ *and* $a_{j_2j_20} < 0$,

$$(a_{j_1j_11}, a_{j_2j_21}) = \underbrace{([2(n_{21} - n_{11})]^{\text{th}}\,\text{IC:IC}, [(2m_{11}), (2n_{21})]^{\text{th}}\,\text{LS})}_{([2(n_{21}-n_{11})],[(2m_{11}),(2n_{21})])\text{-increasing concave lower-saddle}}. \tag{1.135}$$

The equilibrium is a ($[2(n_{21} - n_{11})]^{\text{th}}$ IC:IC, $[(2m_{11}), (2n_{21})]^{\text{th}}$ LS)*-increasing-concave lower-saddle at* $(x^*_{j_1}, x^*_{j_2}) = (a_{j_1j_11}, a_{j_2j_21})$.

(ii) Consider

$$\begin{aligned}\dot{x}_{j_2} &= a_{j_2j_20}\prod_{s=1}^{q_2}(x_{j_2} - a_{j_2j_2s})^{n_{j_2s}}\\ \dot{x}_{j_1} &= a_{j_1j_10}\prod_{r=1}^{p_1}(x_{j_1} - a_{j_1j_1r})^{m_{j_1r}}\prod_{u=1}^{q_1}(x_{j_2} - a_{j_1j_2u})^{n_{j_1u}},\end{aligned} \tag{1.136}$$

where

$$\sum_{r=1}^{p_1} m_{j_1r} = m_1, \sum_{u=1}^{q_1} n_{j_1u} = n_1; \sum_{s=1}^{q_2} n_{j_2s} = n_2. \tag{1.137}$$

(ii_1) For $a_{j_1j_2u} \neq a_{j_2j_2s}$ ($u = 1, 2, \ldots, q_1$; $s = 1, 2, \ldots, q_2$), the first integral manifold is

$$
\begin{aligned}
&\sum_{r_1=1}^{p_1}\sum_{l_1=1}^{m_{j_1r_1}} \frac{A_{l_1m_{j_1r_1}}}{l_1-1}\left[\frac{1}{(x_{j_1}-a_{j_1j_1r_1})^{l_1-1}} - \frac{1}{(x_{j_10}-a_{j_1j_1r_1})^{l_1-1}}\right] \\
&= \frac{a_{j_1j_10}}{a_{j_2j_20}}\Bigg\{\sum_{\beta_0=n_2}^{n_1} b_{j_1j_2\beta_0} \sum_{\substack{\alpha_1+\beta_1=\beta_0-n_{j_2s_1}\\ \alpha_1\in[0,\beta_0-n_2],\\ \beta_1\in[n_2-n_{j_2s_1},\beta_0-n_{j_2s_1}]}} \sum_{\substack{\alpha_2+\beta_2=\beta_1-\Sigma_{i=1}^2 n_{j_2s_i}\\ \alpha_2\in[0,\beta_1-n_2]\\ \beta_2\in[n_2-\Sigma_{i=1}^2 n_{j_2s_i},\beta_1-\Sigma_{i=1}^2 n_{j_2s_i}]}} \cdots \\
&\sum_{\substack{\alpha_{q_2-1}+\beta_{q_2-1}=\beta_{q_2-2}-\Sigma_{i=1}^{q_2-1} n_{j_2s_i}\\ \alpha_{q_2-1}\in[0,\beta_{q_2-2}-n_2]\\ \beta_{q_2-1}\in[n_2-\Sigma_{i=1}^{q_2-1} n_{j_2s_i},\beta_{p-2}-\Sigma_{i=1}^{q_2-1} n_{j_2s_i}]}} \prod_{j=1}^{q_2-1} \frac{(\beta_{j-1}-\Sigma_{i=1}^{j} n_{j_2s_i})!}{\alpha_j!\beta_j!}(a_{j_2j_2s_{j+1}}-a_{j_2j_2s_j})^{\alpha_j} \\
&\times \frac{1}{\beta_{q_2-1}-n_{j_2s_{q_2}}+1}\left[(x_{j_2}-a_{j_2j_2s_\rho})^{\beta_{q_2-1}-n_{j_2s_{q_2}}+1} - (x_{j_20}-a_{j_2j_2s_\rho})^{\beta_{q_2-1}-n_{j_2s_{q_2}}+1}\right] \\
&- \sum_{\beta_0=n_2}^{n_1} b_{j_1j_2\beta_0} \sum_{k=2}^{q_2-1}\Bigg[\sum_{\substack{\alpha_1+\beta_1=\beta_0-n_{j_2s_1}\\ \alpha_1\in[0,\beta_0-n_2),\\ \beta_1\in(n_2-n_{j_2s_1},\beta_0-n_{j_2s_1}]}} \sum_{\substack{\alpha_2+\beta_2=\beta_1-\Sigma_{i=1}^2 n_{j_2s_i}\\ \alpha_2\in[0,\beta_1-n_2]\\ \beta_2\in(n_2-\Sigma_{i=1}^2 n_{j_2s_i},\beta_1-\Sigma_{i=1}^2 n_{j_2s_i})}} \cdots \\
&\times\Bigg(\sum_{\substack{\alpha_{k-1}+\beta_{k-1}=\beta_{k-2}-\Sigma_{i=1}^{k-1} n_{j_2s_i}\\ \alpha_{k-1}\in(\beta_{k-2}-n_2+n_j2s_k,\beta_{k-2}-\Sigma_{i=1}^{k-1} n_{j2si}]\\ \beta_{k-1}\in[0,n_2-\Sigma_{i=1}^{k} n_{j_2s_i})}} \prod_{j=1}^{k-1} \frac{(\beta_{j-1}-\Sigma_{i=1}^{j} n_{j_2s_i})!}{\alpha_j!\beta_j!}(a_{j_2j_2s_{j+1}}-a_{j_2j_2s_j})^{\alpha_j}\quad, \\
&\times \sum_{\rho=k}^{q_2}\sum_{l_\rho=1}^{\bar{n}_{j_2s_\rho}} \frac{A^{(s_k)}_{l_\rho\bar{n}_{j_2s_\rho}}}{l_\rho-1}\left[\frac{1}{(x_{j_2}-a_{j_2j_2s_\rho})^{l_\rho-1}} - \frac{1}{(x_{j_20}-a_{j_2j_2s_\rho})^{l_\rho-1}}\right] \\
&+ \sum_{\substack{\alpha_{k-1}+\beta_{k-1}=\beta_{k-2}-\Sigma_{i=1}^{k-1} n_{j2si}\\ \alpha_{k-1}\in(\beta_{k-2}-n_2,\beta_{k-2}-n_2+n_{j_2s_k}]\\ \beta_{k-1}\in[n_2-\Sigma_{i=1}^{k} n_{j_2s_i},n_2-\Sigma_{i=1}^{k-1} n_{j2si})}} \prod_{j=1}^{k-1} \frac{(\beta_{j-1}-\Sigma_{i=1}^{j} n_{j_2s_i})!}{\alpha_j!\beta_j!}(a_{j_2j_2s_{j+1}}-a_{j_2j_2s_j})^{\alpha_j} \\
&\times \sum_{\rho=k+1}^{q_2}\sum_{l_\rho=1}^{\bar{n}_{j_2s_\rho}} \frac{B^{(s_k)}_{l_\rho\bar{n}_{j_2s_\rho}}}{l_\rho-1}\left[\frac{1}{(x_{j_2}-a_{j_2j_2s_\rho})^{l_\rho-1}} - \frac{1}{(x_{j_20}-a_{j_2j_2s_\rho})^{l_\rho-1}}\right]\Bigg)\Bigg] \\
&-\Bigg(\sum_{\beta_0=0}^{n_{j_2s_1}-1} b_{j_1j_2\beta_0}\sum_{\rho=1}^{q_2}\sum_{l_\rho=1}^{\bar{n}_{j_2s_\rho}} \frac{A^{(s_1)}_{l_\rho\bar{n}_{j_2s_\rho}}}{l_\rho-1}\left[\frac{1}{(x_{j_2}-a_{j_2j_2s_\rho})^{l_\rho-1}} - \frac{1}{(x_{j_20}-a_{j_2j_2s_\rho})^{l_\rho-1}}\right] \\
&+ \sum_{\beta_0=n_{j_2s_1}}^{n_2-1} b_{j_1j_2\beta_0}\sum_{\rho=2}^{q_2}\sum_{l_\rho=1}^{\bar{n}_{j_2s_\rho}} \frac{B^{(s_1)}_{l_\rho\bar{n}_{j_2s_\rho}}}{l_\rho-1}\left[\frac{1}{(x_{j_2}-a_{j_2j_2s_\rho})^{l_\rho-1}} - \frac{1}{(x_{j_20}-a_{j_2j_2s_\rho})^{l_\rho-1}}\right]\Bigg)\Bigg\},
\end{aligned}
\tag{1.138}
$$

where

$$
\begin{aligned}
A_{l_1 m_{j_1 r_1}} &= \frac{1}{(m_{j_1 r_1}-l_1)!}\frac{d^{m_{j_1 r_1}-l_1}}{dx_{j_1}^{m_{j_1 r_1}-l_1}}\frac{1}{\prod_{\sigma=2}^{p_1}\left(x_{j_1}-a_{j_1 j_1 r_\sigma}\right)^{m_{j_1 r\sigma}}}\Big|_{x_{j_1}^*=a_{j_1 j_1 r_1}},\\
A_{m_{j_1 r_1} m_{j_1 r_1}} &= \frac{1}{\prod_{\sigma=2}^{p_1}\left(a_{j_1 j_1 r_1}-a_{j_1 j_1 r\sigma}\right)^{m_{j_1 r\sigma}}};\\
A^{(s_k)}_{l_\rho \overline{n}_{j_2 s_\rho}} &= \frac{1}{(\overline{n}_{j_2 s_\rho}-l_\rho)!}\frac{d^{\overline{n}_{j_2 s_\rho}-l_\rho}}{dx_{j_2}^{\overline{n}_{j_2 s_\rho}-l_\rho}}\frac{1}{\prod_{\sigma=k,\sigma\neq\rho}^{q_2}\left(x_{j2}-a_{j2j2s\sigma}\right)^{n_{j2s\sigma}}}\Big|_{x_{j_2}^*=a_{j_2 j_2 s_\rho}}\\
&(l_\rho = 1,2,\ldots,\overline{n}_{j2sp}, = k, k+1,\ldots,q_2),\\
A^{(s_k)}_{\overline{n}_{j_2 s_\rho}\overline{n}_{j_2 s_\rho}} &= \frac{1}{\prod_{\sigma=k,\sigma\neq\rho}^{q_2}(a_{j_2 j_2 s_\rho}-a_{j_s j_s s\sigma})^{n_{j2js\sigma}}},\\
B^{(s_k)}_{l_\rho \overline{n}_{j_2 s_\rho}} &= \frac{1}{(\overline{n}_{j_2 s_\rho}-l_\rho)!}\frac{d^{\overline{n}_{j_2 s_\rho}-l_\rho}}{dx_{j_2}^{\overline{n}_{j_2 s_\rho}-l_\rho}}\frac{(x_{j_2}-a_{j_2 j_2 s_k})^{\beta_{k-1}-\sum_{i=1}^{k} n_{j_2 s_i}}}{\prod_{\sigma=k+1,\sigma\neq\rho}^{q_2}(x_{j_2}-a_{j_2 j_2 s_\sigma})^{n_{j2s\sigma}}}\Big|_{x_{j_2}^*=a_{j_2 j_2 s_\rho}}\\
&(l_\rho = 1,2,\ldots,\overline{n}_{j_2 s_\rho}, \rho = k+1,\ldots,q_2),\\
B^{(s_k)}_{\overline{n}_{j_2 s_\rho}\overline{n}_{j_2 s_\rho}} &= \frac{(a_{j2j2sp}-a_{j2j2sk})^{\beta_{k-1}-\sum_{i=1}^{k} n_{j2si}}}{\prod_{\sigma=k+1,\sigma\neq\rho}^{q_2}(a_{j_2 j_2 s_p}-aj_s j_s s_\sigma)^{nj_2 s_\sigma}},\\
&\{\overline{n}_{j_2 s_\rho}|\rho = k,k+1,\ldots,q_2\} = \{\sum_{i=1}^{k} n_{j_2 s_i}-\beta_{k-1}, n_{j_2 s_{k+1}},\ldots,n_{j_2 s_{q2}}\},\\
&\text{for } k\in\{1,2,\ldots,q_2-1\};
\end{aligned}
\tag{1.139}
$$

and

$$
\begin{aligned}
b_{j_1 j_2 \beta_0} &= \sum_{t_1,l_1=0}^{t_1+l_1=n_{j_1 u_1}}\cdots\sum_{t_{q_1},l_{q_1}=0}^{t_{q_1}+l_{q_1}=n_{j_1 u_{q_1}}}\prod_{k=1}^{q_1}\frac{n_{j_1 u_k}!}{t_k!l_k!}(a_{j_2 j_2 s_1}-a_{j_1 j_2 u_k})^{t_k}\delta_{\beta_0}^{\Sigma_{k=1}^{q_1} l_k},\\
&\frac{1}{\beta_{q_2-1}-n_{j_2 q_2}+1}\left[(x_{j_2}-a_{j_2 j_2 s_{q_2}})^{\beta_{q_2-1}-n_{j_2 q_2}+1}-(x_{j_2 0}-a_{j_2 j_2 s_{q_2}})^{\beta_{q_2-1}-n_{j_2 q_2}+1}\right]\\
&= \ln\frac{|x_{j_2}-a_{j_2 j_2 s_{q_2}}|}{|x_{j_2 0}-a_{j_2 j_2 s_{q_2}}|},\ \text{for } \beta_{q_2-1}-n_{j_2 q_2}+1=0;\\
&\frac{1}{l_k-1}\left[\frac{1}{(x_{j_2}-a_{j_2 j_2 s_k})^{l_k-1}}-\frac{1}{(x_{j_2 0}-a_{j_2 j_2 s_k})^{l_k-1}}\right] = -\ln\frac{|x_{j_2}-a_{j_2 j_2 s_k}|}{|x_{j_2 0}-a_{j_2 j_2 s_k}|},\\
&\text{for } l_k = 1\ (k=1,2,\ldots,q_2-1);\\
&\frac{1}{l_1-1}\left[\frac{1}{(x_{j_1}-a_{j_1 j_1 r_1})^{l_1-1}}-\frac{1}{(x_{j_1 0}-a_{j_1 j_1 r_1})^{l_1-1}}\right] = -\ln\frac{|x_{j_1}-a_{j_1 j_1 r_1}|}{|x_{j_1 0}-a_{j_1 j_1 r_1}|},\\
&\text{for } l_1 = 1.
\end{aligned}
\tag{1.140}
$$

The network matrices of singular equilibriums and 1-dimensional flows are defined as

$$\begin{aligned}
&\bigcup_{\eta=j_1}^{j_2}\bigcup_{r=1}^{p_1}\bigcup_{l=1}^{q_1+q_2}(a_{j_1j_1r}, c_{\eta j_2 l})\\
&=\bigcup_{r=1}^{p_1}\bigcup_{u=1}^{q_1}\underbrace{(a_{j_1j_1r}, a_{j_1j_2u})}_{([m_{j_1r},n_{j_1u}]\text{XX,xF})}\oplus\bigcup_{r=1}^{p_1}\bigcup_{s=1}^{q_2}\underbrace{(a_{j_1j_1r}, a_{j_2j_2s})}_{(m_{j_1r},n_{j_2s})\text{-XX}},\\
&\{c_{\eta j_2 l}|\eta\in\{j_1,j_2\}, l=1,2,\ldots,q_1+q_2\}\\
&=\text{sort}\{a_{j_1j_2u}, a_{j_2j_2s}|u=1,2,\ldots q_1; s=1,2,\ldots,q_2\},\\
&\text{for } c_{\eta_1 j_2 l_1}<c_{\eta_2 j_2 l_2} \text{ for } l_1<l_2,\ \eta_1,\eta_2\in\{j_1,j_2\};
\end{aligned}\tag{1.141}$$

where

$$\begin{aligned}
&\bigcup_{r=1}^{p_1}\bigcup_{u=1}^{q_1}\underbrace{(a_{j_1j_1r}, a_{j_1j_2u})}_{([m_{j_1r},n_{j_1u}]\text{XX,xF})}\\
&=\begin{bmatrix}
\underbrace{(a_{j_1j_1 1}, a_{j_1j_2q_1})}_{([m_{j_11},n_{j_1q_1}]\text{XX,xF})} & \cdots & \underbrace{(a_{j_1j_1(p_1-1)}, a_{j_1j_2q_1})}_{([m_{1(p_1-1)},n_{j_1q_1}]\text{XX,xF})} & \underbrace{(a_{j_1j_1p_1}, a_{j_1j_2q_1})}_{([m_{j_1p_1},n_{j_1q_1}]\text{XX,xF})}\\
\underbrace{(a_{j_1j_1 1}, a_{j_1j_2(q_1-1)})}_{([m_{j_11},n_{j_1(q_1-1)}]\text{XX,xF})} & \cdots & \underbrace{(a_{j_1j_1(p_1-1)}, a_{j_1j_2(q_1-1})}_{([m_{j_1(p_1-1)},n_{j_1(q_1-1)}]\text{XX,xF})} & \underbrace{(a_{j_1j_1p_1}, a_{j_1j_2(q_1-1)})}_{([m_{j_1p_1},n_{j_1(q_1-1)}]\text{XX,xF})}\\
\vdots & \vdots & \vdots & \vdots\\
\underbrace{(a_{j_1j_1 1}, a_{j_1j_2 1})}_{([m_{j_11},n_{j_11}]\text{XX,xF})} & \cdots & \underbrace{(a_{j_1j_1(p_1-1)}, a_{j_1j_2 1})}_{([m_{j_1(p_1-1)},n_{j_11}]\text{XX,xF})} & \underbrace{(a_{j_1j_1p_1}, a_{j_1j_2 1})}_{([m_{j_1p_1},n_{j_11}]\text{XX,xF})}
\end{bmatrix}_{q_1\times p_1},
\end{aligned}\tag{1.142}$$

and

$$\begin{aligned}
&\bigcup_{r=1}^{p_1}\bigcup_{s=1}^{q_2}\underbrace{(a_{j_1j_1r}, a_{j_2j_2s})}_{(m_{j_2r},n_{j_2s})\text{-XX}}\\
&=\begin{bmatrix}
\underbrace{(a_{j_1j_1 1}, a_{j_2j_2q_2})}_{(m_{j_11},n_{j_2q_1})\text{-YY}} & \cdots & \underbrace{(a_{j_1j_1(p_1-1)}, a_{j_2j_2q_2})}_{(m_{j_1(p_1-1)},n_{(2)q_1})\text{-YY}} & \underbrace{(a_{j_1j_1p_1}, a_{j_2j_2q_2})}_{(m_{j_1p_1},n_{j_2q_2})\text{-YY}}\\
\underbrace{(a_{j_1j_1 1}, a_{j_2j_2(q_1-1)})}_{(m_{j_11},n_{j_2(q_1-1)})\text{-YY}} & \cdots & \underbrace{(a_{j_1j_1(p_1-1)}, a_{j_2j_2(q_2-1)})}_{(m_{j_1(p_1-1)},n_{q_1-1})\text{-YY}} & \underbrace{(a_{j_1j_1p_1}, a_{j_2j_2(q_2-1)})}_{([m_{j_1p_1},n_{j_2(q_2-1)})\text{-YY}}\\
\vdots & \vdots & \vdots & \vdots\\
\underbrace{(a_{j_1j_1 1}, a_{j_2j_2 1})}_{(m_{j_11},n_{j_21})\text{-YY}} & \cdots & \underbrace{(a_{j_1j_1(p_1-1)}, a_{j_2j_2 1})}_{(m_{j_1(p_1-1)},n_{21})\text{-YY}} & \underbrace{(a_{j_1j_1p_1}, a_{j_2j_2 1})}_{(m_{j_1p_1},n_{j_21})\text{-YY}}
\end{bmatrix}_{q_2\times p_1}.
\end{aligned}\tag{1.143}$$

(ii_{1a}) There is an equilibrium for 1-dimensional flow as

$$(x_{j_1}^*, x_{j_2}^*) = (a_{j_1j_1r_1}, a_{j_1j_2u_1}). \tag{1.144}$$

(ii_{1a1}) For $m_{j_1r_1} = 2m_{j_1r_1}1 + 1$ and $n_{j_1u_1} = 2n_{j_1u_1}1 + 1$, the equilibrium of $(x_{j_1}^*, x_{j_2}^*) = (a_{j_1j_1r_1}, a_{j_1j_2u_1})$ has the following properties.

- *For $a_{j_1j_10}\prod_{r_2=1,r_2\neq r_1}^{p_1}(a_{j_1j_1r_1} - a_{j_1j_1r_2})^{m_{j_1r_2}}\prod_{u_2=1,u_2\neq u_1}^{q_1}(a_{j_1j_2u_1} - a_{j_1j_2u_2})^{n_{j_1u_2}} > 0$ and $a_{j_2j_20}\prod_{s=1}^{q_2}(a_{j_1j_2u_1} - a_{j_2j_2s})^{n_{j_2s}} > 0$,*

$$(a_{j_1j_1r_1}, a_{j_1j_2u_1}) = \underbrace{([(2m_{j_1r_1}1{+}1), (2n_{j_1u_1}1 + 1)]\text{DP:UP, pF})}_{[(2m_{j_1r_1}1+1),(2n_{j_1u_1}1+1)]\text{-hyperbolic flow }(+)} \tag{1.145}$$

  The flow at $(x_{j_1}^*, x_{j_2}^*){=}(a_{j_1j_1r_1}, a_{j_1j_2u_1})$ is a ([$(2m_{j_1r_1}1{+}1)$, $(2n_{j_1u_1}1{+}1)$]DP:UP, pF)-positive hyperbolic flow.
- *For $a_{j_1j_10}\prod_{r_2=1,r_2\neq r_1}^{p_1}(a_{j_1j_1r_1} - a_{j_1j_1r_2})^{m_{j_1r_2}}\prod_{u_2=1,u_2\neq u_1}^{q_1}(a_{j_1j_2u_1} - a_{j_1j_2u_2})^{n_{j_1u_2}} < 0$ and $a_{j_2j_20}\prod_{s=1}^{q_2}(a_{j_1j_2u_1} - a_{j_2j_2s})^{n_{j_2s}} > 0$,*

$$(a_{j_1j_1r_1}, a_{j_1j_2u_1}) = \underbrace{([(2m_{j_1r_1}1{+}1), (2n_{j_1u_1}1 + 1)]\text{UP:DP, pF})}_{[(2m_{j_1r_1}1+1),(2n_{j_1u_1}1+1)]\text{-hyperbolic-secant flow }(+)} \tag{1.146}$$

  The flow at $(x_{j_1}^*, x_{j_2}^*){=}(a_{j_1j_1r_1}, a_{j_1j_2u_1})$ is a ([$(2m_{j_1r_1}1{+}1)$, $(2n_{j_1u_1}1{+}1)$]UP:DP, pF)-positive hyperbolic-secant flow.
- *For $a_{j_1j_10}\prod_{r_2=1,r_2\neq r_1}^{p_1}(a_{j_1j_1r_1} - a_{j_1j_1r_2})^{m_{j_1r_2}}\prod_{u_2=1,u_2\neq u_1}^{q_1}(a_{j_1j_2u_1} - a_{j_1j_2u_2})^{n_{j_1u_2}} > 0$ and $a_{j_2j_20}\prod_{s=1}^{q_2}(a_{j_1j_2u_1} - a_{j_2j_2s})^{n_{j_2s}} < 0$,*

$$(a_{j_1j_1r_1}, a_{j_1j_2u_1}) = \underbrace{([(2m_{j_1r_1}1{+}1), (2n_{j_1u_1}1 + 1)]\text{UP:DP, nF})}_{[(2m_{j_1r_1}1+1),(2n_{j_1u_1}1+1)]\text{-hyperbolic-secant flow }(-)} \tag{1.147}$$

  The flow at $(x_{j_1}^*,x_{j_2}^*) = (a_{j_1j_1r_1},a_{j_1j_2u_1})$ is a ([$(2m_{j_1r_1}1{+}1),(2n_{j_1u_1}1{+}1)$]UP:DP,nF)-*negative hyperbolic-secant flow.*
- *For $a_{j_1j_10}\prod_{r_2=1,r_2\neq r_1}^{p_1}(a_{j_1j_1r_1} - a_{j_1j_1r_2})^{m_{j_1r_2}}\prod_{u_2=1,u_2\neq u_1}^{q_1}(a_{j_1j_2u_1} - a_{j_1j_2u_2})^{n_{j_1u_2}} < 0$ and $a_{j_2j_20}\prod_{s=1}^{q_2}(a_{j_1j_2u_1} - a_{j_2j_2s})^{n_{j_2s}} < 0$,*

$$(a_{j_1j_1r_1}, a_{j_1j_2u_1}) = \underbrace{([(2m_{j_1r_1}1{+}1), (2n_{j_1u_1}1 + 1)]\text{DP:UP, nF})}_{[(2m_{j_1r_1}1+1),(2n_{j_1u_1}1+1)]\text{-hyperbolic flow }(-)} \tag{1.148}$$

  The flow at $(x_{j_1}^*,x_{j_2}^*) = (a_{j_1j_1r_1},a_{j_1j_2u_1})$ is a ([$(2m_{j_1r_1}1{+}1),(2n_{j_1u_1}1{+}1)$]DP:UP,pF)-*negative hyperbolic flow.*

(ii_{1a2}) For $m_{j_1r_1} = 2m_{j_1r_1}1$ and $n_{j_1u_1} = 2n_{j_1u_1}1 + 1$, the equilibrium of $(x_{j_1}^*, x_{j_2}^*) = (a_{j_1j_1r_1}, a_{j_1j_2u_1})$ for the x_{j_1}-direction has the following properties.

- *For* $a_{j_1j_10}\prod_{r_2=1,r_2\neq r_1}^{p_1}(a_{j_1j_1r_1}-a_{j_1j_1r_2})^{m_{j_1r_2}}\prod_{u_2=1,u_2\neq u_1}^{q_1}(a_{j_1j_2u_1}-a_{j_1j_2u_2})^{n_{j_1u_2}}>0$ *and* $a_{j_2j_20}\prod_{s=1}^{q_2}(a_{j_1j_2u_1}-a_{j_2j_2s})^{n_{j_2s}}>0$,

$$(a_{j_1j_1r_1},a_{j_1j_2u_1})=\underbrace{([(2m_{j_1r_1}1),(2n_{j_1u_1}1+1)]\text{UP:UP, pF})}_{[(2m_{j_1r_1}1),(2n_{j_1u_1}1+1)]\text{-hyperbolic-secant-to-hyperbolic flow }(+)}. \tag{1.149}$$

The flow at $(x_{j_1}^*,x_{j_2}^*)=(a_{j_1j_1r_1},a_{j_1j_2u_1})$ *is a* $([(2m_{j_1r_1}1),(2n_{j_1u_1}1+1)]$UP:UP, pF)*-positive hyperbolic-secant to hyperbolic flow.*

- *For* $a_{j_1j_10}\prod_{r_2=1,r_2\neq r_1}^{p_1}(a_{j_1j_1r_1}-a_{j_1j_1r_2})^{m_{j_1r_2}}\prod_{u_2=1,u_2\neq u_1}^{q_1}(a_{j_1j_2u_1}-a_{j_1j_2u_2})^{n_{j_1u_2}}<0$ *and* $a_{j_2j_20}\prod_{s=1}^{q_2}(a_{j_1j_2u_1}-a_{j_2j_2s})^{n_{j_2s}}>0$,

$$(a_{j_1j_1r_1},a_{j_1j_2u_1})=\underbrace{([(2m_{j_1r_1}1),(2n_{j_1u_1}1+1)]\text{DP:DP, pF})}_{[(2m_{j_1r_1}1),(2n_{j_1u_1}1+1)]\text{-hyperbolic-to-hyperbolic-secant flow }(+)}. \tag{1.150}$$

The flow at $(x_{j_1}^*,x_{j_2}^*)=(a_{j_1j_1r_1},a_{j_1j_2u_1})$ *is a* $([(2m_{j_1r_1}1),(2n_{j_1u_1}1+1)]$DP:DP, pF)*-positive hyperbolic to hyperbolic-secant flow.*

- *For* $a_{j_1j_10}\prod_{r_2=1,r_2\neq r_1}^{p_1}(a_{j_1j_1r_1}-a_{j_1j_1r_2})^{m_{j_1r_2}}\prod_{u_2=1,u_2\neq u_1}^{q_1}(a_{j_1j_2u_1}-a_{j_1j_2u_2})^{n_{j_1u_2}}>0$ *and* $a_{j_2j_20}\prod_{s=1}^{q_2}(a_{j_1j_2u_1}-a_{j_2j_2s})^{n_{j_2s}}<0$,

$$(a_{j_1j_1r_1},a_{j_1j_2u_1})=\underbrace{([(2m_{j_1r_1}1),(2n_{j_1u_1}1+1)]\text{DP:DP, nF})}_{[(2m_{j_1r_1}1),(2n_{j_1u_1}1+1)]\text{-hyperbolic-to-hyperbolic-secant flow }(-)}. \tag{1.151}$$

The flow at $(x_{j_1}^*,x_{j_2}^*)=(a_{j_1j_1r_1},a_{j_1j_2u_1})$ *is a* $([(2m_{j_1r_1}1),(2n_{j_1u_1}1+1)]$DP:DP, nF)*-negative hyperbolic-to-hyperbolic-secant flow.*

- *For* $a_{j_1j_10}\prod_{r_2=1,r_2\neq r_1}^{p_1}(a_{j_1j_1r_1}-a_{j_1j_1r_2})^{m_{j_1r_2}}\prod_{u_2=1,u_2\neq u_1}^{q_1}(a_{j_1j_2u_1}-a_{j_1j_2u_2})^{n_{j_1u_2}}<0$ *and* $a_{j_2j_20}\prod_{s=1}^{q_2}(a_{j_1j_2u_1}-a_{j_2j_2s})^{n_{j_2s}}<0$,

$$(a_{j_1j_1r_1},a_{j_1j_2u_1})=\underbrace{([(2m_{j_1r_1}1),(2n_{j_1u_1}1+1)]\text{UP:UP, nF})}_{[(2m_{j_1r_1}1),(2n_{j_1u_1}1+1)]\text{-hyperbolic-secant-to-hyperbolic flow }(-)}. \tag{1.152}$$

The flow at $(x_{j_1}^*,x_{j_2}^*)=(a_{j_1j_1r_1},a_{j_1j_2u_1})$ *is a* $([(2m_{j_1r_1}1),(2n_{j_1u_1}1+1)]$UP:UP, nF)*-negative hyperbolic-secant-to-hyperbolic flow.*

(ii$_{1a3}$) For $m_{j_1r_1}=2m_{j_1r_1}1+1$ *and* $n_{j_1u_1}=2n_{j_1u_1}1$, *the equilibrium of* $(x_{j_1}^*,x_{j_2}^*)=(a_{j_1j_1r_1},a_{j_1j_2u_1})$ *for the* x_{j_1}*-direction has the following properties.*

- *For* $a_{j_1j_10}\prod_{r_2=1,r_2\neq r_1}^{p_1}(a_{j_1j_1r_1}-a_{j_1j_1r_2})^{m_{j_1r_2}}\prod_{u_2=1,u_2\neq u_1}^{q_1}(a_{j_1j_2u_1}-a_{j_1j_2u_2})^{n_{j_1u_2}}>0$ *and* $a_{j_2j_20}\prod_{s=1}^{q_2}(a_{j_1j_2u_1}-a_{j_2j_2s})^{n_{j_2s}}>0$,

$$(a_{j_1j_1r_1},a_{j_1j_2u_1})=\underbrace{([(2m_{j_1r_1}1+1),(2n_{j_1u_1}1)]_{\text{DI:II}}\text{SO, pF})}_{[(2m_{j_1r_1}1+1),(2n_{j_1u_1}1)]\text{-decreasing-to-increasing inflection source flow }(+)}. \tag{1.153}$$

The flow at $(x_{j_1}^*,x_{j_2}^*)=(a_{j_1j_1r_1},a_{j_1j_2u_1})$ *is a* $([(2m_{j_1r_1}1+1),(2n_{j_1u_1}1)]_{\text{DI:II}}$SO, pF)*-positive decreasing-to-increasing inflection source flow.*

- *For* $a_{j_1j_10}\prod_{r_2=1,r_2\neq r_1}^{p_1}(a_{j_1j_1r_1}-a_{j_1j_1r_2})^{m_{j_1r_2}}\prod_{u_2=1,u_2\neq u_1}^{q_1}(a_{j_1j_2u_1}-a_{j_1j_2u_2})^{n_{j_1u_2}}<0$ *and* $a_{j_2j_20}\prod_{s=1}^{q_2}(a_{j_1j_2u_1}-a_{j_2j_2s})^{n_{j_2s}}>0$,

$$(a_{j_1j_1r_1},a_{j_1j_2u_1})=\underbrace{([(2m_{j_1r_1}1+1),(2n_{j_1u_1}1)]_{\text{II:DI}}\text{SI, pF})}_{[(2m_{j_1r_1}1+1),(2n_{j_1u_1}1)]\text{-increasing-to-decreasing inflection sink flow }(+)}. \quad (1.154)$$

The flow at $(x_{j_1}^*,x_{j_2}^*)=(a_{j_1j_1r_1},a_{j_1j_2u_1})$ *is a* $([(2m_{j_1r_1}1+1),(2n_{j_1u_1}1)]_{\text{II:DI}}\text{SI, pF})$-*positive increasing-to-decreasing inflection source flow.*

- *For* $a_{j_1j_10}\prod_{r_2=1,r_2\neq r_1}^{p_1}(a_{j_1j_1r_1}-a_{j_1j_1r_2})^{m_{j_1r_2}}\prod_{u_2=1,u_2\neq u_1}^{q_1}(a_{j_1j_2u_1}-a_{j_1j_2u_2})^{n_{j_1u_2}}>0$ *and* $a_{j_2j_20}\prod_{s=1}^{q_2}(a_{j_1j_2u_1}-a_{j_2j_2s})^{n_{j_2s}}<0$,

$$(a_{j_1j_1r_1},a_{j_1j_2u_1})=\underbrace{([(2m_{j_1r_1}1+1),(2n_{j_1u_1}1)]_{\text{II:DI}}\text{SO, nF})}_{[(2m_{j_1r_1}1+1),(2n_{j_1u_1}1)]\text{-increasing-to-decreasing inflection source flow }(-)}. \quad (1.155)$$

The flow at $(x_{j_1}^*,x_{j_2}^*)=(a_{j_1j_1r_1},a_{j_1j_2u_1})$ *is a* $([(2m_{j_1r_1}1+1),(2n_{j_1u_1}1)]_{\text{II:DI}}\text{SO, nF})$-*negative increasing-to-decreasing inflection source flow.*

- *For* $a_{j_1j_10}\prod_{r_2=1,r_2\neq r_1}^{p_1}(a_{j_1j_1r_1}-a_{j_1j_1r_2})^{m_{j_1r_2}}\prod_{u_2=1,u_2\neq u_1}^{q_1}(a_{j_1j_2u_1}-a_{j_1j_2u_2})^{n_{j_1u_2}}<0$ *and* $a_{j_2j_20}\prod_{s=1}^{q_2}(a_{j_1j_2u_1}-a_{j_2j_2s})^{n_{j_2s}}<0$,

$$(a_{j_1j_1r_1},a_{j_1j_2u_1})=\underbrace{([(2m_{j_1r_1}1+1),(2n_{j_1u_1}1)]_{\text{DI:II}}\text{SI, nF})}_{[(2m_{j_1r_1}1+1),(2n_{j_1u_1}1)]\text{-decreasing-to-increasing inflection sink flow }(-)}. \quad (1.156)$$

The flow at $(x_{j_1}^*,x_{j_2}^*)=(a_{j_1j_1r_1},a_{j_1j_2u_1})$ *is a* $([(2m_{j_1r_1}1+1),(2n_{j_1u_1}1)]_{\text{DI:II}}\text{SI, nF})$-*negative decreasing-to-increasing inflection sink flow.*

(ii$_{1a4}$) For $m_{j_1r_1}=2m_{j_1r_1}1$ *and* $n_{j_1u_1}=2n_{j_1u_1}1$, *the equilibrium of* $(x_{j_1}^*,x_{j_2}^*)=(a_{j_1j_1r_1},a_{j_1j_2u_1})$ *for the* x_{j_1}*-direction has the following properties.*

- *For* $a_{j_1j_10}\prod_{r_2=1,r_2\neq r_1}^{p_1}(a_{j_1j_1r_1}-a_{j_1j_1r_2})^{m_{j_1r_2}}\prod_{u_2=1,u_2\neq u_1}^{q_1}(a_{j_1j_2u_1}-a_{j_1j_2u_2})^{n_{j_1u_2}}>0$ *and* $a_{j_2j_20}\prod_{s=1}^{q_2}(a_{j_1j_2u_1}-a_{j_2j_2s})^{n_{j_2s}}>0$,

$$(a_{j_1j_1r_1},a_{j_1j_2u_1})=\underbrace{([(2m_{j_1r_1}1),(2n_{j_1u_1}1)]_{\text{II:II}}\text{US, pF})}_{[(2m_{j_1r_1}1),(2n_{j_1u_1}1)]\text{-increasing-inflection upper-saddle flow }(+)}. \quad (1.157)$$

The flow at $(x_{j_1}^*,x_{j_2}^*)=(a_{j_1j_1r_1},a_{j_1j_2u_1})$ *is a* $([(2m_{j_1r_1}1),(2n_{j_1u_1}1)]_{\text{II:II}}\text{US, pF})$-*positive increasing-inflection upper-saddle flow.*

- *For* $a_{j_1j_10}\prod_{r_2=1,r_2\neq r_1}^{p_1}(a_{j_1j_1r_1}-a_{j_1j_1r_2})^{m_{j_1r_2}}\prod_{u_2=1,u_2\neq u_1}^{q_1}(a_{j_1j_2u_1}-a_{j_1j_2u_2})^{n_{j_1u_2}}<0$ *and* $a_{j_2j_20}\prod_{s=1}^{q_2}(a_{j_1j_2u_1}-a_{j_2j_2s})^{n_{j_2s}}>0$,

$$(a_{j_1j_1r_1},a_{j_1j_2u_1})=\underbrace{([(2m_{j_1r_1}1),(2n_{j_1u_1}1)]_{\text{DI:DI}}\text{LS, pF})}_{[(2m_{j_1r_1}1),(2n_{j_1u_1}1)]\text{-decreasing-inflection lower-saddle flow }(+)}. \quad (1.158)$$

The flow at $(x_{j_1}^*, x_{j_2}^*) = (a_{j_1j_1r_1}, a_{j_1j_2u_1})$ *is a* $([(2m_{j_1r_1}1), (2n_{j_1u_1}1)]_{\text{DI:DI}}\text{LS, pF})$-*positive decreasing-inflection lower-saddle flow.*

- *For* $a_{j_1j_10}\prod_{r_2=1,r_2\neq r_1}^{p_1}(a_{j_1j_1r_1}-a_{j_1j_1r_2})^{m_{j_1r_2}}\prod_{u_2=1,u_2\neq u_1}^{q_1}(a_{j_1j_2u_1}-a_{j_1j_2u_2})^{n_{j_1u_2}}>0$ *and* $a_{j_2j_20}\prod_{s=1}^{q_2}(a_{j_1j_2u_1}-a_{j_2j_2s})^{n_{j_2s}}<0$,

$$(a_{j_1j_1r_1}, a_{j_1j_2u_1}) = \underbrace{([(2m_{j_1r_1}1), (2n_{j_1u_1}1)]_{\text{DI:DI}}\text{US, nF})}_{[(2m_{j_1r_1}1),(2n_{j_1u_1}1)]\text{-decreasing-inflection upper-saddle flow }(-)}. \tag{1.159}$$

The flow at $(x_{j_1}^*, x_{j_2}^*) = (a_{j_1j_1r_1}, a_{j_1j_2u_1})$ *is a* $([(2m_{j_1r_1}1), (2n_{j_1u_1}1)]_{\text{DI:DI}}\text{US, nF})$-*negative decreasing-inflection upper-saddle flow.*

- *For* $a_{j_1j_10}\prod_{r_2=1,r_2\neq r_1}^{p_1}(a_{j_1j_1r_1}-a_{j_1j_1r_2})^{m_{j_1r_2}}\prod_{u_2=1,u_2\neq u_1}^{q_1}(a_{j_1j_2u_1}-a_{j_1j_2u_2})^{n_{j_1u_2}}<0$ *and* $a_{j_2j_20}\prod_{s=1}^{q_2}(a_{j_1j_2u_1}-a_{j_2j_2s})^{n_{j_2s}}<0$,

$$(a_{j_1j_1r_1}, a_{j_1j_2u_1}) = \underbrace{([(2m_{j_1r_1}1), (2n_{j_1u_1}1)]_{\text{II:II}}\text{LS, nF})}_{[(2m_{j_1r_1}1),(2n_{j_1u_1}1)]\text{-increasing-inflection lower-saddle flow }(-)}. \tag{1.160}$$

The flow at $(x_{j_1}^*, x_{j_2}^*) = (a_{j_1j_1r_1}, a_{j_1j_2u_1})$ *is a* $([(2m_{j_1r_1}1), (2n_{j_1u_1}1)]_{\text{II:II}}\text{LS, nF})$-*negative increasing-inflection lower-saddle flow.*

(ii_{1b}) *There is an equilibrium for self-univariate equilibrium as*

$$(x_{j_1}^*, x_{j_2}^*) = (a_{j_1j_1r_1}, a_{j_2j_2s_1}). \tag{1.161}$$

(ii_{1b1}) *For* $m_{j_1r_1}=2m_{j_1r_1}1+1$ *and* $n_{j_2s_1}=2n_{j_2s_1}1+1$, *the equilibrium of* $(x_{j_1}^*, x_{j_2}^*) = (a_{j_1j_1r_1}, a_{j_2j_2s_1})$ *has the following properties.*

- *For* $a_{j_1j_10}\prod_{r_2=1,r_2\neq r_1}^{p_1}(a_{j_1j_1r_1}-a_{j_1j_1r_2})^{m_{j_1r_2}}\prod_{u=1}^{q_1}(a_{j_2j_2s_1}-a_{j_1j_2u})^{n_{j_1u}}>0$ *and* $a_{j_2j_20}\prod_{s_2=1,s_2\neq s_1}^{q_2}(a_{j_2j_2s_1}-a_{j_2j_2s_2})^{n_{j_2s_2}}>0$,

$$(a_{j_1j_1r_1}, a_{j_2j_2s_1}) = \underbrace{((2m_{j_1r_1}1+1)^{\text{th}}\text{ SO}, (2n_{j_2s_1}1+1)^{\text{th}}\text{ SO})}_{((2m_{j_1r_1}1+1),(2n_{j_2s_1}1+1))\text{-source}}. \tag{1.162}$$

The flow at $(x_{j_1}^*, x_{j_2}^*) = (a_{j_1j_1r_1}, a_{j_2j_2s_1})$ *is a* $((2m_{j_1r_1}1+1)^{\text{th}}\text{SO}, (2n_{j_2s_1}1+1)^{\text{th}}\text{ SO})$-*source.*

- *For* $a_{j_1j_10}\prod_{r_2=1,r_2\neq r_1}^{p_1}(a_{j_1j_1r_1}-a_{j_1j_1r_2})^{m_{j_1r_2}}\prod_{u=1}^{q_1}(a_{j_2j_2s_1}-a_{j_1j_2u})^{n_{j_1u}}<0$ *and* $a_{j_2j_20}\prod_{s_2=1,s_2\neq s_1}^{q_2}(a_{j_2j_2s_1}-a_{j_2j_2s_2})^{n_{j_2s_2}}>0$,

$$(a_{j_1j_1r_1}, a_{j_2j_2s_1}) = \underbrace{((2m_{j_1r_1}1+1)^{\text{th}}\text{ SI}, (2n_{j_2s_1}1+1)^{\text{th}}\text{ SO})}_{((2m_{j_1r_1}1+1),(2n_{j_2s_1}1+1))\text{-saddle}}. \tag{1.163}$$

The flow at $(x_{j_1}^*, x_{j_2}^*) = (a_{j_1j_1r_1}, a_{j_2j_2s_1})$ *is a* $((2m_{j_1r_1}1+1)^{\text{th}}\text{SI}, (2n_{j_2s_1}1+1)^{\text{th}}\text{ SO})$-*saddle.*

- *For* $a_{j_1j_10}\prod_{r_2=1,r_2\neq r_1}^{p_1}(a_{j_1j_1r_1}-a_{j_1j_1r_2})^{m_{j_1r_2}}\prod_{u=1}^{q_1}(a_{j_2j_2s_1}-a_{j_1j_2u})^{n_{j_1u}}>0$ *and* $a_{j_2j_20}\prod_{s_2=1,s_2\neq s_1}^{q_2}(a_{j_2j_2s_1}-a_{j_2j_2s_2})^{n_{j_2s_2}}<0$,

$$(a_{j_1j_1r_1},a_{j_2j_2s_1})=\underbrace{((2m_{j_1r_1}1+1)^{\text{th}}\,\text{SO},(2n_{j_2s_1}1+1)^{\text{th}}\,\text{SI})}_{((2m_{j_1r_1}1+1),(2n_{j_2s_1}1+1))\text{-saddle}}. \tag{1.164}$$

The flow at $(x_{j_1}^*,x_{j_2}^*)=(a_{j_1j_1r_1},a_{j_2j_2s_1})$ *is a* $((2m_{j_1r_1}1+1)^{\text{th}}\text{SO},(2n_{j_2s_1}1+1)^{\text{th}}\,\text{SI})$*-saddle.*

- *For* $a_{j_1j_10}\prod_{r_2=1,r_2\neq r_1}^{p_1}(a_{j_1j_1r_1}-a_{j_1j_1r_2})^{m_{j_1r_2}}\prod_{u=1}^{q_1}(a_{j_2j_2s_1}-a_{j_1j_2u})^{n_{j_1u}}<0$ *and* $a_{j_2j_20}\prod_{s_2=1,s_2\neq s_1}^{q_2}(a_{j_2j_2s_1}-a_{j_2j_2s_2})^{n_{j_2s_2}}<0$,

$$(a_{j_1j_1r_1},a_{j_2j_2s_1})=\underbrace{((2m_{j_1r_1}1+1)^{\text{th}}\,\text{SI},(2n_{j_2s_1}1+1)^{\text{th}}\,\text{SI})}_{((2m_{j_1r_1}1+1),(2n_{j_2s_1}1+1))\text{-sink}}. \tag{1.165}$$

The flow at $(x_{j_1}^*,x_{j_2}^*)=(a_{j_1j_1r_1},a_{j_2j_2s_1})$ *is a* $((2m_{j_1r_1}1+1)^{\text{th}}\text{SI},(2n_{j_2s_1}1+1)^{\text{th}}\text{SI})$*-sink.*

(ii_{1b2}) For $m_{j_1r_1}=2m_{j_1r_1}1$ *and* $n_{j_2s_1}=2n_{j_2s_1}1+1$, *the equilibrium of* $(x_{j_1}^*,x_{j_2}^*)=(a_{j_1j_1r_1},a_{j_2j_2s_1})$ *has the following properties.*

- *For* $a_{j_1j_10}\prod_{r_2=1,r_2\neq r_1}^{p_1}(a_{j_1j_1r_1}-a_{j_1j_1r_2})^{m_{j_1r_2}}\prod_{u=1}^{q_1}(a_{j_2j_2s_1}-a_{j_1j_2u})^{n_{j_1u}}>0$ *and* $a_{j_2j_20}\prod_{s_2=1,s_2\neq s_1}^{q_2}(a_{j_2j_2s_1}-a_{j_2j_2s_2})^{n_{j_2s_2}}>0$,

$$(a_{j_1j_1r_1},a_{j_2j_2s_1})=\underbrace{((2m_{j_1r_1}1)^{\text{th}}\,\text{US},(2n_{j_ss_1}1+1)^{\text{th}}\,\text{SO})}_{((2m_{j_1r_1}1),(2n_{j_2s_1}1+1))\text{-upper-saddle source}}. \tag{1.166}$$

The flow at $(x_{j_1}^*,x_{j_2}^*)=(a_{j_1j_1r_1},a_{j_2j_2s_1})$ *is a* $((2m_{j_1r_1}1)^{\text{th}}\,\text{US},(2n_{j_ss_1}1+1)^{\text{th}}\,\text{SO})$*-upper-saddle source.*

- *For* $a_{j_1j_10}\prod_{r_2=1,r_2\neq r_1}^{p_1}(a_{j_1j_1r_1}-a_{j_1j_1r_2})^{m_{j_1r_2}}\prod_{u=1}^{q_1}(a_{j_2j_2s_1}-a_{j_1j_2u})^{n_{j_1u}}<0$ *and* $a_{j_2j_20}\prod_{s_2=1,s_2\neq s_1}^{q_2}(a_{j_2j_2s_1}-a_{j_2j_2s_2})^{n_{j_2s_2}}>0$,

$$(a_{j_1j_1r_1},a_{j_2j_2s_1})=\underbrace{((2m_{j_1r_1}1)^{\text{th}}\,\text{LS},(2n_{j_2s_1}1+1)^{\text{th}}\,\text{SO})}_{((2m_{j_1r_1}1),(2n_{j_2s_1}1+1))\text{-lower-saddle source}}. \tag{1.167}$$

The flow at $(x_{j_1}^*,x_{j_2}^*)=(a_{j_1j_1r_1},a_{j_2j_2s_1})$ *is a* $((2m_{j_1r_1}1)^{\text{th}}\,\text{LS},(2n_{j_2s_1}1+1)^{\text{th}}\,\text{SO})$*-lower-saddle saddle.*

- *For* $a_{j_1j_10}\prod_{r_2=1,r_2\neq r_1}^{p_1}(a_{j_1j_1r_1}-a_{j_1j_1r_2})^{m_{j_1r_2}}\prod_{u=1}^{q_1}(a_{j_2j_2s_1}-a_{j_1j_2u})^{n_{j_1u}}>0$ *and* $a_{j_2j_20}\prod_{s_2=1,s_2\neq s_1}^{q_2}(a_{j_2j_2s_1}-a_{j_2j_2s_2})^{n_{j_2s_2}}<0$,

$$(a_{j_1j_1r_1},a_{j_2j_2s_1})=\underbrace{((2m_{j_1r_1}1)^{\text{th}}\,\text{US},(2n_{j_2s_1}1+1)^{\text{th}}\,\text{SI})}_{((2m_{j_1r_1}1),(2n_{j_2s_1}1+1))\text{-upper-saddle sink}}. \tag{1.168}$$

The flow at $(x_{j_1}^*, x_{j_2}^*) = (a_{j_1j_1r_1}, a_{j_2j_2s_1})$ *is a* $((2m_{j_1r_1}1)^{\text{th}}\,\text{US}, (2n_{j_2s_1}1+1)^{\text{th}}\,\text{SI})$*-upper-saddle sink.*

- *For* $a_{j_1j_10}\prod_{r_2=1,r_2\neq r_1}^{p_1}(a_{j_1j_1r_1}-a_{j_1j_1r_2})^{m_{j_1r_2}}\prod_{u=1}^{q_1}(a_{j_2j_2s_1}-a_{j_1j_2u})^{n_{j_1u}}<0$ *and* $a_{j_2j_20}\prod_{s_2=1,s_2\neq s_1}^{q_2}(a_{j_2j_2s_1}-a_{j_2j_2s_2})^{n_{j_2s_2}}<0$,

$$(a_{j_1j_1r_1}, a_{j_2j_2s_1}) = \underbrace{((2m_{j_1r_1}1)^{\text{th}}\,\text{LS}, (2n_{j_2s_1}1+1)^{\text{th}}\,\text{SI})}_{((2m_{j_1r_1}1),(2n_{j_2s_1}1+1))\text{-lower-saddle sink}}. \tag{1.169}$$

The flow at $(x_{j_1}^*, x_{j_2}^*) = (a_{j_1j_1r_1}, a_{j_2j_2s_1})$ *is a* $((2m_{j_1r_1}1)^{\text{th}}\,\text{LS}, (2n_{j_2s_1}1+1)^{\text{th}}\,\text{SI})$*-lower-saddle sink.*

(ii_{1b3}) For $m_{j_1r_1} = 2m_{j_1r_1}1+1$ *and* $n_{j_2s_1} = 2n_{j_2s_1}1$*, the equilibrium of* $(x_{j_1}^*, x_{j_2}^*) = (a_{j_1j_1r_1}, a_{j_2j_2s_1})$ *has the following properties.*

- *For* $a_{j_1j_10}\prod_{r_2=1,r_2\neq r_1}^{p_1}(a_{j_1j_1r_1}-a_{j_1j_1r_2})^{m_{j_1r_2}}\prod_{u=1}^{q_1}(a_{j_2j_2s_1}-a_{j_1j_2u})^{n_{j_1u}}>0$ *and* $a_{j_2j_20}\prod_{s_2=1,s_2\neq s_1}^{q_2}(a_{j_2j_2s_1}-a_{j_2j_2s_2})^{n_{j_2s_2}}>0$,

$$(a_{j_1j_1r_1}, a_{j_2j_2s_1}) = \underbrace{((2m_{j_1r_1}1+1)^{\text{th}}\,\text{SO}, (2n_{j_2s_1}1)^{\text{th}}\,\text{US})}_{((2m_{j_1r_1}1+1),(2n_{j_2s_1}1))\text{-upper-saddle source}}. \tag{1.170}$$

The flow at $(x_{j_1}^*, x_{j_2}^*) = (a_{j_1j_1r_1}, a_{j_2j_2s_1})$ *is a* $((2m_{j_1r_1}1+1)^{\text{th}}\,\text{SO}, (2n_{j_2s_1}1)^{\text{th}}\,\text{US})$*-upper-saddle source.*

- *For* $a_{j_1j_10}\prod_{r_2=1,r_2\neq r_1}^{p_1}(a_{j_1j_1r_1}-a_{j_1j_1r_2})^{m_{j_1r_2}}\prod_{u=1}^{q_1}(a_{j_2j_2s_1}-a_{j_1j_2u})^{n_{j_1u}}<0$ *and* $a_{j_2j_20}\prod_{s_2=1,s_2\neq s_1}^{q_2}(a_{j_2j_2s_1}-a_{j_2j_2s_2})^{n_{j_2s_2}}>0$,

$$(a_{j_1j_1r_1}, a_{j_2j_2s_1}) = \underbrace{((2m_{j_1r_1}1+1)^{\text{th}}\,\text{SI}, (2n_{j_2s_1}1)^{\text{th}}\,\text{US})}_{((2m_{j_1r_1}1+1),(2n_{j_2s_1}1))\text{-upper-saddle sink}}. \tag{1.171}$$

The flow at $(x_{j_1}^*, x_{j_2}^*) = (a_{j_1j_1r_1}, a_{j_2j_2s_1})$ *is a* $((2m_{j_1r_1}1+1)^{\text{th}}\,\text{SI}, (2n_{j_2s_1}1)^{\text{th}}\,\text{US})$*-upper-saddle sink.*

- *For* $a_{j_1j_10}\prod_{r_2=1,r_2\neq r_1}^{p_1}(a_{j_1j_1r_1}-a_{j_1j_1r_2})^{m_{j_1r_2}}\prod_{u=1}^{q_1}(a_{j_2j_2s_1}-a_{j_1j_2u})^{n_{j_1u}}>0$ *and* $a_{j_2j_20}\prod_{s_2=1,s_2\neq s_1}^{q_2}(a_{j_2j_2s_1}-a_{j_2j_2s_2})^{n_{j_2s_2}}<0$,

$$(a_{j_1j_1r_1}, a_{j_2j_2s_1}) = \underbrace{((2m_{j_1r_1}1+1)^{\text{th}}\,\text{SO}, (2n_{j_2s_1}1)^{\text{th}}\,\text{LS})}_{((2m_{j_1r_1}1+1),(2n_{j_2s_1}1))\text{-lower-saddle source}}. \tag{1.172}$$

The flow at $(x_{j_1}^*, x_{j_2}^*) = (a_{j_1j_1r_1}, a_{j_2j_2s_1})$ *is a* $((2m_{j_1r_1}1+1)^{\text{th}}\,\text{SO}, (2n_{j_2s_1}1)^{\text{th}}\,\text{LS})$*-lower-saddle source.*

- *For* $a_{j_1j_10}\prod_{r_2=1,r_2\neq r_1}^{p_1}(a_{j_1j_1r_1}-a_{j_1j_1r_2})^{m_{j_1r_2}}\prod_{u=1}^{q_1}(a_{j_2j_2s_1}-a_{j_1j_2u})^{n_{j_1u}}<0$ *and* $a_{j_2j_20}\prod_{s_2=1,s_2\neq s_1}^{q_2}(a_{j_2j_2s_1}-a_{j_2j_2s_2})^{n_{j_2s_2}}<0$,

$$(a_{j_1j_1r_1}, a_{j_2j_2s_1}) = \underbrace{((2m_{j_1r_11}+1)^{\text{th}}\,\text{SI}, (2n_{j_2s_11})^{\text{th}}\,\text{LS})}_{((2m_{j_1r_11}+1),(2n_{j_2s_11}))\text{-lower-saddle sink}}. \tag{1.173}$$

The flow at $(x_{j_1}^*, x_{j_2}^*) = (a_{j_1j_1r_1}, a_{j_2j_2s_1})$ *is a* $((2m_{j_1r_11}+1)^{\text{th}}\,\text{SI}, (2n_{j_2s_11})^{\text{th}}\,\text{LS})$ *-lower-saddle sink.*

(ii_{1b4}) For $m_{j_1r_1} = 2m_{j_1r_11}$ *and* $n_{j_2s_1} = 2n_{j_2s_11}$, *the equilibrium of* $(x_{j_1}^*, x_{j_2}^*) = (a_{j_1j_1r_1}, a_{j_2j_2s_1})$ *has the following properties.*

- *For* $a_{j_1j_10}\prod_{r_2=1,r_2\neq r_1}^{p_1}(a_{j_1j_1r_1}-a_{j_1j_1r_2})^{m_{j_1r_2}}\prod_{u=1}^{q_1}(a_{j_2j_2s_1}-a_{j_1j_2u})^{n_{j_1u}}>0$ *and* $a_{j_2j_20}\prod_{s_2=1,s_2\neq s_1}^{q_2}(a_{j_2j_2s_1}-a_{j_2j_2s_2})^{n_{j_2s_2}}>0$,

$$(a_{j_1j_1r_1}, a_{j_2j_2s_1}) = \underbrace{((2m_{j_1r_11})^{\text{th}}\,\text{US}, (2n_{j_2s_11})^{\text{th}}\,\text{US})}_{((2m_{j_1r_11}),(2n_{j_2s_11}))\text{-double-saddle}}. \tag{1.174}$$

  The flow at $(x_{j_1}^*, x_{j_2}^*) = (a_{j_1j_1r_1}, a_{j_2j_2s_1})$ *is a* $((2m_{j_1r_11})^{\text{th}}\,\text{US}, (2n_{j_2s_11})^{\text{th}}\,\text{US})$*-double-saddle.*
- *For* $a_{j_1j_10}\prod_{r_2=1,r_2\neq r_1}^{p_1}(a_{j_1j_1r_1}-a_{j_1j_1r_2})^{m_{j_1r_2}}\prod_{u=1}^{q_1}(a_{j_2j_2s_1}-a_{j_1j_2u})^{n_{j_1u}}<0$ *and* $a_{j_2j_20}\prod_{s_2=1,s_2\neq s_1}^{q_2}(a_{j_2j_2s_1}-a_{j_2j_2s_2})^{n_{j_2s_2}}>0$,

$$(a_{j_1j_1r_1}, a_{j_2j_2s_1}) = \underbrace{((2m_{j_1r_11})^{\text{th}}\,\text{LS}, (2n_{j_2s_11})^{\text{th}}\,\text{US})}_{((2m_{j_1r_11}),(2n_{j_2s_11}))\text{-double-saddle}}. \tag{1.175}$$

  The flow at $(x_{j_1}^*, x_{j_2}^*) = (a_{j_1j_1r_1}, a_{j_2j_2s_1})$ *is a* $((2m_{j_1r_11})^{\text{th}}\,\text{LS}, (2n_{j_2s_11})^{\text{th}}\,\text{US})$*-double-saddle.*
- *For* $a_{j_1j_10}\prod_{r_2=1,r_2\neq r_1}^{p_1}(a_{j_1j_1r_1}-a_{j_1j_1r_2})^{m_{j_1r_2}}\prod_{u=1}^{q_1}(a_{j_2j_2s_1}-a_{j_1j_2u})^{n_{j_1u}}>0$ *and* $a_{j_2j_20}\prod_{s_2=1,s_2\neq s_1}^{q_2}(a_{j_2j_2s_1}-a_{j_2j_2s_2})^{n_{j_2s_2}}<0$,

$$(a_{j_1j_1r_1}, a_{j_2j_2s_1}) = \underbrace{((2m_{j_1r_11})^{\text{th}}\,\text{US}, (2n_{j_2s_11})^{\text{th}}\,\text{LS})}_{((2m_{j_1r_11}),(2n_{j_2s_11}))\text{-double-saddle}}. \tag{1.176}$$

  The flow at $(x_{j_1}^*, x_{j_2}^*) = (a_{j_1j_1r_1}, a_{j_2j_2s_1})$ *is a* $((2m_{j_1r_11})^{\text{th}}\,\text{US}, (2n_{j_2s_11})^{\text{th}}\,\text{LS})$*-double-saddle.*
- *For* $a_{j_1j_10}\prod_{r_2=1,r_2\neq r_1}^{p_1}(a_{j_1j_1r_1}-a_{j_1j_1r_2})^{m_{j_1r_2}}\prod_{u=1}^{q_1}(a_{j_2j_2s_1}-a_{j_1j_2u})^{n_{j_1u}}<0$ *and* $a_{j_2j_20}\prod_{s_2=1,s_2\neq s_1}^{q_2}(a_{j_2j_2s_1}-a_{j_2j_2s_2})^{n_{j_2s_2}}<0$,

$$(a_{j_1j_1r_1}, a_{j_2j_2s_1}) = \underbrace{((2m_{j_1r_11})^{\text{th}}\,\text{LS}, (2n_{j_2s_11})^{\text{th}}\,\text{LS})}_{((2m_{j_1r_11}),(2n_{j_2s_11}))\text{-double-saddle}}. \tag{1.177}$$

  The flow at $(x_{j_1}^*, x_{j_2}^*) = (a_{j_1j_1r_1}, a_{j_2j_2s_1})$ *is a* $((2m_{j_1r_11})^{\text{th}}\,\text{LS}, (2n_{j_2s_11})^{\text{th}}\,\text{LS})$*-double-saddle.*

(ii₂) For $a_{j_1j_2u_1} = a_{j_2j_2s_1}$ $(u_1 \in \{1, 2, \ldots, q_1\};\ s \in \{1, 2, \ldots, q_2\})$

$$\begin{aligned}
\dot{x}_{j_2} &= a_{j_2j_20}(x_{j_2} - a_{j_2j_2s_1})^{n_{j_2s_1}} \prod\nolimits_{s_1=1, s_2\neq s_1}^{q_2} (x_{j_2} - a_{j_2j_2s_2})^{n_{j_2s_2}},\\
\dot{x}_{j_1} &= a_{j_1j_10}(x_{j_2} - a_{j_1j_2u_1})^{n_{j_1u_1}} \prod\nolimits_{r=1}^{p_1} (x_{j_1} - a_{j_1j_1r})^{m_{j_1r}} \prod\nolimits_{u_1=1, u_2\neq u_1}^{q_1} (x_{j_2} - a_{j_1j_2u_2})^{n_{j_1u_2}},
\end{aligned} \tag{1.178}$$

The first integral manifold is

$$\begin{aligned}
&\sum_{r_1=1}^{p_1}\sum_{l_1=1}^{m_{j_1r_1}} \frac{A_{l_1m_{j_1r_1}}}{l_1-1}\left[\frac{1}{(x_{j_1}-a_{j_1j_1r_1})^{l_1-1}} - \frac{1}{(x_{j_10}-a_{j_1j_1r_1})^{l_1-1}}\right]\\
&= \frac{a_{j_1j_10}}{a_{j_2j_20}}\left\{\sum_{\beta_0=n_2}^{n_1} b_{j_1j_2\beta_0} \sum_{\substack{\alpha_1\in[0,\beta_0-n_2],\\ \beta_1\in[n_2-n_{j_2s_1},\beta_0-n_{j_2s_1}]}}^{\alpha_1+\beta_1=\beta_0-n_{j_2s_1}} \ \sum_{\substack{\alpha_2\in[0,\beta_1-n_2]\\ \beta_2\in[n_2-\Sigma_{i=1}^2 n_{j_2s_i},\beta_1-\Sigma_{i=1}^2 n_{j_2s_i}]}}^{\alpha_2+\beta_2=\beta_1-\Sigma_{i=1}^2 n_{j_2s_i}} \cdots\right.\\
&\sum_{\substack{\alpha_{p-1}\in[0,\beta_{q_2-2}-n_2]\\ \beta_{q_2-1}\in[n_2-\Sigma_{i=1}^{q_2-1} n_{j_2s_i},\beta_{q_2-2}-\Sigma_{i=1}^{q_2-1} n_{j_2s_i}]}}^{\alpha_{q_2-1}+\beta_{q_2-1}=\beta_{q_2-2}-\Sigma_{i=1}^{q_2-1} n_{j_2s_i}} \ \prod_{j=1}^{q_2-1} \frac{(\beta_{j-1}-\Sigma_{i=1}^{j} n_{j_2s_i})!}{\alpha_j!\beta_j!}(a_{j_2j_2s_{j+1}} - a_{j_2j_2s_j})^{\alpha_j}\\
&\times \frac{1}{\beta_{q_2-1}-n_{j_2r_{q_2}}+1}\left[(x_{j_2}-a_{j_2j_2s_\rho})^{\beta_{q_2-1}-n_{j_2r_{q_2}}+1} - (x_{j_20}-a_{j_2j_2s_\rho})^{\beta_{q_2-1}-n_{j_2r_{q_2}}+1}\right]\\
&- \sum_{\beta_0=q_2}^{q_1} b_{j_1j_2\beta_0} \sum_{k=2}^{q_2-1}\left[\sum_{\substack{\alpha_1\in[0,\beta_0-n_2),\\ \beta_1\in(n_2-n_{j_2s_1},\beta_0-n_{j_2s_1}]}}^{\alpha_1+\beta_1=\beta_0-n_{j_2s_1}} \ \sum_{\substack{\alpha_2\in[0,\beta_1-n_2]\\ \beta_2\in(n_2-\Sigma_{i=1}^2 n_{j_2s_i},\beta_1-\Sigma_{i=1}^2 n_{j_2s_i})}}^{\alpha_2+\beta_2=\beta_1-\Sigma_{i=1}^2 n_{j_2s_i}} \cdots\right.\\
&\times\left(\sum_{\substack{\alpha_{k-1}\in(\beta_{k-2}-n_2+n_{j_2k},\beta_{k-2}-\Sigma_{i=1}^{k-1} n_{j_2s_i}]\\ \beta_{k-1}\in[0,n_2-\Sigma_{i=1}^{k} n_{j_2s_i})}}^{\alpha_{k-1}+\beta_{k-1}=\beta_{k-2}-\Sigma_{i=1}^{k-1} n_{j_2s_i}} \ \prod_{j=1}^{k-1} \frac{(\beta_{k-2}-\Sigma_{i=1}^{j} n_{j_2s_i})!}{\alpha_j!\beta_j!}(a_{j_2j_2s_j+1} - a_{j_2j_2s_j})^{\alpha_j}\right.\\
&\times \sum_{\rho=k}^{q_2}\sum_{l_\rho=1}^{n_{j_2s_\rho}} \frac{A_{l_\rho n_{j_2s_\rho}}^{(s_k)}}{l_\rho-1}\left[\frac{1}{(x_{j_2}-a_{j_2j_2s_\rho})^{l_\rho-1}} - \frac{1}{(x_{j_20}-a_{j_2j_2s_\rho})^{l_\rho-1}}\right]\\
&+ \sum_{\substack{\alpha_{k-1}\in(\beta_{k-2}-n_2,\beta_{k-2}-n_2+n_{j_2s_k}]\\ \beta_{k-1}\in[n_2-\Sigma_{i=1}^{k} n_{j_2s_i},n_2-\Sigma_{i=1}^{k-1} n_{j_2s_i})}}^{\alpha_{k-1}+\beta_{k-1}=\beta_{k-2}-\Sigma_{i=1}^{k-1} n_{j_2s_i}} \ \prod_{j=1}^{k-1} \frac{(\beta_{j-1}-\Sigma_{i=1}^{j} n_{j_2s_i})!}{\alpha_j!\beta_j!}(a_{j_2j_2s_j+1} - a_{j_2j_2s_j})^{\alpha j}\\
&\left.\left.\left.\times \sum_{\rho=k+1}^{q_2}\sum_{l_\rho=1}^{m'_{s_\rho}} \frac{B_{l_\rho\bar{n}_{j_2s_\rho}}^{(s_k)}}{l_\rho-1}\left[\frac{1}{(x_{j_2}-a_{j_2j_2s_\rho})^{l_\rho-1}} - \frac{1}{(x_{j_20}-a_{j_2j_2s_\rho})^{l_\rho-1}}\right]\right)\right] + \Pi(u_1)\right\},
\end{aligned} \tag{1.179}$$

where

$$\Pi(u_1) = -\left(\sum_{\beta_0=0}^{n_{j_2s_1}-1} b_{j_1j_2\beta_0} \sum_{\rho=1}^{q_2}\sum_{l_\rho=1}^{\bar{n}_{j_2s_\rho}} \frac{A^{(s_1)}_{l_\rho\bar{n}_{j_2s_\rho}}}{l_\rho - 1}\left[\frac{1}{(x_{j_2} - a_{j_2j_2s_\rho})^{l_\rho-1}} - \frac{1}{(x_{j_20} - a_{j_2j_2s_\rho})^{l_\rho-1}}\right]\right.$$
$$\left. + \sum_{\beta_0=n_{j_2s_1}}^{n_2-1} b_{j_1j_2\beta_0} \sum_{\rho=2}^{q_2}\sum_{l_\rho=1}^{\bar{n}_{j_2s_\rho}} \frac{B^{(s_1)}_{l_\rho\bar{n}_{j_2s_\rho}}}{l_\rho - 1}\left[\frac{1}{(x_{j_2} - a_{j_2j_2s_\rho})^{l_\rho-1}} - \frac{1}{(x_{j_20} - a_{j_2j_2s_\rho})^{l_\rho-1}}\right]\right)$$

for $n_{j_1u_1} < n_{j_2s_1}$;

$$\Pi(u_1) = -\sum_{\beta_0=n_{j_2s_1}}^{n_2-1} b_{j_1j_2\beta_0} \sum_{\rho=2}^{q_2}\sum_{l_\rho=1}^{\bar{n}_{j_2s_\rho}} \frac{B^{(s_1)}_{l_\rho\bar{n}_{j_2s_\rho}}}{l_\rho - 1}\left[\frac{1}{(x_{j_2} - a_{j_2j_2s_\rho})^{l_\rho-1}} - \frac{1}{(x_{j_20} - a_{j_2j_2s_\rho})^{l_\rho-1}}\right],$$

for $n_{j_1u_1} = n_{j_2s_1}$;

$$\Pi(u_1) = -\sum_{\beta_0=n_{j_1u_1}}^{n_2-1} b_{j_1j_2\beta_0} \sum_{\rho=2}^{q_2}\sum_{l_\rho=1}^{\bar{n}_{j_2s_\rho}} \frac{B^{(s_1)}_{l_\rho\bar{n}_{j_2s_\rho}}}{l_\rho - 1}\left[\frac{1}{(x_{j_2} - a_{j_2j_2s_\rho})^{l_\rho-1}} - \frac{1}{(x_{j_20} - a_{j_2j_2s_\rho})^{l_\rho-1}}\right],$$

for $n_{j_1u_1} > n_{j_2s_1}$. (1.180)

and

$$b_{j_1j_2\beta_0} = \sum_{t_2,l_2=0}^{t_2+l_2=n_{j_1u_2}} \cdots \sum_{t_{q_1},l_{q_1}=0}^{t_{q_1}+l_{q_1}=n_{j_1u_{q_1}}} \prod_{k=2}^{q_1} \frac{n_{j_1u_k}!}{t_k!l_k!}(a_{j_2j_2s_1} - a_{j_1j_2u_k})^{t_k}\delta_{\beta_0}^{\Sigma_{k=2}^{q_1} l_k + n_{j_1u_1}},$$

$$\frac{1}{\beta_{q_2-1} - n_{j_2q_2} + 1}\left[(x_{j_2} - a_{j_2j_2s_{q_2}})^{\beta_{q_2-1}-n_{j_2q_2}+1} - (x_{j_20} - a_{j_2j_2s_{q_2}})^{\beta_{q_2-1}-n_{j_2q_2}+1}\right]$$
$$= \ln\frac{|x_{j_2} - a_{j_2j_2s_{q_2}}|}{|x_{j_20} - a_{j_2j_2s_{q_2}}|}, \text{ for } \beta_{q_2-1} - n_{j_2q_2} + 1 = 0;$$
$$\frac{1}{l_k - 1}\left[\frac{1}{(x_{j_2} - a_{j_2j_2s_k})^{l_k-1}} - \frac{1}{(x_{j_20} - a_{j_2j_2s_k})^{l_k-1}}\right] = -\ln\frac{|x_{j_2} - a_{j_2j_2s_k}|}{|x_{j_20} - a_{j_2j_2s_k}|},$$
$$\text{for } l_k = 1\ (k = 1, 2, \ldots, q_2 - 1);$$
$$\frac{1}{l_1 - 1}\left[\frac{1}{(x_{j_1} - a_{j_1j_1r_1})^{l_1-1}} - \frac{1}{(x_{j_10} - a_{j_1j_1r_1})^{l_1-1}}\right] = -\ln\frac{|x_{j_1} - a_{j_1j_1r_1}|}{|x_{j_10} - a_{j_1j_1r_1}|},$$
$$\text{for } l_1 = 1.$$
(1.181)

Define a network matrix of hybrid singular 1-dimensional flows and singular equilibriums plus a singular infinite-equilibrium as

$$\text{NM} = \underbrace{(a_{j_1j_1r_1}, a_{j_2j_2s_1})}_{([(n_{j_1u_1}-n_{j_2s_1})]^{\text{th}}\text{XX},[m_{j_1r_1},n_{j_2s_1}]^{\text{th}}\text{YY})} \oplus \bigcup_{r_2=1,r_2\neq r_1}^{p_1} \underbrace{(a_{j_1j_1r_2}, a_{j_2j_2s_1})}_{([(n_{j_1u_1}-n_{j_2s_1})]^{\text{th}}\text{XX},[m_{j_1r_2},n_{j_2s_1}]^{\text{th}}\text{YY})}$$
$$\oplus \bigcup_{r=1}^{p_1} \bigcup_{u_2=1,u_2\neq u_1}^{q_1} \underbrace{(a_{j_1j_1r}, a_{j_1j_2u_2})}_{([m_{j_1r},n_{j_1u_2}]\text{ZZ,xx})} \oplus \bigcup_{r=1}^{p_1} \bigcup_{s_2=1,s_2\neq s_1}^{q_2} \underbrace{(a_{j_1j_1r}, a_{j_2j_2s_2})}_{(m_{j_1r},n_{j_2s_2})^{\text{th}}\text{-UU}}. \tag{1.182}$$

(ii_{2a}) The infinite-equilibriums and the switching equilibriums for $n_{j_2s_1} \leq n_{j_1u_1}$ are presented.

(ii_{2a1}) For $n_{j_2s_1} \leq n_{j_1u_1}$, 16 types of infinite-equilibriums exist at $x_{j_2}^* = a_{j_2j_2s_1} = a_{j_1j_2u_1}$.

(ii_{2a1a}) For $n_{j_1u_1} = 2n_{j_1u_1 1} + 1$ and $n_{j_2s_1} = 2n_{j_2s_1 1} + 1$, the infinite-equilibrium of $x_{j_2}^* = a_{j_1j_2u_1} = a_{j_2j_2s_1}$ with $\overline{x}_{j_1} \neq a_{j_1j_1r_1}$ has the following properties.

- *For $a_{j_1j_10}\prod_{r=1}^{p_1}(\overline{x}_{j_1} - a_{j_1j_1r})^{m_{j_1r}}\prod_{u_2=1,u_2\neq u_1}^{q_1}(a_{j_1j_2u_1} - a_{j_1j_2u_2})^{n_{j_1u_2}} > 0$ and $a_{j_2j_20}\prod_{s_2=1,s_2\neq s_1}^{q_2}(a_{j_2j_2s_1} - a_{j_2j_2s_2})^{n_{j_2s_2}} > 0$,*

$$(\overline{x}_{j_1}, a_{j_2j_2s_1}) = \underbrace{([2(n_{j_1u_1 1} - n_{j_2s_1 1})]^{\text{th}}\,\text{II}, (2n_{j_2s_1 1}+1)^{\text{th}}\,\text{SO})}_{([2(n_{j_1u_1 1}-n_{j_2s_1 1})],(2n_{j_2s_1 1}+1))\text{-increasing-inflection source}}. \tag{1.183}$$

The infinite-equilibrium is a ([2($n_{j_1u_1 1} - n_{j_2s_1 1}$)]th II, ($2n_{j_2s_1 1} + 1$)th SO)-increasing-inflection source at $x_{j_2}^* = a_{j_1j_2u_1} = a_{j_2j_2s_1}$.

- *For $a_{j_1j_10}\prod_{r=1}^{p_1}(\overline{x}_{j_1} - a_{j_1j_1r})^{m_{j_1r}}\prod_{u_2=1,u_2\neq u_1}^{q_1}(a_{j_1j_2u_1} - a_{j_1j_2u_2})^{n_{j_1u_2}} < 0$ and $a_{j_2j_20}\prod_{s_2=1,s_2\neq s_1}^{q_2}(a_{j_2j_2s_1} - a_{j_2j_2s_2})^{n_{j_2s_2}} > 0$,*

$$(\overline{x}_{j_1}, a_{j_2j_2s_1}) = \underbrace{([2(n_{j_1u_1 1} - n_{j_2s_1 1})]^{\text{th}}\,\text{DI}, (2n_{j_2s_1 1}+1)^{\text{th}}\,\text{SO})}_{([2(n_{j_1u_1 1}-n_{j_2s_1 1})],(2n_{j_2s_1 1}+1))\text{-decreasing-inflection source}}. \tag{1.184}$$

The infinite-equilibrium is a ([2($n_{j_1u_1 1} - n_{j_2s_1 1}$)]th DI, ($2n_{j_2s_1 1} + 1$)th SO)-decreasing-inflection source at $x_{j_2}^* = a_{j_1j_2u_1} = a_{j_2j_2s_1}$.

- *For $a_{j_1j_10}\prod_{r=1}^{p_1}(\overline{x}_{j_1} - a_{j_1j_1r})^{m_{j_1r}}\prod_{u_2=1,u_2\neq u_1}^{q_1}(a_{j_1j_2u_1} - a_{j_1j_2u_2})^{n_{j_1u_2}} > 0$ and $a_{j_2j_20}\prod_{s_2=1,s_2\neq s_1}^{q_2}(a_{j_2j_2s_1} - a_{j_2j_2s_2})^{n_{j_2s_2}} < 0$,*

$$(\overline{x}_{j_1}, a_{j_2j_2s_1}) = \underbrace{([2(n_{j_1u_1 1} - n_{j_2s_1 1})]^{\text{th}}\,\text{DI}, (2n_{j_2s_1 1}+1)^{\text{th}}\,\text{SI})}_{([2(n_{j_1u_1 1}-n_{j_2s_1 1})],(2n_{j_2s_1 1}+1))\text{-decreasing-inflection sink}}. \tag{1.185}$$

The infinite-equilibrium is a ([2($n_{j_1u_1 1} - n_{j_2s_1 1}$)]th DI, ($2n_{j_2s_1 1} + 1$)]th SI)-decreasing-inflection sink at $x_{j_2}^* = a_{j_1j_2u_1} = a_{j_2j_2s_1}$.

- *For $a_{j_1j_10}\prod_{r=1}^{p_1}(\overline{x}_{j_1} - a_{j_1j_1r})^{m_{j_1r}}\prod_{u_2=1,u_2\neq u_1}^{q_1}(a_{j_1j_2u_1} - a_{j_1j_2u_2})^{n_{j_1u_2}} < 0$ and $a_{j_2j_20}\prod_{s_2=1,s_2\neq s_1}^{q_2}(a_{j_2j_2s_1} - a_{j_2j_2s_2})^{n_{j_2s_2}} < 0$,*

$$(\overline{x}_{j_1}, a_{j_2j_2s_1}) = \underbrace{([2(n_{j_1u_1 1} - n_{j_2s_1 1})]^{\text{th}}\,\text{II}, (2n_{j_2s_1 1} + 1)^{\text{th}}\,\text{SI})}_{([2(n_{j_1u_1 1} - n_{j_2s_1 1})],(2n_{j_2s_1 1}+1))\text{-increasing-inflection sink}}. \tag{1.186}$$

The infinite-equilibrium is a $([2(n_{j_1u_1 1} - n_{j_2s_1 1})]^{\text{th}}\,\text{II}, (2n_{j_2s_1 1} + 1)^{\text{th}}\,\text{SI})$*-increasing-inflection sink at* $x_{j_2}^* = a_{j_1j_2u_1} = a_{j_2j_2s_1}$.

(ii_{2a1b}*) For* $n_{j_1u_1} = 2n_{j_1u_1 1}$ *and* $n_{j_2s_1} = 2n_{j_2s_1 1} + 1$*, the infinite-equilibrium of* $x_{j_2}^* = a_{j_1j_2u_1} = a_{j_2j_2s_1}$ *with* $\overline{x}_{j_1} \neq a_{j_1j_1r_1}$ *has the following properties.*

- *For* $a_{j_1j_10}\prod_{r=1}^{p_1}(\overline{x}_{j_1} - a_{j_1j_1r})^{m_{j_1r}}\prod_{u_2=1,u_2\neq u_1}^{q_1}(a_{j_1j_2u_1} - a_{j_1j_2u_2})^{n_{j_1u_2}} > 0$ *and* $a_{j_2j_20}\prod_{s_2=1,s_2\neq s_1}^{q_2}(a_{j_2j_2s_1} - a_{j_2j_2s_2})^{n_{j_2s_2}} > 0$,

$$(\overline{x}_{j_1}, a_{j_2j_2s_1}) = \underbrace{([2(n_{j_1u_1 1} - n_{j_2s_1 1}) - 1]^{\text{th}}\,\text{UP}, (2n_{j_2s_1 1} + 1)^{\text{th}}\,\text{SO})}_{([2(n_{j_1u_1 1} - n_{j_2s_1 1})-1],(2n_{j_2s_1 1}+1))\text{-up-parabola source}}. \tag{1.187}$$

The infinite-equilibrium is a $([2(n_{j_1u_1 1} - n_{j_2s_1 1}) - 1]^{\text{th}}\,\text{UP}, (2n_{j_2s_1 1} + 1)^{\text{th}}\,\text{SO})$*-up parabola source at* $x_{j_2}^* = a_{j_1j_2u_1} = a_{j_2j_2s_1}$.

- *For* $a_{j_1j_10}\prod_{r=1}^{p_1}(\overline{x}_{j_1} - a_{j_1j_1r})^{m_{j_1r}}\prod_{u_2=1,u_2\neq u_1}^{q_1}(a_{j_1j_2u_1} - a_{j_1j_2u_2})^{n_{j_1u_2}} < 0$ *and* $a_{j_2j_20}\prod_{s_2=1,s_2\neq s_1}^{q_2}(a_{j_2j_2s_1} - a_{j_2j_2s_2})^{n_{j_2s_2}} > 0$,

$$(\overline{x}_{j_1}, a_{j_2j_2s_1}) = \underbrace{([2(n_{j_1u_1 1} - n_{j_2s_1 1}) - 1]^{\text{th}}\,\text{DP}, (2n_{j_2s_1 1} + 1)^{\text{th}}\,\text{SO})}_{([2(n_{j_1u_1 1} - n_{j_2s_1 1})-1],(2n_{j_2s_1 1}+1))\text{-down-parabola source}}. \tag{1.188}$$

The infinite-equilibrium is a $([2(n_{j_1u_1 1} - n_{j_2s_1 1}) - 1]^{\text{th}}\,\text{DP}, (2n_{j_2s_1 1} + 1)^{\text{th}}\,\text{SO})$*-down parabola source at* $x_{j_2}^* = a_{j_1j_2u_1} = a_{j_2j_2s_1}$.

- *For* $a_{j_1j_10}\prod_{r=1}^{p_1}(\overline{x}_{j_1} - a_{j_1j_1r})^{m_{j_1r}}\prod_{u_2=1,u_2\neq u_1}^{q_1}(a_{j_1j_2u_1} - a_{j_1j_2u_2})^{n_{j_1u_2}} > 0$ *and* $a_{j_2j_20}\prod_{s_2=1,s_2\neq s_1}^{q_2}(a_{j_2j_2s_1} - a_{j_2j_2s_2})^{n_{j_2s_2}} < 0$,

$$(\overline{x}_{j_1}, a_{j_2j_2s_1}) = \underbrace{([2(n_{j_1u_1 1} - n_{j_2s_1 1}) - 1]^{\text{th}}\,\text{DP}, (2n_{j_2s_1 1} + 1)^{\text{th}}\,\text{SI})}_{([2(n_{j_1u_1 1} - n_{j_2s_1 1})-1],(2n_{j_2s_1 1}+1))\text{-down-parabola sink}}. \tag{1.189}$$

The infinite-equilibrium is a $([2(n_{j_1u_1 1} - n_{j_2s_1 1}) - 1]^{\text{th}}\,\text{DP}, (2n_{j_2s_1 1} + 1)^{\text{th}}\,\text{SI})$*-down parabola sink at* $x_{j_2}^* = a_{j_1j_2u_1} = a_{j_2j_2s_1}$.

- *For* $a_{j_1j_10}\prod_{r=1}^{p_1}(\overline{x}_{j_1} - a_{j_1j_1r})^{m_{j_1r}}\prod_{u_2=1,u_2\neq u_1}^{q_1}(a_{j_1j_2u_1} - a_{j_1j_2u_2})^{n_{j_1u_2}} < 0$ *and* $a_{j_2j_20}\prod_{s_2=1,s_2\neq s_1}^{q_2}(a_{j_2j_2s_1} - a_{j_2j_2s_2})^{n_{j_2s_2}} < 0$,

$$(\overline{x}_{j_1}, a_{j_2j_2s_1}) = \underbrace{([2(n_{j_1u_1 1} - n_{j_2s_1 1}) - 1]^{\text{th}}\,\text{UP}, (2n_{j_2s_1 1} + 1)^{\text{th}}\,\text{SI})}_{([2(n_{j_1u_1 1} - n_{j_2s_1 1})-1],(2n_{j_2s_1 1}+1))\text{-down-parabola sink}}. \tag{1.190}$$

The infinite-equilibrium is a $([2(n_{j_1u_1 1} - n_{j_2s_1 1}) - 1]^{\text{th}}\,\text{UP}, (2n_{j_2s_1 1} + 1)^{\text{th}}\,\text{SI})$*-up-parabola sink at* $x_{j_2}^* = a_{j_1j_2u_1} = a_{j_2j_2s_1}$.

(ii_{2a1c}) For $n_{j_1u_1} = 2n_{j_1u_1 1} + 1$ and $n_{j_2s_1} = 2n_{j_2s_1 1}$, the infinite-equilibrium of $x^*_{j_2} = a_{j_1j_2u_1} = a_{j_2j_2s_1}$ with $\bar{x}_{j_1} \neq a_{j_1j_1r_1}$ has the following properties.

- *For* $a_{j_1j_10}\prod_{r=1}^{p_1}(\bar{x}_{j_1} - a_{j_1j_1r})^{m_{j_1r}}\prod_{u_2=1,u_2\neq u_1}^{q_1}(a_{j_1j_2u_1} - a_{j_1j_2u_2})^{n_{j_1u_2}} > 0$ *and* $a_{j_2j_20}\prod_{s_2=1,s_2\neq s_1}^{q_2}(a_{j_2j_2s_1} - a_{j_2j_2s_2})^{n_{j_2s_2}} > 0$,

$$(\bar{x}_{j_1}, a_{j_2j_2s_1}) = \underbrace{([2(n_{j_1u_1 1} - n_{j_2s_1 1}) + 1]^{\text{th}}\,\text{UP},\, (2n_{j_2s_1 1})^{\text{th}}\,\text{US})}_{([2(n_{j_1u_1 1}-n_{j_2s_1 1})+1],(2n_{j_2s_1 1}))\text{-up-parabola upper-saddle}}. \tag{1.191}$$

The infinite-equilibrium is a $([2(n_{j_1u_1 1}-n_{j_2s_1 1})+1]^{\text{th}}\text{UP},(2n_{j_2s_1 1})^{\text{th}}\text{US})$*-up-parabola upper-saddle at* $x^*_{j_2} = a_{j_1j_2u_1} = a_{j_2j_2s_1}$.

- *For* $a_{j_1j_10}\prod_{r=1}^{p_1}(\bar{x}_{j_1} - a_{j_1j_1r})^{m_{j_1r}}\prod_{u_2=1,u_2\neq u_1}^{q_1}(a_{j_1j_2u_1} - a_{j_1j_2u_2})^{n_{j_1u_2}} < 0$ *and* $a_{j_2j_20}\prod_{s_2=1,s_2\neq s_1}^{q_2}(a_{j_2j_2s_1} - a_{j_2j_2s_2})^{n_{j_2s_2}} > 0$,

$$(\bar{x}_{j_1}, a_{j_2j_2s_1}) = \underbrace{([2(n_{j_1u_1 1} - n_{j_2s_1 1}) + 1]^{\text{th}}\,\text{DP},\, (2n_{j_2s_1 1})^{\text{th}}\,\text{US})}_{([2(n_{j_1u_1 1}-n_{j_2s_1 1})+1],(2n_{j_2s_1 1}))\text{-down-parabola upper-saddle}}. \tag{1.192}$$

The infinite-equilibrium is a $([2(n_{j_1u_1 1}-n_{j_2s_1 1})+1]^{\text{th}}\text{DP},(2n_{j_2s_1 1})^{\text{th}}\text{US})$*-down-parabola upper-saddle at* $x^*_{j_2} = a_{j_1j_2u_1} = a_{j_2j_2s_1}$.

- *For* $a_{j_1j_10}\prod_{r=1}^{p_1}(\bar{x}_{j_1} - a_{j_1j_1r})^{m_{j_1r}}\prod_{u_2=1,u_2\neq u_1}^{q_1}(a_{j_1j_2u_1} - a_{j_1j_2u_2})^{n_{j_1u_2}} > 0$ *and* $a_{j_2j_20}\prod_{s_2=1,s_2\neq s_1}^{q_2}(a_{j_2j_2s_1} - a_{j_2j_2s_2})^{n_{j_2s_2}} < 0$,

$$(\bar{x}_{j_1}, a_{j_2j_2s_1}) = \underbrace{([2(n_{j_1u_1 1} - n_{j_2s_1 1}) + 1]^{\text{th}}\,\text{DP},\, (2n_{j_2s_1 1})^{\text{th}}\,\text{LS})}_{([2(n_{j_1u_1 1}-n_{j_2s_1 1})+1],(2n_{j_2s_1 1}))\text{-down-parabola lower-saddle}}. \tag{1.193}$$

The infinite-equilibrium is a $([2(n_{j_1u_1 1}-n_{j_2s_1 1})+1]^{\text{th}}\text{DP},(2n_{j_2s_1 1})^{\text{th}}\text{LS})$*-down-parabola lower-saddle* at $x^*_{j1} = a_{j1j1r1} = a_{j2j1s1}$.

- *For* $a_{j_1j_10}\prod_{r=1}^{p_1}(\bar{x}_{j_1} - a_{j_1j_1r})^{m_{j_1r}}\prod_{u_2=1,u_2\neq u_1}^{q_1}(a_{j_1j_2u_1} - a_{j_1j_2u_2})^{n_{j_1u_2}} < 0$ *and* $a_{j_2j_20}\prod_{s_2=1,s_2\neq s_1}^{q_2}(a_{j_2j_2s_1} - a_{j_2j_2s_2})^{n_{j_2s_2}} < 0$,

$$(\bar{x}_{j_1}, a_{j_2j_2s_1}) = \underbrace{([2(n_{j_1u_1 1} - n_{j_2s_1 1}) + 1]^{\text{th}}\,\text{UP},\, (2n_{j_2s_1 1})^{\text{th}}\,\text{LS})}_{([2(n_{j_1u_1 1}-n_{j_2s_1 1})+1],(2n_{j_2s_1 1}))\text{-up-parabola lower-saddle}}. \tag{1.194}$$

The infinite-equilibrium is a $([2(n_{j_1u_1 1}-n_{j_2s_1 1})+1]^{\text{th}}\text{UP},(2n_{j_2s_1 1})^{\text{th}}\text{LS})$*-up-parabola lower-saddle at* $x^*_{j_2} = a_{j_1j_2u_1} = a_{j_2j_2s_1}$.

(ii_{2a1d}) For $n_{j_1u_1} = 2n_{j_1u_1 1}$ and $n_{j_2s_1} = 2n_{j_2s_1 1}$, the infinite-equilibrium of $x^*_{j_2} = a_{j_1j_2u_1} = a_{j_2j_2s_1}$ with $\bar{x}_{j_1} \neq a_{j_1j_1r_1}$ has the following properties.

- *For* $a_{j_1j_10}\prod_{r=1}^{p_1}(\bar{x}_{j_1} - a_{j_1j_1r})^{m_{j_1r}}\prod_{u_2=1,u_2\neq u_1}^{q_1}(a_{j_1j_2u_1} - a_{j_1j_2u_2})^{n_{j_1u_2}} > 0$ *and* $a_{j_2j_20}\prod_{s_2=1,s_2\neq s_1}^{q_2}(a_{j_2j_2s_1} - a_{j_2j_2s_2})^{n_{j_2s_2}} > 0$,

$$(\overline{x}_{j_1}, a_{j_2j_2s_1}) = \underbrace{([2(n_{j_1u_1}1 - n_{j_2s_1}1)]^{\text{th}}\,\text{II}, (2n_{j_2s_1}1)]^{\text{th}}\,\text{US})}_{([2(n_{j_1u_1}1-n_{j_2s_1}1)],(2n_{j_2s_1}1)])\text{-increasing inflection upper-saddle}}. \tag{1.195}$$

The infinite-equilibrium is a $([2(n_{j_1u_1}1-n_{j_2s_1}1)]^{\text{th}}\text{II}, (2n_{j_2s_1}1)^{\text{th}}\text{US})$*-increasing inflection upper-saddle at* $x^*_{j_2} = a_{j_1j_2u_1} = a_{j_2j_2s_1}$.

- *For* $a_{j_1j_10}\prod_{r=1}^{p_1}(\overline{x}_{j_1} - a_{j_1j_1r})^{m_{j_1r}}\prod_{u_2=1,u_2\neq u_1}^{q_1}(a_{j_1j_2u_1} - a_{j_1j_2u_2})^{n_{j_1u_2}} < 0$ *and* $a_{j_2j_20}\prod_{s_2=1,s_2\neq s_1}^{q_2}(a_{j_2j_2s_1} - a_{j_2j_2s_2})^{n_{j_2s_2}} > 0$,

$$(\overline{x}_{j_1}, a_{j_2j_2s_1}) = \underbrace{([2(n_{j_1u_1}1 - n_{j_2s_1}1)]^{\text{th}}\,\text{DI}, (2n_{j_2s_1}1)^{\text{th}}\,\text{US})}_{([2(n_{j_1u_1}1-n_{j_2s_1}1)],(2n_{j_2s_1}1))\text{-decreasing inflection upper-saddle}}. \tag{1.196}$$

The infinite-equilibrium is a $([2(n_{j_1u_1}1-n_{j_2s_1}1)]^{\text{th}}\text{DI}, (2n_{j_2s_1}1)^{\text{th}}\text{US})$*-decreasing inflection upper-saddle at* $x^*_{j_2} = a_{j_1j_2u_1} = a_{j_2j_2s_1}$.

- *For* $a_{j_1j_10}\prod_{r=1}^{p_1}(\overline{x}_{j_1} - a_{j_1j_1r})^{m_{j_1r}}\prod_{u_2=1,u_2\neq u_1}^{q_1}(a_{j_1j_2u_1} - a_{j_1j_2u_2})^{n_{j_1u_2}} > 0$ *and* $a_{j_2j_20}\prod_{s_2=1,s_2\neq s_1}^{q_2}(a_{j_2j_2s_1} - a_{j_2j_2s_2})^{n_{j_2s_2}} < 0$,

$$(\overline{x}_{j_1}, a_{j_2j_2s_1}) = \underbrace{([2(n_{j_1u_1}1 - n_{j_2s_1}1)]^{\text{th}}\,\text{DI}, (2n_{j_2s_1}1)^{\text{th}}\,\text{LS})}_{([2(n_{j_1u_1}1-n_{j_2s_1}1)],(2n_{j_2s_1}1)])\text{-decreasing inflection lower-saddle}}. \tag{1.197}$$

The infinite-equilibrium is a $([2(n_{j_1u_1}1-n_{j_2s_1}1)]^{\text{th}}\text{DI}, (2n_{j_2s_1}1)^{\text{th}}\text{LS})$*-decreasing inflection lower-saddle at* $x^*_{j_2} = a_{j_1j_2u_1} = a_{j_2j_2s_1}$.

- *For* $a_{j_1j_10}\prod_{r=1}^{p_1}(\overline{x}_{j_1} - a_{j_1j_1r})^{m_{j_1r}}\prod_{u_2=1,u_2\neq u_1}^{q_1}(a_{j_1j_2u_1} - a_{j_1j_2u_2})^{n_{j_1u_2}} < 0$ *and* $a_{j_2j_20}\prod_{s_2=1,s_2\neq s_1}^{q_2}(a_{j_2j_2s_1} - a_{j_2j_2s_2})^{n_{j_2s_2}} < 0$,

$$(\overline{x}_{j_1}, a_{j_2j_2s_1}) = \underbrace{([2(n_{j_1u_1}1 - n_{j_2s_1}1)]^{\text{th}}\,\text{II}, (2n_{j_2s_1}1)^{\text{th}}\,\text{LS})}_{([2(n_{j_1u_1}1-n_{j_2s_1}1)],(2n_{j_2s_1}1)])\text{-increasing inflection lower-saddle}}. \tag{1.198}$$

The infinite-equilibrium is a $([2(n_{j_1u_1}1-n_{j_2s_1}1)]^{\text{th}}\text{II}, (2n_{j_2s_1}1)^{\text{th}}\text{LS})$*-increasing inflection lower-saddle at* $x^*_{j_2} = a_{j_1j_2u_1} = a_{j_2j_2s_1}$.

(ii$_{2a2}$) There are 32 types of equilibrium $(x^*_{j_1}, x^*_{j_2}) = (a_{j_1j_1r_1}, a_{j_2j_2s_1})$ *on the infinite-equilibrium of* $x^*_{j_2} = a_{j_1j_2u_1} = a_{j_2j_2s_1}$.

(ii$_{2a2a}$) For $m_{j_1r_1} = 2m_{j_1r_1}1 + 1, n_{j_1u_1} = 2n_{j_1u_1}1 + 1$ *and* $n_{j_2s_1} = 2n_{j_2s_1}1 + 1$, *the equilibrium of* $(x^*_{j_1}, x^*_{j_2}) = (a_{j_1j_1r_1}, a_{j_2j_2s_1})$ *on the infinite-equilibrium* $x^*_{j_2} = a_{j_1j_2u_1} = a_{j_2j_2s_1}$ *has the following properties.*

- *For* $a_{j_1j_10}\prod_{r_2=1,r_2\neq r_1}^{p_1}(a_{j_1j_1r_1} - a_{j_1j_1r_2})^{m_{j_1r_2}}\prod_{u_2=1,u_2\neq u_1}^{q_1}(a_{j_1j_2u_1} - a_{j_1j_2u_2})^{n_{j_1u_2}} > 0$ *and* $a_{j_2j_20}\prod_{s_2=1,s_2\neq s_1}^{q_2}(a_{j_2j_2s_1} - a_{j_2j_2s_2})^{n_{j_2s_2}} > 0$,

$$\begin{aligned}&(a_{j_1j_1r_1}, a_{j_2j_2s_1})\\ &= \underbrace{([2(n_{j_1u_1}1 - n_{j_2s_1}1)]^{\text{th}}\,\text{DI:II}, [(2m_{j_1r_1}1 + 1), (2n_{j_2s_1}1 + 1)]^{\text{th}}\,\text{SO})}_{([2(n_{j_1u_1}1-n_{j_2s_1}1)],[(2m_{j_1r_1}1+1),(2n_{j_2s_1}1+1)])\text{-decreasing-to-increasing-inflection source}}.\end{aligned} \tag{1.199}$$

The equilibrium is a $([2(n_{j_1u_1 1}-n_{j_2s_1 1})]^{\text{th}}$ DI:II, $[(2m_{j_1r_1 1}+1),(2n_{j_2s_1 1}+1)]^{\text{th}}$ SO)-*decreasing-to-increasing-inflection source at* $(x_{j_1}^*, x_{j_2}^*) = (a_{j_1j_1r_1}, a_{j_2j_2s_1})$.

- *For* $a_{j_1j_10}\prod_{r_2=1,r_2\neq r_1}^{p_1}(a_{j_1j_1r_1}-a_{j_1j_1r_2})^{m_{j_1r_2}}\prod_{u_2=1,u_2\neq u_1}^{q_1}(a_{j_1j_2u_1}-a_{j_1j_2u_2})^{n_{j_1u_2}}<0$ *and* $a_{j_2j_20}\prod_{s_2=1,s_2\neq s_1}^{q_2}(a_{j_2j_2s_1}-a_{j_2j_2s_2})^{n_{j_2s_2}}>0$,

$$(a_{j_1j_1r_1}, a_{j_2j_2s_1}) = \underbrace{([2(n_{j_1u_1 1}-n_{j_2s_1 1})]^{\text{th}}\ \text{II:DI},\ [(2m_{j_1r_1 1}+1),(2n_{j_2s_1 1}+1)]^{\text{th}}\ \text{SO})}_{([2(n_{j_1u_1 1}-n_{j_2s_1 1})],[(2m_{j_1r_1 1}+1),(2n_{j_2s_1 1}+1)])\text{-increasing-to-decreasing-inflection source}}. \quad (1.200)$$

The equilibrium is a $([2(n_{j_1u_1 1}-n_{j_2s_1 1})]^{\text{th}}$ II:DI, $[(2m_{j_1r_1 1}+1),(2n_{j_2s_1 1}+1)]^{\text{th}}$ SO)-*increasing-to-decreasing-inflection source at* $(x_{j_1}^*, x_{j_2}^*) = (a_{j_1j_1r_1}, a_{j_2j_2s_1})$.

- *For* $a_{j_1j_10}\prod_{r_2=1,r_2\neq r_1}^{p_1}(a_{j_1j_1r_1}-a_{j_1j_1r_2})^{m_{j_1r_2}}\prod_{u_2=1,u_2\neq u_1}^{q_1}(a_{j_1j_2u_1}-a_{j_1j_2u_2})^{n_{j_1u_2}}>0$ *and* $a_{j_2j_20}\prod_{s_2=1,s_2\neq s_1}^{q_2}(a_{j_2j_2s_1}-a_{j_2j_2s_2})^{n_{j_2s_2}}<0$,

$$(a_{j_1j_1r_1}, a_{j_2j_2s_1}) = \underbrace{([2(n_{j_1u_1 1}-n_{j_2s_1 1})]^{\text{th}}\ \text{II:DI},\ [(2m_{j_1r_1 1}+1),(2n_{j_2s_1 1}+1)]^{\text{th}}\ \text{SI})}_{([2(n_{j_1u_1 1}-n_{j_2s_1 1})],[(2m_{j_1r_1 1}+1),(2n_{j_2s_1 1}+1)])\text{-increasing-to-decreasing-inflection sink}}. \quad (1.201)$$

The equilibrium is a $([2(n_{j_1u_1 1}-n_{j_2s_1 1})]^{\text{th}}$ II:DI, $[(2m_{j_1r_1 1}+1),(2n_{j_2s_1 1}+1)]^{\text{th}}$ SI)-*increasing-to-decreasing-inflection sink at* $(x_{j_1}^*, x_{j_2}^*) = (a_{j_1j_1r_1}, a_{j_2j_2s_1})$.

- *For* $a_{j_1j_10}\prod_{r_2=1,r_2\neq r_1}^{p_1}(a_{j_1j_1r_1}-a_{j_1j_1r_2})^{m_{j_1r_2}}\prod_{u_2=1,u_2\neq u_1}^{q_1}(a_{j_1j_2u_1}-a_{j_1j_2u_2})^{n_{j_1u_2}}<0$ *and* $a_{j_2j_20}\prod_{s_2=1,s_2\neq s_1}^{q_2}(a_{j_2j_2s_1}-a_{j_2j_2s_2})^{n_{j_2s_2}}<0$,

$$(a_{j_1j_1r_1}, a_{j_2j_2s_1}) = \underbrace{([2(n_{j_1u_1 1}-n_{j_2s_1 1})]^{\text{th}}\ \text{DI:II},\ [(2m_{j_1r_1 1}+1),(2n_{j_2s_1 1}+1)]^{\text{th}}\ \text{SI})}_{([2(n_{j_1u_1 1}-n_{j_2s_1 1})],[(2m_{j_1r_1 1}+1),(2n_{j_2s_1 1}+1)])\text{-decreasing-to-increasing-inflection sink}}. \quad (1.202)$$

The equilibrium is a $([2(n_{j_1u_1 1}-n_{j_2s_1 1})]^{\text{th}}$ DI:II, $[(2m_{j_1r_1 1}+1),(2n_{j_2s_1 1}+1)]^{\text{th}}$ SI)-*decreasing-to-increasing-inflection sink at* $(x_{j_1}^*, x_{j_2}^*) = (a_{j_1j_1r_1}, a_{j_2j_2s_1})$.

(ii$_{2a2b}$) For $m_{j_1r_1}=2m_{j_1r_1 1}+1$, $n_{j_1u_1}=2n_{j_1u_1 1}$ *and* $n_{j_2s_1}=2n_{j_2s_1 1}+1$, *the equilibrium of* $(x_{j_1}^*, x_{j_2}^*) = (a_{j_1j_1r_1}, a_{j_2j_2s_1})$ *on the infinite-equilibrium* $x_{j_2}^* = a_{j_1j_2u_1} = a_{j_2j_2s_1}$ *has the following properties.*

- *For* $a_{j_1j_10}\prod_{r_2=1,r_2\neq r_1}^{p_1}(a_{j_1j_1r_1}-a_{j_1j_1r_2})^{m_{j_1r_2}}\prod_{u_2=1,u_2\neq u_1}^{q_1}(a_{j_1j_2u_1}-a_{j_1j_2u_2})^{n_{j_1u_2}}>0$ *and* $a_{j_2j_20}\prod_{s_2=1,s_2\neq s_1}^{q_2}(a_{j_2j_2s_1}-a_{j_2j_2s_2})^{n_{j_2s_2}}>0$,

$$(a_{j_1j_1r_1}, a_{j_2j_2s_1}) = \underbrace{([2(n_{j_1u_1 1}-n_{j_2s_1 1})-1]^{\text{th}}\ \text{DP:UP},\ [(2m_{j_1r_1 1}+1),(2n_{j_2s_1 1}+1)]^{\text{th}}\ \text{SO})}_{([2(n_{j_1u_1 1}-n_{j_2s_1 1})-1],[(2m_{j_1r_1 1}+1),(2n_{j_2s_1 1}+1)])\text{-hyperbolic source}}. \quad (1.203)$$

The equilibrium is a $([2(n_{j_1u_1 1}-n_{j_2s_1 1})-1]^{\text{th}}$ DP:UP,$[(2m_{j_1r_1 1}+1),(2n_{j_2s_1 1}+1)]^{\text{th}}$ SO)- *hyperbolic source at* $x_{j_2}^* = a_{j_1j_2u_1} = a_{j_2j_2s_1}$.

- *For* $a_{j_1j_10}\prod_{r_2=1,r_2\neq r_1}^{p_1}(a_{j_1j_1r_1}-a_{j_1j_1r_2})^{m_{j_1r_2}}\prod_{u_2=1,u_2\neq u_1}^{q_1}(a_{j_1j_2u_1}-a_{j_1j_2u_2})^{n_{j_1u_2}}<0$ *and* $a_{j_2j_20}\prod_{s_2=1,s_2\neq s_1}^{q_2}(a_{j_2j_2s_1}-a_{j_2j_2s_2})^{n_{j_2s_2}}>0$,

$$(a_{j_1j_1r_1},a_{j_2j_2s_1}) = \underbrace{([2(n_{j_1u_1 1}-n_{j_2s_1 1})-1]^{\text{th}}\text{ UP:DP, }[(2m_{j_1r_1 1}+1),(2n_{j_2s_1 1}+1)]^{\text{th}}\text{ SO})}_{([2(n_{j_1u_1 1}-n_{j_2s_1 1})-1],[(2m_{j_1r_1 1}+1),(2n_{j_2s_1 1}+1)])\text{-hyperbolic-secant source}}. \tag{1.204}$$

The equilibrium is a $([2(n_{j_1u_1 1}-n_{j_2s_1 1})-1]^{\text{th}}$ UP:DP,$[(2m_{j_1r_1 1}+1),(2n_{j_2s_1 1}+1)]^{\text{th}}$ SO)-*hyperbolic-secant source at* $(x_{j_1}^*,x_{j_2}^*) = (a_{j_1j_1r_1},a_{j_2j_2s_1})$.

- *For* $a_{j_1j_10}\prod_{r_2=1,r_2\neq r_1}^{p_1}(a_{j_1j_1r_1}-a_{j_1j_1r_2})^{m_{j_1r_2}}\prod_{u_2=1,u_2\neq u_1}^{q_1}(a_{j_1j_2u_1}-a_{j_1j_2u_2})^{n_{j_1u_2}}>0$ *and* $a_{j_2j_20}\prod_{s_2=1,s_2\neq s_1}^{q_2}(a_{j_2j_2s_1}-a_{j_2j_2s_2})^{n_{j_2s_2}}<0$,

$$(a_{j_1j_1r_1},a_{j_2j_2s_1}) = \underbrace{([2(n_{j_1u_1 1}-n_{j_2s_1 1})-1]^{\text{th}}\text{ UP:DP, }[(2m_{j_1r_1 1}+1),(2n_{j_2s_1 1}+1)]^{\text{th}}\text{ SI})}_{([2(n_{j_1u_1 1}-n_{j_2s_1 1})-1],[(2m_{j_1r_1 1}+1),(2n_{j_2s_1 1}+1)])\text{-hyperbolic-secant sink}}. \tag{1.205}$$

The equilibrium is a $([2(n_{j_1u_1 1}-n_{j_2s_1 1})-1]^{\text{th}}$ UP:DP,$[(2m_{j_1r_1 1}+1),(2n_{j_2s_1 1}+1)]^{\text{th}}$ SI)- *hyperbolic-secant sink at* $(x_{j_1}^*,x_{j_2}^*) = (a_{j_1j_1r_1},a_{j_2j_2s_1})$.

- *For* $a_{j_1j_10}\prod_{r_2=1,r_2\neq r_1}^{p_1}(a_{j_1j_1r_1}-a_{j_1j_1r_2})^{m_{j_1r_2}}\prod_{u_2=1,u_2\neq u_1}^{q_1}(a_{j_1j_2u_1}-a_{j_1j_2u_2})^{n_{j_1u_2}}<0$ *and* $a_{j_2j_20}\prod_{s_2=1,s_2\neq s_1}^{q_2}(a_{j_2j_2s_1}-a_{j_2j_2s_2})^{n_{j_2s_2}}<0$,

$$(a_{j_1j_1r_1},a_{j_2j_2s_1}) = \underbrace{([2(n_{j_1u_1 1}-n_{j_2s_1 1})-1]^{\text{th}}\text{ DP:UP, }[(2m_{j_1r_1 1}+1),(2n_{j_2s_1 1}+1)]^{\text{th}}\text{ SI})}_{([2(n_{j_1u_1 1}-n_{j_2s_1 1})-1],[(2m_{j_1r_1 1}+1),(2n_{j_2s_1 1}+1)])\text{-hyperbolic sink}}. \tag{1.206}$$

The equilibrium is a $([2(n_{j_1u_1 1}-n_{j_2s_1 1})-1]^{\text{th}}$ DP:UP,$[(2m_{j_1r_1 1}+1),(2n_{j_2s_1 1}+1)]^{\text{th}}$ SI)- *hyperbolic sink at* $(x_{j_1}^*,x_{j_2}^*) = (a_{j_1j_1r_1},a_{j_2j_2s_1})$.

(ii_{2a2c}) For $m_{j_1r_1}=2m_{j_1r_1 1}+1$, $n_{j_1u_1}=2n_{j_1u_1 1}+1$ *and* $n_{j_2s_1}=2n_{j_2s_1 1}$, *the equilibrium of* $(x_{j_1}^*,x_{j_2}^*) = (a_{j_1j_1u_1},a_{j_2j_2s_1})$ *on the infinite-equilibrium* $x_{j_2}^* = a_{j_1j_2u_1} = a_{j_2j_2s_1}$ *has the following properties.*

- *For* $a_{j_1j_10}\prod_{r_2=1,r_2\neq r_1}^{p_1}(a_{j_1j_1r_1}-a_{j_1j_1r_2})^{m_{j_1r_2}}\prod_{u_2=1,u_2\neq u_1}^{q_1}(a_{j_1j_2u_1}-a_{j_1j_2u_2})^{n_{j_1u_2}}>0$ *and* $a_{j_2j_20}\prod_{s_2=1,s_2\neq s_1}^{q_2}(a_{j_2j_2s_1}-a_{j_2j_2s_2})^{n_{j_2s_2}}>0$,

$$(a_{j_1j_1r_1}, a_{j_2j_2s_1})$$
$$= \underbrace{([2(n_{j_1u_1}1 - n_{j_2s_1}1) + 1]^{\text{th}}\,\text{DP:UP}, [(2m_{j_1r_1}1 + 1), (2n_{j_2s_1}1)]^{\text{th}}\,\text{US})}_{([2(n_{j_1u_1}1-n_{j_2s_1}1)+1],[(2m_{j_1r_1}1+1),(2n_{j_2s_1}1)])\text{-hyperbolic upper-saddle}}. \quad (1.207)$$

The equilibrium is a $([2(n_{j_1u_1}1-n_{j_2s_1}1)+1]^{\text{th}}$ DP:UP,$[(2m_{j_1r_1}1+1),(2n_{j_2s_1}1)]^{\text{th}}$ US)- *hyperbolic upper-saddle at* $(x^*_{j_1}, x^*_{j_2}) = (a_{j_1j_1r_1}, a_{j_2j_2s_1})$.

- *For* $a_{j_1j_10}\prod_{r_2=1,r_2\neq r_1}^{p_1}(a_{j_1j_1r_1} - a_{j_1j_1r_2})^{m_{j_1r_2}}\prod_{u_2=1,u_2\neq u_1}^{q_1}(a_{j_1j_2u_1} - a_{j_1j_2u_2})^{n_{j_1u_2}} < 0$ *and* $a_{j_2j_20}\prod_{s_2=1,s_2\neq s_1}^{q_2}(a_{j_2j_2s_1} - a_{j_2j_2s_2})^{n_{j_2s_2}} > 0$,

$$(a_{j_1j_1r_1}, a_{j_2j_2s_1})$$
$$= \underbrace{([2(n_{j_1u_1}1 - n_{j_2s_1}1) + 1]^{\text{th}}\,\text{UP:DP}, [(2m_{j_1r_1}1 + 1), (2n_{j_2s_1}1)]^{\text{th}}\,\text{US})}_{([2(n_{j_1u_1}1-n_{j_2s_1}1)+1],[(2m_{j_1r_1}1+1),(2n_{j_2s_1}1)])\text{-hyperbolic-secant upper-saddle}}. \quad (1.208)$$

The equilibrium is a $([2(n_{j_1u_1}1-n_{j_2s_1}1)+1]^{\text{th}}$ UP:DP,$[(2m_{j_1r_1}1+1),(2n_{j_2s_1}1)]^{\text{th}}$ US)- *hyperbolic-secant upper-saddle at* $(x^*_{j_1}, x^*_{j_2}) = (a_{j_1j_1r_1}, a_{j_2j_2s_1})$.

- *For* $a_{j_1j_10}\prod_{r_2=1,r_2\neq r_1}^{p_1}(a_{j_1j_1r_1} - a_{j_1j_1r_2})^{m_{j_1r_2}}\prod_{u_2=1,u_2\neq u_1}^{q_1}(a_{j_1j_2u_1} - a_{j_1j_2u_2})^{n_{j_1u_2}} > 0$ *and* $a_{j_2j_20}\prod_{s_2=1,s_2\neq s_1}^{q_2}(a_{j_2j_2s_1} - a_{j_2j_2s_2})^{n_{j_2s_2}} < 0$,

$$(a_{j_1j_1r_1}, a_{j_2j_2s_1})$$
$$= \underbrace{([2(n_{j_1u_1}1 - n_{j_2s_1}1) + 1]^{\text{th}}\,\text{UP:DP}, [(2m_{j_1r_1}1 + 1), (2n_{j_2s_1}1)]^{\text{th}}\,\text{LS})}_{([2(n_{j_1u_1}1-n_{j_2s_1}1)+1],[(2m_{j_1r_1}1+1),(2n_{j_2s_1}1)])\text{-hyperbolic-secant lower-saddle}}. \quad (1.209)$$

The equilibrium is a $([2(n_{j_1u_1}1-n_{j_2s_1}1)+1]^{\text{th}}$ UP:DP,$[(2m_{j_1r_1}1+1),(2n_{j_2s_1}1)]^{\text{th}}$ LS)-*hyperbolic-secant lower-saddle at* $(x^*_{j_1}, x^*_{j_2}) = (a_{j_1j_1r_1}, a_{j_2j_2s_1})$.

- *For* $a_{j_1j_10}\prod_{r_2=1,r_2\neq r_1}^{p_1}(a_{j_1j_1r_1} - a_{j_1j_1r_2})^{m_{j_1r_2}}\prod_{u_2=1,u_2\neq u_1}^{q_1}(a_{j_1j_2u_1} - a_{j_1j_2u_2})^{n_{j_1u_2}} < 0$ *and* $a_{j_2j_20}\prod_{s_2=1,s_2\neq s_1}^{q_2}(a_{j_2j_2s_1} - a_{j_2j_2s_2})^{n_{j_2s_2}} < 0$,

$$(a_{j_1j_1r_1}, a_{j_2j_2s_1})$$
$$= \underbrace{([2(n_{j_1u_1}1 - n_{j_2s_1}1) + 1]^{\text{th}}\,\text{DP:UP}, [(2m_{j_1r_1}1 + 1), (2n_{j_2s_1}1)]^{\text{th}}\,\text{LS})}_{([2(n_{j_1u_1}1-n_{j_2s_1}1)+1],[(2m_{j_1r_1}1+1),(2n_{j_2s_1}1)])\text{-hyperbolic lower-saddle}}. \quad (1.210)$$

The equilibrium is a $([2(n_{j_1u_1}1-n_{j_2s_1}1)-1]^{\text{th}}$ DP:UP,$[(2m_{j_1r_1}1+1),(2n_{j_2s_1}1)]^{\text{th}}$ LS)-*hyperbolic lower-saddle at* $(x^*_{j_1}, x^*_{j_2}) = (a_{j_1j_1r_1}, a_{j_2j_2s_1})$.

(ii_{2a2d}) For $m_{j_1r_1} = 2m_{j_1r_1}1 + 1$, $n_{j_1u_1} = 2n_{j_1u_1}1$ *and* $n_{j_2s_1} = 2n_{j_2s_1}1$, *the equilibrium of* $(x^*_{j_1}, x^*_{j_2}) = (a_{j_1j_1r_1}, a_{j_2j_2s_1})$ *on the infinite-equilibrium* $x^*_{j_2} = a_{j_1j_2u_1} = a_{j_2j_2s_1}$ *has the following properties.*

- *For* $a_{j_1j_10}\prod_{r_2=1,r_2\neq r_1}^{p_1}(a_{j_1j_1r_1} - a_{j_1j_1r_2})^{m_{j_1r_2}}\prod_{u_2=1,u_2\neq u_1}^{q_1}(a_{j_1j_2u_1} - a_{j_1j_2u_2})^{n_{j_1u_2}} > 0$ *and* $a_{j_2j_20}\prod_{s_2=1,s_2\neq s_1}^{q_2}(a_{j_2j_2s_1} - a_{j_2j_2s_2})^{n_{j_2s_2}} > 0$,

$$
\begin{aligned}
&(a_{j_1j_1r_1}, a_{j_2j_2s_1}) \\
&= \underbrace{([2(n_{j_1u_1}1 - n_{j_2s_1}1)]^{\text{th}}\ \text{DI:II}, [(2m_{j_1r_1}1+1), (2n_{j_2s_1}1)]^{\text{th}}\ \text{US})}_{([2(n_{j_1u_1}1-n_{j_2s_1}1)],[(2m_{j_1r_1}1+1),(2n_{j_2s_1}1)])\text{-decreasing-to-increasing inflection upper-saddle}}.
\end{aligned} \tag{1.211}
$$

The equilibrium is a $([2(n_{j_1u_1}1 - n_{j_2s_1}1)]^{\text{th}}$ DI:II, $[(2m_{j_1r_1}1+1), (2n_{j_2s_1}1)]^{\text{th}}$ US)-*decreasing-to-increasing inflection upper-saddle at* $(x^*_{j_1}, x^*_{j_2}) = (a_{j_1j_1r_1}, a_{j_2j_2s_1})$.

- *For* $a_{j_1j_10}\prod_{r_2=1,r_2\neq r_1}^{p_1}(a_{j_1j_1r_1} - a_{j_1j_1r_2})^{m_{j_1r_2}}\prod_{u_2=1,u_2\neq u_1}^{q_1}(a_{j_1j_2u_1} - a_{j_1j_2u_2})^{n_{j_1u_2}} < 0$ *and* $a_{j_2j_20}\prod_{s_2=1,s_2\neq s_1}^{q_2}(a_{j_2j_2s_1} - a_{j_2j_2s_2})^{n_{j_2s_2}} > 0$,

$$
\begin{aligned}
&(a_{j_1j_1r_1}, a_{j_2j_2s_1}) \\
&= \underbrace{([2(n_{j_1u_1}1 - n_{j_2s_1}1)]^{\text{th}}\ \text{II:DI}, [(2m_{j_1r_1}1+1), (2n_{j_2s_1}1)]^{\text{th}}\ \text{US})}_{([2(n_{j_1u_1}1-n_{j_2s_1}1)],[(2m_{j_1r_1}1+1),(2n_{j_2s_1}1)])\text{-increasing-to-decreasing inflection upper-saddle}}.
\end{aligned} \tag{1.212}
$$

The equilibrium is a $([2(n_{j_1u_1}1 - n_{j_2s_1}1)]^{\text{th}}$ II:DI, $[(2m_{j_1r_1}1+1), (2n_{j_2s_1}1)]^{\text{th}}$ US)-*increasing-to-decreasing inflection upper-saddle at* $(x^*_{j_1}, x^*_{j_2}) = (a_{j_1j_1r_1}, a_{j_2j_2s_1})$.

- *For* $a_{j_1j_10}\prod_{r_2=1,r_2\neq r_1}^{p_1}(a_{j_1j_1r_1} - a_{j_1j_1r_2})^{m_{j_1r_2}}\prod_{u_2=1,u_2\neq u_1}^{q_1}(a_{j_1j_2u_1} - a_{j_1j_2u_2})^{n_{j_1u_2}} > 0$ *and* $a_{j_2j_20}\prod_{s_2=1,s_2\neq s_1}^{q_2}(a_{j_2j_2s_1} - a_{j_2j_2s_2})^{n_{j_2s_2}} < 0$,

$$
\begin{aligned}
&(a_{j_1j_1r_1}, a_{j_2j_2s_1}) \\
&= \underbrace{([2(n_{j_1u_1}1 - n_{j_2s_1}1)]^{\text{th}}\ \text{II:DI}, [(2m_{j_1r_1}1+1), (2n_{j_2s_1}1)]^{\text{th}}\ \text{LS})}_{([2(n_{j_1u_1}1-n_{j_2s_1}1)],[(2m_{j_1r_1}1+1),(2n_{j_2s_1}1)])\text{-increasing-to-decreasing inflection lower-saddle}}.
\end{aligned} \tag{1.213}
$$

The equilibrium is a $([2(n_{j_1u_1}1 - n_{j_2s_1}1)]^{\text{th}}$ II:DI, $[(2m_{j_1r_1}1+1), (2n_{j_2s_1}1)]^{\text{th}}$ LS)-*increasing-to-decreasing inflection lower-saddle at* $(x^*_{j_1}, x^*_{j_2}) = (a_{j_1j_1r_1}, a_{j_2j_2s_1})$.

- *For* $a_{j_1j_10}\prod_{r_2=1,r_2\neq r_1}^{p_1}(a_{j_1j_1r_1} - a_{j_1j_1r_2})^{m_{j_1r_2}}\prod_{u_2=1,u_2\neq u_1}^{q_1}(a_{j_1j_2u_1} - a_{j_1j_2u_2})^{n_{j_1u_2}} < 0$ *and* $a_{j_2j_20}\prod_{s_2=1,s_2\neq s_1}^{q_2}(a_{j_2j_2s_1} - a_{j_2j_2s_2})^{n_{j_2s_2}} < 0$,

$$
\begin{aligned}
&(a_{j_1j_1r_1}, a_{j_2j_2s_1}) \\
&= \underbrace{([2(n_{j_1u_1}1 - n_{j_2s_1}1)]^{\text{th}}\ \text{DI:II}, [(2m_{j_1r_1}1+1), (2n_{j_2s_1}1)]^{\text{th}}\ \text{LS})}_{([2(n_{j_1u_1}1-n_{j_2s_1}1)],[(2m_{j_1r_1}1+1),(2n_{j_2s_1}1)])\text{-decreasing-to-increasing inflection lower-saddle}}.
\end{aligned} \tag{1.214}
$$

The equilibrium is a $([2(n_{j_1u_1}1 - n_{j_2s_1}1)]^{\text{th}}$ DI:II, $[(2m_{j_1r_1}1+1), (2n_{j_2s_1}1)]^{\text{th}}$ LS)-*decreasing-to-increasing inflection lower-saddle at* $(x^*_{j_1}, x^*_{j_2}) = (a_{j_1j_1r_1}, a_{j_2j_2s_1})$.

(ii_{2a2e}) For $m_{j_1r_1} = 2m_{j_1r_1}1, n_{j_1u_1} = 2n_{j_1u_1}1 + 1$ *and* $n_{j_2s_1} = 2n_{j_2s_1}1 + 1$, *the equilibrium of* $(x^*_{j_1}, x^*_{j_2}) = (a_{j_1j_1r_1}, a_{j_2j_2s_1})$ *on the infinite-equilibrium* $x^*_{j_2} = a_{j_1j_2u_1} = a_{j_2j_2s_1}$ *has the following properties.*

- *For* $a_{j_1j_10}\prod_{r_2=1,r_2\neq r_1}^{p_1}(a_{j_1j_1r_1} - a_{j_1j_1r_2})^{m_{j_1r_2}}\prod_{u_2=1,u_2\neq u_1}^{q_1}(a_{j_1j_2u_1} - a_{j_1j_2u_2})^{n_{j_1u_2}} > 0$ *and* $a_{j_2j_20}\prod_{s_2=1,s_2\neq s_1}^{q_2}(a_{j_2j_2s_1} - a_{j_2j_2s_2})^{n_{j_2s_2}} > 0$,

$$(a_{j_1j_1r_1}, a_{j_2j_2s_1}) = \underbrace{([2(n_{j_1u_1}1 - n_{j_2s_1}1)]^{\text{th}} \text{ II:II}, [(2m_{j_1r_1}1), (2n_{j_2s_1}1 + 1)]^{\text{th}} \text{ SO})}_{([2(n_{j_1u_1}1-n_{j_2s_1}1)],[(2m_{j_1r_1}1),(2n_{j_2s_1}1+1)])\text{-increasing-inflection source}}. \tag{1.215}$$

The equilibrium is a ($[2(n_{j_1u_1}1 - n_{j_2s_1}1)]^{\text{th}}$ II:II, $[(2m_{j_1r_1}1), (2n_{j_2s_1}1 + 1)]^{\text{th}}$ SO)*-increasing-inflection source at* $(x_{j_1}^*, x_{j_2}^*) = (a_{j_1j_1r_1}, a_{j_2j_2s_1})$.

- *For* $a_{j_1j_10}\prod_{r_2=1,r_2\neq r_1}^{p_1}(a_{j_1j_1r_1} - a_{j_1j_1r_2})^{m_{j_1r_2}}\prod_{u_2=1,u_2\neq u_1}^{q_1}(a_{j_1j_2u_1} - a_{j_1j_2u_2})^{n_{j_1u_2}} < 0$ *and* $a_{j_2j_20}\prod_{s_2=1,s_2\neq s_1}^{q_2}(a_{j_2j_2s_1} - a_{j_2j_2s_2})^{n_{j_2s_2}} > 0$,

$$(a_{j_1j_1r_1}, a_{j_2j_2s_1}) = \underbrace{([2(n_{j_1u_1}1 - n_{j_2s_1}1)]^{\text{th}} \text{ DI:DI}, [(2m_{j_1r_1}1), (2n_{j_2s_1}1 + 1)]^{\text{th}} \text{ SO})}_{([2(n_{j_1u_1}1-n_{j_2s_1}1)],[(2m_{j_1r_1}1),(2n_{j_2s_1}1+1)])\text{-decreasing-inflection source}}. \tag{1.216}$$

The equilibrium is a ($[2(n_{j_1u_1}1 - n_{j_2s_1}1)]^{\text{th}}$ DI:DI, $[(2m_{j_1r_1}1), (2n_{j_2s_1}1 + 1)]^{\text{th}}$ SO)*-decreasing-inflection source at* $(x_{j_1}^*, x_{j_2}^*) = (a_{j_1j_1r_1}, a_{j_2j_2s_1})$.

- *For* $a_{j_1j_10}\prod_{r_2=1,r_2\neq r_1}^{p_1}(a_{j_1j_1r_1} - a_{j_1j_1r_2})^{m_{j_1r_2}}\prod_{u_2=1,u_2\neq u_1}^{q_1}(a_{j_1j_2u_1} - a_{j_1j_2u_2})^{n_{j_1u_2}} > 0$ *and* $a_{j_2j_20}\prod_{s_2=1,s_2\neq s_1}^{q_2}(a_{j_2j_2s_1} - a_{j_2j_2s_2})^{n_{j_2s_2}} < 0$,

$$(a_{j_1j_1r_1}, a_{j_2j_2s_1}) = \underbrace{([2(n_{j_1u_1}1 - n_{j_2s_1}1)]^{\text{th}} \text{ DI:DI}, [(2m_{j_1r_1}1), (2n_{j_2s_1}1 + 1)]^{\text{th}} \text{ SI})}_{([2(n_{j_1u_1}1-n_{j_2s_1}1)],[(2m_{j_1r_1}1),(2n_{j_2s_1}1+1)])\text{-decreasing-inflection sink}}. \tag{1.217}$$

The equilibrium is a ($[2(n_{j_1u_1}1 - n_{j_2s_1}1)]^{\text{th}}$ DI:DI, $[(2m_{j_1r_1}1), (2n_{j_2s_1}1 + 1)]^{\text{th}}$ SI)*−decreasing-inflection sink at* $(x_{j_1}^*, x_{j_2}^*) = (a_{j_1j_1r_1}, a_{j_2j_2s_1})$.

- *For* $a_{j_1j_10}\prod_{r_2=1,r_2\neq r_1}^{p_1}(a_{j_1j_1r_1} - a_{j_1j_1r_2})^{m_{j_1r_2}}\prod_{u_2=1,u_2\neq u_1}^{q_1}(a_{j_1j_2u_1} - a_{j_1j_2u_2})^{n_{j_1u_2}} < 0$ *and* $a_{j_2j_20}\prod_{s_2=1,s_2\neq s_1}^{q_2}(a_{j_2j_2s_1} - a_{j_2j_2s_2})^{n_{j_2s_2}} < 0$,

$$(a_{j_1j_1r_1}, a_{j_2j_2s_1}) = \underbrace{([2(n_{j_1u_1}1 - n_{j_2s_1}1)]^{\text{th}} \text{ II:II}, [(2m_{j_1r_1}1), (2n_{j_2s_1}1 + 1)]^{\text{th}} \text{ SI}])}_{([2(n_{j_1u_1}1-n_{j_2s_1}1)],[(2m_{j_1r_1}1),(2n_{j_2s_1}1+1)])\text{-increasing-inflection sink}}. \tag{1.218}$$

The equilibrium is a ($[2(n_{j_1u_1}1 - n_{j_2s_1}1)]^{\text{th}}$ II:II, $[(2m_{j_1r_1}1), (2n_{j_2s_1}1 + 1)]^{\text{th}}$ SI)*-increasing-inflection sink at* $(x_{j_1}^*, x_{j_2}^*) = (a_{j_1j_1r_1}, a_{j_2j_2s_1})$.

(ii_{2a2f}) For $m_{j_1r_1} = 2m_{j_1r_1}1$, $n_{j_1u_1} = 2n_{j_1u_1}1$ *and* $n_{j_2s_1} = 2n_{j_2s_1}1 + 1$, *the equilibrium of* $(x_{j_1}^*, x_{j_2}^*) = (a_{j_1j_1r_1}, a_{j_2j_2s_1})$ *on the infinite-equilibrium* $x_{j_2}^* = a_{j_1j_2u_1} = a_{j_2j_2s_1}$ *has the following properties*

- *For* $a_{j_1j_10}\prod_{r_2=1,r_2\neq r_1}^{p_1}(a_{j_1j_1r_1} - a_{j_1j_1r_2})^{m_{j_1r_2}}\prod_{u_2=1,u_2\neq u_1}^{q_1}(a_{j_1j_2u_1} - a_{j_1j_2u_2})^{n_{j_1u_2}} > 0$ *and* $a_{j_2j_20}\prod_{s_2=1,s_2\neq s_1}^{q_2}(a_{j_2j_2s_1} - a_{j_2j_2s_2})^{n_{j_2s_2}} > 0$,

$$(a_{j_1j_1r_1}, a_{j_2j_2s_1}) = \underbrace{([2(n_{j_1u_1 1} - n_{j_2s_1 1}) - 1]^{\text{th}}\ \text{UP:UP}, [(2m_{j_1r_1 1}), (2n_{j_2s_1 1} + 1)]^{\text{th}}\ \text{SO})}_{([2(n_{j_1u_1 1}-n_{j_2s_1 1})-1],[(2m_{j_1r_1 1}),(2n_{j_2s_1 1}+1)])\text{-up-parabola source}}. \tag{1.219}$$

The equilibrium is a $([2(n_{j_1u_1 1}-n_{j_2s_1 1})-1]^{\text{th}}$ UP:UP,$[(2m_{j_1r_1 1}),(2n_{j_2s_1 1}+1)]^{\text{th}}$ SO)*-up-parabola source at* $(x_{j_1}^*, x_{j_2}^*) = (a_{j_1j_1r_1}, a_{j_2j_2s_1})$.

- *For* $a_{j_1j_10}\prod_{r_2=1,r_2\neq r_1}^{p_1}(a_{j_1j_1r_1} - a_{j_1j_1r_2})^{m_{j_1r_2}}\prod_{u_2=1,u_2\neq u_1}^{q_1}(a_{j_1j_2u_1} - a_{j_1j_2u_2})^{n_{j_1u_2}} < 0$ *and* $a_{j_2j_20}\prod_{s_2=1,s_2\neq s_1}^{q_2}(a_{j_2j_2s_1} - a_{j_2j_2s_2})^{n_{j_2s_2}} > 0$,

$$(a_{j_1j_1r_1}, a_{j_2j_2s_1}) = \underbrace{([2(n_{j_1u_1 1} - n_{j_2s_1 1}) - 1]^{\text{th}}\ \text{DP:DP}, [(2m_{j_1r_1 1}), (2n_{j_2s_1 1} + 1)]^{\text{th}}\ \text{SO})}_{([2(n_{j_1u_1 1}-n_{j_2s_1 1})-1],[(2m_{j_1r_1 1}),(2n_{j_2s_1 1}+1)])\text{-down-parabola source}}. \tag{1.220}$$

The equilibrium is a $([2(n_{j_1u_1 1}-n_{j_2s_1 1})-1]^{\text{th}}$ DP:DP,$[(2m_{j_1r_1 1}),(2n_{j_2s_1 1}+1)]^{\text{th}}$ SO)*-down-parabola source at* $(x_{j_1}^*, x_{j_2}^*) = (a_{j_1j_1r_1}, a_{j_2j_2s_1})$.

- *For* $a_{j_1j_10}\prod_{r_2=1,r_2\neq r_1}^{p_1}(a_{j_1j_1r_1} - a_{j_1j_1r_2})^{m_{j_1r_2}}\prod_{u_2=1,u_2\neq u_1}^{q_1}(a_{j_1j_2u_1} - a_{j_1j_2u_2})^{n_{j_1u_2}} > 0$ *and* $a_{j_2j_20}\prod_{s_2=1,s_2\neq s_1}^{q_2}(a_{j_2j_2s_1} - a_{j_2j_2s_2})^{n_{j_2s_2}} < 0$,

$$(a_{j_1j_1r_1}, a_{j_2j_2s_1}) = \underbrace{([2(n_{j_1u_1 1} - n_{j_2s_1 1}) - 1]^{\text{th}}\ \text{DP:DP}, [(2m_{j_1r_1 1}), (2n_{j_2s_1 1} + 1)]^{\text{th}}\ \text{SI})}_{([2(n_{j_1u_1 1}-n_{j_2s_1 1})-1],[(2m_{j_1r_1 1}),(2n_{j_2s_1 1}+1)])\text{-down-parabola sink}}. \tag{1.221}$$

The equilibrium is a $([2(n_{j_1u_1 1}-n_{j_2s_1 1})-1]^{\text{th}}$ DP:DP,$[(2m_{j_1r_1 1}),(2n_{j_2s_1 1}+1)]^{\text{th}}$ SI)*-down-parabola sink at* $(x_{j_1}^*, x_{j_2}^*) = (a_{j_1j_1r_1}, a_{j_2j_2s_1})$.

- *For* $a_{j_1j_10}\prod_{r_2=1,r_2\neq r_1}^{p_1}(a_{j_1j_1r_1} - a_{j_1j_1r_2})^{m_{j_1r_2}}\prod_{u_2=1,u_2\neq u_1}^{q_1}(a_{j_1j_2u_1} - a_{j_1j_2u_2})^{n_{j_1u_2}} < 0$ *and* $a_{j_2j_20}\prod_{s_2=1,s_2\neq s_1}^{q_2}(a_{j_2j_2s_1} - a_{j_2j_2s_2})^{n_{j_2s_2}} < 0$,

$$(a_{j_1j_1r_1}, a_{j_2j_2s_1}) = \underbrace{([2(n_{j_1u_1 1} - n_{j_2s_1 1}) - 1]^{\text{th}}\ \text{UP:UP}, [(2m_{j_1r_1 1}), (2n_{j_2s_1 1} + 1)]^{\text{th}}\ \text{SI})}_{([2(n_{j_1u_1 1}-n_{j_2s_1 1})-1],[(2m_{j_1r_1 1}),(2n_{j_2s_1 1}+1)])\text{-up-parabola sink}}. \tag{1.222}$$

The equilibrium is a $([2(n_{j_1u_1 1}-n_{j_2s_1 1})-1]^{\text{th}}$ UP:UP,$[(2m_{j_1r_1 1}),(2n_{j_2s_1 1}+1)]^{\text{th}}$ SI)*-up-parabola sink at* $(x_{j_1}^*, x_{j_2}^*) = (a_{j_1j_1r_1}, a_{j_2j_2s_1})$.

(ii$_{2a2g}$) For $m_{j_1r_1} = 2m_{j_1r_1 1}+1$, $n_{j_1u_1} = 2n_{j_1u_1 1}+1$ *and* $n_{j_2s_1} = 2n_{j_2s_1 1}$, *the equilibrium of* $(x_{j_1}^*, x_{j_2}^*) = (a_{j_1j_1r_1}, a_{j_2j_2s_1})$ *on the infinite-equilibrium* $x_{j_2}^* = a_{j_1j_2u_1} = a_{j_2j_2s_1}$ *has the following properties.*

- *For* $a_{j_1j_10}\prod_{r_2=1,r_2\neq r_1}^{p_1}(a_{j_1j_1r_1} - a_{j_1j_1r_2})^{m_{j_1r_2}}\prod_{u_2=1,u_2\neq u_1}^{q_1}(a_{j_1j_2u_1} - a_{j_1j_2u_2})^{n_{j_1u_2}} > 0$ *and* $a_{j_2j_20}\prod_{s_2=1,s_2\neq s_1}^{q_2}(a_{j_2j_2s_1} - a_{j_2j_2s_2})^{n_{j_2s_2}} > 0$,

$$(a_{j_1j_1r_1}, a_{j_2j_2s_1}) = \underbrace{([2(n_{j_1u_1 1} - n_{j_2s_1 1}) + 1]^{\text{th}} \text{ UP:UP}, [(2m_{j_1r_1 1}), (2n_{j_2s_1 1})]^{\text{th}} \text{ US})}_{([2(n_{j_1u_1 1}-n_{j_2s_1 1})+1],[(2m_{j_1r_1 1}),(2n_{j_2s_1 1})])\text{-up-parabola upper-saddle}}. \tag{1.223}$$

The equilibrium is a $([2(n_{j_1u_1 1}-n_{j_2s_1 1})+1]^{\text{th}}$ UP:UP, $[(2m_{j_1r_1 1}), (2n_{j_2s_1 1})]^{\text{th}}$ US)-*up-parabola upper-saddle at* $(x_{j_1}^*, x_{j_2}^*) = (a_{j_1j_1r_1}, a_{j_2j_2s_1})$.

- *For* $a_{j_1j_10}\prod_{r_2=1,r_2\neq r_1}^{p_1}(a_{j_1j_1r_1} - a_{j_1j_1r_2})^{m_{j_1r_2}}\prod_{u_2=1,u_2\neq u_1}^{q_1}(a_{j_1j_2u_1} - a_{j_1j_2u_2})^{n_{j_1u_2}} < 0$ *and* $a_{j_2j_20}\prod_{s_2=1,s_2\neq s_1}^{q_2}(a_{j_2j_2s_1} - a_{j_2j_2s_2})^{n_{j_2s_2}} > 0$,

$$(a_{j_1j_1r_1}, a_{j_2j_2s_1}) = \underbrace{([2(n_{j_1u_1 1} - n_{j_2s_1 1}) + 1]^{\text{th}} \text{ DP:DP}, [(2m_{j_1r_1 1}), (2n_{j_2s_1 1})]^{\text{th}} \text{ US})}_{([2(n_{j_1u_1 1}-n_{j_2s_1 1})+1],[(2m_{j_1r_1 1}),(2n_{j_2s_1 1})])\text{-down-parabola upper-saddle}}. \tag{1.224}$$

The equilibrium is a $([2(n_{j_1u_1 1}-n_{j_2s_1 1})+1]^{\text{th}}$ DP:DP, $[(2m_{j_1r_1 1}), (2n_{j_2s_1 1})]^{\text{th}}$ US)-*down-parabola upper-saddle at* $(x_{j_1}^*, x_{j_2}^*) = (a_{j_1j_1r_1}, a_{j_2j_2s_1})$.

- *For* $a_{j_1j_10}\prod_{r_2=1,r_2\neq r_1}^{p_1}(a_{j_1j_1r_1} - a_{j_1j_1r_2})^{m_{j_1r_2}}\prod_{u_2=1,u_2\neq u_1}^{q_1}(a_{j_1j_2u_1} - a_{j_1j_2u_2})^{n_{j_1u_2}} > 0$ *and* $a_{j_2j_20}\prod_{s_2=1,s_2\neq s_1}^{q_2}(a_{j_2j_2s_1} - a_{j_2j_2s_2})^{n_{j_2s_2}} < 0$,

$$(a_{j_1j_1r_1}, a_{j_2j_2s_1}) = \underbrace{([2(n_{j_1u_1 1} - n_{j_2s_1 1}) + 1]^{\text{th}} \text{ DP:DP}, [(2m_{j_1r_1 1}), (2n_{j_2s_1 1})]^{\text{th}} \text{ LS})}_{([2(n_{j_1u_1 1}-n_{j_2s_1 1})+1],[(2m_{j_1r_1 1}),(2n_{j_2s_1 1})])\text{-down-parabola lower-saddle}}. \tag{1.225}$$

The equilibrium is a $([2(n_{j_1u_1 1}-n_{j_2s_1 1})+1]^{\text{th}}$ DP:DP, $[(2m_{j_1r_1 1}), (2n_{j_2s_1 1})]^{\text{th}}$ LS)-*down-parabola lower-saddle at* $(x_{j_1}^*, x_{j_2}^*) = (a_{j_1j_1r_1}, a_{j_2j_2s_1})$.

- *For* $a_{j_1j_10}\prod_{r_2=1,r_2\neq r_1}^{p_1}(a_{j_1j_1r_1} - a_{j_1j_1r_2})^{m_{j_1r_2}}\prod_{u_2=1,u_2\neq u_1}^{q_1}(a_{j_1j_2u_1} - a_{j_1j_2u_2})^{n_{j_1u_2}} < 0$ *and* $a_{j_2j_20}\prod_{s_2=1,s_2\neq s_1}^{q_2}(a_{j_2j_2s_1} - a_{j_2j_2s_2})^{n_{j_2s_2}} < 0$,

$$(a_{j_1j_1r_1}, a_{j_2j_2s_1}) = \underbrace{([2(n_{j_1u_1 1} - n_{j_2s_1 1}) + 1]^{\text{th}} \text{ UP:UP}, [(2m_{j_1r_1 1}), (2n_{j_2s_1 1})]^{\text{th}} \text{ LS})}_{([2(n_{j_1u_1 1}-n_{j_2s_1 1})+1],[(2m_{j_1r_1 1}),(2n_{j_2s_1 1})])\text{-up-parabola lower-saddle}}. \tag{1.226}$$

The equilibrium is a $([2(n_{j_1u_1 1}-n_{j_2s_1 1})+1]^{\text{th}}$ UP:UP, $[(2m_{j_1r_1 1}), (2n_{j_2s_1 1})]^{\text{th}}$ LS)-*up-parabola lower-saddle at* $(x_{j_1}^*, x_{j_2}^*) = (a_{j_1j_1r_1}, a_{j_2j_2s_1})$.

(ii_{2a2h}) *For* $m_{j_1r_1} = 2m_{j_1r_1 1}, n_{j_1u_1} = 2n_{j_1u_1 1}$ *and* $n_{j_2s_1} = 2n_{j_2s_1 1}$, *the equilibrium of* $(x_{j_1}^*, x_{j_2}^*) = (a_{j_1j_1r_1}, a_{j_2j_2s_1})$ *on the infinite-equilibrium* $x_{j_2}^* = a_{j_1j_2u_1} = a_{j_2j_2s_1}$ *has the following properties.*

- *For* $a_{j_1j_10}\prod_{r_2=1,r_2\neq r_1}^{p_1}(a_{j_1j_1r_1} - a_{j_1j_1r_2})^{m_{j_1r_2}}\prod_{u_2=1,u_2\neq u_1}^{q_1}(a_{j_1j_2u_1} - a_{j_1j_2u_2})^{n_{j_1u_2}} > 0$ *and* $a_{j_2j_20}\prod_{s_2=1,s_2\neq s_1}^{q_2}(a_{j_2j_2s_1} - a_{j_2j_2s_2})^{n_{j_2s_2}} > 0$,

$$(a_{j_1j_1r_1}, a_{j_2j_2s_1}) = \underbrace{([2(n_{j_1u_1 1} - n_{j_2s_1 1})]^{\text{th}}\ \text{II:II}, [(2m_{j_1r_1 1}), (2n_{j_2s_1 1})]^{\text{th}}\ \text{US})}_{([2(n_{j_1u_1 1}-n_{j_2s_1 1})],[(2m_{j_1r_1 1}),(2n_{j_2s_1 1})])\text{-increasing inflection upper-saddle}}. \tag{1.227}$$

The equilibrium is a $([2(n_{j_1u_1 1} - n_{j_2s_1 1})]^{\text{th}}\ \text{II:II}, [(2m_{j_1r_1 1}), (2n_{j_2s_1 1})]^{\text{th}}\ \text{US})$*-increasing inflection upper-saddle at* $(x_{j_1}^*, x_{j_2}^*) = (a_{j_1j_1r_1}, a_{j_2j_2s_1})$.

- *For* $a_{j_1j_1 0}\prod_{r_2=1,r_2\neq r_1}^{p_1}(a_{j_1j_1r_1} - a_{j_1j_1r_2})^{m_{j_1r_2}}\prod_{u_2=1,u_2\neq u_1}^{q_1}(a_{j_1j_2u_1} - a_{j_1j_2u_2})^{n_{j_1u_2}} < 0$ *and* $a_{j_2j_2 0}\prod_{s_2=1,s_2\neq s_1}^{q_2}(a_{j_2j_2s_1} - a_{j_2j_2s_2})^{n_{j_2s_2}} > 0$,

$$(a_{j_1j_1r_1}, a_{j_2j_2s_1}) = \underbrace{([2(n_{j_1u_1 1} - n_{j_2s_1 1})]^{\text{th}}\ \text{DI:DI}, [(2m_{j_1r_1 1}), (2n_{j_2s_1 1})]^{\text{th}}\ \text{US})}_{([2(n_{j_1u_1 1}-n_{j_2s_1 1})],[(2m_{j_1r_1 1}),(2n_{j_2s_1 1})])\text{-decreasing inflection upper-saddle}}. \tag{1.228}$$

The equilibrium is a $([2(n_{j_1u_1 1} - n_{j_2s_1 1})]^{\text{th}}\ \text{DI:DI}, [(2m_{j_1r_1 1}), (2n_{j_2s_1 1})]^{\text{th}}\ \text{US})$*-decreasing inflection upper-saddle at* $(x_{j_1}^*, x_{j_2}^*) = (a_{j_1j_1r_1}, a_{j_2j_2s_1})$.

- *For* $a_{j_1j_1 0}\prod_{r_2=1,r_2\neq r_1}^{p_1}(a_{j_1j_1r_1} - a_{j_1j_1r_2})^{m_{j_1r_2}}\prod_{u_2=1,u_2\neq u_1}^{q_1}(a_{j_1j_2u_1} - a_{j_1j_2u_2})^{n_{j_1u_2}} > 0$ *and* $a_{j_2j_2 0}\prod_{s_2=1,s_2\neq s_1}^{q_2}(a_{j_2j_2s_1} - a_{j_2j_2s_2})^{n_{j_2s_2}} < 0$,

$$(a_{j_1j_1r_1}, a_{j_2j_2s_1}) = \underbrace{([2(n_{j_1u_1 1} - n_{j_2s_1 1})]^{\text{th}}\ \text{DI:DI}, [(2m_{j_1r_1 1}), (2n_{j_2s_1 1})]^{\text{th}}\ \text{LS})}_{([2(n_{j_1u_1 1}-n_{j_2s_1 1})],[(2m_{j_1r_1 1}),(2n_{j_2s_1 1})])\text{-decreasing inflection lower-saddle}}. \tag{1.229}$$

The equilibrium is a $([2(n_{j_1u_1 1} - n_{j_2s_1 1})]^{\text{th}}\ \text{DI:DI}, [(2m_{j_1r_1 1}), (2n_{j_2s_1 1})]^{\text{th}}\ \text{LS})$*-decreasing inflection lower-saddle at* $(x_{j_1}^*, x_{j_2}^*) = (a_{j_1j_1r_1}, a_{j_2j_2s_1})$.

- *For* $a_{j_1j_1 0}\prod_{r_2=1,r_2\neq r_1}^{p_1}(a_{j_1j_1r_1} - a_{j_1j_1r_2})^{m_{j_1r_2}}\prod_{u_2=1,u_2\neq u_1}^{q_1}(a_{j_1j_2u_1} - a_{j_1j_2u_2})^{n_{j_1u_2}} < 0$ *and* $a_{j_2j_2 0}\prod_{s_2=1,s_2\neq s_1}^{q_2}(a_{j_2j_2s_1} - a_{j_2j_2s_2})^{n_{j_2s_2}} < 0$,

$$(a_{j_1j_1r_1}, a_{j_2j_2s_1}) = \underbrace{([2(n_{j_1u_1 1} - n_{j_2s_1 1})]^{\text{th}}\ \text{II:II}, [(2m_{j_1r_1 1}), (2n_{j_2s_1 1})]^{\text{th}}\ \text{LS})}_{([2(n_{j_1u_1 1}-n_{j_2s_1 1})],[(2m_{j_1r_1 1}),(2n_{j_2s_1 1})])\text{-increasing inflection lower-saddle}}. \tag{1.230}$$

The equilibrium is a $([2(n_{j_1u_1 1} - n_{j_2s_1 1})]^{\text{th}}\ \text{II:II}, [(2m_{j_1r_1 1}), (2n_{j_2s_1 1})]^{\text{th}}\ \text{LS})$*-increasing inflection lower-saddle at* $(x_{j_1}^*, x_{j_2}^*) = (a_{j_1j_1r_1}, a_{j_2j_2s_1})$.

(ii_{2b}) The infinite-equilibriums and switching bifurcations for $n_{j_2s_1} > n_{j_1u_1}$ *are presented.*

(ii_{2b1}) For $n_{j_2s_1} > n_{j_1u_1}$, *16 types of infinite-equilibriums exist at* $x_{j_2}^* = a_{j_1j_1r_1} = a_{j_2j_2s_1}$.

(ii_{2b1a}) For $n_{j_1u_1} = 2n_{j_1u_1 1} + 1$ *and* $n_{j_2s_1} = 2n_{j_2s_1 1} + 1$, *the infinite-equilibrium of* $x_{j_2}^* = a_{j_1j_2u_1} = a_{j_2j_2s_1}$ *with* $\overline{x}_{j_1} \neq a_{j_1j_1r_1}$ *has the following properties.*

- *For* $a_{j_1j_10}\prod_{r=1}^{p_1}(\overline{x}_{j_1}-a_{j_1j_1r})^{m_{j_1r}}\prod_{u_2=1,u_2\neq u_1}^{q_1}(a_{j_1j_2u_1}-a_{j_1j_2u_2})^{n_{j_1u_2}}>0$ *and* $a_{j_2j_20}\prod_{s_2=1,s_2\neq s_1}^{q_2}(a_{j_2j_2s_1}-a_{j_2j_2s_2})^{n_{j_2s_2}}>0$,

$$(\overline{x}_{j_1},a_{j_2j_2s_1})=\underbrace{([2(n_{j_2s_11}-n_{j_1u_11})]^{\text{th}}\,\text{IC},(2n_{j_2s_11}+1)^{\text{th}}\,\text{SO})}_{([2(n_{j_2s_11}-n_{j_1u_11})],(2n_{j_2s_11}+1))\text{-increasing-concve source}}. \tag{1.231}$$

The infinite-equilibrium is a $([2(n_{j_2s_11}-n_{j_1u_11})]^{\text{th}}\,\text{IC},(2n_{j_2s_11}+1)^{\text{th}}\,\text{SO})$*-increasing-concave source at* $x_{j_2}^*=a_{j_1j_2u_1}=a_{j_2j_2s_1}$.

- *For* $a_{j_1j_10}\prod_{r=1}^{p_1}(\overline{x}_{j_1}-a_{j_1j_1r})^{m_{j_1r}}\prod_{u_2=1,u_2\neq u_1}^{q_1}(a_{j_1j_2u_1}-a_{j_1j_2u_2})^{n_{j_1u_2}}<0$ *and* $a_{j_2j_20}\prod_{s_2=1,s_2\neq s_1}^{q_2}(a_{j_2j_2s_1}-a_{j_2j_2s_2})^{n_{j_2s_2}}>0$,

$$(\overline{x}_{j_1},a_{j_2j_2s_1})=\underbrace{([2(n_{j_2s_11}-n_{j_1u_11})]^{\text{th}}\,\text{DC},(2n_{j_2s_11}+1)^{\text{th}}\,\text{SO})}_{([2(n_{j_2s_11}-n_{j_1u_11})],(2n_{j_2s_11}+1))\text{-decreasing-concave source}}. \tag{1.232}$$

The infinite-equilibrium is a $([2(n_{j_2s_11}-n_{j_1u_11})]^{\text{th}}\,\text{DC},(2n_{j_2s_11}+1)^{\text{th}}\,\text{SO})$*-decreasing-concave source at* $x_{j_2}^*=a_{j_1j_2u_1}=a_{j_2j_2s_1}$.

- *For* $a_{j_1j_10}\prod_{r=1}^{p_1}(\overline{x}_{j_1}-a_{j_1j_1r})^{m_{j_1r}}\prod_{u_2=1,u_2\neq u_1}^{q_1}(a_{j_1j_2u_1}-a_{j_1j_2u_2})^{n_{j_1u_2}}>0$ *and* $a_{j_2j_20}\prod_{s_2=1,s_2\neq s_1}^{q_2}(a_{j_2j_2s_1}-a_{j_2j_2s_2})^{n_{j_2s_2}}<0$,

$$(\overline{x}_{j_1},a_{j_2j_2s_1})=\underbrace{([2(n_{j_2s_11}-n_{j_1u_11})]^{\text{th}}\,\text{DC},(2n_{j_2s_11}+1)^{\text{th}}\,\text{SI})}_{([2(n_{j_2s_11}-n_{j_1u_11})],(2n_{j_2s_11}+1))\text{-decreasing-concave sink}}. \tag{1.233}$$

The infinite-equilibrium is a $([2(n_{j_2s_11}-n_{j_1u_11})]^{\text{th}}\,\text{DC},(2n_{j_2s_11}+1)]^{\text{th}}\,\text{SI})$*-decreasing-concave sink at* $x_{j_2}^*=a_{j_1j_2u_1}=a_{j_2j_2s_1}$.

- *For* $a_{j_1j_10}\prod_{r=1}^{p_1}(\overline{x}_{j_1}-a_{j_1j_1r})^{m_{j_1r}}\prod_{u_2=1,u_2\neq u_1}^{q_1}(a_{j_1j_2u_1}-a_{j_1j_2u_2})^{n_{j_1u_2}}<0$ *and* $a_{j_2j_20}\prod_{s_2=1,s_2\neq s_1}^{q_2}(a_{j_2j_2s_1}-a_{j_2j_2s_2})^{n_{j_2s_2}}<0$,

$$(\overline{x}_{j_1},a_{j_2j_2s_1})=\underbrace{([2(n_{j_2s_11}-n_{j_1u_11})]^{\text{th}}\,\text{IC},(2n_{j_2s_11}+1)^{\text{th}}\,\text{SI})}_{([2(n_{j_2s_11}-n_{j_1u_11})],(2n_{j_2s_11}+1))\text{-increasing-concave sink}}. \tag{1.234}$$

The infinite-equilibrium is a $([2(n_{j_1u_11}-n_{j_2s_11})]^{\text{th}}\,\text{II},(2n_{j_2s_11}+1)^{\text{th}}\,\text{SI})$*-increasing-concave sink at* $x_{j_2}^*=a_{j_1j_2u_1}=a_{j_2j_2s_1}$.

(ii_{2b1b}) For $n_{j_1u_1}=2n_{j_1u_11}$ *and* $n_{j_2s_1}=2n_{j_2s_11}+1$*, the infinite-equilibrium of* $x_{j_2}^*=a_{j_1j_2u_1}=a_{j_2j_2s_1}$ *with* $\overline{x}_{j_1}\neq a_{j_1j_1r_1}$ *has the following properties.*

- *For* $a_{j_1j_10}\prod_{r=1}^{p_1}(\overline{x}_{j_1}-a_{j_1j_1r})^{m_{j_1r}}\prod_{u_2=1,u_2\neq u_1}^{q_1}(a_{j_1j_2u_1}-a_{j_1j_2u_2})^{n_{j_1u_2}}>0$ *and* $a_{j_2j_20}\prod_{s_2=1,s_2\neq s_1}^{q_2}(a_{j_2j_2s_1}-a_{j_2j_2s_2})^{n_{j_2s_2}}>0$,

$$(\overline{x}_{j_1},a_{j_2j_2s_1})=\underbrace{([2(n_{j_2s_11}-n_{j_1u_11})+1]^{\text{th}}\,\text{DA},(2n_{j_2s_11}+1)^{\text{th}}\,\text{SO})}_{([2(n_{j_2s_11}-n_{j_1u_11})+1],(2n_{j_2s_11}+1))\text{-down-asymptote source}}. \tag{1.235}$$

The infinite-equilibrium is a ($[2(n_{j_2s_11}-n_{j_1u_11})+1]^{\text{th}}$ UP, $(2n_{j_2s_11}+1)^{\text{th}}$ SO)-*down-asymptote source at* $x_{j_2}^* = a_{j_1j_2u_1} = a_{j_2j_2s_1}$.

- *For* $a_{j_1j_10}\prod_{r=1}^{p_1}(\overline{x}_{j_1}-a_{j_1j_1r})^{m_{j_1r}}\prod_{u_2=1,u_2\neq u_1}^{q_1}(a_{j_1j_2u_1}-a_{j_1j_2u_2})^{n_{j_1u_2}} < 0$ *and* $a_{j_2j_20}\prod_{s_2=1,s_2\neq s_1}^{q_2}(a_{j_2j_2s_1}-a_{j_2j_2s_2})^{n_{j_2s_2}} > 0$,

$$(\overline{x}_{j_1}, a_{j_2j_2s_1}) = \underbrace{([2(n_{j_2s_11}-n_{j_1u_11})+1]^{\text{th}}\,\text{UA}, (2n_{j_2s_11}+1)^{\text{th}}\,\text{SO})}_{([2(n_{j_2s_11}-n_{j_1u_11})+1],(2n_{j_2s_11}+1))\text{-up-asymptote source}}. \tag{1.236}$$

The infinite-equilibrium is a ($[2(n_{j_2s_11}-n_{j_1u_11})+1]^{\text{th}}$ UA, $(2n_{j_2s_11}+1)^{\text{th}}$ SO)-*up asymptote source at* $x_{j_2}^* = a_{j_1j_2u_1} = a_{j_2j_2s_1}$.

- *For* $a_{j_1j_10}\prod_{r=1}^{p_1}(\overline{x}_{j_1}-a_{j_1j_1r})^{m_{j_1r}}\prod_{u_2=1,u_2\neq u_1}^{q_1}(a_{j_1j_2u_1}-a_{j_1j_2u_2})^{n_{j_1u_2}} > 0$ *and* $a_{j_2j_20}\prod_{s_2=1,s_2\neq s_1}^{q_2}(a_{j_2j_2s_1}-a_{j_2j_2s_2})^{n_{j_2s_2}} < 0$,

$$(\overline{x}_{j_1}, a_{j_2j_2s_1}) = \underbrace{([2(n_{j_2s_11}-n_{j_1u_11})+1]^{\text{th}}\,\text{UA}, (2n_{j_2s_11}+1)^{\text{th}}\,\text{SI})}_{([2(n_{j_2s_11}-n_{j_1u_11})+1],(2n_{j_2s_11}+1))\text{-up-asymtote sink}}. \tag{1.237}$$

The infinite-equilibrium is a ($[2(n_{j_2s_11}-n_{j_1u_11})+1]^{\text{th}}$ UA, $(2n_{j_2s_11}+1)^{\text{th}}$ SI)-*up-asymptote sink at* $x_{j_2}^* = a_{j_1j_2u_1} = a_{j_2j_2s_1}$.

- *For* $a_{j_1j_10}\prod_{r=1}^{p_1}(\overline{x}_{j_1}-a_{j_1j_1r})^{m_{j_1r}}\prod_{u_2=1,u_2\neq u_1}^{q_1}(a_{j_1j_2u_1}-a_{j_1j_2u_2})^{n_{j_1u_2}} < 0$ *and* $a_{j_2j_20}\prod_{s_2=1,s_2\neq s_1}^{q_2}(a_{j_2j_2s_1}-a_{j_2j_2s_2})^{n_{j_2s_2}} < 0$,

$$(\overline{x}_{j_1}, a_{j_2j_2s_1}) = \underbrace{([2(n_{j_2s_11}-n_{j_1u_11})+1]^{\text{th}}\,\text{DA}, (2n_{j_2s_11}+1)^{\text{th}}\,\text{SI})}_{([2(n_{j_2s_11}-n_{j_1u_11})+1],(2n_{j_2s_11}+1))\text{-down-asymptote sink}}. \tag{1.238}$$

The infinite-equilibrium is a ($[2(n_{j_2s_11}-n_{j_1u_11})+1]^{\text{th}}$ DA, $(2n_{j_2s_11}+1)^{\text{th}}$ SI)-*down-asymptote sink at* $x_{j_2}^* = a_{j_1j_2u_1} = a_{j_2j_2s_1}$.

(ii_{2b1c}) *For* $n_{j_1u_1} = 2n_{j_1u_11}+1$ *and* $n_{j_2s_1} = 2n_{j_2s_11}$, *the infinite-equilibrium of* $x_{j_2}^* = a_{j_1j_2u_1} = a_{j_2j_2s_1}$ *with* $\overline{x}_{j_1} \neq a_{j_1j_1r_1}$ *has the following properties.*

- *For* $a_{j_1j_10}\prod_{r=1}^{p_1}(\overline{x}_{j_1}-a_{j_1j_1r})^{m_{j_1r}}\prod_{u_2=1,u_2\neq u_1}^{q_1}(a_{j_1j_2u_1}-a_{j_1j_2u_2})^{n_{j_1u_2}} > 0$ *and* $a_{j_2j_20}\prod_{s_2=1,s_2\neq s_1}^{q_2}(a_{j_2j_2s_1}-a_{j_2j_2s_2})^{n_{j_2s_2}} > 0$,

$$(\overline{x}_{j_1}, a_{j_2j_2s_1}) = \underbrace{([2(n_{j_2s_11}-n_{j_1u_11})-1]^{\text{th}}\,\text{DA}, (2n_{j_2s_11})^{\text{th}}\,\text{US})}_{([2(n_{j_2s_11}-n_{j_1u_11})-1],(2n_{j_2s_11}))\text{-down-asymptote upper-saddle}}. \tag{1.239}$$

The infinite-equilibrium is a ($[2(n_{j_2s_11}-n_{j_1u_11})-1]^{\text{th}}$ DA, $(2n_{j_2s_11})^{\text{th}}$ US)-*down-asymptote upper-saddle at* $x_{j_2}^* = a_{j_1j_2u_1} = a_{j_2j_2s_1}$.

- *For* $a_{j_1j_10}\prod_{r=1}^{p_1}(\overline{x}_{j_1}-a_{j_1j_1r})^{m_{j_1r}}\prod_{u_2=1,u_2\neq u_1}^{q_1}(a_{j_1j_2u_1}-a_{j_1j_2u_2})^{n_{j_1u_2}} < 0$ *and* $a_{j_2j_20}\prod_{s_2=1,s_2\neq s_1}^{q_2}(a_{j_2j_2s_1}-a_{j_2j_2s_2})^{n_{j_2s_2}} > 0$,

$$(\overline{x}_{j_1}, a_{j_2j_2s_1}) = \underbrace{([2(n_{j_2s_1}1 - n_{j_1u_1}1) - 1]^{\text{th}}\,\text{UA}, (2n_{j_2s_1}1)^{\text{th}}\,\text{US})}_{([2(n_{j_2s_1}1 - n_{j_1u_1}1)-1],(2n_{j_2s_1}1))\text{-up-asymptote upper-saddle}}. \tag{1.240}$$

The infinite-equilibrium is a $([2(n_{j_2s_1}1-n_{j_1u_1}1)-1]^{\text{th}}\text{UA}, (2n_{j_2s_1}1)^{\text{th}}\text{US})$*-up-asymptote upper-saddle at* $x^*_{j_2} = a_{j_1j_2u_1} = a_{j_2j_2s_1}$.

- *For* $a_{j_1j_10}\prod_{r=1}^{p_1}(\overline{x}_{j_1} - a_{j_1j_1r})^{m_{j_1r}}\prod_{u_2=1,u_2\neq u_1}^{q_1}(a_{j_1j_2u_1} - a_{j_1j_2u_2})^{n_{j_1u_2}} > 0$ *and* $a_{j_2j_20}\prod_{s_2=1,s_2\neq s_1}^{q_2}(a_{j_2j_2s_1} - a_{j_2j_2s_2})^{n_{j_2s_2}} < 0$,

$$(\overline{x}_{j_1}, a_{j_2j_2s_1}) = \underbrace{([2(n_{j_2s_1}1 - n_{j_1u_1}1) - 1]^{\text{th}}\,\text{UA}, (2n_{j_2s_1}1)^{\text{th}}\,\text{LS})}_{([2(n_{j_2s_1}1 - n_{j_1u_1}1)-1],(2n_{j_2s_1}1))\text{-up-asymptote lower-saddle}}. \tag{1.241}$$

The infinite-equilibrium is a $([2(n_{j_2s_1}1-n_{j_1u_1}1)-1]^{\text{th}}\text{UA}, (2n_{j_2s_1}1)^{\text{th}}\text{LS})$*-up-asymptote lower-saddle at* $x^*_{j_2} = a_{j_1j_2u_1} = a_{j_2j_2s_1}$.

- *For* $a_{j_1j_10}\prod_{r=1}^{p_1}(\overline{x}_{j_1} - a_{j_1j_1r})^{m_{j_1r}}\prod_{u_2=1,u_2\neq u_1}^{q_1}(a_{j_1j_2u_1} - a_{j_1j_2u_2})^{n_{j_1u_2}} < 0$ *and* $a_{j_2j_20}\prod_{s_2=1,s_2\neq s_1}^{q_2}(a_{j_2j_2s_1} - a_{j_2j_2s_2})^{n_{j_2s_2}} < 0$,

$$(\overline{x}_{j_1}, a_{j_2j_2s_1}) = \underbrace{([2(n_{j_2s_1}1 - n_{j_1u_1}1) - 1]^{\text{th}}\,\text{DA}, (2n_{j_2s_1}1)^{\text{th}}\,\text{LS})}_{([2(n_{j_2s_1}1 - n_{j_1u_1}1)-1],(2n_{j_2s_1}1))\text{-down-asymprote lower-saddle}}. \tag{1.242}$$

The infinite-equilibrium is a $([2(n_{j_2s_1}1-n_{j_1u_1}1)-1]^{\text{th}}\text{DA}, (2n_{j_2s_1}1)^{\text{th}}\text{LS})$*-down-asymptote lower-saddle at* $x^*_{j_2} = a_{j_1j_2u_1} = a_{j_2j_2s_1}$.

(ii$_{2b1d}$*) For* $n_{j_1u_1} = 2n_{j_1u_1}1$ *and* $n_{j_2s_1} = 2n_{j_2s_1}1$*, the infinite-equilibrium of* $x^*_{j_2} = a_{j_1j_2u_1} = a_{j_2j_2s_1}$ *with* $\overline{x}_{j_1} \neq a_{j_1j_1r_1}$ *has the following properties.*

- *For* $a_{j_1j_10}\prod_{r=1}^{p_1}(\overline{x}_{j_1} - a_{j_1j_1r})^{m_{j_1r}}\prod_{u_2=1,u_2\neq u_1}^{q_1}(a_{j_1j_2u_1} - a_{j_1j_2u_2})^{n_{j_1u_2}} > 0$ *and* $a_{j_2j_20}\prod_{s_2=1,s_2\neq s_1}^{q_2}(a_{j_2j_2s_1} - a_{j_2j_2s_2})^{n_{j_2s_2}} > 0$,

$$(\overline{x}_{j_1}, a_{j_2j_2s_1}) = \underbrace{([2(n_{j_2s_1}1 - n_{j_1u_1}1)]^{\text{th}}\,\text{IC}, (2n_{j_2s_1}1)^{\text{th}}\,\text{US})}_{([2(n_{j_2s_1}1 - n_{j_1u_1}1)],(2n_{j_2s_1}1))\text{-increasing-concave upper-saddle}}. \tag{1.243}$$

The infinite-equilibrium is a $([2(n_{j_2s_1}1-n_{j_1u_1}1)]^{\text{th}}\text{IC}, (2n_{j_2s_1}1)^{\text{th}}\text{US})$*-increasing concave upper-saddle at* $x^*_{j_2} = a_{j_1j_2u_1} = a_{j_2j_2s_1}$.

- *For* $a_{j_1j_10}\prod_{r=1}^{p_1}(\overline{x}_{j_1} - a_{j_1j_1r})^{m_{j_1r}}\prod_{u_2=1,u_2\neq u_1}^{q_1}(a_{j_1j_2u_1} - a_{j_1j_2u_2})^{n_{j_1u_2}} < 0$ *and* $a_{j_2j_20}\prod_{s_2=1,s_2\neq s_1}^{q_2}(a_{j_2j_2s_1} - a_{j_2j_2s_2})^{n_{j_2s_2}} > 0$,

$$(\overline{x}_{j_1}, a_{j_2j_2s_1}) = \underbrace{([2(n_{j_2s_1}1 - n_{j_1u_1}1)]^{\text{th}}\,\text{DC}, (2n_{j_2s_1}1)^{\text{th}}\,\text{US})}_{([2(n_{j_2s_1}1 - n_{j_1u_1}1)],(2n_{j_2s_1}1))\text{-decreasing concave upper-saddle}}. \tag{1.244}$$

The infinite-equilibrium is a $([2(n_{j_2s_1}1-n_{j_1u_1}1)]^{\text{th}}\text{DC}, (2n_{j_2s_1}1)^{\text{th}}\text{US})$*-decreasing concave upper-saddle at* $x^*_{j_2} = a_{j_1j_2u_1} = a_{j_2j_2s_1}$.

- *For* $a_{j_1j_10}\prod_{r=1}^{p_1}(\overline{x}_{j_1}-a_{j_1j_1r})^{m_{j_1r}}\prod_{u_2=1,u_2\neq u_1}^{q_1}(a_{j_1j_2u_1}-a_{j_1j_2u_2})^{n_{j_1u_2}}>0$ *and* $a_{j_2j_20}\prod_{s_2=1,s_2\neq s_1}^{q_2}(a_{j_2j_2s_1}-a_{j_2j_2s_2})^{n_{j_2s_2}}<0$,

$$(\overline{x}_{j_1},a_{j_2j_2s_1})=\underbrace{([2(n_{j_2s_11}-n_{j_1u_11})]^{\text{th}}\,\text{DC},\,(2n_{j_2s_11})^{\text{th}}\,\text{US})}_{([2(n_{j_2s_11}-n_{j_1u_11})],(2n_{j_2s_11}))\text{-decreasing concave upper-saddle}}\,. \tag{1.245}$$

The infinite-equilibrium is a $([2(n_{j_2s_11}-n_{j_1u_11})]^{\text{th}}\text{DC},(2n_{j_2s_11})^{\text{th}}\text{LS})$*-decreasing concave lower-saddle at* $x_{j_2}^*=a_{j_1j_2u_1}=a_{j_2j_2s_1}$.

- *For* $a_{j_1j_10}\prod_{r=1}^{p_1}(\overline{x}_{j_1}-a_{j_1j_1r})^{m_{j_1r}}\prod_{u_2=1,u_2\neq u_1}^{q_1}(a_{j_1j_2u_1}-a_{j_1j_2u_2})^{n_{j_1u_2}}<0$ *and* $a_{j_2j_20}\prod_{s_2=1,s_2\neq s_1}^{q_2}(a_{j_2j_2s_1}-a_{j_2j_2s_2})^{n_{j_2s_2}}<0$,

$$(\overline{x}_{j_1},a_{j_2j_2s_1})=\underbrace{([2(n_{j_2s_11}-n_{j_1u_11})]^{\text{th}}\,\text{IC},\,(2n_{j_2s_11})^{\text{th}}\,\text{LS})}_{([2(n_{j_2s_11}-n_{j_1u_11})],(2n_{j_2s_11}))\text{-increasing inflection lower-saddle}}\,. \tag{1.246}$$

The infinite-equilibrium is a $([2(n_{j_2s_11}-n_{j_1u_11})]^{\text{th}}\text{IC},(2n_{j_2s_11})^{\text{th}}\text{LS})$*-increasing concave lower-saddle at* $x_{j_2}^*=a_{j_1j_2u_1}=a_{j_2j_2s_1}$.

(ii_{2b2}) There are 32 types of equilibrium $(x_{j_1}^*,x_{j_2}^*)=(a_{j_1j_1r_1},a_{j_2j_2s_1})$ *on the infinite-equilibrium of* $x_{j_2}^*=a_{j_1j_2u_1}=a_{j_2j_2s_1}$.

(ii_{2b2a}) For $m_{j_1r_1}=2m_{j_1r_11}+1, n_{j_1u_1}=2n_{j_1u_11}+1$ *and* $n_{j_2s_1}=2n_{j_2s_11}+1$*, the equilibrium of* $(x_{j_1}^*,x_{j_2}^*)=(a_{j_1j_1r_1},a_{j_2j_2s_1})$ *on the infinite-equilibrium* $x_{j_2}^*=a_{j_1j_2u_1}=a_{j_2j_2s_1}$ *has the following properties.*

- *For* $a_{j_1j_10}\prod_{r_2=1,r_2\neq r_1}^{p_1}(a_{j_1j_1r_1}-a_{j_1j_1r_2})^{m_{j_1r_2}}\prod_{u_2=1,u_2\neq u_1}^{q_1}(a_{j_1j_2u_1}-a_{j_1j_2u_2})^{n_{j_1u_2}}>0$ *and* $a_{j_2j_20}\prod_{s_2=1,s_2\neq s_1}^{q_2}(a_{j_2j_2s_1}-a_{j_2j_2s_2})^{n_{j_2s_2}}>0$,

$$(a_{j_1j_1r_1},a_{j_2j_2s_1})=\underbrace{([2(n_{j_2s_11}-n_{j_1u_11})]^{\text{th}}\,\text{DC:IC},\,[(2m_{j_1r_11}+1),(2n_{j_2s_11}+1)]^{\text{th}}\,\text{SO})}_{([2(n_{j_2s_11}-n_{j_1u_11})],[(2m_{j_1r_11}+1),(2n_{j_2s_11}+1)])\text{-decreasing-to-increasing concave source}}\,. \tag{1.247}$$

The equilibrium is a $([2(n_{j_2s_11}-n_{j_1u_11})]^{\text{th}}\,\text{DC:IC},[(2m_{j_1r_11}+1),(2n_{j_2s_11}+1)]^{\text{th}}$ SO)*-decreasing-to-increasing-concave source at* $(x_{j_1}^*,x_{j_2}^*)=(a_{j_1j_1r_1},a_{j_2j_2s_1})$.

- *For* $a_{j_1j_10}\prod_{r_2=1,r_2\neq r_1}^{p_1}(a_{j_1j_1r_1}-a_{j_1j_1r_2})^{m_{j_1r_2}}\prod_{u_2=1,u_2\neq u_1}^{q_1}(a_{j_1j_2u_1}-a_{j_1j_2u_2})^{n_{j_1u_2}}<0$ *and* $a_{j_2j_20}\prod_{s_2=1,s_2\neq s_1}^{q_2}(a_{j_2j_2s_1}-a_{j_2j_2s_2})^{n_{j_2s_2}}>0$,

$$(a_{j_1j_1r_1},a_{j_2j_2s_1})=\underbrace{([2(n_{j_2s_11}-n_{j_1u_11})]^{\text{th}}\,\text{IC:DC},\,[(2m_{j_1r_11}+1),(2n_{j_2s_11}+1)]^{\text{th}}\,\text{SO})}_{([2(n_{j_2s_11}-n_{j_1u_11})],[(2m_{j_1r_11}+1),(2n_{j_2s_11}+1)])\text{-increasing-to-decreasing concave source}}\,. \tag{1.248}$$

The equilibrium is a $[2(n_{j_2s_11}-n_{j_1u_11})]^{\text{th}}\,\text{IC:DC},[(2m_{j_1r_11}+1),(2n_{j_2s_11}+1)]^{\text{th}}$ SO)*-increasing-to-decreasing-concave source at* $(x_{j_1}^*,x_{j_2}^*)=(a_{j_1j_1r_1},a_{j_2j_2s_1})$.

- *For* $a_{j_1j_10}\prod_{r_2=1,r_2\neq r_1}^{p_1}(a_{j_1j_1r_1}-a_{j_1j_1r_2})^{m_{j_1r_2}}\prod_{u_2=1,u_2\neq u_1}^{q_1}(a_{j_1j_2u_1}-a_{j_1j_2u_2})^{n_{j_1u_2}}>0$ *and* $a_{j_2j_20}\prod_{s_2=1,s_2\neq s_1}^{q_2}(a_{j_2j_2s_1}-a_{j_2j_2s_2})^{n_{j_2s_2}}<0$,

$$(a_{j_1j_1r_1},a_{j_2j_2s_1})=\underbrace{([2(n_{j_2s_1}1-n_{j_1u_1}1)]^{\text{th}}\text{ IC:DC},[(2m_{j_1r_1}1+1),(2n_{j_2s_1}1+1)]^{\text{th}}\text{ SI})}_{([2(n_{j_2s_1}1-n_{j_1u_1}1)],[(2m_{j_1r_1}1+1),(2n_{j_2s_1}1+1)])\text{-increasing-to-decreasing concave sink}}. \tag{1.249}$$

The equilibrium is a $([2(n_{j_2s_1}1-n_{j_1u_1}1)]^{\text{th}}\text{ IC:DC},[(2m_{j_1r_1}1+1),(2n_{j_2s_1}1+1)]^{\text{th}}$ SI)*-increasing-to-decreasing-concave sink at* $(x_{j_1}^*,x_{j_2}^*)=(a_{j_1j_1r_1},a_{j_2j_2s_1})$.

- *For* $a_{j_1j_10}\prod_{r_2=1,r_2\neq r_1}^{p_1}(a_{j_1j_1r_1}-a_{j_1j_1r_2})^{m_{j_1r_2}}\prod_{u_2=1,u_2\neq u_1}^{q_1}(a_{j_1j_2u_1}-a_{j_1j_2u_2})^{n_{j_1u_2}}<0$ *and* $a_{j_2j_20}\prod_{s_2=1,s_2\neq s_1}^{q_2}(a_{j_2j_2s_1}-a_{j_2j_2s_2})^{n_{j_2s_2}}<0$,

$$(a_{j_1j_1r_1},a_{j_2j_2s_1})=\underbrace{([2(n_{j_2s_1}1-n_{j_1u_1}1)]^{\text{th}}\text{ IC:DC},[(2m_{j_1r_1}1+1),(2n_{j_2s_1}1+1)]^{\text{th}}\text{ SI})}_{([2(n_{j_2s_1}1-n_{j_1u_1}1)],[(2m_{j_1r_1}1+1),(2n_{j_2s_1}1+1)])\text{-increasing-to-decreasing concave sink}}. \tag{1.250}$$

The equilibrium is a $([2(n_{j_2s_1}1-n_{j_1u_1}1)]^{\text{th}}\text{ IC:DC},[(2m_{j_1r_1}1+1),(2n_{j_2s_1}1+1)]^{\text{th}}$ SI)*-decreasing-to-increasing-concave sink at* $(x_{j_1}^*,x_{j_2}^*)=(a_{j_1j_1r_1},a_{j_2j_2s_1})$.

(ii$_{2b2b}$*) For* $m_{j_1r_1}=2m_{j_1r_1}1+1$, $n_{j_1u_1}=2n_{j_1u_1}1$ *and* $n_{j_2s_1}=2n_{j_2s_1}1+1$, *the equilibrium of* $(x_{j_1}^*,x_{j_2}^*)=(a_{j_1j_1r_1},a_{j_2j_2s_1})$ *on the infinite-equilibrium* $x_{j_2}^*=a_{j_1j_2u_1}=a_{j_2j_2s_1}$ *has the following properties.*

- *For* $a_{j_1j_10}\prod_{r_2=1,r_2\neq r_1}^{p_1}(a_{j_1j_1r_1}-a_{j_1j_1r_2})^{m_{j_1r_2}}\prod_{u_2=1,u_2\neq u_1}^{q_1}(a_{j_1j_2u_1}-a_{j_1j_2u_2})^{n_{j_1u_2}}>0$ *and* $a_{j_2j_20}\prod_{s_2=1,s_2\neq s_1}^{q_2}(a_{j_2j_2s_1}-a_{j_2j_2s_2})^{n_{j_2s_2}}>0$,

$$\begin{aligned}&(a_{j_1j_1r_1},a_{j_2j_2s_1})\\&=\underbrace{([2(n_{j_2s_1}1-n_{j_1u_1}1)+1]^{\text{th}}\text{ UA:DA},[(2m_{j_1r_1}1+1),(2n_{j_2s_1}1+1)]^{\text{th}}\text{ SO})}_{([2(n_{j_2s_1}1-n_{j_1u_1}1)+1],[(2m_{j_1r_1}1+1),(2n_{j_2s_1}1+1)])\text{-up-to-down asymptote source}}.\end{aligned} \tag{1.251}$$

The equilibrium is a $([2(n_{j_2s_1}1-n_{j_1u_1}1)+1]^{\text{th}}\text{ UA:DA},[(2m_{j_1r_1}1+1),(2n_{j_2s_1}1+1)]^{\text{th}}$ SO)*-up-to-down asymptote source at* $(x_{j_1}^*,x_{j_2}^*)=(a_{j_1j_1r_1},a_{j_2j_2s_1})$.

- *For* $a_{j_1j_10}\prod_{r_2=1,r_2\neq r_1}^{p_1}(a_{j_1j_1r_1}-a_{j_1j_1r_2})^{m_{j_1r_2}}\prod_{u_2=1,u_2\neq u_1}^{q_1}(a_{j_1j_2u_1}-a_{j_1j_2u_2})^{n_{j_1u_2}}<0$ *and* $a_{j_2j_20}\prod_{s_2=1,s_2\neq s_1}^{q_2}(a_{j_2j_2s_1}-a_{j_2j_2s_2})^{n_{j_2s_2}}>0$,

$$\begin{aligned}&(a_{j_1j_1r_1},a_{j_2j_2s_1})\\&=\underbrace{([2(n_{j_2s_1}1-n_{j_1u_1}1)+1]^{\text{th}}\text{ DA:UA},[(2m_{j_1r_1}1+1),(2n_{j_2s_1}1+1)]^{\text{th}}\text{ SO})}_{([2(n_{j_2s_1}1-n_{j_1u_1}1)+1],[(2m_{j_1r_1}1+1),(2n_{j_2s_1}1+1)])\text{-down-to-up asymptote source}}.\end{aligned} \tag{1.252}$$

The equilibrium is a $([2(n_{j_2s_1}1-n_{j_1u_1}1)+1]^{\text{th}}$ DA:UA,$[(2m_{j_1r_1}1+1),(2n_{j_2s_1}1+1)]^{\text{th}}$ SO)*-down-to-up asymptote source at* $(x_{j_1}^*, x_{j_2}^*) = (a_{j_1j_1r_1}, a_{j_2j_2s_1})$.

- *For* $a_{j_1j_10}\prod_{r_2=1,r_2\neq r_1}^{p_1}(a_{j_1j_1r_1}-a_{j_1j_1r_2})^{m_{j_1r_2}}\prod_{u_2=1,u_2\neq u_1}^{q_1}(a_{j_1j_2u_1}-a_{j_1j_2u_2})^{n_{j_1u_2}} > 0$ *and* $a_{j_2j_20}\prod_{s_2=1,s_2\neq s_1}^{q_2}(a_{j_2j_2s_1}-a_{j_2j_2s_2})^{n_{j_2s_2}} < 0$,

$$
\begin{aligned}
&(a_{j_1j_1r_1}, a_{j_2j_2s_1})\\
&= \underbrace{([2(n_{j_2s_1}1-n_{j_1u_1}1)+1]^{\text{th}}\,\text{DA:UA},\,[(2m_{j_1r_1}1+1),(2n_{j_2s_1}1+1)]^{\text{th}}\,\text{SI})}_{([2(n_{j_2s_1}1-n_{j_1u_1}1)+1],[(2m_{j_1r_1}1+1),(2n_{j_2s_1}1+1)])\text{-down-to-up asymptote sink}}.
\end{aligned}
\tag{1.253}
$$

The equilibrium is a $([2(n_{j_2s_1}1-n_{j_1u_1}1)+1]^{\text{th}}$ DA:UA,$[(2m_{j_1r_1}1+1),(2n_{j_2s_1}1+1)]^{\text{th}}$ SI)*-down-to-up asymptote sink at* $(x_{j_1}^*, x_{j_2}^*) = (a_{j_1j_1r_1}, a_{j_2j_2s_1})$.

- *For* $a_{j_1j_10}\prod_{r_2=1,r_2\neq r_1}^{p_1}(a_{j_1j_1r_1}-a_{j_1j_1r_2})^{m_{j_1r_2}}\prod_{u_2=1,u_2\neq u_1}^{q_1}(a_{j_1j_2u_1}-a_{j_1j_2u_2})^{n_{j_1u_2}} < 0$ *and* $a_{j_2j_20}\prod_{s_2=1,s_2\neq s_1}^{q_2}(a_{j_2j_2s_1}-a_{j_2j_2s_2})^{n_{j_2s_2}} < 0$,

$$
\begin{aligned}
&(a_{j_1j_1r_1}, a_{j_2j_2s_1})\\
&= \underbrace{([2(n_{j_2s_1}1-n_{j_1u_1}1)+1]^{\text{th}}\,\text{UA:DA},\,[(2m_{j_1r_1}1+1),(2n_{j_2s_1}1+1)]^{\text{th}}\,\text{SI})}_{([2(n_{j_2s_1}1-n_{j_1u_1}1)+1],[(2m_{j_1r_1}1+1),(2n_{j_2s_1}1+1)])\text{-up-to-down asymptote sink}}.
\end{aligned}
\tag{1.254}
$$

The equilibrium is a $([2(n_{j_2s_1}1-n_{j_1u_1}1)+1]^{\text{th}}$ UA:DA,$[(2m_{j_1r_1}1+1),(2n_{j_2s_1}1+1)]^{\text{th}}$ SI)*-up-to-down asymptote sink at* $(x_{j_1}^*, x_{j_2}^*) = (a_{j_1j_1r_1}, a_{j_2j_2s_1})$.

(ii_{2b2c}*) For* $m_{j_1r_1} = 2m_{j_1r_1}1+1$, $n_{j_1u_1} = 2n_{j_1u_1}1+1$ *and* $n_{j_2s_1} = 2n_{j_2s_1}1$, *the equilibrium of* $(x_{j_1}^*, x_{j_2}^*) = (a_{j_1j_1r_1}, a_{j_2j_2s_1})$ *on the infinite-equilibrium* $x_{j_2}^* = a_{j_1j_2u_1} = a_{j_2j_2s_1}$ *has the following properties.*

- *For* $a_{j_1j_10}\prod_{r_2=1,r_2\neq r_1}^{p_1}(a_{j_1j_1r_1}-a_{j_1j_1r_2})^{m_{j_1r_2}}\prod_{u_2=1,u_2\neq u_1}^{q_1}(a_{j_1j_2u_1}-a_{j_1j_2u_2})^{n_{j_1u_2}} > 0$ *and* $a_{j_2j_20}\prod_{s_2=1,s_2\neq s_1}^{q_2}(a_{j_2j_2s_1}-a_{j_2j_2s_2})^{n_{j_2s_2}} > 0$,

$$
\begin{aligned}
&(a_{j_1j_1r_1}, a_{j_2j_2s_1})\\
&= \underbrace{([2(n_{j_2s_1}1-n_{j_1u_1}1)-1]^{\text{th}}\,\text{UA:DA},\,[(2m_{j_1r_1}1+1),(2n_{j_2s_1}1)]^{\text{th}}\,\text{US})}_{([2(n_{j_2s_1}1-n_{j_1u_1}1)-1],[(2m_{j_1r_1}1+1),(2n_{j_2s_1}1)])\text{-up-to-down asymptote upper-saddle}}.
\end{aligned}
\tag{1.255}
$$

The equilibrium is a $([2(n_{j_2s_1}1-n_{j_1u_1}1)-1]^{\text{th}}$ UA:DA,$[(2m_{j_1r_1}1+1),(2n_{j_2s_1}1)]^{\text{th}}$ US)*-up-to-down asymptote upper-saddle at* $(x_{j_1}^*, x_{j_2}^*) = (a_{j_1j_1r_1}, a_{j_2j_2s_1})$.

- *For* $a_{j_1j_10}\prod_{r_2=1,r_2\neq r_1}^{p_1}(a_{j_1j_1r_1}-a_{j_1j_1r_2})^{m_{j_1r_2}}\prod_{u_2=1,u_2\neq u_1}^{q_1}(a_{j_1j_2u_1}-a_{j_1j_2u_2})^{n_{j_1u_2}} < 0$ *and* $a_{j_2j_20}\prod_{s_2=1,s_2\neq s_1}^{q_2}(a_{j_2j_2s_1}-a_{j_2j_2s_2})^{n_{j_2s_2}} > 0$,

$$
\begin{aligned}
&(a_{j_1j_1r_1}, a_{j_2j_2s_1})\\
&= \underbrace{([2(n_{j_2s_1}1-n_{j_1u_1}1)-1]^{\text{th}}\,\text{DA:UA},\,[(2m_{j_1r_1}1+1),(2n_{j_2s_1}1)]^{\text{th}}\,\text{US})}_{([2(n_{j_2s_1}1-n_{j_1u_1}1)-1],[(2m_{j_1r_1}1+1),(2n_{j_2s_1}1)])\text{-down-to-up asymptote upper-saddle}}.
\end{aligned}
\tag{1.256}
$$

The equilibrium is a $([2(n_{j_2s_1}1-n_{j_1u_1}1)-1]^{\text{th}}$ DA:UA,$[(2m_{j_1r_1}1+1),(2n_{j_2s_1}1)]^{\text{th}}$ US)*-down-to-up asymptote upper-saddle at* $(x^*_{j_1}, x^*_{j_2}) = (a_{j_1j_1r_1}, a_{j_2j_2s_1})$.

- *For* $a_{j_1j_1 0}\prod_{r_2=1,r_2\neq r_1}^{p_1}(a_{j_1j_1r_1}-a_{j_1j_1r_2})^{m_{j_1r_2}}\prod_{u_2=1,u_2\neq u_1}^{q_1}(a_{j_1j_2u_1}-a_{j_1j_2u_2})^{n_{j_1u_2}}>0$ *and* $a_{j_2j_2 0}\prod_{s_2=1,s_2\neq s_1}^{q_2}(a_{j_2j_2s_1}-a_{j_2j_2s_2})^{n_{j_2s_2}}<0$,

$$\begin{aligned}&(a_{j_1j_1r_1}, a_{j_2j_2s_1})\\&=\underbrace{([2(n_{j_2s_1}1-n_{j_1u_1}1)-1]^{\text{th}}\,\text{DA:UA},\,[(2m_{j_1r_1}1+1),(2n_{j_2s_1}1)]^{\text{th}}\,\text{LS})}_{([2(n_{j_2s_1}1-n_{j_1u_1}1)-1],[(2m_{j_1r_1}1+1),(2n_{j_2s_1}1)])\text{-down-to-up asymptote lower-saddle}}.\end{aligned}\tag{1.257}$$

The equilibrium is a $([2(n_{j_2s_1}1-n_{j_1u_1}1)-1]^{\text{th}}$ DA:UA,$[(2m_{j_1r_1}1+1),(2n_{j_2s_1}1)]^{\text{th}}$ LS)*-down-to-up asymptote lower-saddle at* $(x^*_{j_1}, x^*_{j_2}) = (a_{j_1j_1r_1}, a_{j_2j_2s_1})$.

- *For* $a_{j_1j_1 0}\prod_{r_2=1,r_2\neq r_1}^{p_1}(a_{j_1j_1r_1}-a_{j_1j_1r_2})^{m_{j_1r_2}}\prod_{u_2=1,u_2\neq u_1}^{q_1}(a_{j_1j_2u_1}-a_{j_1j_2u_2})^{n_{j_1u_2}}<0$ *and* $a_{j_2j_2 0}\prod_{s_2=1,s_2\neq s_1}^{q_2}(a_{j_2j_2s_1}-a_{j_2j_2s_2})^{n_{j_2s_2}}<0$,

$$\begin{aligned}&(a_{j_1j_1r_1}, a_{j_2j_2s_1})\\&=\underbrace{([2(n_{j_2s_1}1-n_{j_1u_1}1)-1]^{\text{th}}\,\text{UA:DA},\,[(2m_{j_1r_1}1+1),(2n_{j_2s_1}1)]^{\text{th}}\,\text{LS})}_{([2(n_{j_2s_1}1-n_{j_1u_1}1)-1],[(2m_{j_1r_1}1+1),(2n_{j_2s_1}1)])\text{-up-to-down asymptote lower-saddle}}.\end{aligned}\tag{1.258}$$

The equilibrium is a $([2(n_{j_2s_1}1-n_{j_1u_1}1)-1]^{\text{th}}$ UA:DA,$[(2m_{j_1r_1}1+1),(2n_{j_2s_1}1)]^{\text{th}}$ LS)*-up-to-down asymptote lower-saddle at* $(x^*_{j_1}, x^*_{j_2}) = (a_{j_1j_1r_1}, a_{j_2j_2s_1})$.

(ii$_{2b2d}$) For $m_{j_1r_1}=2m_{j_1r_1}1+1$, $n_{j_1u_1}=2n_{j_1u_1}1$ *and* $n_{j_2s_1}=2n_{j_2s_1}1$, *the equilibrium of* $(x^*_{j_1}, x^*_{j_2}) = (a_{j_1j_1r_1}, a_{j_2j_2s_1})$ *on the infinite-equilibrium* $x^*_{j_2}=a_{j_1j_2u_1}=a_{j_2j_2s_1}$ *has the following properties.*

- *For* $a_{j_1j_1 0}\prod_{r_2=1,r_2\neq r_1}^{p_1}(a_{j_1j_1r_1}-a_{j_1j_1r_2})^{m_{j_1r_2}}\prod_{u_2=1,u_2\neq u_1}^{q_1}(a_{j_1j_2u_1}-a_{j_1j_2u_2})^{n_{j_1u_2}}>0$ *and* $a_{j_2j_2 0}\prod_{s_2=1,s_2\neq s_1}^{q_2}(a_{j_2j_2s_1}-a_{j_2j_2s_2})^{n_{j_2s_2}}>0$,

$$\begin{aligned}&(a_{j_1j_1r_1}, a_{j_2j_2s_1})\\&=\underbrace{([2(n_{j_2s_1}1-n_{j_1u_1}1)]^{\text{th}}\,\text{DC:IC},\,[(2m_{j_1r_1}1+1),(2n_{j_2s_1}1)]^{\text{th}}\,\text{US})}_{([2(n_{j_2s_1}1-n_{j_1u_1}1)],[(2m_{j_1r_1}1+1),(2n_{j_2s_1}1)])\text{-decreasing-to-increasing concave upper-saddle}}.\end{aligned}\tag{1.259}$$

The equilibrium is a $([2(n_{j_2s_1}1-n_{j_1u_1}1)]^{\text{th}}$ DC:IC,$[(2m_{j_1r_1}1+1),(2n_{j_2s_1}1)]^{\text{th}}$US)*-decreasing-to-increasing concave upper-saddle at* $(x^*_{j_1}, x^*_{j_2}) = (a_{j_1j_1r_1}, a_{j_2j_2s_1})$.

- *For* $a_{j_1j_1 0}\prod_{r_2=1,r_2\neq r_1}^{p_1}(a_{j_1j_1r_1}-a_{j_1j_1r_2})^{m_{j_1r_2}}\prod_{u_2=1,u_2\neq u_1}^{q_1}(a_{j_1j_2u_1}-a_{j_1j_2u_2})^{n_{j_1u_2}}<0$ *and* $a_{j_2j_2 0}\prod_{s_2=1,s_2\neq s_1}^{q_2}(a_{j_2j_2s_1}-a_{j_2j_2s_2})^{n_{j_2s_2}}>0$,

$$\begin{aligned}&(a_{j_1j_1r_1}, a_{j_2j_2s_1})\\&=\underbrace{([2(n_{j_2s_1}1-n_{j_1u_1}1)]^{\text{th}}\,\text{IC:DC},\,[(2m_{j_1r_1}1+1),(2n_{j_2s_1}1)]^{\text{th}}\,\text{US})}_{([2(n_{j_2s_1}1-n_{j_1u_1}1)],[(2m_{j_1r_1}1+1),(2n_{j_2s_1}1)])\text{-decreasing-to-increasing concave upper-saddle}}.\end{aligned}\tag{1.260}$$

The equilibrium is a $([2(n_{j_1u_1 1}-n_{j_2s_1 1})]^{\text{th}}$ IC:DC, $[(2m_{j_1r_1 1}+1), (2n_{j_2s_1 1})]^{\text{th}}$ US)-*increasing-to-decreasing concave upper-saddle at* $(x^*_{j_1}, x^*_{j_2}) = (a_{j_1j_1r_1}, a_{j_2j_2s_1})$.

- *For* $a_{j_1j_10}\prod_{r_2=1,r_2\neq r_1}^{p_1}(a_{j_1j_1r_1}-a_{j_1j_1r_2})^{m_{j_1r_2}}\prod_{u_2=1,u_2\neq u_1}^{q_1}(a_{j_1j_2u_1}-a_{j_1j_2u_2})^{n_{j_1u_2}}>0$ *and* $a_{j_2j_20}\prod_{s_2=1,s_2\neq s_1}^{q_2}(a_{j_2j_2s_1}-a_{j_2j_2s_2})^{n_{j_2s_2}}<0$,

$$(a_{j_1j_1r_1}, a_{j_2j_2s_1}) = \underbrace{([2(n_{j_2s_1 1}-n_{j_1u_1 1})]^{\text{th}}\ \text{IC:DC},\ [(2m_{j_1r_1 1}+1),(2n_{j_2s_1 1})]^{\text{th}}\ \text{LS})}_{([2(n_{j_2s_1 1}-n_{j_1u_1 1})],[(2m_{j_1r_1 1}+1),(2n_{j_2s_1 1})])\text{-decreasing-to-increasing concave lower-saddle}}. \quad (1.261)$$

The equilibrium is a $([2(n_{j_2s_1 1}-n_{j_1u_1 1})]^{\text{th}}$ IC:DC, $[(2m_{j_1r_1 1}+1), (2n_{j_2s_1 1})]^{\text{th}}$ LS)-*increasing-to-decreasing concave lower-saddle at* $(x^*_{j_1}, x^*_{j_2}) = (a_{j_1j_1r_1}, a_{j_2j_2s_1})$.

- *For* $a_{j_1j_10}\prod_{r_2=1,r_2\neq r_1}^{p_1}(a_{j_1j_1r_1}-a_{j_1j_1r_2})^{m_{j_1r_2}}\prod_{u_2=1,u_2\neq u_1}^{q_1}(a_{j_1j_2u_1}-a_{j_1j_2u_2})^{n_{j_1u_2}}<0$ *and* $a_{j_2j_20}\prod_{s_2=1,s_2\neq s_1}^{q_2}(a_{j_2j_2s_1}-a_{j_2j_2s_2})^{n_{j_2s_2}}<0$,

$$(a_{j_1j_1r_1}, a_{j_2j_2s_1}) = \underbrace{([2(n_{j_2s_1 1}-n_{j_1u_1 1})]^{\text{th}}\ \text{DC:IC},\ [(2m_{j_1r_1 1}+1),(2n_{j_2s_1 1})]^{\text{th}}\ \text{LS})}_{([2(n_{j_2s_1 1}-n_{j_1u_1 1})],[(2m_{j_1r_1 1}+1),(2n_{j_2s_1 1})])\text{-decreasing-to-increasing concave lower-saddle}}. \quad (1.262)$$

The equilibrium is a $([2(n_{j_1u_1 1}-n_{j_2s_1 1})]^{\text{th}}$ DC:IC, $[(2m_{j_1r_1 1}+1), (2n_{j_2s_1 1})]^{\text{th}}$ LS)-*decreasing-to-increasing concave lower-saddle at* $(x^*_{j_1}, x^*_{j_2}) = (a_{j_1j_1r_1}, a_{j_2j_2s_1})$.

(ii$_{2b2e}$) For $m_{j_1r_1}=2m_{j_1r_1 1}$, $n_{j_1u_1}=2n_{j_1u_1 1}+1$ *and* $n_{j_2s_1}=2n_{j_2s_1 1}+1$, *the equilibrium of* $(x^*_{j_1}, x^*_{j_2}) = (a_{j_1j_1r_1}, a_{j_2j_2s_1})$ *on the infinite-equilibrium* $x^*_{j_2}=a_{j_1j_2u_1}=a_{j_2j_2s_1}$ *has the following properties.*

- *For* $a_{j_1j_10}\prod_{r_2=1,r_2\neq r_1}^{p_1}(a_{j_1j_1r_1}-a_{j_1j_1r_2})^{m_{j_1r_2}}\prod_{u_2=1,u_2\neq u_1}^{q_1}(a_{j_1j_2u_1}-a_{j_1j_2u_2})^{n_{j_1u_2}}>0$ *and* $a_{j_2j_20}\prod_{s_2=1,s_2\neq s_1}^{q_2}(a_{j_2j_2s_1}-a_{j_2j_2s_2})^{n_{j_2s_2}}>0$,

$$(a_{j_1j_1r_1}, a_{j_2j_2s_1}) = \underbrace{([2(n_{j_2s_1 1}-n_{j_1u_1 1})]^{\text{th}}\ \text{IC:IC},\ [(2m_{j_1r_1 1}),(2n_{j_2s_1 1}+1)]^{\text{th}}\ \text{SO})}_{([2(n_{j_2s_1 1}-n_{j_1u_1 1})],[(2m_{j_1r_1 1}),(2n_{j_2s_1 1}+1)])\text{-increasing concave source}}. \quad (1.263)$$

The equilibrium is a $([2(n_{j_2s_1 1}-n_{j_1u_1 1})]^{\text{th}}$ IC:IC, $[(2m_{j_1r_1 1}), (2n_{j_2s_1 1}+1)]^{\text{th}}$ SO)-*increasing-concave source at* $(x^*_{j_1}, x^*_{j_2}) = (a_{j_1j_1r_1}, a_{j_2j_2s_1})$.

- *For* $a_{j_1j_10}\prod_{r_2=1,r_2\neq r_1}^{p_1}(a_{j_1j_1r_1}-a_{j_1j_1r_2})^{m_{j_1r_2}}\prod_{u_2=1,u_2\neq u_1}^{q_1}(a_{j_1j_2u_1}-a_{j_1j_2u_2})^{n_{j_1u_2}}<0$ *and* $a_{j_2j_20}\prod_{s_2=1,s_2\neq s_1}^{q_2}(a_{j_2j_2s_1}-a_{j_2j_2s_2})^{n_{j_2s_2}}>0$,

$$(a_{j_1j_1r_1}, a_{j_2j_2s_1}) = \underbrace{([2(n_{j_2s_1 1}-n_{j_1u_1 1})]^{\text{th}}\ \text{DC:DC},\ [(2m_{j_1r_1 1}),(2n_{j_2s_1 1}+1)]^{\text{th}}\ \text{SO})}_{([2(n_{j_2s_1 1}-n_{j_1u_1 1})],[(2m_{j_1r_1 1}),(2n_{j_2s_1 1}+1)])\text{-decreasing concave source}}. \quad (1.264)$$

The equilibrium is a $([2(n_{j_2s_1}1-n_{j_1u_1}1)]^{\text{th}}$ DC:DC,$[(2m_{j_1r_1}1),(2n_{j_2s_1}1+1)]^{\text{th}}$ SO)-*decreasing-concave source at* $(x_{j_1}^*, x_{j_2}^*) = (a_{j_1j_1r_1}, a_{j_2j_2s_1})$.

- *For* $a_{j_1j_10}\prod_{r_2=1,r_2\neq r_1}^{p_1}(a_{j_1j_1r_1}-a_{j_1j_1r_2})^{m_{j_1r_2}}\prod_{u_2=1,u_2\neq u_1}^{q_1}(a_{j_1j_2u_1}-a_{j_1j_2u_2})^{n_{j_1u_2}} > 0$ *and* $a_{j_2j_20}\prod_{s_2=1,s_2\neq s_1}^{q_2}(a_{j_2j_2s_1}-a_{j_2j_2s_2})^{n_{j_2s_2}} < 0$,

$$(a_{j_1j_1r_1}, a_{j_2j_2s_1}) = \underbrace{([2(n_{j_2s_1}1-n_{j_1u_1}1)]^{\text{th}}\ \text{DC:DC},\ [(2m_{j_1r_1}1),(2n_{j_2s_1}1+1)]^{\text{th}}\ \text{SI})}_{([2(n_{j_2s_1}1-n_{j_1u_1}1)],[(2m_{j_1r_1}1),(2n_{j_2s_1}1+1)])\text{-decreasing concave sink}}. \tag{1.265}$$

The equilibrium is a $([2(n_{j_2s_1}1-n_{j_1u_1}1)]^{\text{th}}$ DC:DC, $[(2m_{j_1r_1}1),(2n_{j_2s_1}1+1)]^{\text{th}}$ SI)-*decreasing-concave sink at* $(x_{j_1}^*, x_{j_2}^*) = (a_{j_1j_1r_1}, a_{j_2j_2s_1})$.

- *For* $a_{j_1j_10}\prod_{r_2=1,r_2\neq r_1}^{p_1}(a_{j_1j_1r_1}-a_{j_1j_1r_2})^{m_{j_1r_2}}\prod_{u_2=1,u_2\neq u_1}^{q_1}(a_{j_1j_2u_1}-a_{j_1j_2u_2})^{n_{j_1u_2}} < 0$ *and* $a_{j_2j_20}\prod_{s_2=1,s_2\neq s_1}^{q_2}(a_{j_2j_2s_1}-a_{j_2j_2s_2})^{n_{j_2s_2}} < 0$,

$$(a_{j_1j_1r_1}, a_{j_2j_2s_1}) = \underbrace{([2(n_{j_2s_1}1-n_{j_1u_1}1)]^{\text{th}}\ \text{IC:IC},\ [(2m_{j_1r_1}1),(2n_{j_2s_1}1+1)]^{\text{th}}\ \text{SI})}_{([2(n_{j_2s_1}1-n_{j_1u_1}1)],[(2m_{j_1r_1}1),(2n_{j_2s_1}1+1)])\text{-increasing concave sink}}. \tag{1.266}$$

The equilibrium is a $([2(n_{j_1u_1}1-n_{j_2s_1}1)]^{\text{th}}$ IC:IC, $[(2m_{j_1r_1}1),(2n_{j_2s_1}1+1)]^{\text{th}}$ SI)-*increasing-concave sink at* $(x_{j_1}^*, x_{j_2}^*) = (a_{j_1j_1r_1}, a_{j_2j_2s_1})$.

(ii_{2b2f}*) For* $m_{j_1r_1} = 2m_{j_1r_1}1$, $n_{j_1u_1} = 2n_{j_1u_1}1$ *and* $n_{j_2s_1} = 2n_{j_2s_1}1+1$, *the equilibrium of* $(x_{j_1}^*, x_{j_2}^*) = (a_{j_1j_1r_1}, a_{j_2j_2s_1})$ *on the infinite-equilibrium* $x_{j_2}^* = a_{j_1j_2u_1} = a_{j_2j_2s_1}$ *has the following properties.*

- *For* $a_{j_1j_10}\prod_{r_2=1,r_2\neq r_1}^{p_1}(a_{j_1j_1r_1}-a_{j_1j_1r_2})^{m_{j_1r_2}}\prod_{u_2=1,u_2\neq u_1}^{q_1}(a_{j_1j_2u_1}-a_{j_1j_2u_2})^{n_{j_1u_2}} > 0$ *and* $a_{j_2j_20}\prod_{s_2=1,s_2\neq s_1}^{q_2}(a_{j_2j_2s_1}-a_{j_2j_2s_2})^{n_{j_2s_2}} > 0$,

$$(a_{j_1j_1r_1}, a_{j_2j_2s_1}) = \underbrace{([2(n_{j_2s_1}1-n_{j_1u_1}1)+1]^{\text{th}}\ \text{DA:DA},\ [(2m_{j_1r_1}1),(2n_{j_2s_1}1+1)]^{\text{th}}\ \text{SO})}_{([2(n_{j_2s_1}1-n_{j_1u_1}1)+1],[(2m_{j_1r_1}1),(2n_{j_2s_1}1+1)])\text{-down-asymptote source}}. \tag{1.267}$$

The equilibrium is a $([2(n_{j_2s_1}1-n_{j_1u_1}1)+1]^{\text{th}}$ DA:DA,$[(2m_{j_1r_1}1),(2n_{j_2s_1}1+1)]^{\text{th}}$ SO)-*down-asymptote source at* $(x_{j_1}^*, x_{j_2}^*) = (a_{j_1j_1r_1}, a_{j_2j_2s_1})$.

- *For* $a_{j_1j_10}\prod_{r_2=1,r_2\neq r_1}^{p_1}(a_{j_1j_1r_1}-a_{j_1j_1r_2})^{m_{j_1r_2}}\prod_{u_2=1,u_2\neq u_1}^{q_1}(a_{j_1j_2u_1}-a_{j_1j_2u_2})^{n_{j_1u_2}} < 0$ *and* $a_{j_2j_20}\prod_{s_2=1,s_2\neq s_1}^{q_2}(a_{j_2j_2s_1}-a_{j_2j_2s_2})^{n_{j_2s_2}} > 0$,

$$(a_{j_1j_1r_1}, a_{j_2j_2s_1}) = \underbrace{([2(n_{j_2s_1}1-n_{j_1u_1}1)+1]^{\text{th}}\ \text{UA:UA},\ [(2m_{j_1r_1}1),(2n_{j_2s_1}1+1)]^{\text{th}}\ \text{SO})}_{([2(n_{j_2s_1}1-n_{j_1u_1}1)+1],[(2m_{j_1r_1}1),(2n_{j_2s_1}1+1)])\text{-up-asymptote source}}. \tag{1.268}$$

The equilibrium is a ($[2(n_{j_2s_1 1}-n_{j_1u_1 1})+1]^{\text{th}}$ UA:UA,$[(2m_{j_1r_1 1}),(2n_{j_2s_1 1}+1)]^{\text{th}}$ SO)*-up-asymptote source at* $(x^*_{j_1}, x^*_{j_2}) = (a_{j_1j_1r_1}, a_{j_2j_2s_1})$.

- *For* $a_{j_1j_1 0}\prod_{r_2=1,r_2\neq r_1}^{p_1}(a_{j_1j_1r_1}-a_{j_1j_1r_2})^{m_{j_1r_2}}\prod_{u_2=1,u_2\neq u_1}^{q_1}(a_{j_1j_2u_1}-a_{j_1j_2u_2})^{n_{j_1u_2}} > 0$ *and* $a_{j_2j_2 0}\prod_{s_2=1,s_2\neq s_1}^{q_2}(a_{j_2j_2s_1}-a_{j_2j_2s_2})^{n_{j_2s_2}} < 0$,

$$(a_{j_1j_1r_1}, a_{j_2j_2s_1}) = \underbrace{([2(n_{j_2s_1 1}-n_{j_1u_1 1})+1]^{\text{th}}\ \text{UA:UA}, [(2m_{j_1r_1 1}), (2n_{j_2s_1 1}+1)]^{\text{th}}\ \text{SI})}_{([2(n_{j_2s_1 1}-n_{j_1u_1 1})+1],[(2m_{j_1r_1 1}),(2n_{j_2s_1 1}+1)])\text{-up-asymptote sink}}. \tag{1.269}$$

The equilibrium is a ($[2(n_{j_2s_1 1}-n_{j_1u_1 1})+1]^{\text{th}}$ UA:UA,$[(2m_{j_1r_1 1}),(2n_{j_2s_1 1}+1)]^{\text{th}}$ SI)*-up-asymptote sink at* $(x^*_{j_1}, x^*_{j_2}) = (a_{j_1j_1r_1}, a_{j_2j_2s_1})$.

- *For* $a_{j_1j_1 0}\prod_{r_2=1,r_2\neq r_1}^{p_1}(a_{j_1j_1r_1}-a_{j_1j_1r_2})^{m_{j_1r_2}}\prod_{u_2=1,u_2\neq u_1}^{q_1}(a_{j_1j_2u_1}-a_{j_1j_2u_2})^{n_{j_1u_2}} < 0$ *and* $a_{j_2j_2 0}\prod_{s_2=1,s_2\neq s_1}^{q_2}(a_{j_2j_2s_1}-a_{j_2j_2s_2})^{n_{j_2s_2}} < 0$,

$$(a_{j_1j_1r_1}, a_{j_2j_2s_1}) = \underbrace{([2(n_{j_2s_1 1}-n_{j_1u_1 1})+1]^{\text{th}}\ \text{DA:DA}, [(2m_{j_1r_1 1}), (2n_{j_2s_1 1}+1)]^{\text{th}}\ \text{SI})}_{([2(n_{j_2s_1 1}-n_{j_1u_1 1})+1],[(2m_{j_1r_1 1}),(2n_{j_2s_1 1}+1)])\text{-down-asymptote sink}}. \tag{1.270}$$

The equilibrium is a ($[2(n_{j_2s_1 1}-n_{j_1u_1 1})+1]^{\text{th}}$ DA:DA,$[(2m_{j_1r_1 1}),(2n_{j_2s_1 1}+1)]^{\text{th}}$ SI)*-down-asymptote sink at* $(x^*_{j_1}, x^*_{j_2}) = (a_{j_1j_1r_1}, a_{j_2j_2s_1})$.

(ii_{2b2g}) *For* $m_{j_1r_1} = 2m_{j_1r_1 1}$, $n_{j_1u_1} = 2n_{j_1u_1 1}+1$ *and* $n_{j_2s_1} = 2n_{j_2s_1 1}$, *the equilibrium of* $(x^*_{j_1}, x^*_{j_2}) = (a_{j_1j_1r_1}, a_{j_2j_2s_1})$ *on the infinite-equilibrium* $x^*_{j_2} = a_{j_1j_2u_1} = a_{j_2j_2s_1}$ *has the following properties.*

- *For* $a_{j_1j_1 0}\prod_{r_2=1,r_2\neq r_1}^{p_1}(a_{j_1j_1r_1}-a_{j_1j_1r_2})^{m_{j_1r_2}}\prod_{u_2=1,u_2\neq u_1}^{q_1}(a_{j_1j_2u_1}-a_{j_1j_2u_2})^{n_{j_1u_2}} > 0$ *and* $a_{j_2j_2 0}\prod_{s_2=1,s_2\neq s_1}^{q_2}(a_{j_2j_2s_1}-a_{j_2j_2s_2})^{n_{j_2s_2}} > 0$,

$$(a_{j_1j_1r_1}, a_{j_2j_2s_1}) = \underbrace{([2(n_{j_2s_1 1}-n_{j_1u_1 1})-1]^{\text{th}}\ \text{DA:DA}, [(2m_{j_1r_1 1}), (2n_{j_2s_1 1})]^{\text{th}}\ \text{US})}_{([2(n_{j_2s_1 1}-n_{j_1u_1 1})-1],[(2m_{j_1r_1 1}),(2n_{j_2s_1 1})])\text{-down-asymptote upper-saddle}}. \tag{1.271}$$

The equilibrium is a ($[2(n_{j_2s_1 1}-n_{j_1u_1 1})-1]^{\text{th}}$ DA:DA, $[(2m_{j_1r_1 1}), (2n_{j_2s_1 1})]^{\text{th}}$ US)*-down-asymptote uppersaddle at* $(x^*_{j_1}, x^*_{j_2}) = (a_{j_1j_1r_1}, a_{j_2j_2s_1})$.

- *For* $a_{j_1j_1 0}\prod_{r_2=1,r_2\neq r_1}^{p_1}(a_{j_1j_1r_1}-a_{j_1j_1r_2})^{m_{j_1r_2}}\prod_{u_2=1,u_2\neq u_1}^{q_1}(a_{j_1j_2u_1}-a_{j_1j_2u_2})^{n_{j_1u_2}} < 0$ *and* $a_{j_2j_2 0}\prod_{s_2=1,s_2\neq s_1}^{q_2}(a_{j_2j_2s_1}-a_{j_2j_2s_2})^{n_{j_2s_2}} > 0$,

$$(a_{j_1j_1r_1}, a_{j_2j_2s_1}) = \underbrace{([2(n_{j_2s_1 1}-n_{j_1u_1 1})-1]^{\text{th}}\ \text{UA:UA}, [(2m_{j_1r_1 1}), (2n_{j_2s_1 1})]^{\text{th}}\ \text{US})}_{([2(n_{j_2s_1 1}-n_{j_1u_1 1})-1],[(2m_{j_1r_1 1}),(2n_{j_2s_1 1})])\text{-up-asymptote upper-saddle}}. \tag{1.272}$$

The equilibrium is a $([2(n_{j_2s_1}1-n_{j_1u_1}1)-1]^{\text{th}}$ UA:UA, $[(2m_{j_1r_1}1),(2n_{j_2s_1}1)]^{\text{th}}$ US)-*up-asymptote upper-saddle at* $(x_{j_1}^*,x_{j_2}^*)=(a_{j_1j_1r_1},a_{j_2j_2s_1})$.

- *For* $a_{j_1j_10}\prod_{r_2=1,r_2\neq r_1}^{p_1}(a_{j_1j_1r_1}-a_{j_1j_1r_2})^{m_{j_1r_2}}\prod_{u_2=1,u_2\neq u_1}^{q_1}(a_{j_1j_2u_1}-a_{j_1j_2u_2})^{n_{j_1u_2}}>0$ *and* $a_{j_2j_20}\prod_{s_2=1,s_2\neq s_1}^{q_2}(a_{j_2j_2s_1}-a_{j_2j_2s_2})^{n_{j_2s_2}}<0$,

$$(a_{j_1j_1r_1},a_{j_2j_2s_1})=\underbrace{([2(n_{j_2s_1}1-n_{j_1u_1}1)-1]^{\text{th}}\text{ UA:UA},[(2m_{j_1r_1}1),(2n_{j_2s_1}1)]^{\text{th}}\text{ LS})}_{([2(n_{j_2s_1}1-n_{j_1u_1}1)-1],[(2m_{j_1r_1}1),(2n_{j_2s_1}1)])\text{-up-asymptote lower-saddle}}. \tag{1.273}$$

The equilibrium is a $([2(n_{j_2s_1}1-n_{j_1u_1}1)-1]^{\text{th}}$ UA:UA, $[(2m_{j_1r_1}1),(2n_{j_2s_1}1)]^{\text{th}}$ LS)-*up-asymptote lower-saddle at* $(x_{j_1}^*,x_{j_2}^*)=(a_{j_1j_1r_1},a_{j_2j_2s_1})$.

- *For* $a_{j_1j_10}\prod_{r_2=1,r_2\neq r_1}^{p_1}(a_{j_1j_1r_1}-a_{j_1j_1r_2})^{m_{j_1r_2}}\prod_{u_2=1,u_2\neq u_1}^{q_1}(a_{j_1j_2u_1}-a_{j_1j_2u_2})^{n_{j_1u_2}}<0$ *and* $a_{j_2j_20}\prod_{s_2=1,s_2\neq s_1}^{q_2}(a_{j_2j_2s_1}-a_{j_2j_2s_2})^{n_{j_2s_2}}<0$,

$$(a_{j_1j_1r_1},a_{j_2j_2s_1})=\underbrace{([2(n_{j_2s_1}1-n_{j_1u_1}1)-1]^{\text{th}}\text{ DA:DA},[(2m_{j_1r_1}1),(2n_{j_2s_1}1)]^{\text{th}}\text{ LS})}_{([2(n_{j_2s_1}1-n_{j_1u_1}1)-1],[(2m_{j_1r_1}1),(2n_{j_2s_1}1)])\text{-down-asymptote lower-saddle}}. \tag{1.274}$$

The equilibrium is a $([2(n_{j_2s_1}1-n_{j_1u_1}1)-1]^{\text{th}}$ DA:DA, $[(2m_{j_1r_1}1),(2n_{j_2s_1}1)]^{\text{th}}$ LS)-*down-asymptote lower-saddle at* $(x_{j_1}^*,x_{j_2}^*)=(a_{j_1j_1r_1},a_{j_2j_2s_1})$.

(ii$_{2b2h}$) For $m_{j_1r_1}=2m_{j_1r_1}1$, $n_{j_1u_1}=2n_{j_1u_1}1$ *and* $n_{j_2s_1}=2n_{j_2s_1}1$, *the equilibrium of* $(x_{j_1}^*,x_{j_2}^*)=(a_{j_1j_1r_1},a_{j_2j_2s_1})$ *on the infinite-equilibrium* $x_{j_2}^*=a_{j_1j_2u_1}=a_{j_2j_2s_1}$ *has the following properties.*

- *For* $a_{j_1j_10}\prod_{r_2=1,r_2\neq r_1}^{p_1}(a_{j_1j_1r_1}-a_{j_1j_1r_2})^{m_{j_1r_2}}\prod_{u_2=1,u_2\neq u_1}^{q_1}(a_{j_1j_2u_1}-a_{j_1j_2u_2})^{n_{j_1u_2}}>0$ *and* $a_{j_2j_20}\prod_{s_2=1,s_2\neq s_1}^{q_2}(a_{j_2j_2s_1}-a_{j_2j_2s_2})^{n_{j_2s_2}}>0$,

$$(a_{j_1j_1r_1},a_{j_2j_2s_1})=\underbrace{([2(n_{j_2s_1}1-n_{j_1u_1}1)]^{\text{th}}\text{ IC:IC},[(2m_{j_1r_1}1),(2n_{j_2s_1}1)]^{\text{th}}\text{ US})}_{([2(n_{j_2s_1}1-n_{j_1u_1}1)],[(2m_{j_1r_1}1),(2n_{j_2s_1}1)])\text{-increasing-concave upper-saddle}}. \tag{1.275}$$

The equilibrium is a $([2(n_{j_2s_1}1-n_{j_1u_1}1)]^{\text{th}}$ IC:IC, $[(2m_{j_1r_1}1),(2n_{j_2s_1}1)]^{\text{th}}$ US)-*increasing concave upper-saddle at* $(x_{j_1}^*,x_{j_2}^*)=(a_{j_1j_1r_1},a_{j_2j_2s_1})$.

- *For* $a_{j_1j_10}\prod_{r_2=1,r_2\neq r_1}^{p_1}(a_{j_1j_1r_1}-a_{j_1j_1r_2})^{m_{j_1r_2}}\prod_{u_2=1,u_2\neq u_1}^{q_1}(a_{j_1j_2u_1}-a_{j_1j_2u_2})^{n_{j_1u_2}}<0$ *and* $a_{j_2j_20}\prod_{s_2=1,s_2\neq s_1}^{q_2}(a_{j_2j_2s_1}-a_{j_2j_2s_2})^{n_{j_2s_2}}>0$,

$$(a_{j_1j_1r_1},a_{j_2j_2s_1})=\underbrace{([2(n_{j_2s_1}1-n_{j_1u_1}1)]^{\text{th}}\text{ DC:DC},[(2m_{j_1r_1}1),(2n_{j_2s_1}1)]^{\text{th}}\text{ US})}_{([2(n_{j_2s_1}1-n_{j_1u_1}1)],[(2m_{j_1r_1}1),(2n_{j_2s_1}1)])\text{-decreasing-concave upper-saddle}}. \tag{1.276}$$

The equilibrium is a $([2(n_{j_2s_1}1 - n_{j_1u_1}1)]^{\text{th}}\,\text{DC:DC}, [(2m_{j_1r_1}1), (2n_{j_2s_1}1)]^{\text{th}}\,\text{US})$*-decreasing concave upper-saddle at* $(x_{j_1}^*, x_{j_2}^*) = (a_{j_1j_1r_1}, a_{j_2j_2s_1})$.

- *For* $a_{j_1j_10}\prod_{r_2=1,r_2\neq r_1}^{p_1}(a_{j_1j_1r_1} - a_{j_1j_1r_2})^{m_{j_1r_2}}\prod_{u_2=1,u_2\neq u_1}^{q_1}(a_{j_1j_2u_1} - a_{j_1j_2u_2})^{n_{j_1u_2}} > 0$ *and* $a_{j_2j_20}\prod_{s_2=1,s_2\neq s_1}^{q_2}(a_{j_2j_2s_1} - a_{j_2j_2s_2})^{n_{j_2s_2}} < 0$,

$$(a_{j_1j_1r_1}, a_{j_2j_2s_1}) = \underbrace{([2(n_{j_2s_1}1 - n_{j_1u_1}1)]^{\text{th}}\,\text{DC:DC}, [(2m_{j_1r_1}1), (2n_{j_2s_1}1)]^{\text{th}}\,\text{LS})}_{([2(n_{j_2s_1}1-n_{j_1u_1}1)],[(2m_{j_1r_1}1),(2n_{j_2s_1}1)])\text{-decreasing-concave lower-saddle}}. \tag{1.277}$$

The equilibrium is a $([2(n_{j_2s_1}1 - n_{j_1u_1}1)]^{\text{th}}\,\text{DC:DC}, [(2m_{j_1r_1}1), (2n_{j_2s_1}1)]^{\text{th}}\,\text{LS})$*-decreasing concave lower-saddle at* $(x_{j_1}^*, x_{j_2}^*) = (a_{j_1j_1r_1}, a_{j_2j_2s_1})$.

- *For* $a_{j_1j_10}\prod_{r_2=1,r_2\neq r_1}^{p_1}(a_{j_1j_1r_1} - a_{j_1j_1r_2})^{m_{j_1r_2}}\prod_{u_2=1,u_2\neq u_1}^{q_1}(a_{j_1j_2u_1} - a_{j_1j_2u_2})^{n_{j_1u_2}} < 0$ *and* $a_{j_2j_20}\prod_{s_2=1,s_2\neq s_1}^{q_2}(a_{j_2j_2s_1} - a_{j_2j_2s_2})^{n_{j_2s_2}} < 0$,

$$(a_{j_1j_1r_1}, a_{j_2j_2s_1}) = \underbrace{([2(n_{j_2s_1}1 - n_{j_1u_1}1)]^{\text{th}}\,\text{IC:IC}, [(2m_{j_1r_1}1), (2n_{j_2s_1}1)]^{\text{th}}\,\text{LS})}_{([2(n_{j_2s_1}1-n_{j_1u_1}1)],[(2m_{j_1r_1}1),(2n_{j_2s_1}1)])\text{-increasing-concave lower-saddle}}. \tag{1.278}$$

The equilibrium is a $([2(n_{j_1u_1}1 - n_{j_2s_1}1)]^{\text{th}}\,\text{IC:IC}, [(2m_{j_1r_1}1), (2n_{j_2s_1}1)]^{\text{th}}\,\text{LS})$*-increasing concave lower-saddle at* $(x_{j_1}^*, x_{j_2}^*) = (a_{j_1j_1r_1}, a_{j_2j_2s_1})$.

(iii) Consider

$$\begin{aligned}
\dot{x}_{j_2} &= a_{j_2j_20}\prod_{s=1}^{n_2}(x_{j_2} - a_{j_2j_2s}),\\
\dot{x}_{j_1} &= a_{j_1j_10}\prod_{r=1}^{m_1}(x_{j_1} - a_{j_1j_1r})\prod_{u=1}^{n_1}(x_{j_2} - a_{j_1j_2u}).
\end{aligned} \tag{1.279}$$

The first integral manifold is

$$\begin{aligned}
\sum_{r_1=1}^{m_1} A_{r_1}\ln\frac{|x_{j_1} - a_{j_1j_1r_1}|}{|x_{j_10} - a_{j_1j_1r_1}|} &= \frac{a_{j_1j_10}}{a_{j_2j_20}}\Bigg\{\sum_{\beta_0=n_2}^{n_1} b_{j_1j_2\beta_0}\sum_{\substack{\alpha_1\in[0,\beta_0-n_2],\\ \beta_1\in[n_2-1,\beta_0-1]}}^{\alpha_1+\beta_1=\beta_0-1}\;\sum_{\substack{\alpha_2\in[0,\beta_1-n_2]\\ \beta_2\in[n_2-2,\beta_1-2]}}^{\alpha_2+\beta_2=\beta_1-2}\cdots\\
&\sum_{\substack{\alpha_{n_2-1}\in[0,\beta_{n_2-2}-n_2]\\ \beta_{n_2-1}\in[1,\beta_{n_2-2}-n_2+1]}}^{\alpha_{n_2-1}+\beta_{n_2-1}=\beta_{n_2-2}-n_2+1}\;\prod_{j=1}^{n_2-1}\frac{(\beta_{j-1}-j)!}{\alpha_j!\beta_j!}(a_{j_2j_2s_{j+1}} - a_{j_2j_2s_j})^{\alpha_j}\\
&\times\frac{1}{\beta_{n_2-1}}\left[(x_{j_2} - a_{j_2j_2n_2})^{\beta_{n_2-1}} - (x_{j_20} - a_{j_2j_2n_2})^{\beta_{n_2-1}}\right]
\end{aligned}$$

$$
\begin{aligned}
& - \sum_{\beta_0=n_2}^{n_1} b_{j_1 j_2 \beta_0} \sum_{k=2}^{n_2-1} \Bigg[\sum_{\substack{\alpha_1+\beta_1=\beta_0-1 \\ \alpha_1\in[0,\beta_0-n_2), \\ \beta_1\in(n_2-1,\beta_0-1]}} \sum_{\substack{\alpha_2+\beta_2=\beta_1-2 \\ \alpha_2\in[0,\beta_1-n_2] \\ \beta_2\in(n_2-2,\beta_1-2)}} \cdots \\
& \times \Bigg(\sum_{\substack{\alpha_{k-1}+\beta_{k-1}=\beta_{k-2}-k+1 \\ \alpha_{k-1}\in(\beta_{k-2}-n_2+1,\beta_{k-2}-k+1] \\ \beta_{k-1}\in[0,n_2-k)}} \prod_{j=1}^{k-1} \frac{(\beta_{j-1}-j)!}{\alpha_j!\beta_j!} (a_{j_2 j_2 s_{j+1}} - a_{j_2 j_2 s_j})^{\alpha_j} \\
& \times \sum_{\rho=k}^{n_2} A_{\rho 1}^{(s_k)} \ln \frac{|x_{j_2} - a_{j_2 j_2 s_\rho}|}{|x_{j_2 0} - a_{j_2 j_2 s_\rho}|} \\
& + \sum_{\substack{\alpha_{k-1}+\beta_{k-1}=\beta_{k-2}-k+1 \\ \alpha_{k-1}\in(\beta_{k-2}-n_2,\beta_{k-2}-n_2+1] \\ \beta_{k-1}\in[n_2-k,n_2-k+1)}} \prod_{j=1}^{k-1} \frac{(\beta_{j-1}-j)!}{\alpha_j!\beta_j!} (a_{j_2 j_2 s_{j+1}} - a_{j_2 j_2 s_j})^{\alpha_j} \\
& \times \sum_{\rho=k+1}^{n_2} B_{\rho 1}^{(s_k)} \ln \frac{|x_{j_2} - a_{j_2 j_2 s_\rho}|}{|x_{j_2 0} - a_{j_2 j_2 s_\rho}|} \Bigg) \Bigg] + b_{j_1 j_2 0} \sum_{\rho=1}^{n_2} A_{\rho 1}^{(s_1)} \ln \frac{|x_{j_2} - a_{j_2 j_2 s_\rho}|}{|x_{j_2 0} - a_{j_2 j_2 s_\rho}|} \\
& + \sum_{\beta_0=1}^{n_2-1} b_{j_1 j_2 \beta_0} \sum_{\rho=2}^{n_2} B_{\rho 1}^{(s_1)} \frac{|x_{j_2} - a_{j_2 j_2 s_\rho}|}{|x_{j_2 0} - a_{j_2 j_2 s_\rho}|} \Bigg\};
\end{aligned}
\tag{1.280}
$$

where

$$
\begin{aligned}
& A_{r_1} = \frac{1}{\prod_{r_2=1,r_2\neq r_1}^{m_1} (a_{j_1 j_1 r_1} - a_{j_1 j_1 r_\sigma})}; \\
& b_{j_1 j_2 \beta_0} = \sum_{t_1,l_1=0}^{t_1+l_1=1} \cdots \sum_{t_q,l_q=0}^{t_q+l_q=1} \prod_{k=1}^{n_1} (a_{j_2 j_2 s_1} - a_{j_1 j_2 u_k})^{t_k} \delta_{\beta_0}^{\Sigma_{r=1}^{n_1} l_r}; \\
& \lim_{\beta_{n_2-1}\to 0} \frac{1}{\beta_{n_2-1}} \left[(x_{j_2} - a_{j_2 j_2 s_{n_2}})^{\beta_{n_2-1}} - (x_{j_2 0} - a_{j_2 j_2 s_{n_2}})^{\beta_{n_2-1}} \right] = \ln \frac{|x_{j_2} - a_{j_2 j_2 s_{n_2}}|}{|x_{j_2 0} - a_{j_2 j_2 s_{n_2}}|}; \\
& A_{\rho 1}^{(s_k)} = \frac{1}{\prod_{\sigma=k,\sigma\neq\rho}^{n_2} (a_{j_2 j_2 s_\rho} - a_{j_2 j_2 s_\sigma})}, \text{ for } \rho = \{k, k+1, \ldots, n_2\}, \\
& B_{\rho 1}^{(s_k)} = \frac{(a_{j_2 j_2 s_\rho} - a_{j_2 j_2 s_k})^{\beta_{k-1}-k}}{\prod_{\sigma=k+1,\sigma\neq\rho}^{n_2} (a_{j_2 j_2 s_\rho} - a_{j_2 j_2 s_\sigma})}, \text{ for } \rho = \{k+1, k+2, \ldots, n_2\}; \\
& k \in \{1, 2, \ldots, n_2 - 1\}.
\end{aligned}
\tag{1.281}
$$

(iii$_1$) For $a_{j_1 j_2 u} \neq a_{j_2 j_2 s}$ ($s = 1, 2, \ldots, n_2; u = 1, 2, \ldots, n_1$), the simple 1-dimensional flows and simple equilibriums exist.

(iii$_{1a}$) The equilibrium of $(x_{j_1}^*, x_{j_2}^*) = (a_{j_1 j_1 r_1}, a_{j_1 j_2 u_1})$ ($r_1 = 1, 2, \ldots, m_1; u_1 = 1, 2, \ldots, n_1$) for the -direction has the following properties.

- *For* $a_{j_1j_10}\prod_{r_2=1,r_2\neq r_1}^{m_1}(a_{j_1j_1r_1}-a_{j_1j_1r_2})\prod_{u_2=1,u_2\neq u_1}^{n_1}(a_{j_1j_2u_2}-a_{j_1j_2u_2})>0$ *and* $a_{j_2j_20}\prod_{s=1}^{n_2}(a_{j_1j_2u_1}-a_{j_2j_2s})>0$,

$$(a_{j_1j_1r_1},a_{j_1j_2u_1})=\underbrace{(\text{DP:UP, pF})}_{\text{hyperbolic flow }(+)}. \tag{1.282}$$

- For $a_{j_1j_10}\prod_{r_2=1,r_2\neq r_1}^{m_1}(a_{j_1j_1r_1}-a_{j_1j_1r_2})\prod_{u_2=1,u_2\neq u_1}^{n_1}(a_{j_1j_2u_2}-a_{j_1j_2u_2})<0$ *and* $a_{j_2j_20}\prod_{s=1}^{n_2}(a_{j_1j_2u_1}-a_{j_2j_2s})>0$,

$$(a_{j_1j_1r_1},a_{j_1j_2u_1})=\underbrace{(\text{UP:DP, pF})}_{\text{hyperbolic-secant flow }(+)}. \tag{1.283}$$

- For $a_{j_1j_10}\prod_{r_2=1,r_2\neq r_1}^{m_1}(a_{j_1j_1r_1}-a_{j_1j_1r_2})\prod_{u_2=1,u_2\neq u_1}^{n_1}(a_{j_1j_2u_2}-a_{j_1j_2u_2})>0$ *and* $a_{j_2j_20}\prod_{s=1}^{n_2}(a_{j_1j_2u_1}-a_{j_2j_2s})<0$,

$$(a_{j_1j_1r_1},a_{j_1j_2u_1})=\underbrace{(\text{UP:DP, nF})}_{\text{hyperbolic-secant flow }(-)}. \tag{1.284}$$

- For $a_{j_1j_10}\prod_{r_2=1,r_2\neq r_1}^{m_1}(a_{j_1j_1r_1}-a_{j_1j_1r_2})\prod_{u_2=1,u_2\neq u_1}^{n_1}(a_{j_1j_2u_2}-a_{j_1j_2u_2})<0$ *and* $a_{j_2j_20}\prod_{s=1}^{n_2}(a_{j_1j_2u_1}-a_{j_2j_2s})<0$,

$$(a_{j_1j_1r_1},a_{j_1j_2u_1})=\underbrace{(\text{DP:UP, nF})}_{\text{hyperbolic flow }(-)}. \tag{1.285}$$

(iii$_{Ib}$) The equilibrium of $(x_{j_1}^*,x_{j_2}^*)=(a_{j_1j_1r_1},a_{j_2j_2s_1})$ $(r_1=1,2,\ldots,m_1;\,s_1=1,2,\ldots,n_2)$ *for the* x_{j_1}*-direction has the following properties.*

- For $a_{j_1j_10}\prod_{r_2=1,r_2\neq r_1}^{m_1}(a_{j_1j_1r_1}-a_{j_1j_1r_2})\prod_{u=1}^{n_1}(a_{j_2j_2s_1}-a_{j_1j_2u})>0$ and $a_{j_2j_20}\prod_{s_2=1,s_2\neq s_1}^{n_2}(a_{j_2j_2s_1}-a_{j_2j_2s_2})>0$,

$$(a_{j_1j_1r_1},a_{j_2j_2s_1})=\underbrace{(\text{SO, SO})}_{\text{source}}. \tag{1.286}$$

- For $a_{j_1j_10}\prod_{r_2=1,r_2\neq r_1}^{m_1}(a_{j_1j_1r_1}-a_{j_1j_1r_2})\prod_{u=1}^{n_1}(a_{j_2j_2s_1}-a_{j_1j_2u})<0$ and $a_{j_2j_20}\prod_{s_2=1,s_2\neq s_1}^{n_2}(a_{j_2j_2s_1}-a_{j_2j_2s_2})>0$,

$$(a_{j_1j_1r_1},a_{j_2j_2s_1})=\underbrace{(\text{SI, SO})}_{\text{saddle}}. \tag{1.287}$$

- For $a_{j_1j_10}\prod_{r_2=1,r_2\neq r_1}^{m_1}(a_{j_1j_1r_1}-a_{j_1j_1r_2})\prod_{u=1}^{n_1}(a_{j_2j_2s_1}-a_{j_1j_2u}) > 0$ and $a_{j_2j_20}\prod_{s_2=1,s_2\neq s_1}^{n_2}(a_{j_2j_2s_1}-a_{j_2j_2s_2}) < 0$,

$$(a_{j_1j_1r_1}, a_{j_2j_2s_1}) = \underbrace{(\text{SO, SI})}_{\text{saddle}}. \tag{1.288}$$

- For $a_{j_1j_10}\prod_{r_2=1,r_2\neq r_1}^{m_1}(a_{j_1j_1r_1}-a_{j_1j_1r_2})\prod_{u=1}^{n_1}(a_{j_2j_2s_1}-a_{j_1j_2u}) < 0$ and $a_{j_2j_20}\prod_{s_2=1,s_2\neq s_1}^{n_2}(a_{j_2j_2s_1}-a_{j_2j_2s_2}) < 0$,

$$(a_{j_1j_1r_1}, a_{j_2j_2s_1}) = \underbrace{(\text{SI, SI})}_{\text{sink}}. \tag{1.289}$$

Define

$$\begin{aligned}
&\bigcup_{r=1}^{m_1}\bigcup_{l=1}^{n_1+n_2}(a_{j_1j_1r}, c_{j_2j_2l})\\
&=\bigcup_{r=1}^{m_1}\bigcup_{u=1}^{n_1}\underbrace{(a_{j_1j_1r}, a_{j_1j_2u})}_{\text{(XX,xF)}}\oplus\bigcup_{r=1}^{m_1}\bigcup_{s=1}^{n_2}\underbrace{(a_{j_1j_1r}, a_{j_2j_2s})}_{\text{YY}},\\
&\{c_{j_2j_2l}|l=1,2,\ldots,n_1+n_2\}\\
&=\text{sort}\{a_{j_1j_2u}, a_{j_2j_2s}|u=1,2,\ldots,n_1; s=1,2,\ldots,n_2\},\\
&\text{for}\, c_{j_2j_2l_1} < c_{j_2j_2l_2} \text{ for } (l_1 < l_2);
\end{aligned} \tag{1.290}$$

where

$$\begin{aligned}
&\bigcup_{r=1}^{m_1}\bigcup_{u=1}^{n_1}\underbrace{(a_{j_1j_1r}, a_{j_1j_2u})}_{\text{XX}}\\
&=\begin{bmatrix}
(a_{j_1j_11}, a_{j_1j_2n_1}) & \cdots & (a_{j_1j_1(m_1-1)}, a_{j_1j_2n_1}) & (a_{j_1j_1m_1}, a_{j_1j_2n_1})\\
(a_{j_1j_11}, a_{j_1j_2(n_1-1)}) & \cdots & (a_{j_1j_1(m_1-1)}, a_{j_1j_2(n_1-1)}) & (a_{j_1j_1m_1}, a_{j_1j_2(n_1-1)})\\
\vdots & \vdots & \vdots & \vdots\\
(a_{j_1j_11}, a_{j_1j_21}) & \cdots & (a_{j_1j_1(m_1-1)}, a_{j_1j_21}) & (a_{j_1j_1m_1}, a_{j_1j_21})
\end{bmatrix}_{n_1\times m_1}.
\end{aligned} \tag{1.291}$$

and

$$\begin{aligned}
&\bigcup_{r=1}^{m_1}\bigcup_{s=1}^{n_2}\underbrace{(a_{j_1j_1r}, a_{j_2j_2s})}_{\text{YY}}\\
&=\begin{bmatrix}
(a_{j_1j_11}, a_{j_2j_2n_2}) & \cdots & (a_{j_1j_1(m_1-1)}, a_{j_2j_2n_2}) & (a_{j_1j_1m_1}, a_{j_2j_2n_2})\\
(a_{j_1j_11}, a_{j_2j_2(n_2-1)}) & \cdots & (a_{j_1j_1(m_1-1)}, a_{j_2j_2(n_2-1)}) & (a_{j_1j_1m_1}, a_{j_2j_2(n_2-1)})\\
\vdots & \vdots & \vdots & \vdots\\
(a_{j_1j_11}, a_{j_2j_21}) & \cdots & (a_{j_1j_1(m_1-1)}, a_{j_2j_21}) & (a_{j_1j_1m_1}, a_{j_2j_21})
\end{bmatrix}_{n_2\times m_1}.
\end{aligned} \tag{1.292}$$

(iii_2) For $a_{j_1j_2u_1} = a_{j_2j_2s_1}$, the dynamical system becomes

$$\begin{aligned}\dot{x}_{j_1} &= a_{j_1j_10}(x_{j_2} - a_{j_1j_2u_1})\prod\nolimits_{r=1}^{m}(x_{j_1} - a_{j_1j_1r})\prod\nolimits_{u_2=1,u_2\neq u_1}^{n_1}(x_{j_2} - a_{j_1j_2u_2}),\\ \dot{x}_{j_2} &= a_{j_2j_20}(x_{j_2} - a_{j_2j_2s_1})\prod\nolimits_{s_2=1,s_2\neq s_1}^{n_2}(x_{j_2} - a_{j_2j_2s_2}).\end{aligned} \tag{1.293}$$

The first integral manifolds can be determined from Eqs. (1.280) through the equivalent dynamical system as

$$\begin{aligned}\dot{x}_{j_1} &= a_{j_1j_10}\prod\nolimits_{r=1}^{m}(x_{j_1} - a_{j_1j_1r})\prod\nolimits_{u_2=1,u_2\neq u_1}^{n_1}(x_{j_2} - a_{j_1j_2u_2}),\\ \dot{x}_{j_2} &= a_{j_2j_20}\prod\nolimits_{s_2=1,s_2\neq s_1}^{n_2}(x_{j_2} - a_{j_2j_2s_2}).\\ n_1 &\Rightarrow n_1 - 1,\, u_{j+1} \Rightarrow u_j\ (j = 1, 2, \ldots, n_1 - 1);\\ n_2 &\Rightarrow n_2 - 1,\, s_{i+1} \Rightarrow s_i\ (i = 1, 2, \ldots, n_2 - 1).\end{aligned} \tag{1.294}$$

Define a network matrix of 1-dimentional flows and equilibriums with an infinite-equilibrium as

$$\begin{aligned}\text{NM} = {}&\underbrace{(a_{j_1j_1r_1}, a_{j_2j_2s_1})}_{\text{(XX, YY)}} \oplus \bigcup\nolimits_{r_2=1,r_2\neq r_1}^{m_1}\underbrace{(a_{j_1j_1r_2}, a_{j_2j_2s_1})}_{\text{(XX,YY)}}\\ &\oplus \bigcup\nolimits_{r=1}^{m_1}\bigcup\nolimits_{u_2=1,u_2\neq u_1}^{n_1}\underbrace{(a_{j_1j_1r}, a_{j_1j_2u_2})}_{\text{(XX,XF)}} \oplus \bigcup\nolimits_{r=1}^{m_1}\bigcup\nolimits_{s_2=1,s_2\neq s_1}^{n_2}\underbrace{(a_{j_1j_1r}, a_{j_2j_2s_2})}_{\text{ZZ}},\end{aligned} \tag{1.295}$$

(iii_{2a}) The infinite-equilibrium of $x_{j_2}^* = a_{j_1j_2u_1} = a_{j_2j_2s_1}$ with $\overline{x}_{j_1} \neq a_{j_1j_1r_1}$ has the following properties.

- *For* $a_{j_1j_10}\prod_{r=1}^{m_1}(\overline{x}_{j_1} - a_{j_1j_1r})\prod_{u_2=1,u_2\neq u_1}^{n_1}(a_{j_1j_2u_1} - a_{j_1j_2u_2}) > 0$ and $a_{j_2j_20}\prod_{s_2=1,s_2\neq s_1}^{n_2}(a_{j_2j_2s_1} - a_{j_2j_2s_2}) > 0$,

$$(\overline{x}_{j_1}, a_{j_2j_2s_1}) = \underbrace{\text{(II, SO)}}_{\text{increasing-inflection source}}. \tag{1.296}$$

The infinite-equilibrium is a (II, SO)*-increasing-inflection source at* $x_{j_2}^* = a_{j_1j_2u_1} = a_{j_2j_2s_1}$.

- *For* $a_{j_1j_10}\prod_{r=1}^{m_1}(\overline{x}_{j_1}-a_{j_1j_1r})\prod_{u_2=1,u_2\neq u_1}^{n_1}(a_{j_1j_2u_1}-a_{j_1j_2u_2}) < 0$ and $a_{j_2j_20}\prod_{s_2=1,s_2\neq s_1}^{n_2}(a_{j_2j_2s_1}-a_{j_2j_2s_2}) > 0$,

$$(\overline{x}_{j_1}, a_{j_2j_2s_1}) = \underbrace{\text{(DI, SO)}}_{\text{decreasing-inflection source}}. \tag{1.297}$$

The infinite-equilibrium is a (DI, SO)*-decreasing-inflection source at* $x^*_{j_2} = a_{j_1j_2u_1} = a_{j_2j_2s_1}$.

- *For* $a_{j_1j_10}\prod_{r=1}^{m_1}(\overline{x}_{j_1}-a_{j_1j_1r})\prod_{u_2=1,u_2\neq u_1}^{n_1}(a_{j_1j_2u_1}-a_{j_1j_2u_2}) > 0$ and $a_{j_2j_20}\prod_{s_2=1,s_2\neq s_1}^{n_2}(a_{j_2j_2s_1}-a_{j_2j_2s_2}) < 0$,

$$(\overline{x}_{j_1}, a_{j_2j_2s_1}) = \underbrace{\text{(DI, SI)}}_{\text{decreasing-inflection sink}}. \tag{1.298}$$

The infinite-equilibrium is a (DI, SI)*-decreasing-inflection sink at* $x^*_{j_2} = a_{j_1j_2u_1} = a_{j_2j_2s_1}$.

- *For* $a_{j_1j_10}\prod_{r=1}^{m_1}(\overline{x}_{j_1}-a_{j_1j_1r})\prod_{u_2=1,u_2\neq u_1}^{n_1}(a_{j_1j_2u_1}-a_{j_1j_2u_2}) < 0$ and $a_{j_2j_20}\prod_{s_2=1,s_2\neq s_1}^{n_2}(a_{j_2j_2s_1}-a_{j_2j_2s_2}) < 0$,

$$(\overline{x}_{j_1}, a_{j_2j_2s_1}) = \underbrace{\text{(II, SI)}}_{\text{increasing-inflection sink}}. \tag{1.299}$$

The infinite-equilibrium is a (II, SI)*-increasing-inflection sink at* $x^*_{j_2} = a_{j_1j_2u_1} = a_{j_2j_2s_1}$.

(iii_{2b}) The equilibrium of $(x^*_{j_1}, x^*_{j_2}) = (a_{j_1j_1r_1}, a_{j_2j_2s_1})$ *on the infinite-equilibrium of* $x^*_{j_2} = a_{j_1j_2u_1} = a_{j_2j_2s_1}$ *has the following properties.*

- *For* $a_{j_1j_10}\prod_{r_2=1,r_2\neq r_1}^{m_1}(a_{j_1j_1r_1}-a_{j_1j_1r_2})\prod_{u_2=1,u_2\neq u_1}^{n_2}(a_{j_1j_2u_1}-a_{j_1j_2u_2}) > 0$ *and* $a_{j_2j_20}\prod_{s_2=1,s_2\neq s_1}^{n_2}(a_{j_2j_2s_1}-a_{j_2j_2s_2}) > 0$,

$$(a_{j_1j_1r_1}, a_{j_2j_2s_1}) = \underbrace{\text{(DI:II, SO)}}_{\text{decreasing-to-increasing-inflection source}}. \tag{1.300}$$

The equilibrium is a (DI:II, SO)*-decreasing-to-increasing-inflection source at* $(x^*_{j_1}, x^*_{j_2}) = (a_{j_1j_1r_1}, a_{j_2j_2s_1})$.

- *For* $a_{j_1j_10}\prod_{r_2=1,r_2\neq r_1}^{m_1}(a_{j_1j_1r_1}-a_{j_1j_1r_2})\prod_{u_2=1,u_2\neq u_1}^{n_2}(a_{j_1j_2u_1}-a_{j_1j_2u_2}) < 0$ *and* $a_{j_2j_20}\prod_{s_2=1,s_2\neq s_1}^{n_2}(a_{j_2j_2s_1}-a_{j_2j_2s_2}) > 0$,

$$(a_{j_1j_1r_1}, a_{j_2j_2s_1}) = \underbrace{\text{(II:DI, SO)}}_{\text{increasing-to-decreasing-inflection source}}. \tag{1.301}$$

The equilibrium is an (II:DI, SO)-*increasing-to-decreasing-inflection source at* $(x_{j_1}^*, x_{j_2}^*) = (a_{j_1j_1r_1}, a_{j_2j_2s_1})$.

- *For* $a_{j_1j_10}\prod_{r_2=1,r_2\neq r_1}^{m_1}(a_{j_1j_1r_1}-a_{j_1j_1r_2})\prod_{u_2=1,u_2\neq u_1}^{n_2}(a_{j_1j_2u_1}-a_{j_1j_2u_2}) > 0$ *and* $a_{j_2j_20}\prod_{s_2=1,s_2\neq s_1}^{n_2}(a_{j_2j_2s_1}-a_{j_2j_2s_2}) < 0$,

$$(a_{j_1j_1r_1}, a_{j_2j_2s_1}) = \underbrace{(\text{II:DI, SI})}_{\text{increasing-to-decreasing-inflection sink}}. \tag{1.302}$$

The equilibrium is an (II:DI, SO)-*increasing-to-decreasing-inflection sink at* $(x_{j_1}^*, x_{j_2}^*) = (a_{j_1j_1r_1}, a_{j_2j_2s_1})$.

- *For* $a_{j_1j_10}\prod_{r_2=1,r_2\neq r_1}^{m_1}(a_{j_1j_1r_1}-a_{j_1j_1r_2})\prod_{u_2=1,u_2\neq u_1}^{n_2}(a_{j_1j_2u_1}-a_{j_1j_2u_2}) < 0$ *and* $a_{j_2j_20}\prod_{s_2=1,s_2\neq s_1}^{n_2}(a_{j_2j_2s_1}-a_{j_2j_2s_2}) < 0$,

$$(a_{j_1j_1r_1}, a_{j_2j_2s_1}) = \underbrace{(\text{DI:II, SI})}_{\text{decreasing-to-increasing-inflection sink}}. \tag{1.303}$$

The equilibrium is a (DI:II, SI)-*decreasing-to-increasing-inflection sink at* $(x_{j_1}^*, x_{j_2}^*) = (a_{j_1j_1r_1}, a_{j_2j_2s_1})$.

(iv) The 1-dimensional flow at $(x_{j_1}^*, x_{j_2}^*) = (a_{j_1j_11}, a_{j_1j_21})$ *with* $a_{j_2j_21} \neq a_{j_1j_21}$ *has four classes of 1-diemsnional flow bifurcations with* $\dot{x}_{j_2}^* = a_{j_2j_20}(a_{j_1j_21} - a_{j_2j_21})^{n_2}$.

(iv$_1$) For $m_1 = 2m_{11}+1$ *and* $n_1 = 2n_{11}+1$, *the equilibrium of* $(x_{j_1}^*, x_{j_2}^*) = (a_{j_1j_11}, a_{j_1j_21})$ *for the* x_{j_1}*-direction has the following bifurcation properties with*

$$\begin{aligned}
&2m_{11}+1-\sum_{r_1=1}^{p_1} m_{j_1r_1}^{(1)} = 2\sum_{\rho_1=1}^{l_1} m_{\rho_1},\\
&2n_{11}+1-\sum_{u_1=1}^{q_1} n_{j_1u_1}^{(1)} = 2\sum_{\sigma_1=1}^{l_2} n_{\sigma_1};\\
&\sum_{r_2=1}^{\bar{p}_1} m_{j_1r_2}^{(2)} = 2m_{11}+1,\ \sum_{u_2=1}^{\bar{q}_1} n_{j_1u_2}^{(2)} = 2n_{11}+1;\\
&\sum_{r_i=1}^{p_1} m_{j_1r_i}^{(i)} = 2m_{11}+1,\ \sum_{u_i=1}^{q_1} n_{j_1u_i}^{(2)} = 2n_{11}+1,\ \text{ for } i=1,2;
\end{aligned} \tag{1.304}$$

l_1, l_2*-quadratic polynomials are without real roots, and the* ρ_1^{th} *and* σ_1^{th}*-quadratic polynomials are with powers* m_{ρ_1} *and* n_{σ_1}, *respectively.*

(iv$_{1a}$) For $a_{j_1j_10} > 0$ *and* $a_{j_2j_20}(a_{j_1j_21} - a_{j_2j_21})^{n_2} > 0$, *there is a* ([$(2m_{11}+1)$, $(2n_{11}+1)$] DP:UP, pF)-*positive hyperbolic flow as*

$$(a_{j_1j_11}, a_{j_1j_21}) = \underbrace{([(2m_{11}+1), (2n_{11}+1)]\text{DP:UP, pF})}_{[(2m_{11}+1),(2n_{11}+1)]\text{-hyperbolic flow }(+)}. \tag{1.305}$$

There are three following ([(2m_{11}+1), (2n_{11} + 1)]DP:UP, pF)*-positive hyperbolic flow appearing and switching bifurcations.*

(iv_{1a1}) The ([(2m_{11}+1), (2n_{11} + 1)]DP:UP, pF)*-positive hyperbolic flow appearing bifurcation is from a* (DP:UP, pF)*-positive hyperbolic-flow to a* (2m_{11}+1) × (2n_{11}+1)*-hyperbolic and hyperbolic-secant flow network as*

$$\begin{aligned}\underbrace{(a_{j_1j_11}, a_{j_1j_21})}_{\text{(DP:UP, pF)-flow}} &\rightleftharpoons \underbrace{([(2m_{11}+1), (2n_{11}+1)]\text{DP:UP, pF})}_{[(2m_{11}+1),(2n_{11}+1)]\text{-hyperbolic flow }(+)}\\ &\rightleftharpoons \bigcup_{r=1}^{2m_{11}+1}\bigcup_{u=1}^{2n_{11}+1}\underbrace{(a_{j_1j_1r}, a_{j_1j_2u})}_{\text{YY}},\end{aligned} \tag{1.306}$$

where

$$\bigcup_{r=1}^{2m_{11}+1}\bigcup_{u=1}^{2n_{11}+1}\underbrace{(a_{j_1j_1r}, a_{j_1j_2u})}_{\text{YY}} = \begin{bmatrix} \underbrace{(\text{DP:UP, pF})}_{\text{hyperbolic flow }(+)} & \cdots & \underbrace{(\text{UP:DP,pF})}_{\text{hyperbolic-secant flow }(+)} & \underbrace{(\text{DP:UP, pF})}_{\text{hyperbolic flow }(+)}\\ \underbrace{(\text{UP:DP, pF})}_{\text{hyperbolic-secant flow }(+)} & \cdots & \underbrace{(\text{DP:UP, pF})}_{\text{hyperbolic flow }(+)} & \underbrace{(\text{UP:DP, pF})}_{\text{hyperbolic-secant flow }(+)}\\ \vdots & \vdots & \vdots & \vdots\\ \underbrace{(\text{DP:UP, pF})}_{\text{hyperbolic flow }(+)} & \cdots & \underbrace{(\text{UP:DP, pF})}_{\text{hyperbolic-secant flow }(+)} & \underbrace{(\text{DP:UP, pF})}_{\text{hyperbolic flow }(+)} \end{bmatrix}_{(2n_{11}+1)\times(2m_{11}+1)} \tag{1.307}$$

(iv_{1a2}) The ([(2m_{11}+1), (2n_{11} + 1)]DP:UP, pF)*-positive hyperbolic flow appearing bifurcation is from a* $q_1 \times p_1$ *to* $\overline{q}_1 \times \overline{p}_1$ *1-dimensional flow networks as*

$$\begin{aligned}\bigcup_{r_1=1}^{p_1}\bigcup_{u_1=1}^{q_1}\underbrace{(a_{j_1j_1r_1}, a_{j_1j_2u_1})}_{([m_{j_1r_1},n_{j_1u_1}]\text{-XX,pF})} &\rightleftharpoons \underbrace{([(2m_{11}+1), (2n_{11}+1)]\text{DP:UP, pF})}_{[(2m_{11}+1),(2n_{11}+1)]\text{-hyperbolic flow }(+)}\\ &\rightleftharpoons \bigcup_{r_2=1}^{\overline{p}_1}\bigcup_{u_2=1}^{\overline{q}_1}\underbrace{(a_{j_1j_1r_2}, a_{j_1j_2u_2})}_{([m_{j_1r_2},n_{j_1u_2}]\text{-YY,pF})}.\end{aligned} \tag{1.308}$$

(iv_{1a3}) The ([(2m_{11}+1), (2n_{11} + 1)]DP:UP, pF)*-positive hyperbolic flow switching bifurcation is for the switching of two networks of* $q_1 \times p_1$ *1-dimensional flows as*

$$\bigcup_{r_1=1}^{p_1}\bigcup_{u_1=1}^{q_1}\underbrace{(a_{j_1j_1r_1},a_{j_1j_2u_1})}_{([m_{j_1r_1},n_{j_1u_1}]\text{-XX,pF})}\rightleftharpoons\underbrace{([(2m_{11}+1),(2n_{11}+1)]\text{DP:UP, pF})}_{[(2m_{11}+1),(2n_{11}+1)]\text{-hyperbolic flow }(+)}$$
$$\rightleftharpoons\bigcup_{r_2=1}^{p_1}\bigcup_{u_2=1}^{q_1}\underbrace{(a_{j_1j_1r_2},a_{j_1j_2u_2})}_{([m_{j_1r_2},n_{j_1u_2}]\text{-YY,pF})}. \tag{1.309}$$

(iv_{1b}) For $a_{j_1j_10}<0$ and $a_{j_2j_20}(a_{j_1j_21}-a_{j_2j_21})^{n_2}>0$, there is a ([(2$m_{11}$+1), (2$n_{11}$+1)] UP:DP, pF)*-positive hyperbolic-secant flow as*

$$(a_{j_1j_11},a_{j_1j_21})=\underbrace{([(2m_{11}+1),(2n_{11}+1)]\text{UP:DP, pF})}_{[(2m_{11}+1),(2n_{11}+1)]\text{-hyperbolic-secant flow }(+)}. \tag{1.310}$$

There are three following ([(2m_{11}+1), (2n_{11} + 1)]UP:DP, pF)*-positive hyperbolic-secant flow appearing and switching bifurcations.*

(iv_{1b1}) The ([(2m_{11}+1), (2n_{11} + 1)]UP:DP, pF)*-positive hyperbolic-secant flow appearing bifurcation is from a* (UP:DP,pF)*-positive hyperbolic-secant flow to a* $(2m_{11}+1)\times(2n_{11}+1)$*-hyperbolic and hyperbolic-secant flow network as*

$$\underbrace{(a_{j_1j_11},a_{j_1j_21})}_{\text{(UP:DP, pF)-flow}}\rightleftharpoons\underbrace{([(2m_{11}+1),(2n_{11}+1)]\text{UP:DP, pF})}_{[(2m_{11}+1),(2n_{11}+1)]\text{-hyperbolic-secant flow }(+)}$$
$$\rightleftharpoons\bigcup_{r=1}^{2m_{11}+1}\bigcup_{u=1}^{2n_{11}+1}\underbrace{(a_{j_1j_1r},a_{j_1j_2u})}_{\text{YY}}, \tag{1.311}$$

where

$$\bigcup_{r=1}^{2m_{11}+1}\bigcup_{u=1}^{2n_{11}+1}\underbrace{(a_{j_1j_1r},a_{j_1j_2u})}_{\text{YY}}$$
$$=\begin{bmatrix}\underbrace{\text{(UP:DP, pF)}}_{\text{hyperbolic-secant flow }(+)} & \cdots & \underbrace{\text{(DP:UP, pF)}}_{\text{hyperbolic flow }(+)} & \underbrace{\text{(UP:DP, pF)}}_{\text{hyperbolic-secant flow }(+)}\\ \underbrace{\text{(DP:UP, pF)}}_{\text{hyperbolic flow }(+)} & \cdots & \underbrace{\text{(UP:DP, pF)}}_{\text{hyperbolic-secant flow }(+)} & \underbrace{\text{(DP:UP, pF)}}_{\text{hyperbolic flow }(+)}\\ \vdots & \vdots & \vdots & \vdots\\ \underbrace{\text{(UP:DP, pF)}}_{\text{hyperbolic-secant flow }(+)} & \cdots & \underbrace{\text{(DP:UP, pF)}}_{\text{hyperbolic flow }(+)} & \underbrace{\text{(UP:DP, pF)}}_{\text{hyperbolic-secant flow }(+)}\end{bmatrix}_{(2n_{11}+1)\times(2m_{11}+1)}. \tag{1.312}$$

(iv_{1b2}) *The* ([(2m_{11}+1), (2n_{11} + 1)]UP:DP, pF)*-positive hyperbolic-secant flow appearing bifurcation is from a* $q_1 \times p_1$ *to* $\overline{q}_1 \times \overline{p}_1$ *1-dimensional flow networks as*

$$\bigcup_{r_1=1}^{p_1}\bigcup_{u_1=1}^{q_1}\underbrace{(a_{j_1j_1r_1}, a_{j_1j_2u_1})}_{([m_{j_1r_1}, n_{j_1u_1}]\text{-XX,pF})} \rightleftharpoons \underbrace{([(2m_{11}+1), (2n_{11}+1)]\text{UP:DP, pF})}_{[(2m_{11}+1),(2n_{11}+1)]\text{-hyperbolic-secant flow }(+)}$$
$$\rightleftharpoons \bigcup_{r_2=1}^{\overline{p}_1}\bigcup_{u_2=1}^{\overline{q}_1}\underbrace{(a_{j_1j_1r_2}, a_{j_1j_2u_2})}_{([m_{j_1r_2}, n_{j_1u_2}]\text{-YY,pF})}. \tag{1.313}$$

(iv_{1b3}) *The* ([(2m_{11}+1), (2n_{11} + 1)]UP:DP, pF)*-positive hyperbolic-secant flow switching bifurcation is for the switching of two networks of* $q_1 \times p_1$ *1-dimensional flows as*

$$\bigcup_{r_1=1}^{p_1}\bigcup_{u_1=1}^{q_1}\underbrace{(a_{j_1j_1r_1}, a_{j_1j_2u_1})}_{([m_{j_1r_1}, n_{j_1u_1}]\text{-XX,pF})} \rightleftharpoons \underbrace{([(2m_{11}+1), (2n_{11}+1)]\text{UP:DP, pF})}_{[(2m_{11}+1),(2n_{11}+1)]\text{-hyperbolic flow }(+)}$$
$$\rightleftharpoons \bigcup_{r_2=1}^{p_1}\bigcup_{u_2=1}^{q_1}\underbrace{(a_{j_1j_1r_2}, a_{j_1j_2u_2})}_{([m_{j_1r_2}, n_{j_1u_2}]\text{-YY,pF})}. \tag{1.314}$$

(iv_{1c}) *For* $a_{j_1j_10} > 0$ *and* $a_{j_2j_20}(a_{j_1j_21} - a_{j_2j_21})^{n_2} < 0$, *there is a* ([(2$m_{11}$+1), (2$n_{11}$+1)] UP:DP, nF)*-negative hyperbolic-secant flow as*

$$(a_{j_1j_11}, a_{j_1j_21}) = \underbrace{([(2m_{11}+1), (2n_{11}+1)]\text{UP:DP, nF})}_{[(2m_{11}+1),(2n_{11}+1)]\text{-hyperbolic-secant flow }(-)}. \tag{1.315}$$

There are three following ([(2m_{11}+1), (2n_{11} + 1)]UP:DP, nF)*-negative hyperbolic-secant flow appearing and switching bifurcations.*

(iv_{1c1}) *The* ([(2m_{11}+1), (2n_{11} + 1)]UP:DP, nF)*-negative hyperbolic-secant flow appearing bifurcation is from a* (UP:DP, nF)*-negative hyperbolic-secant flow to a* (2m_{11}+1) × (2n_{11}+1)*-hyperbolic and hyperbolic-secant flow network as*

$$\underbrace{(a_{j_1j_11}, a_{j_1j_21})}_{(\text{UP:DP, nF})\text{-flow}} \rightleftharpoons \underbrace{([(2m_{11}+1), (2n_{11}+1)]\text{UP:DP, nF})}_{[(2m_{11}+1),(2n_{11}+1)]\text{-hyperbolic-secant flow }(-)}$$
$$\rightleftharpoons \bigcup_{r=1}^{2m_{11}+1}\bigcup_{u=1}^{2n_{11}+1}\underbrace{(a_{j_1j_1r}, a_{j_1j_2u})}_{\text{YY}}, \tag{1.316}$$

where

$$\bigcup_{r=1}^{2m_{11}+1}\bigcup_{u=1}^{2n_{11}+1}\underbrace{(a_{j_1j_1r},a_{j_1j_2u})}_{\text{YY}}$$
$$=\begin{bmatrix}\underbrace{(\text{UP:DP, nF})}_{\text{hyperbolic-secant flow }(-)} & \cdots & \underbrace{(\text{DP:UP, nF})}_{\text{hyperbolic flow }(-)} & \underbrace{(\text{UP:DP, nF})}_{\text{hyperbolic-secant flow }(-)}\\ \underbrace{(\text{DP:UP, nF})}_{\text{hyperbolic flow }(-)} & \cdots & \underbrace{(\text{UP:DP, nF})}_{\text{hyperbolic-secant flow }(-)} & \underbrace{(\text{DP:UP, nF})}_{\text{hyperbolic flow }(-)}\\ \vdots & \vdots & \vdots & \vdots\\ \underbrace{(\text{UP:DP, nF})}_{\text{hyperbolic-secant flow }(-)} & \cdots & \underbrace{(\text{DP:UP, nF})}_{\text{hyperbolic flow }(-)} & \underbrace{(\text{UP:DP, nF})}_{\text{hyperbolic-secant flow }(-)}\end{bmatrix}_{(2n_{11}+1)\times(2m_{11}+1)}. \tag{1.317}$$

(iv_{1c2}) The ([($2m_{11}$+1), ($2n_{11}$ + 1)]UP:DP, nF)*-negative hyperbolic-secant flow appearing bifurcation is from a* $q_1 \times p_1$ *to* $\overline{q}_1 \times \overline{p}_1$ *1-dimensional flow networks as*

$$\bigcup_{r_1=1}^{p_1}\bigcup_{u_1=1}^{q_1}\underbrace{(a_{j_1j_1r_1},a_{j_1j_2u_1})}_{([m_{j_1r_1},n_{j_1u_1}]\text{-XX,nF})}\rightleftharpoons\underbrace{([(2m_{11}+1),(2n_{11}+1)]\text{UP:DP, nF})}_{[(2m_{11}+1),(2n_{11}+1)]\text{-hyperbolic-secant flow }(-)}$$
$$\rightleftharpoons\bigcup_{r_2=1}^{\overline{p}_1}\bigcup_{u_2=1}^{\overline{q}_1}\underbrace{(a_{j_1j_1r_2},a_{j_1j_2u_2})}_{([m_{j_1r_2},n_{j_1u_2}]\text{-YY,nF})}. \tag{1.318}$$

(iv_{1c3}) The ([($2m_{11}$+1), ($2n_{11}$ + 1)]UP:DP, nF)*-negative hyperbolic-secant flow switching bifurcation is for the switching of two networks of* $q_1 \times p_1$ *1-dimensional flows as*

$$\bigcup_{r_1=1}^{p_1}\bigcup_{u_1=1}^{q_1}\underbrace{(a_{j_1j_1r_1},a_{j_1j_2u_1})}_{([m_{j_1r_1},n_{j_1u_1}]\text{-XX,nF})}\rightleftharpoons\underbrace{([(2m_{11}+1),(2n_{11}+1)]\text{UP:DP, nF})}_{[(2m_{11}+1),(2n_{11}+1)]\text{-hyperbolic flow }(-)}$$
$$\rightleftharpoons\bigcup_{r_2=1}^{p_1}\bigcup_{u_2=1}^{q_1}\underbrace{(a_{j_1j_1r_2},a_{j_1j_2u_2})}_{([m_{j_1r_2},n_{j_1u_2}]\text{-YY,nF})}. \tag{1.319}$$

(iv_{1d}) For $a_{j_1j_10} < 0$ *and* $a_{j_2j_20}(a_{j_1j_21} - a_{j_2j_21})^{n_2} < 0$*, there is a* ([($2m_{11}$+1), ($2n_{11}$+1)] DP:UP, nF)*-negative hyperbolic flow as*

$$(a_{j_1j_11},a_{j_1j_21})=\underbrace{([(2m_{11}+1),(2n_{11}+1)]\text{DP:UP, nF})}_{[(2m_{11}+1),(2n_{11}+1)]\text{-hyperbolic flow }(-)}. \tag{1.320}$$

There are three following ([($2m_{11}$+1), ($2n_{11}$ + 1)]DP:UP, nF)*-negative hyperbolic flow appearing and switching bifurcations.*

(iv_{1d1}) The $([(2m_{11}+1),(2n_{11}+1)]\text{DP:UP, nF})$*-negative hyperbolic flow appearing bifurcation is from a* (DP:UP, nF)*-negative hyperbolic-secant flow to a* $(2m_{11}+1)\times(2n_{11}+1)$*-hyperbolic and hyperbolic-secant flow network as*

$$\begin{aligned}\underbrace{(a_{j_1j_11},a_{j_1j_21})}_{\text{(DP:UP, nF)-flow}} &\rightleftharpoons \underbrace{([(2m_{11}+1),(2n_{11}+1)]\text{DP:UP, nF})}_{[(2m_{11}+1),(2n_{11}+1)]\text{-hyperbolic flow }(-)}\\ &\rightleftharpoons \bigcup\nolimits_{r=1}^{2m_{11}+1}\bigcup\nolimits_{u=1}^{2n_{11}+1}\underbrace{(a_{j_1j_1r},a_{j_1j_2u})}_{\text{YY}},\end{aligned}\tag{1.321}$$

where

$$\bigcup\nolimits_{r=1}^{2m_{11}+1}\bigcup\nolimits_{u=1}^{2n_{11}+1}\underbrace{(a_{j_1j_1r},a_{j_1j_2u})}_{\text{YY}}$$
$$=\begin{bmatrix}\underbrace{\text{(UP:DP, nF)}}_{\text{hyperbolic-secant flow }(-)} & \cdots & \underbrace{\text{(DP:UP, nF)}}_{\text{hyperbolic flow }(-)} & \underbrace{\text{(UP:DP, nF)}}_{\text{hyperbolic-secant flow }(-)}\\ \underbrace{\text{(DP:UP, nF)}}_{\text{hyperbolic flow }(-)} & \cdots & \underbrace{\text{(UP:DP, nF)}}_{\text{hyperbolic-secant flow }(-)} & \underbrace{\text{(DP:UP, nF)}}_{\text{hyperbolic flow }(-)}\\ \vdots & \vdots & \vdots & \vdots\\ \underbrace{\text{(UP:DP, nF)}}_{\text{hyperbolic-secant flow }(-)} & \cdots & \underbrace{\text{(DP:UP, nF)}}_{\text{hyperbolic flow }(-)} & \underbrace{\text{(UP:DP, nF)}}_{\text{hyperbolic-secant flow }(-)}\end{bmatrix}_{(2n_{11}+1)\times(2m_{11}+1)}.\tag{1.322}$$

(iv_{1d2}) The $([(2m_{11}+1),(2n_{11}+1)]\text{DP:UP, nF})$*-negative hyperbolic flow appearing bifurcation is from a* $q_1\times p_1$ *to* $\overline{q}_1\times\overline{p}_1$ *1-dimensional flow networks as*

$$\begin{aligned}\bigcup\nolimits_{r_1=1}^{p_1}\bigcup\nolimits_{u_1=1}^{q_1}\underbrace{(a_{j_1j_1r_1},a_{j_1j_2u_1})}_{([m_{j_1r_1},n_{j_1u_1}]\text{-XX,nF})} &\rightleftharpoons \underbrace{([(2m_{11}+1),(2n_{11}+1)]\text{DP:UP, nF})}_{[(2m_{11}+1),(2n_{11}+1)]\text{-hyperbolic flow }(-)}\\ &\rightleftharpoons \bigcup\nolimits_{r_2=1}^{\overline{p}_1}\bigcup\nolimits_{u_2=1}^{\overline{q}_1}\underbrace{(a_{j_1j_1r_2},a_{j_1j_2u_2})}_{([m_{j_1r_2},n_{j_1u_2}]\text{-YY,nF})}.\end{aligned}\tag{1.323}$$

(iv_{1d3}) The $([(2m_{11}+1),(2n_{11}+1)]\text{DP:UP, nF})$*-negative hyperbolic flow switching bifurcation is for the switching of two networks of* $q_1\times p_1$ *1-dimensional flows as*

$$\begin{aligned}\bigcup\nolimits_{r_1=1}^{p_1}\bigcup\nolimits_{u_1=1}^{q_1}\underbrace{(a_{j_1j_1r_1},a_{j_1j_2u_1})}_{([m_{j_1r_1},n_{j_1u_1}]\text{-XX,nF})} &\rightleftharpoons \underbrace{([(2m_{11}+1),(2n_{11}+1)]\text{DP:UP, nF})}_{[(2m_{11}+1),(2n_{11}+1)]\text{-hyperbolic flow }(-)}\\ &\rightleftharpoons \bigcup\nolimits_{r_2=1}^{p_1}\bigcup\nolimits_{u_2=1}^{q_1}\underbrace{(a_{j_1j_1r_2},a_{j_1j_2u_2})}_{([m_{j_1r_2},n_{j_1u_2}]\text{-YY,nF})}.\end{aligned}\tag{1.324}$$

(iv_2) For $m_1 = 2m_{11}$ and $n_1 = 2n_{11}+1$, the equilibrium of $(x_{j_1}^*, x_{j_2}^*) = (a_{j_1j_11}, a_{j_1j_21})$ for the x_{j_1}-direction has the following bifurcation properties with

$$\begin{aligned}
&2m_{11} - \sum_{r_1=1}^{p_1} m_{j_1r_1}^{(1)} = 2\sum_{\rho_1=1}^{l_1} m_{\rho_1},\\
&2n_{11} + 1 - \sum_{u_1=1}^{q_1} n_{j_1u_1}^{(1)} = 2\sum_{\sigma_1=1}^{l_2} n_{\sigma_1};\\
&\sum_{r_2=1}^{\overline{p}_1} m_{j_1r_2}^{(2)} = 2m_{11}, \sum_{u_2=1}^{\overline{q}_1} n_{j_1u_2}^{(2)} = 2n_{11}+1;\\
&\sum_{r_i=1}^{p_1} m_{j_1r_i}^{(i)} = 2m_{11}, \sum_{u_i=1}^{q_1} n_{j_1u_i}^{(2)} = 2n_{11}+1, \text{ for } i=1,2;
\end{aligned}\tag{1.325}$$

l_1, l_2-quadratic polynomials are without real roots, and the ρ_1^{th} and σ_1^{th}-quadratic polynomials are with powers m_{ρ_1} and n_{σ_1}, respectively.

(iv_{2a}) For $a_{j_1j_10} > 0$ and $a_{j_2j_20}(a_{j_1j_21} - a_{j_2j_21})^{n_2} > 0$, there is a ([(2$m_{11}$), (2$n_{11}$ + 1)]UP:UP, pF)*-positive hyperbolic-secant-to-hyperbolic flow as*

$$(a_{j_1j_11}, a_{j_1j_21}) = \underbrace{([(2m_{11}), (2n_{11}+1)]\text{UP:UP, pF})}_{[(2m_{11}),(2n_{11}+1)]\text{-hyperbolic-secant-to-hyperbolic flow }(+)}. \tag{1.326}$$

There are three following ([(2m_{11}), (2n_{11} + 1)]UP:UP, pF)*-positive hyperbolic-secant-to-hyperbolic flow appearing and switching bifurcations.*

(iv_{2a1}) The ([(2m_{11}),(2n_{11}+1)]UP:UP,pF)*-positive hyperbolic-secant-to-hyperbolic flow appearing bifurcation is from a* (UP,pF)*-positive up-parabola flow to a* $(2m_{11}) \times (2n_{11}+1)$*-hyperbolic and hyperbolic-secant flow network as*

$$\begin{aligned}
\underbrace{(a_{j_1j_21}, \dot{x}_{j_2})}_{(\text{UP, pF})\text{-flow}} &\rightleftharpoons \underbrace{([(2m_{11}), (2n_{11}+1)]\text{UP:UP, pF})}_{[(2m_{11}),(2n_{11}+1)]\text{-hyperbolic-secant-to-hyperbolic flow }(+)}\\
&\rightleftharpoons \bigcup_{r=1}^{2m_{11}} \bigcup_{u=1}^{2n_{11}+1} \underbrace{(a_{j_1j_1r}, a_{j_1j_2u})}_{\text{YY}},
\end{aligned}\tag{1.327}$$

where

$$\begin{aligned}
&\bigcup_{r=1}^{2m_{11}} \bigcup_{u=1}^{2n_{11}+1} \underbrace{(a_{j_1j_1r}, a_{j_1j_2u})}_{\text{YY}}\\
&= \begin{bmatrix}
\underbrace{(\text{UP:DP, pF})}_{\text{hyperbolic-secant flow }(+)} & \cdots & \underbrace{(\text{UP:DP, pF})}_{\text{hyperbolic-secant flow }(+)} & \underbrace{(\text{DP:UP, pF})}_{\text{hyperbolic flow }(+)}\\
\underbrace{(\text{DP:UP, pF})}_{\text{hyperbolic flow }(+)} & \cdots & \underbrace{(\text{DP:UP, pF})}_{\text{hyperbolic flow }(+)} & \underbrace{(\text{UP:UP, pF})}_{\text{hyperbolic-secant flow }(+)}\\
\vdots & \vdots & \vdots & \vdots\\
\underbrace{(\text{UP:DP, pF})}_{\text{hyperbolic-secant flow }(+)} & \cdots & \underbrace{(\text{UP:DP, pF})}_{\text{hyperbolic-secant flow }(+)} & \underbrace{(\text{DP:UP, pF})}_{\text{hyperbolic flow }(+)}
\end{bmatrix}_{(2n_{11}+1)\times(2m_{11})}.
\end{aligned}\tag{1.328}$$

(iv_{2a2}) The $([(2m_{11}),(2n_{11}+1)]\text{UP:UP,pF})$*-positive hyperbolic-secant-to-hyperbolic flow appearing bifurcation is from a* $q_1 \times p_1$ *to* $\overline{q}_1 \times \overline{p}_1$ *1-dimensional flow networks as*

$$\bigcup_{r_1=1}^{p_1}\bigcup_{u_1=1}^{q_1}\underbrace{(a_{j_1j_1r_1},a_{j_1j_2u_1})}_{([m_{j_1r_1},n_{j_1u_1}]\text{-XX, pF})} \rightleftharpoons \underbrace{([(2m_{11}),(2n_{11}+1)]\text{UP:UP, pF})}_{[(2m_{11}),(2n_{11}+1)]\text{-hyperbolic-secant-to-hyperbolic flow }(+)} \rightleftharpoons \bigcup_{r_2=1}^{\overline{p}_1}\bigcup_{u_2=1}^{\overline{q}_1}\underbrace{(a_{j_1j_1r_2},a_{j_1j_2u_2})}_{([m_{j_1r_2},n_{j_1u_2}]\text{-YY, pF})}. \quad (1.329)$$

(iv_{2a3}) The $([(2m_{11}),(2n_{11}+1)]\text{UP:UP,pF})$*-positive hyperbolic-secant-to-hyperbolic flow switching bifurcation is for the switching of two networks of* $q_1 \times p_1$ *1-dimensional flows as*

$$\bigcup_{r_1=1}^{p_1}\bigcup_{u_1=1}^{q_1}\underbrace{(a_{j_1j_1r_1},a_{j_1j_2u_1})}_{([m_{j_1r_1},n_{j_1u_1}]\text{-XX, pF})} \rightleftharpoons \underbrace{([(2m_{11}),(2n_{11}+1)]\text{UP:UP, pF})}_{[(2m_{11}),(2n_{11}+1)]\text{-hyperbolic-secant-to-hyperbolic flow }(+)} \rightleftharpoons \bigcup_{r_2=1}^{p_1}\bigcup_{u_2=1}^{q_1}\underbrace{(a_{j_1j_1r_2},a_{j_1j_2u_2})}_{([m_{j_1r_2},n_{j_1u_2}]\text{-YY, pF})}. \quad (1.330)$$

(iv_{2b}) For $a_{j_1j_10} < 0$ *and* $a_{j_2j_20}(a_{j_1j_21}-a_{j_2j_21})^{n_2} > 0$*, there is a* $([(2m_{11}),(2n_{11}+1)]\text{DP:DP, pF})$*-positive hyperbolic-to-hyperbolic-secant flow as*

$$(a_{j_1j_11},a_{j_1j_21}) = \underbrace{([(2m_{11}),(2n_{11}+1)]\text{DP:DP, pF})}_{[(2m_{11}),(2n_{11}+1)]\text{-hyerpbolic-to-hyperbolic-secant flow }(+)}. \quad (1.331)$$

There are three following $([(2m_{11}),(2n_{11}+1)]\text{DP:DP, pF})$*-positive hyperbolic-to-hyperbolic-secant flow appearing and switching bifurcations.*

(iv_{2b1}) The $([(2m_{11}),(2n_{11}+1)]\text{DP:DP, pF})$*-positive hyperbolic-to-hyperbolic-secant flow appearing bifurcation is from a* (DP, pF)*-positive down-parabola flow to a* $(2m_{11}) \times (2n_{11}+1)$*-hyperbolic and hyperbolic-secant flow network as*

$$\underbrace{(a_{j_1j_21},\dot{x}_{j_2})}_{(\text{DP, pF})\text{-flow}} \rightleftharpoons \underbrace{([(2m_{11}),(2n_{11}+1)]\text{DP:DP, pF})}_{[(2m_{11}),(2n_{11}+1)]\text{-hyerpbolic-to-hyperbolic-secant flow }(+)} \rightleftharpoons \bigcup_{r=1}^{2m_{11}}\bigcup_{u=1}^{2n_{11}+1}\underbrace{(a_{j_1j_1r},a_{j_1j_2u})}_{\text{YY}}, \quad (1.332)$$

where

$$\bigcup_{r=1}^{2m_{11}}\bigcup_{u=1}^{2n_{11}+1}\underbrace{(a_{j_1j_1r},a_{j_1j_2u})}_{\text{YY}}$$

$$=\begin{bmatrix}\underbrace{(\text{DP:UP, pF})}_{\text{hyperbolic flow }(+)} & \cdots & \underbrace{(\text{DP:UP, pF})}_{\text{hyperbolic flow }(+)} & \underbrace{(\text{UP:DP, pF})}_{\text{hyperbolic-secant flow }(+)}\\ \underbrace{(\text{UP:DP, pF})}_{\text{hyperbolic-secant flow }(+)} & \cdots & \underbrace{(\text{UP:DP, pF})}_{\text{hyperbolic-secant flow }(+)} & \underbrace{(\text{DP:UP, pF})}_{\text{hyperbolic flow }(+)}\\ \vdots & \vdots & \vdots & \vdots\\ \underbrace{(\text{DP:UP, pF})}_{\text{hyperbolic flow }(+)} & \cdots & \underbrace{(\text{DP:UP, pF})}_{\text{hyperbolic flow }(+)} & \underbrace{(\text{UP:DP, pF})}_{\text{hyperbolic-secant flow }(+)}\end{bmatrix}_{(2n_{11}+1)\times(2m_{11})}. \tag{1.333}$$

(iv_{2b2}) The ([$(2m_{11}), (2n_{11}+1)$]DP:DP, pF)*-positive hyperbolic-to-hyperbolic-secant flow appearing bifurcation is from a $q_1\times p_1$ to $\overline{q}_1\times\overline{p}_1$ 1-dimensional flow networks as*

$$\bigcup_{r_1=1}^{p_1}\bigcup_{u_1=1}^{q_1}\underbrace{(a_{j_1j_1r_1},a_{j_1j_2u_1})}_{([m_{j_1r_1},n_{j_1u_1}]\text{-XX, pF})}\rightleftharpoons\underbrace{([(2m_{11}),(2n_{11}+1)]\text{DP:DP, pF})}_{[(2m_{11}),(2n_{11}+1)]\text{-hyerpbolic-to-hyperbolic-secant flow }(+)}$$
$$\rightleftharpoons\bigcup_{r_2=1}^{\overline{p}_1}\bigcup_{u_2=1}^{\overline{q}_1}\underbrace{(a_{j_1j_1r_2},a_{j_1j_2u_2})}_{([m_{j_1r_2},n_{j_1u_2}]\text{-YY, pF})}. \tag{1.334}$$

(iv_{2b3}) The ([$(2m_{11}),(2n_{11}+1)$]DP:DP, pF)*-positive hyperbolic-to-hyperbolic-secant flow switching bifurcation is for the switching of two networks of $q_1\times p_1$ 1-dimensional flows as*

$$\bigcup_{r_1=1}^{p_1}\bigcup_{u_1=1}^{q_1}\underbrace{(a_{j_1j_1r_1},a_{j_1j_2u_1})}_{([m_{j_1r_1},n_{j_1u_1}]\text{-XX, pF})}\rightleftharpoons\underbrace{([(2m_{11}),(2n_{11}+1)]\text{DP:DP, pF})}_{[(2m_{11}),(2n_{11}+1)]\text{-hyerpbolic-to-hyperbolic-secant flow }(+)}$$
$$\rightleftharpoons\bigcup_{r_2=1}^{p_1}\bigcup_{u_2=1}^{q_1}\underbrace{(a_{j_1j_1r_2},a_{j_1j_2u_2})}_{([m_{j_1r_2},n_{j_1u_2}]\text{-YY, pF})}. \tag{1.335}$$

(iv_{2c}) For $a_{j_1j_10}>0$ and $a_{j_2j_20}(a_{j_1j_21}-a_{j_2j_21})^{n_2}<0$, there is a ([$(2m_{11}),(2n_{11}+1)$]DP:DP, nF)*-negative hyperbolic-to-hyperbolic-secant flow as*

$$(a_{j_1j_11},a_{j_1j_21})=\underbrace{([(2m_{11}),(2n_{11}+1)]\text{DP:DP, nF})}_{[(2m_{11}),(2n_{11}+1)]\text{-hyperbolic-to-hyperbolic-secant flow }(-)}. \tag{1.336}$$

There are three following ([$(2m_{11}),(2n_{11}+1)$]DP:DP, nF)*-negative hyperbolic-to-hyperbolic-secant flow appearing and switching bifurcations* .

(iv_{2c1}) The ([(2m_{11}), (2n_{11} + 1)]DP:DP, nF)*-negative hyperbolic-to-hyperbolic-secant flow appearing bifurcation is from a* (DP, nF)*-negative down-parabola flow to a* $(2m_1) \times (2n_1{+}1)$*-hyperbolic and hyperbolic-secant flow network as*

$$\underbrace{(a_{j_1j_21}, \dot{x}_{j_2})}_{\text{(DP, nF)-flow}} \rightleftharpoons \underbrace{([(2m_{11}), (2n_{11}+1)]\text{DP:DP, nF})}_{[(2m_{11}),(2n_{11}+1)]\text{-hyperbolic-to-hyperbolic-secant flow }(-)} \rightleftharpoons \bigcup_{r=1}^{2m_{11}} \bigcup_{u=1}^{2n_{11}+1} \underbrace{(a_{j_1j_1r}, a_{j_1j_2u})}_{\text{YY}}, \tag{1.337}$$

where

$$\bigcup_{r=1}^{2m_{11}} \bigcup_{u=1}^{2n_{11}+1} \underbrace{(a_{j_1j_1r}, a_{j_1j_2u})}_{\text{YY}} = \begin{bmatrix} \underbrace{(\text{DP:UP, nF})}_{\text{hyperbolic flow }(-)} & \cdots & \underbrace{(\text{DP:UP, nF})}_{\text{hyperbolic flow }(-)} & \underbrace{(\text{UP:UP, nF})}_{\text{hyperbolic-secant flow }(-)} \\ \underbrace{(\text{UP:UP, nF})}_{\text{hyperbolic-secant flow }(-)} & \cdots & \underbrace{(\text{UP:UP, nF})}_{\text{hyperbolic-secant flow }(-)} & \underbrace{(\text{DP:UP, nF})}_{\text{hyperbolic flow }(-)} \\ \vdots & \vdots & \vdots & \vdots \\ \underbrace{(\text{DP:UP, nF})}_{\text{hyperbolic flow }(-)} & \cdots & \underbrace{(\text{DP:UP, nF})}_{\text{hyperbolic flow }(-)} & \underbrace{(\text{UP:UP, nF})}_{\text{hyperbolic-secant flow }(-)} \end{bmatrix}_{(2n_{11}+1)\times(2m_{11})}. \tag{1.338}$$

(iv_{2c2}) The ([(2m_{11}), (2n_{11} + 1)]DP:DP, nF)*-negative hyperbolic-to-hyperbolic-secant flow appearing bifurcation is from a* $q_1 \times p_1$ *to* $\overline{q}_1 \times \overline{p}_1$ *1-dimensional flow networks as*

$$\bigcup_{r_1=1}^{p_1} \bigcup_{u_1=1}^{q_1} \underbrace{(a_{j_1j_1r_1}, a_{j_1j_2u_1})}_{([m_{j_1r_1}, n_{j_1u_1}]\text{-XX, nF})} \rightleftharpoons \underbrace{([(2m_{11}), (2n_{11}+1)]\text{DP:DP, nF})}_{[(2m_{11}),(2n_{11}+1)]\text{-hyperbolic-to-hyperbolic-secant flow }(-)} \rightleftharpoons \bigcup_{r_2=1}^{\overline{p}_1} \bigcup_{u_2=1}^{\overline{q}_1} \underbrace{(a_{j_1j_1r_2}, a_{j_1j_2u_2})}_{([m_{j_1r_2}, n_{j_1u_2}]\text{-YY, nF})}. \tag{1.339}$$

(iv_{2c3}) The ([(2m_{11}), (2n_{11} + 1)]DP:DP, nF)*-negative hyperbolic-to-hyperbolic-secant flow switching bifurcation is for the switching of two networks of* $q_1 \times p_1$ *1-dimensional flows as*

$$\bigcup_{r_1=1}^{p_1} \bigcup_{u_1=1}^{q_1} \underbrace{(a_{j_1j_1r_1}, a_{j_1j_2u_1})}_{([m_{j_1r_1}, n_{j_1u_1}]\text{-XX, nF})} \rightleftharpoons \underbrace{([(2m_{11}), (2n_{11}+1)]\text{DP:DP, nF})}_{[(2m_{11}),(2n_{11}+1)]\text{-hyperbolic-to-hyperbolic-secant flow }(-)} \rightleftharpoons \bigcup_{r_2=1}^{p_1} \bigcup_{u_2=1}^{q_1} \underbrace{(a_{j_1j_1r_2}, a_{j_1j_2u_2})}_{([m_{j_1r_2}, n_{j_1u_2}]\text{-YY, nF})}. \tag{1.340}$$

(iv$_{2d}$) For $a_{j_1j_10}<0$ *and* $a_{j_2j_20}(a_{j_1j_21}-a_{j_2j_21})^{n_2}<0$*, there is a* ([(2$m_{11}$), (2$n_{11}$ + 1)]UP:UP, nF)*-negative hyperbolic-secant-to-hyperbolic flow as*

$$(a_{j_1j_11},a_{j_1j_21})=\underbrace{([(2m_{11}),(2n_{11}+1)]\text{UP:UP, nF})}_{[(2m_{11}),(2n_{11}+1)]\text{-hyperbolic-secant-to-hyperbolic flow }(-)}. \tag{1.341}$$

There are three following ([(2m_{11}), (2n_{11} + 1)]UP:UP, nF)*-negative hyperbolic-secant-to-hyperbolic flow appearing and switching bifurcations.*

(iv$_{2d1}$) The ([(2m_{11}), (2n_{11}+1)]UP:UP, nF)*-negative hyperbolic-secant-to-hyperbolic flow appearing bifurcation is from a* (UP, nF)*-negative up-parabola flow to a* (2m_{11}) × (2n_{11}+1)*-hyperbolic and hyperbolic-secant flow network as*

$$\begin{aligned}\underbrace{(a_{j_1j_21},\dot{x}_{j_2})}_{\text{(UP, nF)-flow}}&\rightleftharpoons\underbrace{([(2m_{11}),(2n_{11}+1)]\text{UP:UP, nF})}_{[(2m_{11}),(2n_{11}+1)]\text{-hyperbolic-secant-to-hyperbolic flow }(-)}\\&\rightleftharpoons\bigcup_{r=1}^{2m_{11}}\bigcup_{u=1}^{2n_{11}+1}\underbrace{(a_{j_1j_1r},a_{j_1j_2u})}_{\text{YY}},\end{aligned} \tag{1.342}$$

where

$$\bigcup_{r=1}^{2m_{11}}\bigcup_{u=1}^{2n_{11}+1}\underbrace{(a_{j_1j_1r},a_{j_1j_2u})}_{\text{YY}}$$
$$=\begin{bmatrix}\underbrace{\text{(UP:UP, nF)}}_{\text{hyperbolic-secant flow }(-)} & \cdots & \underbrace{\text{(DP:UP, nF)}}_{\text{hyperbolic flow }(-)} & \underbrace{\text{(UP:UP, nF)}}_{\text{hyperbolic-secant flow }(-)}\\ \underbrace{\text{(DP:UP, nF)}}_{\text{hyperbolic flow }(-)} & \cdots & \underbrace{\text{(UP:UP, nF)}}_{\text{hyperbolic-secant flow }(-)} & \underbrace{\text{(DP:UP, nF)}}_{\text{hyperbolic flow }(-)}\\ \vdots & \vdots & \vdots & \vdots\\ \underbrace{\text{(UP:UP, nF)}}_{\text{hyperbolic-secant flow }(-)} & \cdots & \underbrace{\text{(DP:UP, nF)}}_{\text{hyperbolic flow }(-)} & \underbrace{\text{(UP:UP, nF)}}_{\text{hyperbolic-secant flow }(-)}\end{bmatrix}_{(2n_{11}+1)\times(2m_{11})}. \tag{1.343}$$

(iv$_{2d2}$) The ([(2m_{11}), (2n_{11}+1)]UP:UP, nF)*-negative hyperbolic-secant-to-hyperbolic flow appearing bifurcation is from a* $q_1\times p_1$ *to* $\overline{q}_1\times\overline{p}_1$ *1-dimensional flow networks as*

$$\begin{aligned}\bigcup_{r_1=1}^{p_1}\bigcup_{u_1=1}^{q_1}\underbrace{(a_{j_1j_1r_1},a_{j_1j_2u_1})}_{([m_{j_1r_1},n_{j_1u_1}]\text{-XX, nF})}&\rightleftharpoons\underbrace{([(2m_{11}),(2n_{11}+1)]\text{UP:UP, nF})}_{[(2m_{11}),(2n_{11}+1)]\text{-hyperbolic-secant-to-hyperbolic flow }(-)}\\&\rightleftharpoons\bigcup_{r_2=1}^{\overline{p}_1}\bigcup_{u_2=1}^{\overline{q}_1}\underbrace{(a_{j_1j_1r_2},a_{j_1j_2u_2})}_{([m_{j_1r_2},n_{j_1u_2}]\text{-YY, nF})}.\end{aligned} \tag{1.344}$$

(iv_{2d3}) The ([(2m_{11}), (2n_{11}+1)]UP:UP, nF)*-negative hyperbolic-secant-to-hyperbolic flow switching bifurcation is for the switching of two networks of* $q_1 \times p_1$ *1-dimensional flows as*

$$\bigcup\nolimits_{r_1=1}^{p_1}\bigcup\nolimits_{u_1=1}^{q_1}\underbrace{(a_{j_1j_1r_1}, a_{j_1j_2u_1})}_{([m_{j_1r_1}, n_{j_1u_1}]\text{-XX, nF})} \rightleftharpoons \underbrace{([(2m_{11}), (2n_{11}+1)]\text{UP:UP, nF})}_{[(2m_{11}),(2n_{11}+1)]\text{-hyperbolic-secant-to-hyperbolic flow }(-)}$$
$$\rightleftharpoons \bigcup\nolimits_{r_2=1}^{p_1}\bigcup\nolimits_{u_2=1}^{q_1}\underbrace{(a_{j_1j_1r_2}, a_{j_1j_2u_2})}_{([m_{j_1r_2}, n_{j_1u_2}]\text{-YY, nF})}. \tag{1.345}$$

(iv_3) For $m_1 = 2m_{11}+1$ *and* $n_1 = 2n_{11}$*, the equilibrium of* $(x^*_{j_1}, x^*_{j_2}) = (a_{j_1j_11}, a_{j_1j_21})$ *for the* x_{j_1}*-direction has the following bifurcation properties with*

$$\begin{aligned}
&2m_{11}+1-\sum_{r_1=1}^{p_1} m^{(1)}_{j_1r_1} = 2\sum_{\rho_1=1}^{l_1} m_{\rho_1},\\
&2n_{11}-\sum_{u_1=1}^{q_1} n^{(1)}_{j_1u_1} = 2\sum_{\sigma_1=1}^{l_2} n_{\sigma_1};\\
&\sum_{r_2=1}^{\bar{p}_1} m^{(2)}_{j_1r_2} = 2m_{11}+1, \sum_{u_2=1}^{\bar{q}_1} n^{(2)}_{j_1u_2} = 2n_{11};\\
&\sum_{r_i=1}^{p_1} m^{(i)}_{j_1r_i} = 2m_{11}+1, \sum\nolimits_{u_i=1}^{q_1} n^{(2)}_{j_1u_i} = 2n_{11}, \text{ for } i=1,2;
\end{aligned} \tag{1.346}$$

l_1, l_2*-quadratic polynomials are without real roots, and the* ρ_1^{th} *and* σ_1^{th}*-quadratic polynomials are with powers* m_{ρ_1} *and* n_{σ_1}*, respectively.*

(iv_{3a}) For $a_{j_1j_10} > 0$ *and* $a_{j_2j_20}(a_{j_1j_21} - a_{j_2j_21})^{n_2} > 0$*, there is a* ([(2$m_{11}$ + 1), (2n_{11})]$_{\text{DI:II}}$SO, pF)*-positive inflection-source flow as*

$$(a_{j_1j_11}, a_{j_1j_21}) = \underbrace{([(2m_{11}+1), (2n_{11})]_{\text{DI:II}}\text{SO, pF})}_{[(2m_{11}+1),(2n_{11})]\text{-inflection-source flow }(+)}. \tag{1.347}$$

There are three following ([(2m_{11}+1), (2n_{11})]$_{\text{DI:II}}$SO, pF)*-positive inflection-source flow appearing and switching bifurcations.*

(iv_{3a1}) The ([(2m_{11}+1), (2n_{11})]$_{\text{DI:II}}$SO, pF)*-positive inflection-source flow appearing bifurcation is from an* (SO, pF)*-positive inflection-source flow to a* (2m_{11} + 1) × (2n_{11})*-hyperbolic and hyperbolic-secant flow network as*

$$\underbrace{(a_{j_1j_11}, \dot{x}_{j_2})}_{\text{(SO, pF)-flow}} \rightleftharpoons \underbrace{([(2m_{11}+1), (2n_{11})]_{\text{DI:II}}\text{SO, pF})}_{[(2m_{11}+1),(2n_{11})]\text{-infelction source flow }(+)}$$
$$\rightleftharpoons \bigcup\nolimits_{r=1}^{2m_{11}+1}\bigcup\nolimits_{u=1}^{2n_{11}}\underbrace{(a_{j_1j_1r}, a_{j_1j_2u})}_{\text{YY}}, \tag{1.348}$$

where

$$\bigcup_{r=1}^{2m_{11}+1}\bigcup_{u=1}^{2n_{11}}\underbrace{(a_{j_1j_1r},a_{j_1j_2u})}_{\text{YY}}$$
$$=\begin{bmatrix}\underbrace{\text{(DP:UP, pF)}}_{\text{hyperbolic flow }(+)} & \cdots & \underbrace{\text{(UP:DP, pF)}}_{\text{hyperbolic-secant flow }(+)} & \underbrace{\text{(DP:UP, pF)}}_{\text{hyperbolic flow }(+)}\\ \underbrace{\text{(UP:DP, pF)}}_{\text{hyperbolic-secant flow }(+)} & \cdots & \underbrace{\text{(DP:UP, pF)}}_{\text{hyperbolic flow }(+)} & \underbrace{\text{(UP:DP, pF)}}_{\text{hyperbolic-secant flow }(+)}\\ \vdots & \vdots & \vdots & \vdots\\ \underbrace{\text{(UP:DP, pF)}}_{\text{hyperbolic-secant flow }(+)} & \cdots & \underbrace{\text{(DP:UP, pF)}}_{\text{hyperbolic flow }(+)} & \underbrace{\text{(UP:DP, pF)}}_{\text{hyperbolic-secant flow }(+)}\end{bmatrix}_{(2n_{11})\times(2m_{11}+1)}. \tag{1.349}$$

(iv_{3a2}) The $([(2m_{11}+1),(2n_{11})]_{\text{DI:II}}\text{SO, pF})$*-positive decreasing-to-increasing inflection source flow appearing bifurcation is from a* $q_1\times p_1$ *to* $\overline{q}_1\times\overline{p}_1$ *1-dimensional flow networks as*

$$\begin{aligned}\bigcup_{r_1=1}^{p_1}\bigcup_{u_1=1}^{q_1}\underbrace{(a_{j_1j_1r_1},a_{j_1j_2u_1})}_{([m^{(1)}_{j_1r_1},n^{(1)}_{j_1u_1}]\text{-YY, pF})} &\rightleftharpoons \underbrace{([(2m_{11}+1),(2n_{11})]_{\text{DI:II}}\text{SO, pF})}_{[(2m_{11}+1),(2n_{11})]\text{-inflection-source flow }(+)}\\ &\rightleftharpoons \bigcup_{r_2=1}^{\overline{p}_1}\bigcup_{u_2=1}^{\overline{q}_1}\underbrace{(a_{j_1j_1r_2},a_{j_1j_2u_2})}_{([m^{(1)}_{j_1r_2},n^{(1)}_{j_1u_2}]\text{-YY, pF})}.\end{aligned} \tag{1.350}$$

(iv_{3a3}) The $([(2m_{11}+1),(2n_{11})]_{\text{DI:II}}\text{SO, pF})$*-positive inflection-source flow switching bifurcation is for the switching of two networks of* $q_1\times p_1$ *1-dimensional flows as*

$$\begin{aligned}\bigcup_{r_1=1}^{p_1}\bigcup_{u_1=1}^{q_1}\underbrace{(a_{j_1j_1r_1},a_{j_1j_2u_1})}_{([m_{j_1r_1},n_{j_1u_1}]\text{-XX, pF})} &\rightleftharpoons \underbrace{([(2m_{11}+1),(2n_{11})]_{\text{DI:II}}\text{SO, pF})}_{[(2m_{11}+1),(2n_{11})]\text{-inflection-source flow }(+)}\\ &\rightleftharpoons \bigcup_{r_2=1}^{p_1}\bigcup_{u_2=1}^{q_1}\underbrace{(a_{j_1j_1r_2},a_{j_1j_2u_2})}_{([m_{j_1r_2},n_{j_1u_2}]\text{-YY,pF})}.\end{aligned} \tag{1.351}$$

(iv_{3b}) For $a_{j_1j_10}<0$ *and* $a_{j_2j_20}(a_{j_1j_21}-a_{j_2j_21})^{n_2}>0$, *there is a* $([(2m_{11}+1),(2n_{11})]_{\text{II:DI}}\text{SI, pF})$*-positive inflection-sink flow as*

$$(a_{j_1j_11},a_{j_1j_21})=\underbrace{([(2m_{11}+1),(2n_{11})]_{\text{II:DI}}\text{SI, pF})}_{[(2m_{11}+1),(2n_{11})]\text{-inflection-sink flow }(+)}. \tag{1.352}$$

There are three following ([(2m_{11}+1), (2n_{11})]$_{\text{II:DI}}$SI, pF)*-positive increasing-to-decreasing inflection sink flow appearing and switching bifurcations.*

(iv$_{3b1}$) The ([(2m_{11}+1), (2n_{11})]$_{\text{II:DI}}$SI, pF)*-positive increasing-to-decreasing inflection sink flow appearing bifurcation is from an* (SI, pF)*-positive sink flow to a* $(2m_{11}+1)\times(2n_{11})$*-hyperbolic and hyperbolic-secant flow network as*

$$\underbrace{(a_{j_1j_1 1}, \dot{x}_{j_2})}_{\text{(SI, pF)-flow}} \rightleftharpoons \underbrace{([(2m_{11}+1), (2n_{11})]_{\text{II:DI}}\text{SI, pF})}_{[(2m_{11}+1),(2n_{11})]\text{-increasing-to-dereasing inflection sink flow }(+)}$$
$$\rightleftharpoons \bigcup\nolimits_{r=1}^{2m_{11}+1}\bigcup\nolimits_{u=1}^{2n_{11}} \underbrace{(a_{j_1j_1r}, a_{j_1j_2u})}_{\text{YY}}, \tag{1.353}$$

where

$$\bigcup\nolimits_{r=1}^{2m_{11}+1}\bigcup\nolimits_{u=1}^{2n_{11}} \underbrace{(a_{j_1j_1r}, a_{j_1j_2u})}_{\text{YY}}$$
$$= \begin{bmatrix} \underbrace{\text{(UP:DP,pF)}}_{\text{hyperbolic-secant flow }(+)} & \cdots & \underbrace{\text{(DP:UP, pF)}}_{\text{hyperbolic flow }(+)} & \underbrace{\text{(UP:DP, pF)}}_{\text{hyperbolic-secant flow }(+)} \\ \underbrace{\text{(DP:UP, pF)}}_{\text{hyperbolic flow }(+)} & \cdots & \underbrace{\text{(UP:DP, pF)}}_{\text{hyperbolic-secant flow }(+)} & \underbrace{\text{(DP:UP, pF)}}_{\text{hyperbolic flow }(+)} \\ \vdots & \vdots & \vdots & \vdots \\ \underbrace{\text{(DP:UP, pF)}}_{\text{hyperbolic flow }(+)} & \cdots & \underbrace{\text{(UP:DP, pF)}}_{\text{hyperbolic-secant flow }(+)} & \underbrace{\text{(DP:UP, pF)}}_{\text{hyperbolic flow }(+)} \end{bmatrix}_{(2n_{11})\times(2m_{11}+1)}. \tag{1.354}$$

(iv$_{3b2}$) The ([(2m_{11}+1), (2n_{11})]$_{\text{II:DI}}$SI, pF)*-positive inflection-sink flow appearing bifurcation is from a* $q_1 \times p_1$ *to* $\overline{q}_1 \times \overline{p}_1$ *1-dimensional flow networks as*

$$\bigcup\nolimits_{r_1=1}^{p_1}\bigcup\nolimits_{u_1=1}^{q_1} \underbrace{(a_{j_1j_1r_1}, a_{j_1j_2u_1})}_{([m_{j_1r_1}, n_{j_1u_1}]\text{-XX, pF})} \rightleftharpoons \underbrace{([(2m_{11}+1), (2n_{11})]_{\text{II:DI}}\text{SI, pF})}_{[(2m_{11}+1),(2n_{11})]\text{-inflection-sink flow }(+)}$$
$$\rightleftharpoons \bigcup\nolimits_{r_2=1}^{\overline{p}_1}\bigcup\nolimits_{u_2=1}^{\overline{q}_1} \underbrace{(a_{j_1j_1r_2}, a_{j_1j_2u_2})}_{([m_{j_1r_2}, n_{j_1u_2}]\text{-YY, pF})}. \tag{1.355}$$

(iv$_{3b3}$) The ([(2m_{11}+1), (2n_{11})]$_{\text{II:DI}}$SI, pF)*-positive inflection-sink flow switching bifurcation is for the switching of two networks of* $q_1 \times p_1$ *1-dimensional flows as*

$$\bigcup\nolimits_{r_1=1}^{p_1}\bigcup\nolimits_{u_1=1}^{q_1} \underbrace{(a_{j_1j_1r_1}, a_{j_1j_2u_1})}_{([m_{j_1r_1}, n_{j_1u_1}]\text{-XX, pF})} \rightleftharpoons \underbrace{([(2m_{11}+1), (2n_{11})]_{\text{II:DI}}\text{SI, pF})}_{[(2m_{11}+1),(2n_{11})]\text{-inflection-sink flow }(+)}$$
$$\rightleftharpoons \bigcup\nolimits_{r_2=1}^{p_1}\bigcup\nolimits_{u_2=1}^{q_1} \underbrace{(a_{j_1j_1r_2}, a_{j_1j_2u_2})}_{([m_{j_1r_2}, n_{j_1u_2}]\text{-YY, pF})}. \tag{1.356}$$

(iv_{3c}) For $a_{j_1j_10} > 0$ and $a_{j_2j_20}(a_{j_1j_21} - a_{j_2j_21})^{n_2} < 0$, there is a $([(2m_{11}+1), (2n_{11})]_{\text{II:DI}}\text{SO, nF})$*-negative inflection-source flow as*

$$(a_{j_1j_11}, a_{j_1j_21}) = \underbrace{([(2m_{11}+1), (2n_{11})]_{\text{II:DI}}\text{SO, nF})}_{[(2m_{11}+1),(2n_{11})]\text{-inflection-source flow }(-)}. \tag{1.357}$$

There are three following $([(2m_{11}+1), (2n_{11})]_{\text{II:DI}}\text{SO, nF})$*-negative inflection-source flow appearing and switching bifurcations.*

(iv_{3c1}) The $([(2m_{11}+1),(2n_{11})]_{\text{II:DI}}\text{SO,nF})$*-negative inflection-source flow appearing bifurcation is from an* (SO, nF)*-negative sink flow to a* $(2m_{11}+1) \times (2n_{11})$*-hyperbolic and hyperbolic-secant flow network as*

$$\begin{aligned}\underbrace{(a_{j_1j_11}, \dot{x}_{j_2})}_{\text{(SO, nF)-flow}} &\rightleftharpoons \underbrace{([(2m_{11}+1), (2n_{11})]_{\text{II:DI}}\text{SO, nF})}_{[(2m_{11}+1),(2n_{11})]\text{-inflection-source flow }(-)}\\ &\rightleftharpoons \bigcup\nolimits_{r=1}^{2m_{11}+1} \bigcup\nolimits_{u=1}^{2n_{11}} \underbrace{(a_{j_1j_1r}, a_{j_1j_2u})}_{\text{YY}},\end{aligned} \tag{1.358}$$

where

$$\bigcup\nolimits_{r=1}^{2m_{11}+1} \bigcup\nolimits_{u=1}^{2n_{11}} \underbrace{(a_{j_1j_1r}, a_{j_1j_2u})}_{\text{YY}}$$
$$= \begin{bmatrix} \underbrace{\text{(UP:DP, nF)}}_{\text{hyperbolic-secant flow }(-)} & \cdots & \underbrace{\text{(DP:UP, nF)}}_{\text{hyperbolic flow }(-)} & \underbrace{\text{(UP:DP, nF)}}_{\text{hyperbolic-secant flow }(-)} \\ \underbrace{\text{(DP:UP, nF)}}_{\text{hyperbolic flow }(-)} & \cdots & \underbrace{\text{(UP:DP, nF)}}_{\text{hyperbolic-secant flow }(-)} & \underbrace{\text{(DP:UP, nF)}}_{\text{hyperbolic flow }(-)} \\ \vdots & \vdots & \vdots & \vdots \\ \underbrace{\text{(UP:DP, nF)}}_{\text{hyperbolic-secant flow }(-)} & \cdots & \underbrace{\text{(DP:UP, nF)}}_{\text{hyperbolic flow }(-)} & \underbrace{\text{(UP:DP, nF)}}_{\text{hyperbolic-secant flow }(-)} \end{bmatrix}_{(2n_{11})\times(2m_{11}+1)}. \tag{1.359}$$

(iv_{3c2}) The $([(2m_{11}+1),(2n_{11})]_{\text{II:DI}}\text{SO,nF})$*-negative inflection-source flow appearing bifurcation is from a* $q_1 \times p_1$ *to* $\overline{q}_1 \times \overline{p}_1$ *1-dimensional flow networks as*

$$\begin{aligned}\bigcup\nolimits_{r_1=1}^{p_1} \bigcup\nolimits_{u_1=1}^{q_1} \underbrace{(a_{j_1j_1r_1}, a_{j_1j_2u_1})}_{([m_{j_1r_1},n_{j_1u_1}]\text{-XX, nF})} &\rightleftharpoons \underbrace{([(2m_{11}+1), (2n_{11})]_{\text{II:DI}}\text{SO, nF})}_{[(2m_{11}+1),(2n_{11})]\text{-inflection-source flow }(-)}\\ &\rightleftharpoons \bigcup\nolimits_{r_2=1}^{\overline{p}_1} \bigcup\nolimits_{u_2=1}^{\overline{q}_1} \underbrace{(a_{j_1j_1r_2}, a_{j_1j_2u_2})}_{([m_{j_1r_2},n_{j_1u_2}]\text{-YY, nF})}.\end{aligned} \tag{1.360}$$

(iv_{3c3}) The $([(2m_{11}+1),(2n_{11})]_{\text{II:DI}}\text{SO, nF})$*-negative inflection-source flow switching bifurcation is for the switching of two networks of* $q_1 \times p_1$ *1-dimensional flows as*

$$\bigcup_{r_1=1}^{p_1}\bigcup_{u_1=1}^{q_1}\underbrace{(a_{j_1j_1r_1},a_{j_1j_2u_1})}_{([m_{j_1r_1},n_{j_1u_1}]\text{-XX, nF})} \rightleftharpoons \underbrace{([(2m_{11}+1),(2n_{11})]_{\text{II:DI}}\text{SO, nF})}_{[(2m_{11}+1),(2n_{11})]\text{-inflection-source flow }(-)}$$
$$\rightleftharpoons \bigcup_{r_2=1}^{p_1}\bigcup_{u_2=1}^{q_1}\underbrace{(a_{j_1j_1r_2},a_{j_1j_2u_2})}_{([m_{j_1r_2},n_{j_1u_2}]\text{-YY, nF})}. \tag{1.361}$$

(iv_{3d}) For $a_{j_1j_10}<0$ *and* $a_{j_2j_20}(a_{j_1j_21}-a_{j_2j_21})^{n_2}<0$*, there is a* $([(2m_{11}+1),(2n_{11})]_{\text{DI:II}}\text{SI, nF})$*-negative inflection-sink flow as*

$$(a_{j_1j_11},a_{j_1j_21}) = \underbrace{([(2m_{11}+1),(2n_{11})]_{\text{DI:II}}\text{SI, nF})}_{[(2m_{11}+1),(2n_{11})]\text{-inflection-sink flow }(-)}. \tag{1.362}$$

There are three following $([(2m_{11}+1),(2n_{11})]_{\text{DI:II}}\text{SI, nF})$ *-negative inflection-sink flow appearing and switching bifurcations.*

(iv_{3d1}) The $([(2m_{11}+1),(2n_{11})]_{\text{DI:II}}\text{SI, nF})$*-negative inflection-sink flow appearing bifurcation is from an (SI, nF)-negative sink flow to a* $(2m_{11}+1)\times(2n_{11})$*-hyperbolic and hyperbolic-secant flow network as*

$$\underbrace{(a_{j_1j_11},a_{j_1j_21})}_{\text{(SI, nF)-flow}} \rightleftharpoons \underbrace{([(2m_{11}+1),(2n_{11})]_{\text{DI:II}}\text{SI, nF})}_{[(2m_{11}+1),(2n_{11})]\text{-inflection-sink flow }(-)}$$
$$\rightleftharpoons \bigcup_{r=1}^{2m_{11}+1}\bigcup_{u=1}^{2n_{11}}\underbrace{(a_{j_1j_1r},a_{j_1j_2u})}_{\text{YY}}, \tag{1.363}$$

where

$$\bigcup_{r=1}^{2m_{11}+1}\bigcup_{u=1}^{2n_{11}}\underbrace{(a_{j_1j_1r},a_{j_1j_2u})}_{\text{YY}}$$
$$= \begin{bmatrix} \underbrace{\text{(UP:DP, nF)}}_{\text{hyperbolic-secant flow }(-)} & \cdots & \underbrace{\text{(DP:UP, nF)}}_{\text{hyperbolic flow }(-)} & \underbrace{\text{(UP:DP, nF)}}_{\text{hyperbolic-secant flow }(-)} \\ \underbrace{\text{(DP:UP, nF)}}_{\text{hyperbolic flow }(-)} & \cdots & \underbrace{\text{(UP:DP, nF)}}_{\text{hyperbolic-secant flow }(-)} & \underbrace{\text{(DP:UP, nF)}}_{\text{hyperbolic flow }(-)} \\ \vdots & \vdots & \vdots & \vdots \\ \underbrace{\text{(DP:UP, nF)}}_{\text{hyperbolic flow }(-)} & \cdots & \underbrace{\text{(UP:DP, nF)}}_{\text{hyperbolic-secant flow }(-)} & \underbrace{\text{(DP:UP, nF)}}_{\text{hyperbolic flow }(-)} \end{bmatrix}_{(2n_{11})\times(2m_{11}+1)}. \tag{1.364}$$

(iv_{3d2}) The ([(2m_{11}+1), (2n_{11})]$_{\mathrm{DI:II}}$SI, nF)*-negative inflection-sink flow appearing bifurcation is from a* $q_1 \times p_1$ *to* $\overline{q}_1 \times \overline{p}_1$ *1-dimensional flow networks as*

$$\bigcup\nolimits_{r_1=1}^{p_1}\bigcup\nolimits_{u_1=1}^{q_1}\underbrace{(a_{j_1j_1r_1},a_{j_1j_2u_1})}_{([m_{j_1r_1},n_{j_1u_1}]\text{-XX, nF})}\rightleftharpoons\underbrace{([(2m_{11}+1),(2n_{11})]_{\mathrm{DI:II}}\mathrm{SI,\ nF})}_{[(2m_{11}+1),(2n_{11})]\text{-inflection-sink flow }(-)}$$
$$\rightleftharpoons\bigcup\nolimits_{r_2=1}^{\overline{p}_1}\bigcup\nolimits_{u_2=1}^{\overline{q}_1}\underbrace{(a_{j_1j_1r_2},a_{j_1j_2u_2})}_{([m_{j_1r_2},n_{j_1u_2}]\text{-YY, nF})}. \tag{1.365}$$

(iv_{3d3}) The ([(2m_{11}+1), (2n_{11})]$_{\mathrm{DI:II}}$SI, nF)*-negative inflection-sink flow switching bifurcation is for the switching of two networks of* $q_1 \times p_1$ *1-dimensional flows as*

$$\bigcup\nolimits_{r_1=1}^{p_1}\bigcup\nolimits_{u_1=1}^{q_1}\underbrace{(a_{j_1j_1r_1},a_{j_1j_2u_1})}_{([m_{j_1r_1},n_{j_1u_1}]\text{-XX, nF})}\rightleftharpoons\underbrace{([(2m_{11}+1),(2n_{11})]_{\mathrm{DI:II}}\mathrm{SI,\ nF})}_{[(2m_{11}+1),(2n_{11})]\text{-inflection-sink flow }(-)}$$
$$\rightleftharpoons\bigcup\nolimits_{r_2=1}^{p_1}\bigcup\nolimits_{u_2=1}^{q_1}\underbrace{(a_{j_1j_1r_2},a_{j_1j_2u_2})}_{([m_{j_1r_2},n_{j_1u_2}]\text{-XX, nF})}. \tag{1.366}$$

(iv_4) For $m_1 = 2m_{11}$ *and* $n_1 = 2n_{11}$*, the equilibrium of* $(x_{j_1}^*, x_{j_2}^*) = (a_{j_1j_1 1}, a_{j_1j_2 1})$ *for the* x_{j_1}*-direction has the following bifurcation properties with*

$$\begin{aligned}
&2m_{11}-\sum_{r_1=1}^{p_1}m_{j_1r_1}^{(1)}=2\sum_{\rho_1=1}^{l_1}m_{\rho_1},\\
&2n_{11}-\sum_{u_1=1}^{q_1}n_{j_1u_1}^{(1)}=2\sum_{\sigma_1=1}^{l_2}n_{\sigma_1};\\
&\sum_{r_2=1}^{\overline{p}_1}m_{j_1r_2}^{(2)}=2m_{11},\ \sum_{u_2=1}^{\overline{q}_1}n_{j_1u_2}^{(2)}=2n_{11};\\
&\sum_{r_i=1}^{p_1}m_{j_1r_i}^{(i)}=2m_{11},\ \sum_{u_i=1}^{q_1}n_{j_1u_i}^{(2)}=2n_{11},\ \text{for } i=1,2;
\end{aligned} \tag{1.367}$$

l_1, l_2*-quadratic polynomials are without real roots, and the* ρ_1^{th} *and* σ_1^{th}*-quadratic polynomials are with powers* m_{ρ_1} *and* n_{σ_1}*, respectively.*

(iv_{4a}) For $a_{j_1j_1 0}>0$ *and* $a_{j_2j_2 0}(a_{j_1j_2 1}-a_{j_2j_2 1})^{n_2}>0$*, there is a* ([(2$m_{11}$),(2$n_{11}$)]$_{\mathrm{II:II}}$US,pF)*-positive increasing-inflection upper-saddle flow as*

$$(a_{j_1j_1 1},a_{j_1j_2 1})=\underbrace{([(2m_{11}),(2n_{11})]_{\mathrm{II:II}}\mathrm{US,\ pF})}_{[(2m_{11}),(2n_{11})]\text{-increasing inflection upper-saddle flow }(+)}. \tag{1.368}$$

There are three following ([(2m_{11}), (2n_{11})]$_{\rm II:II}$US, pF)*-positive increasing-inflection upper-saddle flow appearing and switching bifurcations.*

(iv$_{4a1}$) The ([(2m_{11}), (2n_{11})]$_{\rm II:II}$US, pF)*-positive increasing-inflection upper-saddle flow appearing bifurcation is from a* (pF, pF)*-flow to a* (2m_{11}) × (2n_{11})*-hyperbolic and hyperbolic-secant flow network as*

$$\underbrace{(\dot{x}_{j_1}, \dot{x}_{j_2})}_{\text{(pF, pF)-flow}} \rightleftharpoons \underbrace{([(2m_{11}), (2n_{11})]_{\rm II:II}\text{US, pF})}_{[(2m_{11}),(2n_{11})]\text{-increasing inflection upper-saddle flow }(+)}$$
$$\rightleftharpoons \bigcup_{r=1}^{2m_{11}} \bigcup_{u=1}^{2n_{11}} \underbrace{(a_{j_1 j_1 r}, a_{j_1 j_2 u})}_{\text{YY}}, \quad (1.369)$$

where

$$\bigcup_{r=1}^{2m_{11}} \bigcup_{u=1}^{2n_{11}} \underbrace{(a_{j_1 j_1 r}, a_{j_1 j_2 u})}_{\text{YY}}$$
$$= \begin{bmatrix} \underbrace{\text{(UP:DP, pF)}}_{\text{hyperbolic-secant flow }(+)} & \cdots & \underbrace{\text{(UP:DP, pF)}}_{\text{hyperbolic-secant flow }(+)} & \underbrace{\text{(DP:UP, pF)}}_{\text{hyperbolic flow }(+)} \\ \underbrace{\text{(DP:UP, pF)}}_{\text{hyperbolic flow }(+)} & \cdots & \underbrace{\text{(DP:UP, pF)}}_{\text{hyperbolic flow }(+)} & \underbrace{\text{(UP:DP, pF)}}_{\text{hyperbolic-secant flow }(+)} \\ \vdots & \vdots & \vdots & \vdots \\ \underbrace{\text{(DP:UP, pF)}}_{\text{hyperbolic flow }(+)} & \cdots & \underbrace{\text{(DP:UP, pF)}}_{\text{hyperbolic flow }(+)} & \underbrace{\text{(UP:DP, pF)}}_{\text{hyperbolic-secant flow }(+)} \end{bmatrix}_{(2n_{11})\times(2m_{11})} . \quad (1.370)$$

(iv$_{4a2}$) The ([(2m_{11}), (2n_{11})]$_{\rm II:II}$US, pF)*-positive increasing-inflection upper-saddle flow appearing bifurcation is from a* $q_1 \times p_1$ *to* $\overline{q}_1 \times \overline{p}_1$ *1-dimensional flow networks as*

$$\bigcup_{r_1=1}^{p_1} \bigcup_{u_1=1}^{q_1} \underbrace{(a_{j_1 j_1 r_1}, a_{j_1 j_2 u_1})}_{([m_{j_1 r_1}, n_{j_1 u_1}]\text{-XX, pF})} \rightleftharpoons \underbrace{([(2m_{11}), (2n_{11})]_{\rm II:II}\text{US, pF})}_{[(2m_{11}),(2n_{11})]\text{-increasing inflection upper-saddle flow }(+)}$$
$$\rightleftharpoons \bigcup_{r_2=1}^{\overline{p}_1} \bigcup_{u_2=1}^{\overline{q}_1} \underbrace{(a_{j_1 j_1 r_2}, a_{j_1 j_2 u_2})}_{([m_{j_1 r_2}, n_{j_1 u_2}]\text{-YY, pF})} . \quad (1.371)$$

(iv$_{4a3}$) The ([(2m_{11}), (2n_{11})]$_{\rm II:II}$US, pF)*-positive increasing-inflection upper-saddle flow switching bifurcation is for the switching of two networks of* $q_1 \times p_1$ *1-dimensional flows as*

$$\bigcup_{r_1=1}^{p_1}\bigcup_{u_1=1}^{q_1}\underbrace{(a_{j_1j_1r_1},a_{j_1j_2u_1})}_{([m_{j_1r_1},n_{j_1u_1}]\text{-XX, pF})}\rightleftharpoons\underbrace{([(2m_{11}),(2n_{11})]_{\text{II:II}}\text{US, pF})}_{[(2m_{11}),(2n_{11})]\text{-increasing inflection upper-saddle flow }(+)}$$
$$\rightleftharpoons\bigcup_{r_2=1}^{p_1}\bigcup_{u_2=1}^{q_1}\underbrace{(a_{j_1j_1r_2},a_{j_1j_2u_2})}_{([m_{j_1r_2},n_{j_1u_2}]\text{-YY, pF})}. \tag{1.372}$$

(iv$_{4b}$) For $a_{j_1j_10}<0$ and $a_{j_2j_20}(a_{j_1j_21}-a_{j_2j_21})^{n_2}>0$, there is a ([(2$m_{11}$), (2$n_{11}$)]$_{\text{DI:DI}}$LS, pF)*-positive decreasing-inflection lower-saddle flow as*

$$(a_{j_1j_11},a_{j_1j_21})=\underbrace{([(2m_{11}),(2n_{11})]_{\text{DI:DI}}\text{LS, pF})}_{[(2m_{11}),(2n_{11})]\text{-decreasing-inflection lower-saddle flow }(+)}. \tag{1.373}$$

There are three following ([(2m_{11}), (2n_{11})]$_{\text{DI:DI}}$LS, pF)*-positive decreasing-inflection lower-saddle flow appearing and switching bifurcations.*

(iv$_{4b1}$) The ([(2m_{11}), (2n_{11})]$_{\text{DI:DI}}$LS, pF)*-positive decreasing-inflection lower-saddle flow appearing bifurcation is from an* (nF, pF)*-flow to a* (2m_{11}) × (2n_{11})*-decreasing-inflection lower-saddle flow network as*

$$\underbrace{(\dot{x}_{j_1},\dot{x}_{j_2})}_{\text{(nF, pF)-flow}}\rightleftharpoons\underbrace{([(2m_{11}),(2n_{11})]_{\text{DI:DI}}\text{LS, pF})}_{[(2m_{11}),(2n_{11})]\text{-dereasing-inflection lower-saddle flow }(+)}$$
$$\rightleftharpoons\bigcup_{r=1}^{2m_{11}}\bigcup_{u=1}^{2n_{11}}\underbrace{(a_{j_1j_1r},a_{j_1j_2u})}_{\text{YY}}, \tag{1.374}$$

where

$$\bigcup_{r=1}^{2m_{11}}\bigcup_{u=1}^{2n_{11}}\underbrace{(a_{j_1j_1r},a_{j_1j_2u})}_{\text{YY}}$$
$$=\begin{bmatrix}\underbrace{\text{(DP:UP, pF)}}_{\text{hyperbolic flow }(+)} & \cdots & \underbrace{\text{(DP:UP, pF)}}_{\text{hyperbolic flow }(+)} & \underbrace{\text{(UP:DP, pF)}}_{\text{hyperbolic-secant flow }(+)}\\ \underbrace{\text{(UP:DP, pF)}}_{\text{hyperbolic-secant flow }(+)} & \cdots & \underbrace{\text{(UP:DP, pF)}}_{\text{hyperbolic-secant flow }(+)} & \underbrace{\text{(DP:UP, pF)}}_{\text{hyperbolic flow }(+)}\\ \vdots & \vdots & \vdots & \vdots\\ \underbrace{\text{(UP:DP, pF)}}_{\text{hyperbolic-secant flow }(+)} & \cdots & \underbrace{\text{(UP:DP, pF)}}_{\text{hyperbolic-secant flow }(+)} & \underbrace{\text{(DP:UP, pF)}}_{\text{hyperbolic flow }(+)}\end{bmatrix}_{(2n_{11})\times(2m_{11})}. \tag{1.375}$$

(iv$_{4b2}$) The ([(2m_{11}), (2n_{11})]$_{\text{DI:DI}}$LS, pF)*-positive decreasing-inflection lower-saddle flow appearing bifurcation is from a $q_1\times p_1$ to $\overline{q}_1\times\overline{p}_1$ 1-dimensional flow networks as*

$$\bigcup_{r_1=1}^{p_1}\bigcup_{u_1=1}^{q_1}\underbrace{(a_{j_1j_1r_1},a_{j_1j_2u_1})}_{([m_{j_1r_1},n_{j_1u_1}]\text{-XX, pF})} \rightleftharpoons \underbrace{([(2m_{11}),(2n_{11})]_{\text{DI:DI}}\text{LS, pF})}_{[(2m_{11}),(2n_{11})]\text{-dereasing-inflection lower-saddle flow }(+)}$$
$$\rightleftharpoons \bigcup_{r_2=1}^{\bar{p}_1}\bigcup_{u_2=1}^{\bar{q}_1}\underbrace{(a_{j_1j_1r_2},a_{j_1j_2u_2})}_{([m_{j_1r_2},n_{j_1u_2}]\text{-YY, pF}))}. \tag{1.376}$$

(iv$_{4b3}$) The $([(2m_{11}),(2n_{11})]_{\text{DI:DI}}\text{LS, pF})$*-positive decreasing-inflection lower-saddle flow switching bifurcation is for the switching of two networks of* $q_1 \times p_1$ *1-dimensional flows as*

$$\bigcup_{r_1=1}^{p_1}\bigcup_{u_1=1}^{q_1}\underbrace{(a_{j_1j_1r_1},a_{j_1j_2u_1})}_{([m_{j_1r_1},n_{j_1u_1}]\text{-XX, pF})} \rightleftharpoons \underbrace{([(2m_{11}),(2n_{11})]_{\text{DI:DI}}\text{LS, pF})}_{[(2m_{11}),(2n_{11})]\text{-dereasing-inflection lower-saddle flow }(+)}$$
$$\rightleftharpoons \bigcup_{r_2=1}^{p_1}\bigcup_{u_2=1}^{q_1}\underbrace{(a_{j_1j_1r_2},a_{j_1j_2u_2})}_{([m_{j_1r_2},n_{j_1u_2}]\text{-YY, pF})}. \tag{1.377}$$

(iv$_{4c}$) For $a_{j_1j_10} > 0$ *and* $a_{j_2j_20}(a_{j_1j_21} - a_{j_2j_21})^{n_2} < 0$*, there is a* $([(2m_{11}),(2n_{11})]_{\text{DI:DI}}\text{US, nF})$*-negative decreasing-inflection upper-saddle flow as*

$$(a_{j_1j_11},a_{j_1j_21}) = \underbrace{([(2m_{11}),(2n_{11})]_{\text{DI:DI}}\text{US, nF})}_{[(2m_{11}),(2n_{11})]\text{-dereasing-inflection upper-saddle flow }(-)}. \tag{1.378}$$

There are three following $([(2m_{11}),(2n_{11})]_{\text{DI:DI}}\text{US, nF})$ *-negative decreasing-inflection upper-saddle flow appearing and switching bifurcations.*

(iv$_{4c1}$) The $([(2m_{11}),(2n_{11})]_{\text{DI:DI}}\text{US, nF})$*-negative decreasing-inflection upper-saddle flow appearing bifurcation is from a* (pF, nF)*-flow to a* $(2m_{11}) \times (2n_{11})$*-decreasing-inflection upper-saddle flow network as*

$$\underbrace{(\dot{x}_{j_1},\dot{x}_{j_2})}_{\text{(pF, nF)-flow}} \rightleftharpoons \underbrace{([(2m_{11}),(2n_{11})]_{\text{DI:DI}}\text{US, nF})}_{[(2m_{11}),(2n_{11})]\text{-dereasing-inflection upper-saddle flow }(-)}$$
$$\rightleftharpoons \bigcup_{r=1}^{2m_{11}+1}\bigcup_{u=1}^{2n_{11}}\underbrace{(a_{j_1j_1r},a_{j_1j_2u})}_{\text{YY}}, \tag{1.379}$$

where

$$\bigcup_{r=1}^{2m_{11}} \bigcup_{u=1}^{2n_{11}} \underbrace{(a_{j_1j_1r}, a_{j_1j_2u})}_{\text{YY}}$$

$$= \begin{bmatrix} \underbrace{\text{(DP:UP, nF)}}_{\text{hyperbolic flow }(-)} & \cdots & \underbrace{\text{(DP:UP, nF)}}_{\text{hyperbolic flow }(-)} & \underbrace{\text{(UP:DP, nF)}}_{\text{hyperbolic-secant flow }(-)} \\ \underbrace{\text{(UP:DP, nF)}}_{\text{hyperbolic-secant flow }(-)} & \cdots & \underbrace{\text{(UP:DP, nF)}}_{\text{hyperbolic-secant flow }(-)} & \underbrace{\text{(DP:UP, nF)}}_{\text{hyperbolic flow }(-)} \\ \vdots & \vdots & \vdots & \vdots \\ \underbrace{\text{(UP:DP, nF)}}_{\text{hyperbolic-secant flow }(-)} & \cdots & \underbrace{\text{(UP:DP, nF)}}_{\text{hyperbolic-secant flow }(-)} & \underbrace{\text{(DP:UP, nF)}}_{\text{hyperbolic flow }(-)} \end{bmatrix}_{(2n_{11})\times(2m_{11})} . \tag{1.380}$$

(iv_{4c2}) The ($[(2m_{11}), (2n_{11})]_{\text{DI:DI}}$US, nF)*-negative decreasing-inflection upper-saddle flow appearing bifurcation is from a* $q_1 \times p_1$ *to* $\overline{q}_1 \times \overline{p}_1$ *1-dimensional flow networks as*

$$\bigcup_{r_1=1}^{p_1} \bigcup_{u_1=1}^{q_1} \underbrace{(a_{j_1j_1r_1}, a_{j_1j_2u_1})}_{([m_{j_1r_1}, n_{j_1u_1}]\text{-XX, nF})} \rightleftharpoons \underbrace{([(2m_{11}), (2n_{11})]_{\text{DI:DI}}\text{US, nF})}_{[(2m_{11}),(2n_{11})]\text{-dereasing-inflection upper-saddle flow }(-)}$$
$$\rightleftharpoons \bigcup_{r_2=1}^{\overline{p}_1} \bigcup_{u_2=1}^{\overline{q}_1} \underbrace{(a_{j_1j_1r_2}, a_{j_1j_2u_2})}_{([m_{j_1r_2}, n_{j_1u_2}]\text{-YY, nF})} . \tag{1.381}$$

(iv_{4c3}) The ($[(2m_{11}), (2n_{11})]_{\text{DI:DI}}$US, nF)*-negative decreasing-inflection upper-saddle flow switching bifurcation is for the switching of two networks of* $q_1 \times p_1$ *1-dimensional flows as*

$$\bigcup_{r_1=1}^{p_1} \bigcup_{u_1=1}^{q_1} \underbrace{(a_{j_1j_1r_1}, a_{j_1j_2u_1})}_{([m_{j_1r_1}, n_{j_1u_1}]\text{-XX, nF})} \rightleftharpoons \underbrace{([(2m_{11}), (2n_{11})]_{\text{DI:DI}}\text{US, nF})}_{[(2m_{11}),(2n_{11})]\text{-dereasing-inflection upper-saddle flow }(-)}$$
$$\rightleftharpoons \bigcup_{r_2=1}^{p_1} \bigcup_{u_2=1}^{q_1} \underbrace{(a_{j_1j_1r_2}, a_{j_1j_2u_2})}_{([m_{j_1r_2}, n_{j_1u_2}]\text{-XX, nF})} . \tag{1.382}$$

(iv_{4d}) For $a_{j_1j_10} < 0$ *and* $a_{j_2j_20}(a_{j_1j_21} - a_{j_2j_21})^{n_2} < 0$*, there is a* ($[(2m_{11}), (2n_{11})]_{\text{II:II}}$LS, nF)*-negative increasing inflection lower-saddle flow as*

$$(a_{j_1j_11}, a_{j_1j_21}) = \underbrace{([(2m_{11}), (2n_{11})]_{\text{II:II}}\text{LS, nF})}_{[(2m_{11}),(2n_{11})]\text{-increasing-inflection lower-saddle flow }(-)} . \tag{1.383}$$

There are three following ($[(2m_{11}), (2n_{11})]_{\text{II:II}}$LS, nF)*-negative increasing-inflection lower-saddle flow appearing and switching bifurcations.*

(iv_{4d1}) The $([(2m_{11}),(2n_{11})]_{\text{II:II}}\text{LS, nF})$*-negative increasing-inflection lower-saddle flow appearing bifurcation is from an* (nF, nF)*-flow to a* $(2m_{11})\times(2n_{11})$*-hyperbolic and hyperbolic-secant flow network as*

$$\underbrace{(\dot{x}_{j_1},\dot{x}_{j_2})}_{\text{(nF, nF)-flow}} \rightleftharpoons \underbrace{([(2m_{11}),(2n_{11})]_{\text{II:II}}\text{LS, nF})}_{[(2m_{11}),(2n_{11})]\text{-increasing-inflection lower-saddle flow }(-)}$$
$$\rightleftharpoons \bigcup\nolimits_{r=1}^{2m_{11}}\bigcup\nolimits_{u=1}^{2n_{11}}\underbrace{(a_{j_1j_1r},a_{j_1j_2u})}_{\text{YY}}, \tag{1.384}$$

where

$$\bigcup\nolimits_{r=1}^{2m_{11}}\bigcup\nolimits_{u=1}^{2n_{11}}\underbrace{(a_{j_1j_1r},a_{j_1j_2u})}_{\text{YY}}$$
$$=\begin{bmatrix} \underbrace{(\text{DP:UP, nF})}_{\text{hyperbolic flow }(-)} & \cdots & \underbrace{(\text{DP:UP, nF})}_{\text{hyperbolic flow }(-)} & \underbrace{(\text{UP:DP, nF})}_{\text{hyperbolic-secant flow }(-)} \\ \underbrace{(\text{UP:DP, nF})}_{\text{hyperbolic-secant flow }(-)} & \cdots & \underbrace{(\text{UP:DP, nF})}_{\text{hyperbolic-secant flow }(-)} & \underbrace{(\text{DP:UP, nF})}_{\text{hyperbolic flow }(-)} \\ \vdots & \vdots & \vdots & \vdots \\ \underbrace{(\text{UP:DP, nF})}_{\text{hyperbolic-secant flow }(-)} & \cdots & \underbrace{(\text{UP:DP, nF})}_{\text{hyperbolic-secant flow }(-)} & \underbrace{(\text{DP:UP, nF})}_{\text{hyperbolic flow }(-)} \end{bmatrix}_{(2n_{11})\times(2m_{11})}. \tag{1.385}$$

(iv_{4d2}) The $([(2m_{11}),(2n_{11})]_{\text{II:II}}\text{LS, nF})$*-negative increasing-inflection lower-saddle flow appearing bifurcation is from a* $q_1\times p_1$ *to* $\overline{q}_1\times\overline{p}_1$ *1-dimensional flow networks as*

$$\bigcup\nolimits_{r_1=1}^{p_1}\bigcup\nolimits_{u_1=1}^{q_1}\underbrace{(a_{j_1j_1r_1},a_{j_1j_2u_1})}_{([m_{j_1r_1},n_{j_1u_1}]\text{-XX, nF})} \rightleftharpoons \underbrace{([(2m_{11}),(2n_{11})]_{\text{II:II}}\text{LS, nF})}_{[(2m_{11}),(2n_{11})]\text{-increasing-inflection lower-saddle flow }(-)}$$
$$\rightleftharpoons \bigcup\nolimits_{r_2=1}^{\overline{p}_1}\bigcup\nolimits_{u_2=1}^{\overline{q}_1}\underbrace{(a_{j_1j_1r_2},a_{j_1j_2u_2})}_{([m_{j_1r_2},n_{j_1u_2}]\text{-YY, nF}))}. \tag{1.386}$$

(iv_{4d3}) The $([(2m_{11}),(2n_{11})]_{\text{II:II}}\text{LS, nF})$ *-negative increasing-inflection lower-saddle flow switching bifurcation is for the switching of two networks of* $q_1\times p_1$ *1-dimensional flows as*

$$\bigcup\nolimits_{r_1=1}^{p_1}\bigcup\nolimits_{u_1=1}^{q_1}\underbrace{(a_{j_1j_1r_1},a_{j_1j_2u_1})}_{([m_{j_1r_1},n_{j_1u_1}]\text{-XX, nF})} \rightleftharpoons \underbrace{([(2m_{11}),(2n_{11})]_{\text{II:II}}\text{LS, nF})}_{[(2m_{11}),(2n_{11})]\text{-increasing-inflection lower-saddle flow }(-)}$$
$$\rightleftharpoons \bigcup\nolimits_{r_2=1}^{p_1}\bigcup\nolimits_{u_2=1}^{q_1}\underbrace{(a_{j_1j_1r_2},a_{j_1j_2u_2})}_{([m_{j_1r_2},n_{j_1u_2}]\text{-YY, nF})}. \tag{1.387}$$

(v) The equilibrium of $(x_{j_1}^*, x_{j_2}^*) = (a_{j_1j_1 1}, a_{j_2j_2 1})$ *with* $a_{j_2j_2 1} \neq a_{j_1j_2 1}$ *has four classes of equilibrium bifurcations.*

(v_I) For $m_1 = 2m_{11} + 1$ *and* $n_2 = 2n_{21} + 1$*, the equilibrium of* $(x_{j_1}^*, x_{j_2}^*) = (a_{j_1j_1 1}, a_{j_1j_2 1})$ *for the* x_{j_1}*-direction has the following bifurcation properties with*

$$\begin{aligned}
&2m_{11} + 1 - \sum_{r_1=1}^{p_1} m_{j_1 r_1}^{(1)} = 2\sum_{\rho_1=1}^{l_1} m_{\rho_1},\\
&2n_{21} + 1 - \sum_{s_1=1}^{q_2} n_{j_2 s_1}^{(1)} = 2\sum_{\sigma_1=1}^{l_2} n_{\sigma_1};\\
&\sum_{r_2=1}^{\bar{p}_1} m_{j_1 r_2}^{(2)} = 2m_{11} + 1, \sum_{s_2=1}^{\bar{q}_2} n_{j_2 s_2}^{(2)} = 2n_{21} + 1;\\
&\sum_{r_i=1}^{p_1} m_{j_1 r_i}^{(i)} = 2m_{11} + 1, \sum_{s_i=1}^{q_2} n_{j_2 s_i}^{(2)} = 2n_{21} + 1, \text{ for } i = 1, 2;
\end{aligned} \tag{1.388}$$

l_1, l_2*-quadratic polynomials are without real roots, and the* ρ_1^{th} *and* σ_1^{th}*-quadratic polynomials are with powers* m_{ρ_1} *and* n_{σ_1}*, respectively.*

(v_{Ia}) For $a_{j_1j_1 0}(a_{j_2j_2 1} - a_{j_1j_2 1})^{n_1} > 0$ *and* $a_{j_2j_2 0} > 0$*, there is a* $((2m_{11} + 1), (2n_{21} + 1))$*-source equilibrium at* $(x_{j_1}^*, x_{j_2}^*) = (a_{j_1j_1 1}, a_{j_2j_2 1})$ *as*

$$(a_{j_1j_1 1}, a_{j_2j_2 1}) = \underbrace{((2m_{11}{+}1)^{\text{th}}\,\text{SO}, (2n_{21}{+}1)^{\text{th}}\,\text{SO})}_{((2m_{11}+1),(2n_{21}+1))\text{-source}}. \tag{1.389}$$

The following three $((2m_{11}{+}1)^{\text{th}}\,\text{SO}, (2n_{21}{+}1)^{\text{th}}\,\text{SO})$*-source appearing and switching bifurcations as follows.*

(v_{Ia1}) The $((2m_{11}{+}1)^{\text{th}}\,\text{SO}, (2n_{21}{+}1)^{\text{th}}\,\text{SO})$*-source appearing bifurcation is from an* (SO, SO)*-source to a*$(2m_{11} + 1) \times (2n_{21} + 1)$*-sink, source, and saddles network as*

$$\begin{aligned}
\underbrace{(a_{j_1j_1 1}, a_{j_2j_2 1})}_{(\text{SO, SO})\text{-source}} &\rightleftharpoons \underbrace{((2m_{11}{+}1)^{\text{th}}\,\text{SO}, (2n_{21}{+}1)^{\text{th}}\,\text{SO})}_{((2m_{11}+1),(2n_{21}+1))\text{-source}}\\
&\rightleftharpoons \bigcup_{r=1}^{2m_{11}+1} \bigcup_{s=1}^{2n_{21}+1} \underbrace{(a_{j_1j_1 r}, a_{j_2j_2 s})}_{\text{XX}},
\end{aligned} \tag{1.390}$$

where

$$\bigcup_{r=1}^{2m_{11}+1}\bigcup_{s=1}^{2n_{21}+1}\underbrace{(a_{j_1j_1r},a_{j_2j_2s})}_{XX}$$

$$=\begin{bmatrix}\underbrace{(SO, SO)}_{source} & \cdots & \underbrace{(SI, SO)}_{saddle} & \underbrace{(SO, SO)}_{source}\\ \underbrace{(SO, SI)}_{saddle} & \cdots & \underbrace{(SI, SI)}_{sink} & \underbrace{(SO, SI)}_{saddle}\\ \vdots & \vdots & \vdots & \vdots\\ \underbrace{(SO, SO)}_{source} & \cdots & \underbrace{(SI, SO)}_{saddle} & \underbrace{(SO, SO)}_{source}\end{bmatrix}_{(2n_{21}+1)\times(2m_{11}+1)}. \tag{1.391}$$

(v_{1a2}) *The* $((2m_{11}+1)^{\text{th}}$ SO, $(2n_{21}+1)^{\text{th}}$ SO)*-source appearing bifurcation is from a* $q_2\times p_1$ *to* $\overline{q}_2\times\overline{p}_1$ *sink, source, and saddle networks as*

$$\bigcup_{r_1=1}^{p_1}\bigcup_{s_1=1}^{q_2}\underbrace{(a_{j_1j_1r_1},a_{j_2j_2s_1})}_{(m_{j_1r_1},n_{j_2s_1})\text{-}XX}\rightleftharpoons\underbrace{((2m_{11}+1)^{\text{th}}\text{ SO},(2n_{21}+1)^{\text{th}}\text{ SO})}_{((2m_{11}+1),(2n_{21}+1))\text{-source}}$$
$$\rightleftharpoons\bigcup_{r_2=1}^{\overline{p}_1}\bigcup_{s_2=1}^{\overline{q}_2}\underbrace{(a_{j_1j_1r_2},a_{j_2j_2s_2})}_{(m_{j_1r_2},n_{j_2s_2})\text{-}YY}. \tag{1.392}$$

(v_{1a3}) *The* $((2m_{11}+1)^{\text{th}}$ SO, $(2n_{21}+1)^{\text{th}}$ SO)*-source switching bifurcation is for the switching of two networks of* $q_2\times p_1$ *sink, source, and saddles as*

$$\bigcup_{r_1=1}^{p_1}\bigcup_{s_1=1}^{q_2}\underbrace{(a_{j_1j_1r_1},a_{j_2j_2s_1})}_{(m_{j_1r_1},n_{j_2s_1})\text{-}XX}\rightleftharpoons\underbrace{((2m_{11}+1)^{\text{th}}\text{ SO},(2n_{21}+1)^{\text{th}}\text{ SO})}_{((2m_{11}+1),(2n_{21}+1))\text{-source}}$$
$$\rightleftharpoons\bigcup_{r_2=1}^{p_1}\bigcup_{s_2=1}^{q_2}\underbrace{(a_{j_1j_1r_2},a_{j_2j_2s_2})}_{(m_{j_1r_2},n_{j_2s_2})\text{-}YY}. \tag{1.393}$$

(v_{1b}) *For* $a_{j_1j_1 0}(a_{j_2j_2 1}-a_{j_1j_2 1})^{n_1}<0$ *and* $a_{j_2j_2 0}>0$, *there is a* $((2m_{11}+1),(2n_{21}+1))$ *-saddle equilibrium at* $(x_{j_1}^*,x_{j_2}^*)=(a_{j_1j_1 1},a_{j_2j_2 1})$ *as*

$$(a_{j_1j_1 1},a_{j_2j_2 1})=\underbrace{((2m_{11}+1)^{\text{th}}\text{ SI},(2n_{21}+1)^{\text{th}}\text{ SO})}_{((2m_{11}+1),(2n_{21}+1))\text{-saddle}}. \tag{1.394}$$

The following three $((2m_{11}+1)^{\text{th}}$ SI, $(2n_{21}+1)^{\text{th}}$ SO)*-saddle appearing and switching bifurcations as follows.*

(v_{1b1}) *The* $((2m_{11}+1)^{\text{th}}$ SI, $(2n_{21}+1)^{\text{th}}$ SO)*-saddle appearing bifurcation is from an* (SI, SO)*-saddle to a*$(2m_{11}+1)\times(2n_{21}+1)$*-sink, source, and saddles network as*

$$\underbrace{(a_{j_1j_11}, a_{j_2j_21})}_{\text{(SI, SO)-saddle}} \rightleftharpoons \underbrace{((2m_{11}+1)^{\text{th}}\text{ SI}, (2n_{21}+1)^{\text{th}}\text{ SO})}_{((2m_{11}+1),(2n_{21}+1))\text{-saddle}}$$
$$\rightleftharpoons \bigcup\nolimits_{r=1}^{2m_{11}+1}\bigcup\nolimits_{s=1}^{2n_{21}+1}\underbrace{(a_{j_1j_1r}, a_{j_2j_2s})}_{\text{XX}}, \tag{1.395}$$

where

$$\bigcup\nolimits_{r=1}^{2m_{11}+1}\bigcup\nolimits_{s=1}^{2n_{21}+1}\underbrace{(a_{j_1j_1r}, a_{j_2j_2s})}_{\text{XX}}$$
$$= \begin{bmatrix} \underbrace{\text{(SI, SO)}}_{\text{saddle}} & \cdots & \underbrace{\text{(SO, SO)}}_{\text{source}} & \underbrace{\text{(SI, SO)}}_{\text{saddle}} \\ \underbrace{\text{(SI, SI)}}_{\text{sink}} & \cdots & \underbrace{\text{(SO, SI)}}_{\text{saddle}} & \underbrace{\text{(SI, SI)}}_{\text{sink}} \\ \vdots & \vdots & \vdots & \vdots \\ \underbrace{\text{(SI, SO)}}_{\text{saddle}} & \cdots & \underbrace{\text{(SO, SO)}}_{\text{source}} & \underbrace{\text{(SI, SO)}}_{\text{saddle}} \end{bmatrix}_{(2n_{21}+1)\times(2m_{11}+1)}. \tag{1.396}$$

(v_{1b2}) *The* $((2m_{11}+1)^{\text{th}}\text{ SI}, (2n_{21}+1)^{\text{th}}\text{ SO})$*-saddle appearing bifurcation is from a* $q_2 \times p_1$ *to* $\overline{q}_2 \times \overline{p}_1$ *sink, source, and saddle networks as*

$$\bigcup\nolimits_{r_1=1}^{p_1}\bigcup\nolimits_{s_1=1}^{q_2}\underbrace{(a_{j_1j_1r_1}, a_{j_2j_2s_1})}_{(m_{j_1r_1},n_{j_2s_1})\text{-XX}} \rightleftharpoons \underbrace{((2m_{11}+1)^{\text{th}}\text{ SI}, (2n_{21}+1)^{\text{th}}\text{ SO})}_{((2m_{11}+1),(2n_{21}+1))\text{-saddle}}$$
$$\rightleftharpoons \bigcup\nolimits_{r_2=1}^{\overline{p}_1}\bigcup\nolimits_{s_2=1}^{\overline{q}_2}\underbrace{(a_{j_1j_1r_2}, a_{j_2j_2s_2})}_{(m_{j_1r_2},n_{j_2s_2})\text{-YY}}. \tag{1.397}$$

(v_{1b3}) *The* $((2m_{11}+1)^{\text{th}}\text{ SI}, (2n_{21}+1)^{\text{th}}\text{ SO})$*-saddle switching bifurcation is for the switching of two networks of* $q_2 \times p_1$ *sink, source, and saddles as*

$$\bigcup\nolimits_{r_1=1}^{p_1}\bigcup\nolimits_{s_1=1}^{q_2}\underbrace{(a_{j_1j_1r_1}, a_{j_2j_2s_1})}_{(m_{j_1r_1},n_{j_2s_1})\text{-XX}} \rightleftharpoons \underbrace{((2m_{11}+1)^{\text{th}}\text{ SI}, (2n_{21}+1)^{\text{th}}\text{ SO})}_{((2m_{11}+1),(2n_{21}+1))\text{-saddle}}$$
$$\rightleftharpoons \bigcup\nolimits_{r_2=1}^{p_1}\bigcup\nolimits_{s_2=1}^{q_2}\underbrace{(a_{j_1j_1r_2}, a_{j_2j_2s_2})}_{(m_{j_1r_2},n_{j_2s_2})\text{-YY}}. \tag{1.398}$$

(v_{1c}) *For* $a_{j_1j_10}(a_{j_2j_21} - a_{j_1j_21})^{n_1} > 0$ *and* $a_{j_2j_20} < 0$, *there is a* $((2m_{11}+1), (2n_{21}+1))$*-saddle equilibrium at* $(x^*_{j_1}, x^*_{j_2}) = (a_{j_1j_11}, a_{j_2j_21})$ *as*

$$(a_{j_1j_11}, a_{j_2j_21}) = \underbrace{((2m_{11}+1)^{\text{th}}\text{ SO}, (2n_{21}+1)^{\text{th}}\text{ SI})}_{((2m_{11}+1),(2n_{21}+1))\text{-saddle}}. \tag{1.399}$$

The following three $((2m_{11}+1)^{\text{th}}\,\text{SO}, (2n_{21}+1)^{\text{th}}\,\text{SI})$*-saddle appearing and switching bifurcations as follows.*

(v_{1c1}) *The* $((2m_{11}+1)^{\text{th}}\,\text{SO}, (2n_{21}+1)^{\text{th}}\,\text{SI})$*-saddle appearing bifurcation is from an* (SO, SI)*-saddle to a* $(2m_{11}+1)\times(2n_{21}+1)$*-sink, source, and saddles network as*

$$\underbrace{(a_{j_1j_11}, a_{j_2j_21})}_{\text{(SO, SI)-saddle}} \rightleftharpoons \underbrace{((2m_{11}+1)^{\text{th}}\,\text{SO}, (2n_{21}+1)^{\text{th}}\,\text{SI})}_{((2m_{11}+1),(2n_{21}+1))\text{-saddle}} \rightleftharpoons \bigcup_{r=1}^{2m_{11}+1}\bigcup_{s=1}^{2n_{21}+1}\underbrace{(a_{j_1j_1r}, a_{j_2j_2s})}_{\text{XX}}, \tag{1.400}$$

where

$$\bigcup_{r=1}^{2m_{11}+1}\bigcup_{s=1}^{2n_{21}+1}\underbrace{(a_{j_1j_1r}, a_{j_2j_2s})}_{\text{XX}} = \begin{bmatrix} \underbrace{\text{(SO, SI)}}_{\text{saddle}} & \cdots & \underbrace{\text{(SI, SI)}}_{\text{sink}} & \underbrace{\text{(SO, SI)}}_{\text{saddle}} \\ \underbrace{\text{(SO, SO)}}_{\text{source}} & \cdots & \underbrace{\text{(SI, SO)}}_{\text{saddle}} & \underbrace{\text{(SO, SO)}}_{\text{source}} \\ \vdots & \vdots & \vdots & \vdots \\ \underbrace{\text{(SO, SI)}}_{\text{saddle}} & \cdots & \underbrace{\text{(SI, SI)}}_{\text{sink}} & \underbrace{\text{(SO, SI)}}_{\text{saddle}} \end{bmatrix}_{(2n_{21}+1)\times(2m_{11}+1)}. \tag{1.401}$$

(v_{1c2}) *The* $((2m_{11}+1)^{\text{th}}\,\text{SO}, (2n_{21}+1)^{\text{th}}\,\text{SI})$*-saddle appearing bifurcation is from a* $q_2\times p_1$ *to* $\overline{q}_2\times\overline{p}_1$ *sink, source, and saddle networks as*

$$\bigcup_{r_1=1}^{p_1}\bigcup_{s_1=1}^{q_2}\underbrace{(a_{j_1j_1r_1}, a_{j_2j_2s_1})}_{(m_{j_1r_1},n_{j_2s_1})\text{-XX}} \rightleftharpoons \underbrace{((2m_{11}+1)^{\text{th}}\,\text{SO}, (2n_{21}+1)^{\text{th}}\,\text{SI})}_{((2m_{11}+1),(2n_{21}+1))\text{-saddle}} \rightleftharpoons \bigcup_{r_2=1}^{\overline{p}_1}\bigcup_{s_2=1}^{\overline{q}_2}\underbrace{(a_{j_1j_1r_2}, a_{j_2j_2s_2})}_{(m_{j_1r_2},n_{j_2s_2})\text{-YY}}. \tag{1.402}$$

(v_{1c3}) *The* $((2m_{11}+1)^{\text{th}}\,\text{SO}, (2n_{21}+1)^{\text{th}}\,\text{SI})$*-saddle switching bifurcation is for the switching of two networks of* $q_2\times p_1$ *sink, source, and saddles as*

$$\bigcup_{r_1=1}^{p_1}\bigcup_{s_1=1}^{q_2}\underbrace{(a_{j_1j_1r_1}, a_{j_2j_2s_1})}_{(m_{j_1r_1},n_{j_2s_1})\text{-XX}} \rightleftharpoons \underbrace{((2m_{11}+1)^{\text{th}}\,\text{SO}, (2n_{21}+1)^{\text{th}}\,\text{SI})}_{((2m_{11}+1),(2n_{21}+1))\text{-saddle}} \rightleftharpoons \bigcup_{r_2=1}^{p_1}\bigcup_{s_2=1}^{q_2}\underbrace{(a_{j_1j_1r_2}, a_{j_2j_2s_2})}_{(m_{j_1r_2},n_{j_2s_2})\text{-YY}}. \tag{1.403}$$

(v_{1d}) For $a_{j_1 j_1 0}(a_{j_2 j_2 1} - a_{j_1 j_2 1})^{n_1} < 0$ and $a_{j_2 j_2 0} < 0$, there is a $((2m_{11}+1), (2n_{21}+1))$-sink equilibrium at $(x^*_{j_1}, x^*_{j_2}) = (a_{j_1 j_1 1}, a_{j_2 j_2 1})$ as

$$(a_{j_1 j_1 1}, a_{j_2 j_2 1}) = \underbrace{((2m_{11}+1)^{\text{th}}\,\text{SI}, (2n_{21}+1)^{\text{th}}\,\text{SI})}_{((2m_{11}+1),(2n_{21}+1))\text{-sink}}. \tag{1.404}$$

The following three $((2m_{11}+1)^{\text{th}}$ SI, $(2n_{21}+1)^{\text{th}}$ SI)-sink appearing and switching bifurcations as follows.

(v_{1d1}) The $((2m_{11}+1)^{\text{th}}$ SI, $(2n_{21}+1)^{\text{th}}$ SI)-sink appearing bifurcation is from an (SI, SI)-sink to a $(2m_{11}+1) \times (2n_{21}+1)$-sink, source, and saddles network as

$$\begin{aligned}\underbrace{(a_{j_1 j_1 1}, a_{j_2 j_2 1})}_{(\text{SI, SI})\text{-sink}} &\rightleftharpoons \underbrace{((2m_{11}+1)^{\text{th}}\,\text{SI}, (2n_{21}+1)^{\text{th}}\,\text{SI})}_{((2m_{11}+1),(2n_{21}+1))\text{-sink}} \\ &\rightleftharpoons \bigcup\nolimits_{r=1}^{2m_{11}+1} \bigcup\nolimits_{s=1}^{2n_{21}+1} \underbrace{(a_{j_1 j_1 r}, a_{j_2 j_2 s})}_{\text{XX}},\end{aligned} \tag{1.405}$$

where

$$\begin{aligned}&\bigcup\nolimits_{r=1}^{2m_{11}+1} \bigcup\nolimits_{s=1}^{2n_{21}+1} \underbrace{(a_{j_1 j_1 r}, a_{j_2 j_2 s})}_{\text{XX}} \\ &= \begin{bmatrix} \underbrace{(\text{SI, SI})}_{\text{sink}} & \cdots & \underbrace{(\text{SO, SI})}_{\text{saddle}} & \underbrace{(\text{SI, SI})}_{\text{sink}} \\ \underbrace{(\text{SI, SO})}_{\text{saddle}} & \cdots & \underbrace{(\text{SO, SO})}_{\text{source}} & \underbrace{(\text{SI, SO})}_{\text{saddle}} \\ \vdots & \vdots & \vdots & \vdots \\ \underbrace{(\text{SI, SI})}_{\text{sink}} & \cdots & \underbrace{(\text{SO, SI})}_{\text{saddle}} & \underbrace{(\text{SI, SI})}_{\text{sink}} \end{bmatrix}_{(2n_{21}+1)\times(2m_{11}+1)}.\end{aligned} \tag{1.406}$$

(v_{1d2}) The $((2m_{11}+1)^{\text{th}}$ SI, $(2n_{21}+1)^{\text{th}}$ SI)-sink appearing bifurcation is from a $q_2 \times p_1$ to $\bar{q}_2 \times \bar{p}_1$ sink, source, and saddle networks as

$$\begin{aligned}\bigcup\nolimits_{r_1=1}^{p_1} \bigcup\nolimits_{s_1=1}^{q_2} \underbrace{(a_{j_1 j_1 r_1}, a_{j_2 j_2 s_1})}_{(m_{j_1 r_1}, n_{j_2 s_1})\text{-XX}} &\rightleftharpoons \underbrace{((2m_{11}+1)^{\text{th}}\,\text{SI}, (2n_{21}+1)^{\text{th}}\,\text{SI})}_{((2m_{11}+1),(2n_{21}+1))\text{-sink}} \\ &\rightleftharpoons \bigcup\nolimits_{r_2=1}^{\bar{p}_1} \bigcup\nolimits_{s_2=1}^{\bar{q}_2} \underbrace{(a_{j_1 j_1 r_2}, a_{j_2 j_2 s_2})}_{(m_{j_1 r_2}, n_{j_2 s_2})\text{-YY}}.\end{aligned} \tag{1.407}$$

(v_{1d3}) The $((2m_{11}+1)^{\text{th}}$ SI, $(2n_{21}+1)^{\text{th}}$ SI)-sink switching bifurcation is for the switching of two networks of $q_2 \times p_1$ sink, source, and saddles as

$$\bigcup_{r_1=1}^{p_1}\bigcup_{s_1=1}^{q_2}\underbrace{(a_{j_1j_1r_1},a_{j_2j_2s_1})}_{(m_{j_1r_1},n_{j_2s_1})\text{-XX}} \rightleftharpoons \underbrace{((2m_{11}+1)^{\text{th}}\,\text{SI},\,(2n_{21}+1)^{\text{th}}\,\text{SI})}_{((2m_{11}+1),(2n_{21}+1))\text{-sink}}$$
$$\rightleftharpoons \bigcup_{r_2=1}^{p_1}\bigcup_{s_2=1}^{q_2}\underbrace{(a_{j_1j_1r_2},a_{j_2j_2s_2})}_{(m_{j_1r_2},n_{j_2s_2})\text{-YY}}. \tag{1.408}$$

(v_2) For $m_1 = 2m_{11}+1$ and $n_2 = 2n_{21}$, the equilibrium of $(x^*_{j_1}, x^*_{j_2}) = (a_{j_1j_11}, a_{j_1j_21})$ for the x_{j_1}-direction has the following bifurcation properties with

$$2m_{11}+1-\sum_{r_1=1}^{p_1} m^{(1)}_{j_1r_1} = 2\sum_{\rho_1=1}^{l_1} m_{\rho_1},$$
$$2n_{21}-\sum_{s_1=1}^{q_2} n^{(1)}_{j_2s_1} = 2\sum_{\sigma_1=1}^{l_2} n_{\sigma_1};$$
$$\sum_{r_2=1}^{\bar{p}_1} m^{(2)}_{j_1r_2} = 2m_{11}+1,\ \sum_{s_2=1}^{\bar{q}_2} n^{(2)}_{j_2s_2} = 2n_{21};$$
$$\sum_{r_i=1}^{p_1} m^{(i)}_{j_1r_i} = 2m_{11}+1,\ \sum_{s_i=1}^{q_2} n^{(2)}_{j_2s_i} = 2n_{21},\ \text{for } i=1,2; \tag{1.409}$$

l_1, l_2-quadratic polynomials are without real roots, and the ρ_1^{th} and σ_1^{th}-quadratic polynomials are with powers m_{ρ_1} and n_{σ_1}, respectively.

(v_{2a}) For $a_{j_1j_10}(a_{j_2j_21}-a_{j_1j_21})^{n_1} > 0$ and $a_{j_2j_20} > 0$, there is a $((2m_{11}+1),(2n_{21}))$-upper-saddle source equilibrium at $(x^*_{j_1}, x^*_{j_2}) = (a_{j_1j_11}, a_{j_2j_21})$ as

$$(a_{j_1j_11},a_{j_2j_21}) = \underbrace{((2m_{11}+1)^{\text{th}}\,\text{SO},\,(2n_{21})^{\text{th}}\,\text{US})}_{((2m_{11}+1),(2n_{21}))\text{-upper-saddle-source}}. \tag{1.410}$$

The following three $((2m_{11}+1)^{\text{th}}$ SO, $(2n_{21})^{\text{th}}$ US)-upper-saddle-source appearing and switching bifurcations as follows .

(v_{2a1}) The $((2m_{11}+1)^{\text{th}}$ SO, $(2n_{21})^{\text{th}}$ US)-upper-saddle-source appearing bifurcation is from an (SO, pF)-positive source flow to a $(2m_{11}+1)\times(2n_{21})$-sink, source, and saddles network as

$$\underbrace{(a_{j_1j_11},\dot{x}_{j_2})}_{\text{(SO, pF) - flow}} \rightleftharpoons \underbrace{((2m_{11}+1)^{\text{th}}\text{SO},\,(2n_{21})^{\text{th}}\text{US})}_{((2m_{11}+1),(2n_{21}))\text{-upper-saddle source}}$$
$$\rightleftharpoons \bigcup_{r=1}^{2m_{11}+1}\bigcup_{s=1}^{2n_{21}}\underbrace{(a_{j_1j_1r},a_{j_2j_2s})}_{\text{XX}}, \tag{1.411}$$

where

$$\bigcup_{r=1}^{2m_{11}+1} \bigcup_{s=1}^{2n_{21}} \underbrace{(a_{j_1j_1r}, a_{j_2j_2s})}_{\text{XX}}$$
$$= \begin{bmatrix} \underbrace{(\text{SO, SO})}_{\text{source}} & \cdots & \underbrace{(\text{SI, SO})}_{\text{saddle}} & \underbrace{(\text{SO, SO})}_{\text{source}} \\ \underbrace{(\text{SO, SI})}_{\text{saddle}} & \cdots & \underbrace{(\text{SI, SI})}_{\text{sink}} & \underbrace{(\text{SO, SI})}_{\text{saddle}} \\ \vdots & \vdots & \vdots & \vdots \\ \underbrace{(\text{SO, SI})}_{\text{saddle}} & \cdots & \underbrace{(\text{SI, SI})}_{\text{sink}} & \underbrace{(\text{SO, SI})}_{\text{saddle}} \end{bmatrix}_{(2n_{21})\times(2m_{11}+1)}. \tag{1.412}$$

(v_{2a2}) *The* $((2m_{11}+1)^{\text{th}}\ \text{SO}, (2n_{21})^{\text{th}}\ \text{US})$*-upper-saddle-source appearing bifurcation is from a* $q_2 \times p_1$ *to* $\overline{q}_2 \times \overline{p}_1$ *sink, source, and saddle networks as*

$$\bigcup_{r_1=1}^{p_1} \bigcup_{s_1=1}^{q_2} \underbrace{(a_{j_1j_1r_1}, a_{j_2j_2s_1})}_{(m_{j_1r_1}, n_{j_2s_1})\text{-XX}} \rightleftharpoons \underbrace{((2m_{11}+1)^{\text{th}}\ \text{SO}, (2n_{21})^{\text{th}}\ \text{US})}_{((2m_{11}+1),(2n_{21}))\text{-upper-saddle-source}}$$
$$\rightleftharpoons \bigcup_{r_2=1}^{\overline{p}_1} \bigcup_{s_2=1}^{\overline{q}_2} \underbrace{(a_{j_1j_1r_2}, a_{j_2j_2s_2})}_{(m_{j_1r_2}, n_{j_2s_2})\text{-YY}}. \tag{1.413}$$

(v_{2a3}) *The* $((2m_{11}+1)^{\text{th}}\ \text{SO}, (2n_{21})^{\text{th}}\ \text{US})$*-upper-saddle-source switching bifurcation is for the switching of two networks of* $q_2 \times p_1$ *sink, source, and saddles as*

$$\bigcup_{r_1=1}^{p_1} \bigcup_{s_1=1}^{q_2} \underbrace{(a_{j_1j_1r_1}, a_{j_2j_2s_1})}_{(m_{j_1r_1}, n_{j_2s_1})\text{-XX}} \rightleftharpoons \underbrace{((2m_{11}+1)^{\text{th}}\ \text{SO}, (2n_{21})^{\text{th}}\ \text{US})}_{((2m_{11}+1),(2n_{21}))\text{-upper-saddle-source}}$$
$$\rightleftharpoons \bigcup_{r_2=1}^{p_1} \bigcup_{s_2=1}^{q_2} \underbrace{(a_{j_1j_1r_2}, a_{j_2j_2s_2})}_{(m_{j_1r_2}, n_{j_2s_2})\text{-YY}}. \tag{1.414}$$

(v_{2b}) *For* $a_{j_1j_10}(a_{j_2j_21} - a_{j_1j_21})^{n_1} < 0$ *and* $a_{j_2j_20} > 0$*, there is a* $((2m_{11}+1), (2n_{21}))$*-upper-saddle-sink equilibrium at* $(x_{j_1}^*, x_{j_2}^*) = (a_{j_1j_11}, a_{j_2j_21})$ *as*

$$(a_{j_1j_11}, a_{j_2j_21}) = \underbrace{((2m_{11}+1)^{\text{th}}\ \text{SI}, (2n_{21})^{\text{th}}\ \text{US})}_{((2m_{11}+1),(2n_{21}))\text{-upper-saddle sink}}. \tag{1.415}$$

The following three $((2m_{11}+1)^{\text{th}}\ \text{SI}, (2n_{21})^{\text{th}}\ \text{US})$*-upper-saddle-sink appearing and switching bifurcations as follows.*

(v_{2b1}) The $((2m_{11}+1)^{\text{th}}$ SI, $(2n_{21})^{\text{th}}$ US)*-upper-saddle-sink appearing bifurcation is from an* (SI, pF)*-sink flow to a* $(2m_{11}+1)\times(2n_{21})$*-sink, source, and saddles network as*

$$\underbrace{(a_{j_1j_1 1},\dot{x}_{j_2})}_{\text{(SI, pF)-flow}} \rightleftharpoons \underbrace{((2m_{11}+1)^{\text{th}}\text{ SI}, (2n_{21})^{\text{th}}\text{ US})}_{((2m_{11}+1),(2n_{21}))\text{-upper-saddle-sink}}$$
$$\rightleftharpoons \bigcup\nolimits_{r=1}^{2m_{11}+1}\bigcup\nolimits_{s=1}^{2n_{21}}\underbrace{(a_{j_1j_1r},a_{j_2j_2s})}_{\text{XX}}, \tag{1.416}$$

where

$$\bigcup\nolimits_{r=1}^{2m_{11}+1}\bigcup\nolimits_{s=1}^{2n_{21}}\underbrace{(a_{j_1j_1r},a_{j_2j_2s})}_{\text{XX}}$$
$$= \begin{bmatrix} \underbrace{(\text{SI, SO})}_{\text{saddle}} & \cdots & \underbrace{(\text{SO, SO})}_{\text{source}} & \underbrace{(\text{SI, SO})}_{\text{saddle}} \\ \underbrace{(\text{SI, SI})}_{\text{sink}} & \cdots & \underbrace{(\text{SO, SI})}_{\text{saddle}} & \underbrace{(\text{SI, SI})}_{\text{sink}} \\ \vdots & \vdots & \vdots & \vdots \\ \underbrace{(\text{SI, SI})}_{\text{sink}} & \cdots & \underbrace{(\text{SO, SI})}_{\text{saddle}} & \underbrace{(\text{SI, SI})}_{\text{sink}} \end{bmatrix}_{(2n_{21})\times(2m_{11}+1)}. \tag{1.417}$$

(v_{2b2}) The $((2m_{11}+1)^{\text{th}}$ SI, $(2n_{21})^{\text{th}}$ US)*-upper-saddle-sink appearing bifurcation is from a* $q_2\times p_1$ *to* $\overline{q}_2\times\overline{p}_1$ *sink, source, and saddle networks as*

$$\bigcup\nolimits_{r_1=1}^{p_1}\bigcup\nolimits_{s_1=1}^{q_2}\underbrace{(a_{j_1j_1r_1},a_{j_2j_2s_1})}_{(m_{j_1r_1},n_{j_2s_1})\text{-XX}} \rightleftharpoons \underbrace{((2m_{11}+1)^{\text{th}}\text{ SI}, (2n_{21})^{\text{th}}\text{ US})}_{((2m_{11}+1),(2n_{21}))\text{-upper-saddle-sink}}$$
$$\rightleftharpoons \bigcup\nolimits_{r_2=1}^{\overline{p}_1}\bigcup\nolimits_{s_2=1}^{\overline{q}_2}\underbrace{(a_{j_1j_1r_2},a_{j_2j_2s_2})}_{(m_{j_1r_2},n_{j_2s_2})\text{-YY}}. \tag{1.418}$$

(v_{2b3}) The $((2m_{11}+1)^{\text{th}}$ SI, $(2n_{21})^{\text{th}}$ US)*-upper-saddle-sink switching bifurcation is for the switching of two networks of* $q_2\times p_1$ *sink, source, and saddles as*

$$\bigcup\nolimits_{r_1=1}^{p_1}\bigcup\nolimits_{s_1=1}^{q_2}\underbrace{(a_{j_1j_1r_1},a_{j_2j_2s_1})}_{(m_{j_1r_1},n_{j_2s_1})\text{-XX}} \rightleftharpoons \underbrace{((2m_{11}+1)^{\text{th}}\text{ SI}, (2n_{21})^{\text{th}}\text{ US})}_{((2m_{11}+1),(2n_{21}))\text{-upper-saddle-sink}}$$
$$\rightleftharpoons \bigcup\nolimits_{r_2=1}^{p_1}\bigcup\nolimits_{s_2=1}^{q_2}\underbrace{(a_{j_1j_1r_2},a_{j_2j_2s_2})}_{(m_{j_1r_2},n_{j_2s_2})\text{-YY}}. \tag{1.419}$$

(v_{2c}) For $a_{j_1j_10}(a_{j_2j_21} - a_{j_1j_21})^{n_1} > 0$ and $a_{j_2j_20} < 0$, there is a $((2m_{11}+1),(2n_{21}))$-lower-saddle-source equilibrium at $(x^*_{j_1}, x^*_{j_2}) = (a_{j_1j_11}, a_{j_2j_21})$ as

$$(a_{j_1j_11}, a_{j_2j_21}) = \underbrace{((2m_{11}+1)^{\text{th}}\,\text{SO},\,(2n_{21})^{\text{th}}\,\text{LS})}_{((2m_{11}+1),(2n_{21}))\text{-lower-saddle-source}}. \tag{1.420}$$

The following three $((2m_{11}+1)^{\text{th}}$ SO, $(2n_{21})^{\text{th}}$ LS)*-lower-saddle-source appearing and switching bifurcations as follows.*

(v_{2c1}) The $((2m_{11}+1)^{\text{th}}$ SO, $(2n_{21})^{\text{th}}$ LS) *-lower-saddle-source appearing bifurcation is from an* (SO, nF)*-negative source flow to a* $(2m_{11}+1)\times(2n_{21})$*-sink, source, and saddles network as*

$$\begin{aligned}\underbrace{(a_{j_1j_11}, \dot{x}_{j_2})}_{(\text{SO, nF})\text{-flow}} &\rightleftharpoons \underbrace{((2m_{11}+1)^{\text{th}}\,\text{SO},\,(2n_{21})^{\text{th}}\,\text{LS})}_{((2m_{11}+1),(2n_{21}))\text{-lower-saddle-source}}\\ &\rightleftharpoons \bigcup\nolimits_{r=1}^{2m_{11}+1}\bigcup\nolimits_{s=1}^{2n_{21}}\underbrace{(a_{j_1j_1r}, a_{j_2j_2s})}_{\text{XX}},\end{aligned} \tag{1.421}$$

where

$$\bigcup\nolimits_{r=1}^{2m_{11}+1}\bigcup\nolimits_{s=1}^{2n_{21}+1}\underbrace{(a_{j_1j_1r}, a_{j_2j_2s})}_{\text{XX}} = \begin{bmatrix}\underbrace{(\text{SO, SI})}_{\text{saddle}} & \cdots & \underbrace{(\text{SI, SI})}_{\text{sink}} & \underbrace{(\text{SO, SI})}_{\text{saddle}}\\ \underbrace{(\text{SO, SO})}_{\text{source}} & \cdots & \underbrace{(\text{SI, SO})}_{\text{saddle}} & \underbrace{(\text{SO, SO})}_{\text{source}}\\ \vdots & \vdots & \vdots & \vdots\\ \underbrace{(\text{SO, SO})}_{\text{source}} & \cdots & \underbrace{(\text{SI, SO})}_{\text{saddle}} & \underbrace{(\text{SO, SO})}_{\text{source}}\end{bmatrix}_{(2n_{21})\times(2m_{11}+1)}. \tag{1.422}$$

(v_{2c2}) The $((2m_{11}+1)^{\text{th}}$ SO, $(2n_{21})^{\text{th}}$ LS)*-lower-saddle-source appearing bifurcation is from a $q_2 \times p_1$ to $\overline{q}_2 \times \overline{p}_1$ sink, source, and saddle networks as*

$$\begin{aligned}\bigcup\nolimits_{r_1=1}^{p_1}\bigcup\nolimits_{s_1=1}^{q_2}\underbrace{(a_{j_1j_1r_1}, a_{j_2j_2s_1})}_{(m_{j_1r_1},n_{j_2s_1})\text{-XX}} &\rightleftharpoons \underbrace{((2m_{11}+1)^{\text{th}}\,\text{SO},\,(2n_{21})^{\text{th}}\,\text{LS})}_{((2m_{11}+1),(2n_{21}))\text{-lower-saddle-source}}\\ &\rightleftharpoons \bigcup\nolimits_{r_2=1}^{\overline{p}_1}\bigcup\nolimits_{s_2=1}^{\overline{q}_2}\underbrace{(a_{j_1j_1r_2}, a_{j_2j_2s_2})}_{(m_{j_1r_2},n_{j_2s_2})\text{-YY}}.\end{aligned} \tag{1.423}$$

(v_{2c3}) The $((2m_{11}+1)^{\text{th}}\,\text{SO},\,(2n_{21})^{\text{th}}\,\text{LS})$*-lower-saddle-source switching bifurcation is for the switching of two networks of* $q_2 \times p_1$ *sink, source, and saddles as*

$$\bigcup_{r_1=1}^{p_1}\bigcup_{s_1=1}^{q_2}\underbrace{(a_{j_1j_1r_1},a_{j_2j_2s_1})}_{(m_{j_1r_1},n_{j_2s_1})\text{-XX}} \rightleftharpoons \underbrace{((2m_{11}+1)^{\text{th}}\,\text{SO},\,(2n_{21})^{\text{th}}\,\text{LS})}_{((2m_{11}+1),(2n_{21}))\text{-lower-saddle-source}}$$
$$\rightleftharpoons \bigcup_{r_2=1}^{p_1}\bigcup_{s_2=1}^{q_2}\underbrace{(a_{j_1j_1r_2},a_{j_2j_2s_2})}_{(m_{j_1r_2},n_{j_2s_2})\text{-YY}}. \tag{1.424}$$

(v_{2d}) For $a_{j_1j_10}(a_{j_2j_21}-a_{j_1j_21})^{n_1}<0$ *and* $a_{j_2j_20}<0$*, there is a* $((2m_{11}+1),(2n_{21}))$*-sink equilibrium at* $(x_{j_1}^*,x_{j_2}^*)=(a_{j_1j_11},a_{j_2j_21})$ *as*

$$(a_{j_1j_11},a_{j_2j_21})=\underbrace{((2m_{11}+1)^{\text{th}}\,\text{SI},\,(2n_{21})^{\text{th}}\,\text{LS})}_{((2m_{11}+1),(2n_{21}))\text{-lower-saddle-sink}}. \tag{1.425}$$

The following three $((2m_{11}+1)^{\text{th}}\,\text{SI},\,(2n_{21})^{\text{th}}\,\text{LS})$*-lower-saddle-sink appearing and switching bifurcations as follows.*

(v_{2dI}) The $((2m_{11}+1)^{\text{th}}\,\text{SI},\,(2n_{21})^{\text{th}}\,\text{LS})$*-lower-saddle-sink appearing bifurcation is from an* (SI, nF)*-negative sink to a* $(2m_{11}+1)\times(2n_{21})$*-sink, source, and saddles network as*

$$\underbrace{(a_{j_1j_11},\dot{x}_{j_2})}_{(\text{SI, nF})\text{-flow}} \rightleftharpoons \underbrace{((2m_{11}+1)^{\text{th}}\,\text{SI},\,(2n_{21})^{\text{th}}\,\text{LS})}_{((2m_{11}+1),(2n_{21}))\text{-lower-saddle-sink}}$$
$$\rightleftharpoons \bigcup_{r=1}^{2m_{11}+1}\bigcup_{s=1}^{2n_{21}}\underbrace{(a_{j_1j_1r},a_{j_2j_2s})}_{\text{XX}}, \tag{1.426}$$

where

$$\bigcup_{r=1}^{2m_{11}+1}\bigcup_{s=1}^{2n_{21}}\underbrace{(a_{j_1j_1r},a_{j_2j_2s})}_{\text{XX}}$$
$$=\begin{bmatrix}\underbrace{(\text{SI, SI})}_{\text{sink}} & \cdots & \underbrace{(\text{SO, SI})}_{\text{saddle}} & \underbrace{(\text{SI, SI})}_{\text{sink}}\\ \underbrace{(\text{SI, SO})}_{\text{saddle}} & \cdots & \underbrace{(\text{SO, SO})}_{\text{source}} & \underbrace{(\text{SI, SO})}_{\text{saddle}}\\ \vdots & \vdots & \vdots & \vdots\\ \underbrace{(\text{SI, SO})}_{\text{saddle}} & \cdots & \underbrace{(\text{SO, SO})}_{\text{source}} & \underbrace{(\text{SI, SO})}_{\text{saddle}}\end{bmatrix}_{(2n_{21})\times(2m_{11}+1)}. \tag{1.427}$$

(v_{2d2}) The $((2m_{11}+1)^{\text{th}}\,\text{SI},\,(2n_{21})^{\text{th}}\,\text{LS})$*-lower-saddle-sink appearing bifurcation is from a* $q_2 \times p_1$ *to* $\overline{q}_2 \times \overline{p}_1$ *sink, source, and saddle networks as*

$$\bigcup_{r_1=1}^{p_1}\bigcup_{s_1=1}^{q_2}\underbrace{(a_{j_1j_1r_1},a_{j_2j_2s_1})}_{(m_{j_1r_1},n_{j_2s_1})\text{-XX}} \rightleftharpoons \underbrace{((2m_{11}+1)^{\text{th}}\,\text{SI},\,(2n_{21})^{\text{th}}\,\text{LS})}_{((2m_{11}+1),(2n_{21}))\text{-lower-saddle-sink}} \rightleftharpoons \bigcup_{r_2=1}^{\overline{p}_1}\bigcup_{s_2=1}^{\overline{q}_2}\underbrace{(a_{j_1j_1r_2},a_{j_2j_2s_2})}_{(m_{j_1r_2},n_{j_2s_2})\text{-YY}}. \tag{1.428}$$

(v_{2d3}) The $((2m_{11}+1)^{\text{th}}\,\text{SI},\,(2n_{21})^{\text{th}}\,\text{LS})$*-lower-saddle-sink switching bifurcation is for the switching of two networks of* $q_2 \times p_1$ *sink, source, and saddles as*

$$\bigcup_{r_1=1}^{p_1}\bigcup_{s_1=1}^{q_2}\underbrace{(a_{j_1j_1r_1},a_{j_2j_2s_1})}_{(m_{j_1r_1},n_{j_2s_1})\text{-XX}} \rightleftharpoons \underbrace{((2m_{11}+1)^{\text{th}}\,\text{SI},\,(2n_{21})^{\text{th}}\,\text{LS})}_{((2m_{11}+1),(2n_{21}))\text{-lower-saddle-sink}} \rightleftharpoons \bigcup_{r_2=1}^{p_1}\bigcup_{s_2=1}^{q_2}\underbrace{(a_{j_1j_1r_2},a_{j_2j_2s_2})}_{(m_{j_1r_2},n_{j_2s_2})\text{-YY}}. \tag{1.429}$$

(v_3) For $m_1 = 2m_{11}$ *and* $n_2 = 2n_{21} + 1$*, the equilibrium of* $(x^*_{j_1}, x^*_{j_2}) = (a_{j_1j_1 1}, a_{j_2j_2 1})$ *has the following bifurcation properties with*

$$\begin{aligned}
&2m_{11} - \sum_{r_1=1}^{p_1} m^{(1)}_{j_1r_1} = 2\sum_{\rho_1=1}^{l_1} m_{\rho_1},\\
&2n_{21} + 1 - \sum_{s_1=1}^{q_2} n^{(1)}_{j_2s_1} = 2\sum_{\sigma_1=1}^{l_2} n_{\sigma_1};\\
&\sum_{r_2=1}^{\overline{p}_1} m^{(2)}_{j_1r_2} = 2m_{11},\ \sum_{s_2=1}^{\overline{q}_2} n^{(2)}_{j_2s_2} = 2n_{21} + 1;\\
&\sum_{r_i=1}^{p_1} m^{(i)}_{j_1r_i} = 2m_{11},\ \sum_{s_i=1}^{q_2} n^{(2)}_{j_2s_i} = 2n_{21} + 1,\ \text{for } i = 1, 2;
\end{aligned} \tag{1.430}$$

l_1, l_2-quadratic polynomials are without real roots, and the ρ_1^{th} *and* σ_1^{th}*-quadratic polynomials are with powers* m_{ρ_1} *and* n_{σ_1}*, respectively.*

(v_{3a}) For $a_{j_1j_1 0}(a_{j_2j_2 1} - a_{j_1j_2 1})^{n_1} > 0$ *and* $a_{j_2j_2 0} > 0$*, there is a* $((2m_{11}),(2n_{21} + 1))$*-upper-saddle-source equilibrium at* $(x^*_{j_1}, x^*_{j_2}) = (a_{j_1j_1 1}, a_{j_2j_2 1})$ *as*

$$(a_{j_1j_1 1}, a_{j_2j_2 1}) = \underbrace{((2m_{11})^{\text{th}}\,\text{US},\,(2n_{21}+1)^{\text{th}}\,\text{SO})}_{((2m_{11}),(2n_{21}+1))\text{-upper-saddle-source}}. \tag{1.431}$$

The following three $((2m_{11})^{\text{th}}\,\text{US}, (2n_{21}+1)^{\text{th}}\,\text{SO})$-*upper-saddle-source appearing and switching bifurcations as follows.*

(v_{3a1}) *The* $((2m_{11})^{\text{th}}\,\text{US}, (2n_{21}+1)^{\text{th}}\,\text{SO})$-*upper-saddle-source appearing bifurcation is from a* (pF, SO)-*positive source flow to a* $(2m_{11}) \times (2n_{21}+1)$-*sink, source, and saddles network as*

$$\begin{aligned}
\underbrace{(\dot{x}_{j_1}, a_{j_2j_21})}_{\text{(pF, SO)-flow}} &\rightleftharpoons \underbrace{((2m_{11})^{\text{th}}\,\text{US}, (2n_{21}+1)^{\text{th}}\,\text{SO})}_{((2m_{11}),(2n_{21}+1))\text{-upper-saddle-source}} \\
&\rightleftharpoons \bigcup\nolimits_{r=1}^{2m_{11}} \bigcup\nolimits_{s=1}^{2n_{21}+1} \underbrace{(a_{j_1j_1r}, a_{j_2j_2s})}_{\text{XX}},
\end{aligned} \tag{1.432}$$

where

$$\begin{aligned}
&\bigcup\nolimits_{r=1}^{2m_{11}} \bigcup\nolimits_{s=1}^{2n_{21}+1} \underbrace{(a_{j_1j_1r}, a_{j_2j_2s})}_{\text{XX}} \\
&= \begin{bmatrix}
\underbrace{(\text{SI, SO})}_{\text{saddle}} & \cdots & \underbrace{(\text{SI, SO})}_{\text{saddle}} & \underbrace{(\text{SO, SO})}_{\text{source}} \\
\underbrace{(\text{SI, SI})}_{\text{sink}} & \cdots & \underbrace{(\text{SI, SI})}_{\text{sink}} & \underbrace{(\text{SO, SI})}_{\text{saddle}} \\
\vdots & \vdots & \vdots & \vdots \\
\underbrace{(\text{SI, SO})}_{\text{saddle}} & \cdots & \underbrace{(\text{SI, SO})}_{\text{saddle}} & \underbrace{(\text{SO, SO})}_{\text{source}}
\end{bmatrix}_{(2n_{21}+1)\times(2m_{11})}.
\end{aligned} \tag{1.433}$$

(v_{3a2}) *The* $((2m_{11})^{\text{th}}\,\text{US}, (2n_{21}+1)^{\text{th}}\,\text{SO})$-*upper-saddle-source appearing bifurcation is from a* $q_2 \times p_1$ *to* $\overline{q}_2 \times \overline{p}_1$ *sink, source, and saddle networks as*

$$\begin{aligned}
\bigcup\nolimits_{r_1=1}^{p_1} \bigcup\nolimits_{s_1=1}^{q_2} \underbrace{(a_{j_1j_1r_1}, a_{j_2j_2s_1})}_{(m_{j_1r_1}, n_{j_2s_1})\text{-XX}} &\rightleftharpoons \underbrace{((2m_{11})^{\text{th}}\,\text{US}, (2n_{21}+1)^{\text{th}}\,\text{SO})}_{((2m_{11}),(2n_{21}+1))\text{-upper-saddle-source}} \\
&\rightleftharpoons \bigcup\nolimits_{r_2=1}^{\overline{p}_1} \bigcup\nolimits_{s_2=1}^{\overline{q}_2} \underbrace{(a_{j_1j_1r_2}, a_{j_2j_2s_2})}_{(m_{j_1r_2}, n_{j_2s_2})\text{-YY}}.
\end{aligned} \tag{1.434}$$

(v_{3a3}) *The* $((2m_{11})^{\text{th}}\,\text{US}, (2n_{21}+1)^{\text{th}}\,\text{SO})$-*upper-saddle-source switching bifurcation is for the switching of two networks of* $q_2 \times p_1$ *sink, source, and saddles as*

$$\begin{aligned}
\bigcup\nolimits_{r_1=1}^{p_1} \bigcup\nolimits_{s_1=1}^{q_2} \underbrace{(a_{j_1j_1r_1}, a_{j_2j_2s_1})}_{(m_{j_1r_1}, n_{j_2s_1})\text{-XX}} &\rightleftharpoons \underbrace{((2m_{11})^{\text{th}}\,\text{US}, (2n_{21}+1)^{\text{th}}\,\text{SO})}_{((2m_{11}),(2n_{21}+1))\text{-upper-saddle-source}} \\
&\rightleftharpoons \bigcup\nolimits_{r_2=1}^{p_1} \bigcup\nolimits_{s_2=1}^{q_2} \underbrace{(a_{j_1j_1r_2}, a_{j_2j_2s_2})}_{(m_{j_1r_2}, n_{j_2s_2})\text{-YY}}.
\end{aligned} \tag{1.435}$$

(v_{3b}) *For* $a_{j_1j_10}(a_{j_2j_21}-a_{j_1j_21})^{n_1}<0$ *and* $a_{j_2j_20}>0$, *there is a* $((2m_{11}),(2n_{21}+1))$*-lower-saddle-source equilibrium at* $(x^*_{j_1},x^*_{j_2})=(a_{j_1j_11},a_{j_2j_21})$ *as*

$$(a_{j_1j_11},a_{j_2j_21})=\underbrace{((2m_{11})^{\text{th}}\,\text{LS},\,(2n_{21}+1)^{\text{th}}\,\text{SO})}_{((2m_{11}),(2n_{21}+1))\text{-lower-saddle-source}}. \tag{1.436}$$

The following three $((2m_{11})^{\text{th}}\,\text{LS},\,(2n_{21}+1)^{\text{th}}\,\text{SO})$*-lower-saddle-source appearing and switching bifurcations as follows.*

(v_{3b1}) *The* $((2m_{11})^{\text{th}}\,\text{LS},\,(2n_{21}+1)^{\text{th}}\,\text{SO})$*-lower-saddle-source appearing bifurcation is from an* (nF, SO)*-negative source flow to a* $(2m_{11})\times(2n_{21}+1)$*-sink, source, and saddles network as*

$$\begin{aligned}\underbrace{(\dot{x}_{j_1},a_{j_2j_21})}_{\text{(nF,SO)-flow}} &\rightleftharpoons \underbrace{((2m_{11})^{\text{th}}\,\text{LS},\,(2n_{21}+1)^{\text{th}}\,\text{SO})}_{((2m_{11}),(2n_{21}+1))\text{-lower-saddle-source}}\\ &\rightleftharpoons \bigcup\nolimits_{r=1}^{2m_{11}}\bigcup\nolimits_{s=1}^{2n_{21}+1}\underbrace{(a_{j_1j_1r},a_{j_2j_2s})}_{\text{XX}},\end{aligned} \tag{1.437}$$

where

$$\begin{aligned}&\bigcup\nolimits_{r=1}^{2m_{11}}\bigcup\nolimits_{s=1}^{2n_{21}+1}\underbrace{(a_{j_1j_1r},a_{j_2j_2s})}_{\text{XX}}\\ &=\begin{bmatrix}\underbrace{(\text{SO, SO})}_{\text{source}} & \cdots & \underbrace{(\text{SO, SO})}_{\text{source}} & \underbrace{(\text{SI, SO})}_{\text{saddle}}\\ \underbrace{(\text{SO, SI})}_{\text{saddle}} & \cdots & \underbrace{(\text{SO, SI})}_{\text{saddle}} & \underbrace{(\text{SI, SI})}_{\text{sink}}\\ \vdots & \vdots & \vdots & \vdots\\ \underbrace{(\text{SO, SO})}_{\text{source}} & \cdots & \underbrace{(\text{SO, SO})}_{\text{source}} & \underbrace{(\text{SI, SO})}_{\text{saddle}}\end{bmatrix}_{(2n_{21}+1)\times(2m_{11})}.\end{aligned} \tag{1.438}$$

(v_{3b2}) *The* $((2m_{11})^{\text{th}}\,\text{LS},\,(2n_{21}+1)^{\text{th}}\,\text{SO})$*-lower-saddle-source appearing bifurcation is from a* $q_2\times p_1$ *to* $\overline{q}_2\times\overline{p}_1$ *sink, source, and saddle networks as*

$$\begin{aligned}\bigcup\nolimits_{r_1=1}^{p_1}\bigcup\nolimits_{s_1=1}^{q_2}\underbrace{(a_{j_1j_1r_1},a_{j_2j_2s_1})}_{(m_{j_1r_1},n_{j_2s_1})\text{-XX}} &\rightleftharpoons \underbrace{((2m_{11})^{\text{th}}\,\text{LS},\,(2n_{21}+1)^{\text{th}}\,\text{SO})}_{((2m_{11}),(2n_{21}+1))\text{-lower-saddle-source}}\\ &\rightleftharpoons \bigcup\nolimits_{r_2=1}^{\overline{p}_1}\bigcup\nolimits_{s_2=1}^{\overline{q}_2}\underbrace{(a_{j_1j_1r_2},a_{j_2j_2s_2})}_{(m_{j_1r_2},n_{j_2s_2})\text{-YY}}.\end{aligned} \tag{1.439}$$

(v_{3b3}) The $((2m_{11})^{\text{th}}\,\text{LS}, (2n_{21}+1)^{\text{th}}\,\text{SO})$ *-lower-saddle-source switching bifurcation is for the switching of two networks of* $q_2 \times p_1$ *sink, source, and saddles as*

$$\bigcup_{r_1=1}^{p_1}\bigcup_{s_1=1}^{q_2}\underbrace{(a_{j_1j_1r_1}, a_{j_2j_2s_1})}_{(m_{j_1r_1}, n_{j_2s_1})\text{-XX}} \rightleftharpoons \underbrace{((2m_{11})^{\text{th}}\,\text{LS}, (2n_{21}+1)^{\text{th}}\,\text{SO})}_{((2m_{11}),(2n_{21}+1))\text{-lower-saddle-source}}$$
$$\rightleftharpoons \bigcup_{r_2=1}^{p_1}\bigcup_{s_2=1}^{q_2}\underbrace{(a_{j_1j_1r_2}, a_{j_2j_2s_2})}_{(m_{j_1r_2}, n_{j_2s_2})\text{-YY}}. \tag{1.440}$$

(v_{3c}) For $a_{j_1j_10}(a_{j_2j_21} - a_{j_1j_21})^{n_1} > 0$ *and* $a_{j_2j_20} < 0$*, there is a* $((2m_{11}), (2n_{21}+1))$ *-upper-saddle-sink equilibrium at* $(x^*_{j_1}, x^*_{j_2}) = (a_{j_1j_11}, a_{j_2j_21})$ *as*

$$(a_{j_1j_11}, a_{j_2j_21}) = \underbrace{((2m_{11})^{\text{th}}\,\text{US}, (2n_{21}+1)^{\text{th}}\,\text{SI})}_{((2m_{11}),(2n_{21}+1))\text{-upper-saddle-sink}}. \tag{1.441}$$

The following three $((2m_{11})^{\text{th}}\,\text{US}, (2n_{21}+1)^{\text{th}}\,\text{SI})$*-upper-saddle-sink appearing and switching bifurcations as follows.*

(v_{3c1}) The $((2m_{11})^{\text{th}}\,\text{US}, (2n_{21}+1)^{\text{th}}\,\text{SI})$*-upper-saddle-sink appearing bifurcation is from a* (pF, SI)*-negative sink flow to a* $(2m_{11}) \times (2n_{21}+1)$*-sink, source, and saddles network as*

$$\underbrace{(\dot{x}_{j_1}, a_{j_2j_21})}_{(\text{pF, SI})\text{-flow}} \rightleftharpoons \underbrace{((2m_{11})^{\text{th}}\,\text{US}, (2n_{21}+1)^{\text{th}}\,\text{SI})}_{((2m_{11}),(2n_{21}+1))\text{-upper-saddle-sink}}$$
$$\rightleftharpoons \bigcup_{r=1}^{2m_{11}}\bigcup_{s=1}^{2n_{21}+1}\underbrace{(a_{j_1j_1r}, a_{j_2j_2s})}_{\text{XX}}, \tag{1.442}$$

where

$$\bigcup_{r=1}^{2m_{11}}\bigcup_{s=1}^{2n_{21}+1}\underbrace{(a_{j_1j_1r}, a_{j_2j_2s})}_{\text{XX}}$$
$$= \begin{bmatrix} \underbrace{(\text{SI, SI})}_{\text{sink}} & \cdots & \underbrace{(\text{SI, SI})}_{\text{sink}} & \underbrace{(\text{SO, SI})}_{\text{saddle}} \\ \underbrace{(\text{SI, SO})}_{\text{saddle}} & \cdots & \underbrace{(\text{SI, SO})}_{\text{saddle}} & \underbrace{(\text{SO, SO})}_{\text{source}} \\ \vdots & \vdots & \vdots & \vdots \\ \underbrace{(\text{SI, SI})}_{\text{sink}} & \cdots & \underbrace{(\text{SI, SI})}_{\text{sink}} & \underbrace{(\text{SO, SI})}_{\text{saddle}} \end{bmatrix}_{(2n_{21}+1)\times(2m_{11})}. \tag{1.443}$$

(v_{3c2}) *The* $((2m_{11})^{\text{th}}\,\text{US}, (2n_{21}+1)^{\text{th}}\,\text{SI})$-*upper-saddle-sink appearing bifurcation is from a* $q_2 \times p_1$ *to* $\overline{q}_2 \times \overline{p}_1$ *sink, source, and saddle networks as*

$$\begin{aligned}\bigcup\nolimits_{r_1=1}^{p_1}\bigcup\nolimits_{s_1=1}^{q_2}\underbrace{(a_{j_1j_1r_1}, a_{j_2j_2s_1})}_{(m_{j_1r_1}, n_{j_2s_1})\text{-XX}} &\rightleftharpoons \underbrace{((2m_{11})^{\text{th}}\,\text{US}, (2n_{21}+1)^{\text{th}}\,\text{SI})}_{((2m_{11}),(2n_{21}+1))\text{-upper-saddle-sink}}\\ &\rightleftharpoons \bigcup\nolimits_{r_2=1}^{\overline{p}_1}\bigcup\nolimits_{s_2=1}^{\overline{q}_2}\underbrace{(a_{j_1j_1r_2}, a_{j_2j_2s_2})}_{(m_{j_1r_2}, n_{j_2s_2})\text{-YY}}.\end{aligned} \tag{1.444}$$

(v_{3c3}) *The* $((2m_{11})^{\text{th}}\,\text{US}, (2n_{21}+1)^{\text{th}}\,\text{SI})$-*upper-saddle-sink switching bifurcation is for the switching of two networks of* $q_2 \times p_1$ *sink, source, and saddles as*

$$\begin{aligned}\bigcup\nolimits_{r_1=1}^{p_1}\bigcup\nolimits_{s_1=1}^{q_2}\underbrace{(a_{j_1j_1r_1}, a_{j_2j_2s_1})}_{(m_{j_1r_1}, n_{j_2s_1})\text{-XX}} &\rightleftharpoons \underbrace{((2m_{11})^{\text{th}}\,\text{US}, (2n_{21}+1)^{\text{th}}\,\text{SI})}_{((2m_{11}),(2n_{21}+1))\text{-upper-saddle-sink}}\\ &\rightleftharpoons \bigcup\nolimits_{r_2=1}^{p_1}\bigcup\nolimits_{s_2=1}^{q_2}\underbrace{(a_{j_1j_1r_2}, a_{j_2j_2s_2})}_{(m_{j_1r_2}, n_{j_2s_2})\text{-YY}}.\end{aligned} \tag{1.445}$$

(v_{3d}) *For* $a_{j_1j_10}(a_{j_2j_21}-a_{j_1j_21})^{n_1} < 0$ *and* $a_{j_2j_20} < 0$, *there is a* $((2m_{11}), (2n_{21}+1))$-*sink equilibrium at* $(x_{j_1}^*, x_{j_2}^*) = (a_{j_1j_11}, a_{j_2j_21})$ *as*

$$(a_{j_1j_11}, a_{j_2j_21}) = \underbrace{((2m_{11})^{\text{th}}\,\text{LS}, (2n_{21}+1)^{\text{th}}\,\text{SI})}_{((2m_{11}),(2n_{21}+1))\text{-lower-saddle-sink}}. \tag{1.446}$$

The following three $((2m_{11})^{\text{th}}\,\text{LS}, (2n_{21}+1)^{\text{th}}\,\text{SI})$-*lower-saddle-sink appearing and switching bifurcations as follows.*

(v_{3d1}) *The* $((2m_{11})^{\text{th}}\,\text{LS}, (2n_{21}+1)^{\text{th}}\,\text{SI})$-*lower-saddle-sink appearing bifurcation is from an* (nF, SI)-*negative sink flow to a*$(2m_{11}) \times (2n_{21}+1)$-*sink, source, and saddles network as*

$$\underbrace{(\dot{x}_{j_1}, a_{j_2j_21})}_{\text{(nF, SI)-flow}} \rightleftharpoons \underbrace{((2m_{11})^{\text{th}}\,\text{LS}, (2n_{21}+1)^{\text{th}}\,\text{SI})}_{((2m_{11}),(2n_{21}+1))\text{-lower-saddle-sink}} \rightleftharpoons \bigcup\nolimits_{r=1}^{2m_{11}}\bigcup\nolimits_{s=1}^{2n_{21}+1}\underbrace{(a_{j_1j_1r}, a_{j_2j_2s})}_{\text{XX}}, \tag{1.447}$$

where

$$\bigcup_{r=1}^{2m_{11}}\bigcup_{s=1}^{2n_{21}+1}\underbrace{(a_{j_1j_1r},a_{j_2j_2s})}_{\text{XX}}$$
$$=\begin{bmatrix}\underbrace{(\text{SO, SI})}_{\text{saddle}} & \cdots & \underbrace{(\text{SO, SI})}_{\text{saddle}} & \underbrace{(\text{SI, SI})}_{\text{sink}}\\ \underbrace{(\text{SO, SO})}_{\text{source}} & \cdots & \underbrace{(\text{SO, SO})}_{\text{source}} & \underbrace{(\text{SI, SO})}_{\text{saddle}}\\ \vdots & \vdots & \vdots & \vdots\\ \underbrace{(\text{SO, SI})}_{\text{saddle}} & \cdots & \underbrace{(\text{SO, SI})}_{\text{saddle}} & \underbrace{(\text{SI, SI})}_{\text{sink}}\end{bmatrix}_{(2n_{21}+1)\times(2m_{11})}. \tag{1.448}$$

(v_{3d2}) The $((2m_{11})^{\text{th}}\,\text{LS},\,(2n_{21}+1)^{\text{th}}\,\text{SI})$*-lower-saddle-sink appearing bifurcation is from a* $q_2\times p_1$ *to* $\overline{q}_2\times\overline{p}_1$ *sink, source, and saddle networks as*

$$\bigcup_{r_1=1}^{p_1}\bigcup_{s_1=1}^{q_2}\underbrace{(a_{j_1j_1r_1},a_{j_2j_2s_1})}_{(m_{j_1r_1},n_{j_2s_1})\text{-XX}}\rightleftharpoons\underbrace{((2m_{11})^{\text{th}}\,\text{LS},\,(2n_{21}+1)^{\text{th}}\,\text{SI})}_{((2m_{11}),(2n_{21}+1))\text{-lower-saddle-sink}}$$
$$\rightleftharpoons\bigcup_{r_2=1}^{\overline{p}_1}\bigcup_{s_2=1}^{\overline{q}_2}\underbrace{(a_{j_1j_1r_2},a_{j_2j_2s_2})}_{(m_{j_1r_2},n_{j_2s_2})\text{-YY}}. \tag{1.449}$$

(v_{3d3}) The $((2m_{11})^{\text{th}}\,\text{LS},\,(2n_{21}+1)^{\text{th}}\,\text{SI})$*-lower-saddle-sink switching bifurcation is for the switching of two networks of* $q_2\times p_1$ *sink, source, and saddles as*

$$\bigcup_{r_1=1}^{p_1}\bigcup_{s_1=1}^{q_2}\underbrace{(a_{j_1j_1r_1},a_{j_2j_2s_1})}_{(m_{j_1r_1},n_{j_2s_1})\text{-XX}}\rightleftharpoons\underbrace{((2m_{11})^{\text{th}}\,\text{LS},\,(2n_{21}+1)^{\text{th}}\,\text{SI})}_{((2m_{11}),(2n_{21}+1))\text{-lower-saddle-sink}}$$
$$\rightleftharpoons\bigcup_{r_2=1}^{p_1}\bigcup_{s_2=1}^{q_2}\underbrace{(a_{j_1j_1r_2},a_{j_2j_2s_2})}_{(m_{j_1r_2},n_{j_2s_2})\text{-YY}}. \tag{1.450}$$

(v_4) For $m_1=2m_{11}$ *and* $n_2=2n_{21}$*, the equilibrium of* $(x_{j_1}^*,x_{j_2}^*)=(a_{j_1j_11},a_{j_2j_21})$ *has the following bifurcation properties with*

$$2m_{11}-\sum_{r_1=1}^{p_1}m_{j_1r_1}^{(1)}=2\sum_{\rho_1=1}^{l_1}m_{\rho_1},$$
$$2n_{21}-\sum_{s_1=1}^{q_2}n_{j_2s_1}^{(1)}=2\sum_{\sigma_1=1}^{l_2}n_{\sigma_1};$$
$$\sum_{r_2=1}^{\overline{p}_1}m_{j_1r_2}^{(2)}=2m_{11},\ \sum_{s_2=1}^{\overline{q}_2}n_{j_2s_2}^{(2)}=2n_{21};$$

$$\sum_{r_i=1}^{p_1} m_{j_1 r_i}^{(i)} = 2m_{11}, \sum_{s_i=1}^{q_2} n_{j_2 s_i}^{(2)} = 2n_{21}, \text{ for } i = 1, 2; \tag{1.451}$$

l_1, l_2-quadratic polynomials are without real roots, and the ρ_1^{th} and σ_1^{th}-quadratic polynomials are with powers m_{ρ_1} and n_{σ_1}, respectively.

(v_{4a}) For $a_{j_1 j_1 0}(a_{j_2 j_2 1} - a_{j_1 j_2 1})^{n_1} > 0$ and $a_{j_2 j_2 0} > 0$, there is a $((2m_{11}), (2n_{21}))$-double-saddle equilibrium at $(x_{j_1}^*, x_{j_2}^*) = (a_{j_1 j_1 1}, a_{j_2 j_2 1})$ as

$$(a_{j_1 j_1 1}, a_{j_2 j_2 1}) = \underbrace{((2m_{11})^{\text{th}}\,\text{US}, (2n_{21})^{\text{th}}\,\text{US})}_{((2m_{11}),(2n_{21}))\text{-double-saddle}}. \tag{1.452}$$

The following three $((2m_{11})^{\text{th}}\,\text{US}, (2n_{21})^{\text{th}}\,\text{US})$*-double-saddle appearing and switching bifurcations as follows.*

(v_{4a1}) The $((2m_{11})^{\text{th}}\,\text{US}, (2n_{21})^{\text{th}}\,\text{US})$*-double-saddle appearing bifurcation is from a* (pF, pF)*-flow to a $(2m_{11}) \times (2n_{21})$-sink, source, and saddles network as*

$$\begin{aligned}\underbrace{(\dot{x}_{j_1}, \dot{x}_{j_2})}_{(\text{pF, pF})\text{-flow}} &\rightleftharpoons \underbrace{((2m_{11})^{\text{th}}\,\text{US}, (2n_{21})^{\text{th}}\,\text{US})}_{((2m_{11}),(2n_{21}))\text{-double-saddle}} \\ &\rightleftharpoons \bigcup\nolimits_{r=1}^{2m_{11}} \bigcup\nolimits_{s=1}^{2n_{21}} \underbrace{(a_{j_1 j_1 r}, a_{j_2 j_2 s})}_{\text{XX}},\end{aligned} \tag{1.453}$$

where

$$\begin{aligned}&\bigcup\nolimits_{r=1}^{2m_{11}} \bigcup\nolimits_{s=1}^{2n_{21}+1} \underbrace{(a_{j_1 j_1 r}, a_{j_2 j_2 s})}_{\text{XX}} \\ &= \begin{bmatrix} \underbrace{(\text{SI, SO})}_{\text{saddle}} & \cdots & \underbrace{(\text{SI, SO})}_{\text{saddle}} & \underbrace{(\text{SO, SO})}_{\text{source}} \\ \underbrace{(\text{SI, SI})}_{\text{sink}} & \cdots & \underbrace{(\text{SI, SI})}_{\text{sink}} & \underbrace{(\text{SO, SI})}_{\text{saddle}} \\ \vdots & \vdots & \vdots & \vdots \\ \underbrace{(\text{SI, SI})}_{\text{sink}} & \cdots & \underbrace{(\text{SI, SI})}_{\text{sink}} & \underbrace{(\text{SO, SI})}_{\text{saddle}} \end{bmatrix}_{(2n_{21})\times(2m_{11})}.\end{aligned} \tag{1.454}$$

(v_{4a2}) The $((2m_{11})^{\text{th}}\,\text{US}, (2n_{21})^{\text{th}}\,\text{US})$*-double-saddle appearing bifurcation is from a $q_2 \times p_1$ to $\overline{q}_2 \times \overline{p}_1$ sink, source, and saddle networks as*

$$\begin{aligned}\bigcup\nolimits_{r_1=1}^{p_1} \bigcup\nolimits_{s_1=1}^{q_2} \underbrace{(a_{j_1 j_1 r_1}, a_{j_2 j_2 s_1})}_{(m_{j_1 r_1}, n_{j_2 s_1})\text{-XX}} &\rightleftharpoons \underbrace{((2m_{11})^{\text{th}}\,\text{US}, (2n_{21})^{\text{th}}\,\text{US})}_{((2m_{11}),(2n_{21}))\text{-double-saddle}} \\ &\rightleftharpoons \bigcup\nolimits_{r_2=1}^{\overline{p}_1} \bigcup\nolimits_{s_2=1}^{\overline{q}_2} \underbrace{(a_{j_1 j_1 r_2}, a_{j_2 j_2 s_2})}_{(m_{j_1 r_2}, n_{j_2 s_2})\text{-YY}}.\end{aligned} \tag{1.455}$$

$(v_{4a3}$) The $((2m_{11})^{\text{th}}\,\text{US}, (2n_{21})^{\text{th}}\,\text{US})$*-double-saddle switching bifurcation is for the switching of two networks of* $q_2 \times p_1$ *sink, source, and saddles as*

$$\bigcup_{r_1=1}^{p_1}\bigcup_{s_1=1}^{q_2}\underbrace{(a_{j_1j_1r_1}, a_{j_2j_2s_1})}_{(m_{j_1r_1}, n_{j_2s_1})\text{-XX}} \rightleftharpoons \underbrace{((2m_{11})^{\text{th}}\,\text{US}, (2n_{21})^{\text{th}}\,\text{US})}_{((2m_{11}),(2n_{21}))\text{-double-saddle}}$$
$$\rightleftharpoons \bigcup_{r_2=1}^{p_1}\bigcup_{s_2=1}^{q_2}\underbrace{(a_{j_1j_1r_2}, a_{j_2j_2s_2})}_{(m_{j_1r_2}, n_{j_2s_2})\text{-YY}}. \tag{1.456}$$

(v_{4b}) For $a_{j_1j_10}(a_{j_2j_21} - a_{j_1j_21})^{n_1} < 0$ *and* $a_{j_2j_20} > 0$*, there is a* $((2m_{11}), (2n_{21}))$*-double-saddle equilibrium at* $(x_{j_1}^*, x_{j_2}^*) = (a_{j_1j_11}, a_{j_2j_21})$ *as*

$$(a_{j_1j_11}, a_{j_2j_21}) = \underbrace{((2m_{11})^{\text{th}}\,\text{LS}, (2n_{21})^{\text{th}}\,\text{US})}_{((2m_{11}),(2n_{21}))\text{-double-saddle}}. \tag{1.457}$$

The following three $((2m_{11})^{\text{th}}\,\text{LS}, (2n_{21})^{\text{th}}\,\text{US})$*-double-saddle appearing and switching bifurcations as follows.*

$(v_{4b1}$) The $((2m_{11})^{\text{th}}\,\text{LS}, (2n_{21})^{\text{th}}\,\text{US})$*-double-saddle appearing bifurcation is from an* (nF, pF)*-flow to a* $(2m_{11}) \times (2n_{21})$*-sink, source, and saddles network as*

$$\underbrace{(\dot{x}_{j_1}, \dot{x}_{j_2})}_{\text{(nF, pF)-flow}} \rightleftharpoons \underbrace{((2m_{11})^{\text{th}}\,\text{LS}, (2n_{21})^{\text{th}}\,\text{US})}_{((2m_{11}),(2n_{21}))\text{-double-saddle}}$$
$$\rightleftharpoons \bigcup_{r=1}^{2m_{11}}\bigcup_{s=1}^{2n_{21}}\underbrace{(a_{j_1j_1r}, a_{j_2j_2s})}_{\text{XX}}, \tag{1.458}$$

where

$$\bigcup_{r=1}^{2m_{11}}\bigcup_{s=1}^{2n_{21}}\underbrace{(a_{j_1j_1r}, a_{j_2j_2s})}_{\text{XX}}$$
$$= \begin{bmatrix} \underbrace{(\text{SO, SO})}_{\text{source}} & \cdots & \underbrace{(\text{SO, SO})}_{\text{source}} & \underbrace{(\text{SI, SO})}_{\text{saddle}} \\ \underbrace{(\text{SO, SI})}_{\text{saddle}} & \cdots & \underbrace{(\text{SO, SI})}_{\text{saddle}} & \underbrace{(\text{SI, SI})}_{\text{sink}} \\ \vdots & \vdots & \vdots & \vdots \\ \underbrace{(\text{SO, SI})}_{\text{saddle}} & \cdots & \underbrace{(\text{SO, SI})}_{\text{saddle}} & \underbrace{(\text{SI, SI})}_{\text{sink}} \end{bmatrix}_{(2n_{21})\times(2m_{11})}. \tag{1.459}$$

(v_{4b2}) The $((2m_{11})^{\text{th}}\,\text{LS}, (2n_{21})^{\text{th}}\,\text{US})$*-double-saddle appearing bifurcation is from a* $q_2 \times p_1$ *to* $\overline{q}_2 \times \overline{p}_1$ *sink, source, and saddle networks as*

$$\bigcup_{r_1=1}^{p_1}\bigcup_{s_1=1}^{q_2}\underbrace{(a_{j_1j_1r_1},a_{j_2j_2s_1})}_{(m_{j_1r_1},n_{j_2s_1})\text{-XX}} \rightleftharpoons \underbrace{((2m_{11})^{\text{th}}\,\text{LS},(2n_{21})^{\text{th}}\,\text{US})}_{((2m_{11}),(2n_{21}))\text{-double-saddle}}$$
$$\rightleftharpoons \bigcup_{r_2=1}^{\bar{p}_1}\bigcup_{s_2=1}^{\bar{q}_2}\underbrace{(a_{j_1j_1r_2},a_{j_2j_2s_2})}_{(m_{j_1r_2},n_{j_2s_2})\text{-YY}}. \tag{1.460}$$

(v_{4b3}) The $((2m_{11})^{\text{th}}\,\text{LS},(2n_{21})^{\text{th}}\,\text{US})$*-double-saddleswitching bifurcation is for the switching of two networks of* $q_2\times p_1$ *sink, source, and saddles as*

$$\bigcup_{r_1=1}^{p_1}\bigcup_{s_1=1}^{q_2}\underbrace{(a_{j_1j_1r_1},a_{j_2j_2s_1})}_{(m_{j_1r_1},n_{j_2s_1})\text{-XX}} \rightleftharpoons \underbrace{((2m_{11})^{\text{th}}\,\text{LS},(2n_{21})^{\text{th}}\,\text{US})}_{((2m_{11}),(2n_{21}))\text{-double-saddle}}$$
$$\rightleftharpoons \bigcup_{r_2=1}^{p_1}\bigcup_{s_2=1}^{q_2}\underbrace{(a_{j_1j_1r_2},a_{j_2j_2s_2})}_{(m_{j_1r_2},n_{j_2s_2})\text{-YY}}. \tag{1.461}$$

(v_{4c}) For $a_{j_1j_1 0}(a_{j_2j_2 1}-a_{j_1j_2 1})^{n_1}>0$ *and* $a_{j_2j_2 0}<0$*, there is a* $((2m_{11}),(2n_{21}))$ *-double-saddle equilibrium at* $(x_{j_1}^*,x_{j_2}^*)=(a_{j_1j_1 1},a_{j_2j_2 1})$ *as*

$$(a_{j_1j_1 1},a_{j_2j_2 1})=\underbrace{((2m_{11})^{\text{th}}\,\text{LS},(2n_{21})^{\text{th}}\,\text{US})}_{((2m_{11}),(2n_{21}))\text{-double-saddle}}. \tag{1.462}$$

The following three $((2m_{11})^{\text{th}}\,\text{US},(2n_{21})^{\text{th}}\,\text{LS})$*-double-saddle appearing and switching bifurcations as follows.*

(v_{4c1}) The $((2m_{11})^{\text{th}}\,\text{US},(2n_{21})^{\text{th}}\,\text{LS})$*-double-saddle appearing bifurcation is from a* (pF, nF)*-flow to a* $(2m_{11})\times(2n_{21})$*-sink, source, and saddles network as*

$$\underbrace{(\dot{x}_{j_1},\dot{x}_{j_2})}_{(\text{pF,nF})\text{-flow}} \rightleftharpoons \underbrace{((2m_{11})^{\text{th}}\,\text{LS},(2n_{21})^{\text{th}}\,\text{US})}_{((2m_{11}),(2n_{21}))\text{-double-saddle}}$$
$$\rightleftharpoons \bigcup_{r=1}^{2m_{11}}\bigcup_{s=1}^{2n_{21}}\underbrace{(a_{j_1j_1r},a_{j_2j_2s})}_{\text{XX}}, \tag{1.463}$$

where

$$\bigcup_{r=1}^{2m_{11}}\bigcup_{s=1}^{2n_{21}}\underbrace{(a_{j_1j_1r},a_{j_2j_2s})}_{\text{XX}}$$
$$=\begin{bmatrix}\underbrace{(\text{SI, SI})}_{\text{sink}} & \cdots & \underbrace{(\text{SI, SI})}_{\text{sink}} & \underbrace{(\text{SO, SI})}_{\text{saddle}}\\ \underbrace{(\text{SI, SO})}_{\text{saddle}} & \cdots & \underbrace{(\text{SI, SO})}_{\text{saddle}} & \underbrace{(\text{SO, SO})}_{\text{source}}\\ \vdots & \vdots & \vdots & \vdots\\ \underbrace{(\text{SI, SO})}_{\text{saddle}} & \cdots & \underbrace{(\text{SI, SO})}_{\text{saddle}} & \underbrace{(\text{SO, SO})}_{\text{source}}\end{bmatrix}_{(2n_{21}+1)\times(2m_{11})}. \tag{1.464}$$

(v_{4c2}) The $((2m_{11})^{\text{th}}\,\text{US}, (2n_{21})^{\text{th}}\,\text{LS})$*-double-saddle appearing bifurcation is from a* $q_2 \times p_1$ *to* $\overline{q}_2 \times \overline{p}_1$ *sink, source, and saddle networks as*

$$\bigcup_{r_1=1}^{p_1}\bigcup_{s_1=1}^{q_2}\underbrace{(a_{j_1j_1r_1}, a_{j_2j_2s_1})}_{(m_{j_1r_1}, n_{j_2s_1})\text{-XX}} \rightleftharpoons \underbrace{((2m_{11})^{\text{th}}\,\text{LS}, (2n_{21})^{\text{th}}\,\text{US})}_{((2m_{11}),(2n_{21}))\text{-double-saddle}}$$
$$\rightleftharpoons \bigcup_{r_2=1}^{\overline{p}_1}\bigcup_{s_2=1}^{\overline{q}_2}\underbrace{(a_{j_1j_1r_2}, a_{j_2j_2s_2})}_{(m_{j_1r_2}, n_{j_2s_2})\text{-YY}}. \tag{1.465}$$

(v_{4c3}) The $((2m_{11})^{\text{th}}\,\text{US}, (2n_{21})^{\text{th}}\,\text{LS})$*-double-saddle switching bifurcation is for the switching of two networks of* $q_2 \times p_1$ *sink, source, and saddles as*

$$\bigcup_{r_1=1}^{p_1}\bigcup_{s_1=1}^{q_2}\underbrace{(a_{j_1j_1r_1}, a_{j_2j_2s_1})}_{(m_{j_1r_1}, n_{j_2s_1})\text{-XX}} \rightleftharpoons \underbrace{((2m_{11})^{\text{th}}\,\text{LS}, (2n_{21})^{\text{th}}\,\text{US})}_{((2m_{11}),(2n_{21}))\text{-double-saddle}}$$
$$\rightleftharpoons \bigcup_{r_2=1}^{p_1}\bigcup_{s_2=1}^{q_2}\underbrace{(a_{j_1j_1r_2}, a_{j_2j_2s_2})}_{(m_{j_1r_2}, n_{j_2s_2})\text{-YY}}. \tag{1.466}$$

(v_{4d}) For $a_{j_1j_10}(a_{j_2j_21} - a_{j_1j_21})^{n_1} < 0$ *and* $a_{j_2j_20} < 0$*, there is a* $((2m_{11}), (2n_{21}))$*-double-saddle equilibrium at* $(x^*_{j_1}, x^*_{j_2}) = (a_{j_1j_11}, a_{j_2j_21})$ *as*

$$(a_{j_1j_11}, a_{j_2j_21}) = \underbrace{((2m_{11})^{\text{th}}\,\text{LS}, (2n_{21})^{\text{th}}\,\text{LS})}_{((2m_{11}),(2n_{21}+1))\text{-double-saddle}}. \tag{1.467}$$

The following three $((2m_{11})^{\text{th}}\,\text{LS}, (2n_{21})^{\text{th}}\,\text{LS})$*-double-saddle appearing and switching bifurcations as follows.*

(v_{4d1}) The $((2m_{11})^{\text{th}}\,\text{LS}, (2n_{21})^{\text{th}}\,\text{LS})$*-double-saddle appearing bifurcation is from an* (nF, nF)*-flow to a* $(2m_{11}) \times (2n_{21})$*-sink, source, and saddles network as*

$$\underbrace{(\dot{x}_{j_1}, \dot{x}_{j_2})}_{(\text{nF, nF})\text{-flow}} \rightleftharpoons \underbrace{((2m_{11})^{\text{th}}\,\text{LS}, (2n_{21})^{\text{th}}\,\text{LS})}_{((2m_{11}),(2n_{21}))\text{-double-saddle}}$$
$$\rightleftharpoons \bigcup_{r=1}^{2m_{11}}\bigcup_{s=1}^{2n_{21}}\underbrace{(a_{j_1j_1r}, a_{j_2j_2s})}_{\text{XX}}, \tag{1.468}$$

where

$$\bigcup_{r=1}^{2m_{11}} \bigcup_{s=1}^{2n_{21}} \underbrace{(a_{j_1j_1r}, a_{j_2j_2s})}_{\text{XX}}$$
$$= \begin{bmatrix} \underbrace{(\text{SO, SI})}_{\text{saddle}} & \cdots & \underbrace{(\text{SO, SI})}_{\text{saddle}} & \underbrace{(\text{SI, SI})}_{\text{sink}} \\ \underbrace{(\text{SO, SO})}_{\text{source}} & \cdots & \underbrace{(\text{SO, SO})}_{\text{source}} & \underbrace{(\text{SI, SO})}_{\text{saddle}} \\ \vdots & \vdots & \vdots & \vdots \\ \underbrace{(\text{SO, SO})}_{\text{source}} & \cdots & \underbrace{(\text{SO, SO})}_{\text{source}} & \underbrace{(\text{SI, SO})}_{\text{saddle}} \end{bmatrix}_{(2n_{21})\times(2m_{11})} . \tag{1.469}$$

(v_{4d2}) The $((2m_{11})^{\text{th}}$ LS, $(2n_{21})^{\text{th}}$ LS)*-double-saddle appearing bifurcation is from a* $q_2 \times p_1$ *to* $\overline{q}_2 \times \overline{p}_1$ *sink, source, and saddle networks as*

$$\bigcup_{r_1=1}^{p_1} \bigcup_{s_1=1}^{q_2} \underbrace{(a_{j_1j_1r_1}, a_{j_2j_2s_1})}_{(m_{j_1r_1}, n_{j_2s_1})\text{-XX}} \rightleftharpoons \underbrace{((2m_{11})^{\text{th}}\,\text{LS}, (2n_{21})^{\text{th}}\,\text{LS})}_{((2m_{11}),(2n_{21}))\text{-double-saddle}}$$
$$\rightleftharpoons \bigcup_{r_2=1}^{\overline{p}_1} \bigcup_{s_2=1}^{\overline{q}_2} \underbrace{(a_{j_1j_1r_2}, a_{j_2j_2s_2})}_{(m_{j_1r_2}, n_{j_2s_2})\text{-YY}}. \tag{1.470}$$

(v_{4d3}) The $((2m_{11})^{\text{th}}$ LS, $(2n_{21})^{\text{th}}$ LS)*-double-saddle switching bifurcation is for the switching of two networks of* $q_2 \times p_1$ *sink, source, and saddles as*

$$\bigcup_{r_1=1}^{p_1} \bigcup_{s_1=1}^{q_2} \underbrace{(a_{j_1j_1r_1}, a_{j_2j_2s_1})}_{(m_{j_1r_1}, n_{j_2s_1})\text{-XX}} \rightleftharpoons \underbrace{((2m_{11})^{\text{th}}\,\text{LS}, (2n_{21})^{\text{th}}\,\text{LS})}_{((2m_{11}),(2n_{21}))\text{-double-saddle}}$$
$$\rightleftharpoons \bigcup_{r_2=1}^{p_1} \bigcup_{s_2=1}^{q_2} \underbrace{(a_{j_1j_1r_2}, a_{j_2j_2s_2})}_{(m_{j_1r_2}, n_{j_2s_2})\text{-YY}}. \tag{1.471}$$

(vi) For $a_{j_2j_21} = a_{j_1j_21}$*, there are 64 types of singular infinite-equilibriums at* $x_{j_2}^* = a_{j_2j_21} = a_{j_1j_21}$*. For* $n_1 \geq n_2$*, 32 types of singular infinite-equilibriums are*

$$(a_{j_1j_11}, a_{j_2j_21}) = \underbrace{([n_1 - n_2]^{\text{th}}\,\text{XX}, [m_1, n_2]^{\text{th}}\,\text{YY})}_{([n_1-n_2],[m_1,n_2])\text{-XXYY}}, \tag{1.472}$$

where

$$\begin{aligned}
&\mathrm{XX} \in \{\mathrm{DI{:}II, II{:}DI, II{:}DI, DI{:}II}\},\\
&\mathrm{YY} \in \{\mathrm{SO, SO, SI, SI}\};\\
&\mathrm{XX} \in \{\mathrm{DP{:}UP, UP{:}DP, UP{:}DP, DP{:}UP}\},\\
&\mathrm{YY} \in \{\mathrm{SO, SO, SI, SI}\};\\
&\mathrm{XX} \in \{\mathrm{DP{:}UP, UP{:}DP, UP{:}DP, DP{:}UP}\},\\
&\mathrm{YY} \in \{\mathrm{US, US, LS, LS}\};\\
&\mathrm{XX} \in \{\mathrm{DI{:}II, II{:}DI, II{:}DI, DI{:}II}\},\\
&\mathrm{YY} \in \{\mathrm{US, US, LS, LS}\};\\
&\mathrm{XX} \in \{\mathrm{II{:}II, DI{:}DI, DI{:}DI, II{:}II}\},\\
&\mathrm{YY} \in \{\mathrm{SO, SO, SI, SI}\};\\
&\mathrm{XX} \in \{\mathrm{UP{:}UP, DP{:}DP, DP{:}DP, UP{:}UP}\},\\
&\mathrm{YY} \in \{\mathrm{SO, SO, SI, SI}\};\\
&\mathrm{XX} \in \{\mathrm{UP{:}UP, DP{:}DP, DP{:}DP, UP{:}UP}\},\\
&\mathrm{YY} \in \{\mathrm{US, US, LS, LS}\};\\
&\mathrm{XX} \in \{\mathrm{II{:}II, DI{:}DI, DI{:}DI, II{:}II}\},\\
&\mathrm{YY} \in \{\mathrm{US, US, LS, LS}\}.
\end{aligned} \tag{1.473}$$

For $n_1 < n_2$, 32 types of singular infinite-equilibriums are

$$(a_{j_1 j_1 1}, a_{j_2 j_2 1}) = \underbrace{([n_2 - n_1]^{\mathrm{th}}\,\mathrm{ZZ}, [m_1, n_2]^{\mathrm{th}}\,\mathrm{YY})}_{([n_2-n_1],[m_1,n_2])\text{-ZZYY}}, \tag{1.474}$$

where

$$\begin{aligned}
&\mathrm{ZZ} \in \{\mathrm{DC{:}\ IC, IC{:}\ DC, IC{:}\ DC, DC{:}\ IC,}\}\\
&\mathrm{YY} \in \{\mathrm{SO, SO, SI, SI;}\}\\
&\mathrm{XX} \in \{\mathrm{UA{:}DA, DA{:}UA, DA{:}UA, UA{:}DA,}\}\\
&\mathrm{YY} \in \{\mathrm{SO, SO, SI, SI;}\}\\
&\mathrm{XX} \in \{\mathrm{UA{:}DA, DA{:}UA, DA{:}UA, UA{:}DA,}\}\\
&\mathrm{YY} \in \{\mathrm{US, US, LS, LS;}\}\\
&\mathrm{ZZ} \in \{\mathrm{DC{:}\ IC, IC{:}\ DC, IC{:}\ DC, DC{:}\ IC,}\}\\
&\mathrm{YY} \in \{\mathrm{US, US, LS, LS;}\}\\
&\mathrm{ZZ} \in \{\mathrm{IC{:}IC, DC{:}DC, DC{:}DC, IC{:}IC,}\}\\
&\mathrm{YY} \in \{\mathrm{SO, SO, SI, SI;}\}\\
&\mathrm{XX} \in \{\mathrm{DA{:}DA, UA{:}UA, UA{:}UA, DA{:}DA,}\}\\
&\mathrm{YY} \in \{\mathrm{SO, SO, SI, SI;}\}\\
&\mathrm{XX} \in \{\mathrm{DA{:}DA, UA{:}UA, UA{:}UA, DA{:}DA,}\}\\
&\mathrm{YY} \in \{\mathrm{US, US, LS, LS;}\}\\
&\mathrm{ZZ} \in \{\mathrm{IC{:}IC, DC{:}DC, DC{:}DC, IC{:}IC,}\}\\
&\mathrm{YY} \in \{\mathrm{US, US, LS, LS}\}.
\end{aligned} \tag{1.475}$$

(vi_1) For $n_2 < n_1$, the singularity in the x_{j_1}-direction more than singularity in the x_{j_2}-direction, and the following three ($[n_1 - n_2]^{\text{th}}$ XX, $[m_1, n_2]^{\text{th}}$ YY)-XXYY *switching bifurcations are given as follows.*

(vi_{1a}) The ($[n_1 - n_2]^{\text{th}}$ XX, $[m_1, n_2]^{\text{th}}$ YY)-XXYY *switching bifurcation is from a singular equilibrium with a singular 1-dimensional flow switching to a singular 1-dimensional flow with a singular equilibrium.*

$$\left\{\begin{array}{c}\underbrace{(a_{j_1j_1}1, a_{j_2j_2}1)}_{(m_1,n_2)-\text{YY}_1}\\ \underbrace{(a_{j_1j_1}1, a_{j_1j_2}1)}_{((m_1,n_1)^{\text{th}}\text{XX}_1,\text{xF})}\end{array}\right\} \rightleftharpoons \underbrace{([n_1 - n_2]^{\text{th}}\,\text{XX}, [m_1, n_2]^{\text{th}}\,\text{YY})}_{([n_1-n_2],[m_1,n_2])\text{-XXYY}} \rightleftharpoons \left\{\begin{array}{c}\underbrace{(a_{j_1j_1}1, a_{j_1j_2}1)}_{((m_1,n_1)^{\text{th}}\text{XX}_2,\text{xF})}\\ \underbrace{(a_{j_1j_1}1, a_{j_2j_2}1)}_{(m_1,n_2)\text{-YY}_2}\end{array}\right\} \tag{1.476}$$

with

$$\begin{array}{l}\text{XX}_1, \text{XX}_2 \in \{\text{DP:UP, UP:DP; UP:UP, DP:DP; }_{\text{II:DI}}\text{SO},_{\text{DI:II}}\text{SO},\\ {}_{\text{II:DI}}\text{SI},_{\text{DI:II}}\text{SI; }_{\text{II:II}}\text{US},_{\text{II:II}}\text{LS},_{\text{DI:DI}}\text{US},_{\text{DI:DI}}\text{LS;}\}\\ \text{xF}_1, \text{xF}_2 \in \{\text{pF, nF;}\}\\ \text{YY}_1,\text{YY}_2 \in \{\text{sink, source, saddles; upper (lower)-saddle-source,}\\ \text{upper (lower)-saddle-sink; double-saddles}\}.\end{array} \tag{1.477}$$

(vi_{1b}) The ($[n_1 - n_2]^{\text{th}}$ XX, $[m_1, n_2]^{\text{th}}$ YY)-XXYY *switching bifurcation is for two $m_1 \times (n_1 + n_2)$ networks of non-singular equilibrium and hyperbolic and hyperbolic-secant flows.*

$$\begin{array}{l}\bigcup\nolimits_{r=1}^{m_1}\bigcup\nolimits_{u_1=1}^{n_1}\underbrace{(a_{j_1j_1r}, a_{j_1j_2u_1})}_{(\text{XX}_1,\text{xF}_1)} \oplus \bigcup\nolimits_{r=1}^{m_1}\bigcup\nolimits_{s_1=1}^{n_2}\underbrace{(a_{j_1j_1r}, a_{j_2j_2s_1})}_{\text{YY}_1}\\ \rightleftharpoons \underbrace{([n_1 - n_2]^{\text{th}}\,\text{XX}, [m_1, n_2]^{\text{th}}\,\text{YY})}_{([n_1-n_2],[m_1,n_2])\text{-XXYY}}\\ \rightleftharpoons \bigcup\nolimits_{r=1}^{m_1}\bigcup\nolimits_{u_2=1}^{n_1}\underbrace{(a_{j_1j_1r}, a_{j_1j_2u_1})}_{(\text{XX}_2,\text{xF}_2)} \oplus \bigcup\nolimits_{r=1}^{m_1}\bigcup\nolimits_{s_2=1}^{n_1}\underbrace{(a_{j_1j_1r}, a_{j_2j_2s_2})}_{\text{YY}_2},\end{array} \tag{1.478}$$

with

$$\begin{array}{l}\text{YY}_1, \text{YY}_2 \in \{\text{sink, source, saddle}\};\\ \text{xF}_1, \text{xF}_2 \in \{\text{pF, nF}\};\\ \text{XX}_1, \text{XX}_2 \in \{\text{hyperbolic flow, hypersecant flow}\}.\end{array} \tag{1.479}$$

(vi_{1c}) The $([n_1-n_2]^{\text{th}}\,\text{XX}, [m_1,n_2]^{\text{th}}\,\text{YY})$-XXYY *switching bifurcation is for two* $p_1 \times (q_1+q_2)$ *networks of singular equilibriums and singular 1-dimensional flows.*

$$\begin{aligned}
&\bigcup\nolimits_{r=1}^{p_1}\bigcup\nolimits_{u_1=1}^{q_1}\underbrace{(a_{j_1j_1r}, a_{j_1j_2u_1})}_{([m_{j_1r}, n_{j_1u_1}]\text{XX}_1,\text{xF}_1)} \oplus \bigcup\nolimits_{r=1}^{p_1}\bigcup\nolimits_{s_1=1}^{q_2}\underbrace{(a_{j_1j_1r}, a_{j_2j_2s_1})}_{(m_{j_1r}, n_{j_2s_1})\text{-YY}_1}\\
&\rightleftharpoons \underbrace{([n_1-n_2]^{\text{th}}\,\text{XX}, [m_1,n_2]^{\text{th}}\,\text{YY})}_{([n_1-n_2],[m_1,n_2])\text{-XXYY}}\\
&\rightleftharpoons \bigcup\nolimits_{r=1}^{p_1}\bigcup\nolimits_{u_2=1}^{q_1}\underbrace{(a_{j_1j_1r}, a_{j_1j_2u_2})}_{([m_{j_1r}, n_{j_1u_2}]\text{XX}_1,\text{xF}_1)} \oplus \bigcup\nolimits_{r=1}^{p_1}\bigcup\nolimits_{s_2=1}^{q_2}\underbrace{(a_{j_1j_1r}, a_{j_2j_2s_2})}_{(m_{j_1r}, n_{j_2s_2})\text{-YY}_1}
\end{aligned} \tag{1.480}$$

with

$$\begin{aligned}
&\text{XX}_1, \text{XX}_2 \in \{\text{DP:UP, UP:DP; UP:UP, DP:DP;}_{\text{II:DI}}\text{SO}, {}_{\text{DI:II}}\text{SO},\\
&{}_{\text{II:DI}}\text{SI},{}_{\text{DI:II}}\text{SI}; {}_{\text{II:II}}\text{US},{}_{\text{II:II}}\text{LS},{}_{\text{DI:DI}}\text{US},{}_{\text{DI:DI}}\text{LS}\};\\
&\text{xF}_1, \text{xF}_2 \in \{\text{pF, nF}\};\\
&\text{YY}_1, \text{YY}_2 \in \{\text{sink, source, saddles; upper (lower)-saddle-source,}\\
&\ \text{upper (lower)-saddle-sink; double-saddles}\};
\end{aligned} \tag{1.481}$$

and

$$m_1 = \sum_{r_i=1}^{p_1} m_{j_1r_i},\ n_1 = \sum_{u_i=1}^{q_1} n_{j_1u_i},\ n_2 = \sum_{s_i=1}^{q_1} n_{j_2s_i};\ i=1,2. \tag{1.482}$$

(vi_2) For $n_1 < n_2$, *the singularity in the* x_{j_2}*-direction more than singularity in* x_{j_1}*-direction, and the following three* $([n_2-n_1]^{\text{th}}\,\text{ZZ}, [m_1,n_2]^{\text{th}}\,\text{YY})$-ZZYY *switching bifurcations are given as follows.*

(vi_{2a}) The $([n_2-n_1]^{\text{th}}\,\text{ZZ}, [m_1,n_2]^{\text{th}}\,\text{YY})$-ZZYY *switching bifurcation is for a singular equilibrium with a singular 1-dimensional flow switching to a singular 1-dimensional flow with a singular equilibrium.*

$$\left\{\begin{matrix}\underbrace{(a_{j_1j_1 1}, a_{j_2j_2 1})}_{(m_1,n_2)\text{-YY}_1}\\ \underbrace{(a_{j_1j_1 1}, a_{j_1j_2 1})}_{((m_1,n_1)^{\text{th}}\,\text{XX}_1,\text{xF})}\end{matrix}\right\} \rightleftharpoons \underbrace{([n_2-n_1]^{\text{th}}\,\text{ZZ}, [m_1,n_2]^{\text{th}}\,\text{YY})}_{([n_2-n_1],[m_1,n_2])\text{-ZZYY}} \rightleftharpoons \left\{\begin{matrix}\underbrace{(a_{j_1j_1 1}, a_{j_1j_2 1})}_{((m_1,n_1)^{\text{th}}\,\text{XX}_2,\text{xF})}\\ \underbrace{(a_{j_1j_1 1}, a_{j_2j_2 1})}_{(m_1,n_2)\text{-YY}_2}\end{matrix}\right\} \tag{1.483}$$

with

$$\text{XX}_1, \text{XX}_2 \in \{\text{DP:UP, UP:DP; UP:UP, DP:DP; }{}_{\text{II:DI}}\text{SO},{}_{\text{DI:II}}\text{SO},$$

$$
{}_{\mathrm{II:DI}}\mathrm{SI},{}_{\mathrm{DI:II}}\mathrm{SI};\ {}_{\mathrm{II:II}}\mathrm{US},{}_{\mathrm{II:II}}\mathrm{LS},{}_{\mathrm{DI:DI}}\mathrm{US},{}_{\mathrm{DI:DI}}\mathrm{LS}\};\\
\mathrm{xF}_1,\mathrm{xF}_2 \in \{\mathrm{pF},\ \mathrm{nF}\};\\
\mathrm{YY}_1,\mathrm{YY}_2 \in \{\text{sink, source, saddles; upper (lower)-saddle-source,}\\
\text{upper (lower)-saddle-sink; double-saddles}\}. \tag{1.484}
$$

(vi_{2b}) The ($[n_2-n_1]^{\text{th}}$ ZZ, $[m_1, n_2]^{\text{th}}$ YY)-ZZYY *switching bifurcation is for two $m_1 \times (n_1+n_2)$ networks of non-singular equilibrium and hyperbolic and hyperbolic-secant flows.*

$$
\begin{aligned}
&\bigcup_{r=1}^{m_1}\bigcup_{u_1=1}^{n_1}\underbrace{(a_{j_1j_1r},a_{j_1j_2u_1})}_{(\mathrm{XX}_1,\mathrm{xF}_1)}\oplus\bigcup_{r=1}^{m_1}\bigcup_{s_1=1}^{n_2}\underbrace{(a_{j_1j_1r},a_{j_2j_2s_1})}_{\mathrm{YY}_1}\\
&\rightleftharpoons\underbrace{([n_2-n_1]^{\text{th}}\ \mathrm{ZZ},\ [m_1,n_2]^{\text{th}}\ \mathrm{YY})}_{([n_2-n_1],[m_1,n_2])\text{-ZZYY}}\\
&\rightleftharpoons\bigcup_{r=1}^{m_1}\bigcup_{u_2=1}^{n_1}\underbrace{(a_{j_1j_1r},a_{j_1j_2u_1})}_{(XX_2,\mathrm{xF}_2)}\oplus\bigcup_{r=1}^{m_1}\bigcup_{s_2=1}^{n_1}\underbrace{(a_{j_1j_1r},a_{j_2j_2s_2})}_{\mathrm{YY}_2},
\end{aligned}\tag{1.485}
$$

with

$$
\mathrm{YY}_1,\mathrm{YY}_2 \in \{\text{sink, source, saddle}\};\\
\mathrm{xF}_1,\mathrm{xF}_2 \in \{\mathrm{pF},\ \mathrm{nF}\};\\
\mathrm{XX}_1,\mathrm{XX}_2 \in \{\text{hyperbolic flow, hyperbolic-secant flow}\}. \tag{1.486}
$$

(vi_{2c}) The ($[n_2-n_1]^{\text{th}}$ ZZ, $[m_1, n_2]^{\text{th}}$ YY)-ZZYY *switching bifurcation is for two $p_1 \times (q_1+q_2)$ networks of singular equilibriums and singular 1-dimensional flows.*

$$
\begin{aligned}
&\bigcup_{r=1}^{p_1}\bigcup_{u_1=1}^{q_1}\underbrace{(a_{j_1j_1r},a_{j_1j_2u_1})}_{([m_{j_1r},n_{j_1u_1}]\mathrm{XX}_1,\mathrm{xF}_1)}\oplus\bigcup_{r=1}^{p_1}\bigcup_{s_1=1}^{q_2}\underbrace{(a_{j_1j_1r},a_{j_2j_2s_1})}_{(\mathrm{m}_{j_1r},n_{j_2s_1})\text{-}\mathrm{YY}_1}\\
&\rightleftharpoons\underbrace{([n_2-n_1]^{\text{th}}\ \mathrm{ZZ},\ [m_1,n_2]^{\text{th}}\ \mathrm{YY})}_{([n_2-n_1],[m_1,n_2])\text{-ZZYY}}\\
&\rightleftharpoons\bigcup_{r=1}^{p_1}\bigcup_{u_2=1}^{q_1}\underbrace{(a_{j_1j_1r},a_{j_1j_2u_2})}_{([m_{j_1r},n_{j_1u_2}]XX_1,\mathrm{xF}_2)}\oplus\bigcup_{r=1}^{p_1}\bigcup_{s_2=1}^{q_2}\underbrace{(a_{j_1j_1r},a_{j_2j_2s_2})}_{(m_{j_1r},n_{j_2s_2})\text{-}\mathrm{YY}_2}
\end{aligned}\tag{1.487}
$$

with

$$
\mathrm{XX}_1,\mathrm{XX}_2 \in \{\mathrm{DP{:}UP},\ \mathrm{UP{:}DP};\ \mathrm{UP{:}UP},\ \mathrm{DP{:}DP};\ {}_{\mathrm{II:DI}}\mathrm{SO},{}_{\mathrm{DI:II}}\mathrm{SO},\\
{}_{\mathrm{II:DI}}\mathrm{SI},{}_{\mathrm{DI:II}}\mathrm{SI};\ {}_{\mathrm{II:II}}\mathrm{US},{}_{\mathrm{II:II}}\mathrm{LS},{}_{\mathrm{DI:DI}}\mathrm{US},{}_{\mathrm{DI:DI}}\mathrm{LS}\};\\
\mathrm{xF}_1,\mathrm{xF}_2 \in \{\mathrm{pF},\ \mathrm{nF}\};
$$

$$\mathrm{YY}_1, \mathrm{YY}_2 \in \{\text{sink, source, saddles; upper (lower)-saddle-source, upper (lower)-saddle-sink; double-saddle}\}. \tag{1.488}$$

and

$$m_1 = \sum_{r_i=1}^{p_1} m_{j_1 r_i}, n_1 = \sum_{u_i=1}^{q_1} n_{j_1 u_i}, n_2 = \sum_{s_i=1}^{q_1} n_{j_2 s_i};\ i = 1, 2. \tag{1.489}$$

Chapter 2
Proof of Theorem 1.1

In this chapter, Theorem 2.1 is proved. The theorem about the singular equilibriums and 1-diemsnional flows with infinite-equilibriums in the self and product polynomial systems is proved first. The first integral manifolds are obtained for singular and nonsingular sinks, saddles, sources, saddle-sources and saddle-sinks, double-saddles. The theorem about hybrid networks of singular equilibriums and 1-diemsnional flows in the self and product polynomial systems are proved, and the corresponding infinite-equilibriums are discussed. The theorem about hybrid network of simple equilibriums and 1-diemsnional flows with infinite-equilibriums in the self and product polynomial systems is proved. The theorem about appearing and switching of simple and singular 1-dimensional flows and equilibriums is proved. Finally, the theorem about the switching of networks of simple and singular equilibriums and flows is also proved.

2.1 Singular Equilibriums and Infinite-Equilibriums

In this section, the theorem about the singular equilibriums and 1-dimensional flows in the self and product polynomial systems will be proved.

2.1.1 Singular Equilibriums and 1-Dimensional Flows

(i) Consider a two-dimensional, nonlinear dynamical system as

$$\begin{aligned}\dot{x}_{j_2} &= a_{j_2j_20}(x_{j_2} - a_{j_2j_21})^{n_2},\\ \dot{x}_{j_1} &= a_{j_1j_10}(x_{j_1} - a_{j_1j_11})^{m_1}(x_{j_2} - a_{j_1j_21})^{n_1}.\end{aligned}$$

A. C. J. Luo, *Two-Dimensional Self and Product Polynomial Systems*,
https://doi.org/10.1007/978-981-96-5483-3_2

(i_1) The foregoing equation becomes

$$\frac{dx_{j_1}}{dx_{j_2}}=\frac{a_{j_1j_10}(x_{j_1}-a_{j_1j_11})^{m_1}(x_{j_2}-a_{j_1j_21})^{n_1}}{a_{j_2j_20}(x_{j_2}-a_{j_2j_21})^{n_2}},$$

or

$$\begin{aligned}\frac{dx_{j_1}}{(x_{j_1}-a_{j_1j_11})^{m_1}}&=\frac{a_{j_1j_10}}{a_{j_2j_20}}\frac{(x_{j_2}-a_{j_1j_21})^{n_1}}{(x_{j_2}-a_{j_2j_21})^{n_2}}dx_{j_2}\\&=\frac{a_{j_1j_10}}{a_{j_2j_20}}\sum_{\alpha,\beta=0}^{\alpha+\beta=n_1}\frac{n_1!}{\alpha!\beta!}(a_{j_2j_21}-a_{j_1j_21})^{\alpha}(x_{j_2}-a_{j_2j_21})^{\beta-n_2}dx_{j_2}.\end{aligned}$$

With an initial condition (x_{j_10},x_{j_20}) art time $t=t_0$, integration of the foregoing equation gives the first integral manifold as

$$\begin{aligned}&-\frac{1}{m_1-1}\left[\frac{1}{(x_{j_1}-a_{j_1j_11})^{m_1-1}}-\frac{1}{(x_{j_10}-a_{j_1j_11})^{m_1-1}}\right]\\&=\frac{a_{j_1j_10}}{a_{j_2j_20}}\sum_{\alpha,\beta=0}^{\alpha+\beta=n_1}\frac{n_1!}{\alpha!\beta!}(a_{j_2j_21}-a_{j_1j_21})^{\alpha}\\&\times\frac{1}{\beta-n_2+1}\left[(x_{j_2}-a_{j_2j_21})^{\beta-n_2+1}-(x_{j_20}-a_{j_2j_21})^{\beta-n_2+1}\right],\\&\text{for } m_1\neq 1,\ \beta-n_2+1\neq 0;\end{aligned}$$

and

$$\begin{aligned}&-\frac{1}{m_1-1}\left[\frac{1}{(x_{j_1}-a_{j_1j_11})^{m_1-1}}-\frac{1}{(x_{j_10}-a_{j_1j_11})^{m_1-1}}\right]=\ln\frac{|x_{j_1}-a_{j_1j_11}|}{|x_{j_10}-a_{j_1j_11}|},\\&\text{for } m_1=1;\\&\frac{1}{\beta-n_2+1}\left[(x_{j_2}-a_{j_2j_21})^{\beta-n_2+1}-(x_{j_20}-a_{j_2j_21})^{\beta-n_2+1}\right]=\ln\frac{|x_{j_2}-a_{j_2j_21}|}{|x_{j_20}-a_{j_2j_21}|},\\&\text{for } \beta-n_2+1=0.\end{aligned}$$

(i_{1a}) At the equilibrium of $x^*_{j_2}=a_{j_1j_21}$ and $\overline{x}_{j_1}\neq a_{j_1j_11}$, in phase space,

$$\begin{aligned}&\left.\frac{dx_{j_1}}{dx_{j_2}}\right|_{x^*_{j_2}=a_{j_1j_21}}=\left.\frac{a_{j_1j_10}(\overline{x}_{j_1}-a_{j_1j_11})^{m_1}(x_{j_2}-a_{j_1j_21})^{n_1}}{a_{j_2j_20}(x_{j_2}-a_{j_2j_21})^{n_2}}\right|_{x^*_{j_2}=a_{j_1j_21}}=0,\\&\vdots\\&\left.\frac{d^{n_1}x_{j_1}}{dx^{n_1}_{j_2}}\right|_{x^*_{j_2}=a_{j_1j_21}}=0.\end{aligned}$$

If

$$\frac{d^{n_1+1}x_{j_1}}{dx_{j_2}^{n_1+1}}\bigg|_{x_{j_2}^*=a_{j_1j_2 1}} = \frac{a_{j_1j_1 0}(\overline{x}_{j_1} - a_{j_1j_1 1})^{m_1} n_1!}{a_{j_2j_2 0}(a_{j_1j_2 1} - a_{j_2j_2 1})^{n_2}} > 0,$$

the $(2n_{11}+1)^{\text{th}}$ -order up-parabola flows exists for $n_1 = 2n_{11}+1$, and the $(2n_{11})^{\text{th}}$ -order increasing-inflection flow exists for $n_1 = 2n_{11}$. If

$$\frac{d^{n_1+1}x_{j_1}}{dx_{j_2}^{n_1+1}}\bigg|_{x_{j_2}^*=a_{j_1j_2 1}} = \frac{a_{j_1j_1 0}(\overline{x}_{j_1} - a_{j_1j_1 1})^{m_1} n_1!}{a_{j_2j_2 0}(a_{j_1j_2 1} - a_{j_2j_2 1})^{n_2}} < 0,$$

the $(2n_{11}+1)^{\text{th}}$-order down-parabola flows exists for $n_1 = 2n_{11}+1$, and the $(2n_{11})^{\text{th}}$ -order decreasing-inflection flow exists for $n_1 = 2n_{11}$.

At the equilibrium of $x_{j_1}^* = a_{j_1j_1 1}$ and $\overline{x}_{j_2} \neq a_{j_1j_2 1}, a_{j_2j_2 1}$, the variational equation is

$$\begin{aligned}\dot{x}_{j_2} &= a_{j_2j_2 0}(\overline{x}_{j_2} - a_{j_2j_2 1})^{n_2},\\ \Delta\dot{x}_{j_1} &= a_{j_1j_1 0}(\overline{x}_{j_2} - a_{j_1j_2 1})^{n_1}(\Delta x_{j_1})^{m_1}.\end{aligned}$$

The flows at $x_{j_1}^* = a_{j_1j_1 1}$ and $\overline{x}_{j_2} \neq a_{j_1j_2 1}$ are the $(2m_{11}+1)^{\text{th}}$-order source and sink for $a_{j_1j_1 0}(\overline{x}_{j_2} - a_{j_1j_2 1})^{n_1} > 0$ and $a_{j_1j_1 0}(\overline{x}_{j_2} - a_{j_1j_2 1})^{n_1} < 0$ for $m_1 = 2m_{11}+1$. The flows at $x_{j_1}^* = a_{j_1j_1 1}$ and $\overline{x}_{j_2} \neq a_{j_1j_2 1}, a_{j_2j_2 1}$, are the $(2m_{11})^{\text{th}}$-order upper-saddle and lower-saddle for $a_{j_1j_1 0}(\overline{x}_{j_2} - a_{j_1j_2 1})^{n_1} > 0$ and $a_{j_1j_1 0}(\overline{x}_{j_2} - a_{j_1j_2 1})^{n_1} < 0$ for $m_1 = 2m_{11}$.

Let $\overline{x}_{j_1} = a_{j_1j_1 1}$ and $\overline{x}_{j_2} = a_{j_1j_2 1} \neq a_{j_2j_2 1}$. Therefore, from the above discussion, there are four cases($\text{i}_{1\text{a}1}$) $m_1 = 2m_{11}+1$ and $n_1 = 2n_{11}+1$, ($\text{i}_{1\text{a}2}$) $m_1 = 2m_{11}$ and $n_1 = 2n_{11}+1$,($\text{i}_{1\text{a}3}$) $m_1 = 2m_{11}+1$ and $n_1 = 2n_{11}$,($\text{i}_{1\text{a}4}$) $m_1 = 2m_{11}$ and $n_1 = 2n_{11}$.

($\text{i}_{1\text{a}1}$) For $m_1 = 2m_{11}+1$ and $n_1 = 2n_{11}+1$, the equilibrium of $(x_{j_1}^*, x_{j_2}^*) = (a_{j_1j_1 1}, a_{j_1j_2 1})$ has the following properties as in Eqs. (1.4)–(1.7).

- For $a_{j_1j_1 0} > 0$ and $a_{j_2j_2 0}(a_{j_1j_2 1} - a_{j_2j_2 1})^{n_2} > 0$,

$$(a_{j_1j_1 1}, a_{j_1j_2 1}) = \underbrace{([(2m_{11}{+}1), (2n_{11}+1)]\text{DP:UP, pF})}_{[(2m_{11}+1),(2n_{11}+1)]\text{-hyperbolic flow }(+)}.$$

The flow at $(x_{j_1}^*, x_{j_2}^*) = (a_{j_1j_1 1}, a_{j_1j_2 1})$ is a ([$(2m_{11}+1)$, $(2n_{11}+1)$]DP:UP, pF)-positive hyperbolic flow.

- For $a_{j_1j_1 0} < 0$ and $a_{j_2j_2 0}(a_{j_1j_2 1} - a_{j_2j_2 1})^{n_2} > 0$,

$$(a_{j_1j_1 1}, a_{j_1j_2 1}) = \underbrace{([(2m_{11}{+}1), (2n_{11}+1)]\text{UP:DP, pF})}_{[(2m_{11}+1),(2n_{11}+1)]\text{-hyperbolic-secant flow }(+)}.$$

The flow at $(x_{j_1}^*, x_{j_2}^*) = (a_{j_1j_1 1}, a_{j_1j_2 1})$ is a ([$(2m_{11}+1)$, $(2n_{11}+1)$]UP:DP, pF)-positive hyperbolic-secant flow.

- For $a_{j_1j_10} > 0$ and $a_{j_2j_20}(a_{j_1j_21} - a_{j_2j_21})^{n_2} < 0$,

$$(a_{j_1j_11}, a_{j_1j_21}) = \underbrace{([(2m_{11}+1), (2n_{11}+1)]\text{DP:UP,nF})}_{[(2m_{11}+1),(2n_{11}+1)]\text{-hyperbolic flow }(-)}.$$

The flow at $(x^*_{j_1}, x^*_{j_2}) = (a_{j_1j_11}, a_{j_1j_21})$ is a ([(2m_{11}+1), (2n_{11} + 1)]UP:DP, nF)-negative hyperbolic-secant flow.

- For $a_{j_1j_10} < 0$ and $a_{j_2j_20}(a_{j_1j_21} - a_{j_2j_21})^{n_2} < 0$,

$$(a_{j_1j_11}, a_{j_1j_21}) = \underbrace{([(2m_{11}+1), (2n_{11}+1)]\text{DP:UP,nF})}_{[(2m_{11}+1),(2n_{11}+1)]\text{-hyperbolic flow }(-)}.$$

The flow at $(x^*_{j_1}, x^*_{j_2}) = (a_{j_1j_11}, a_{j_1j_21})$ is a ([(2m_{11}+ 1), (2n_{11} + 1)]UP:UP, pF)-negative hyperbolic flow.

(i_{1a2}) For $m_1 = 2m_{11}$ and $n_1 = 2n_{11} + 1$, the equilibrium of $(x^*_{j_1}, x^*_{j_2}) = (a_{j_1j_11}, a_{j_1j_21})$ for the x_{j_1}-direction has the following properties as in Eqs. (1.8)–(1.11).

- For $a_{j_1j_10} > 0$ and $a_{j_2j_20}(a_{j_1j_21} - a_{j_2j_21})^{n_2} > 0$,

$$(a_{j_1j_11}, a_{j_1j_21}) = \underbrace{([(2m_{11}), (2n_{11}+1)]\text{UP:UP, pF})}_{[(2m_{11}),(2n_{11}+1)]\text{-hyperbolic-secant-to-hyperbolic flow }(+)}.$$

The flow at $(x^*_{j_1}, x^*_{j_2}) = (a_{j_1j_11}, a_{j_1j_21})$ is a ([(2m_{11}), (2n_{11} + 1)]UP:UP, pF)-positive hyperbolic-secant to hyperbolic flow.

- For $a_{j_1j_10} < 0$ and $a_{j_2j_20}(a_{j_1j_21} - a_{j_2j_21})^{n_2} > 0$,

$$(a_{j_1j_11}, a_{j_1j_21}) = \underbrace{([(2m_{11}), (2n_{11}+1)]\text{DP:DP, pF})}_{[(2m_{11}),(2n_{11}+1)]\text{-hyperbolic-to-hyperbolic-secant flow }(+)}.$$

The flow at $(x^*_{j_1}, x^*_{j_2}) = (a_{j_1j_11}, a_{j_1j_21})$ is a ([(2m_{11}), (2n_{11} + 1)]DP:DP, pF)-positive hyperbolic to hyperbolic-secant flow.

- For $a_{j_1j_10} > 0$ and $a_{j_2j_20}(a_{j_1j_21} - a_{j_2j_21})^{n_2} < 0$,

$$(a_{j_1j_11}, a_{j_1j_21}) = \underbrace{([(2m_{11}), (2n_{11}+1)]\text{DP:DP, nF})}_{[(2m_{11}),(2n_{11}+1)]\text{-hyperbolic-to-hyperbolic-secant flow }(-)}.$$

The flow at $(x^*_{j_1}, x^*_{j_2}) = (a_{j_1j_11}, a_{j_1j_21})$ is a ([(2m_{11}), (2n_{11} + 1)]DP:DP, nF)-negative hyperbolic-to-hyperbolic-secant flow.

- For $a_{j_1j_10} < 0$ and $a_{j_2j_20}(a_{j_1j_21} - a_{j_2j_21})^{n_2} < 0$,

$$(a_{j_1j_11}, a_{j_1j_21}) = \underbrace{([(2m_{11}), (2n_{11}+1)]\text{UP:UP, nF})}_{[(2m_{11}),(2n_{11}+1)]\text{-hyperbolic-secant-to-hyperbolic flow }(-)}.$$

The flow at $(x_{j_1}^*, x_{j_2}^*) = (a_{j_1 j_1 1}, a_{j_1 j_2 1})$ is a ([(2m_{11}), (2n_{11} + 1)]UP:UP, nF)-negative hyperbolic-secant-to-hyperbolic flow.

(i_{1a3}) For $m_1 = 2m_{11} + 1$ and $n_1 = 2n_{11}$, the equilibrium of $(x_{j_1}^*, x_{j_2}^*) = (a_{j_1 j_1 1}, a_{j_1 j_2 1})$ for the x_{j_1}-direction has the following properties as in Eqs. (1.12)–(1.15).

- For $a_{j_1 j_1 0} > 0$ and $a_{j_2 j_2 0}(a_{j_1 j_2 1} - a_{j_2 j_2 1})^{n_2} > 0$,

$$(a_{j_1 j_1 1}, a_{j_1 j_2 1}) = \underbrace{([(2m_{11}+1), (2n_{11})]_{\mathrm{DI:II}}\mathrm{SO,\ pF})}_{[(2m_{11}+1),(2n_{11})]\text{-decreasing-to-increasing inflection source flow }(+)}.$$

The flow at $(x_{j_1}^*, x_{j_2}^*) = (a_{j_1 j_1 1}, a_{j_1 j_2 1})$ is a $([(2m_{11}+1), (2n_{11})]_{\mathrm{DI:II}}\mathrm{SO, pF})$-positive decreasing-to-increasing inflection source flow.

- For $a_{j_1 j_1 0} < 0$ and $a_{j_2 j_2 0}(a_{j_1 j_2 1} - a_{j_2 j_2 1})^{n_2} > 0$,

$$(a_{j_1 j_1 1}, a_{j_1 j_2 1}) = \underbrace{([(2m_{11}+1), (2n_{11})]_{\mathrm{II:DI}}\mathrm{SI,\ pF})}_{[(2m_{11}+1),(2n_{11})]\text{-increasing-to-decreasing inflection sink flow }(+)}.$$

The flow at $(x_{j_1}^*, x_{j_2}^*) = (a_{j_1 j_1 1}, a_{j_1 j_2 1})$ is a $([(2m_{11}+1), (2n_{11})]_{\mathrm{II:DI}}\mathrm{SI, pF})$-positive increasing-to-decreasing inflection source flow.

- For $a_{j_1 j_1 0} > 0$ and $a_{j_2 j_2 0}(a_{j_1 j_2 1} - a_{j_2 j_2 1})^{n_2} < 0$,

$$(a_{j_1 j_1 1}, a_{j_1 j_2 1}) = \underbrace{([(2m_{11}+1), (2n_{11})]_{\mathrm{II:DI}}\mathrm{SO,nF})}_{[(2m_{11}+1),(2n_{11})]\text{-increasing-to-decreasing inflection source flow }(-)}.$$

The flow at $(x_{j_1}^*, x_{j_2}^*) = (a_{j_1 j_1 1}, a_{j_1 j_2 1})$ is a $([(2m_{11}+1), (2n_{11})]_{\mathrm{II:DI}}\mathrm{SO,nF})$-negative increasing-to-decreasing inflection source flow.

- For $a_{j_1 j_1 0} < 0$ and $a_{j_2 j_2 0}(a_{j_1 j_2 1} - a_{j_2 j_2 1})^{n_2} < 0$,

$$(a_{j_1 j_1 1}, a_{j_1 j_2 1}) = \underbrace{([(2m_{11}+1), (2n_{11})]_{\mathrm{DI:II}}\mathrm{SI,nF})}_{[(2m_{11}+1),(2n_{11})]\text{-decreasing-to-increasing inflection sink flow }(-)}.$$

(i_{1a4}) For $m_1 = 2m_{11}$ and $n_1 = 2n_{11}$, the equilibrium of $(x_{j_1}^*, x_{j_2}^*) = (a_{j_1 j_1 1}, a_{j_1 j_2 1})$ for the x_{j_1}-direction has the following properties as in Eqs. (1.16)–(1.19).

- For $a_{j_1 j_1 0} > 0$ and $a_{j_2 j_2 0}(a_{j_1 j_2 1} - a_{j_2 j_2 1})^{n_2} > 0$,

$$(a_{j_1 j_1 1}, a_{j_1 j_2 1}) = \underbrace{([(2m_{11}), (2n_{11})]_{\mathrm{II:II}}\mathrm{US,pF})}_{[(2m_{11}),(2n_{11})]\text{-increasing-inflection upper-saddle flow }(+)}.$$

- For $a_{j_1 j_1 0} < 0$ and $a_{j_2 j_2 0}(a_{j_1 j_2 1} - a_{j_2 j_2 1})^{n_2} > 0$,

$$(a_{j_1 j_1 1}, a_{j_1 j_2 1}) = \underbrace{([(2m_{11}), (2n_{11})]_{\mathrm{DI:DI}}\mathrm{LS,pF})}_{[(2m_{11}),(2n_{11})]\text{-decreasing-inflection lower-saddle flow }(+)}.$$

The flow at $(x_{j_1}^*, x_{j_2}^*) = (a_{j_1 j_1 1}, a_{j_1 j_2 1})$ is a ([(2m_1), (2n_1)]DI:DILS,pF)-positive decreasing-inflection lower-saddle flow.

- For $a_{j_1 j_1 0} > 0$ and $a_{j_2 j_2 0}(a_{j_1 j_2 1} - a_{j_2 j_2 1})^{n_2} < 0$,

$$(a_{j_1 j_1 1}, a_{j_1 j_2 1}) = \underbrace{([(2m_{11}), (2n_{11})]_{\text{DI:DI}}\text{US, nF})}_{[(2m_{11}),(2n_{11})]\text{-decreasing-inflection upper-saddle flow }(-)}.$$

The flow at $(x_{j_1}^*, x_{j_2}^*) = (a_{j_1 j_1 1}, a_{j_1 j_2 1})$ is a ([(2m_{11}), (2n_{11})]DI:DIUS, nF)-negative decreasing-inflection upper-saddle flow.

- For $a_{j_1 j_1 0} < 0$ and $a_{j_2 j_2 0}(a_{j_1 j_2 1} - a_{j_2 j_2 1})^{n_2} < 0$,

$$(a_{j_1 j_1 1}, a_{j_1 j_2 1}) = \underbrace{([(2m_{11}), (2n_{11})]_{\text{II:II}}\text{LS, nF})}_{[(2m_{11}),(2n_{11})]\text{-increasing-inflection lower-saddle flow }(-)}.$$

The flow at $(x_{j_1}^*, x_{j_2}^*) = (a_{j_1 j_1 1}, a_{j_1 j_2 1})$ is a ([(2m_{11}), (2n_{11})]II:IILS, nF)-negative increasing-inflection lower-saddle flow.

(i_{1b}) There is an equilibrium for self-univariate equilibrium as

$$(x_{j_1}^*, x_{j_2}^*) = (a_{j_1 j_1 1}, a_{j_2 j_2 1}).$$

The variational equation at such an equilibrium with $a_{j_2 j_2 1} \neq a_{j_1 j_2 1}$ is

$$\begin{aligned}\Delta\dot{x}_{j_2} &= a_{j_2 j_2 0}(\Delta x_{j_2})^{n_2},\\ \Delta\dot{x}_{j_1} &= a_{j_1 j_1 0}(\Delta x_{j_1})^{m_1}(a_{j_2 j_2 1} - a_{j_1 j_2 1})^{n_1}.\end{aligned}$$

For $m_1 = 2m_{11} + 1$, flows at equilibrium $(x_{j_1}^*, x_{j_2}^*) = (a_{j_1 j_1 1}, a_{j_2 j_2 1})$ in the x_{j_1}-direction are $(2m_{11} + 1)^{\text{th}}$-source and sink flows for $a_{j_1 j_1 0}(a_{j_2 j_2 1} - a_{j_1 j_2 1})^{n_1} > 0$ and $a_{j_1 j_1 0}(a_{j_2 j_2 1} - a_{j_1 j_2 1})^{n_1} < 0$, respectively. For $m_1 = 2m_{11}$, flows at equilibrium $(x_{j_1}^*, x_{j_2}^*) = (a_{j_1 j_1 1}, a_{j_2 j_2 1})$ in the x_{j_1}-direction are $(2m_{11})^{\text{th}}$- upper-saddle and lower-saddle for $a_{j_1 j_1 0}(a_{j_2 j_2 1} - a_{j_1 j_2 1})^{n_1} > 0$ and $a_{j_1 j_1 0}(a_{j_2 j_2 1} - a_{j_1 j_2 1})^{n_1} < 0$, respectively.

For $n_2 = 2n_{21}+1$, flows at equilibrium $(x_{j_1}^*, x_{j_2}^*) = (a_{j_1 j_1 1}, a_{j_2 j_2 1})$ in the x_{j_2}-direction are $(2n_{21} + 1)^{\text{th}}$-source and sink flows for $a_{j_2 j_2 0} > 0$ and $a_{j_2 j_2 0} < 0$, respectively. For $n_2 = 2n_{21}$, flows at equilibrium $(x_{j_1}^*, x_{j_2}^*) = (a_{j_1 j_1 1}, a_{j_2 j_2 1})$ in the x_{j_2}-direction are $(2n_{21})^{\text{th}}$-upper-saddle and lower-saddle for $a_{j_2 j_2 0} > 0$ and $a_{j_2 j_2 0} < 0$, respectively. Thus there are sixteen types of equilibriums.

(i_{1b1}) For $m_1 = 2m_{11} + 1$ and $n_2 = 2n_{21} + 1$,the equilibrium of $(x_{j_1}^*, x_{j_2}^*) = (a_{j_1 j_1 1}, a_{j_2 j_2 1})$ has the following properties as in Eqs. (1.21)–(1.24).

- For $a_{j_1 j_1 0}(a_{j_2 j_2 1} - a_{j_1 j_2 1})^{n_1} > 0$ and $a_{j_2 j_2 0} > 0$,

$$(a_{j_1 j_1 1}, a_{j_2 j_2 1}) = \underbrace{((2m_{11}+1)^{\text{th}}\text{SO}, (2n_{21}+1)^{\text{th}}\text{SO})}_{((2m_{11}+1),(2n_{21}+1))\text{-source}}.$$

The flow at $(x^*_{j_1}, x^*_{j_2}) = (a_{j_1j_11}, a_{j_2j_21})$ is a $((2m_{11}+1)^{\text{th}}\text{SO}, (2n_{21}+1)^{\text{th}}\text{SO})$-source.

- For $a_{j_1j_10}(a_{j_2j_21} - a_{j_1j_21})^{n_1} < 0$ and $a_{j_2j_20} > 0$,

$$(a_{j_1j_11}, a_{j_2j_21}) = \underbrace{((2m_{11}+1)^{\text{th}}\text{SI}, (2n_{21}+1)^{\text{th}}\text{SO})}_{((2m_{11}+1),(2n_{21}+1))\text{-saddle}}.$$

The flow at $(x^*_{j_1}, x^*_{j_2}) = (a_{j_1j_11}, a_{j_2j_21})$ is a $((2m_{11}+1)^{\text{th}}\text{SI}, (2n_{21}+1)^{\text{th}}\text{SO})$-saddle.

- For $a_{j_1j_10}(a_{j_2j_21} - a_{j_1j_21})^{n_1} > 0$ and $a_{j_2j_20} < 0$,

$$(a_{j_1j_11}, a_{j_2j_21}) = \underbrace{((2m_{11}+1)^{\text{th}}\text{SO}, (2n_{21}+1)^{\text{th}}\text{SI})}_{((2m_{11}+1),(2n_{21}+1))\text{-saddle}}.$$

The flow at $(x^*_{j_1}, x^*_{j_2}) = (a_{j_1j_11}, a_{j_2j_21})$ is a $((2m_{11}+1)^{\text{th}}\text{SO}, (2n_{21}+1)^{\text{th}}\text{SI})$-saddle.

- For $a_{j_1j_10}(a_{j_2j_21} - a_{j_1j_21})^{n_1} < 0$ and $a_{j_2j_20} < 0$,

$$(a_{j_1j_11}, a_{j_2j_21}) = \underbrace{((2m_{11}+1)^{\text{th}}\text{SI}, (2n_{21}+1)^{\text{th}}\text{SI})}_{((2m_{11}+1),(2n_{21}+1))\text{-sink}}.$$

The flow at $(x^*_{j_1}, x^*_{j_2}) = (a_{j_1j_11}, a_{j_2j_21})$ is a $((2m_{11}+1)^{\text{th}}\text{SI}, (2n_{21}+1)^{\text{th}}\text{SI})$-sink.

(i_{1b2}) For $m_1 = 2m_{11}$ and $n_2 = 2n_{21} + 1$, the equilibrium of $(x^*_{j_1}, x^*_{j_2}) = (a_{j_1j_11}, a_{j_2j_21})$ has the following properties as in Eqs. (1.25)–(1.28).

- For $a_{j_1j_10}(a_{j_2j_21} - a_{j_1j_21})^{n_1} > 0$ and $a_{j_2j_20} > 0$,

$$(a_{j_1j_11}, a_{j_2j_21}) = \underbrace{((2m_{11})^{\text{th}}\text{US}, (2n_{21}+1)^{\text{th}}\text{SO})}_{((2m_{11}),(2n_{21}+1))\text{-upper-saddle source}}.$$

The flow at $(x^*_{j_1}, x^*_{j_2}) = (a_{j_1j_11}, a_{j_2j_21})$ is a $((2m_{11})^{\text{th}}\text{US}, (2n_{21}+1)^{\text{th}}\text{SO})$-upper-saddle source.

- For $a_{j_1j_10}(a_{j_2j_21} - a_{j_1j_21})^{n_1} < 0$ and $a_{j_2j_20} > 0$,

$$(a_{j_1j_11}, a_{j_2j_21}) = \underbrace{((2m_{11})^{\text{th}}\text{LS}, (2n_{21}+1)^{\text{th}}\text{SO})}_{((2m_{11}),(2n_{21}+1))\text{-lower-saddle source}}.$$

The flow at $(x^*_{j_1}, x^*_{j_2}) = (a_{j_1j_11}, a_{j_2j_21})$ is a $((2m_{11})^{\text{th}}\text{LS}, (2n_{21}+1)^{\text{th}}\text{SO})$-lower-saddle source.

- For $a_{j_1j_10}(a_{j_2j_21} - a_{j_1j_21})^{n_1} > 0$ and $a_{j_2j_20} < 0$,

$$(a_{j_1j_11}, a_{j_2j_21}) = \underbrace{((2m_{11})^{\text{th}}\text{US}, (2n_{21}+1)^{\text{th}}\text{SI})}_{((2m_{11}),(2n_{21}+1))\text{-upper-saddle sink}}.$$

The flow at $(x^*_{j_1}, x^*_{j_2}) = (a_{j_1j_11}, a_{j_2j_21})$ is a $((2m_{11})^{\text{th}}\text{US}, (2n_{21}+1)^{\text{th}}\text{SI})$-upper-saddle sink.

- For $a_{j_1j_10}(a_{j_2j_21} - a_{j_1j_21})^{n_1} < 0$ and $a_{j_2j_20} < 0$,

$$(a_{j_1j_11}, a_{j_2j_21}) = \underbrace{((2m_{11})^{\text{th}}\text{LS}, (2n_{21}+1)^{\text{th}}\text{SI})}_{((2m_{11}),(2n_{21}+1))\text{-lower-saddle sink}}.$$

The flow at $(x^*_{j_1}, x^*_{j_2}) = (a_{j_1j_11}, a_{j_2j_21})$ is a $((2m_{11})^{\text{th}}\text{LS}, (2n_{21}+1)^{\text{th}}\text{SI})$-lower-saddle sink.

(i_{1b3}) For $m_1 = 2m_{11} + 1$ and $n_2 = 2n_{21}$, the equilibrium of $(x^*_{j_1}, x^*_{j_2}) = (a_{j_1j_11}, a_{j_2j_21})$ has the following properties as in Eqs. (1.29)–(1.32).

- For $a_{j_1j_10}(a_{j_2j_21} - a_{j_1j_21})^{n_1} > 0$ and $a_{j_2j_20} > 0$,

$$(a_{j_1j_11}, a_{j_2j_21}) = \underbrace{((2m_{11}+1)^{\text{th}}\text{SO}, (2n_{21})^{\text{th}}\text{US})}_{((2m_{11}+1),(2n_{21}))\text{-upper-saddle source}}.$$

The flow at $(x^*_{j_1}, x^*_{j_2}) = (a_{j_1j_11}, a_{j_2j_21})$ is a $((2m_{11} + 1)^{\text{th}}\text{SO}, (2n_{21})^{\text{th}}\text{US})$-upper-saddle source.

- For $a_{j_1j_10}(a_{j_2j_21} - a_{j_1j_21})^{n_1} < 0$ and $a_{j_2j_20} > 0$,

$$(a_{j_1j_11}, a_{j_2j_21}) = \underbrace{((2m_{11}+1)^{\text{th}}\text{SI}, (2n_{21})^{\text{th}}\text{US})}_{((2m_{11}+1),(2n_{21}))\text{-upper-saddle sink}}.$$

The flow at $(x^*_{j_1}, x^*_{j_2}) = (a_{j_1j_11}, a_{j_2j_21})$ is a $((2m_{11}+1)^{\text{th}}\text{SI}, (2n_{21})^{\text{th}}\text{US})$-upper-saddle sink.

- For $a_{j_1j_10}(a_{j_2j_21} - a_{j_1j_21})^{n_1} > 0$ and $a_{j_2j_20} < 0$,

$$(a_{j_1j_11}, a_{j_2j_21}) = \underbrace{((2m_{11}+1)^{\text{th}}\text{SO}, (2n_{21})^{\text{th}}\text{LS})}_{((2m_{11}+1),(2n_{21}))\text{-lower-saddle source}}.$$

The flow at $(x^*_{j_1}, x^*_{j_2}) = (a_{j_1j_11}, a_{j_2j_21})$ is a $((2m_{11}+1)^{\text{th}}\text{SO}, (2n_{21})^{\text{th}}\text{LS})$-lower-saddle source.

- For $a_{j_1j_10}(a_{j_2j_21} - a_{j_1j_21})^{n_1} < 0$ and $a_{j_2j_20} < 0$,

$$(a_{j_1j_11}, a_{j_2j_21}) = \underbrace{((2m_{11}+1)^{\text{th}}\text{SI}, (2n_{21})^{\text{th}}\text{LS})}_{((2m_{11}+1),(2n_{21}))\text{-lower-saddle sink}}.$$

The flow at $(x^*_{j_1}, x^*_{j_2}) = (a_{j_1j_11}, a_{j_2j_21})$ is a $((2m_{11}+1)^{\text{th}}\text{SI}, (2n_{21})^{\text{th}}\text{LS})$-lower-saddle sink.

(i_{1b4}) For $m_1 = 2m_{11}$ and $n_2 = 2n_{21}$, the equilibrium of $(x^*_{j_1}, x^*_{j_2}) = (a_{j_1j_11}, a_{j_2j_21})$ has the following properties as in Eqs. (1.33)–(1.36).

- For $a_{j_1j_10}(a_{j_2j_21} - a_{j_1j_21})^{n_1} > 0$ and $a_{j_2j_20} > 0$,

$$(a_{j_1j_11}, a_{j_2j_21}) = \underbrace{((2m_{11})^{\text{th}}\text{US}, (2n_{21})^{\text{th}}\text{US})}_{((2m_{11}),(2n_{21}))\text{-double-saddle}}.$$

The flow at $(x_{j_1}^*, x_{j_2}^*) = (a_{j_1j_11}, a_{j_2j_21})$ is a $((2m_{11})^{\text{th}}\text{US}, (2n_{21})^{\text{th}}\text{US})$-double-saddle.

- For $a_{j_1j_10}(a_{j_2j_21} - a_{j_1j_21})^{n_1} < 0$ and $a_{j_2j_20} > 0$,

$$(a_{j_1j_11}, a_{j_2j_21}) = \underbrace{((2m_{11})^{\text{th}}\text{LS}, (2n_{21})^{\text{th}}\text{US})}_{((2m_{11}),(2n_{21}))\text{-double-saddle}}.$$

The flow at $(x_{j_1}^*, x_{j_2}^*) = (a_{j_1j_11}, a_{j_2j_21})$ is a $((2m_{11})^{\text{th}}\text{LS}, (2n_{21})^{\text{th}}\text{US})$-double-saddle.

- For $a_{j_1j_10}(a_{j_2j_21} - a_{j_1j_21})^{n_1} > 0$ and $a_{j_2j_20} < 0$,

$$(a_{j_1j_11}, a_{j_2j_21}) = \underbrace{((2m_{11})^{\text{th}}\text{US}, (2n_{21})^{\text{th}}\text{LS})}_{((2m_{11}),(2n_{21}))\text{-double-saddle}}.$$

The flow at $(x_{j_1}^*, x_{j_2}^*) = (a_{j_1j_11}, a_{j_2j_21})$ is a $((2m_{11})^{\text{th}}\text{US}, (2n_{21})^{\text{th}}\text{LS})$-double-saddle.

- For $a_{j_1j_10}(a_{j_2j_21} - a_{j_1j_21})^{n_1} < 0$ and $a_{j_2j_20} < 0$,

$$(a_{j_1j_11}, a_{j_2j_21}) = \underbrace{((2m_{11})^{\text{th}}\text{LS}, (2n_{21})^{\text{th}}\text{LS})}_{((2m_{11}),(2n_{21}))\text{-double-saddle}}.$$

The flow at $(x_{j_1}^*, x_{j_2}^*) = (a_{j_1j_11}, a_{j_2j_21})$ is a $((2m_{11})^{\text{th}}\text{US}, (2n_{21})^{\text{th}}\text{LS})$-double-saddle.

2.1.2 *Infinite-Equilibriums*

(i_2) For $a_{j_2j_21} = a_{j_1j_21}$, the dynamical equation gives

$$\frac{dx_{j_1}}{dx_{j_2}} = \frac{a_{j_1j_10}(x_{j_1} - a_{j_1j_11})^{m_1}(x_{j_2} - a_{j_2j_21})^{n_1-n_2}}{a_{j_2j_20}},$$

or

$$\frac{dx_{j_1}}{(x_{j_1} - a_{j_1j_11})^{m_1}} = \frac{a_{j_1j_10}}{a_{j_2j_20}}(x_{j_2} - a_{j_2j_21})^{n_1-n_2}dx_{j_2}.$$

With an initial condition $(x_{j_1 0}, x_{j_2 0})$ art time $t = t_0$, integration of the foregoing equation gives the first integral manifold as

$$
\begin{aligned}
&-\frac{1}{m_1-1}\left[\frac{1}{(x_{j_1}-a_{j_1j_11})^{m_1-1}}-\frac{1}{(x_{j_10}-a_{j_1j_11})^{m_1-1}}\right]\\
&=\frac{a_{j_1j_10}}{a_{j_2j_20}}\frac{1}{n_1-n_2+1}[(x_{j_2}-a_{j_2j_21})^{n_1-n_2+1}-(x_{j_20}-a_{j_2j_21})^{n_1-n_2+1}],\\
&\text{for } m_1\neq 1,\ n_2-n_1\neq 1;
\end{aligned}
$$

and

$$
\begin{aligned}
&-\frac{1}{m_1-1}\left[\frac{1}{(x_{j_1}-a_{j_1j_11})^{m_1-1}}-\frac{1}{(x_{j_10}-a_{j_1j_11})^{m_1-1}}\right]=\ln\frac{|x_{j_1}-a_{j_1j_11}|}{|x_{j_10}-a_{j_1j_11}|},\\
&\text{for } m_1=1;
\end{aligned}
$$

$$
\begin{aligned}
&\frac{1}{n_1-n_2+1}\left[(x_{j_2}-a_{j_2j_21})^{n_1-n_2+1}-(x_{j_20}-a_{j_2j_21})^{n_1-n_2+1}\right]=\ln\frac{|x_{j_2}-a_{j_2j_21}|}{|x_{j_20}-a_{j_2j_21}|},\\
&\text{for } n_1-n_2+1=0.
\end{aligned}
$$

(i_{2a}) For $n_1 \geq n_2$, the first integral manifold is

$$
\begin{aligned}
&-\frac{1}{m_1-1}[\frac{1}{(x_{j_1}-a_{j_1j_11})^{m_1-1}}-\frac{1}{(x_{j_10}-a_{j_1j_11})^{m_1-1}}]\\
&=\frac{a_{j_1j_10}}{a_{j_2j_20}}\frac{1}{n_1-n_2+1}[(x_{j_2}-a_{j_1j_21})^{n_1-n_2+1}-(x_{j_20}-a_{j_1j_21})^{n_1-n_2+1}],\\
&\text{for } m_1\neq 1;\\
&-\frac{1}{m_1-1}\left[\frac{1}{(x_{j_1}-a_{j_1j_11})^{m_1-1}}-\frac{1}{(x_{j_10}-a_{j_1j_11})^{m_1-1}}\right]=\ln\frac{|x_{j_1}-a_{j_1j_11}|}{|x_{j_10}-a_{j_1j_11}|},\\
&\text{for } m_1=1.
\end{aligned}
$$

(i_{2a1}) For $x^*_{j_2} = a_{j_2j_21} = a_{j_1j_21}$ and $\overline{x}_{j_1} \neq a_{j_1j_11}$, the foregoing equation becomes

$$
\begin{aligned}
&\left.\frac{dx_{j_1}}{dx_{j_2}}\right|_{x^*_{j_2}=a_{j_1j_21}}=\left.\frac{a_{j_1j_10}(\overline{x}_{j_1}-a_{j_1j_11})^{m_1}(x_{j_2}-a_{j_1j_21})^{n_1-n_2}}{a_{j_2j_20}}\right|_{x^*_{j_2}=a_{j_1j_21}}=0,\\
&\vdots\\
&\left.\frac{d^{(n_1-n_2)}x_{j_1}}{dx_{j_2}^{(n_1-n_2)}}\right|_{x^*_{j_2}=a_{j_1j_21}}=0.
\end{aligned}
$$

If

$$\left.\frac{d^{(n_1-n_2)+1}x_{j_1}}{dx_{j_2}^{(n_1-n_2)+1}}\right|_{x_{j_2}^*=a_{j_1j_21}} = \frac{a_{j_1j_10}(n_1-n_2)!(\overline{x}_{j_1}-a_{j_1j_11})^{m_1}}{a_{j_2j_20}} > 0,$$

the flows at $x_{j_2}^* = a_{j_2j_21} = a_{j_1j_21}$ are $[2(n_{11}-n_{21}) \pm 1]^{\text{th}}$-order up-parabola flow for $(n_1 = 2n_{11}+1, n_2 = 2n_{21}$ and $n_1 = 2n_{11}, n_2 = 2n_{21}+1)$, and $[2(n_{11}-n_{21})]^{\text{th}}$-order increasing-inflection flow for $(n_1 = 2n_{11}+1, n_2 = 2n_{21}+1$ and $n_1 = 2n_{11}, n_2 = 2n_{21})$. If

$$\left.\frac{d^{(n_1-n_2)+1}x_{j_1}}{dx_{j_2}^{(n_1-n_2)+1}}\right|_{x_{j_2}^*=a_{j_1j_21}} = \frac{a_{j_1j_10}(n_1-n_2)!(\overline{x}_{j_1}-a_{j_1j_11})^{m_1}}{a_{j_2j_20}} < 0,$$

the flows at $x_{j_2}^* = a_{j_2j_21} = a_{j_1j_21}$ are $[2(n_{11}-n_{21}) \pm 1]^{\text{th}}$-order down-parabola flow for $(n_1 = 2n_{11}+1, n_2 = 2n_{21}$ and $n_1 = 2n_{11}, n_2 = 2n_{21}+1)$, and $[2(n_{11}-n_{21})]^{\text{th}}$-order decreasing-inflection flow for $(n_1 = 2n_{11}+1, n_2 = 2n_{21}+1$ and $n_1 = 2n_{11}, n_2 = 2n_{21})$.

The variational equation at $x_{j_2}^* = a_{j_2j_21} = a_{j_1j_21}$ is

$$\Delta\dot{x}_{j_2} = a_{j_2j_20}(\Delta x_{j_2})^{n_2}.$$

The flows at $x_{j_2}^* = a_{j_2j_21} = a_{j_1j_21}$ are $(2n_{21}+1)^{\text{th}}$ source and sink flows for $a_{j_2j_20} > 0$ and $a_{j_2j_20} < 0$, respectively. The flows at $x_{j_2}^* = a_{j_2j_21} = a_{j_1j_21}$ are $(2n_{21})^{\text{th}}$-upper-saddle and lower-saddle flows for $a_{j_2j_20} > 0$ and $a_{j_2j_20} < 0$, respectively.

At $x_{j_1}^* = a_{j_1j_11}$, the variational equation for $\overline{x}_{j_2} \neq a_{j_1j_21} = a_{j_2j_21}$ is

$$\Delta\dot{x}_{j_1} = a_{j_1j_10}(\overline{x}_{j_2} - a_{j_1j_21})^{n_1}(\Delta x_{j_1})^{m_1}.$$

The flows at $x_{j_1}^* = a_{j_1j_11}$ and $\overline{x}_{j_2} \neq a_{j_1j_21} = a_{j_2j_21}$ are $(2m_{11}+1)^{\text{th}}$-source and sink flows for $a_{j_1j_10}(\overline{x}_{j_2} - a_{j_1j_21})^{n_1} > 0$ and $a_{j_1j_10}(\overline{x}_{j_2} - a_{j_1j_21})^{n_1} < 0$, respectively. The flows at $x_{j_1}^* = a_{j_1j_11}$ and $\overline{x}_{j_2} \neq a_{j_1j_21} = a_{j_2j_21}$ are $(2m_{11})^{\text{th}}$-upper saddle and lower-saddle flows for $a_{j_1j_10}(\overline{x}_{j_2} - a_{j_1j_21})^{n_1} > 0$ and $a_{j_1j_10}(\overline{x}_{j_2} - a_{j_1j_21})^{n_1} < 0$, respectively.

Thus, based on the above analysis, the flows at $x_{j_2}^* = a_{j_2j_21} = a_{j_1j_21}$ and on both sides of $x_{j_1}^* = a_{j_1j_11}$ are $[\text{XX}, (2n_{21}+1)]^{\text{th}}$ source and sink flows for $a_{j_2j_20} > 0$ and $a_{j_2j_20} < 0$, respectively. The flows at $x_{j_2}^* = a_{j_2j_21} = a_{j_1j_21}$ are $[\text{XX}, (2n_{21})]^{\text{th}}$-upper-saddle and lower-saddle flows for $a_{j_2j_20} > 0$ and $a_{j_2j_20} < 0$, respectively. Notice that $\text{XX} \in (2m_{11}+1), (2m_{11})$.

Therefore, there are 16 types of infinite-equilibriums at $x_{j_2}^* = a_{j_2j_21} = a_{j_1j_21}$.

(i_{2a1a}) *For* $n_1 = 2n_{11}+1$ *and* $n_2 = 2n_{21}+1$, *the infinite-equilibrium of* $x_{j_2}^* = a_{j_2j_21} = a_{j_1j_21}$ *with* $\overline{x}_{j_1} \neq a_{j_1j_11}$ *has the following properties* as in Eqs. (1.39)–(1.42).

- *For* $a_{j_1j_10}(\overline{x}_{j_1}-a_{j_1j_11})^{m_1}>0$ and $a_{j_2j_20}>0$,

$$(\overline{x}_{j_1},a_{j_2j_21})=\underbrace{([2(n_{11}-n_{21})]^{\text{th}}\text{II},(2n_{21}+1)^{\text{th}}\text{SO})}_{([2(n_{11}-n_{21})],(2n_{21}+1))\text{-increasing inflection source}}.$$

The infinite-equilibrium is a $([2(n_{11}-n_{21})]^{\text{th}}\text{II},(2n_{21}+1)^{\text{th}}\text{SO})$-increasing-inflection source at $x_{j_2}^*=a_{j_2j_21}=a_{j_1j_21}$.

- *For* $a_{j_1j_10}(\overline{x}_{j_1}-a_{j_1j_11})^{m_1}<0$ and $a_{j_2j_20}>0$,

$$(\overline{x}_{j_1},a_{j_2j_21})=\underbrace{([2(n_{11}-n_{21})]^{\text{th}}\text{DI},(2n_{21}+1)^{\text{th}}\text{SO})}_{([2(n_{11}-n_{21})],(2n_{21}+1))\text{-decreasing inflection source}}.$$

The infinite-equilibrium is a $([2(n_{11}-n_{21})]^{\text{th}}\text{DI},(2n_{21}+1)^{\text{th}}\text{SO})$-decreasing-inflection source at $x_{j_2}^*=a_{j_2j_21}=a_{j_1j_21}$.

- *For* $a_{j_1j_10}(\overline{x}_{j_1}-a_{j_1j_11})^{m_1}>0$ and $a_{j_2j_20}<0$,

$$(\overline{x}_{j_1},a_{j_2j_21})=\underbrace{([2(n_{11}-n_{21})]^{\text{th}}\text{DI},(2n_{21}+1)^{\text{th}}\text{SI})}_{([2(n_{11}-n_{21})],(2n_{21}+1))\text{-decreasing inflection sink}}.$$

The infinite-equilibrium is a $([2(n_{11}-n_{21})]^{\text{th}}\text{DI},(2n_{21}+1)^{\text{th}}\text{SI})$-decreasing-inflection sink at $x_{j_2}^*=a_{j_2j_21}=a_{j_1j_21}$.

- *For* $a_{j_1j_10}(\overline{x}_{j_1}-a_{j_1j_11})^{m_1}<0$ and $a_{j_2j_20}<0$,

$$(\overline{x}_{j_1},a_{j_2j_21})=\underbrace{([2(n_{11}-n_{21})]^{\text{th}}\text{II},(2n_{21}+1)^{\text{th}}\text{SI})}_{([2(n_{11}-n_{21})],(2n_{21}+1))\text{-increasing inflection sink}}.$$

The infinite-equilibrium is a $([2(n_{11}-n_{21})]^{\text{th}}\text{II},(2n_{21}+1)^{\text{th}}\text{SI})$-increasing-inflection sink at $x_{j_2}^*=a_{j_2j_21}=a_{j_1j_21}$.

(i_{2a1b}) *For* $n_1=2n_{11}$ *and* $n_2=2n_{21}+1$, *the infinite-equilibrium of* $x_{j_2}^*=a_{j_2j_21}=a_{j_1j_21}$ *with* $\overline{x}_{j_1}\neq a_{j_1j_11}\neq a_{j_2j_11}$ *has the following properties* as in Eqs. (1.43)–(1.46).

- *For* $a_{j_1j_10}(\overline{x}_{j_1}-a_{j_1j_11})^{m_1}>0$ and $a_{j_2j_20}>0$,

$$(\overline{x}_{j_1},a_{j_2j_21})=\underbrace{([2(n_{11}-n_{21})-1]^{\text{th}}\text{UP},(2n_{21}+1)^{\text{th}}\text{SO})}_{([2(n_{11}-n_{21})-1],(2n_{21}+1))\text{-up-parabola source}}.$$

The infinite-equilibrium is a-up-parabola source at $x_{j_2}^*=a_{j_2j_21}=a_{j_1j_21}$.

- *For* $a_{j_1j_10}(\overline{x}_{j_1}-a_{j_1j_11})^{m_1}<0$ and $a_{j_2j_20}>0$,

$$(\overline{x}_{j_1},a_{j_2j_21})=\underbrace{([2(n_{11}-n_{21})-1]^{\text{th}}\text{DP},(2n_{21}+1)^{\text{th}}\text{SO})}_{([2(n_{11}-n_{21})-1],(2n_{21}+1))\text{-down-parabola source}}.$$

The infinite-equilibrium is a $([2(n_{11}-n_{21})-1]^{\text{th}}\text{DP}, (2n_{21}+1)^{\text{th}}\text{SO})$-down-parabola source at $x_{j_2}^* = a_{j_2j_21} = a_{j_1j_21}$.

- *For* $a_{j_1j_10}(\overline{x}_{j_1} - a_{j_1j_11})^{m_1} > 0$ and $a_{j_2j_20} < 0$,

$$(\overline{x}_{j_1}, a_{j_2j_21}) = \underbrace{([2(n_{11}-n_{21})-1]^{\text{th}}\text{DP}, (2n_{21}+1)^{\text{th}}\text{SI})}_{([2(n_{11}-n_{21})-1],(2n_{21}+1))\text{-down-parabola sink}}.$$

The infinite-equilibrium is a $([2(n_{11}-n_{21})-1]^{\text{th}}\text{DP}, (2n_{21}+1)^{\text{th}}\text{SI})$-down-parabola sink at $x_{j_2}^* = a_{j_2j_21} = a_{j_1j_21}$.

- *For* $a_{j_1j_10}(\overline{x}_{j_1} - a_{j_1j_11})^{m_1} < 0$ and $a_{j_2j_20} < 0$,

$$(\overline{x}_{j_1}, a_{j_2j_21}) = \underbrace{([2(n_{11}-n_{21})-1]^{\text{th}}\text{UP}, (2n_{21}+1)^{\text{th}}\text{SI})}_{([2(n_{11}-n_{21})-1],(2n_{21}+1))\text{-up-parabola sink}}.$$

The infinite-equilibrium is a $([2(n_{11}-n_{21})-1]^{\text{th}}\text{UP}, (2n_{21}+1)^{\text{th}}\text{SI})$-up-parabola sink at $x_{j_2}^* = a_{j_2j_21} = a_{j_1j_21}$ *with* $\overline{x}_{j_1} \neq a_{j_1j_11} \neq a_{j_2j_11}$.

($\text{i}_{2\text{a}1\text{c}}$) *For* $n_1 = 2n_{11}+1$ *and* $n_2 = 2n_{21}$, *the infinite-equilibrium of* $x_{j_2}^* = a_{j_2j_21} = a_{j_1j_21}$ *with* $\overline{x}_{j_1} \neq a_{j_1j_11}$ *has the following properties* as in Eqs. (1.47)–(1.50).

- *For* $a_{j_1j_10}(\overline{x}_{j_1} - a_{j_1j_11})^{m_1} > 0$ and $a_{j_2j_20} > 0$,

$$(\overline{x}_{j_1}, a_{j_2j_21}) = \underbrace{([2(n_{11}-n_{21})+1]^{\text{th}}\text{UP}, (2n_{21})^{\text{th}}\text{US})}_{([2(n_{11}-n_{21})+1],(2n_{21}))\text{-up-parabola upper-saddle}}.$$

The infinite-equilibrium is a $([2(n_{11}-n_{21})+1]^{\text{th}}\text{UP}, (2n_{21})^{\text{th}}\text{US})$-up-parabola upper-saddle at $x_{j_2}^* = a_{j_2j_21} = a_{j_1j_21}$.

- *For* $a_{j_1j_10}(\overline{x}_{j_1} - a_{j_1j_11})^{m_1} < 0$ and $a_{j_2j_20} > 0$,

$$(\overline{x}_{j_1}, a_{j_2j_21}) = \underbrace{([2(n_{11}-n_{21})+1]^{\text{th}}\text{DP}, (2n_{21})^{\text{th}}\text{US})}_{([2(n_{11}-n_{21})+1],(2n_{21}))\text{-down-parabola upper-saddle}}.$$

The infinite-equilibrium is a $([2(n_{11}-n_{21})+1]^{\text{th}}\text{DP}, (2n_{21})^{\text{th}}\text{US})$-down-parabola upper-saddle at $x_{j_2}^* = a_{j_2j_21} = a_{j_1j_21}$.

- *For* $a_{j_1j_10}(\overline{x}_{j_1} - a_{j_1j_11})^{m_1} > 0$ and $a_{j_2j_20} < 0$,

$$(\overline{x}_{j_1}, a_{j_2j_21}) = \underbrace{([2(n_{11}-n_{21})+1]^{\text{th}}\text{DP}, (2n_{21})^{\text{th}}\text{LS})}_{([2(n_{11}-n_{21})+1],(2n_{21}))\text{-down-parabola lower-saddle}}.$$

The infinite-equilibrium is a $([2(n_{11}-n_{21})+1]^{\text{th}}\text{DP}, (2n_{21})^{\text{th}}\text{LS})$-down-parabola lower-saddle at $x_{j_2}^* = a_{j_2j_21} = a_{j_1j_21}$ *with* $\overline{x}_{j_1} \neq a_{j_1j_11} \neq a_{j_2j_11}$.

- *For* $a_{j_1j_10}(\overline{x}_{j_1} - a_{j_1j_11})^{m_1} < 0$ and $a_{j_2j_20} < 0$,

$$(\overline{x}_{j_1}, a_{j_2j_2 1}) = \underbrace{([2(n_{11} - n_{21}) + 1]^{\text{th}}\text{UP}, (2n_{21})^{\text{th}}\text{LS})}_{([2(n_{11}-n_{21})+1],(2n_{21}))\text{-up-parabola lower-saddle}}.$$

The infinite-equilibrium is a $([2(n_{11} - n_{21}) + 1]^{\text{th}}\text{UP}, (2n_{21})^{\text{th}}\text{LS})$-up-parabola lower-saddle at $x^*_{j_2} = a_{j_2j_2 1} = a_{j_1j_2 1}$.

($\mathrm{i}_{2\mathrm{a}1\mathrm{d}}$) *For* $n_1 = 2n_{11}$ *and* $n_2 = 2n_{21}$, *the infinite-equilibrium of* $x^*_{j_2} = a_{j_2j_2 1} = a_{j_1j_2 1}$ *with* $\overline{x}_{j_1} \neq a_{j_1j_1 1} \neq a_{j_2j_1 1}$ *has the following properties* as in Eqs. (1.51)–(1.54).

- *For* $a_{j_1j_1 0}(\overline{x}_{j_1} - a_{j_1j_1 1})^{m_1} > 0$ and $a_{j_2j_2 0} > 0$,

$$(\overline{x}_{j_1}, a_{j_2j_2 1}) = \underbrace{([2(n_{11} - n_{21})]^{\text{th}}\text{II}, (2n_{21})^{\text{th}}\text{US})}_{([2(n_{11}-n_{21})],(2n_{21}))\text{-increasing-inflection upper-saddle}}.$$

  The infinite-equilibrium is a $([2(n_{11} - n_{21})]^{\text{th}}\text{II}, (2n_{21})^{\text{th}}\text{US})$-increasing-inflection upper-saddle at $x^*_{j_2} = a_{j_2j_2 1} = a_{j_1j_2 1}$.
- *For* $a_{j_1j_1 0}(\overline{x}_{j_1} - a_{j_1j_1 1})^{m_1} < 0$ and $a_{j_2j_2 0} > 0$,

$$(\overline{x}_{j_1}, a_{j_2j_2 1}) = \underbrace{([2(n_{11} - n_{21})]^{\text{th}}\text{DI}, (2n_{21})^{\text{th}}\text{US})}_{([2(n_{11}-n_{21})],(2n_{21}))\text{-decreasing inflection upper-saddle}}.$$

  The infinite-equilibrium is a $([2(n_{11} - n_{21})]^{\text{th}}\text{DI}, (2n_{21})^{\text{th}}\text{US})$-decreasing-inflection upper-saddle at $x^*_{j_2} = a_{j_2j_2 1} = a_{j_1j_2 1}$.
- *For* $a_{j_1j_1 0}(\overline{x}_{j_1} - a_{j_1j_1 1})^{m_1} > 0$ and $a_{j_2j_2 0} < 0$,

$$(\overline{x}_{j_1}, a_{j_2j_2 1}) = \underbrace{([2(n_{11} - n_{21})]^{\text{th}}\text{DI}, (2n_{21})^{\text{th}}\text{LS})}_{([2(n_{11}-n_{21})],(2n_{21}))\text{-decreasing inflection lower-saddle}}.$$

  The infinite-equilibrium is a $([2(n_{11} - n_{21})]^{\text{th}}\text{DI}, (2n_{21})^{\text{th}}\text{LS})$-decreasing-inflection lower-saddle at $x^*_{j_2} = a_{j_2j_2 1} = a_{j_1j_2 1}$.
- *For* $a_{j_1j_1 0}(\overline{x}_{j_1} - a_{j_1j_1 1})^{m_1} < 0$ and $a_{j_2j_2 0} < 0$,

$$(\overline{x}_{j_1}, a_{j_2j_2 1}) = \underbrace{([2(n_{11} - n_{21})]^{\text{th}}\text{II}, (2n_{21})^{\text{th}}\text{LS})}_{([2(n_{11}-n_{21})],(2n_{21}))\text{-increasing inflection lower-saddle}}.$$

  The infinite-equilibrium is a $([2(n_{11} - n_{21})]^{\text{th}}\text{II}, (2n_{21})^{\text{th}}\text{LS})$-increasing-inflection lower-saddle at $x^*_{j_2} = a_{j_2j_2 1} = a_{j_1j_2 1}$.

($\mathrm{i}_{2\mathrm{a}2}$) On both sides of $\overline{x}_{j_2} = a_{j_1j_1 1}$, the inflection and parabola flows will be switched for $m_1 = (2m_{11} + 1)$, and the inflection and parabola flows will not be switched for $m_1 = 2m_{11}$.

There are 32 types of equilibrium $(x^*_{j_2}, x^*_{j_2}) = (a_{j_1j_1 1}, a_{j_2j_2 1})$ on the infinite-equilibrium of $x^*_{j_2} = a_{j_2j_2 1} = a_{j_1j_2 1}$.

(i_{2a2a}) *For* $m_1 = 2m_{11} + 1$, $n_1 = 2n_{11} + 1$, *and* $n_2 = 2n_{21} + 1$, *the equilibrium* of $(x^*_{j_1}, x^*_{j_2}) = (a_{j_1j_11}, a_{j_2j_21})$ on the infinite-equilibrium of $x^*_{j_2} = a_{j_2j_21} = a_{j_1j_21}$ *has the following properties* as in Eqs. (1.55)–(1.58).

- *For* $a_{j_1j_10} > 0$ and $a_{j_2j_20} > 0$,

$$(a_{j_1j_11}, a_{j_2j_21}) = \underbrace{([2(n_{11} - n_{21})]^{\text{th}}\text{DI:II},[(2m_{11}+1),(2n_{21}+1)]^{\text{th}}\text{SO})}_{([2(n_{11}-n_{21})],[(2m_{11}+1),(2n_{21}+1)])\text{-decreasing-to-increasing inflection source}}.$$

The equilibrium is a $([2(n_{11} - n_{21})]^{\text{th}}\text{DI:II},[(2m_{11}+1),(2n_{21}+1)]^{\text{th}}\text{SO})$-decreasing-to-increasing-inflection source at $(x^*_{j_1}, x^*_{j_2}) = (a_{j_1j_11}, a_{j_2j_21})$.

- *For* $a_{j_1j_10} < 0$ and $a_{j_2j_20} > 0$,

$$(a_{j_1j_11}, a_{j_2j_21}) = \underbrace{([2(n_{11} - n_{21})]^{\text{th}}\text{II:DI},[(2m_{11}+1),(2n_{21}+1)]^{\text{th}}\text{SO})}_{([2(n_{11}-n_{21})],[(2m_{11}+1),(2n_{21}+1)])\text{-increasing-to-decreasing inflection source}}.$$

The equilibrium is a $([2(n_{11} - n_{21})]^{\text{th}}\text{II:DI},[(2m_{11}+1),(2n_{21}+1)]^{\text{th}}\text{SO})$-increasing-to-decreasing-inflection source at $(x^*_{j_1}, x^*_{j_2}) = (a_{j_1j_11}, a_{j_2j_21})$.

- *For* $a_{j_1j_10} > 0$ and $a_{j_2j_20} < 0$,

$$(a_{j_1j_11}, a_{j_2j_21}) = \underbrace{([2(n_{11} - n_{21})]^{\text{th}}\text{II:DI},[(2m_{11}+1),(2n_{21}+1)]^{\text{th}}\text{SI})}_{([2(n_{11}-n_{21})],[(2m_{11}+1),(2n_{21}+1)])\text{-increasing-to-decreasing inflection sink}}.$$

The equilibrium is a $([2(n_{11} - n_{21})]^{\text{th}}\text{II:DI},[(2m_{11}+1),(2n_{21}+1)]^{\text{th}}\text{SI})$-increasing-to-decreasing-inflection sink at $(x^*_{j_1}, x^*_{j_2}) = (a_{j_1j_11}, a_{j_2j_21})$.

- *For* $a_{j_1j_10} < 0$ and $a_{j_2j_20} < 0$,

$$(a_{j_1j_11}, a_{j_2j_21}) = \underbrace{([2(n_{11} - n_{21})]^{\text{th}}\text{DI:II},[(2m_{11}+1),(2n_{21}+1)]^{\text{th}}\text{SI})}_{([2(n_{11}-n_{21})],[(2m_{11}+1),(2n_{21}+1)])\text{-decreasing-to-increasing inflection sink}}.$$

The equilibrium is a $([2(n_{11} - n_{21})]^{\text{th}}\text{DI:II},[(2m_{11}+1),(2n_{21}+1)]^{\text{th}}\text{SI})$-decreasing-to-increasing-inflection sink at $(x^*_{j_1}, x^*_{j_2}) = (a_{j_1j_11}, a_{j_2j_21})$.

(i_{2a2b}) For $m_1 = 2m_{11} + 1$, $n_1 = 2n_{11}$, and $n_2 = 2n_{21} + 1$, the equilibrium of $(x^*_{j_1}, x^*_{j_2}) = (a_{j_1j_11}, a_{j_2j_21})$ on the infinite-equilibrium of $x^*_{j_2} = a_{j_2j_21} = a_{j_1j_21}$ has the following properties as in Eqs. (1.59)–(1.62).

- *For* $a_{j_1j_10} > 0$ and $a_{j_2j_20} > 0$,

$$(a_{j_1j_11}, a_{j_2j_21}) = \underbrace{([2(n_{11} - n_{21}) - 1]^{\text{th}}\text{DP:UP},[(2m_{11}+1),(2n_{21}+1)]^{\text{th}}\text{SO})}_{([2(n_{11}-n_{21})-1],[(2m_{11}+1),(2n_{21}+1)])\text{-hyperbolic source}}.$$

The equilibrium is a $([2(n_{11}-n_{21})-1]^{\text{th}}\text{DP:UP},[(2m_{11}+1),(2n_{21}+1)]^{\text{th}}\ \text{SO})$-hyperbolic source at $(x_{j_1}^*, x_{j_2}^*) = (a_{j_1j_1 1}, a_{j_2j_2 1})$.

- *For* $a_{j_1j_1 0} < 0$ and $a_{j_2j_2 0} > 0$,

$$(a_{j_1j_1 1}, a_{j_2j_2 1}) = \underbrace{([2(n_{11}-n_{21})-1]^{\text{th}}\text{UP:DP},[(2m_{11}+1),(2n_{21}+1)]^{\text{th}}\text{SO})}_{([2(n_{11}-n_{21})-1],[(2m_{11}+1),(2n_{21}+1)])\text{-hyperbolic-secant source}}.$$

The equilibrium is a $([2(n_{11}-n_{21})-1]^{\text{th}}\text{UP:DP},[(2m_{11}+1),(2n_{21}+1)]^{\text{th}}\text{SO})$-hyperbolic-secant source at $(x_{j_1}^*, x_{j_2}^*) = (a_{j_1j_1 1}, a_{j_2j_2 1})$.

- *For* $a_{j_1j_1 0} > 0$ and $a_{j_2j_2 0} < 0$,

$$(a_{j_1j_1 1}, a_{j_2j_2 1}) = \underbrace{([2(n_{11}-n_{21})-1]^{\text{th}}\text{UP:DP},[(2m_{11}+1),(2n_{21}+1)]^{\text{th}}\text{SI})}_{([2(n_{11}-n_{21})-1],[(2m_{11}+1),(2n_{21}+1)])\text{-hyperbolic-secant sink}}.$$

The equilibrium is a $([2(n_{11}-n_{21})-1]^{\text{th}}\text{UP:DP},[(2m_{11}+1),(2n_{21}+1)]^{\text{th}}\text{SI})$-hyperbolic-secant sink at $(x_{j_1}^*, x_{j_2}^*) = (a_{j_1j_1 1}, a_{j_2j_2 1})$.

- *For* $a_{j_1j_1 0} < 0$ and $a_{j_2j_2 0} < 0$,

$$(a_{j_1j_1 1}, a_{j_2j_2 1}) = \underbrace{([2(n_{11}-n_{21})-1]^{\text{th}}\text{DP:UP},[(2m_{11}+1),(2n_{21}+1)]^{\text{th}}\text{SI})}_{([2(n_{11}-n_{21})-1],[(2m_{11}+1),(2n_{21}+1)])\text{-hyperbolic sink}}.$$

The equilibrium is a $([2(n_{11}-n_{21})-1]^{\text{th}}\text{DP:UP},[(2m_{11}+1),(2n_{21}+1)]^{\text{th}}\text{SI})$-hyperbolic sink at $(x_{j_1}^*, x_{j_2}^*) = (a_{j_1j_1 1}, a_{j_2j_2 1})$.

(i_{2a2c}) For $m_1 = 2m_{11}+1, n_1 = 2n_{11}+1$, and $n_2 = 2n_{21}$, the equilibrium of $(x_{j_1}^*, x_{j_2}^*) = (a_{j_1j_1 1}, a_{j_2j_2 1})$ on the infinite-equilibrium of $x_{j_2}^* = a_{j_2j_2 1} = a_{j_1j_2 1}$ has the following properties as in Eqs. (1.63)–(1.66).

- *For* $a_{j_1j_1 0} > 0$ and $a_{j_2j_2 0} > 0$,

$$(a_{j_1j_1 1}, a_{j_2j_2 1}) = \underbrace{([2(n_{11}-n_{21})+1]^{\text{th}}\text{DP:UP},[(2m_{11}+1),(2n_{21})]^{\text{th}}\text{US})}_{([2(n_{11}-n_{21})+1],[(2m_{11}+1),(2n_{21})])\text{-hyperbolic upper-saddle}}.$$

The equilibrium is a $([2(n_{11}-n_{21})+1]^{\text{th}}\text{DP:UP},[(2m_{11}+1),(2n_{21})]^{\text{th}}\text{US})$-hyperbolic upper-saddle at $(x_{j_1}^*, x_{j_2}^*) = (a_{j_1j_1 1}, a_{j_2j_2 1})$.

- *For* $a_{j_1j_1 0} < 0$ and $a_{j_2j_2 0} > 0$,

$$(a_{j_1j_1 1}, a_{j_2j_2 1}) = \underbrace{([2(n_{11}-n_{21})+1]^{\text{th}}\text{UP:DP},[(2m_{11}+1),(2n_{21})]^{\text{th}}\text{US})}_{([2(n_{11}-n_{21})+1],[(2m_{11}+1),(2n_{21})])\text{-hyperbolic-secant upper-saddle}}.$$

The equilibrium is a $([2(n_{11}-n_{21})+1]^{\text{th}}\text{UP:DP},[(2m_{11}+1),(2n_{21})]^{\text{th}}\text{LS})$-hyperbolic secant upper-saddle at $(x^*_{j_1},x^*_{j_2})=(a_{j_1j_1 1},a_{j_2j_2 1})$.

- *For* $a_{j_1j_1 0}>0$ and $a_{j_2j_2 0}<0$,

$$(a_{j_1j_1 1},a_{j_2j_2 1})=\underbrace{([2(n_{11}-n_{21})+1]^{\text{th}}\text{UP:DP},[(2m_{11}+1),(2n_{21})]^{\text{th}}\text{LS})}_{([2(n_{11}-n_{21})+1],[(2m_{11}+1),(2n_{21})])\text{-hyperbolic-secant lower-saddle}}.$$

The equilibrium is a $([2(n_{11}-n_{21})+1]^{\text{th}}\text{UP:DP},[(2m_{11}+1),(2n_{21})]^{\text{th}}\text{LS})$-hyperbolic-secant lower-saddle at $(x^*_{j_1},x^*_{j_2})=(a_{j_1j_1 1},a_{j_2j_2 1})$.

- *For* $a_{j_1j_1 0}<0$ and $a_{j_2j_2 0}<0$,

$$(a_{j_1j_1 1},a_{j_2j_2 1})=\underbrace{([2(n_{11}-n_{21})+1]^{\text{th}}\text{DP:UP},[(2m_{11}+1),(2n_{21})]^{\text{th}}\text{LS})}_{([2(n_{11}-n_{21})+1],[(2m_{11}+1),(2n_{21})])\text{-hyperbolic lower-saddle}}.$$

The equilibrium is a $([2(n_{11}-n_{21})+1]^{\text{th}}\text{DP:UP},[(2m_{11}+1),(2n_{21})]^{\text{th}}\text{LS})$-hyperbolic lower-saddle at $(x^*_{j_1},x^*_{j_2})=(a_{j_1j_1 1},a_{j_2j_2 1})$.

($\text{i}_{2\text{a}2\text{d}}$) For $m_1=2m_{11}+1,n_1=2n_{11}$, and $n_2=2n_{21}$, the equilibrium of $(x^*_{j_2},x^*_{j_2})=(a_{j_1j_1 1},a_{j_2j_2 1})$ on the infinite-equilibrium of $x^*_{j_2}=a_{j_2j_2 1}=a_{j_1j_2 1}$ has the following properties as in Eqs. (1.67)–(1.70).

- For $a_{j_1j_1 0}>0$ and $a_{j_2j_2 0}>0$,

$$(a_{j_1j_1 1},a_{j_2j_2 1})=\underbrace{([2(n_{11}-n_{21})]^{\text{th}}\text{DI:II},[(2m_{11}+1),(2n_{21})]^{\text{th}}\text{US})}_{([2(n_{11}-n_{21})],[(2m_{11}+1),(2n_{21})])\text{-decreasing-to-increasing inflection upper-saddle}}.$$

The equilibrium is a $([2(n_{11}-n_{21})]^{\text{th}}\text{DI:II},[(2m_{11}+1),(2n_{21})]^{\text{th}}\text{US})$-decreasing-to-increasing-inflection upper-saddle at $(x^*_{j_1},x^*_{j_2})=(a_{j_1j_1 1},a_{j_2j_2 1})$.

- For $a_{j_1j_1 0}<0$ and $a_{j_2j_2 0}>0$,

$$(a_{j_1j_1 1},a_{j_2j_2 1})=\underbrace{([2(n_{11}-n_{21})]^{\text{th}}\text{II:DI},[(2m_{11}+1),(2n_{21})]^{\text{th}}\text{US})}_{([2(n_{11}-n_{21})],[(2m_{11}+1),(2n_{21})])\text{-increasing-to-decreasing inflection upper-saddle}}.$$

The equilibrium is a $([2(n_{11}-n_{21})]^{\text{th}}\text{II:DI},[(2m_{11}+1),(2n_{21})]^{\text{th}}\text{US})$-increasing-to-decreasing-inflection upper-saddle at $(x^*_{j_1},x^*_{j_2})=(a_{j_1j_1 1},a_{j_2j_2 1})$.

- For $a_{j_1j_1 0}>0$ and $a_{j_2j_2 0}<0$,

$$(a_{j_1j_1 1},a_{j_2j_2 1})=\underbrace{([2(n_{11}-n_{21})]^{\text{th}}\text{II:DI},[(2m_{11}+1),(2n_{21})]^{\text{th}}\text{LS})}_{([2(n_{11}-n_{21})],[(2m_{11}+1),(2n_{21})])\text{-increasing-to-decreasing inflection lower-saddle}}.$$

The equilibrium is a $([2(n_{11}-n_{21})]^{\text{th}}\text{II:DI},[(2m_{11}+1),(2n_{21})]^{\text{th}}\text{LS})$-increasing-to-decreasing-inflection lower-saddle at $(x_{j_1}^*, x_{j_2}^*) = (a_{j_1j_1 1}, a_{j_2j_2 1})$.

- For $a_{j_1j_1 0} < 0$ and $a_{j_2j_2 0} < 0$,

$$(a_{j_1j_1 1}, a_{j_2j_2 1}) = \underbrace{([2(n_{11}-n_{21})]^{\text{th}}\text{DI:II},[(2m_{11}+1),(2n_{21})]^{\text{th}}\text{LS})}_{([2(n_{11}-n_{21})],[(2m_{11}+1),(2n_{21})])\text{-decreasing-to-increasing inflection lower-saddle}}.$$

The equilibrium is a $([2(n_{11}-n_{21})]^{\text{th}}\text{DI:II},[(2m_{11}+1),(2n_{21})]^{\text{th}}\text{LS})$-decreasing-to-increasing-inflection lower-saddle at $(x_{j_1}^*, x_{j_2}^*) = (a_{j_1j_1 1}, a_{j_2j_2 1})$.

(i_{2a2e}) For $m_1 = 2m_{11}$, $n_1 = 2n_{11}+1$, and $n_2 = 2n_{21}+1$, the equilibrium of $(x_{j_1}^*, x_{j_2}^*) = (a_{j_1j_1 1}, a_{j_2j_2 1})$ on the infinite-equilibrium of $x_{j_2}^* = a_{j_2j_2 1} = a_{j_1j_2 1}$ has the following properties as in Eqs. (1.71)–(1.74)

- For $a_{j_1j_1 0} > 0$ and $a_{j_2j_2 0} > 0$,

$$(a_{j_1j_1 1}, a_{j_2j_2 1}) = \underbrace{([2(n_{11}-n_{21})]^{\text{th}}\text{II:II},[(2m_{11}),(2n_{21}+1)]^{\text{th}}\text{SO})}_{([2(n_{11}-n_{21})],[(2m_{11}),(2n_{21}+1)])\text{-increasing inflection source}}.$$

The equilibrium is a $([2(n_{11}-n_{21})]^{\text{th}}\text{II:II},[(2m_{11}),(2n_{21}+1)]^{\text{th}}\text{SO})$-increasing-inflection source at $(x_{j_1}^*, x_{j_2}^*) = (a_{j_1j_1 1}, a_{j_2j_2 1})$.

- For $a_{j_1j_1 0} < 0$ and $a_{j_2j_2 0} > 0$,

$$(a_{j_1j_1 1}, a_{j_2j_2 1}) = \underbrace{([2(n_{11}-n_{21})]^{\text{th}}\text{DI:DI},[(2m_{11}),(2n_{21}+1)]^{\text{th}}\text{SO})}_{([2(n_{11}-n_{21})],[(2m_{11}),(2n_{21}+1)])\text{-decreasing inflection source}}.$$

The equilibrium is a $([2(n_{11}-n_{21})]^{\text{th}}\text{DI:DI},[(2m_{11}),(2n_{21}+1)]^{\text{th}}\text{SO})$-decreasing-inflection source at $(x_{j_1}^*, x_{j_2}^*) = (a_{j_1j_1 1}, a_{j_2j_2 1})$.

- For $a_{j_1j_1 0} > 0$ and $a_{j_2j_2 0} < 0$,

$$(a_{j_1j_1 1}, a_{j_2j_2 1}) = \underbrace{([2(n_{11}-n_{21})]^{\text{th}}\text{DI:DI},[(2m_{11}),(2n_{21}+1)]^{\text{th}}\text{SI})}_{([2(n_{11}-n_{21})],[(2m_{11}),(2n_{21}+1)])\text{-decreasing inflection sink}}.$$

The equilibrium is a $([2(n_{11}-n_{21})]^{\text{th}}\text{DI:DI},[(2m_{11}),(2n_{21}+1)]^{\text{th}}\text{SI})$-decreasing-inflection sink at $(x_{j_1}^*, x_{j_2}^*) = (a_{j_1j_1 1}, a_{j_2j_2 1})$.

- For $a_{j_1j_1 0} < 0$ and $a_{j_2j_2 0} < 0$,

$$(a_{j_1j_1 1}, a_{j_2j_2 1}) = \underbrace{([2(n_{11}-n_{21})]^{\text{th}}\text{II:II},[(2m_{11}),(2n_{21}+1)]^{\text{th}}\text{SI})}_{([2(n_{11}-n_{21})],[(2m_{11}),(2n_{21}+1)])\text{-increasing inflection sink}}.$$

The equilibrium is a ([2$(n_{11}-n_{21})$]thII:II,[$(2m_{11}),(2n_{21}+1)$]thSI)-increasing-inflection source at $(x^*_{j_1}, x^*_{j_2}) = (a_{j_1j_1 1}, a_{j_2j_2 1})$.

($\text{i}_{2\text{a}2\text{f}}$) For $m_1 = 2m_{11}$, $n_1 = 2n_{11}$, and $n_2 = 2n_{21}+1$, the equilibrium of $(x^*_{j_1}, x^*_{j_2}) = (a_{j_1j_1 1}, a_{j_2j_2 1})$ on the infinite-equilibrium of $x^*_{j_2} = a_{j_2j_2 1} = a_{j_1j_2 1}$ has the following properties as in Eqs. (1.75)–(1.78).

- For $a_{j_1j_1 0} > 0$ and $a_{j_2j_2 0} > 0$,

$$(a_{j_1j_1 1}, a_{j_2j_2 1}) = \underbrace{([2(n_{11}-n_{21})-1]^{\text{th}}\text{UP:UP},[(2m_{11}),(2n_{21}+1)]^{\text{th}}\text{SO})}_{([2(n_{11}-n_{21})-1],[(2m_{11}),(2n_{21}+1)])\text{-up-parabola source}}.$$

The equilibrium is a ([2$(n_{11}-n_{21})-1$]thUP:UP,[$(2m_{11}),(2n_{21}+1)$]thSO)-up-parabola source at $(x^*_{j_1}, x^*_{j_2}) = (a_{j_1j_1 1}, a_{j_2j_2 1})$.

- For $a_{j_1j_1 0} < 0$ and $a_{j_2j_2 0} > 0$,

$$(a_{j_1j_1 1}, a_{j_2j_2 1}) = \underbrace{([2(n_{11}-n_{21})-1]^{\text{th}}\text{DP:DP},[(2m_{11}),(2n_{21}+1)]^{\text{th}}\text{SO})}_{([2(n_{11}-n_{21})-1],[(2m_{11}),(2n_{21}+1)])\text{-down-parabola source}}.$$

The equilibrium is a ([2$(n_{11}-n_{21})-1$]thDP:DP,[$(2m_{11}),(2n_{21}+1)$]thSO)-down-parabola source at $(x^*_{j_1}, x^*_{j_2}) = (a_{j_1j_1 1}, a_{j_2j_2 1})$.

- For $a_{j_1j_1 0} > 0$ and $a_{j_2j_2 0} < 0$,

$$(a_{j_1j_1 1}, a_{j_2j_2 1}) = \underbrace{([2(n_{11}-n_{21})-1]^{\text{th}}\text{DP:DP},[(2m_{11}),(2n_{21}+1)]^{\text{th}}\text{SI})}_{([2(n_{11}-n_{21})-1],[(2m_{11}),(2n_{21}+1)])\text{-down-parabola sink}}.$$

The equilibrium is a ([2$(n_{11}-n_{21})-1$]thDP:DP,[$(2m_{11}),(2n_{21}+1)$]thSI)-down-parabola sink at $(x^*_{j_1}, x^*_{j_2}) = (a_{j_1j_1 1}, a_{j_2j_2 1})$.

- For $a_{j_1j_1 0} < 0$ and $a_{j_2j_2 0} < 0$,

$$(a_{j_1j_1 1}, a_{j_2j_2 1}) = \underbrace{([2(n_{11}-n_{21})-1]^{\text{th}}\text{UP:UP},[(2m_{11}),(2n_{21}+1)]^{\text{th}}\text{SI})}_{([2(n_{11}-n_{21})-1],[(2m_{11}),(2n_{21}+1)])\text{-up-parabola sink}}.$$

The equilibrium is a ([2$(n_{11}-n_{21})-1$]thUP:UP,[$(2m_{11}),(2n_{21}+1)$]thSI)-up-parabola sink at $(x^*_{j_1}, x^*_{j_2}) = (a_{j_1j_1 1}, a_{j_2j_2 1})$.

($\text{i}_{2\text{a}2\text{g}}$) For $m_1 = 2m_{11}$, $n_1 = 2n_{11}+1$, and $n_2 = 2n_{21}$, the equilibrium of $(x^*_{j_1}, x^*_{j_2}) = (a_{j_1j_1 1}, a_{j_2j_2 1})$ on the infinite-equilibrium of $x^*_{j_2} = a_{j_2j_2 1} = a_{j_1j_2 1}$ has the following properties as in Eqs. (1.79)–(1.82).

- For $a_{j_1j_1 0} > 0$ and $a_{j_2j_2 0} > 0$,

$$(a_{j_1j_1 1}, a_{j_2j_2 1}) = \underbrace{([2(n_{11}-n_{21})+1]^{\text{th}}\text{UP:UP},[(2m_{11}),(2n_{21})]^{\text{th}}\text{US})}_{([2(n_{11}-n_{21})+1],[(2m_{11}),(2n_{21}+1)])\text{-up-parabola upper-saddle}}.$$

The equilibrium is a ($[2(n_{11}-n_{21})+1]^{\text{th}}$UP:UP,$[(2m_{11}),(2n_{21})]^{\text{th}}$US)-up-parabola upper-saddle at $(x^*_{j_2},x^*_{j_2})=(a_{j_1j_1 1},a_{j_2j_2 1})$.

- For $a_{j_1j_1 0}<0$ and $a_{j_2j_2 0}>0$,

$$(a_{j_1j_1 1},a_{j_2j_2 1})=\underbrace{([2(n_{11}-n_{21})+1]^{\text{th}}\text{DP:DP},[(2m_{11}),(2n_{21})]^{\text{th}}\text{US})}_{([2(n_{11}-n_{21})+1],[(2m_{11}),(2n_{21}+1)])\text{-down-parabola upper-saddle}}.$$

The equilibrium is a ($[2(n_{11}-n_{21})+1]^{\text{th}}$DP:DP,$[(2m_{11}),(2n_{21})]^{\text{th}}$US)-down-parabola upper-saddle at $(x^*_{j_1},x^*_{j_2})=(a_{j_1j_1 1},a_{j_2j_2 1})$.

- For $a_{j_1j_1 0}>0$ and $a_{j_2j_2 0}<0$,

$$(a_{j_1j_1 1},a_{j_2j_2 1})=\underbrace{([2(n_{11}-n_{21})+1]^{\text{th}}\text{DP:DP},[(2m_{11}),(2n_{21})]^{\text{th}}\text{LS})}_{([2(n_{11}-n_{21})+1],[(2m_{11}),(2n_{21})])\text{-down-parabola lower-saddle}}.$$

The equilibrium is a ($[2(n_{11}-n_{21})+1]^{\text{th}}$DP:DP,$[(2m_{11}),(2n_{21})]^{\text{th}}$LS)-down-parabola lower-saddle at $(x^*_{j_1},x^*_{j_2})=(a_{j_1j_1 1},a_{j_2j_2 1})$.

- For $a_{j_1j_1 0}<0$ and $a_{j_2j_2 0}<0$,

$$(a_{j_1j_1 1},a_{j_2j_2 1})=\underbrace{([2(n_{11}-n_{21})+1]^{\text{th}}\text{UP:UP},[(2m_{11})^{\text{th}},(2n_{21})]\text{LS})}_{([2(n_{11}-n_{21})+1],[(2m_{11}),(2n_{21})])\text{-up-parabola lower-saddle}}.$$

The equilibrium is a ($[2(n_{11}-n_{21})+1]^{\text{th}}$UP:UP,$[(2m_{11}),(2n_{21})]^{\text{th}}$LS)-up-parabola lower-saddle at $(x^*_{j_1},x^*_{j_2})=(a_{j_1j_1 1},a_{j_2j_2 1})$.

(i_{2a2h}) For $m_1=2m_{11}$,$n_1=2n_{11}$, and $n_2=2n_{21}$, the equilibrium of $(x^*_{j_1},x^*_{j_2})=(a_{j_1j_1 1},a_{j_2j_2 1})$ on the infinite-equilibrium of $x^*_{j_2}=a_{j_2j_2 1}=a_{j_1j_2 1}$ has the following properties as in Eqs. (1.83)–(1.86).

- For $a_{j_1j_1 0}>0$ and $a_{j_2j_2 0}>0$,

$$(a_{j_1j_1 1},a_{j_2j_2 1})=\underbrace{([2(n_{11}-n_{21})]^{\text{th}}\text{II:II},[(2m_{11}),(2n_{21})]^{\text{th}}\text{US})}_{([2(n_{11}-n_{21})],[(2m_{11}),(2n_{21})])\text{-increasing inflection upper-saddle}}.$$

The equilibrium is a ($[2(n_{11}-n_{21})]^{\text{th}}$II:II,$[(2m_{11}),(2n_{21})]^{\text{th}}$US)-increasing-inflection upper-saddle at $(x^*_{j_1},x^*_{j_2})=(a_{j_1j_1 1},a_{j_2j_2 1})$.

- For $a_{j_1j_1 0}<0$ and $a_{j_2j_2 0}>0$,

$$(a_{j_1j_1 1},a_{j_2j_2 1})=\underbrace{([2(n_{11}-n_{21})]^{\text{th}}\text{DI:DI},[(2m_{11}),(2n_{21})]^{\text{th}}\text{US})}_{([2(n_{11}-n_{21})],[(2m_{11}),(2n_{21})]),\text{-decreasing-inflection upper-saddle}}.$$

The equilibrium is a ($[2(n_{11}-n_{21})]^{\text{th}}$DI:DI,$[(2m_{11}),(2n_{21})]^{\text{th}}$US)-decreasing-inflection upper-saddle at $(x^*_{j_1},x^*_{j_2})=(a_{j_1j_1 1},a_{j_2j_2 1})$.

- For $a_{j_1j_1 0}>0$ and $a_{j_2j_2 0}<0$,

$$(a_{j_1j_11}, a_{j_2j_21}) = \underbrace{([2(n_{11} - n_{21})]^{\text{th}}\text{DI:DI},[(2m_{11}),(2n_{21})]^{\text{th}}\text{LS})}_{([2(n_{11}-n_{21})],[(2m_{11}+1),(2n_{21})])\text{-decreasing inflection lower-saddle}}.$$

The equilibrium is a $([2(n_{11} - n_{21})]^{\text{th}}\text{DI:DI},[(2m_{11}),(2n_{21})]^{\text{th}}\text{LS})$-decreasing-inflection lower-saddle at $(x_{j_1}^*, x_{j_2}^*) = (a_{j_1j_11}, a_{j_2j_21})$.

- For $a_{j_1j_10} < 0$ and $a_{j_2j_20} < 0$,

$$(a_{j_1j_11}, a_{j_2j_21}) = \underbrace{([2(n_{11} - n_{21})]^{\text{th}}\text{II:II},[(2m_{11}),(2n_{21})]^{\text{th}}\text{LS})}_{([2(n_{11}-n_{21})],[(2m_{11}),(2n_{21})])\text{-increasing inflection lower-saddle}}.$$

The equilibrium is a $([2(n_{11} - n_{21})]^{\text{th}}\text{II:II},[(2m_{11}),(2n_{21})]^{\text{th}}\text{LS})$-increasing-inflection lower-saddle at $(x_{j_1}^*, x_{j_2}^*) = (a_{j_1j_11}, a_{j_2j_21})$.

(i_{2b}) For $n_1 < n_2$, the first integral manifold is

$$\frac{1}{m_1 - 1}\left[\frac{1}{(x_{j_1} - a_{j_1j_11})^{m_1-1}} - \frac{1}{(x_{j_10} - a_{j_1j_11})^{m_1-1}}\right]$$
$$= \frac{a_{j_1j_10}}{a_{j_2j_20}}\frac{1}{n_2 - n_1 - 1}\left[\frac{1}{(x_{j_2} - a_{j_2j_21})^{n_2-n_1-1}} - \frac{1}{(x_{j_20} - a_{j_2j_21})^{n_2-n_1-1}}\right],$$

for $m_1 \neq 1, n_2 - n_1 \neq 1$;

$$\frac{1}{m_1 - 1}\left[\frac{1}{(x_{j_1} - a_{j_1j_11})^{m_1-1}} - \frac{1}{(x_{j_10} - a_{j_1j_11})^{m_1-1}}\right] = -\ln\frac{|x_{j_1} - a_{j_1j_11}|}{|x_{j_10} - a_{j_1j_11}|},$$

for $m_1 = 1$;

$$\frac{1}{n_2 - n_1 - 1}\left[\frac{1}{(x_{j_2} - a_{j_2j_21})^{n_2-n_1-1}} - \frac{1}{(x_{j_20} - a_{j_2j_21})^{n_2-n_1-1}}\right] = -\ln\frac{|x_{j_2} - a_{j_2j_21}|}{|x_{j_20} - a_{j_2j_21}|},$$

for $n_2 - n_1 - 1 = 0$.

There are 32 types of infinite-equilibriums at $x_{j_2}^* = a_{j_2j_21} = a_{j_1j_21}$.

For $x_{j_2}^* = a_{j_2j_21} = a_{j_1j_21}$ and $\overline{x}_{j_1} \neq a_{j_1j_11}$, if

$$\left.\frac{dx_{j_1}}{dx_{j_2}}\right|_{x_{j_2}^*=a_{j_2j_21}\pm\varepsilon} = \lim_{\varepsilon\to 0}\frac{a_{j_1j_10}(\overline{x}_{j_1} - a_{j_1j_11})^{m_1}}{a_{j_2j_20}(\pm\varepsilon)^{n_2-n_1}} = +\infty,$$

the flow at $x_{j_2}^* = a_{j_2j_21} = a_{j_1j_21}$ are a $[2(n_{21} - n_{11})]^{\text{th}}$-order increasing-concave flows for $n_1 = 2n_{11} + 1, n_2 = 2n_{21} + 1$ and $n_1 = 2n_{11}, n_2 = 2n_{21}$), if

$$\left.\frac{dx_{j_1}}{dx_{j_2}}\right|_{x_{j_2}^*=a_{j_2j_21}\pm\varepsilon} = \lim_{\varepsilon\to 0}\frac{a_{j_1j_10}(\overline{x}_{j_1} - a_{j_1j_11})^{m_1}}{a_{j_2j_20}(\pm\varepsilon)^{n_2-n_1}} = -\infty,$$

the flow at $x_{j_2}^* = a_{j_2j_21} = a_{j_1j_21}$ is a $[2(n_{21} - n_{11})]^{\text{th}}$-order decreasing-concave flow for $n_1 = 2n_{11} + 1, n_2 = 2n_{21} + 1$ and $n_1 = 2n_{11}, n_2 = 2n_{21}$).

For $x_{j_2}^* = a_{j_2j_21} = a_{j_1j_21}$ and $\overline{x}_{j_1} \neq a_{j_1j_11}$, if

$$\left.\frac{dx_{j_1}}{dx_{j_2}}\right|_{x^*_{j_2}=a_{j_2j_21}} = \lim_{x^*_{j_2}\to a_{j_2j_21}\pm} \frac{a_{j_1j_10}(\overline{x}_{j_1} - a_{j_1j_11})^{m_1}}{a_{j_2j_20}(x_{j_2} - a_{j_2j_21})^{n_2-n_1}} = \pm\infty,$$

the flow at $x^*_{j_2} = a_{j_2j_21} = a_{j_1j_21}$ is a $[2(n_{21} - n_{11}) \pm 1]^{\text{th}}$-order up-asymptote flow for $n_1 = 2n_{11}, n_2 = 2n_{21} + 1$ and $n_1 = 2n_{11} + 1, n_2 = 2n_{21}$), if

$$\left.\frac{dx_{j_1}}{dx_{j_2}}\right|_{x^*_{j_2}=a_{j_2j_21}\pm\varepsilon} = \lim_{\varepsilon\to 0} \frac{a_{j_1j_10}(\overline{x}_{j_1} - a_{j_1j_11})^{m_1}}{a_{j_2j_20}(\pm\varepsilon)^{n_2-n_1}} = \mp\infty,$$

the flow at $x^*_{j_2} = a_{j_2j_21} = a_{j_1j_21}$ is a $[2(n_{21} - n_{11}) \pm 1]^{\text{th}}$-order down-asymptote flow for $n_1 = 2n_{11}, n_2 = 2n_{21} + 1$ and $n_1 = 2n_{11} + 1, n_2 = 2n_{21}$).

The variational equation at $x^*_{j_2} = a_{j_2j_21} = a_{j_1j_21}$ is

$$\Delta\dot{x}_{j_2} = a_{j_2j_20}(\Delta x_{j_2})^{n_2}.$$

The flows at $x^*_{j_2} = a_{j_2j_21} = a_{j_1j_21}$ are $(2n_{21} + 1)^{\text{th}}$ source and sink flows for $a_{j_2j_20} > 0$ and $a_{j_2j_20} < 0$, respectively. The flows at $x^*_{j_2} = a_{j_2j_21} = a_{j_1j_21}$ are $(2n_{21})^{\text{th}}$-upper-saddle and lower-saddle flows for $a_{j_2j_20} > 0$ and $a_{j_2j_20} < 0$, respectively.

At $x^*_{j_1} = a_{j_1j_11}$, the variational equation for $\overline{x}_{j_2} \neq a_{j_1j_21} = a_{j_2j_21}$ is

$$\Delta\dot{x}_{j_1} = a_{j_1j_10}(\overline{x}_{j_2} - a_{j_1j_21})^{n_1}(\Delta x_{j_1})^{m_1}.$$

The flows at $x^*_{j_1} = a_{j_1j_11}$ and $\overline{x}_{j_2} \neq a_{j_1j_21} = a_{j_2j_21}$ are $(2m_{11} + 1)^{\text{th}}$-source and sink flows for $a_{j_1j_10}(\overline{x}_{j_2} - a_{j_1j_21})^{n_1} > 0$ and $a_{j_1j_10}(\overline{x}_{j_2} - a_{j_1j_21})^{n_1} < 0$, respectively. The flows at $x^*_{j_1} = a_{j_1j_11}$ and $\overline{x}_{j_2} \neq a_{j_1j_21} = a_{j_2j_21}$ are $(2m_{11})^{\text{th}}$-upper saddle and lower-saddle flows for $a_{j_1j_10}(\overline{x}_{j_2} - a_{j_1j_21})^{n_1} > 0$ and $a_{j_1j_10}(\overline{x}_{j_2} - a_{j_1j_21})^{n_1} < 0$, respectively.

Thus, based on the above analysis, the flows at $x^*_{j_2} = a_{j_2j_21} = a_{j_1j_21}$ and on both sides of $x^*_{j_1} = a_{j_1j_11}$ are $[XX, (2n_{21} + 1)]^{\text{th}}$ source and sink flows for $a_{j_2j_20} > 0$ and $a_{j_2j_20} < 0$, respectively. The flows at $x^*_{j_2} = a_{j_2j_21} = a_{j_1j_21}$ are $[\text{XX}, (2n_{21})]^{\text{th}}$-upper-saddle and lower-saddle flows for $a_{j_2j_20} > 0$ and $a_{j_2j_20} < 0$, respectively. Notice that $\text{XX} \in (2m_{11}+1), (2m_{11})$.

(i_{2b1}) There are 16 types of infinite-equilibriums at $x^*_{j_2} = a_{j_2j_21} = a_{j_1j_21}$.

(i_{2b1a}) For $n_1 = 2n_{11} + 1$ and $n_2 = 2n_{21} + 1$, the infinite-equilibrium of $x^*_{j_2} = a_{j_2j_21} = a_{j_1j_21}$ and has the following properties as in (1.88)–(1.91).

- For $a_{j_1j_10}(\overline{x}_{j_1} - a_{j_1j_11})^{m_1} > 0$ and $a_{j_2j_20} > 0$,

$$(\overline{x}_{j_1}, a_{j_2j_21}) = \underbrace{([2(n_{21} - n_{11})]^{\text{th}}\text{IC}, (2n_{21} + 1)^{\text{th}}\text{SO})}_{([2(n_{21}-n_{11})],(2n_{21}+1))\text{-increasing concave source}}.$$

  The infinite-equilibrium is a $([2(n_{21} - n_{11})]^{\text{th}}\text{IC}, (2n_{21} + 1)^{\text{th}}\text{SO})$-increasing-concave source at $x^*_{j_2} = a_{j_2j_21} = a_{j_1j_21}$.

- For $a_{j_1j_10}(\overline{x}_{j_1} - a_{j_1j_11})^{m_1} < 0$ and $a_{j_2j_20} > 0$,

$$(\overline{x}_{j_1}, a_{j_2j_21}) = \underbrace{([2(n_{21} - n_{11})]^{\text{th}}\text{DC}, (2n_{21} + 1)^{\text{th}}\text{SO})}_{([2(n_{21}-n_{11})],(2n_{21}+1))\text{-decreasing concave source}}.$$

The infinite-equilibrium is a $([2(n_{21} - n_{11})]^{\text{th}}\text{DC}, (2n_{21} + 1)^{\text{th}}\text{SO})$-decreasing-concave source at $x^*_{j_2} = a_{j_2j_21} = a_{j_1j_21}$.

- For $a_{j_1j_10}(\overline{x}_{j_1} - a_{j_1j_11})^{m_1} > 0$ and $a_{j_2j_20} < 0$,

$$(\overline{x}_{j_1}, a_{j_2j_21}) = \underbrace{([2(n_{21} - n_{11})]^{\text{th}}\text{DC}, (2n_{21} + 1)^{\text{th}}\text{SI})}_{([2(n_{21}-n_{11})],(2n_{21}+1))\text{-decreasing concave sink}}.$$

The infinite-equilibrium is a $([2(n_{21} - n_{11})]^{\text{th}}\text{DC}, (2n_{21} + 1)^{\text{th}}\text{SI})$-decreasing-concave sink at $x^*_{j_2} = a_{j_2j_21} = a_{j_1j_21}$.

- For $a_{j_1j_10}(\overline{x}_{j_1} - a_{j_1j_11})^{m_1} < 0$ and $a_{j_2j_20} < 0$,

$$(\overline{x}_{j_1}, a_{j_2j_21}) = \underbrace{([2(n_{21} - n_{11})]^{\text{th}}\text{IC}, (2n_{21} + 1)^{\text{th}}\text{SI})}_{([2(n_{21}-n_{11})],(2n_{21}+1))\text{-increasing concave sink}}.$$

The infinite-equilibrium is a $([2(n_{21} - n_{11})]^{\text{th}}\text{IC}, (2n_{21} + 1)^{\text{th}}\text{SI})$-increasing-concave sink at $x^*_{j_2} = a_{j_2j_21} = a_{j_1j_21}$.

(i_{2b1b}) For $n_1 = 2n_{11}$ and $n_2 = 2n_{21}+1$, the infinite-equilibrium of $x^*_{j_2} = a_{j_2j_21} = a_{j_1j_21}$ has the following properties and has the following properties as in Eqs. (1.92)–(1.95).

- For $a_{j_1j_10}(\overline{x}_{j_1} - a_{j_1j_11})^{m_1} > 0$ and $a_{j_2j_20} > 0$,

$$(\overline{x}_{j_1}, a_{j_2j_21}) = \underbrace{([2(n_{21} - n_{11}) + 1]^{\text{th}}\text{DA}, (2n_{21} + 1)^{\text{th}}\text{SO})}_{([2(n_{21}-n_{11})+1],(2n_{21}+1))\text{-down-asymptote source}}.$$

The infinite-equilibrium is a $([2(n_{21} - n_{11}) + 1]^{\text{th}}\text{DA}, (2n_{21} + 1)^{\text{th}}\text{SO})$-down-asymptote source at $x^*_{j_2} = a_{j_2j_21} = a_{j_1j_21}$.

- For $a_{j_1j_10}(\overline{x}_{j_1} - a_{j_1j_11})^{m_1} < 0$ and $a_{j_2j_20} > 0$,

$$(\overline{x}_{j_1}, a_{j_2j_21}) = \underbrace{([2(n_{21} - n_{11}) + 1]^{\text{th}}\text{UA}, (2n_{21} + 1)^{\text{th}}\text{SO})}_{([2(n_{21}-n_{11})+1],(2n_{21}+1))\text{-up-asymptote source}}.$$

The infinite-equilibrium is a $([2(n_{21} - n_{11}) + 1]^{\text{th}}\text{UA}, (2n_{21} + 1)^{\text{th}}\text{SO})$-up-asymptote source at $x^*_{j_2} = a_{j_2j_21} = a_{j_1j_21}$.

- For $a_{j_1j_10}(\overline{x}_{j_1} - a_{j_1j_11})^{m_1} > 0$ and $a_{j_2j_20} < 0$,

$$(\overline{x}_{j_1}, a_{j_2j_21}) = \underbrace{([2(n_{21} - n_{11}) + 1]^{\text{th}}\text{UA}, (2n_{21} + 1)^{\text{th}}\text{SI})}_{([2(n_{21}-n_{11})+1],(2n_{21}+1))\text{-up-asymptote sink}}.$$

The infinite-equilibrium is a $([2(n_{21}-n_{11})+1]^{\text{th}}\text{UA}, (2n_{21}+1)^{\text{th}}\text{SI})$-up-asymptote sink at $x_{j_2}^* = a_{j_2j_21} = a_{j_1j_21}$.

- For $a_{j_1j_10}(\overline{x}_{j_1} - a_{j_1j_11})^{m_1} < 0$ and $a_{j_2j_20} < 0$,

$$(\overline{x}_{j_1}, a_{j_2j_21}) = \underbrace{([2(n_{21}-n_{11})+1]^{\text{th}}\text{DA}, (2n_{21}+1)^{\text{th}}\text{SI})}_{([2(n_{21}-n_{11})+1],(2n_{21}+1))\text{-down-asymptote sink}}.$$

The infinite-equilibrium is a $([2(n_{21}-n_{11})+1]^{\text{th}}\text{DA}, (2n_{21}+1)^{\text{th}}\text{SI})$-down-asymptote sink at $x_{j_2}^* = a_{j_2j_21} = a_{j_1j_21}$.

(i_{2b1c}) For $n_1 = 2n_{11}+1$ and $n_2 = 2n_{21}$, the infinite-equilibrium of $x_{j_2}^* = a_{j_2j_21} = a_{j_1j_21}$ and has the following properties as in Eqs. (1.96)–(1.99).

- For $a_{j_1j_10}(\overline{x}_{j_1} - a_{j_1j_11})^{m_1} > 0$ and $a_{j_2j_20} > 0$,

$$(\overline{x}_{j_1}, a_{j_2j_21}) = \underbrace{([2(n_{21}-n_{11})-1]^{\text{th}}\text{DA}, (2n_{21})^{\text{th}}\text{US})}_{([2(n_{21}-n_{11})-1],(2n_{21}))\text{-down-asymptote upper-saddle}}.$$

The infinite-equilibrium is a $([2(n_{21}-n_{11})-1]^{\text{th}}\text{DA}, (2n_{21})^{\text{th}}\text{US})$-down-asymptote upper-saddle at $x_{j_2}^* = a_{j_2j_21} = a_{j_1j_21}$.

- For $a_{j_1j_10}(\overline{x}_{j_1} - a_{j_1j_11})^{m_1} < 0$ and $a_{j_2j_20} > 0$,

$$(\overline{x}_{j_1}, a_{j_2j_21}) = \underbrace{([2(n_{21}-n_{11})-1]^{\text{th}}\text{UA}, (2n_{21})^{\text{th}}\text{US})}_{([2(n_{21}-n_{11})-1],(2n_{21}))\text{-up-asymptote upper-saddle}}.$$

The infinite-equilibrium is a $([2(n_{21}-n_{11})-1]^{\text{th}}\text{UA}, (2n_{21})^{\text{th}}\text{US})$-up-asymptote upper-saddle at $x_{j_2}^* = a_{j_2j_21} = a_{j_1j_21}$.

- For $a_{j_1j_10}(\overline{x}_{j_1} - a_{j_1j_11})^{m_1} > 0$ and $a_{j_2j_20} < 0$,

$$(\overline{x}_{j_1}, a_{j_2j_21}) = \underbrace{([2(n_{21}-n_{11})-1]^{\text{th}}\text{UA}, (2n_{21}+1)^{\text{th}}\text{LS})}_{([2(n_{21}-n_{11})-1],(2n_{21}))\text{-up-asymptote lower-saddle}}.$$

The infinite-equilibrium is a $([2(n_{21}-n_{11})-1]^{\text{th}}\text{UA}, (2n_{21})^{\text{th}}\text{LS})$-up-asymptote lower-saddle at $x_{j_2}^* = a_{j_2j_21} = a_{j_1j_21}$.

- For $a_{j_1j_10}(\overline{x}_{j_1} - a_{j_1j_11})^{m_1} < 0$ and $a_{j_2j_20} < 0$,

$$(\overline{x}_{j_1}, a_{j_2j_21}) = \underbrace{([2(n_{21}-n_{11})-1]^{\text{th}}\text{DA}, (2n_{21})^{\text{th}}\text{LS})}_{([2(n_{21}-n_{11})-1],(2n_{21}))\text{-down-asymptote lower-saddle}}.$$

The infinite-equilibrium is a $([2(n_{21}-n_{11})-1]^{\text{th}}\text{DA}, (2n_{21})^{\text{th}}\text{LS})$-down-asymptote lower-saddle at $x_{j_2}^* = a_{j_2j_21} = a_{j_1j_21}$.

(i_{2b1d}) For $n_1 = 2n_{11}$, and $n_2 = 2n_{21}$, the infinite-equilibrium of $x^*_{j_2} = a_{j_2j_21} = a_{j_1j_21}$ has the following properties and has the following properties as in Eqs. (1.100)–(1.103).

- For $a_{j_1j_10}(\overline{x}_{j_1} - a_{j_1j_11})^{m_1} > 0$ and $a_{j_2j_20} > 0$,

$$(\overline{x}_{j_1}, a_{j_2j_21}) = \underbrace{([2(n_{21} - n_{11})]^{\text{th}}\text{IC}, (2n_{21})^{\text{th}}\text{US})}_{([2(n_{21}-n_{11})],(2n_{21}))\text{-increasing-concave upper-saddle}}.$$

  The infinite-equilibrium is a $([2(n_{21} - n_{11})]^{\text{th}}$IC, $(2n_{21})^{\text{th}}$US)-increasing-concave upper-saddle at $x^*_{j_2} = a_{j_2j_21} = a_{j_1j_21}$.
- For $a_{j_1j_10}(\overline{x}_{j_1} - a_{j_1j_11})^{m_1} < 0$ and $a_{j_2j_20} > 0$,

$$(\overline{x}_{j_1}, a_{j_2j_21}) = \underbrace{([2(n_{21} - n_{11})]^{\text{th}}\text{DC}, (2n_{21})^{\text{th}}\text{US})}_{([2(n_{21}-n_{11})],(2n_{21}))\text{-decreasing concave upper-saddle}}.$$

  The infinite-equilibrium is a $([2(n_{21} - n_{11})]^{\text{th}}$DC, $(2n_{21})^{\text{th}}$US)-decreasing-concave upper-saddle at $x^*_{j_2} = a_{j_2j_21} = a_{j_1j_21}$.
- For $a_{j_1j_10}(\overline{x}_{j_1} - a_{j_1j_11})^{m_1} > 0$ and $a_{j_2j_20} < 0$,

$$(\overline{x}_{j_1}, a_{j_2j_21}) = \underbrace{([2(n_{21} - n_{11})]^{\text{th}}\text{DC}, (2n_{21})^{\text{th}}\text{LS})}_{([2(n_{21}-n_{11})],(2n_{21}+1))\text{-decreasing concave lower-saddle}}.$$

  The infinite-equilibrium is a $([2(n_{21} - n_{11})]^{\text{th}}$DC, $(2n_{21})^{\text{th}}$LS)-decreasing-concave lower-saddle at $x^*_{j_2} = a_{j_2j_21} = a_{j_1j_21}$.
- For $a_{j_1j_10}(\overline{x}_{j_1} - a_{j_1j_11})^{m_1} < 0$ and $a_{j_2j_20} < 0$,

$$(\overline{x}_{j_1}, a_{j_2j_21}) = \underbrace{([2(n_{21} - n_{11})]^{\text{th}}\text{II}, (2n_{21})^{\text{th}}\text{LS})}_{([2(n_{21}-n_{11})],(2n_{21}))\text{-increasing concave lower-saddle}}.$$

  The infinite-equilibrium is a $([2(n_{21}-n_{11})]^{\text{th}}$II, $(2n_{21})^{\text{th}}$LS)-increasing-concave lower-saddle at $x^*_{j_2} = a_{j_2j_21} = a_{j_1j_21}$.

(i_{1b2}) On both sides of $\overline{x}_{j_1} = a_{j_1j_11}$, the inflection and parabola flows will be switched for $m_1 = (2m_{11} + 1)$, and the asymptote and concave flows will not be switched for $m_1 = 2m_{11}$.

Therefore, there are 32 types of equilibrium $(x^*_{j_2}, x^*_{j_2}) = (a_{j_1j_11}, a_{j_2j_21})$ on the infinite-equilibrium of $x^*_{j_2} = a_{j_2j_21} = a_{j_1j_21}$.

(i_{2b2a}) For $m_1 = 2m_{11} + 1, n_1 = 2n_{11} + 1$, and $n_2 = 2n_{21} + 1$, the equilibrium of $(x^*_{j_1}, x^*_{j_2}) = (a_{j_1j_11}, a_{j_2j_21})$ on the equilibrium of $x^*_{j_2} = a_{j_2j_21} = a_{j_1j_21}$ has the following properties as in Eqs. (1.104)–(1.107).

- For $a_{j_1j_10} > 0$ and $a_{j_2j_20} > 0$,

$$(a_{j_1j_11}, a_{j_2j_21}) = \underbrace{([2(n_{21} - n_{11})]^{\text{th}}\text{DC:IC},[(2m_{11}+1),(2n_{21}+1)]^{\text{th}}\text{SO})}_{([2(n_{21}-n_{11})],[(2m_{11}+1),(2n_{21}+1)])\text{-decreasing-to-increasing concave source}}.$$

The equilibrium is a $([2(n_{21}-n_{11})]^{\text{th}}\text{DC:IC},[(2m_{11}+1),(2n_{21}+1)]^{\text{th}}\text{SO})$-decreasing-to-increasing-concave source at $(x^*_{j_1}, x^*_{j_2}) = (a_{j_1j_11}, a_{j_2j_21})$.

- For $a_{j_1j_10} < 0$ and $a_{j_2j_20} > 0$,

$$(a_{j_1j_11}, a_{j_2j_21}) = \underbrace{([2(n_{21} - n_{11})]^{\text{th}}\text{IC:DC}, [(2m_{11}+1),(2n_{21}+1)]^{\text{th}}\text{SO})}_{([2(n_{21}-n_{11})],[(2m_{11}+1),(2n_{21}+1)])\text{-increasing-to-decreasing concave source}}.$$

The equilibrium is a $([2(n_{21} - n_{11})]^{\text{th}}\text{IC:DC}, [(2m_{11}+1),(2n_{21}+1)]^{\text{th}}\text{SO})$-increasing-to-decreasing-concave source at $(x^*_{j_1}, x^*_{j_2}) = (a_{j_1j_11}, a_{j_2j_21})$.

- For $a_{j_1j_10} > 0$ and $a_{j_2j_20} < 0$,

$$(a_{j_1j_11}, a_{j_2j_21}) = \underbrace{([2(n_{21} - n_{11})]^{\text{th}}\text{IC:DC}, [(2m_{11}+1),(2n_{21}+1)]^{\text{th}}\text{SI})}_{([2(n_{21}-n_{11})],[(2m_{11}+1),(2n_{21}+1)])\text{-increasing-to-decreasing concave sink}}.$$

The equilibrium is a $([2(n_{21} - n_{11})]^{\text{th}}\text{IC:DC}, [(2m_{11}+1),(2n_{21}+1)]^{\text{th}}\text{SI})$-increasing-to-decreasing concave sink at $(x^*_{j_1}, x^*_{j_2}) = (a_{j_1j_11}, a_{j_2j_21})$.

- For $a_{j_1j_10} < 0$ and $a_{j_2j_20} < 0$,

$$(a_{j_1j_11}, a_{j_2j_21}) = \underbrace{([2(n_{21} - n_{11})]^{\text{th}}\text{DC:IC},[(2m_{11}+1),(2n_{21}+1)]^{\text{th}}\text{SI})}_{([2(n_{21}-n_{11})],[(2m_{11}+1),(2n_{21}+1)])\text{-decreasing-to-increasing concave sink}}.$$

The equilibrium is a $([2(n_{21} - n_{11})]^{\text{th}}\text{DC:IC},[(2m_{11}+1),(2n_{21}+1)]^{\text{th}}\text{SI})$-decreasing-to-increasing-concave sink at $(x^*_{j_1}, x^*_{j_2}) = (a_{j_1j_11}, a_{j_2j_21})$.

(i_{2b2b}) For $m_1 = 2m_{11}+1$, $n_1 = 2n_{11}$, and $n_2 = 2n_{21}+1$, the equilibrium of $(x^*_{j_1}, x^*_{j_2}) = (a_{j_1j_11}, a_{j_2j_21})$ on the infinite-equilibrium of $x^*_{j_2} = a_{j_2j_21} = a_{j_1j_21}$ has the following properties as in Eqs. (1.108)–(1.111).

- For $a_{j_1j_10} > 0$ and $a_{j_2j_20} > 0$,

$$(a_{j_1j_11}, a_{j_2j_21}) = \underbrace{([2(n_{21} - n_{11}) + 1]^{\text{th}}\text{UA:DA}, [(2m_{11}+1),(2n_{21}+1)]^{\text{th}}\text{SO})}_{([2(n_{21}-n_{11})+1],[(2m_{11}+1),(2n_{21}+1)])\text{-up-to-down asymptote source}}.$$

The equilibrium is a $([2(n_{21}-n_{11})+1]^{\text{th}}\text{UA:DA}, [(2m_{11}+1),(2n_{21}+1)]^{\text{th}}\text{SO})$-up-to-down asymptote source at $(x^*_{j_1}, x^*_{j_2}) = (a_{j_1j_11}, a_{j_2j_21})$.

- For $a_{j_1j_10} < 0$ and $a_{j_2j_20} > 0$,

$$(a_{j_1j_11}, a_{j_2j_21}) = \underbrace{([2(n_{21} - n_{11}) + 1]^{\text{th}}\text{DA:UA}, [(2m_{11}+1),(2n_{21}+1)]^{\text{th}}\text{SO})}_{([2(n_{21}-n_{11})+1],[(2m_{11}+1),(2n_{21}+1)])\text{-down-to-up asymptote source}}.$$

The equilibrium is a $([2(n_{21}-n_{11})+1]^{\text{th}}\text{DA:UA}, [(2m_{11}+1),(2n_{21}+1)]^{\text{th}}\text{SO})$-down-to-up asymptote source at $(x_{j_1}^*, x_{j_2}^*) = (a_{j_1j_11}, a_{j_2j_21})$.

- For $a_{j_1j_10} > 0$ and $a_{j_2j_20} < 0$,

$$(a_{j_1j_11}, a_{j_2j_21}) = \underbrace{([2(n_{21}-n_{11})+1]^{\text{th}}\text{DA:UA}, [(2m_{11}+1),(2n_{21}+1)]^{\text{th}}\text{SI})}_{([2(n_{21}-n_{11})+1],[(2m_{11}+1),(2n_{21}+1)])\text{-down-to-up asymptote sink}}.$$

The equilibrium is a $([2(n_{21}-n_{11})+1]^{\text{th}}\text{DA:UA}, [(2m_{11}+1),(2n_{21}+1)]^{\text{th}}\text{SI})$-down-to-up asymptote sink at $(x_{j_1}^*, x_{j_2}^*) = (a_{j_1j_11}, a_{j_2j_21})$.

- For $a_{j_1j_10} < 0$ and $a_{j_2j_20} < 0$,

$$(a_{j_1j_11}, a_{j_2j_21}) = \underbrace{([2(n_{21}-n_{11})+1]^{\text{th}}\text{UA:DA}, [(2m_{11}+1),(2n_{21}+1)]^{\text{th}}\text{SI})}_{([2(n_{21}-n_{11})+1],[(2m_{11}+1),(2n_{21}+1)])\text{-up-to-down asymptote sink}}.$$

The equilibrium is a $([2(n_{21}-n_{11})+1]^{\text{th}}\text{UA:DA}, [(2m_{11}+1),(2n_{21}+1)]^{\text{th}}\text{SI})$-up-to-down asymptote sink at $(x_{j_1}^*, x_{j_2}^*) = (a_{j_1j_11}, a_{j_2j_21})$.

($\text{i}_{2\text{b}2\text{c}}$) For $m_1 = 2m_{11}+1$, $n_1 = 2n_{11}+1$, and $n_2 = 2n_{21}$, the equilibrium of $(x_{j_1}^*, x_{j_2}^*) = (a_{j_1j_11}, a_{j_2j_21})$ on the infinite-equilibrium of $x_{j_2}^* = a_{j_2j_21} = a_{j_1j_21}$ has the following properties as in Eqs. (1.112)–(1.115).

- For $a_{j_1j_10} > 0$ and $a_{j_2j_20} > 0$,

$$(a_{j_1j_11}, a_{j_2j_21}) = \underbrace{([2(n_{21}-n_{11})-1]^{\text{th}}\text{UA:DA}, [(2m_{11}+1)^{\text{th}}, (2n_{21})]\text{US})}_{([2(n_{21}-n_{11})-1],[(2m_{11}+1),(2n_{21})])\text{-up-to-down asymptote upper-saddle}}.$$

The equilibrium is a $([2(n_{21}-n_{11})-1]^{\text{th}}\text{UA:DA}, [(2m_{11}+1),(2n_{21})]^{\text{th}}\text{US})$-up-to-down asymptote upper-saddle at $(x_{j_1}^*, x_{j_2}^*) = (a_{j_1j_11}, a_{j_2j_21})$.

- For $a_{j_1j_10} < 0$ and $a_{j_2j_20} > 0$,

$$(a_{j_1j_11}, a_{j_2j_21}) = \underbrace{([2(n_{21}-n_{11})-1]^{\text{th}}\text{DA:UA}, [(2m_{11}+1)^{\text{th}}, (2n_{21})]\text{US})}_{([2(n_{21}-n_{11})-1],[(2m_{11}+1),(2n_{21})])\text{-down-to-up asymptote upper-saddle}}.$$

The equilibrium is a $([2(n_{21}-n_{11})-1]^{\text{th}}\text{DA:UA}, [(2m_{11}+1),(2n_{21})]^{\text{th}}\text{US})$-down-to-up asymptote upper-saddle at $(x_{j_1}^*, x_{j_2}^*) = (a_{j_1j_11}, a_{j_2j_21})$.

- For $a_{j_1j_10} > 0$ and $a_{j_2j_20} < 0$,

$$(a_{j_1j_11}, a_{j_2j_21}) = \underbrace{([2(n_{21}-n_{11})-1]^{\text{th}}\text{DA:UA}, [(2m_{11}+1)^{\text{th}}, (2n_{21})]\text{LS})}_{([2(n_{21}-n_{11})-1],[(2m_{11}+1),(2n_{21})])\text{-down-to-up asymptote lower-saddle}}.$$

The equilibrium is a $([2(n_{21}-n_{11})-1]^{\text{th}}\text{DA:UA}, [(2m_{11}+1),(2n_{21})]^{\text{th}}\text{LS})$-down-to-up asymptote lower-saddle at $(x_{j_1}^*, x_{j_2}^*) = (a_{j_1j_11}, a_{j_2j_21})$.

- For $a_{j_1j_10} < 0$ and $a_{j_2j_20} < 0$,

$$(a_{j_1j_11}, a_{j_2j_21}) = \underbrace{([2(n_{21}-n_{11})-1]^{\text{th}}\text{UA:DA}, [(2m_{11}+1)^{\text{th}}, (2n_{21})]\text{LS})}_{([2(n_{21}-n_{11})-1],[(2m_{11}+1),(2n_{21})])\text{-up-to-down asymptote lower-saddle}}.$$

The equilibrium is a ($[2(n_{21}-n_{11})-1]^{\text{th}}$UA:DA, $[(2m_{11}+1),(2n_{21})]^{\text{th}}$LS)-up-to-down asymptote lower-saddle at $(x^*_{j_1}, x^*_{j_2}) = (a_{j_1j_11}, a_{j_2j_21})$.

(i_{2b2d}) For $m_1 = 2m_{11}+1$, $n_1 = 2n_{11}$, and $n_2 = 2n_{21}$, the equilibrium of $(x^*_{j_1}, x^*_{j_2}) = (a_{j_1j_11}, a_{j_2j_21})$ on the infinite-equilibrium of $x^*_{j_2} = a_{j_2j_21} = a_{j_1j_21}$ has the following properties as in Eqs. (1.116)–(1.119).

- For $a_{j_1j_10} > 0$ and $a_{j_2j_20} > 0$,

$$(a_{j_1j_11}, a_{j_2j_21}) = \underbrace{([2(n_{21}-n_{11})]^{\text{th}}\text{DC:IC},[(2m_{11}+1),(2n_{21})]^{\text{th}}\text{US})}_{([2(n_{21}-n_{11})],[(2m_{11}+1),(2n_{21})])\text{-deceasing-to-increasing concave upper-saddle}}.$$

  The equilibrium is a ($[2(n_{21}-n_{11})]^{\text{th}}$DC:IC,$[(2m_{11}+1),(2n_{21})]^{\text{th}}$US)-decreasing-to-increasing-concave upper-saddle at $(x^*_{j_1}, x^*_{j_2}) = (a_{j_1j_11}, a_{j_2j_21})$.
- For $a_{j_1j_10} < 0$ and $a_{j_2j_20} > 0$,

$$(a_{j_1j_11}, a_{j_2j_21}) = \underbrace{([2(n_{21}-n_{11})]^{\text{th}}\text{IC:DC}, [(2m_{11}+1),(2n_{21})]^{\text{th}}\text{US})}_{([2(n_{21}-n_{11})],[(2m_{11}+1),(2n_{21})])\text{-inceasing-to-dccreasing concave upper-saddle}}.$$

  The equilibrium is a ($[2(n_{21}-n_{11})]^{\text{th}}$IC:DC, $[(2m_{11}+1),(2n_{21})]^{\text{th}}$US)-increasing-to-decreasing-concave upper-saddle at $(x^*_{j_1}, x^*_{j_2}) = (a_{j_1j_11}, a_{j_2j_21})$.
- For $a_{j_1j_10} > 0$ and $a_{j_2j_20} < 0$,

$$(a_{j_1j_11}, a_{j_2j_21}) = \underbrace{([2(n_{21}-n_{11})]^{\text{th}}\text{IC:DC}, [(2m_{11}+1),(2n_{21})]^{\text{th}}\text{LS})}_{([2(n_{21}-n_{11})],[(2m_{11}+1),(2n_{21})])\text{-inceasing-to-dccreasing concave lower-saddle}}.$$

  The equilibrium is a ($[2(n_{21}-n_{11})]^{\text{th}}$IC:DC, $[(2m_{11}+1),(2n_{21})]^{\text{th}}$LS)-increasing-to-decreasing-concave lower-saddle at $(x^*_{j_1}, x^*_{j_2}) = (a_{j_1j_11}, a_{j_2j_21})$.
- For $a_{j_1j_10} < 0$ and $a_{j_2j_20} < 0$,

$$(a_{j_1j_11}, a_{j_2j_21}) = \underbrace{([2(n_{21}-n_{11})]^{\text{th}}\text{DC:IC},[(2m_{11}+1),(2n_{21})]^{\text{th}}\text{LS})}_{([2(n_{21}-n_{11})],[(2m_{11}+1),(2n_{21})])\text{-deceasing-to-increasing concave lower-saddle}}.$$

  The equilibrium is a ($[2(n_{21}-n_{11})]^{\text{th}}$DC:IC,$[(2m_{11}+1),(2n_{21})]^{\text{th}}$LS)-decreasing-to-increasing-concave lower-saddle at $(x^*_{j_1}, x^*_{j_2}) = (a_{j_1j_11}, a_{j_2j_21})$.

(i_{2b2e}) For $m_1 = 2m_{11}$, $n_1 = 2n_{11}+1$, and $n_2 = 2n_{21}+1$, the equilibrium of $(x^*_{j_1}, x^*_{j_2}) = (a_{j_1j_11}, a_{j_2j_21})$ on the infinite-equilibrium of $x^*_{j_2} = a_{j_2j_21} = a_{j_1j_21}$ has the following properties as in Eqs. (1.120)–(1.123).

- For $a_{j_1j_10} > 0$ and $a_{j_2j_20} > 0$,

$$(a_{j_1j_11}, a_{j_2j_21}) = \underbrace{([2(n_{21} - n_{11})]^{\text{th}}\text{IC:IC}, [(2m_{11}), (2n_{21} + 1)]^{\text{th}}\text{SO})}_{([2(n_{21}-n_{11})],[(2m_{11}),(2n_{21}+1)])\text{-increasing concave source}}.$$

The equilibrium is a $([2(n_{21}-n_{11})]^{\text{th}}\text{IC:IC}, [(2m_{11}), (2n_{21}+1)]^{\text{th}}\text{SO})$-increasing-concave source at $(x_{j_1}^*, x_{j_2}^*) = (a_{j_1j_11}, a_{j_2j_21})$.

- For $a_{j_1j_10} < 0$ and $a_{j_2j_20} > 0$,

$$(a_{j_1j_11}, a_{j_2j_21}) = \underbrace{([2(n_{21} - n_{11})]^{\text{th}}\text{DC:DC}, [(2m_{11}), (2n_{21} + 1)]^{\text{th}}\text{SO})}_{([2(n_{21}-n_{11})],[(2m_{11}),(2n_{21}+1)])\text{-decreasing concave source}}.$$

The equilibrium is a $([2(n_{21}-n_{11})]^{\text{th}}\text{DC:DC}, [(2m_{11}), (2n_{21}+1)]^{\text{th}}\text{SO})$-decreasing-concave source at $(x_{j_2}^*, x_{j_2}^*) = (a_{j_1j_11}, a_{j_2j_21})$.

- For $a_{j_1j_10} > 0$ and $a_{j_2j_20} < 0$,

$$(a_{j_1j_11}, a_{j_2j_21}) = \underbrace{([2(n_{21} - n_{11})]^{\text{th}}\text{DC:DC}, [(2m_{11}), (2n_{21} + 1)]^{\text{th}}\text{SI})}_{([2(n_{21}-n_{11})],[(2m_{11}),(2n_{21}+1)])\text{-decreasing concave sink}}.$$

The equilibrium is a $([2(n_{21}-n_{11})]^{\text{th}}\text{DC:DC}, [(2m_{11}), (2n_{21}+1)]^{\text{th}}\text{SI})$-decreasing concave sink at $(x_{j_1}^*, x_{j_2}^*) = (a_{j_1j_11}, a_{j_2j_21})$.

- For $a_{j_1j_10} < 0$ and $a_{j_2j_20} < 0$,

$$(a_{j_1j_11}, a_{j_2j_21}) = \underbrace{([2(n_{21} - n_{11})]^{\text{th}}\text{IC:IC}, [(2m_{11}), (2n_{21} + 1)]^{\text{th}}\text{SI})}_{([2(n_{21}-n_{11})],[(2m_{11}),(2n_{21}+1)])\text{-increasing concave sink}}.$$

The equilibrium is a $([2(n_{21}-n_{11})]^{\text{th}}\text{IC:IC}, [(2m_{11}), (2n_{21}+1)]^{\text{th}}\text{SI})$-increasing-concave sink at $(x_{j_1}^*, x_{j_2}^*) = (a_{j_1j_11}, a_{j_2j_21})$.

(i_{2b2f}) For $m_1 = 2m_{11}$, $n_1 = 2n_{11}$, and $n_2 = 2n_{21} + 1$, the equilibrium of $(x_{j_2}^*, x_{j_2}^*) = (a_{j_1j_11}, a_{j_2j_21})$ on the infinite-equilibrium of $x_{j_2}^* = a_{j_2j_21} = a_{j_1j_21}$ has the following properties as in Eqs. (1.124)–(1.127).

- For $a_{j_1j_10} > 0$ and $a_{j_2j_20} > 0$,

$$(a_{j_1j_11}, a_{j_2j_21}) = \underbrace{([2(n_{21} - n_{11}) + 1]^{\text{th}}\text{DA:DA}, [(2m_{11}), (2n_{21} + 1)]^{\text{th}}\text{SO})}_{([2(n_{21}-n_{11})+1],[(2m_{11}),(2n_{21}+1)])\text{-down-asymptote source}}.$$

The equilibrium is a $([2(n_{21} - n_{11}) + 1]^{\text{th}}\text{DA:DA}, [(2m_{11}), (2n_{21} + 1)]^{\text{th}}\text{SO})$-down asymptote source at $(x_{j_1}^*, x_{j_2}^*) = (a_{j_1j_11}, a_{j_2j_21})$.

- For $a_{j_1j_10} < 0$ and $a_{j_2j_20} > 0$,

$$(a_{j_1j_11}, a_{j_2j_21}) = \underbrace{([2(n_{21} - n_{11}) + 1]^{\text{th}}\text{UA:UA}, [(2m_{11}), (2n_{21} + 1)]^{\text{th}}\text{SO})}_{([2(n_{21}-n_{11})+1],[(2m_{11}),(2n_{21}+1)])\text{-up asymptote source}}.$$

The equilibrium is a $([2(n_{21}-n_{11})+1]^{\text{th}}\text{UA:UA}, [(2m_{11}),(2n_{21}+1)]^{\text{th}}\text{SO})$-up-asymptote source at $(x^*_{j_1}, x^*_{j_2}) = (a_{j_1j_1 1}, a_{j_2j_2 1})$.

- For $a_{j_1j_1 0} > 0$ and $a_{j_2j_2 0} < 0$,

$$(a_{j_1j_1 1}, a_{j_2j_2 1}) = \underbrace{([2(n_{21}-n_{11})+1]^{\text{th}}\text{UA:UA}, [(2m_{11}),(2n_{21}+1)]^{\text{th}}\text{SI})}_{([2(n_{21}-n_{11})+1],[(2m_{11}),(2n_{21}+1)])\text{-up asymptote sink}}.$$

The equilibrium is a $([2(n_{21}-n_{11})+1]^{\text{th}}\text{UA:UA}, [(2m_{11}),(2n_{21}+1)]^{\text{th}}\text{SI})$-up asymptote sink at $(x^*_{j_1}, x^*_{j_2}) = (a_{j_1j_1 1}, a_{j_2j_2 1})$.

- For $a_{j_1j_1 0} < 0$ and $a_{j_2j_2 0} < 0$,

$$(a_{j_1j_1 1}, a_{j_2j_2 1}) = \underbrace{([2(n_{21}-n_{11})+1]^{\text{th}}\text{DA:DA}, [(2m_{11}),(2n_{21}+1)]^{\text{th}}\text{SI})}_{([2(n_{21}-n_{11})+1],[(2m_{11}),(2n_{21}+1)])\text{-down asymptote sink}}.$$

The equilibrium is a $([2(n_{21}-n_{11})+1]^{\text{th}}\text{DA:DA}, [(2m_{11}),(2n_{21}+1)]^{\text{th}}\text{SI})$-down-asymptote sink at $(x^*_{j_1}, x^*_{j_2}) = (a_{j_1j_1 1}, a_{j_2j_2 1})$.

(i_{2b2g}) For $m_1 = 2m_{11}$, $n_1 = 2n_{11}+1$, and $n_2 = 2n_{21}$, the equilibrium of $(x^*_{j_1}, x^*_{j_2}) = (a_{j_1j_1 1}, a_{j_2j_2 1})$ on the infinite-equilibrium of $x^*_{j_2} = a_{j_2j_2 1} = a_{j_1j_2 1}$ has the following properties as in Eqs. (1.128)–(1.131).

- For $a_{j_1j_1 0} > 0$ and $a_{j_2j_2 0} > 0$,

$$(a_{j_1j_1 1}, a_{j_2j_2 1}) = \underbrace{([2(n_{21}-n_{11})-1]^{\text{th}}\text{DA:DA}, [(2m_{11})^{\text{th}}, (2n_{21})]\text{US})}_{([2(n_{21}-n_{11})-1],[(2m_{11}),(2n_{21})])\text{-down asymptote upper-saddle}}.$$

The equilibrium is a $([2(n_{21}-n_{11})-1]^{\text{th}}\text{DA:DA}, [(2m_{11}),(2n_{21})]^{\text{th}}\text{US})$-down asymptote upper-saddle at $(x^*_{j_1}, x^*_{j_2}) = (a_{j_1j_1 1}, a_{j_2j_2 1})$.

- For $a_{j_1j_1 0} < 0$ and $a_{j_2j_2 0} > 0$,

$$(a_{j_1j_1 1}, a_{j_2j_2 1}) = \underbrace{([2(n_{21}-n_{11})-1]^{\text{th}}\text{UA:UA}, [(2m_{11})^{\text{th}}, (2n_{21})]\text{US})}_{([2(n_{21}-n_{11})-1],[(2m_{11}),(2n_{21})])\text{-up asymptote upper-saddle}}.$$

The equilibrium is a $([2(n_{21}-n_{11})-1]^{\text{th}}\text{UA:UA}, [(2m_{11}),(2n_{21})]^{\text{th}}\text{US})$-up asymptote upper-saddle at $(x^*_{j_1}, x^*_{j_2}) = (a_{j_1j_1 1}, a_{j_2j_2 1})$.

- For $a_{j_1j_1 0} > 0$ and $a_{j_2j_2 0} < 0$,

$$(a_{j_1j_1 1}, a_{j_2j_2 1}) = \underbrace{([2(n_{21}-n_{11})-1]^{\text{th}}\text{UA:UA}, [(2m_{11}), (2n_{21})]^{\text{th}}\text{LS})}_{([2(n_{21}-n_{11})-1],[(2m_{11}),(2n_{21})])\text{-up-asymptote lower-saddle}}.$$

The equilibrium is a $([2(n_{21}-n_{11})-1]^{\text{th}}\text{UA:UA}, [(2m_{11}),(2n_{21})]^{\text{th}}\text{LS})$ up asymptote lower-saddle at $(x^*_{j_1}, x^*_{j_2}) = (a_{j_1j_1 1}, a_{j_2j_2 1})$.

- For $a_{j_1j_1 0} < 0$ and $a_{j_2j_2 0} < 0$,

$$(a_{j_1j_11}, a_{j_2j_21}) = \underbrace{([2(n_{21}-n_{11})-1]^{\text{th}}\text{DA:DA}, [(2m_{11}), (2n_{21})]^{\text{th}}\text{LS})}_{([2(n_{21}-n_{11})-1],[(2m_{11}),(2n_{21})])\text{-down asymptote lower-saddle}}.$$

The equilibrium is a $([2(n_{21}-n_{11})-1]^{\text{th}}\text{DA:DA}, [(2m_{11}),(2n_{21})]^{\text{th}}\text{LS})$-down-asymptote lower-saddle at $(x_{j_1}^*, x_{j_2}^*) = (a_{j_1j_11}, a_{j_2j_21})$.

(i_{2b2h}) For $m_1 = 2m_{11}$, $n_1 = 2n_{11}$, and $n_2 = 2n_{21}$, the equilibrium of $(x_{j_1}^*, x_{j_2}^*) = (a_{j_1j_11}, a_{j_2j_21})$ on the infinite-equilibrium of $x_{j_2}^* = a_{j_2j_21} = a_{j_1j_21}$ has the following properties as in Eqs. (1.132)–(1.135).

- For $a_{j_1j_10} > 0$ and $a_{j_2j_20} > 0$,

$$(a_{j_1j_11}, a_{j_2j_21}) = \underbrace{([2(n_{21}-n_{11})]^{\text{th}}\text{IC:IC}, [(2m_{11}),(2n_{21})]^{\text{th}}\text{US})}_{([2(n_{21}-n_{11})],[(2m_{11}),(2n_{21})])\text{-increasing concave upper-saddle}}.$$

The equilibrium is a $([2(n_{21}-n_{11})]^{\text{th}}\text{IC:IC}, [(2m_{11}),(2n_{21})]^{\text{th}}\text{US})$-increasing-concave upper-saddle at $(x_{j_1}^*, x_{j_2}^*) = (a_{j_1j_11}, a_{j_2j_21})$.

- For $a_{j_1j_10} < 0$ and $a_{j_2j_20} > 0$,

$$(a_{j_1j_11}, a_{j_2j_21}) = \underbrace{([2(n_{21}-n_{11})]^{\text{th}}\text{DC:DC}, [(2m_{11}),(2n_{21})]^{\text{th}}\text{US})}_{([2(n_{21}-n_{11})],[(2m_{11}),(2n_{21})])\text{-decreasing concave upper-saddle}}.$$

The equilibrium is a $([2(n_{21}-n_{11})]^{\text{th}}\text{DC:DC}, [(2m_{11}),(2n_{21})]^{\text{th}}\text{US})$-decreasing-concave upper-saddle at $(x_{j_1}^*, x_{j_2}^*) = (a_{j_1j_11}, a_{j_2j_21})$.

- For $a_{j_1j_10} > 0$ and $a_{j_2j_20} < 0$,

$$(a_{j_1j_11}, a_{j_2j_21}) = \underbrace{([2(n_{21}-n_{11})]^{\text{th}}\text{DC:DC}, [(2m_{11}),(2n_{21})]^{\text{th}}\text{LS})}_{([2(n_{21}-n_{11})],[(2m_{11}),(2n_{21})])\text{-decreasing concave lower-saddle}}.$$

The equilibrium is a $([2(n_{21}-n_{11})]^{\text{th}}\text{DC:DC}, [(2m_{11}),(2n_{21})]^{\text{th}}\text{LS})$-decreasing-concave lower-saddle at $(x_{j_1}^*, x_{j_2}^*) = (a_{j_1j_11}, a_{j_2j_21})$.

- For $a_{j_1j_10} < 0$ and $a_{j_2j_20} < 0$,

$$(a_{j_1j_11}, a_{j_2j_21}) = \underbrace{([2(n_{21}-n_{11})]^{\text{th}}\text{IC:IC}, [(2m_{11}),(2n_{21})]^{\text{th}}\text{LS})}_{([2(n_{21}-n_{11})],[(2m_{11}),(2n_{21})])\text{-increasing concave lower-saddle}}.$$

The equilibrium is a $([2(n_{21}-n_{11})]^{\text{th}}\text{IC:IC}, [(2m_{11}),(2n_{21})]^{\text{th}}\text{LS})$-increasing-concave lower-saddle at $(x_{j_1}^*, x_{j_2}^*) = (a_{j_1j_11}, a_{j_2j_21})$.

2.2 Hybrid Networks of Singular Equilibriums and Flows

In this section, theorem about the hybrid networks of singular equilibriums and flows will be proved, and the corresponding infinite-equilibriums will be discussed.

(ii) Consider

$$\dot{x}_{j_2} = a_{j_2j_20}\prod_{s=1}^{q_2}(x_{j_2} - a_{j_2j_2s})^{n_{j_2s}}$$
$$\dot{x}_{j_1} = a_{j_1j_10}\prod_{r=1}^{p_1}(x_{j_1} - a_{j_1j_1r})^{m_{j_1r}}\prod_{u=1}^{q_1}(x_{j_2} - a_{j_1j_2u})^{n_{j_1u}},$$

where

$$\sum_{r=1}^{p_1} m_{j_1r} = m_1, \sum_{u=1}^{q_1} n_{j_2u} = n_1; \sum_{s=1}^{q_2} n_{j_2s} = n_2.$$

2.2.1 *Hybrid Networks of Singular Equilibriums and Flows*

(ii$_1$) For $a_{j_1j_2u} \neq a_{j_2j_2s}$ $(u = 1, 2, \cdots, q_1; s = 1, 2, \cdots, q_2)$, consider

$$\frac{dx_{j_1}}{dx_{j_2}} = \frac{a_{j_1j_10}}{a_{j_2j_20}}\frac{\prod_{r=1}^{p_1}(x_{j_1} - a_{j_1j_1r})^{m_{j_1r}}\prod_{u=1}^{q_1}(x_{j_2} - a_{j_1j_2u})^{n_{j_1u}}}{\prod_{s=1}^{q_2}(x_{j_2} - a_{j_2j_2s})^{n_{j_2s}}}.$$

The deformed equation is

$$\frac{dx_{j_1}}{\prod_{r=1}^{p_1}(x_{j_1} - a_{j_1j_1r})^{m_{j_1r}}} = \frac{a_{j_1j_10}\prod_{u=1}^{q_1}(x_{j_2} - a_{j_1j_2u})^{n_{j_1u}}}{a_{j_2j_20}\prod_{s=1}^{q_2}(x_{j_2} - a_{j_2j_2s})^{n_{j_2s}}}dx_{j_2},$$

The expansion of the above equation gives

$$\sum_{r_1=1}^{p_1}\sum_{l_1=1}^{m_{j_1r_1}}\frac{A_{l_1m_{j_1r_1}}}{(x_{j_1} - a_{j_1j_1r_1})^{l_1}}dx_{j_1}$$
$$= \frac{a_{j_1j_10}}{a_{j_2j_20}}\sum_{\beta_0=0}^{n_2}\frac{\sum_{t_1,l_1=0}^{t_1+l_1=n_{j_1u_1}}\cdots\sum_{t_{q_1},l_q=0}^{t_{q_1}+l_{q_1}=n_{j_1u_{q_1}}}\prod_{k=1}^{q_1}\frac{n_{j_1u_k}!}{t_k!l_k!}(a_{j_2j_2s_1} - a_{j_1j_2u_k})^{t_k}\delta_{\beta_0}^{\Sigma_{r=1}^{q_1}l_r}}{(x_{j_2} - a_{j_2j_2s_1})^{n_{j_2s_1}}\prod_{s_2=1,s_2\neq s_1}^{q_2}(x_{j_2} - a_{j_2j_2s_2})^{n_{j_2s_2}}}$$
$$\times(x_{j_2} - a_{j_2j_2s_1})^{\beta_0}dx_{j_2}$$

where

$$A_{l_1 m_{j_1 r_1}} = \frac{1}{(m_{j_1 r_1} - l_1)!} \frac{d^{m_{j_1 r_1} - l_1}}{dx_{j_1}^{m_{j_1 r_1} - l_1}} \frac{1}{\prod_{\sigma=2}^{p_1} (x_{j_1} - a_{j_1 j_1 r_\sigma})^{m_{j_1 r_\sigma}}} \Big|_{x_{j_1}^* = a_{j_1 j_1 r_1}},$$

$$A_{m_{j_1 r_1} m_{j_1 r_1}} = \frac{1}{\prod_{\sigma=2}^{p_1} (a_{j_1 j_1 r_1} - a_{j_1 j_1 r_\sigma})^{m_{j_1 r_\sigma}}}.$$

Let

$$b_{j_1 j_2 \beta_0} = \sum_{t_1, l_1 = 0}^{t_1 + l_1 = n_{j_1 u_1}} \cdots \sum_{t_{q_1}, l_{q_1} = 0}^{t_{q_1} + l_{q_1} = n_{j_1 u_{q_1}}} \prod_{k=1}^{q_1} \frac{n_{j_1 u_k}!}{t_k! l_k!} (a_{j_2 j_2 s_1} - a_{j_1 j_2 u_k})^{t_k} \delta_{\beta_0}^{\Sigma_{r=1}^{q_1} l_r}.$$

We have

$$\sum_{r_1=1}^{p_1} \sum_{l_1=1}^{m_{j_1 r_1}} \frac{A_{l_1 m_{j_1 r_1}}}{(x_{j_1} - a_{j_1 j_1 r_1})^{l_1}} dx_{j_1}$$
$$= \frac{a_{j_1 j_1 0}}{a_{j_2 j_2 0}} \frac{\sum_{\beta_0=0}^{n_2-1} b_{j_1 j_2 \beta_0} (x_{j_2} - a_{j_2 j_2 s_1})^{\beta_0} + \sum_{\beta_0=n_2}^{n_1} b_{j_1 j_2 \beta_0} (x_{j_2} - a_{j_2 j_2 s_1})^{\beta_0}}{(x_{j_2} - a_{j_2 j_2 s_1})^{n_{j_2 s_1}} \prod_{s_2=1, s_2 \neq s_1}^{q_2} (x_{j_2} - a_{j_2 j_2 s_2})^{n_{j_2 s_2}}} dx_{j_2}.$$

The first part is given by

$$\frac{\sum_{\beta_0=0}^{n_2-1} b_{j_1 j_2 \beta_0} (x_{j_2} - a_{j_2 j_2 s_1})^{\beta_0}}{(x_{j_2} - a_{j_2 j_2 s_1})^{n_{j_2 s_1}} \prod_{s_2=1, s_2 \neq s_1}^{q_2} (x_{j_2} - a_{j_2 j_2 s_2})^{n_{j_2 s_2}}}$$
$$= \sum_{\beta_0=0}^{n_2-1} b_{j_1 j_2 \beta_0} \frac{(x_{j_2} - a_{j_2 j_2 s_1})^{\beta_0 - n_{j_2 s_1}}}{\prod_{s_2=1, s_2 \neq s_1}^{q_2} (x_{j_2} - a_{j_2 j_2 s_2})^{n_{j_2 s_2}}}$$
$$= \sum_{\beta_0=0}^{n_{j_2 s_1}-1} b_{j_1 j_2 \beta_0} \sum_{j=1}^{q_2} \sum_{l_j=1}^{\overline{n}_{j_2 s_j}} A_{l_j \overline{n}_{j_2 s_j}}^{(s_1)} \frac{1}{(x_{j_2} - a_{j_2 j_2 s_j})^{l_j}}$$
$$+ \sum_{\beta_0=n_{j_2 s_1}}^{n_2-1} b_{j_1 j_2 \beta_0} \sum_{j=2}^{q_2} \sum_{l_j=1}^{\overline{n}_{j_2 s_j}} B_{l_j \overline{n}_{j_2 s_j}}^{(s_1)} \frac{1}{(x_{j_2} - a_{j_2 j_2 s_j})^{l_j}};$$

where

$$A_{l_\rho \overline{n}_{j_2 s_\rho}}^{(s_1)} = \frac{1}{(\overline{n}_{j_2 s_\rho} - l_\rho)!} \frac{d^{\overline{n}_{j_2 s_\rho} - l_\rho}}{dx_{j_2}^{\overline{n}_{j_2 s_\rho} - l_\rho}} \frac{1}{\prod_{\sigma=1, \sigma \neq \rho}^{q_2} (x_{j_2} - a_{j_2 j_2 s_\sigma})^{n_{j_2 s_\sigma}}} \Big|_{x_{j_2}^* = a_{j_2 j_2 s_\rho}}$$

$$(l_\rho = 1, 2, \cdots, \overline{n}_{j_2 s_\rho}, \rho = 1, 2, \cdots, q_2),$$

$$A^{(s_1)}_{\overline{n}_{j_2 s_\rho} \overline{n}_{j_2 s_\rho}} = \frac{1}{\prod_{\sigma=1,\sigma\neq\rho}^{q_2} (a_{j_2 j_2 s_\rho} - a_{j_s j_s s_\sigma})^{n_{j_2 s_\sigma}}},$$

$$B^{(s_1)}_{l_\rho \overline{n}_{j_2 s_\rho}} = \frac{1}{(\overline{n}_{j_2 s_\rho} - l_\rho)!} \frac{d^{\overline{n}_{j_2 s_\rho} - l_\rho}}{dx_{j_2}^{\overline{n}_{j_2 s_\rho} - l_\rho}} \frac{(x_{j_2} - a_{j_2 j_2 s_1})^{\beta_0 - n_{j_2 s_1}}}{\prod_{\sigma=2,\sigma\neq\rho}^{q_2} (x_{j_2} - a_{j_2 j_2 s_\sigma})^{n_{(2) s_\sigma}}} \Bigg|_{x^*_{j_2} = a_{j_2 j_2 s_\rho}}$$

$$(l_\rho = 1, 2, \cdots, \overline{n}_{(2) s_\rho}, \rho = 2, 3, \cdots, q_2),$$

$$B^{(s_1)}_{\overline{n}_{j_2 s_\rho} \overline{n}_{j_2 s_\rho}} = \frac{(a_{j_2 j_2 s_\rho} - a_{j_2 j_2 s_1})^{\beta_0 - n_{j_2 s_1}}}{\prod_{\sigma=2,\sigma\neq\rho}^{q_2} (a_{j_2 j_2 s_\rho} - a_{j_s j_s s_\sigma})^{n_{j_2 s_\sigma}}}$$

$$\{\overline{n}_{j_2 s_\rho} | \rho = 1, 2, \cdots, q_2\} = \{n_{j_2 s_1} - \beta_0, n_{j_2 s_2}, \cdots, n_{j_2 s_{q_2}}\},$$

The second part is deformed by

$$\frac{\sum\limits_{\beta_0=n_2}^{n_1} b_{j_1 j_2 \beta_0} (x_{j_2} - a_{j_2 j_2 s_1})^{\beta_0}}{(x_{j_2} - a_{j_2 j_2 s_1})^{n_{j_2 s_1}} \prod\limits_{s_2=1, s_2\neq s_1}^{q_2} (x_{j_2} - a_{j_2 j_2 s_2})^{n_{j_2 s_2}}}$$

$$= \sum_{\beta_0=n_2}^{n_1} b_{j_1 j_2 \beta_0} \frac{(x_{j_2} - a_{j_2 j_2 s_1})^{\beta_0 - n_{j_2 s_1}}}{(x_{j_2} - a_{j_2 j_2 s_2})^{n_{j_2 s_2}} \prod\limits_{s_3=1, s_3\neq s_1, s_2}^{q_2} (x_{j_2} - a_{j_2 j_2 s_3})^{n_{j_2 s_3}}}$$

$$= \sum_{\beta_0=n_2}^{n_1} b_{j_1 j_2 \beta_0} \Bigg[\sum_{\substack{\alpha_1 \in [0, \beta_0 - n_2] \\ \beta_1 \in [n_2 - n_{j_2 s_1}, \beta_0 - n_{j_2 s_1}]}}^{\alpha_1 + \beta_1 = \beta_0 - n_{j_2 s_1}} \frac{(\beta_0 - n_{j_2 s_1})!}{\alpha_1! \beta_1!} \frac{(a_{j_2 j_2 s_2} - a_{j_2 j_2 s_1})^{\alpha_1} (x_{j_2} - a_{j_2 j_2 s_2})^{\beta_1 - n_{j_2 s_2}}}{\prod\limits_{s_3=1, s_3\neq s_1, s_2}^{q_2} (x_{j_2} - a_{j_2 j_2 s_3})^{n_{j_2 s_3}}}$$

$$+ \sum_{\substack{\alpha_1 \in [\beta_0 - \Sigma_{i=1}^2 n_{j_2 s_i}, \beta_0 - n_{j_2 s_1}) \\ \beta_1 \in [0, n_{j_2 s_2})}}^{\alpha_1 + \beta_1 = \beta_0 - n_{j_2 s_1}} \frac{(\beta_0 - n_{j_2 s_1})!}{\alpha_1! \beta_1!} \frac{(a_{j_2 j_2 s_2} - a_{j_2 j_2 s_1})^{\alpha_1}}{(x_{j_1} - a_{j_1 j_1 r_2})^{n_{j_2 s_2} - \beta_1} \prod\limits_{s_3=1, s_3\neq s_1, s_2}^{q_2} (x_{j_2} - a_{j_2 j_2 s_3})^{n_{j_2 s_3}}}$$

$$+ \sum_{\substack{\alpha_1 \in (\beta_0 - n_2, \beta_0 - \Sigma_{i=1}^2 n_{j_2 s_i}] \\ \beta_1 \in [n_{j_2 s_2}, n_2 - n_{j_2 s_1})}}^{\alpha_1 + \beta_1 = \beta_0 - n_{j_2 s_1}} \frac{(\beta_0 - n_{j_2 s_1})!}{\alpha_1! \beta_1!} \frac{(a_{j_2 j_2 s_2} - a_{j_2 j_2 s_1})^{\alpha_1} (x_{j_2} - a_{j_2 j_2 s_2})^{\beta_1 - n_{j_2 s_2}}}{\prod\limits_{s_3=1, s_3\neq s_1, s_2}^{q_2} (x_{j_2} - a_{j_2 j_2 s_3})^{n_{j_2 s_3}}} \Bigg]$$

Thus

$$\frac{\sum\limits_{\beta_0=n_2}^{n_1} b_{j_1 j_2 \beta_0} (x_{j_2} - a_{j_2 j_2 s_1})^{\beta_0}}{(x_{j_2} - a_{j_2 j_2 s_1})^{n_{j_2 s_1}} \prod\limits_{s_2=1, s_2\neq s_1}^{q_2} (x_{j_2} - a_{j_2 j_2 s_2})^{n_{j_2 s_2}}} = \sum_{\beta_0=n_2}^{n_1} b_{j_1 j_2 \beta_0}$$

$$\times \Bigg[\sum_{\substack{\alpha_1 \in [0, \alpha - n_2] \\ \beta_1 \in [n_2 - n_{j_2 s_1}, \beta_0 - n_{j_2 s_1}]}}^{\alpha_1 + \beta_1 = \beta_0 - n_{j_2 s_1}} \frac{(\beta_0 - n_{j_2 s_1})!}{\alpha_1 ! \beta_1 !} (a_{j_2 j_2 s_2} - a_{j_2 j_2 s_1})^{\alpha_1} \frac{(x_{j_2} - a_{j_2 j_2 s_2})^{\beta_1 - n_{j_2 s_2}}}{\prod\limits_{s_3 = 1, s_3 \neq s_1, s_2}^{q_2} (x_{j_2} - a_{j_2 j_2 s_3})^{n_{j_2 s_3}}}$$

$$+ \sum_{\substack{\alpha_1 \in [\beta_0 - \Sigma_{i=1}^2 n_{j_2 s_i}, \beta_0 - n_{j_2 s_1}) \\ \beta_1 \in [0, n_{j_2 s_2})}}^{\alpha_1 + \beta_1 = \beta_0 - n_{j_2 s_1}} \frac{(\beta_0 - n_{j_2 s_1})!}{\alpha_1 ! \beta_1 !} (a_{j_2 j_2 s_2} - a_{j_2 j_2 s_1})^{\alpha_1} \sum_{\rho=2}^{q_2} \sum_{l_2=1}^{\overline{n}_{s_\rho}} A_{l_2 \overline{n}_{s_\rho}}^{(s_2)} \frac{1}{(x_{j_2} - a_{j_2 j_2 s_\rho})^{l_2}}$$

$$\sum_{\substack{\alpha_1 \in (\beta_0 - n_2, \beta_0 - \Sigma_{i=1}^2 n_{j_2 s_i}] \\ \beta_1 \in [n_{j_2 s_2}, n_2 - n_{j_2 s_1})}}^{\alpha_1 + \beta_1 = \beta_0 - n_{j_2 s_1}} \frac{(\beta_0 - n_{j_2 s_1})!}{\alpha_1 ! \beta_1 !} (a_{j_2 j_2 s_2} - a_{j_2 j_2 s_1})^{\alpha_1} \sum_{\rho=3}^{q_2} \sum_{l_2=1}^{\overline{n}_{s_\rho}} B_{l_2 \overline{n}_{s_\rho}}^{(s_2)} \frac{1}{(x_{j_2} - a_{j_2 j_2 s_\rho})^{l_2}} \Bigg],$$

where

$$A_{l_\rho \overline{n}_{j_2 s_\rho}}^{(s_2)} = \frac{1}{(\overline{n}_{j_2 s_\rho} - l_\rho)!} \frac{d^{\overline{n}_{j_2 s_\rho} - l_\rho}}{dx_{j_2}^{\overline{n}_{j_2 s_\rho} - l_\rho}} \frac{1}{\prod_{\sigma=2, \sigma \neq \rho}^{q_2} (x_{j_2} - a_{j_2 j_2 s_\sigma})^{n_{j_2 s_\sigma}}} \Bigg|_{x_{j_2}^* = a_{j_2 j_2 s_\rho}}$$

$$(l_\rho = 1, 2, \cdots, \overline{n}_{j_2 s_\rho}, \rho = 2, 3, \cdots, q_2),$$

$$A_{\overline{n}_{j_2 s_\rho} \overline{n}_{j_2 s_\rho}}^{(s_2)} = \frac{1}{\prod_{\sigma=2, \sigma \neq \rho}^{q_2} (a_{j_2 j_2 s_\rho} - a_{j_s j_s s_\sigma})^{n_{j_2 s_\sigma}}},$$

$$B_{l_\rho \overline{n}_{j_2 s_\rho}}^{(s_2)} = \frac{1}{(\overline{n}_{j_2 s_\rho} - l_\rho)!} \frac{d^{\overline{n}_{j_2 s_\rho} - l_\rho}}{dx_{j_2}^{\overline{n}_{j_2 s_\rho} - l_\rho}} \frac{(x_{j_2} - a_{j_2 j_2 s_2})^{\beta_1 - \sum_{i=1}^2 n_{j_2 s_i}}}{\prod_{\sigma=3, \sigma \neq \rho}^{q_2} (x_{j_2} - a_{j_2 j_2 s_\sigma})^{n_{j_2 s_\sigma}}} \Bigg|_{x_{j_2}^* = a_{j_2 j_2 s_\rho}}$$

$$(l_\rho = 1, 2, \cdots, \overline{n}_{j_2 s_\rho}, \rho = 3, 4, \cdots, q_2),$$

$$B_{\overline{n}_{j_2 s_\rho} \overline{n}_{j_2 s_\rho}}^{(s_2)} = \frac{(a_{j_2 j_2 s_\rho} - a_{j_2 j_2 s_2})^{\beta_1 - \sum_{i=1}^2 n_{j_2 s_i}}}{\prod_{\sigma=3, \sigma \neq \rho}^{q_2} (a_{j_2 j_2 s_\rho} - a_{j_s j_s s_\sigma})^{n_{j_2 s_\sigma}}};$$

$$\{\overline{n}_{j_2 s_\rho} | \rho = 2, 3, \cdots, q_2\} = \left\{ \sum\nolimits_{i=1}^{2} n_{j_2 s_i} - \beta_1, n_{j_2 s_3}, \cdots, n_{j_2 s_{q_2}} \right\}.$$

Repeating the same procedure, the following relations are obtained.

$$\frac{(x_{j_2} - a_{j_2 j_2 s_2})^{\beta_1 - n_{j_2 s_2}}}{\prod\limits_{s_3 = 1, s_3 \neq s_1, s_2}^{q_2} (x_{j_2} - a_{j_2 j_2 s_3})^{n_{j_2 s_3}}}$$

$$= \sum_{\substack{\alpha_2 \in [0, \alpha - n_2] \\ \beta_2 \in [n_2 - \Sigma_{i=1}^2 n_{j_2 s_i}, \beta_0 - \Sigma_{i=1}^2 n_{j_2 s_i}]}}^{\alpha_2 + \beta_2 = \beta_0 - \Sigma_{i=1}^2 n_{j_2 s_i}} \prod_{k=1}^{2} \frac{(\beta_{k-1} - \Sigma_{i=1}^k n_{j_2 s_i})!}{\alpha_k ! \beta_k !}$$

$$\times \frac{(a_{j_2 j_2 s_{k+1}} - a_{j_2 j_2 s_k})^{\alpha_k} (x_{j_2} - a_{j_2 j_2 s_2})^{\beta_1 - n_{j_2 s_2}}}{\prod\limits_{s_3 = 1, s_3 \neq s_1, s_2}^{q_2} (x_{j_2} - a_{j_2 j_2 s_3})^{n_{j_2 s_3}}}$$

$$+\sum_{\substack{\alpha_2\in[\beta_0-\Sigma_{i=1}^{3}n_{j_2s_i},\beta_0-\Sigma_{i=1}^{2}n_{j_2s_i})\\ \beta_2\in[0,n_{j_2s_3})}}^{\alpha_2+\beta_2=\beta_0-\Sigma_{i=1}^{2}n_{j_2s_i}}\prod_{k=1}^{2}\frac{(\beta_{k-1}-\Sigma_{i=1}^{k}n_{j_2s_i})!}{\alpha_k!\beta_k!}(a_{j_2j_2s_{k+1}}-a_{j_2j_2s_k})^{\alpha_k}$$

$$\times\sum_{\rho=3}^{q_2}\sum_{l_2=1}^{\overline{n}_{s_\rho}}A^{(s_3)}_{l_2\overline{n}_{s_\rho}}\frac{1}{(x_{j_2}-a_{j_2j_2s_\rho})^{l_2}}+\sum_{\substack{\alpha_2\in(\beta_0-n_2,\beta_0-\Sigma_{i=1}^{3}n_{j_2s_i}]\\ \beta_2\in[n_{j_2s_3},n_2-\Sigma_{i=1}^{2}n_{j_2s_i})}}^{\alpha_2+\beta_2=\beta_0-\Sigma_{i=1}^{2}n_{j_2s_i}}\prod_{k=1}^{2}\frac{(\beta_{k-1}-\Sigma_{i=1}^{k}n_{j_2s_i})!}{\alpha_k!\beta_k!}$$

$$\times(a_{j_2j_2s_{k+1}}-a_{j_2j_2s_k})^{\alpha_k}\sum_{\rho=4}^{q_2}\sum_{l_2=1}^{\overline{n}_{s_\rho}}\frac{B^{(s_3)}_{l_2\overline{n}_{s_\rho}}}{(x_{j_2}-a_{j_2j_2s_\rho})^{l_2}},$$

and

$$A^{(s_3)}_{l_\rho\overline{n}_{j_2s_\rho}}=\frac{1}{(\overline{n}_{j_2s_\rho}-l_\rho)!}\frac{d^{\overline{n}_{j_2s_\rho}-l_\rho}}{dx_{j_2}^{\overline{n}_{j_2s_\rho}-l_\rho}}\frac{1}{\prod_{\sigma=3,\sigma\neq\rho}^{q_2}(x_{j_2}-a_{j_2j_2s_\sigma})^{n_{j_2s_\sigma}}}\Bigg|_{x^*_{j_2}=a_{j_2j_2s_\rho}}$$

$$(l_\rho=1,2,\cdots,\overline{n}_{j_2s_\rho},\ \rho=3,4,\cdots,q_2),$$

$$A^{(s_3)}_{\overline{n}_{j_2s_\rho}\overline{n}_{j_2s_\rho}}=\frac{1}{\prod_{\sigma=3,\sigma\neq\rho}^{q_2}(a_{j_2j_2s_\rho}-a_{j_sj_ss_\sigma})^{n_{j_2s_\sigma}}},$$

$$B^{(s_3)}_{l_\rho\overline{n}_{j_2s_\rho}}=\frac{1}{(\overline{n}_{j_2s_\rho}-l_\rho)!}\frac{d^{\overline{n}_{j_2s_\rho}-l_\rho}}{dx_{j_2}^{\overline{n}_{j_2s_\rho}-l_\rho}}\frac{(x_{j_2}-a_{j_2j_2s_3})^{\beta_1-\sum_{i=1}^{2}n_{j_2s_i}}}{\prod_{\sigma=3,\sigma\neq\rho}^{q_2}(x_{j_2}-a_{j_2j_2s_\sigma})^{n_{j_2s_\sigma}}}\Bigg|_{x^*_{j_2}=a_{j_2j_2s_\rho}}$$

$$(l_\rho=1,2,\cdots,\overline{n}_{j_2s_\rho},\ \rho=4,5,\cdots,q_2),$$

$$B^{(s_3)}_{\overline{n}_{j_2s_\rho}\overline{n}_{j_2s_\rho}}=\frac{(a_{j_2j_2s_\rho}-a_{j_2j_2s_3})^{\beta_1-\sum_{i=1}^{2}n_{j_2s_i}}}{\prod_{\sigma=4,\sigma\neq\rho}^{q_2}(a_{j_2j_2s_\rho}-a_{j_sj_ss_\sigma})^{n_{j_2s_\sigma}}};$$

$$\{\overline{n}_{j_2s_\rho}|\rho=3,4,\cdots,q_2\}=\left\{\sum_{i=1}^{2}n_{j_2s_i}-\beta_1,n_{j_2s_3},\cdots,n_{j_2s_{q_2}}\right\}.$$

Continuing the same procedure until r_{p-1}, the following relation is obtained as

$$\sum_{r_1=1}^{p_1}\sum_{l_1=1}^{m_{j_1r_1}}\frac{A_{l_1m_{j_1r_1}}}{(x_{j_1}-a_{j_1j_1r_1})^{l_1}}dx_{j_1}=\frac{a_{j_1j_10}}{a_{j_2j_20}}\left\{\sum_{\beta_0=n_2}^{n_1}b_{j_1j_2\beta_0}\sum_{\substack{\alpha_1\in[0,\beta_0-n_2],\\ \beta_1\in[n_2-n_{j_2s_1},\beta_0-n_{j_2s_1}]}}^{\alpha_1+\beta_1=\beta_0-n_{j_2s_1}}\right.$$

$$\sum_{\substack{\alpha_2\in[0,\beta_1-n_2]\\ \beta_2\in[n_2-\Sigma_{i=1}^{2}n_{j_2s_i},\beta_1-\Sigma_{i=1}^{2}n_{j_2s_i}]}}^{\alpha_2+\beta_2=\beta_1-\Sigma_{i=1}^{2}n_{j_2s_i}}\cdots\sum_{\substack{\alpha_{q_2-1}\in[0,\beta_{q_2-2}-n_2]\\ \beta_{q_2-1}\in[n_2-\Sigma_{i=1}^{q_2-1}n_{j_2s_i},\beta_{q_2-2}-\Sigma_{i=1}^{q_2-1}n_{j_2s_i}]}}^{\alpha_{q_2-1}+\beta_{q_2-1}=\beta_{q_2-2}-\Sigma_{i=1}^{q_2-1}n_{j_2s_i}}$$

$$\prod_{j=1}^{q_2-1}\frac{(\beta_{j-1}-\Sigma_{i=1}^{j}n_{j_2s_i})!}{\alpha_j!\beta_j!}(a_{j_2j_2s_{j+1}}-a_{j_2j_2s_j})^{\alpha_j}(x_{j_2}-a_{j_2j_2s_{q_2}})^{\beta_{q_2-1}-n_{j_2s_{q_2}}}dx_{j_2}$$

$$+\sum_{\beta_0=q_2}^{q_1}b_{j_2j_2\beta_0}\times\sum_{k=2}^{q_2-1}\Bigg[\sum_{\substack{\alpha_1\in[0,\beta_0-n_2),\\ \beta_1\in(n_2-n_{j_2s_1},\beta_0-n_{j_2s_1}]}}^{\alpha_1+\beta_1=\beta_0-n_{j_2s_1}}\;\sum_{\substack{\alpha_2\in[0,\beta_1-n_2]\\ \beta_2\in(n_2-\Sigma_{i=1}^{2}n_{j_2s_i},\beta_1-\Sigma_{i=1}^{2}n_{j_2s_i})}}^{\alpha_2+\beta_2=\beta_1-\Sigma_{i=1}^{2}n_{j_2s_i}}$$

$$\cdots\left(\sum_{\substack{\alpha_{k-1}\in(\beta_{k-2}-n_2+n_{j_2s_k},\beta_{k-2}-\Sigma_{i=1}^{k-1}n_{j_2s_i}]\\ \beta_{k-1}\in[0,n_2-\Sigma_{i=1}^{k}n_{j_2s_i})}}^{\alpha_{k-1}+\beta_{k-1}=\beta_{k-2}-\Sigma_{i=1}^{k-1}n_{j_2s_i}}\;\prod_{j=1}^{k-1}\frac{(\beta_{j-1}-\Sigma_{i=1}^{j}n_{j_2s_i})!}{\alpha_j!\beta_j!}(a_{j_2j_2s_{j+1}}-a_{j_2j_2s_j})^{\alpha_j}\right.$$

$$\times\sum_{\rho=k}^{q_2}\sum_{l_\rho=1}^{\overline{n}_{j_2s_\rho}}A^{(s_k)}_{l_\rho\overline{n}_{j_2s_\rho}}\frac{dx_{j_2}}{(x_{j_2}-a_{j_2j_2s_\rho})^{l_\rho}}$$

$$+\sum_{\substack{\alpha_{k-1}\in(\beta_{k-2}-n_2,\beta_{k-2}-n_2+n_{j_2s_k}]\\ \beta_{k-1}\in[n_2-\Sigma_{i=1}^{k}n_{j_2s_i},n_2-\Sigma_{i=1}^{k-1}n_{j_2s_i})}}^{\alpha_{k-1}+\beta_{k-1}=\beta_{k-2}-\Sigma_{i=1}^{k-1}n_{j_2s_i}}\;\prod_{j=1}^{k-1}\frac{(\beta_{j-1}-\Sigma_{i=1}^{j}n_{j_2s_i})!}{\alpha_j!\beta_j!}(a_{j_2j_2s_{j+1}}-a_{j_2j_2s_j})^{\alpha_j}$$

$$\left.\times\sum_{\rho=k+1}^{q_2}\sum_{l_\rho=1}^{\overline{n}_{s_\rho}}B^{(s_k)}_{l_\rho\overline{n}_{j_2s_\rho}}\frac{dx_{j_2}}{(x_{j_2}-a_{j_2j_2s_\rho})^{l_\rho-1}}\right)\Bigg]+\left(\sum_{\beta_0=0}^{n_{j_2s_1}-1}b_{j_1j_2\beta_0}\sum_{\rho=1}^{q_2}\sum_{l_\rho=1}^{\overline{n}_{j_2s_\rho}}A^{(s_1)}_{l_\rho\overline{n}_{j_2s_\rho}}\right.$$

$$\left.\left.\frac{dx_{j_2}}{(x_{j_2}-a_{j_2j_2s_\rho})^{l_\rho}}+\sum_{\beta_0=n_{j_2s_1}}^{n_2-1}b_{j_1j_2\beta_0}\sum_{\rho=2}^{q_2}\sum_{l_\rho=1}^{\overline{n}_{j_2s_\rho}}B^{(s_1)}_{l_\rho\overline{n}_{j_2s_\rho}}\frac{dx_{j_2}}{(x_{j_2}-a_{j_2j_2s_\rho})^{l_\rho}}\right)\right\}.$$

With an initial condition at time, the integration of the foregoing equation gives the first integral manifold as

$$\sum_{r_1=1}^{p_1}\sum_{l_1=1}^{m_{j_1r_1}}\frac{A_{l_1m_{j_1r_1}}}{l_1-1}\left[\frac{1}{(x_{j_1}-a_{j_1j_1r_1})^{l_1-1}}-\frac{1}{(x_{j_10}-a_{j_1j_1r_1})^{l_1-1}}\right]$$

$$=\frac{a_{j_1j_10}}{a_{j_2j_20}}\Bigg[\sum_{\beta_0=n_2}^{n_1}b_{j_1j_2\beta_0}\sum_{\substack{\alpha_1\in[0,\beta_0-n_2],\\ \beta_1\in[n_2-n_{j_2s_1},\beta_0-n_{j_2s_1}]}}^{\alpha_1+\beta_1=\beta_0-n_{j_2s_1}}\;\sum_{\substack{\alpha_2\in[0,\beta_1-n_2]\\ \beta_2\in[n_2-\Sigma_{i=1}^{2}n_{j_2s_i},\beta_1-\Sigma_{i=1}^{2}n_{j_2s_i}]}}^{\alpha_2+\beta_2=\beta_1-\Sigma_{i=1}^{2}n_{j_2s_i}}\cdots$$

$$\sum_{\substack{\alpha_{q_2-1}\in[0,\beta_{q_2-2}-n_2]\\ \beta_{q_2-1}\in[n_2-\Sigma_{i=1}^{q_2-1}n_{j_2s_i},\beta_{q_2-2}-\Sigma_{i=1}^{q_2-1}n_{j_2s_i}]}}^{\alpha_{q_2-1}+\beta_{q_2-1}=\beta_{q_2-2}-\Sigma_{i=1}^{q_2-1}n_{j_2s_i}}\;\prod_{j=1}^{q_2-1}\frac{(\beta_{j-1}-\Sigma_{i=1}^{j}n_{j_2s_i})!}{\alpha_j!\beta_j!}(a_{j_2j_2s_{j+1}}-a_{j_2j_2s_j})^{\alpha_j}$$

$$\times\frac{1}{\beta_{q_2-1}-n_{j_2s_{q_2}}+1}\left[(x_{j_2}-a_{j_2j_2s_{q_2}})^{\beta_{q_2-1}-n_{j_2s_{q_2}}+1}-(x_{j_20}-a_{j_2j_2s_{q_2}})^{\beta_{q_2-1}-n_{j_2r_{q_2}}+1}\right]$$

$$
-\sum_{\beta_0=n_2}^{n_1} b_{j_1j_2\beta_0}\sum_{k=2}^{q_2-1}\Bigg(\sum_{\substack{\alpha_1+\beta_1=\beta_0-n_{j_2s_1}\\ \alpha_1\in[0,\beta_0-n_2),\\ \beta_1\in(n_2-n_{j_2s_1},\beta_0-n_{j_2s_1}]}}\ \sum_{\substack{\alpha_2+\beta_2=\beta_1-\Sigma_{i=1}^2 n_{j_2s_i}\\ \alpha_2\in[0,\beta_1-n_2]\\ \beta_2\in(n_2-\Sigma_{i=1}^2 n_{j_2s_i},\beta_1-\Sigma_{i=1}^2 n_{j_2s_i})}}\cdots
$$

$$
\times\Bigg\{\sum_{\substack{\alpha_{k-1}+\beta_{k-1}=\beta_{k-2}-\Sigma_{i=1}^{k-1}n_{j_2s_i}\\ \alpha_{k-1}\in(\beta_{k-2}-n_2+n_{j_2s_k},\beta_{k-2}-\Sigma_{i=1}^{k-1}n_{j_2s_i}]\\ \beta_{k-1}\in[0,n_2-\Sigma_{i=1}^{k}n_{j_2s_i})}}\prod_{j=1}^{k-1}\frac{(\beta_{j-1}-\Sigma_{i=1}^{j}n_{j_2s_i})!}{\alpha_j!\beta_j!}(a_{j_2j_2s_{j+1}}-a_{j_2j_2s_j})^{\alpha_j}
$$

$$
\times\sum_{\rho=k}^{q_2}\sum_{l_\rho=1}^{\overline{n}_{j_2s_\rho}}\frac{A^{(s_k)}_{l_\rho\overline{n}_{j_2s_\rho}}}{l_\rho-1}\left[\frac{1}{(x_{j_2}-a_{j_2j_2s_\rho})^{l_\rho-1}}-\frac{1}{(x_{j_20}-a_{j_2j_2s_\rho})^{l_\rho-1}}\right]
$$

$$
+\sum_{\substack{\alpha_{k-1}+\beta_{k-1}=\beta_{k-2}-\Sigma_{i=1}^{k-1}n_{j_2s_i}\\ \alpha_{k-1}\in(\beta_{k-2}-n_2,\beta_{k-2}-n_2+n_{j_2k}]\\ \beta_{k-1}\in[n_2-\Sigma_{i=1}^{k}n_{j_2s_i},n_2-\Sigma_{i=1}^{k-1}n_{j_2s_i})}}\prod_{j=1}^{k-1}\frac{(\beta_{j-1}-\Sigma_{i=1}^{j}n_{j_2s_i})!}{\alpha_j!\beta_j!}(a_{j_2j_2s_{j+1}}-a_{j_2j_2s_j})^{\alpha_j}
$$

$$
\times\sum_{\rho=k+1}^{q_2}\sum_{l_\rho=1}^{\overline{n}_{j_2s_\rho}}\frac{B^{(s_k)}_{l_\rho\overline{n}_{j_2s_\rho}}}{l_\rho-1}\left[\frac{1}{(x_{j_2}-a_{j_2j_2s_\rho})^{l_\rho-1}}-\frac{1}{(x_{j_20}-a_{j_2j_2s_\rho})^{l_\rho-1}}\right]\Bigg\}\Bigg)
$$

$$
-\Bigg\{\sum_{\beta_0=0}^{n_{j_2s_1}-1} b_{j_1j_2\beta_0}\sum_{\rho=1}^{q_2}\sum_{l_\rho=1}^{\overline{n}_{j_2s_\rho}}\frac{A^{(s_1)}_{l_\rho\overline{n}_{j_2s_\rho}}}{l_\rho-1}\left[\frac{1}{(x_{j_2}-a_{j_2j_2s_\rho})^{l_\rho-1}}-\frac{1}{(x_{j_20}-a_{j_2j_2s_\rho})^{l_\rho-1}}\right]
$$

$$
+\sum_{\beta_0=n_{j_2s_1}}^{n_2-1} b_{j_1j_2\beta_0}\sum_{\rho=2}^{q_2}\sum_{l_\rho=1}^{\overline{n}_{j_2s_\rho}}\frac{B^{(s_1)}_{l_\rho\overline{n}_{j_2s_\rho}}}{l_\rho-1}\left[\frac{1}{(x_{j_2}-a_{j_2j_2s_\rho})^{l_\rho-1}}-\frac{1}{(x_{j_20}-a_{j_2j_2s_\rho})^{l_\rho-1}}\right]\Bigg\}\Bigg],
$$

where

$$
A^{(s_k)}_{l_\rho\overline{n}_{j_2s_\rho}}=\frac{1}{(\overline{n}_{j_2s_\rho}-l_\rho)!}\frac{d^{\overline{n}_{j_2s_\rho}-l_\rho}}{dx_{j_2}^{\overline{n}_{j_2s_\rho}-l_\rho}}\frac{1}{\prod_{\sigma=k,\sigma\neq\rho}^{q_2}(x_{j_2}-a_{j_2j_2s_\sigma})^{n_{j_2s_\sigma}}}\Bigg|_{x^*_{j_2}=a_{j_2j_2s_\rho}}
$$

$$
(l_\rho=1,2,\cdots,\overline{n}_{j_2s_\rho},\ \rho=k,k+1,\cdots,q_2),
$$

$$
A^{(s_k)}_{\overline{n}_{j_2s_\rho}\overline{n}_{j_2s_\rho}}=\frac{1}{\prod_{\sigma=k,\sigma\neq\rho}^{q_2}(a_{j_2j_2s_\rho}-a_{j_sj_ss_\sigma})^{n_{j_2s_\sigma}}},
$$

$$
B^{(s_k)}_{l_\rho\overline{n}_{j_2s_\rho}}=\frac{1}{(\overline{n}_{j_2s_\rho}-l_\rho)!}\frac{d^{\overline{n}_{j_2s_\rho}-l_\rho}}{dx_{j_2}^{\overline{n}_{j_2s_\rho}-l_\rho}}\frac{(x_{j_2}-a_{j_2j_2s_k})^{\beta_{k-1}-\sum_{i=1}^{k}n_{j_2s_i}}}{\prod_{\sigma=k+1,\sigma\neq\rho}^{q_2}(x_{j_2}-a_{j_2j_2s_\sigma})^{n_{j_2s_\sigma}}}\Bigg|_{x^*_{j_2}=a_{j_2j_2s_\rho}}
$$

$$
(l_\rho=1,2,\cdots,\overline{n}_{j_2s_\rho},\ \rho=k+1,\cdots,q_2),
$$

$$
B^{(s_k)}_{\overline{n}_{j_2s_\rho}\overline{n}_{j_2s_\rho}}=\frac{(a_{j_2j_2s_\rho}-a_{j_2j_2s_k})^{\beta_{k-1}-\sum_{i=1}^{k}n_{j_2s_i}}}{\prod_{\sigma=k+1,\sigma\neq\rho}^{q_2}(a_{j_2j_2s_\rho}-a_{j_sj_ss_\sigma})^{n_{j_2s_\sigma}}},
$$

$$
\{\overline{n}_{j_2s_\rho}|\rho=k,k+1,\cdots,q_2\}=\left\{\sum_{i=1}^{k}n_{j_2s_i}-\beta_{k-1},n_{j_2s_{k+1}},\cdots,n_{j_2s_{q_2}}\right\},
$$

for $k\in\{1,2,\cdots,q_2-1\}$;

and

$$
\begin{aligned}
& b_{j_1j_2\beta_0}=\sum_{t_1,l_1=0}^{t_1+l_1=n_{j_1u_1}}\cdots\sum_{t_{q_1},l_{q_1}=0}^{t_{q_1}+l_{q_1}=n_{j_1u_{q_1}}}\prod_{k=1}^{q_1}\frac{n_{j_1u_k}!}{t_k!l_k!}(a_{j_2j_2s_1}-a_{j_1j_2u_k})^{t_k}\delta_{\beta_0}^{\Sigma_{k=1}^{q_1}l_k},\\
& \frac{1}{\beta_{q_2-1}-n_{j_2q_2}+1}[(x_{j_2}-a_{j_2j_2s_{q_2}})^{\beta_{q_2-1}-n_{j_2q_2}+1}-(x_{j_20}-a_{j_2j_2s_{q_2}})^{\beta_{q_2-1}-n_{j_2q_2}+1}]\\
& =\ln\frac{|x_{j_2}-a_{j_2j_2s_{q_2}}|}{|x_{j_20}-a_{j_2j_2s_{q_2}}|},\ \text{for } \beta_{q_2-1}-n_{j_2q_2}+1=0;\\
& \frac{1}{l_k-1}\left[\frac{1}{(x_{j_2}-a_{j_2j_2s_k})^{l_k-1}}-\frac{1}{(x_{j_20}-a_{j_2j_2s_k})^{l_k-1}}\right]=-\ln\frac{|x_{j_2}-a_{j_2j_2s_k}|}{|x_{j_20}-a_{j_2j_2s_k}|},\\
& \text{for } l_k=1\,(k=1,2,\cdots,q_2-1);\\
& \frac{1}{l_1-1}\left[\frac{1}{(x_{j_1}-a_{j_1j_1r_1})^{l_1-1}}-\frac{1}{(x_{j_10}-a_{j_1j_1r_1})^{l_1-1}}\right]=-\ln\frac{|x_{j_1}-a_{j_1j_1r_1}|}{|x_{j_10}-a_{j_1j_1r_1}|},\\
& \text{for } l_1=1.
\end{aligned}
$$

The network matrices of singular equilibriums and 1-dimensional flows are defined as

$$
\begin{aligned}
& \bigcup_{\eta=j_1}^{j_2}\bigcup_{r=1}^{p_1}\bigcup_{l=1}^{q_1+q_2}(a_{j_1j_1r},c_{\eta j_2l})\\
& =\bigcup_{r=1}^{p_1}\bigcup_{u=1}^{q_1}\underbrace{(a_{j_1j_1r},a_{j_1j_2u})}_{([m_{j_1r},n_{j_1u}]-XX,xF)}\oplus\bigcup_{r=1}^{p_1}\bigcup_{s=1}^{q_2}\underbrace{(a_{j_1j_1r},a_{j_2j_2s})}_{(m_{j_1r},n_{j_1s})-XX},\\
& \{c_{\eta j_2l}|\eta\in\{j_1,j_2\},l=1,2,\cdots,q_1+q_2\}\\
& =\text{sort}\{a_{j_1j_2u},a_{j_2j_2s}|u=1,2,\cdots q_1;s=1,2,\cdots,q_2\},\\
& c_{\eta_1j_2l_1}<c_{\eta_2j_2l_2}\ \text{for } l_1<l_2,\eta_1,\eta_2\in\{j_1,j_2\};
\end{aligned}
$$

where

$$
\begin{aligned}
& \bigcup_{r=1}^{p_1}\bigcup_{u=1}^{q_1}\underbrace{(a_{j_1j_1r},a_{j_1j_2u})}_{([m_{j_1r},n_{j_1u}]\text{-XX,xF})}\\
& =\begin{bmatrix}
\underbrace{(a_{j_1j_11},a_{j_1j_2q_1})}_{([m_{j_11},n_{j_1q_1}]\text{XX,xF})} & \cdots & \underbrace{(a_{j_1j_1(p_1-1)},a_{j_1j_2q_1})}_{([m_{j_1(p_1-1)},n_{j_1q_1}]\text{XX,xF})} & \underbrace{(a_{j_1j_1p_1},a_{j_1j_2q_1})}_{([m_{j_1p_1},n_{j_1q_1}]\text{XX,xF})}\\
\underbrace{(a_{j_1j_11},a_{j_1j_2(q_1-1)})}_{([m_{j_11},n_{j_1(q_1-1)}]\text{XX,xF})} & \cdots & \underbrace{(a_{j_1j_1(p_1-1)},a_{j_1j_2(q_1-1)})}_{([m_{j_1(p_1-1)},n_{j_1(q_1-1)}]\text{XX,xF})} & \underbrace{(a_{j_1j_1p_1},a_{j_1j_2(q_1-1)})}_{([m_{j_11},n_{j_1(q_1-1)}]\text{XX,xF})}\\
\vdots & \vdots & \vdots & \vdots\\
\underbrace{(a_{j_1j_11},a_{j_1j_21})}_{([m_{j_11},n_{j_11}]\text{XX,xF})} & \cdots & \underbrace{(a_{j_1j_1(p_1-1)},a_{j_1j_21})}_{([m_{j_1(p_1-1)},n_{j_11}]\text{XX,xF})} & \underbrace{(a_{j_1j_11},a_{j_1j_21})}_{([m_{j_1p_1},n_{j_11}]\text{XX,xF})}
\end{bmatrix}_{q_1\times p_1}
\end{aligned}
$$

and

$$\bigcup_{r=1}^{p_1}\bigcup_{s=1}^{q_2}\underbrace{(a_{j_1j_1r},a_{j_2j_2s})}_{(m_{j_2r},n_{j_2s})\text{-XX}}$$

$$=\begin{bmatrix}\underbrace{(a_{j_1j_1 1},a_{j_2j_2q_2})}_{(m_{j_11},n_{j_2q_1})\text{-YY}} & \cdots & \underbrace{(a_{j_1j_1(p_1-1)},a_{j_2j_2q_2})}_{(m_{j_1(p_1-1)},n_{(2)q_1})\text{-YY}} & \underbrace{(a_{j_1j_1p_1},a_{j_2j_2q_2})}_{(m_{j_1p_1},n_{j_2q_2})\text{-YY}}\\ \underbrace{(a_{j_1j_11},a_{j_2j_2(q_1-1)})}_{(m_{j_11},n_{j_2(q_1-1)})\text{-YY}} & \cdots & \underbrace{(a_{j_1j_1(p_1-1)},a_{j_2j_2(q_2-1)})}_{(m_{j_1(p_1-1)},n_{q_1-1})\text{-YY}} & \underbrace{(a_{j_1j_1p_1},a_{j_2j_2(q_2-1)})}_{(m_{j_1p_1},n_{j_2(q_2-1)})\text{-YY}}\\ \vdots & \vdots & \vdots & \vdots\\ \underbrace{(a_{j_1j_11},a_{j_2j_21})}_{(m_{j_11},n_{j_21})\text{-YY}} & \cdots & \underbrace{(a_{j_1j_1(p_1-1)},a_{j_2j_21})}_{(m_{j_1(p_1-1)},n_{21})\text{-YY}} & \underbrace{(a_{j_1j_11},a_{j_2j_21})}_{(m_{j_1p_1},n_{j_21})\text{-YY}}\end{bmatrix}_{q_2\times P_2}$$

($\mathrm{ii}_{1\mathrm{a}}$) There is an equilibrium for 1-dimensional flow as

$$(x_{j_1}^*,x_{j_2}^*)=(a_{j_1j_1r_1},a_{j_1j_2u_1}).$$

For $a_{j_1j_2u}\neq a_{j_2j_2s}$ $(u=1,2,\cdots,q_1;\,s=1,2,\cdots,q_2)$, consider

$$\frac{dx_{j_1}}{dx_{j_2}}=\frac{a_{j_1j_10}}{a_{j_2j_20}}\frac{\prod_{r=1}^{p_1}(x_{j_1}-a_{j_1j_1r})^{m_{j_1r}}\prod_{u=1}^{q_1}(x_{j_2}-a_{j_1j_2u})^{n_{j_1u}}}{\prod_{s=1}^{q_2}(x_{j_2}-a_{j_2j_2s})^{n_{j_2s}}}.$$

At the equilibrium of $x_{j_2}^*=a_{j_1j_2u_1}$ and $\overline{x}_{j_1}\neq a_{j_1j_1r}(r=1,2,\cdots,p_1)$, in phase space,

$$\begin{aligned}\left.\frac{dx_{j_1}}{dx_{j_2}}\right|_{x_{j_2}^*=a_{j_1j_2u_1}}&=\frac{a_{j_1j_10}}{a_{j_2j_20}}\frac{\prod_{r=1}^{p_1}(\overline{x}_{j_1}-a_{j_1j_1r})^{m_{j_1r}}\prod_{u_2=1,u_2\neq u_1}^{q_1}(x_{j_2}-a_{j_1j_2u_2})^{n_{j_1u_2}}}{\prod_{s=1}^{q_2}(x_{j_2}-a_{j_2j_2s})^{n_{j_2s}}}\\&\quad\times(x_{j_2}-a_{j_1j_2u_1})^{n_{j_1u_1}}\Big|_{x_{j_2}^*=a_{j_1j_2u_1}}=0,\\&\ \vdots\\\left.\frac{d^{n_{j_1u_1}}x_{j_1}}{dx_{j_2}^{n_{j_1u_1}}}\right|_{x_{j_2}^*=a_{j_1j_2u_1}}&=0.\end{aligned}$$

If

$$\begin{aligned}&\left.\frac{d^{n_{j_1u_1}+1}x_{j_1}}{dx_{j_2}^{n_{j_1u_1}+1}}\right|_{x_{j_2}^*=a_{j_1j_2u_1}}\\&=\frac{a_{j_1j_10}}{a_{j_2j_20}}\frac{\prod_{r=1}^{p_1}(\overline{x}_{j_1}-a_{j_1j_1r})^{m_{j_1r}}\prod_{u_2=1,u_2\neq u_1}^{q_1}(a_{j_1j_2u_1}-a_{j_1j_2u_2})^{n_{j_1u_2}}n_{j_1u_1}!}{\prod_{s=1}^{q_2}(a_{j_1j_2u_1}-a_{j_2j_2s})^{n_{j_2s}}}>0,\end{aligned}$$

the $(2n_{j_1u_11}+1)^{\text{th}}$ -order up-parabola flows exists for $n_{j_1u_1}=2n_{j_1u_11}+1$, and the $(2n_{j_1u_11})^{\text{th}}$ -order increasing-inflection flow exists for $n_{j_1u_1}=2n_{j_1u_11}$. If

$$\begin{aligned}&\frac{d^{n_{j_1u_1}+1}x_{j_1}}{dx_{j_2}^{n_{j_1u_1}+!}}\bigg|_{x_{j_2}^*=a_{j_1j_2u_1}}\\&=\frac{a_{j_1j_10}}{a_{j_2j_20}}\frac{\prod_{r=1}^{p_1}(\overline{x}_{j_1}-a_{j_1j_1r})^{m_{j_1r}}\prod_{u_2=1,u_2\neq u_1}^{q_1}(a_{j_1j_2u_1}-a_{j_1j_2u_2})^{n_{j_1u_2}}n_{j_1u_1}!}{\prod_{s=1}^{q_2}(a_{j_1j_2u_1}-a_{j_2j_2s})^{n_{j_2s}}}<0,\end{aligned}$$

the $(2n_{j_1u_11}+1)^{\text{th}}$ -order down-parabola flows exists for $n_{j_1u_1}=2n_{j_1u_11}+1$, and the $(2n_{j_1u_11})^{\text{th}}$ -order decreasing-inflection flow exists for $n_{j_1u_1}=2n_{j_1u_11}$.

Let

$$\dot{x}_{j_2}=a_{j_2j_20}\prod_{s=1}^{q_2}(a_{j_1j_2u_1}-a_{j_2j_2s})^{n_{j_2s}}.$$

For $a_{j_2j_20}\prod_{s=1}^{q_2}(a_{j_1j_2u_1}-a_{j_2j_2s})^{n_{j_2s}}>0$ and $a_{j_2j_20}\prod_{s=1}^{q_2}(a_{j_1j_2u_1}-a_{j_2j_2s})^{n_{j_2s}}<0$, the parabola and inflection flows are positive and negative in the x_{j_2}-direction, respectively.

At $x_{j_1}^*=a_{j_1j_1r_1}$ and $\overline{x}_{j_2}\neq a_{j_1j_2u},a_{j_2j_2s}(u=1,2,\cdots,q_1;s=1,2,\cdots,q_2)$, the variational equation is

$$\begin{aligned}\dot{x}_{j_2}&=a_{j_2j_20}\prod_{s=1}^{q_2}(\overline{x}_{j_2}-a_{j_2j_2s})^{n_{j_2s}},\\\Delta\dot{x}_{j_1}&=a_{j_1j_10}\prod_{r_2=1,r_2\neq r_1}^{p_1}(a_{j_1j_1r_1}-a_{j_1j_1r_2})^{m_{j_1r_2}}\prod_{u=1}^{q_1}(\overline{x}_{j_2}-a_{j_1j_2u})^{n_{j_1u}}(\Delta x_{j_1})^{m_{j_1r_1}}.\end{aligned}$$

For

$$\begin{aligned}&a_{j_1j_10}\prod_{r_2=1,r_2\neq r_1}^{p_1}(a_{j_1j_1r_1}-a_{j_1j_1r_2})^{m_{j_1r_2}}\prod_{u=1}^{q_1}(\overline{x}_{j_2}-a_{j_1j_2u})^{n_{j_1u}}>0\text{ and}\\&a_{j_1j_10}\prod_{r_2=1,r_2\neq r_1}^{p_1}(a_{j_1j_1r_1}-a_{j_1j_1r_2})^{m_{j_1r_2}}\prod_{u=1}^{q_1}(\overline{x}_{j_2}-a_{j_1j_2u})^{n_{j_1u}}<0,\end{aligned}$$

the flows at $x_{j_1}^*=a_{j_1j_1r_1}$ and $\overline{x}_{j_2}\neq a_{j_1j_2u},a_{j_2j_2s}$ are the $(2m_{j_1r_11}+1)^{\text{th}}$-order source and sink for $m_{j_1r_1}=2m_{j_1r_11}+1$; and the flows at $x_{j_1}^*=a_{j_1j_1r_1}$ and $\overline{x}_{j_2}\neq a_{j_1j_2u},a_{j_2j_2s}$, are the $(2m_{j_1r_11})^{\text{th}}$-order upper-saddle and lower-saddle for $m_{j_1r_1}=2m_{j_1r_11}$.

Let $\overline{x}_{j_1}=a_{j_1j_1r_1}$ and $\overline{x}_{j_2}=a_{j_1j_2u_1}\neq a_{j_2j_2s}$. Therefore, from the above discussion, there are four cases: ($\text{ii}_{1\text{a}1}$) $m_{j_1r_1}=2m_{j_1r_11}+1$ and $n_{j_1u_1}=2n_{j_1u_11}+1$, ($\text{ii}_{1\text{a}2}$) $m_{j_1r_1}=2m_{j_1r_11}$ and $n_{j_1u_1}=2n_{j_1u_11}+1$,($\text{ii}_{1\text{a}3}$) $m_{j_1r_1}=2m_{j_1r_11}+1$ and $n_{j_1u_1}=2n_{j_1u_11}$,($\text{ii}_{1\text{a}4}$) $m_{j_1r_1}=2m_{j_1r_11}$ and $n_{j_1u_1}=2n_{j_1u_11}$.

($\text{ii}_{1\text{a}1}$) For $m_{j_1r_1}=2m_{j_1r_11}+1$ and $n_{j_1u_1}=2n_{j_1u_11}+1$, the equilibrium of $(x_{j_1}^*,x_{j_2}^*)=(a_{j_1j_1r_1},a_{j_1j_2u_1})$ has the following properties as in Eqs. (1.145)–(1.148).

- For $a_{j_1j_10}\prod_{r_2=1,r_2\neq r_1}^{p_1}(a_{j_1j_1r_1}-a_{j_1j_1r_2})^{m_{j_1r_2}}\prod_{u_2=1,u_2\neq u_1}^{q_1}(a_{j_1j_2u_1}-a_{j_1j_2u_2})^{n_{j_1u_2}}>0$ and $a_{j_2j_20}\prod_{s=1}^{q_2}(a_{j_1j_2u_1}-a_{j_2j_2s})^{n_{j_2s}}>0$,

$$(a_{j_1j_1r_1}, a_{j_1j_2u_1}) = \underbrace{([(2m_{j_1r_1}1+1), (2n_{j_1u_1}1+1)]\text{DP:UP, pF})}_{[(2m_{j_1r_1}1+1),(2n_{j_1u_1}1+1)]\text{-hyperbolic }(+)}$$

The flow at $(x^*_{j_1}, x^*_{j_2}) = (a_{j_1j_1r_1}, a_{j_1j_2u_1})$is a([(2$m_{j_1r_1}1$+1), (2$n_{j_1u_1}1$ + 1)]DP:UP, pF)-positive hyperbolic flow.

- For $a_{j_1j_10}\prod_{r_2=1,r_2\neq r_1}^{p_1}(a_{j_1j_1r_1}-a_{j_1j_1r_2})^{m_{j_1r_2}}\prod_{u_2=1,u_2\neq u_1}^{q_1}(a_{j_1j_2u_1}-a_{j_1j_2u_2})^{n_{j_1u_2}}<0$ and $a_{j_2j_20}\prod_{s=1}^{q_2}(a_{j_1j_2u_1}-a_{j_2j_2s})^{n_{j_2s}}>0$,

$$(a_{j_1j_1r_1}, a_{j_1j_2u_1}) = \underbrace{([(2m_{j_1r_1}1+1), (2n_{j_1u_1}1+1)]\text{UP:DP, pF})}_{[(2m_{j_1r_1}1+1),(2n_{j_1u_1}1+1)]\text{-hyperbolic-secant flow }(+)}$$

The flow at $(x^*_{j_1}, x^*_{j_2}) = (a_{j_1j_1r_1}, a_{j_1j_2u_1})$is a([(2$m_{j_1r_1}1$+1), (2$n_{j_1u_1}1$ + 1)]UP:DP, pF)-positive hyperbolic-secant flow.

- For $a_{j_1j_10}\prod_{r_2=1,r_2\neq r_1}^{p_1}(a_{j_1j_1r_1}-a_{j_1j_1r_2})^{m_{j_1r_2}}\prod_{u_2=1,u_2\neq u_1}^{q_1}(a_{j_1j_2u_1}-a_{j_1j_2u_2})^{n_{j_1u_2}}>0$ and $a_{j_2j_20}\prod_{s=1}^{q_2}(a_{j_1j_2u_1}-a_{j_2j_2s})^{n_{j_2s}}<0$,

$$(a_{j_1j_1r_1}, a_{j_1j_2u_1}) = \underbrace{([(2m_{j_1r_1}1+1), (2n_{j_1u_1}1+1)]\text{UP:DP, nF})}_{[(2m_{j_1r_1}1+1),(2n_{j_1u_1}1+1)]\text{-hyperbolic-secant flow }(-)}$$

The flow at $(x^*_{j_1}, x^*_{j_2}) = (a_{j_1j_1r_1}, a_{j_1j_2u_1})$is a([(2$m_{j_1r_1}1$+1), (2$n_{j_1u_1}1$ + 1)]UP:DP, nF)-negative hyperbolic-secant flow.

- For $a_{j_1j_10}\prod_{r_2=1,r_2\neq r_1}^{p_1}(a_{j_1j_1r_1}-a_{j_1j_1r_2})^{m_{j_1r_2}}\prod_{u_2=1,u_2\neq u_1}^{q_1}(a_{j_1j_2u_1}-a_{j_1j_2u_2})^{n_{j_1u_2}}<0$ and $a_{j_2j_20}\prod_{s=1}^{q_2}(a_{j_1j_2u_1}-a_{j_2j_2s})^{n_{j_2s}}<0$,

$$(a_{j_1j_1r_1}, a_{j_1j_2u_1}) = \underbrace{([(2m_{j_1r_1}1+1), (2n_{j_1u_1}1+1)]\text{DP:UP,nF})}_{[(2m_{j_1r_1}1+1),(2n_{j_1u_1}1+1)]\text{-hyperbolic }(-)}$$

The flow at $(x^*_{j_1}, x^*_{j_2}) = (a_{j_1j_1r_1}, a_{j_1j_2u_1})$is a([(2$m_{j_1r_1}1$+1), (2$n_{j_1u_1}1$ + 1)]DP:UP, nF)-negative hyperbolic-secant flow.

(ii_{1a2}) For $m_{j_1r_1} = 2m_{j_1r_1}1$ and $n_{j_1u_1} = 2n_{j_1u_1}1+1$, the equilibrium of $(x^*_{j_1}, x^*_{j_2}) = (a_{j_1j_1r_1}, a_{j_1j_2u_1})$ for the x_{j_1}-direction has the following properties as in Eqs. (1.149)–(1.152).

- For $a_{j_1j_10}\prod_{r_2=1,r_2\neq r_1}^{p_1}(a_{j_1j_1r_1}-a_{j_1j_1r_2})^{m_{j_1r_2}}\prod_{u_2=1,u_2\neq u_1}^{q_1}(a_{j_1j_2u_1}-a_{j_1j_2u_2})^{n_{j_1u_2}}>0$ and $a_{j_2j_20}\prod_{s=1}^{q_2}(a_{j_1j_2u_1}-a_{j_2j_2s})^{n_{j_2s}}>0$,

$$(a_{j_1j_1r_1}, a_{j_1j_2u_1}) = \underbrace{([(2m_{j_1r_1}1), (2n_{j_1u_1}1+1)]\text{UP:UP, pF})}_{[(2m_{j_1r_1}1),(2n_{j_1u_1}1+1)]\text{-hyperbolic-secant-to-hyperbolic flow }(+)}.$$

The flow at $(x^*_{j_1}, x^*_{j_2}) = (a_{j_1j_1r_1}, a_{j_1j_2u_1})$ is a ([(2$m_{j_1r_1}1$), (2$n_{j_1u_1}1$ + 1)]UP:UP, pF)-positive hyperbolic-secant to hyperbolic flow.

- For $a_{j_1j_10}\prod_{r_2=1,r_2\neq r_1}^{p_1}(a_{j_1j_1r_1}-a_{j_1j_1r_2})^{m_{j_1r_2}}\prod_{u_2=1,u_2\neq u_1}^{q_1}(a_{j_1j_2u_1}-a_{j_1j_2u_2})^{n_{j_1u_2}}<0$ and $a_{j_2j_20}\prod_{s=1}^{q_2}(a_{j_1j_2u_1}-a_{j_2j_2s})^{n_{j_2s}}>0$,

$$(a_{j_1j_1r_1},a_{j_1j_2u_1})=\underbrace{([(2m_{j_1r_1}1),(2n_{j_1u_1}1+1)]\text{DP:DP, pF})}_{[(2m_{j_1r_1}1),(2n_{j_1u_1}1+1)]\text{-hyperbolic-to-hyperbolic-secant flow }(+)}.$$

The flow at $(x_{j_1}^*,x_{j_2}^*)=(a_{j_1j_1r_1},a_{j_1j_2u_1})$ is a $([(2m_{j_1r_1}1),(2n_{j_1u_1}1+1)]\text{DP:DP, pF})$-positive hyperbolic to hyperbolic-secant flow.

- For $a_{j_1j_10}\prod_{r_2=1,r_2\neq r_1}^{p_1}(a_{j_1j_1r_1}-a_{j_1j_1r_2})^{m_{j_1r_2}}\prod_{u_2=1,u_2\neq u_1}^{q_1}(a_{j_1j_2u_1}-a_{j_1j_2u_2})^{n_{j_1u_2}}>0$ and $a_{j_2j_20}\prod_{s=1}^{q_2}(a_{j_1j_2u_1}-a_{j_2j_2s})^{n_{j_2s}}<0$,

$$(a_{j_1j_1r_1},a_{j_1j_2u_1})=\underbrace{([(2m_{j_1r_1}1),(2n_{j_1u_1}1+1)]\text{DP:DP, nF})}_{[(2m_{j_1r_1}1),(2n_{j_1u_1}1+1)]\text{-hyperbolic-to-hyperbolic-secant flow }(-)}.$$

The flow at $(x_{j_1}^*,x_{j_2}^*)=(a_{j_1j_1r_1},a_{j_1j_2u_1})$ is a $([(2m_{j_1r_1}1),(2n_{j_1u_1}1+1)]\text{DP:DP, nF})$-negative hyperbolic-to-hyperbolic-secant flow.

- For $a_{j_1j_10}\prod_{r_2=1,r_2\neq r_1}^{p_1}(a_{j_1j_1r_1}-a_{j_1j_1r_2})^{m_{j_1r_2}}\prod_{u_2=1,u_2\neq u_1}^{q_1}(a_{j_1j_2u_1}-a_{j_1j_2u_2})^{n_{j_1u_2}}<0$ and $a_{j_2j_20}\prod_{s=1}^{q_2}(a_{j_1j_2u_1}-a_{j_2j_2s})^{n_{j_2s}}<0$,

$$(a_{j_1j_1r_1},a_{j_1j_2u_1})=\underbrace{([(2m_{j_1r_1}1),(2n_{j_1u_1}1+1)]\text{UP:UP, nF})}_{[(2m_{j_1r_1}1),(2n_{j_1u_1}1+1)]\text{-hyperbolic-secant-to-hyperbolic flow }(-)}.$$

The flow at $(x_{j_1}^*,x_{j_2}^*)=(a_{j_1j_1r_1},a_{j_1j_2u_1})$ is a $([(2m_{j_1r_1}1),(2n_{j_1u_1}1+1)]\text{UP:UP, nF})$-negative hyperbolic-secant-to-hyperbolic flow.

($\text{ii}_{1\text{a}3}$) For $m_{j_1r_1}=2m_{j_1r_1}1+1$ and $n_{j_1u_1}=2n_{j_1u_1}1$, the equilibrium of $(x_{j_1}^*,x_{j_2}^*)=(a_{j_1j_1r_1},a_{j_1j_2u_1})$ for the x_{j_1}-direction has the following properties as in Eqs. (1.153)–(1.156).

- For $a_{j_1j_10}\prod_{r_2=1,r_2\neq r_1}^{p_1}(a_{j_1j_1r_1}-a_{j_1j_1r_2})^{m_{j_1r_2}}\prod_{u_2=1,u_2\neq u_1}^{q_1}(a_{j_1j_2u_1}-a_{j_1j_2u_2})^{n_{j_1u_2}}>0$ and $a_{j_2j_20}\prod_{s=1}^{q_2}(a_{j_1j_2u_1}-a_{j_2j_2s})^{n_{j_2s}}>0$,

$$(a_{j_1j_1r_1},a_{j_1j_2u_1})=\underbrace{([(2m_{j_1r_1}1+1),(2n_{j_1u_1}1)]_{\text{DI:II}}\text{SO, pF})}_{[(2m_{j_1r_1}1+1),(2n_{j_1u_1}1)]\text{-decreasing-to-increasing inflection source flow }(+)}.$$

The flow at $(x_{j_1}^*,x_{j_2}^*)=(a_{j_1j_1r_1},a_{j_1j_2u_1})$ is a $([(2m_{j_1r_1}1+1),(2n_{j_1u_1}1)]_{\text{DI:II}}\text{SO, pF})$-positive decreasing-to-increasing inflection source flow.

- For $a_{j_1j_10}\prod_{r_2=1,r_2\neq r_1}^{p_1}(a_{j_1j_1r_1}-a_{j_1j_1r_2})^{m_{j_1r_2}}\prod_{u_2=1,u_2\neq u_1}^{q_1}(a_{j_1j_2u_1}-a_{j_1j_2u_2})^{n_{j_1u_2}}<0$ and $a_{j_2j_20}\prod_{s=1}^{q_2}(a_{j_1j_2u_1}-a_{j_2j_2s})^{n_{j_2s}}>0$,

$$(a_{j_1j_1r_1},a_{j_1j_2u_1})=\underbrace{([(2m_{j_1r_1}1+1),(2n_{j_1u_1}1)]_{\text{II:DI}}\text{SI, pF})}_{[(2m_{j_1r_1}1+1),(2n_{j_1u_1}1)]\text{-increasing-to-decreasing inflection sink flow }(+)}.$$

The flow at $(x_{j_1}^*, x_{j_2}^*) = (a_{j_1j_1r_1}, a_{j_1j_2u_1})$ is a $([(2m_{j_1r_1}1+1), (2n_{j_1u_1}1)]_{\text{II:DI}}\text{SI, pF})$-positive increasing-to-decreasing inflection source flow.

- For $a_{j_1j_10}\prod_{r_2=1,r_2\neq r_1}^{p_1}(a_{j_1j_1r_1}-a_{j_1j_1r_2})^{m_{j_1r_2}}\prod_{u_2=1,u_2\neq u_1}^{q_1}(a_{j_1j_2u_1}-a_{j_1j_2u_2})^{n_{j_1u_2}}>0$ and $a_{j_2j_20}\prod_{s=1}^{q_2}(a_{j_1j_2u_1}-a_{j_2j_2s})^{n_{j_2s}}<0$,

$$(a_{j_1j_1r_1}, a_{j_1j_2u_1}) = \underbrace{([(2m_{j_1r_1}1+1), (2n_{j_1u_1}1)]_{\text{II:DI}}\text{SO,nF})}_{[(2m_{j_1r_1}1+1),(2n_{j_1u_1}1)]\text{-increasing-to-decreasing inflection source flow }(-)}.$$

The flow at $(x_{j_1}^*, x_{j_2}^*) = (a_{j_1j_1r_1}, a_{j_1j_2u_1})$ is a $([(2m_{j_1r_1}1+1), (2n_{j_1u_1}1)]_{\text{II:DI}}\text{SO,nF})$-negative increasing-to-decreasing inflection source flow.

- For $a_{j_1j_10}\prod_{r_2=1,r_2\neq r_1}^{p_1}(a_{j_1j_1r_1}-a_{j_1j_1r_2})^{m_{j_1r_2}}\prod_{u_2=1,u_2\neq u_1}^{q_1}(a_{j_1j_2u_1}-a_{j_1j_2u_2})^{n_{j_1u_2}}<0$ and $a_{j_2j_20}\prod_{s=1}^{q_2}(a_{j_1j_2u_1}-a_{j_2j_2s})^{n_{j_2s}}<0$,

$$(a_{j_1j_1r_1}, a_{j_1j_2u_1}) = \underbrace{([(2m_{j_1r_1}1+1), (2n_{j_1u_1}1)]_{\text{DI:II}}\text{SI,nF})}_{[(2m_{j_1r_1}1+1),(2n_{j_1u_1}1)]\text{-decreasing-to-increasing inflection sink flow }(-)}.$$

The flow at $(x_{j_1}^*, x_{j_2}^*) = (a_{j_1j_1r_1}, a_{j_1j_2u_1})$ is a $([(2m_{j_1r_1}1+1), (2n_{j_1u_1}1)]_{\text{DI:II}}\text{SI,nF})$-negative decreasing-to-increasing inflection sink flow.

(ii_{1a4}) For $m_{j_1r_1} = 2m_{j_1r_1}1$ and $n_{j_1u_1} = 2n_{j_1u_1}1$, the equilibrium of $(x_{j_1}^*, x_{j_2}^*) = (a_{j_1j_1r_1}, a_{j_1j_2u_1})$ for the x_{j_1}-direction has the following properties as in Eqs. (1.157)–(1.160).

- For $a_{j_1j_10}\prod_{r_2=1,r_2\neq r_1}^{p_1}(a_{j_1j_1r_1}-a_{j_1j_1r_2})^{m_{j_1r_2}}\prod_{u_2=1,u_2\neq u_1}^{q_1}(a_{j_1j_2u_1}-a_{j_1j_2u_2})^{n_{j_1u_2}}>0$ and $a_{j_2j_20}\prod_{s=1}^{q_2}(a_{j_1j_2u_1}-a_{j_2j_2s})^{n_{j_2s}}>0$,

$$(a_{j_1j_1r_1}, a_{j_1j_2u_1}) = \underbrace{([(2m_{j_1r_1}1), (2n_{j_1u_1}1)]_{\text{II:II}}\text{US,pF})}_{[(2m_{j_1r_1}1),(2n_{j_1u_1}1)]\text{-increasing-inflection upper-saddle flow }(+)}.$$

The flow at $(x_{j_1}^*, x_{j_2}^*) = (a_{j_1j_1r_1}, a_{j_1j_2u_1})$ is a $([(2m_{j_1r_1}1), (2n_{j_1u_1}1)]_{\text{II:II}}\text{US,pF})$-positive increasing-inflection upper-saddle flow.

- For $a_{j_1j_10}\prod_{r_2=1,r_2\neq r_1}^{p_1}(a_{j_1j_1r_1}-a_{j_1j_1r_2})^{m_{j_1r_2}}\prod_{u_2=1,u_2\neq u_1}^{q_1}(a_{j_1j_2u_1}-a_{j_1j_2u_2})^{n_{j_1u_2}}<0$ and $a_{j_2j_20}\prod_{s=1}^{q_2}(a_{j_1j_2u_1}-a_{j_2j_2s})^{n_{j_2s}}>0$,

$$(a_{j_1j_1r_1}, a_{j_1j_2u_1}) = \underbrace{([(2m_{j_1r_1}1), (2n_{j_1u_1}1)]_{\text{DI:DI}}\text{LS,pF})}_{[(2m_{j_1r_1}1),(2n_{j_1u_1}1)]\text{-decreasing-inflection lower-saddle flow }(+)}.$$

The flow at $(x_{j_1}^*, x_{j_2}^*) = (a_{j_1j_1r_1}, a_{j_1j_2u_1})$ is a $([(2m_{j_1r_1}1), (2n_{j_1u_1}1)]_{\text{DI:DI}}\text{LS,pF})$-positive decreasing-inflection lower-saddle flow.

- For $a_{j_1j_10}\prod_{r_2=1,r_2\neq r_1}^{p_1}(a_{j_1j_1r_1}-a_{j_1j_1r_2})^{m_{j_1r_2}}\prod_{u_2=1,u_2\neq u_1}^{q_1}(a_{j_1j_2u_1}-a_{j_1j_2u_2})^{n_{j_1u_2}}>0$ and $a_{j_2j_20}\prod_{s=1}^{q_2}(a_{j_1j_2u_1}-a_{j_2j_2s})^{n_{j_2s}}<0$,

$$(a_{j_1j_1r_1}, a_{j_1j_2u_1}) = \underbrace{([(2m_{j_1r_1}1), (2n_{j_1u_1}1)]_{\text{DI:DI}}\text{US, nF})}_{[(2m_{j_1r_1}1),(2n_{j_1u_1}1)]\text{-decreasing-inflection upper-saddle flow }(-)}.$$

The flow at $(x_{j_1}^*, x_{j_2}^*) = (a_{j_1j_1r_1}, a_{j_1j_2u_1})$ is a $([(2m_{j_1r_1}1), (2n_{j_1u_1}1)]_{\mathrm{DI:DI}}\mathrm{US}, \mathrm{nF})$-negative decreasing-inflection upper-saddle flow.

- For $a_{j_1j_10}\prod_{r_2=1,r_2\neq r_1}^{p_1}(a_{j_1j_1r_1}-a_{j_1j_1r_2})^{m_{j_1r_2}}\prod_{u_2=1,u_2\neq u_1}^{q_1}(a_{j_1j_2u_1}-a_{j_1j_2u_2})^{n_{j_1u_2}}<0$ and $a_{j_2j_20}\prod_{s=1}^{q_2}(a_{j_1j_2u_1}-a_{j_2j_2s})^{n_{j_2s}}<0$,

$$(a_{j_1j_1r_1}, a_{j_1j_2u_1}) = \underbrace{([(2m_{j_1r_1}1), (2n_{j_1u_1}1)]_{\mathrm{II:II}}\mathrm{LS}, \mathrm{nF})}_{[(2m_{j_1r_1}1),(2n_{j_1u_1}1)]\text{-increasing-inflection lower-saddle flow }(-)}.$$

The flow at $(x_{j_1}^*, x_{j_2}^*) = (a_{j_1j_1r_1}, a_{j_1j_2u_1})$ is a $([(2m_{j_1r_1}1), (2n_{j_1u_1}1)]_{\mathrm{II:II}}\mathrm{LS}, \mathrm{nF})$-negative increasing-inflection lower-saddle flow.

$(\mathrm{ii}_{1\mathrm{b}})$ There is an equilibrium for self-univariate equilibrium as

$$(x_{j_1}^*, x_{j_2}^*) = (a_{j_1j_1r_1}, a_{j_2j_2s_1}).$$

The corresponding variational equations are

$$\Delta\dot{x}_{j_2} = a_{j_2j_20}\prod_{s_2=1,s_2\neq s_1}^{q_2}(a_{j_2j_2s_1}-a_{j_2j_2s_2})^{n_{j_2s_2}}(\Delta x_{j_2})^{n_{j_2s_1}}$$
$$\Delta\dot{x}_{j_1} = a_{j_1j_10}\prod_{r_2=1,r_2\neq r_1}^{p_1}(a_{j_1j_1r_1}-a_{j_1j_1r_2})^{m_{j_1r_2}}\prod_{u=1}^{q_1}(a_{j_2j_2s_1}-a_{j_1j_2u})^{n_{j_1u}}(\Delta x_{j_1})^{m_{j_1r_1}}.$$

For

$$a_{j_1j_10}\prod_{r_2=1,r_2\neq r_1}^{p_1}(a_{j_1j_1r_1}-a_{j_1j_1r_2})^{m_{j_1r_2}}\prod_{u=1}^{q_1}(\overline{x}_{j_2}-a_{j_1j_2u})^{n_{j_1u}}>0 \text{ and}$$
$$a_{j_1j_10}\prod_{r_2=1,r_2\neq r_1}^{p_1}(a_{j_1j_1r_1}-a_{j_1j_1r_2})^{m_{j_1r_2}}\prod_{u=1}^{q_1}(\overline{x}_{j_2}-a_{j_1j_2u})^{n_{j_1u}}<0,$$

the flows at $x_{j_1}^* = a_{j_1j_1r_1}$ and $\overline{x}_{j_2}\neq a_{j_1j_2u}, a_{j_2j_2s}$, are the $(2m_{j_1r_1}1+1)^{\text{th}}$-order source and sink for $m_{j_1r_1}=2m_{j_1r_1}1+1$; and the flows at $x_{j_1}^* = a_{j_1j_1r_1}$ and $\overline{x}_{j_2}\neq a_{j_1j_2u}, a_{j_2j_2s}$, are the $(2m_{j_1r_1}1)^{\text{th}}$-order upper-saddle and lower-saddle for $m_{j_1r_1}=2m_{j_1r_1}1$.

For

$$a_{j_2j_20}\prod_{s_2=1,s_2\neq s_1}^{q_2}(a_{j_2j_2s_1}-a_{j_2j_2s_2})^{n_{j_2s_2}}>0 \text{ and}$$
$$a_{j_2j_20}\prod_{s_2=1,s_2\neq s_1}^{q_2}(a_{j_2j_2s_1}-a_{j_2j_2s_2})^{n_{j_2s_2}}<0,$$

the flows at $x_{j_2}^* = a_{j_2j_2s_1}\neq a_{j_1j_2u}$ and $\overline{x}_{j_1}\neq a_{j_1j_1}1$ are the $(2n_{j_2s_1}1+1)^{\text{th}}$-order source and sink for $n_{j_2s_1}=2n_{j_2s_1}1+1$; and the flows at $x_{j_2}^* = a_{j_2j_2s_1}\neq a_{j_1j_2u}$ and $\overline{x}_{j_1}\neq a_{j_1j_1}1$ are the $(2n_{j_2s_1}1)^{\text{th}}$-order upper-saddle and lower-saddle for $n_{j_2s_1}=2n_{j_2s_1}1$.

$(\mathrm{ii}_{1\mathrm{b}1})$ For $m_{j_1r_1}=2m_{j_1r_1}1+1$ and $n_{j_2s_1}=2n_{j_2s_1}1+1$, the equilibrium of $(x_{j_1}^*, x_{j_2}^*) = (a_{j_1j_1r_1}, a_{j_2j_2s_1})$ has the following properties as in Eqs. (1.162)–(1.165).

- For $a_{j_1j_10}\prod_{r_2=1,r_2\neq r_1}^{p_1}(a_{j_1j_1r_1}-a_{j_1j_1r_2})^{m_{j_1r_2}}\prod_{u=1}^{q_1}(a_{j_2j_2s_1}-a_{j_1j_2u})^{n_{j_1u}}>0$ and $a_{j_2j_20}\prod_{s_2=1,s_2\neq s_1}^{q_2}(a_{j_2j_2s_1}-a_{j_2j_2s_2})^{n_{j_2s_2}}>0$,

$$(a_{j_1j_1r_1}, a_{j_2j_2s_1}) = \underbrace{((2m_{j_1r_1}1 + 1)^{\text{th}}\text{SO},(2n_{j_2s_1}1 + 1)^{\text{th}}\text{SO})}_{((2m_{j_1r_1}1+1),(2n_{j_2s_1}1+1))\text{-source}}.$$

The flow at $(x_{j_1}^*, x_{j_2}^*)=(a_{j_1j_1r_1}, a_{j_2j_2s_1})$ is a $((2m_{j_1r_1}1+1)^{\text{th}}\text{SO},(2n_{j_2s_1}1+1)^{\text{th}}\text{SO})$-source.

- For $a_{j_1j_10}\prod_{r_2=1,r_2\neq r_1}^{p_1}(a_{j_1j_1r_1}-a_{j_1j_1r_2})^{m_{j_1r_2}}\prod_{u=1}^{q_1}(a_{j_2j_2s_1}-a_{j_1j_2u})^{n_{j_1u}}<0$ and $a_{j_2j_20}\prod_{s_2=1,s_2\neq s_1}^{q_2}(a_{j_2j_2s_1}-a_{j_2j_2s_2})^{n_{j_2s_2}}>0$,

$$(a_{j_1j_1r_1}, a_{j_2j_2s_1}) = \underbrace{((2m_{j_1r_1}1 + 1)^{\text{th}}\text{SI},(2n_{j_2s_1}1 + 1)^{\text{th}}\text{SO})}_{((2m_{j_1r_1}1+1),(2n_{j_2s_1}1+1))\text{-saddle}}.$$

The flow at $(x_{j_1}^*, x_{j_2}^*)=(a_{j_1j_1r_1}, a_{j_2j_2s_1})$ is a $((2m_{j_1r_1}1+1)^{\text{th}}\text{SI},(2n_{j_2s_1}1+1)^{\text{th}}\text{SO})$-saddle.

- For $a_{j_1j_10}\prod_{r_2=1,r_2\neq r_1}^{p_1}(a_{j_1j_1r_1}-a_{j_1j_1r_2})^{m_{j_1r_2}}\prod_{u=1}^{q_1}(a_{j_2j_2s_1}-a_{j_1j_2u})^{n_{j_1u}}>0$ and $a_{j_2j_20}\prod_{s_2=1,s_2\neq s_1}^{q_2}(a_{j_2j_2s_1}-a_{j_2j_2s_2})^{n_{j_2s_2}}<0$,

$$(a_{j_1j_1r_1}, a_{j_2j_2s_1}) = \underbrace{((2m_{j_1r_1}1 + 1)^{\text{th}}\text{SO},(2n_{j_2s_1}1 + 1)^{\text{th}}\text{SI})}_{((2m_{j_1r_1}1+1),(2n_{j_2s_1}1+1))\text{-saddle}}.$$

The flow at $(x_{j_1}^*, x_{j_2}^*)=(a_{j_1j_1r_1}, a_{j_2j_2s_1})$ is a $((2m_{j_1r_1}1+1)^{\text{th}}\text{SO},(2n_{j_2s_1}1+1)^{\text{th}}\text{SI})$-saddle.

- For $a_{j_1j_10}\prod_{r_2=1,r_2\neq r_1}^{p_1}(a_{j_1j_1r_1}-a_{j_1j_1r_2})^{m_{j_1r_2}}\prod_{u=1}^{q_1}(a_{j_2j_2s_1}-a_{j_1j_2u})^{n_{j_1u}}<0$ and $a_{j_2j_20}\prod_{s_2=1,s_2\neq s_1}^{q_2}(a_{j_2j_2s_1}-a_{j_2j_2s_2})^{n_{j_2s_2}}<0$,

$$(a_{j_1j_1r_1}, a_{j_2j_2s_1}) = \underbrace{((2m_{j_1r_1}1 + 1)^{\text{th}}\text{SI},(2n_{j_2s_1}1 + 1)^{\text{th}}\text{SI})}_{((2m_{j_1r_1}1+1),(2n_{j_2s_1}1+1))\text{-sink}}.$$

The flow at $(x_{j_1}^*, x_{j_2}^*)=(a_{j_1j_1r_1}, a_{j_2j_2s_1})$ is a $((2m_{j_1r_1}1+1)^{\text{th}}\text{SI},(2n_{j_2s_1}1+1)^{\text{th}}\text{SI})$-sink.

(ii$_{1b2}$) For $m_{j_1r_1} = 2m_{j_1r_1}1{+}1$ and $n_{j_ss_1} = 2n_{j_ss_1}1 + 1$, the equilibrium of $(x_{j_1}^*, x_{j_2}^*) = (a_{j_1j_1}1, a_{j_2j_2}1)$ has the following properties as in Eqs. (1.166)–(1.169).

- For $a_{j_1j_10}\prod_{r_2=1,r_2\neq r_1}^{p_1}(a_{j_1j_1r_1}-a_{j_1j_1r_2})^{m_{j_1r_2}}\prod_{u=1}^{q_1}(a_{j_2j_2s_1}-a_{j_1j_2u})^{n_{j_1u}}>0$ and $a_{j_2j_20}\prod_{s_2=1,s_2\neq s_1}^{q_2}(a_{j_2j_2s_1}-a_{j_2j_2s_2})^{n_{j_2s_2}}>0$,

$$(a_{j_1j_1r_1}, a_{j_2j_2s_1}) = \underbrace{((2m_{j_1r_1}1)^{\text{th}}\text{US},(2n_{j_ss_1}1 + 1)^{\text{th}}\text{SO})}_{((2m_{j_1r_1}1),(2n_{j_2s_1}1+1))\text{-upper-saddle source}}.$$

The flow at $(x_{j_1}^*, x_{j_2}^*) = (a_{j_1j_1r_1}, a_{j_2j_2s_1})$ is a $((2m_{j_1r_1}1)^{\text{th}}\text{US},(2n_{j_ss_1}1+1)^{\text{th}}\text{SO})$-upper-saddle source.

- For $a_{j_1j_10}\prod_{r_2=1,r_2\neq r_1}^{p_1}(a_{j_1j_1r_1}-a_{j_1j_1r_2})^{m_{j_1r_2}}\prod_{u=1}^{q_1}(a_{j_2j_2s_1}-a_{j_1j_2u})^{n_{j_1u}}<0$ and $a_{j_2j_20}\prod_{s_2=1,s_2\neq s_1}^{q_2}(a_{j_2j_2s_1}-a_{j_2j_2s_2})^{n_{j_2s_2}}>0$,

$$(a_{j_1j_1r_1},a_{j_2j_2s_1})=\underbrace{((2m_{j_1r_1}1)^{\text{th}}\text{LS},(2n_{j_2s_1}1+1)^{\text{th}}\text{SO})}_{((2m_{j_1r_1}1),(2n_{j_2s_1}1+1))\text{-lower-saddle source}}.$$

The flow at $(x_{j_1}^*,x_{j_2}^*)=(a_{j_1j_1r_1},a_{j_2j_2s_1})$ is a $((2m_{j_1r_1}1)^{\text{th}}\text{LS},(2n_{j_2s_1}1+1)^{\text{th}}\text{SO})$-lower-saddle saddle.

- For $a_{j_1j_10}\prod_{r_2=1,r_2\neq r_1}^{p_1}(a_{j_1j_1r_1}-a_{j_1j_1r_2})^{m_{j_1r_2}}\prod_{u=1}^{q_1}(a_{j_2j_2s_1}-a_{j_1j_2u})^{n_{j_1u}}>0$ and $a_{j_2j_20}\prod_{s_2=1,s_2\neq s_1}^{q_2}(a_{j_2j_2s_1}-a_{j_2j_2s_2})^{n_{j_2s_2}}<0$,

$$(a_{j_1j_1r_1},a_{j_2j_2s_1})=\underbrace{((2m_{j_1r_1}1)^{\text{th}}\text{US},(2n_{j_2s_1}1+1)^{\text{th}}\text{SI})}_{((2m_{j_1r_1}1),(2n_{j_2s_1}1+1))\text{-upper-saddle sink}}.$$

The flow at $(x_{j_1}^*,x_{j_2}^*)=(a_{j_1j_1r_1},a_{j_2j_2s_1})$ is a $((2m_{j_1r_1}1)^{\text{th}}\text{US},(2n_{j_2s_1}1+1)^{\text{th}}\text{SI})$-upper-saddle sink.

- For $a_{j_1j_10}\prod_{r_2=1,r_2\neq r_1}^{p_1}(a_{j_1j_1r_1}-a_{j_1j_1r_2})^{m_{j_1r_2}}\prod_{u=1}^{q_1}(a_{j_2j_2s_1}-a_{j_1j_2u})^{n_{j_1u}}<0$ and $a_{j_2j_20}\prod_{s_2=1,s_2\neq s_1}^{q_2}(a_{j_2j_2s_1}-a_{j_2j_2s_2})^{n_{j_2s_2}}<0$,

$$(a_{j_1j_1r_1},a_{j_2j_2s_1})=\underbrace{((2m_{j_1r_1}1)^{\text{th}}\text{LS},(2n_{j_2s_1}1+1)^{\text{th}}\text{SI})}_{((2m_{j_1r_1}1),(2n_{j_2s_1}1+1))\text{-lower-saddle sink}}.$$

The flow at $(x_{j_1}^*,x_{j_2}^*)=(a_{j_1j_1r_1},a_{j_2j_2s_1})$ is a $((2m_{j_1r_1}1)^{\text{th}}\text{LS},(2n_{j_2s_1}1+1)^{\text{th}}\text{SI})$-lower-saddle sink.

(ii$_{1b3}$) For $m_{j_1r_1}=2m_{j_1r_1}+1$ and $n_{j_2s_1}=2n_{j_2s_1}1$, the equilibrium of $(x_{j_1}^*,x_{j_2}^*)=(a_{j_1j_1r_1},a_{j_2j_2s_1})$ has the following properties as in Eqs. (1.170)–(1.173).

- For $a_{j_1j_10}\prod_{r_2=1,r_2\neq r_1}^{p_1}(a_{j_1j_1r_1}-a_{j_1j_1r_2})^{m_{j_1r_2}}\prod_{u=1}^{q_1}(a_{j_2j_2s_1}-a_{j_1j_2u})^{n_{j_1u}}>0$ and $a_{j_2j_20}\prod_{s_2=1,s_2\neq s_1}^{q_2}(a_{j_2j_2s_1}-a_{j_2j_2s_2})^{n_{j_2s_2}}>0$,

$$(a_{j_1j_1r_1},a_{j_2j_2s_1})=\underbrace{((2m_{j_1r_1}1+1)^{\text{th}}\text{SO},(2n_{j_2s_1}1)^{\text{th}}\text{US})}_{((2m_{j_1r_1}1+1),(2n_{j_2s_1}1))\text{-upper-saddle source}}.$$

The flow at $(x_{j_1}^*,x_{j_2}^*)=(a_{j_1j_1r_1},a_{j_2j_2s_1})$ is a $((2m_{j_1r_1}1+1)^{\text{th}}\text{SO},(2n_{j_2s_1}1)^{\text{th}}\text{US})$-upper-saddle source.

- For $a_{j_1j_10}\prod_{r_2=1,r_2\neq r_1}^{p_1}(a_{j_1j_1r_1}-a_{j_1j_1r_2})^{m_{j_1r_2}}\prod_{u=1}^{q_1}(a_{j_2j_2s_1}-a_{j_1j_2u})^{n_{j_1u}}<0$ and $a_{j_2j_20}\prod_{s_2=1,s_2\neq s_1}^{q_2}(a_{j_2j_2s_1}-a_{j_2j_2s_2})^{n_{j_2s_2}}>0$,

$$(a_{j_1j_1r_1},a_{j_2j_2s_1})=\underbrace{((2m_{j_1r_1}1+1)^{\text{th}}\text{SI},(2n_{j_2s_1}1)^{\text{th}}\text{US})}_{((2m_{j_1r_1}1+1),(2n_{j_2s_1}1))\text{-upper-saddle sink}}.$$

The flow at $(x_{j_1}^*, x_{j_2}^*) = (a_{j_1j_1r_1}, a_{j_2j_2s_1})$ is a $((2m_{j_1r_1}1+1)^{\text{th}}\text{SI},(2n_{j_2s_1}1)^{\text{th}}\text{US})$-upper-saddle sink.

- For $a_{j_1j_10}\prod_{r_2=1,r_2\neq r_1}^{p_1}(a_{j_1j_1r_1}-a_{j_1j_1r_2})^{m_{j_1r_2}}\prod_{u=1}^{q_1}(a_{j_2j_2s_1}-a_{j_1j_2u})^{n_{j_1u}}>0$ and $a_{j_2j_20}\prod_{s_2=1,s_2\neq s_1}^{q_2}(a_{j_2j_2s_1}-a_{j_2j_2s_2})^{n_{j_2s_2}}<0$,

$$(a_{j_1j_1r_1}, a_{j_2j_2s_1}) = \underbrace{((2m_{j_1r_1}1+1)^{\text{th}}\text{SO},(2n_{j_2s_1}1)^{\text{th}}\text{LS})}_{((2m_{j_1r_1}1+1),(2n_{j_2s_1}1))\text{-lower-saddle source}}.$$

The flow at $(x_{j_1}^*, x_{j_2}^*) = (a_{j_1j_1r_1}, a_{j_2j_2s_1})$ is a $((2m_{j_1r_1}1+1)^{\text{th}}\text{SO},(2n_{j_2s_1}1)^{\text{th}}\text{LS})$-lower-saddle source.

- For $a_{j_1j_10}\prod_{r_2=1,r_2\neq r_1}^{p_1}(a_{j_1j_1r_1}-a_{j_1j_1r_2})^{m_{j_1r_2}}\prod_{u=1}^{q_1}(a_{j_2j_2s_1}-a_{j_1j_2u})^{n_{j_1u}}<0$ and $a_{j_2j_20}\prod_{s_2=1,s_2\neq s_1}^{q_2}(a_{j_2j_2s_1}-a_{j_2j_2s_2})^{n_{j_2s_2}}<0$,

$$(a_{j_1j_1r_1}, a_{j_2j_2s_1}) = \underbrace{((2m_{j_1r_1}1+1)^{\text{th}}\text{SI},(2n_{j_2s_1}1)^{\text{th}}\text{LS})}_{((2m_{j_1r_1}1+1),(2n_{j_2s_1}1))\text{-lower-saddle sink}}.$$

The flow at $(x_{j_1}^*, x_{j_2}^*) = (a_{j_1j_1r_1}, a_{j_2j_2s_1})$ is a $((2m_{j_1r_1}1+1)^{\text{th}}\text{SI},(2n_{j_2s_1}1)^{\text{th}}\text{LS})$-lower-saddle sink.

($\text{ii}_{1\text{b}4}$) For $m_{j_1r_1} = 2m_{j_1r_1}1$ and $n_{j_2s_1} = 2n_{j_2s_1}1$, the equilibrium of $(x_{j_1}^*, x_{j_2}^*) = (a_{j_1j_1r_1}, a_{j_2j_2s_1})$ has the following properties as in Eqs. (1.174)–(1.177).

- For $a_{j_1j_10}\prod_{r_2=1,r_2\neq r_1}^{p_1}(a_{j_1j_1r_1}-a_{j_1j_1r_2})^{m_{j_1r_2}}\prod_{u=1}^{q_1}(a_{j_2j_2s_1}-a_{j_1j_2u})^{n_{j_1u}}>0$ and $a_{j_2j_20}\prod_{s_2=1,s_2\neq s_1}^{q_2}(a_{j_2j_2s_1}-a_{j_2j_2s_2})^{n_{j_2s_2}}>0$,

$$(a_{j_1j_1r_1}, a_{j_2j_2s_1}) = \underbrace{((2m_{j_1r_1}1)^{\text{th}}\text{US},(2n_{j_2s_1}1)^{\text{th}}\text{US})}_{((2m_{j_1r_1}1),(2n_{j_2s_1}1))\text{-double-saddle}}.$$

The flow at $(x_{j_1}^*, x_{j_2}^*)=(a_{j_1j_1r_1}, a_{j_2j_2s_1})$ is a $((2m_{j_1r_1}1)^{\text{th}}\text{US},(2n_{j_2s_1}1)^{\text{th}}\text{US})$-double-saddle.

- For $a_{j_1j_10}\prod_{r_2=1,r_2\neq r_1}^{p_1}(a_{j_1j_1r_1}-a_{j_1j_1r_2})^{m_{j_1r_2}}\prod_{u=1}^{q_1}(a_{j_2j_2s_1}-a_{j_1j_2u})^{n_{j_1u}}<0$ and $a_{j_2j_20}\prod_{s_2=1,s_2\neq s_1}^{q_2}(a_{j_2j_2s_1}-a_{j_2j_2s_2})^{n_{j_2s_2}}>0$,

$$(a_{j_1j_1r_1}, a_{j_2j_2s_1}) = \underbrace{((2m_{j_1r_1}1)^{\text{th}}\text{LS},(2n_{j_2s_1}1)^{\text{th}}\text{US})}_{((2m_{j_1r_1}1),(2n_{j_2s_1}1))\text{-double-saddle}}.$$

The flow at $(x_{j_1}^*, x_{j_2}^*)=(a_{j_1j_1r_1}, a_{j_2j_2s_1})$ is a $((2m_{j_1r_1}1)^{\text{th}}\text{LS},(2n_{j_2s_1}1)^{\text{th}}\text{US})$-double-saddle.

- For $a_{j_1j_10}\prod_{r_2=1,r_2\neq r_1}^{p_1}(a_{j_1j_1r_1}-a_{j_1j_1r_2})^{m_{j_1r_2}}\prod_{u=1}^{q_1}(a_{j_2j_2s_1}-a_{j_1j_2u})^{n_{j_1u}}>0$ and $a_{j_2j_20}\prod_{s_2=1,s_2\neq s_1}^{q_2}(a_{j_2j_2s_1}-a_{j_2j_2s_2})^{n_{j_2s_2}}<0$,

$$(a_{j_1j_1r_1}, a_{j_2j_2s_1}) = \underbrace{((2m_{j_1r_1}1)^{\text{th}}\text{US},(2n_{j_2s_1}1)^{\text{th}}\text{LS})}_{((2m_{j_1r_1}1),(2n_{j_2s_1}1))\text{-double-saddle}}.$$

The flow at $(x_{j_1}^*,x_{j_2}^*)=(a_{j_1j_1r_1},a_{j_2j_2s_1})$ is a $((2m_{j_1r_1}1)^{\text{th}}\text{US},(2n_{j_2s_1}1)^{\text{th}}\text{LS})$-double-saddle.

- For $a_{j_1j_10}\prod_{r_2=1,r_2\neq r_1}^{p_1}(a_{j_1j_1r_1}-a_{j_1j_1r_2})^{m_{j_1r_2}}\prod_{u=1}^{q_1}(a_{j_2j_2s_1}-a_{j_1j_2u})^{n_{j_1u}}<0$ and $a_{j_2j_20}\prod_{s_2=1,s_2\neq s_1}^{q_2}(a_{j_2j_2s_1}-a_{j_2j_2s_2})^{n_{j_2s_2}}<0$,

$$(a_{j_1j_1r_1}, a_{j_2j_2s_1}) = \underbrace{((2m_{j_1r_1}1)^{\text{th}}\text{LS},(2n_{j_2s_1}1)^{\text{th}}\text{LS})}_{((2m_{j_1r_1}1),(2n_{j_2s_1}1))\text{-double-saddle}}.$$

The flow at $(x_{j_1}^*,x_{j_2}^*)=(a_{j_1j_1r_1},a_{j_2j_2s_1})$ is a $((2m_{j_1r_1}1)^{\text{th}}\text{LS},(2n_{j_2s_1}1)^{\text{th}}\text{LS})$-double-saddle.

2.2.2 *Infinite-Equilibriums in the Hybrid Networks*

(ii$_2$) For $a_{j_1j_2u_1}=a_{j_2j_2s_1}$ $(u_1\in\{1,2,\cdots,q_1\};\ s\in\{1,2,\cdots,q_2\})$

$$\begin{aligned}\dot{x}_{j_2} &= a_{j_2j_20}(x_{j_2}-a_{j_2j_2s_1})^{n_{j_2s_1}}\prod_{s_1=1,s_2\neq s_1}^{q_2}(x_{j_2}-a_{j_2j_2s_2})^{n_{j_2s_2}},\\ \dot{x}_{j_1} &= a_{j_1j_10}(x_{j_2}-a_{j_1j_2u_1})^{n_{j_1u_1}}\prod_{r=1}^{p_1}(x_{j_1}-a_{j_1j_1r})^{m_{j_1r}}\prod_{u_1=1,u_2\neq u_1}^{q_1}(x_{j_2}-a_{j_1j_2u_2})^{n_{j_1u_2}}.\end{aligned}$$

Consider

$$\frac{dx_{j_1}}{dx_{j_2}}=\frac{a_{j_1j_10}}{a_{j_2j_20}}\frac{(x_{j_2}-a_{j_1j_2u_1})^{n_{j_1u_1}}\prod_{r=1}^{p_1}(x_{j_1}-a_{j_1j_1r})^{m_{j_1r}}\prod_{u_1=1,u_2\neq u_1}^{q_1}(x_{j_2}-a_{j_1j_2u_2})^{n_{j_1u_2}}}{(x_{j_2}-a_{j_2j_2s_1})^{n_{j_2s_1}}\prod_{s_1=1,s_2\neq s_1}^{q_2}(x_{j_2}-a_{j_2j_2s_2})^{n_{j_2s_2}}}.$$

The deformed equation is

$$\frac{dx_{j_1}}{\prod_{r=1}^{p_1}(x_{j_1}-a_{j_1j_1r})^{m_{j_1r}}}=\frac{a_{j_1j_10}}{a_{j_2j_20}}\frac{(x_{j_2}-a_{j_1j_2u_1})^{n_{j_1u_1}}\prod_{u_1=1,u_2\neq u_1}^{q_1}(x_{j_2}-a_{j_1j_2u_2})^{n_{j_1u_2}}}{(x_{j_2}-a_{j_2j_2s_1})^{n_{j_2s_1}}\prod_{s_1=1,s_2\neq s_1}^{q_2}(x_{j_2}-a_{j_2j_2s_2})^{n_{j_2s_2}}}dx_{j_2},$$

The expansion of the above equation gives

$$\begin{aligned}&\sum_{r_1=1}^{p_1}\sum_{l_1=1}^{m_{j_1r_1}}\frac{A_{l_1m_{j_1r_1}}}{(x_{j_1}-a_{j_1j_1r_1})^{l_1}}dx_{j_1}=\frac{a_{j_1j_10}}{a_{j_2j_20}}\sum_{\beta_0=n_{j_1u_1}}^{n_2}(x_{j_2}-a_{j_2j_2s_1})^{\beta_0}\\ &\times\frac{\sum_{t_2,l_2=0}^{t_2+l_2=n_{j_1u_2}}\cdots\sum_{t_{q_1},l_{q_1}=0}^{t_{q_1}+l_{q_1}=n_{j_1u_{q_1}}}\prod_{k=2}^{q_1}\frac{n_{j_1u_k}!}{t_k!l_k!}(a_{j_2j_2s_1}-a_{j_1j_2u_k})^{t_k}\delta_{\beta_0}^{n_{j_1u_2}+\Sigma_{r=2}^{q_1}l_r}}{(x_{j_2}-a_{j_2j_2s_1})^{n_{j_2s_1}}\prod_{s_2=1,s_2\neq s_1}^{q_2}(x_{j_2}-a_{j_2j_2s_2})^{n_{j_2s_2}}}dx_{j_2},\end{aligned}$$

where

$$A_{l_1 m_{j_1 r_1}} = \frac{1}{(m_{j_1 r_1} - l_1)!} \frac{d^{m_{j_1 r_1} - l_1}}{dx_{j_1}^{m_{j_1 r_1} - l_1}} \frac{1}{\prod_{\sigma=2}^{p_1} (x_{j_1} - a_{j_1 j_1 r_\sigma})^{m_{j_1 r_\sigma}}} \Big|_{x_{j_1}^* = a_{j_1 j_1 r_1}},$$

$$A_{m_{j_1 r_1} m_{j_1 r_1}} = \frac{1}{\prod_{\sigma=2}^{p_1} (a_{j_1 j_1 r_1} - a_{j_1 j_1 r_\sigma})^{m_{j_1 r_\sigma}}}.$$

Let

$$b_{j_1 j_2 \beta_0} = \sum_{t_2, l_2 = 0}^{t_2 + l_2 = n_{j_1 u_2}} \cdots \sum_{t_{q_1}, l_{q_1} = 0}^{t_{q_1} + l_{q_1} = n_{j_1 u_{q_1}}} \prod_{k=2}^{q_1} \frac{n_{j_1 u_k}!}{t_k! l_k!} (a_{j_2 j_2 s_1} - a_{j_1 j_2 u_k})^{t_k} \delta_{\beta_0}^{n_{j_1 u_2} + \Sigma_{r=2}^{q_1} l_r}.$$

We have

$$\sum_{r_1=1}^{p_1} \sum_{l_1=1}^{m_{j_1 r_1}} \frac{A_{l_1 m_{j_1 r_1}}}{(x_{j_1} - a_{j_1 j_1 r_1})^{l_1}} dx_{j_1}$$

$$= \frac{a_{j_1 j_1 0}}{a_{j_2 j_2 0}} \frac{\sum_{\beta_0 = n_{j_1 u_1}}^{n_2 - 1} b_{j_1 j_2 \beta_0} (x_{j_2} - a_{j_2 j_2 s_1})^{\beta_0} + \sum_{\beta_0 = n_2}^{n_1} b_{j_1 j_2 \beta_0} (x_{j_2} - a_{j_2 j_2 s_1})^{\beta_0}}{(x_{j_2} - a_{j_2 j_2 s_1})^{n_{j_2 s_1}} \prod_{s_2 = 1, s_2 \neq s_1}^{q_2} (x_{j_2} - a_{j_2 j_2 s_2})^{n_{j_2 s_2}}} dx_{j_2}.$$

For $n_{j_1 u_1} < n_{j_2 s_1}$, the first part is given by

$$\frac{\sum_{\beta_0 = n_{j_1 u_1}}^{n_2 - 1} b_{j_1 j_2 \beta_0} (x_{j_2} - a_{j_2 j_2 s_1})^{\beta_0}}{(x_{j_2} - a_{j_2 j_2 s_1})^{n_{j_2 s_1}} \prod_{s_2 = 1, s_2 \neq s_1}^{q_2} (x_{j_2} - a_{j_2 j_2 s_2})^{n_{j_2 s_2}}}$$

$$= \sum_{\beta_0 = n_{j_1 u_1}}^{n_2 - 1} b_{j_1 j_2 \beta_0} \frac{(x_{j_2} - a_{j_2 j_2 s_1})^{\beta_0 - n_{j_2 s_1}}}{\prod_{s_2 = 1, s_2 \neq s_1}^{q_2} (x_{j_2} - a_{j_2 j_2 s_2})^{n_{j_2 s_2}}}$$

$$= \sum_{\beta_0 = n_{j_1 u_1}}^{n_{j_2 s_1} - 1} b_{j_1 j_2 \beta_0} \sum_{j=1}^{q_2} \sum_{l_j = 1}^{\bar{n}_{j_2 s_j}} A_{l_j \bar{n}_{j_2 s_j}}^{(s_1)} \frac{1}{(x_{j_2} - a_{j_2 j_2 s_j})^{l_j}}$$

$$+ \sum_{\beta_0 = m_{r_1}}^{n_2 - 1} b_{j_1 j_2 \beta_0} \sum_{j=2}^{q_2} \sum_{l_j = 1}^{\bar{n}_{j_2 s_j}} B_{l_j \bar{n}_{j_2 s_j}}^{(s_1)} \frac{1}{(x_{j_2} - a_{j_2 j_2 s_j})^{l_j}};$$

where

$$A^{(s_1)}_{l_\rho \overline{n}_{j_2 s_\rho}} = \frac{1}{(\overline{n}_{j_2 s_\rho} - l_\rho)!} \frac{d^{\overline{n}_{j_2 s_\rho} - l_\rho}}{dx_{j_2}^{\overline{n}_{j_2 s_\rho} - l_\rho}} \frac{1}{\prod_{\sigma=1,\sigma\neq\rho}^{q_2} (x_{j_2} - a_{j_2 j_2 s_\sigma})^{n_{j_2 s_\sigma}}} \Big|_{x^*_{j_2} = a_{j_2 j_2 s_\rho}}$$

$$(l_\rho = 1, 2, \cdots, \overline{n}_{j_2 s_\rho}, \rho = 1, 2, \cdots, q_2),$$

$$A^{(s_1)}_{\overline{n}_{j_2 s_\rho} \overline{n}_{j_2 s_\rho}} = \frac{1}{\prod_{\sigma=1,\sigma\neq\rho}^{q_2} (a_{j_2 j_2 s_\rho} - a_{j_s j_s s_\sigma})^{n_{j_2 s_\sigma}}},$$

$$B^{(s_1)}_{l_\rho \overline{n}_{j_2 s_\rho}} = \frac{1}{(\overline{n}_{j_2 s_\rho} - l_\rho)!} \frac{d^{\overline{n}_{j_2 s_\rho} - l_\rho}}{dx_{j_2}^{\overline{n}_{j_2 s_\rho} - l_\rho}} \frac{(x_{j_2} - a_{j_2 j_2 s_1})^{\beta_0 - n_{j_2 s_1}}}{\prod_{\sigma=2,\sigma\neq\rho}^{q_2} (x_{j_2} - a_{j_2 j_2 s_\sigma})^{n_{(2) s_\sigma}}} \Big|_{x^*_{j_2} = a_{j_2 j_2 s_\rho}}$$

$$(l_\rho = 1, 2, \cdots, \overline{n}_{(2) s_\rho}, \rho = 2, 3, \cdots, q_2),$$

$$B^{(s_1)}_{\overline{n}_{j_2 s_\rho} \overline{n}_{j_2 s_\rho}} = \frac{(a_{j_2 j_2 s_\rho} - a_{j_2 j_2 s_1})^{\beta_0 - n_{j_2 s_1}}}{\prod_{\sigma=2,\sigma\neq\rho}^{q_2} (a_{j_2 j_2 s_\rho} - a_{j_s j_s s_\sigma})^{n_{j_2 s_\sigma}}}$$

$$\{\overline{n}_{j_2 s_\rho} | \rho = 1, 2, \cdots, q_2\} = \{n_{j_2 s_1} - \beta_0, n_{j_2 s_2}, \cdots, n_{j_2 s_{q_2}}\}.$$

For $n_{j_1 u_1} = n_{j_2 s_1}$, the first part is given by

$$\frac{\sum_{\beta_0 = n_{j_1 u_1}}^{n_2 - 1} b_{j_1 j_2 \beta_0} (x_{j_2} - a_{j_2 j_2 s_1})^{\beta_0}}{(x_{j_2} - a_{j_2 j_2 s_1})^{n_{j_2 s_1}} \prod_{s_2=1, s_2\neq s_1}^{q_2} (x_{j_2} - a_{j_2 j_2 s_j})^{n_{j_2 s_2}}}$$

$$= \sum_{\beta_0 = n_{j_1 u_1}}^{n_2 - 1} b_{j_1 j_2 \beta_0} \frac{(x_{j_2} - a_{j_2 j_2 s_1})^{\beta_0 - n_{j_2 s_1}}}{\prod_{s_2=1, s_2\neq s_1}^{q_2} (x_{j_2} - a_{j_2 j_2 s_j})^{n_{j_2 s_2}}}$$

$$= \sum_{\beta_0 = m_{r_1}}^{n_2 - 1} b_{j_1 j_2 \beta_0} \sum_{j=2}^{q_2} \sum_{l_j=1}^{\overline{n}_{j_2 s_j}} B^{(s_1)}_{l_j \overline{n}_{j_2 s_j}} \frac{1}{(x_{j_2} - a_{j_2 j_2 s_j})^{l_j}}.$$

For $n_{j_1 u_1} > n_{j_2 s_1}$, the first part is given by

$$\frac{\sum_{\beta_0 = n_{j_1 u_1}}^{n_2 - 1} b_{j_1 j_2 \beta_0} (x_{j_2} - a_{j_2 j_2 s_1})^{\beta_0}}{(x_{j_2} - a_{j_2 j_2 s_1})^{n_{j_2 s_1}} \prod_{s_2=1, s_2\neq s_1}^{q_2} (x_{j_2} - a_{j_2 j_2 s_2})^{n_{j_2 s_2}}}$$

$$= \sum_{\beta_0 = n_{j_1 u_1}}^{n_2 - 1} b_{j_1 j_2 \beta_0} \frac{(x_{j_2} - a_{j_2 j_2 s_1})^{\beta_0 - n_{j_2 s_1}}}{\prod_{s_2=1, s_2\neq s_1}^{q_2} (x_{j_2} - a_{j_2 j_2 s_2})^{n_{j_2 s_2}}}$$

$$= \sum_{\beta_0 = n_{j_1 u_1}}^{n_2 - 1} b_{j_1 j_2 \beta_0} \sum_{j=2}^{q_2} \sum_{l_j=1}^{\overline{n}_{j_2 s_j}} B^{(s_1)}_{l_j \overline{n}_{j_2 s_j}} \frac{1}{(x_{j_2} - a_{j_2 j_2 s_j})^{l_j}};$$

Following the proof of Eqs. (1.105)–(1.109) for $a_{j_2j_2s_1} \neq a_{j_1j_2u_1}$, with the above equations, the first integral manifold for $a_{j_2j_2s_1} = a_{j_1j_2u_1}$ is given by

$$
\begin{aligned}
&\sum_{r_1=1}^{p_1}\sum_{l_1=1}^{m_{j_1r_1}} \frac{A_{l_1m_{j_1r_1}}}{l_1-1}\Big[\frac{1}{(x_{j_1}-a_{j_1j_1r_1})^{l_1-1}} - \frac{1}{(x_{j_10}-a_{j_1j_1r_1})^{l_1-1}}\Big] \\
&= \frac{a_{j_1j_10}}{a_{j_2j_20}}\Bigg\{ \sum_{\beta_0=n_2}^{n_1} b_{j_1j_2\beta_0} \sum_{\substack{\alpha_1\in[0,\beta_0-n_2],\\ \beta_1\in[n_2-n_{j_2s_1},\beta_0-n_{(2)r_1}]}}^{\alpha_1+\beta_1=\beta_0-n_{j_2s_1}} \quad \sum_{\substack{\alpha_2\in[0,\beta_1-n_2]\\ \beta_2\in[n_2-\Sigma_{i=1}^{2}n_{j_2s_i},\beta_1-\Sigma_{i=1}^{2}n_{j_2s_i}]}}^{\alpha_2+\beta_2=\beta_1-\Sigma_{i=1}^{2}n_{j_2s_i}} \cdots \\
&\quad \sum_{\substack{\alpha_{p-1}\in[0,\beta_{q_2-2}-n_2]\\ \beta_{q_2-1}\in[n_2-\Sigma_{i=1}^{q_2-1}n_{j_2s_i},\beta_{p-2}-\Sigma_{i=1}^{q_2-1}n_{j_2s_i}]}}^{\alpha_{q_2-1}+\beta_{q_2-1}=\beta_{q_2-2}-\Sigma_{i=1}^{q_2-1}n_{j_2s_i}} \quad \prod_{j=1}^{q_2-1} \frac{(\beta_{j-1}-\Sigma_{i=1}^{j}n_{j_2s_i})!}{\alpha_j!\beta_j!}(a_{j_2j_2s_{j+1}}-a_{j_2j_2s_j})^{\alpha_j} \\
&\quad \times \frac{1}{\beta_{q_2-1}-n_{j_2r_{q_2}}+1}\Big[(x_{j_2}-a_{j_2j_2s_\rho})^{\beta_{q_2-1}-n_{j_2r_{q_2}}+1} - (x_{j_20}-a_{j_2j_2s_\rho})^{\beta_{q_2-1}-n_{j_2r_{q_2}}+1}\Big] \\
&\quad - \sum_{\beta_0=q_2}^{q_1} b_{j_1j_2\beta_0} \sum_{k=2}^{q_2-1} \Bigg[\sum_{\substack{\alpha_1\in[0,\beta_0-n_2),\\ \beta_1\in(n_2-n_{j_2s_1},\beta_0-n_{j_2s_i}]}}^{\alpha_1+\beta_1=\beta_0-n_{j_2s_1}} \quad \sum_{\substack{\alpha_2\in[0,\beta_1-n_2]\\ \beta_2\in(n_2-\Sigma_{i=1}^{2}n_{j_2s_i},\beta_1-\Sigma_{i=1}^{2}n_{j_2s_i})}}^{\alpha_2+\beta_2=\beta_1-\Sigma_{i=1}^{2}n_{j_2s_i}} \cdots \\
&\quad \times \Bigg(\sum_{\substack{\alpha_{k-1}\in(\beta_{k-2}-n_2+n_{j_2k},\beta_{k-2}-\Sigma_{i=1}^{k-1}n_{(2)r_i}]\\ \beta_{k-1}\in[0,n_2-\Sigma_{i=1}^{k}n_{j_2s_i})}}^{\alpha_{k-1}+\beta_{k-1}=\beta_{k-2}-\Sigma_{i=1}^{k-1}n_{j_2s_i}} \quad \prod_{j=1}^{k-1} \frac{(\beta_{j-1}-\Sigma_{i=1}^{j}n_{j_2s_i})!}{\alpha_j!\beta_j!}(a_{j_2j_2s_{j+1}}-a_{j_2j_2s_j})^{\alpha_j} \\
&\quad \times \sum_{\rho=k}^{q_2}\sum_{l_\rho=1}^{n_{j_2s_\rho}} \frac{A^{(s_k)}_{l_\rho n_{j_2s_\rho}}}{l_\rho-1}\Big[\frac{1}{(x_{j_2}-a_{j_2j_2s_\rho})^{l_\rho-1}} - \frac{1}{(x_{j_20}-a_{j_2j_2s_\rho})^{l_\rho-1}}\Big] \\
&\quad + \sum_{\substack{\alpha_{k-1}\in(\beta_{k-2}-n_2,\beta_{k-2}-n_2+n_{j_2k}]\\ \beta_{k-1}\in[n_2-\Sigma_{i=1}^{k}n_{j_2s_i},n_2-\Sigma_{i=1}^{k-1}n_{j_2s_i})}}^{\alpha_{k-1}+\beta_{k-1}=\beta_{k-2}-\Sigma_{i=1}^{k-1}n_{j_2s_i}} \quad \prod_{j=1}^{k-1} \frac{(\beta_{j-1}-\Sigma_{i=1}^{j}n_{j_2s_i})!}{\alpha_j!\beta_j!}(a_{j_2j_2s_{j+1}}-a_{j_2j_2s_j})^{\alpha_j} \\
&\quad \times \sum_{\rho=k+1}^{q_2}\sum_{l_\rho=1}^{m'_{s_\rho}} \frac{B^{(s_k)}_{l_\rho \bar{n}_{j_2s_\rho}}}{l_\rho-1}\Big[\frac{1}{(x_{j_2}-a_{j_2j_2s_\rho})^{l_\rho-1}} - \frac{1}{(x_{j_20}-a_{j_2j_2s_\rho})^{l_\rho-1}}\Big] \Bigg) \Bigg] + \Pi(u_1) \Bigg\},
\end{aligned}
$$

where

$$\Pi(u_1) = -\left(\sum_{\beta_0=0}^{n_{j_2 s_1}-1} b_{j_1 j_2 \beta_0} \sum_{\rho=1}^{q_2} \sum_{l_\rho=1}^{\overline{n}_{j_2 s_\rho}} \frac{A^{(s_1)}_{l_\rho \overline{n}_{j_2 s_\rho}}}{l_\rho - 1} \left[\frac{1}{(x_{j_2} - a_{j_2 j_2 s_\rho})^{l_\rho - 1}} - \frac{1}{(x_{j_2 0} - a_{j_2 j_2 s_\rho})^{l_\rho - 1}} \right] \right.$$
$$\left. + \sum_{\beta_0=n_{j_2 s_1}}^{n_2-1} b_{j_1 j_2 \beta_0} \sum_{\rho=2}^{q_2} \sum_{l_\rho=1}^{\overline{n}_{j_2 s_\rho}} \frac{B^{(s_1)}_{l_\rho \overline{n}_{j_2 s_\rho}}}{l_\rho - 1} \left[\frac{1}{(x_{j_2} - a_{j_2 j_2 s_\rho})^{l_\rho - 1}} - \frac{1}{(x_{j_2 0} - a_{j_2 j_2 s_\rho})^{l_\rho - 1}} \right] \right)$$

for $n_{j_1 u_1} < n_{j_2 s_1}$;

$$\Pi(u_1) = -\sum_{\beta_0=n_{j_2 s_1}}^{n_2-1} b_{j_1 j_2 \beta_0} \sum_{\rho=2}^{q_2} \sum_{l_\rho=1}^{\overline{n}_{j_2 s_\rho}} \frac{B^{(s_1)}_{l_\rho \overline{n}_{j_2 s_\rho}}}{l_\rho - 1} \left[\frac{1}{(x_{j_2} - a_{j_2 j_2 s_\rho})^{l_\rho - 1}} - \frac{1}{(x_{j_2 0} - a_{j_2 j_2 s_\rho})^{l_\rho - 1}} \right],$$

for $n_{j_1 u_1} = n_{j_2 s_1}$;

$$\Pi(u_1) = -\sum_{\beta_0=n_{j_1 u_1}}^{n_2-1} b_{j_1 j_2 \beta_0} \sum_{\rho=2}^{q_2} \sum_{l_\rho=1}^{\overline{n}_{j_2 s_\rho}} \frac{B^{(s_1)}_{l_\rho \overline{n}_{j_2 s_\rho}}}{l_\rho - 1} \left[\frac{1}{(x_{j_2} - a_{j_2 j_2 s_\rho})^{l_\rho - 1}} - \frac{1}{(x_{j_2 0} - a_{j_2 j_2 s_\rho})^{l_\rho - 1}} \right],$$

for $n_{j_1 u_1} > n_{j_2 s_1}$;

and

$$b_{j_1 j_2 \beta_0} = \sum_{t_2, l_2=0}^{t_2+l_2=n_{j_1 u_2}} \cdots \sum_{t_{q_1}, l_{q_1}=0}^{t_{q_1}+l_{q_1}=n_{j_1 u_{q_1}}} \prod_{k=1}^{q_1} \frac{n_{j_1 u_k}!}{t_k! l_k!} (a_{j_2 j_2 s_1} - a_{j_1 j_2 u_k})^{t_k} \delta_{\beta_0}^{n_{j_1 u_1} + \Sigma_{k=2}^{q_1} l_k},$$

$$\frac{1}{\beta_{q_2-1} - n_{j_2 q_2} + 1} [(x_{j_2} - a_{j_2 j_2 s_{q_2}})^{\beta_{q_2-1} - n_{j_2 q_2} + 1} - (x_{j_2 0} - a_{j_2 j_2 s_{q_2}})^{\beta_{q_2-1} - n_{j_2 q_2} + 1}]$$
$$= \ln \frac{|x_{j_2} - a_{j_2 j_2 s_{q_2}}|}{|x_{j_2 0} - a_{j_2 j_2 s_{q_2}}|}, \text{ for } \beta_{q_2-1} - n_{j_2 q_2} + 1 = 0;$$
$$\frac{1}{l_k - 1} \left[\frac{1}{(x_{j_2} - a_{j_2 j_2 s_k})^{l_k - 1}} - \frac{1}{(x_{j_2 0} - a_{j_2 j_2 s_k})^{l_k - 1}} \right] = -\ln \frac{|x_{j_2} - a_{j_2 j_2 s_k}|}{|x_{j_2 0} - a_{j_2 j_2 s_k}|},$$
for $l_k = 1\ (k = 1, 2, \cdots, q_2 - 1)$;
$$\frac{1}{l_1 - 1} \left[\frac{1}{(x_{j_1} - a_{j_1 j_1 r_1})^{l_1 - 1}} - \frac{1}{(x_{j_1 0} - a_{j_1 j_1 r_1})^{l_1 - 1}} \right] = -\ln \frac{|x_{j_1} - a_{j_1 j_1 r_1}|}{|x_{j_1 0} - a_{j_1 j_1 r_1}|},$$
for $l_1 = 1$.

Define a network matrix of hybrid singular 1-dimensional flow and singular equilibriums as

$$\mathrm{NM} = \underbrace{(a_{j_1 j_1 r_1}, a_{j_2 j_2 s_1})}_{([(n_{j_1 u_1} - n_{j_2 s_1})]^{\text{th}}\text{XX}, [m_{j_1 r_1}, n_{j_2 s_1}]^{\text{th}}\text{YY})} \oplus \bigcup_{r_2=1, r_2 \neq r_1}^{p_1} \underbrace{(a_{j_1 j_1 r_2}, a_{j_2 j_2 s_1})}_{([(n_{j_1 u_1} - n_{j_2 s_1})]^{\text{th}}\text{XX}, [m_{j_1 r_2}, n_{j_2 s_1}]^{\text{th}}\text{YY})}$$
$$\oplus \bigcup_{r=1}^{p_1} \bigcup_{u_2=1, u_2 \neq u_1}^{q_1} \underbrace{(a_{j_1 j_1 r}, a_{j_1 j_2 u_2})}_{([m_{j_1 r}, n_{j_1 u_2}]\text{ZZ}, \text{xx})} \oplus \bigcup_{r=1}^{p_1} \bigcup_{s_2=1, s_2 \neq s_1}^{q_2} \underbrace{(a_{j_1 j_1 r}, a_{j_2 j_2 s_2})}_{(m_{j_1 r}, n_{j_2 s_2})^{\text{th}}\text{-UU}}.$$

($\text{ii}_{2\text{a}}$) For $n_{j_2s_1} \le n_{j_1u_1}$, at $x_{j_2}^* = a_{j_2j_2s_1} = a_{j_1j_2u_1}$ and $\overline{x}_{j_1} \ne a_{j_1j_1r}(r = 1, 2, \cdots, p_1)$, in phase space,

$$
\begin{aligned}
&\frac{dx_{j_1}}{dx_{j_2}}\bigg|_{x_{j_2}^*=a_{j_1j_2u_1}} \\
&= \frac{a_{j_1j_10}(x_{j_2} - a_{j_1j_2u_1})^{n_{j_1u_1}-n_{j_2s_1}} \prod_{r=1}^{p_1} (\overline{x}_{j_1} - a_{j_1j_1r})^{m_{j_1r}}}{a_{j_2j_20} \prod_{s_2=1,s_2\ne s_1}^{q_2} (x_{j_2} - a_{j_2j_2s_2})^{n_{j_2s_2}}} \\
&\quad \times \prod_{u_2=1,u_2\ne u_1}^{q_1} (x_{j_2} - a_{j_1j_2u_2})^{n_{j_1u_2}}\bigg|_{x_{j_2}^*=a_{j_1j_2u_1}} \\
&= 0, \\
&\vdots \\
&\frac{d^{(n_{j_1u_1}-n_{j_2s_1})}x_{j_1}}{dx_{j_2}^{(n_{j_1u_1}-n_{j_2s_1})}}\bigg|_{x_{j_2}^*=a_{j_1j_2u_1}} = 0.
\end{aligned}
$$

If

$$
\begin{aligned}
&\frac{d^{(n_{j_1u_1}-n_{j_2s_1})+1}x_{j_1}}{dx_{j_2}^{(n_{j_1u_1}-n_{j_2s_1})+1}}\bigg|_{x_{j_2}^*=a_{j_1j_2u_1}} \\
&= \frac{a_{j_1j_10}(n_{j_1u_1} - n_{j_2s_1})! \prod_{r=1}^{p_1} (\overline{x}_{j_1} - a_{j_1j_1r})^{m_{j_1r}} \prod_{u_2=1,u_2\ne u_1}^{q_1} (a_{j_1j_2u_1} - a_{j_1j_2u_2})^{n_{j_1u_2}}}{a_{j_2j_20} \prod_{s_2=1,s_2\ne s_1}^{q_2} (a_{j_2j_2s_1} - a_{j_2j_2s_2})^{n_{j_2s_2}}} \\
&> 0,
\end{aligned}
$$

the flows at $x_{j_2}^* = a_{j_2j_2s_1} = a_{j_1j_2u_1}$ are $[2(n_{j_1u_11} - n_{j_2s_11}) \pm 1]^{\text{th}}$-order up-parabola flow for ($n_{j_1u_1} = 2n_{j_1u_11} + 1$, $n_{j_2s_1} = 2n_{j_2s_11}$ and $n_{j_1u_1} = 2n_{j_1u_11}$, $n_{j_2s_1} = 2n_{j_2s_11} + 1$), and $[2(n_{j_1u_11} - n_{j_2s_11})]^{\text{th}}$-order increasing-inflection flow for ($n_{j_1u_1} = 2n_{j_1u_11} + 1$, $n_{j_2s_1} = 2n_{j_2s_11} + 1$) and $n_{j_1u_1} = 2n_{j_1u_11}$, $n_{j_2s_1} = 2n_{j_2s_11}$).

If

$$
\begin{aligned}
&\frac{d^{(n_{j_1u_1}-n_{j_2s_1})+1}x_{j_1}}{dx_{j_2}^{(n_{j_1u_1}-n_{j_2s_1})+1}}\bigg|_{x_{j_2}^*=a_{j_1j_2u_1}} \\
&= \frac{a_{j_1j_10}(n_{j_1u_1} - n_{j_2s_1})! \prod_{r=1}^{p_1} (\overline{x}_{j_1} - a_{j_1j_1r})^{m_{j_1s}} \prod_{u_2=1,u_2\ne u_1}^{q_1} (a_{j_1j_2u_1} - a_{j_1j_2u_2})^{n_{j_1u_2}}}{a_{j_2j_20} \prod_{s_2=1,s_2\ne s_1}^{q_2} (a_{j_2j_2s_1} - a_{j_2j_2s_2})^{n_{j_2s_2}}} \\
&< 0,
\end{aligned}
$$

the flows at $x_{j_2}^* = a_{j_2j_2s_1} = a_{j_1j_2u_1}$ are $[2(n_{j_1u_11} - n_{j_2s_11}) \pm 1]^{\text{th}}$-order down-parabola flow for ($n_{j_1u_1} = 2n_{j_1u_11} + 1$, $n_{j_2s_1} = 2n_{j_2s_11}$ and $n_{j_1u_1} = 2n_{j_1u_11}$, $n_{j_2s_1} = 2n_{j_2s_11} + 1$), and $[2(n_{j_1u_11} - n_{j_2s_11})]^{\text{th}}$-order decreasing-inflection flow for ($n_{j_1u_1} = 2n_{j_1u_11} + 1$, $n_{j_2s_1} = 2n_{j_2s_11} + 1$) and $n_{j_1u_1} = 2n_{j_1u_11}$, $n_{j_2s_1} = 2n_{j_2s_11}$).

The variational equation at $x_{j_2}^* = a_{j_2j_2s_1} = a_{j_1j_2u_1}$ is

$$\Delta\dot{x}_{j_2} = a_{j_2j_20}\prod_{s_2=1,s_2\neq s_1}^{q_2}(a_{j_2j_2s_1} - a_{j_2j_2s_2})^{n_{j_2s_2}}(\Delta x_{j_2})^{n_{j_2s_1}}.$$

For

$$a_{j_2j_20}\prod_{s_2=1,s_2\neq s_1}^{q_2}(a_{j_2j_2s_1} - a_{j_2j_2s_2})^{n_{j_2s_2}} > 0 \text{ and}$$
$$a_{j_2j_20}\prod_{s_2=1,s_2\neq s_1}^{q_2}(a_{j_2j_2s_1} - a_{j_2j_2s_2})^{n_{j_2s_2}} < 0,$$

the flows at $x_{j_2}^* = a_{j_2j_2s_1} = a_{j_1j_2u_1}$ are $(2n_{j_2s_1 1} + 1)^{\text{th}}$ source and sink flows for $n_{j_2s_1} = 2n_{j_2s_1 1} + 1$, and $(2n_{j_2s_1 1})^{\text{th}}$-upper-saddle and lower-saddle flows for $n_{j_2s_1} = 2n_{j_2s_1 1}$.

(ii_{2a1}) For $n_{j_2s_1} \leq n_{j_1u_1}$, 16 types of infinite-equilibriums exist at $x_{j_2}^* = a_{j_2j_2s_1} = a_{j_1j_2u_1}$..

(ii_{2a1a}) For $n_{j_1u_1} = 2n_{j_1u_1 1} + 1$ and $n_{j_2s_1} = 2n_{j_2s_1 1} + 1$, the infinite-equilibrium of $x_{j_2}^* = a_{j_1j_2u_1} = a_{j_2j_2s_1}$ with $\overline{x}_{j_1} \neq a_{j_1j_1r_1}$ has the following properties as in Eqs. (1.183)–(1.186).

- For $a_{j_1j_10}\prod_{r=1}^{p_1}(\overline{x}_{j_1} - a_{j_1j_1r})^{m_{j_1r}}\prod_{u_2=1,u_2\neq u_1}^{q_1}(a_{j_1j_2u_1} - a_{j_1j_2u_2})^{n_{j_1u_2}} > 0$ and $a_{j_2j_20}\prod_{s_2=1,s_2\neq s_1}^{q_2}(a_{j_2j_2s_1} - a_{j_2j_2s_2})^{n_{j_2s_2}} > 0$,

$$(\overline{x}_{j_1}, a_{j_2j_2s_1}) = \underbrace{([2(n_{j_1u_1 1} - n_{j_2s_1 1})]^{\text{th}}\text{II}, (2n_{j_2s_1 1} + 1)^{\text{th}}\text{SO})}_{([2(n_{j_1u_1 1} - n_{j_2s_1 1})],(2n_{j_2s_1 1}+1))\text{-increasing-inflection source}}.$$

  The infinite-equilibrium is a $([2(n_{j_1u_1 1} - n_{j_2s_1 1})]^{\text{th}}\text{II}, (2n_{j_2s_1 1} + 1)^{\text{th}}\text{SO})$-increasing-inflection source at $x_{j_2}^* = a_{j_1j_2u_1} = a_{j_2j_2s_1}$.

- For $a_{j_1j_10}\prod_{r=1}^{p_1}(\overline{x}_{j_1} - a_{j_1j_1r})^{m_{j_1r}}\prod_{u_2=1,u_2\neq u_1}^{q_1}(a_{j_1j_2u_1} - a_{j_1j_2u_2})^{n_{j_1u_2}} < 0$ and $a_{j_2j_20}\prod_{s_2=1,s_2\neq s_1}^{q_2}(a_{j_2j_2s_1} - a_{j_2j_2s_2})^{n_{j_2s_2}} > 0$,

$$(\overline{x}_{j_1}, a_{j_2j_2s_1}) = \underbrace{([2(n_{j_1u_1 1} - n_{j_2s_1 1})]^{\text{th}}\text{DI}, (2n_{j_2s_1 1} + 1)^{\text{th}}\text{SO})}_{([2(n_{j_1u_1 1} - n_{j_2s_1 1})],(2n_{j_2s_1 1}+1))\text{-decreasing - inflection source}}.$$

  The infinite-equilibrium is a $([2(n_{j_1u_1 1} - n_{j_2s_1 1})]^{\text{th}}\text{DI}, (2n_{j_2s_1 1} + 1)^{\text{th}}\text{SO})$-decreasing-inflection source at $x_{j_2}^* = a_{j_1j_2u_1} = a_{j_2j_2s_1}$.

- For $a_{j_1j_10}\prod_{r=1}^{p_1}(\overline{x}_{j_1} - a_{j_1j_1r})^{m_{j_1r}}\prod_{u_2=1,u_2\neq u_1}^{q_1}(a_{j_1j_2u_1} - a_{j_1j_2u_2})^{n_{j_1u_2}} > 0$ and $a_{j_2j_20}\prod_{s_2=1,s_2\neq s_1}^{q_2}(a_{j_2j_2s_1} - a_{j_2j_2s_2})^{n_{j_2s_2}} < 0$,

$$(\overline{x}_{j_1}, a_{j_2j_2s_1}) = \underbrace{([2(n_{j_1u_1 1} - n_{j_2s_1 1})]^{\text{th}}\text{DI}, (2n_{j_2s_1 1} + 1)^{\text{th}}\text{SI})}_{([2(n_{j_1u_1 1} - n_{j_2s_1 1})],(2n_{j_2s_1 1}+1))\text{-decreasing - inflection sink}}.$$

The infinite-equilibrium is a $([2(n_{j_1u_1 1}-n_{j_2s_1 1})]^{\text{th}}\text{DI}, (2n_{j_2s_1 1}+1)]^{\text{th}}\text{SI})$-decreasing-inflection sink at $x_{j_2}^*=a_{j_1j_2u_1}=a_{j_2j_2s_1}$.

- For $a_{j_1j_10}\prod_{r=1}^{p_1}(\bar{x}_{j_1}-a_{j_1j_1r})^{m_{j_1r}}\prod_{u_2=1,u_2\neq u_1}^{q_1}(a_{j_1j_2u_1}-a_{j_1j_2u_2})^{n_{j_1u_2}}<0$ and $a_{j_2j_20}\prod_{s_2=1,s_2\neq s_1}^{q_2}(a_{j_2j_2s_1}-a_{j_2j_2s_2})^{n_{j_2s_2}}<0$,

$$(\bar{x}_{j_1},a_{j_2j_2s_1})=\underbrace{([2(n_{j_1u_1 1}-n_{j_2s_1 1})]^{\text{th}}\text{II}, (2n_{j_2s_1 1}+1)^{\text{th}}\text{SI})}_{([2(n_{j_1u_1 1}-n_{j_2s_1 1})],(2n_{j_2s_1 1}+1))\text{-increasing-inflection sink}}.$$

The infinite-equilibrium is a $([2(n_{j_1u_1 1}-n_{j_2s_1 1})]^{\text{th}}\text{II}, (2n_{j_2s_1 1}+1)^{\text{th}}\text{SI})$-increasing-inflection sink at $x_{j_2}^*=a_{j_1j_2u_1}=a_{j_2j_2s_1}$.

($\text{ii}_{2\text{a}1\text{b}}$) For $n_{j_1u_1}=2n_{j_1u_1 1}$ and $n_{j_2s_1}=2n_{j_2s_1 1}+1$, the infinite-equilibrium of $x_{j_2}^*=a_{j_1j_2u_1}=a_{j_2j_2s_1}$ with $\bar{x}_{j_1}\neq a_{j_1j_1r_1}$ has the following properties as in Eqs. (1.187)–(1.190).

- For $a_{j_1j_10}\prod_{r=1}^{p_1}(\bar{x}_{j_1}-a_{j_1j_1r})^{m_{j_1r}}\prod_{u_2=1,u_2\neq u_1}^{q_1}(a_{j_1j_2u_1}-a_{j_1j_2u_2})^{n_{j_1u_2}}>0$ and $a_{j_2j_20}\prod_{s_2=1,s_2\neq s_1}^{q_2}(a_{j_2j_2s_1}-a_{j_2j_2s_2})^{n_{j_2s_2}}>0$,

$$(\bar{x}_{j_1},a_{j_2j_2s_1})=\underbrace{([2(n_{j_1u_1 1}-n_{j_2s_1 1})-1]^{\text{th}}\text{UP}, (2n_{j_2s_1 1}+1)^{\text{th}}\text{SO})}_{([2(n_{j_1u_1 1}-n_{j_2s_1 1})-1],(2n_{j_2s_1 1}+1))\text{-up-parabola source}}.$$

The infinite-equilibrium is a $([2(n_{j_1u_1 1}-n_{j_2s_1 1})-1]^{\text{th}}\text{UP}, (2n_{j_2s_1 1}+1)^{\text{th}}\text{SO})$-up parabola source at $x_{j_2}^*=a_{j_1j_2u_1}=a_{j_2j_2s_1}$.

- For $a_{j_1j_10}\prod_{r=1}^{p_1}(\bar{x}_{j_1}-a_{j_1j_1r})^{m_{j_1r}}\prod_{u_2=1,u_2\neq u_1}^{q_1}(a_{j_1j_2u_1}-a_{j_1j_2u_2})^{n_{j_1u_2}}<0$ and $a_{j_2j_20}\prod_{s_2=1,s_2\neq s_1}^{q_2}(a_{j_2j_2s_1}-a_{j_2j_2s_2})^{n_{j_2s_2}}>0$,

$$(\bar{x}_{j_1},a_{j_2j_2s_1})=\underbrace{([2(n_{j_1u_1 1}-n_{j_2s_1 1})-1]^{\text{th}}\text{DP}, (2n_{j_2s_1 1}+1)^{\text{th}}\text{SO})}_{([2(n_{j_1u_1 1}-n_{j_2s_1 1})-1],(2n_{j_2s_1 1}+1))\text{-down-parabola source}}.$$

The infinite-equilibrium is a $([2(n_{j_1u_1 1}-n_{j_2s_1 1})-1]^{\text{th}}\text{DP}, (2n_{j_2s_1 1}+1)^{\text{th}}\text{SO})$-down parabola source at $x_{j_2}^*=a_{j_1j_2u_1}=a_{j_2j_2s_1}$.

- For $a_{j_1j_10}\prod_{r=1}^{p_1}(\bar{x}_{j_1}-a_{j_1j_1r})^{m_{j_1r}}\prod_{u_2=1,u_2\neq u_1}^{q_1}(a_{j_1j_2u_1}-a_{j_1j_2u_2})^{n_{j_1u_2}}>0$ and $a_{j_2j_20}\prod_{s_2=1,s_2\neq s_1}^{q_2}(a_{j_2j_2s_1}-a_{j_2j_2s_2})^{n_{j_2s_2}}<0$,

$$(\bar{x}_{j_1},a_{j_2j_2s_1})=\underbrace{([2(n_{j_1u_1 1}-n_{j_2s_1 1})-1]^{\text{th}}\text{DP}, (2n_{j_2s_1 1}+1)^{\text{th}}\text{SI})}_{([2(n_{j_1u_1 1}-n_{j_2s_1 1})-1],(2n_{j_2s_1 1}+1))\text{-down-parabola sink}}.$$

The infinite-equilibrium is a $([2(n_{j_1u_1 1}-n_{j_2s_1 1})-1]^{\text{th}}\text{DP}, (2n_{j_2s_1 1}+1)^{\text{th}}\text{SI})$-down parabola sink at $x_{j_2}^*=a_{j_1j_2u_1}=a_{j_2j_2s_1}$.

- For $a_{j_1j_10}\prod_{r=1}^{p_1}(\bar{x}_{j_1}-a_{j_1j_1r})^{m_{j_1r}}\prod_{u_2=1,u_2\neq u_1}^{q_1}(a_{j_1j_2u_1}-a_{j_1j_2u_2})^{n_{j_1u_2}}<0$ and $a_{j_2j_20}\prod_{s_2=1,s_2\neq s_1}^{q_2}(a_{j_2j_2s_1}-a_{j_2j_2s_2})^{n_{j_2s_2}}<0$,

$$(\overline{x}_{j_1}, a_{j_2j_2s_1}) = \underbrace{([2(n_{j_1u_1 1} - n_{j_2s_1 1}) - 1]^{\text{th}}\text{UP}, (2n_{j_2s_1 1} + 1)^{\text{th}}\text{SI})}_{([2(n_{j_1u_1 1} - n_{j_2s_1 1})-1],(2n_{j_2s_1 1}+1))\text{-down-parabola sink}}.$$

The infinite-equilibrium is a $([2(n_{j_1u_1 1}-n_{j_2s_1 1})-1]^{\text{th}}\text{UP}, (2n_{j_2s_1 1}+1)^{\text{th}}\text{SI})$-up-parabola sink at $x^*_{j_2} = a_{j_1j_2u_1} = a_{j_2j_2s_1}$.

($\text{ii}_{2\text{a1c}}$) For $n_{j_1u_1} = 2n_{j_1u_1 1} + 1$ and $n_{j_2s_1} = 2n_{j_2s_1 1}$, the infinite-equilibrium of $x^*_{j_2} = a_{j_1j_2u_1} = a_{j_2j_2s_1}$ with $\overline{x}_{j_1} \neq a_{j_1j_1r_1}$ has the following properties as in Eqs. (1.191)–(1.194).

- For $a_{j_1j_10} \prod_{r=1}^{p_1} (\overline{x}_{j_1} - a_{j_1j_1r})^{m_{j_1r}} \prod_{u_2=1,u_2\neq u_1}^{q_1} (a_{j_1j_2u_1} - a_{j_1j_2u_2})^{n_{j_1u_2}} > 0$ and $a_{j_2j_20} \prod_{s_2=1,s_2\neq s_1}^{q_2} (a_{j_2j_2s_1} - a_{j_2j_2s_2})^{n_{j_2s_2}} > 0$,

$$(\overline{x}_{j_1}, a_{j_2j_2s_1}) = \underbrace{([2(n_{j_1u_1 1} - n_{j_2s_1 1}) + 1]^{\text{th}}\text{UP}, (2n_{j_2s_1 1})^{\text{th}}\text{US})}_{([2(n_{j_1u_1 1} - n_{j_2s_1 1})+1],(2n_{j_2s_1 1}))\text{-up-parabola upper-saddle}}.$$

The infinite-equilibrium is a $([2(n_{j_1u_1 1}-n_{j_2s_1 1})+1]^{\text{th}}\text{UP}, (2n_{j_2s_1 1})^{\text{th}}\text{US})$-up-parabola upper-saddle at $x^*_{j_2} = a_{j_1j_2u_1} = a_{j_2j_2s_1}$.

- For $a_{j_1j_10} \prod_{r=1}^{p_1} (\overline{x}_{j_1} - a_{j_1j_1r})^{m_{j_1r}} \prod_{u_2=1,u_2\neq u_1}^{q_1} (a_{j_1j_2u_1} - a_{j_1j_2u_2})^{n_{j_1u_2}} < 0$ and $a_{j_2j_20} \prod_{s_2=1,s_2\neq s_1}^{q_2} (a_{j_2j_2s_1} - a_{j_2j_2s_2})^{n_{j_2s_2}} > 0$,

$$(\overline{x}_{j_1}, a_{j_2j_2s_1}) = \underbrace{([2(n_{j_1u_1 1} - n_{j_2s_1 1}) + 1]^{\text{th}}\text{DP}, (2n_{j_2s_1 1})^{\text{th}}\text{US})}_{([2(n_{j_1u_1 1} - n_{j_2s_1 1})+1],(2n_{j_2s_1 1}))\text{-down-parabola upper-saddle}}.$$

The infinite-equilibrium is a $([2(n_{j_1u_1 1}-n_{j_2s_1 1})+1]^{\text{th}}\text{DP}, (2n_{j_2s_1 1})^{\text{th}}\text{US})$-down-parabola upper-saddle at $x^*_{j_2} = a_{j_1j_2u_1} = a_{j_2j_2s_1}$.

- For $a_{j_1j_10} \prod_{r=1}^{p_1} (\overline{x}_{j_1} - a_{j_1j_1r})^{m_{j_1r}} \prod_{u_2=1,u_2\neq u_1}^{q_1} (a_{j_1j_2u_1} - a_{j_1j_2u_2})^{n_{j_1u_2}} > 0$ and $a_{j_2j_20} \prod_{s_2=1,s_2\neq s_1}^{q_2} (a_{j_2j_2s_1} - a_{j_2j_2s_2})^{n_{j_2s_2}} < 0$,

$$(\overline{x}_{j_1}, a_{j_2j_2s_1}) = \underbrace{([2(n_{j_1u_1 1} - n_{j_2s_1 1}) + 1]^{\text{th}}\text{DP}, (2n_{j_2s_1 1})^{\text{th}}\text{LS})}_{([2(n_{j_1u_1 1} - n_{j_2s_1 1})+1],(2n_{j_2s_1 1}))\text{-down-parabola lower-saddle}}.$$

The infinite-equilibrium is a $([2(n_{j_1u_1 1}-n_{j_2s_1 1})+1]^{\text{th}}\text{DP}, (2n_{j_2s_1 1})^{\text{th}}\text{LS})$-down-parabola lower-saddle at.

- For $a_{j_1j_10} \prod_{r=1}^{p_1} (\overline{x}_{j_1} - a_{j_1j_1r})^{m_{j_1r}} \prod_{u_2=1,u_2\neq u_1}^{q_1} (a_{j_1j_2u_1} - a_{j_1j_2u_2})^{n_{j_1u_2}} < 0$ and $a_{j_2j_20} \prod_{s_2=1,s_2\neq s_1}^{q_2} (a_{j_2j_2s_1} - a_{j_2j_2s_2})^{n_{j_2s_2}} < 0$,

$$(\overline{x}_{j_1}, a_{j_2j_2s_1}) = \underbrace{([2(n_{j_1u_1 1} - n_{j_2s_1 1}) + 1]^{\text{th}}\text{UP}, (2n_{j_2s_1 1})^{\text{th}}\text{LS})}_{([2(n_{j_1u_1 1} - n_{j_2s_1 1})+1],(2n_{j_2s_1 1}))\text{-up-parabola lower-saddle}}.$$

The infinite-equilibrium is a $([2(n_{j_1u_1 1} - n_{j_2s_1 1}) + 1]^{\text{th}}\text{UP}, (2n_{j_2s_1 1})^{\text{th}}\text{LS})$-up-parabola lower-saddle at $x^*_{j_2} = a_{j_1j_2u_1} = a_{j_2j_2s_1}$.

(ii_{2a1d}) For $n_{j_1u_1} = 2n_{j_1u_1 1}$ and $n_{j_2s_1} = 2n_{j_2s_1 1}$, the infinite-equilibrium of $x^*_{j_2} = a_{j_1j_2u_1} = a_{j_2j_2s_1}$ with $\overline{x}_{j_1} \neq a_{j_1j_1r_1}$ has the following properties as in Eqs. (1.195)–(1.198).

- For $a_{j_1j_10}\prod_{r=1}^{p_1}(\overline{x}_{j_1} - a_{j_1j_1r})^{m_{j_1r}}\prod_{u_2=1,u_2\neq u_1}^{q_1}(a_{j_1j_2u_1} - a_{j_1j_2u_2})^{n_{j_1u_2}} > 0$ and $a_{j_2j_20}\prod_{s_2=1,s_2\neq s_1}^{q_2}(a_{j_2j_2s_1} - a_{j_2j_2s_2})^{n_{j_2s_2}} > 0$,

$$(\overline{x}_{j_1}, a_{j_2j_2s_1}) = \underbrace{([2(n_{j_1u_1 1} - n_{j_2s_1 1})]^{\text{th}}\text{II}, (2n_{j_2s_1 1})]^{\text{th}}\text{US})}_{([2(n_{j_1u_1 1}-n_{j_2s_1 1})],(2n_{j_2s_1 1})])\text{-increasing inflection upper-saddle}}.$$

The infinite-equilibrium is a $([2(n_{j_1u_1 1}-n_{j_2s_1 1})]^{\text{th}}\text{II}, (2n_{j_2s_1 1})^{\text{th}}\text{US})$-increasing inflection upper-saddle at $x^*_{j_2} = a_{j_1j_2u_1} = a_{j_2j_2s_1}$.

- For $a_{j_1j_10}\prod_{r=1}^{p_1}(\overline{x}_{j_1} - a_{j_1j_1r})^{m_{j_1r}}\prod_{u_2=1,u_2\neq u_1}^{q_1}(a_{j_1j_2u_1} - a_{j_1j_2u_2})^{n_{j_1u_2}} > 0$ and $a_{j_2j_20}\prod_{s_2=1,s_2\neq s_1}^{q_2}(a_{j_2j_2s_1} - a_{j_2j_2s_2})^{n_{j_2s_2}} > 0$,

$$(\overline{x}_{j_1}, a_{j_2j_2s_1}) = \underbrace{([2(n_{j_1u_1 1} - n_{j_2s_1 1})]^{\text{th}}\text{DI}, (2n_{j_2s_1 1})^{\text{th}}\text{US})}_{([2(n_{j_1u_1 1}-n_{j_2s_1 1})],(2n_{j_2s_1 1}))\text{-decreasing inflection upper-saddle}}.$$

The infinite-equilibrium is a $([2(n_{j_1u_1 1}-n_{j_2s_1 1})]^{\text{th}}\text{DI}, (2n_{j_2s_1 1})^{\text{th}}\text{US})$-decreasing inflection upper-saddle at $x^*_{j_2} = a_{j_1j_2u_1} = a_{j_2j_2s_1}$.

- For $a_{j_1j_10}\prod_{r=1}^{p_1}(\overline{x}_{j_1} - a_{j_1j_1r})^{m_{j_1r}}\prod_{u_2=1,u_2\neq u_1}^{q_1}(a_{j_1j_2u_1} - a_{j_1j_2u_2})^{n_{j_1u_2}} > 0$ and $a_{j_2j_20}\prod_{s_2=1,s_2\neq s_1}^{q_2}(a_{j_2j_2s_1} - a_{j_2j_2s_2})^{n_{j_2s_2}} < 0$,

$$(\overline{x}_{j_1}, a_{j_2j_2s_1}) = \underbrace{([2(n_{j_1u_1 1} - n_{j_2s_1 1})]^{\text{th}}\text{DI}, (2n_{j_2s_1 1})^{\text{th}}\text{LS})}_{([2(n_{j_1u_1 1}-n_{j_2s_1 1})],(2n_{j_2s_1 1})])\text{-increasing-to-decreasing inflection lower-saddle}}.$$

The infinite-equilibrium is a $([2(n_{j_1u_1 1}-n_{j_2s_1 1})]^{\text{th}}\text{DI}, (2n_{j_2s_1 1})^{\text{th}}\text{LS})$-decreasing inflection lower-saddle at $x^*_{j_2} = a_{j_1j_2u_1} = a_{j_2j_2s_1}$.

- For $a_{j_1j_10}\prod_{r=1}^{p_1}(\overline{x}_{j_1} - a_{j_1j_1r})^{m_{j_1r}}\prod_{u_2=1,u_2\neq u_1}^{q_1}(a_{j_1j_2u_1} - a_{j_1j_2u_2})^{n_{j_1u_2}} < 0$ and $a_{j_2j_20}\prod_{s_2=1,s_2\neq s_1}^{q_2}(a_{j_2j_2s_1} - a_{j_2j_2s_2})^{n_{j_2s_2}} < 0$,

$$(\overline{x}_{j_1}, a_{j_2j_2s_1}) = \underbrace{([2(n_{j_1u_1 1} - n_{j_2s_1 1})]^{\text{th}}\text{II}, (2n_{j_2s_1 1})^{\text{th}}\text{LS})}_{([2(n_{j_1u_1 1}-n_{j_2s_1 1})],(2n_{j_2s_1 1})])\text{-increasing inflection lower-saddle}}.$$

The infinite-equilibrium is a $([2(n_{j_1u_1 1}-n_{j_2s_1 1})]^{\text{th}}\text{II}, (2n_{j_2s_1 1})^{\text{th}}\text{LS})$-increasing inflection lower-saddle at $x^*_{j_2} = a_{j_1j_2u_1} = a_{j_2j_2s_1}$.

(ii_{2a2}) At $x^*_{j_1} = a_{j_1j_1r_1}$, the variational equation for $\overline{x}_{j_2} \neq a_{j_1j_2u}$ ($u = 1, 2, \cdots, q_1$) is

$$\Delta\dot{x}_{j_1} = a_{j_1j_10}\prod_{r_2=1,r_2\neq r_1}^{p_1}(a_{j_1j_1r_1} - a_{j_1j_1r_2})^{m_{j_1r_2}}\prod_{u=1}^{q_1}(\overline{x}_{j_1} - a_{j_1j_2u})^{n_{j_1u}}(\Delta x_{j_1})^{m_{j_1r_1}}.$$

For

$$a_{j_1j_10}\prod\nolimits_{r_2=1,r_2\neq r_1}^{p_1}(a_{j_1j_1r_1}-a_{j_1j_1r_2})^{m_{j_1r_2}}\prod\nolimits_{u=1}^{q_1}(\bar{x}_{j_1}-a_{j_1j_2u})^{n_{j_1u}}>0 \text{ and}$$
$$a_{j_1j_10}\prod\nolimits_{r_2=1,r_2\neq r_1}^{p_1}(a_{j_1j_1r_1}-a_{j_1j_1r_2})^{m_{j_1r_2}}\prod\nolimits_{u=1}^{q_1}(\bar{x}_{j_1}-a_{j_1j_2u})^{n_{j_1u}}<0,$$

the flows at $x^*_{j_1}=a_{j_1j_1r_1}$ and $\bar{x}_{j_2}\neq a_{j_1j_2u}$ are $(2m_{j_1r_1 1}+1)^{\text{th}}$-source and sink flows for $m_{j_1r_1}=2m_{j_1r_1 1}+1$; and $(2m_{11})^{\text{th}}$-upper saddle and lower-saddle flows for $m_{j_1r_1}=2m_{j_1r_1 1}$.

On both sides of $\bar{x}_{j_2}=a_{j_1j_1 1}$, the inflection or parabola flows in case ($\text{ii}_{2\text{a}1}$) will be changed from increasing to decreasing or up to down, and vice versa for $m_{j_1r_1}=(2m_{j_1r_1 1}+1)$, and the inflection or parabola flows will not be changed between increasing and decreasing or between up and down for $m_{j_1r_1}=2m_{j_1r_1 1}$.

Therefore, from ($\text{ii}_{2\text{a}1}$) at $\bar{x}_{j_1}=a_{j_1j_1r_1}$, with the above analysis, there are 32 types of equilibrium $(x^*_{j_1},x^*_{j_2})=(a_{j_1j_1r_1},a_{j_2j_2s_1})$ on the infinite-equilibrium of $x^*_{j_2}=a_{j_1j_2u_1}=a_{j_2j_2s_1}$.

($\text{ii}_{2\text{a}2\text{a}}$) For $m_{j_1r_1}=2m_{j_1r_1 1}+1$, $n_{j_1u_1}=2n_{j_1u_1 1}+1$, and $n_{j_2s_1}=2n_{j_2s_1 1}+1$, the equilibrium of $(x^*_{j_1},x^*_{j_2})=(a_{j_1j_1r_1},a_{j_2j_2s_1})$ on the infinite-equilibrium $x^*_{j_2}=a_{j_1j_2u_1}=a_{j_2j_2s_1}$ has the following properties as in Eqs. (1.199)–(1.202).

- For $a_{j_1j_10}\prod_{r_2=1,r_2\neq r_1}^{p_1}(a_{j_1j_1r_1}-a_{j_1j_1r_2})^{m_{j_1r_2}}\prod_{u_2=1,u_2\neq u_1}^{q_1}(a_{j_1j_2u_1}-a_{j_1j_2u_2})^{n_{j_1u_2}}>0$ and $a_{j_2j_20}\prod_{s_2=1,s_2\neq s_1}^{q_2}(a_{j_2j_2s_1}-a_{j_2j_2s_2})^{n_{j_2s_2}}>0$,

$$\begin{aligned}&(a_{j_1j_1r_1},a_{j_2j_2s_1})\\&=\underbrace{([2(n_{j_1u_1 1}-n_{j_2s_1 1})]^{\text{th}}\text{DI:II},[(2m_{j_1r_1 1}+1),(2n_{j_2s_1 1}+1)]^{\text{th}}\text{SO})}_{([2(n_{j_1u_1 1}-n_{j_2s_1 1})],[(2m_{j_1r_1 1}+1),(2n_{j_2s_1 1}+1)])\text{-decreasing-to-increasing-inflection source}}.\end{aligned}$$

  The equilibrium is a $([2(n_{j_1u_1 1}-n_{j_2s_1 1})]^{\text{th}}\text{DI:II},[(2m_{j_1r_1 1}+1),(2n_{j_2s_1 1}+1)]^{\text{th}}\text{SO})$-decreasing-to-increasing-inflection source at $(x^*_{j_1},x^*_{j_2})=(a_{j_1j_1r_1},a_{j_2j_2s_1})$.
- For $a_{j_1j_10}\prod_{r_2=1,r_2\neq r_1}^{p_1}(a_{j_1j_1r_1}-a_{j_1j_1r_2})^{m_{j_1r_2}}\prod_{u_2=1,u_2\neq u_1}^{q_1}(a_{j_1j_2u_1}-a_{j_1j_2u_2})^{n_{j_1u_2}}<0$ and $a_{j_2j_20}\prod_{s_2=1,s_2\neq s_1}^{q_2}(a_{j_2j_2s_1}-a_{j_2j_2s_2})^{n_{j_2s_2}}>0$,

$$\begin{aligned}&(a_{j_1j_1r_1},a_{j_2j_2s_1})\\&=\underbrace{([2(n_{j_1u_1 1}-n_{j_2s_1 1})]^{\text{th}}\text{II:DI},[(2m_{j_1r_1 1}+1),(2n_{j_2s_1 1}+1)]^{\text{th}}\text{SO})}_{([2(n_{j_1u_1 1}-n_{j_2s_1 1})],[(2m_{j_1r_1 1}+1),(2n_{j_2s_1 1}+1)])\text{-increasing-to-decreasing-inflection source}}.\end{aligned}$$

  The equilibrium is a $([2(n_{j_1u_1 1}-n_{j_2s_1 1})]^{\text{th}}\text{II:DI},[(2m_{j_1r_1 1}+1),(2n_{j_2s_1 1}+1)]^{\text{th}}\text{SO})$-increasing-to-decreasing-inflection source at $(x^*_{j_1},x^*_{j_2})=(a_{j_1j_1r_1},a_{j_2j_2s_1})$.
- For $a_{j_1j_10}\prod_{r_2=1,r_2\neq r_1}^{p_1}(a_{j_1j_1r_1}-a_{j_1j_1r_2})^{m_{j_1r_2}}\prod_{u_2=1,u_2\neq u_1}^{q_1}(a_{j_1j_2u_1}-a_{j_1j_2u_2})^{n_{j_1u_2}}>0$ and $a_{j_2j_20}\prod_{s_2=1,s_2\neq s_1}^{q_2}(a_{j_2j_2s_1}-a_{j_2j_2s_2})^{n_{j_2s_2}}<0$,

$$\begin{aligned}&(a_{j_1j_1r_1}, a_{j_2j_2s_1})\\&= \underbrace{([2(n_{j_1u_1 1}-n_{j_2s_1 1})]^{\rm th}\text{II:DI},[(2m_{j_1r_1 1}+1),(2n_{j_2s_1 1}+1)]^{\rm th}\text{SI})}_{([2(n_{j_1u_1 1}-n_{j_2s_1 1})],[(2m_{j_1r_1 1}+1),(2n_{j_2s_1 1}+1)])\text{-increasing-to-decreasing-inflection sink}}.\end{aligned}$$

- For $a_{j_1j_10}\prod_{r_2=1,r_2\neq r_1}^{p_1}(a_{j_1j_1r_1}-a_{j_1j_1r_2})^{m_{j_1r_2}}\prod_{u_2=1,u_2\neq u_1}^{q_1}(a_{j_1j_2u_1}-a_{j_1j_2u_2})^{n_{j_1u_2}}<0$ and $a_{j_2j_20}\prod_{s_2=1,s_2\neq s_1}^{q_2}(a_{j_2j_2s_1}-a_{j_2j_2s_2})^{n_{j_2s_2}}<0$,

$$\begin{aligned}&(a_{j_1j_1r_1}, a_{j_2j_2s_1})\\&= \underbrace{([2(n_{j_1u_1 1}-n_{j_2s_1 1})]^{\rm th}\text{DI:II},[(2m_{j_1r_1 1}+1),(2n_{j_2s_1 1}+1)]^{\rm th}\text{SI})}_{([2(n_{j_1u_1 1}-n_{j_2s_1 1})],[(2m_{j_1r_1 1}+1),(2n_{j_2s_1 1}+1)])\text{-decreasing-to-increasing-inflection sink}}.\end{aligned}$$

The equilibrium is a $([2(n_{j_1u_1 1}-n_{j_2s_1 1})]^{\rm th}\text{DI:II},[(2m_{j_1r_1 1}+1),(2n_{j_2s_1 1}+1)]^{\rm th}\text{SI})$-decreasing-to-increasing-inflection sink at $(x_{j_1}^*, x_{j_2}^*)=(a_{j_1j_1r_1}, a_{j_2j_2s_1})$.

(ii_{2a2b}) For $m_{j_1r_1}=2m_{j_1r_1 1}+1, n_{j_1u_1}=2n_{j_1u_1 1}$, and $n_{j_2s_1}=2n_{j_2s_1 1}+1$, the equilibrium of $(x_{j_1}^*, x_{j_2}^*)=(a_{j_1j_1r_1}, a_{j_2j_2s_1})$ on the infinite-equilibrium $x_{j_2}^*=a_{j_1j_2u_1}=a_{j_2j_2s_1}$ has the following properties as in Eqs. (1.203)–(1.206).

- For $a_{j_1j_10}\prod_{r_2=1,r_2\neq r_1}^{p_1}(a_{j_1j_1r_1}-a_{j_1j_1r_2})^{m_{j_1r_2}}\prod_{u_2=1,u_2\neq u_1}^{q_1}(a_{j_1j_2u_1}-a_{j_1j_2u_2})^{n_{j_1u_2}}>0$ and $a_{j_2j_20}\prod_{s_2=1,s_2\neq s_1}^{q_2}(a_{j_2j_2s_1}-a_{j_2j_2s_2})^{n_{j_2s_2}}>0$,

$$\begin{aligned}&(a_{j_1j_1r_1}, a_{j_2j_2s_1})\\&= \underbrace{([2(n_{j_1u_1 1}-n_{j_2s_1 1})-1]^{\rm th}\text{DP:UP},[(2m_{j_1r_1 1}+1),(2n_{j_2s_1 1}+1)]^{\rm th}\text{SO})}_{([2(n_{j_1u_1 1}-n_{j_2s_1 1})-1],[(2m_{j_1r_1 1}+1),(2n_{j_2s_1 1}+1)])\text{-hyperbolic source}}.\end{aligned}$$

The equilibrium is a $([2(n_{j_1u_1 1}-n_{j_2s_1 1})-1]^{\rm th}\text{DP:UP},[(2m_{j_1r_1 1}+1),(2n_{j_2s_1 1}+1)]^{\rm th}$ SO)-hyperbolic source at $x_{j_2}^*=a_{j_1j_2u_1}=a_{j_2j_2s_1}$.
- For $a_{j_1j_10}\prod_{r_2=1,r_2\neq r_1}^{p_1}(a_{j_1j_1r_1}-a_{j_1j_1r_2})^{m_{j_1r_2}}\prod_{u_2=1,u_2\neq u_1}^{q_1}(a_{j_1j_2u_1}-a_{j_1j_2u_2})^{n_{j_1u_2}}<0$ and $a_{j_2j_20}\prod_{s_2=1,s_2\neq s_1}^{q_2}(a_{j_2j_2s_1}-a_{j_2j_2s_2})^{n_{j_2s_2}}>0$,

$$\begin{aligned}&(a_{j_1j_1r_1}, a_{j_2j_2s_1})\\&= \underbrace{([2(n_{j_1u_1 1}-n_{j_2s_1 1})-1]^{\rm th}\text{UP:DP},[(2m_{j_1r_1 1}+1),(2n_{j_2s_1 1}+1)]^{\rm th}\text{SO})}_{([2(n_{j_1u_1 1}-n_{j_2s_1 1})-1],[(2m_{j_1r_1 1}+1),(2n_{j_2s_1 1}+1)])\text{-hyperbolic - secant source}}.\end{aligned}$$

The equilibrium is a $([2(n_{j_1u_1 1}-n_{j_2s_1 1})-1]^{\rm th}\text{UP:DP},[(2m_{j_1r_1 1}+1),(2n_{j_2s_1 1}+1)]^{\rm th}$ SO)-hyperbolic-secant source at $(x_{j_1}^*, x_{j_2}^*)=(a_{j_1j_1r_1}, a_{j_2j_2s_1})$.
- For $a_{j_1j_10}\prod_{r_2=1,r_2\neq r_1}^{p_1}(a_{j_1j_1r_1}-a_{j_1j_1r_2})^{m_{j_1r_2}}\prod_{u_2=1,u_2\neq u_1}^{q_1}(a_{j_1j_2u_1}-a_{j_1j_2u_2})^{n_{j_1u_2}}>0$ and $a_{j_2j_20}\prod_{s_2=1,s_2\neq s_1}^{q_2}(a_{j_2j_2s_1}-a_{j_2j_2s_2})^{n_{j_2s_2}}<0$,

$$(a_{j_1j_1r_1}, a_{j_2j_2s_1}) = \underbrace{([2(n_{j_1u_1}1 - n_{j_2s_1}1) - 1]^{\text{th}}\text{UP:DP,}[(2m_{j_1r_1}1 + 1), (2n_{j_2s_1}1 + 1)]^{\text{th}}\text{SI})}_{([2(n_{j_1u_1}1-n_{j_2s_1}1)-1],[(2m_{j_1r_1}1+1),(2n_{j_2s_1}1+1)])\text{-hyperbolic - secant sink}}.$$

The equilibrium is a $([2(n_{j_1u_1}1-n_{j_2s_1}1)-1]^{\text{th}}$UP:DP,$[(2m_{j_1r_1}1+1), (2n_{j_2s_1}1+1)]^{\text{th}}$ SI)-hyperbolic-secant sink at $(x_{j_1}^*, x_{j_2}^*) = (a_{j_1j_1r_1}, a_{j_2j_2s_1})$.

- For $a_{j_1j_10}\prod_{r_2=1,r_2\neq r_1}^{p_1}(a_{j_1j_1r_1} - a_{j_1j_1r_2})^{m_{j_1r_2}}\prod_{u_2=1,u_2\neq u_1}^{q_1}(a_{j_1j_2u_1} - a_{j_1j_2u_2})^{n_{j_1u_2}} < 0$ and $a_{j_2j_20}\prod_{s_2=1,s_2\neq s_1}^{q_2}(a_{j_2j_2s_1} - a_{j_2j_2s_2})^{n_{j_2s_2}} < 0$,

$$(a_{j_1j_1r_1}, a_{j_2j_2s_1}) = \underbrace{([2(n_{j_1u_1}1 - n_{j_2s_1}1) - 1]^{\text{th}}\text{DP:UP,}[(2m_{j_1r_1}1 + 1), (2n_{j_2s_1}1 + 1)]^{\text{th}}\text{SI})}_{([2(n_{j_1u_1}1-n_{j_2s_1}1)-1],[(2m_{j_1r_1}1+1),(2n_{j_2s_1}1+1)])\text{-hyperbolic sink}}.$$

The equilibrium is a $([2(n_{j_1u_1}1-n_{j_2s_1}1)-1]^{\text{th}}$DP:UP,$[(2m_{j_1r_1}1+1), (2n_{j_2s_1}1+1)]^{\text{th}}$ SI)-hyperbolic sink at $(x_{j_1}^*, x_{j_2}^*) = (a_{j_1j_1r_1}, a_{j_2j_2s_1})$.

(ii_{2a2c}) For $m_{j_1r_1} = 2m_{j_1r_1}1+1$, $n_{j_1u_1} = 2n_{j_1u_1}1+1$, and $n_{j_2s_1} = 2n_{j_2s_1}1$, the equilibrium of $(x_{j_1}^*, x_{j_2}^*) = (a_{j_1j_1u_1}, a_{j_2j_2s_1})$ on the infinite-equilibrium $x_{j_2}^* = a_{j_1j_2u_1} = a_{j_2j_2s_1}$ has the following properties as in Eqs. (1.207)–(1.210).

- For $a_{j_1j_10}\prod_{r_2=1,r_2\neq r_1}^{p_1}(a_{j_1j_1r_1} - a_{j_1j_1r_2})^{m_{j_1r_2}}\prod_{u_2=1,u_2\neq u_1}^{q_1}(a_{j_1j_2u_1} - a_{j_1j_2u_2})^{n_{j_1u_2}} > 0$ and $a_{j_2j_20}\prod_{s_2=1,s_2\neq s_1}^{q_2}(a_{j_2j_2s_1} - a_{j_2j_2s_2})^{n_{j_2s_2}} > 0$,

$$(a_{j_1j_1r_1}, a_{j_2j_2s_1}) = \underbrace{([2(n_{j_1u_1}1 - n_{j_2s_1}1) + 1]^{\text{th}}\text{DP:UP,}[(2m_{j_1r_1}1 + 1), (2n_{j_2s_1}1)]^{\text{th}}\text{US})}_{([2(n_{j_1u_1}1-n_{j_2s_1}1)+1],[(2m_{j_1r_1}1+1),(2n_{j_2s_1}1)])\text{-hyperbolic upper-saddle}}.$$

The equilibrium is a $([2(n_{j_1u_1}1 - n_{j_2s_1}1) + 1]^{\text{th}}$DP:UP,$[(2m_{j_1r_1}1 + 1), (2n_{j_2s_1}1)]^{\text{th}}$ US)-hyperbolic upper-saddle at $(x_{j_1}^*, x_{j_2}^*) = (a_{j_1j_1r_1}, a_{j_2j_2s_1})$.

- For $a_{j_1j_10}\prod_{r_2=1,r_2\neq r_1}^{p_1}(a_{j_1j_1r_1} - a_{j_1j_1r_2})^{m_{j_1r_2}}\prod_{u_2=1,u_2\neq u_1}^{q_1}(a_{j_1j_2u_1} - a_{j_1j_2u_2})^{n_{j_1u_2}} < 0$ and $a_{j_2j_20}\prod_{s_2=1,s_2\neq s_1}^{q_2}(a_{j_2j_2s_1} - a_{j_2j_2s_2})^{n_{j_2s_2}} > 0$,

$$(a_{j_1j_1r_1}, a_{j_2j_2s_1}) = \underbrace{([2(n_{j_1u_1}1 - n_{j_2s_1}1) + 1]^{\text{th}}\text{UP:DP,}[(2m_{j_1r_1}1 + 1), (2n_{j_2s_1}1)]^{\text{th}}\text{US})}_{([2(n_{j_1u_1}1-n_{j_2s_1}1)+1],[(2m_{j_1r_1}1+1),(2n_{j_2s_1}1)])\text{-hyperbolic - secant upper-saddle}}.$$

The equilibrium is a $([2(n_{j_1u_1}1 - n_{j_2s_1}1) + 1]^{\text{th}}$UP:DP,$[(2m_{j_1r_1}1 + 1), (2n_{j_2s_1}1)]^{\text{th}}$ US)-hyperbolic-secant upper-saddle at $(x_{j_1}^*, x_{j_2}^*) = (a_{j_1j_1r_1}, a_{j_2j_2s_1})$.

- For $a_{j_1j_10}\prod_{r_2=1,r_2\neq r_1}^{p_1}(a_{j_1j_1r_1} - a_{j_1j_1r_2})^{m_{j_1r_2}}\prod_{u_2=1,u_2\neq u_1}^{q_1}(a_{j_1j_2u_1} - a_{j_1j_2u_2})^{n_{j_1u_2}} > 0$ and $a_{j_2j_20}\prod_{s_2=1,s_2\neq s_1}^{q_2}(a_{j_2j_2s_1} - a_{j_2j_2s_2})^{n_{j_2s_2}} < 0$,

$$(a_{j_1j_1r_1}, a_{j_2j_2s_1}) = \underbrace{([2(n_{j_1u_1}1 - n_{j_2s_1}1) + 1]^{\text{th}}\text{UP:DP},[(2m_{j_1r_1}1 + 1), (2n_{j_2s_1}1)]^{\text{th}}\text{LS})}_{([2(n_{j_1u_1}1-n_{j_2s_1}1)+1],[(2m_{j_1r_1}1+1),(2n_{j_2s_1}1)])\text{-hyperbolic - secant lower-saddle}}.$$

The equilibrium is a $([2(n_{j_1u_1}1 - n_{j_2s_1}1) + 1]^{\text{th}}\text{UP:DP},[(2m_{j_1r_1}1 + 1), (2n_{j_2s_1}1)]^{\text{th}}$ LS)-hyperbolic-secant lower-saddle at $(x^*_{j_1}, x^*_{j_2}) = (a_{j_1j_1r_1}, a_{j_2j_2s_1})$.

- For $a_{j_1j_10}\prod_{r_2=1,r_2\neq r_1}^{p_1}(a_{j_1j_1r_1} - a_{j_1j_1r_2})^{m_{j_1r_2}}\prod_{u_2=1,u_2\neq u_1}^{q_1}(a_{j_1j_2u_1} - a_{j_1j_2u_2})^{n_{j_1u_2}} < 0$ and $a_{j_2j_20}\prod_{s_2=1,s_2\neq s_1}^{q_2}(a_{j_2j_2s_1} - a_{j_2j_2s_2})^{n_{j_2s_2}} < 0$,

$$(a_{j_1j_1r_1}, a_{j_2j_2s_1}) = \underbrace{([2(n_{j_1u_1}1 - n_{j_2s_1}1) + 1]^{\text{th}}\text{DP:UP},[(2m_{j_1r_1}1 + 1), (2n_{j_2s_1}1)]^{\text{th}}\text{LS})}_{([2(n_{j_1u_1}1-n_{j_2s_1}1)+1],[(2m_{j_1r_1}1+1),(2n_{j_2s_1}1)])\text{-hyperbolic lower-saddle}}.$$

The equilibrium is a $([2(n_{j_1u_1}1 - n_{j_2s_1}1) + 1]^{\text{th}}\text{DP:UP},[(2m_{j_1r_1}1 + 1), (2n_{j_2s_1}1)]^{\text{th}}$ LS)-hyperbolic lower-saddle at $(x^*_{j_1}, x^*_{j_2}) = (a_{j_1j_1r_1}, a_{j_2j_2s_1})$.

(ii_{2a2d}) For $m_{j_1r_1} = 2m_{j_1r_1}1 + 1$, $n_{j_1u_1} = 2n_{j_1u_1}1$, and $n_{j_2s_1} = 2n_{j_2s_1}1$, the equilibrium of $(x^*_{j_1}, x^*_{j_2}) = (a_{j_1j_1r_1}, a_{j_2j_2s_1})$ on the infinite-equilibrium $x^*_{j_2} = a_{j_1j_2u_1} = a_{j_2j_2s_1}$ has the following properties as in Eqs. (1.211)–(1.214).

- For $a_{j_1j_10}\prod_{r_2=1,r_2\neq r_1}^{p_1}(a_{j_1j_1r_1} - a_{j_1j_1r_2})^{m_{j_1r_2}}\prod_{u_2=1,u_2\neq u_1}^{q_1}(a_{j_1j_2u_1} - a_{j_1j_2u_2})^{n_{j_1u_2}} > 0$ and $a_{j_2j_20}\prod_{s_2=1,s_2\neq s_1}^{q_2}(a_{j_2j_2s_1} - a_{j_2j_2s_2})^{n_{j_2s_2}} > 0$,

$$(a_{j_1j_1r_1}, a_{j_2j_2s_1}) = \underbrace{([2(n_{j_1u_1}1 - n_{j_2s_1}1)]^{\text{th}}\text{DI:II},[(2m_{j_1r_1}1 + 1), (2n_{j_2s_1}1)]^{\text{th}}\text{US})}_{([2(n_{j_1u_1}1-n_{j_2s_1}1)],[(2m_{j_1r_1}1+1),(2n_{j_2s_1}1)])\text{-decreasing-to-increasing inflection upper-saddle}}.$$

The equilibrium is a $([2(n_{j_1u_1}1 - n_{j_2s_1}1)]^{\text{th}}\text{DI:II},[(2m_{j_1r_1}1 + 1), (2n_{j_2s_1}1)]^{\text{th}}\text{US})$-decreasing-to-increasing inflection upper-saddle at $(x^*_{j_1}, x^*_{j_2}) = (a_{j_1j_1r_1}, a_{j_2j_2s_1})$.

- For $a_{j_1j_10}\prod_{r_2=1,r_2\neq r_1}^{p_1}(a_{j_1j_1r_1} - a_{j_1j_1r_2})^{m_{j_1r_2}}\prod_{u_2=1,u_2\neq u_1}^{q_1}(a_{j_1j_2u_1} - a_{j_1j_2u_2})^{n_{j_1u_2}} < 0$ and $a_{j_2j_20}\prod_{s_2=1,s_2\neq s_1}^{q_2}(a_{j_2j_2s_1} - a_{j_2j_2s_2})^{n_{j_2s_2}} > 0$,

$$(a_{j_1j_1r_1}, a_{j_2j_2s_1}) = \underbrace{([2(n_{j_1u_1}1 - n_{j_2s_1}1)]^{\text{th}}\text{II:DI},[(2m_{j_1r_1}1 + 1), (2n_{j_2s_1}1)]^{\text{th}}\text{US})}_{([2(n_{j_1u_1}1-n_{j_2s_1}1)],[(2m_{j_1r_1}1+1),(2n_{j_2s_1}1)])\text{-increasing-to-decreasing inflection upper-saddle}}.$$

The equilibrium is a $([2(n_{j_1u_1}1 - n_{j_2s_1}1)]^{\text{th}}\text{II:DI},[(2m_{j_1r_1}1 + 1), (2n_{j_2s_1}1)]^{\text{th}}\text{US})$-increasing-to-decreasing inflection upper-saddle at $(x^*_{j_1}, x^*_{j_2}) = (a_{j_1j_1r_1}, a_{j_2j_2s_1})$.

- For $a_{j_1j_10}\prod_{r_2=1,r_2\neq r_1}^{p_1}(a_{j_1j_1r_1} - a_{j_1j_1r_2})^{m_{j_1r_2}}\prod_{u_2=1,u_2\neq u_1}^{q_1}(a_{j_1j_2u_1} - a_{j_1j_2u_2})^{n_{j_1u_2}} > 0$ and $a_{j_2j_20}\prod_{s_2=1,s_2\neq s_1}^{q_2}(a_{j_2j_2s_1} - a_{j_2j_2s_2})^{n_{j_2s_2}} < 0$,

$$(a_{j_1j_1r_1}, a_{j_2j_2s_1}) = \underbrace{([2(n_{j_1u_1}1 - n_{j_2s_1}1)]^{\text{th}}\text{II:DI},[(2m_{j_1r_1}1+1),(2n_{j_2s_1}1)]^{\text{th}}\text{LS})}_{([2(n_{j_1u_1}1-n_{j_2s_1}1)],[(2m_{j_1r_1}1+1),(2n_{j_2s_1}1)])\text{-increasing-to-decreasing inflection lower-saddle}}.$$

The equilibrium is a $([2(n_{j_1u_1}1 - n_{j_2s_1}1)]^{\text{th}}\text{II:DI},[(2m_{j_1r_1}1+1),(2n_{j_2s_1}1)]^{\text{th}}\text{LS})$-increasing-to-decreasing inflection lower-saddle at $(x_{j_1}^*, x_{j_2}^*) = (a_{j_1j_1r_1}, a_{j_2j_2s_1})$.

- For $a_{j_1j_10}\prod_{r_2=1,r_2\neq r_1}^{p_1}(a_{j_1j_1r_1}-a_{j_1j_1r_2})^{m_{j_1r_2}}\prod_{u_2=1,u_2\neq u_1}^{q_1}(a_{j_1j_2u_1}-a_{j_1j_2u_2})^{n_{j_1u_2}}<0$ and $a_{j_2j_20}\prod_{s_2=1,s_2\neq s_1}^{q_2}(a_{j_2j_2s_1}-a_{j_2j_2s_2})^{n_{j_2s_2}}<0$,

$$(a_{j_1j_1r_1}, a_{j_2j_2s_1}) = \underbrace{([2(n_{j_1u_1}1 - n_{j_2s_1}1)]^{\text{th}}\text{DI:II},[(2m_{j_1r_1}1+1),(2n_{j_2s_1}1)]^{\text{th}}\text{LS})}_{([2(n_{j_1u_1}1-n_{j_2s_1}1)],[(2m_{j_1r_1}1+1),(2n_{j_2s_1}1)])\text{-decreasing-to-increasing inflection lower-saddle}}.$$

The equilibrium is a $([2(n_{j_1u_1}1 - n_{j_2s_1}1)]^{\text{th}}\text{DI:II},[(2m_{j_1r_1}1+1),(2n_{j_2s_1}1)]^{\text{th}}\text{LS})$-decreasing-to-increasing inflection lower-saddle at $(x_{j_1}^*, x_{j_2}^*) = (a_{j_1j_1r_1}, a_{j_2j_2s_1})$.

(ii_{2a2e}) For $m_{j_1r_1} = 2m_{j_1r_1}1$, $n_{j_1u_1} = 2n_{j_1u_1}1+1$, and $n_{j_2s_1} = 2n_{j_2s_1}1+1$, the equilibrium of $(x_{j_1}^*, x_{j_2}^*) = (a_{j_1j_1r_1}, a_{j_2j_2s_1})$ on the infinite-equilibrium $x_{j_2}^* = a_{j_1j_2u_1} = a_{j_2j_2s_1}$ has the following properties as in Eqs. (1.215)–(1.218).

- For $a_{j_1j_10}\prod_{r_2=1,r_2\neq r_1}^{p_1}(a_{j_1j_1r_1}-a_{j_1j_1r_2})^{m_{j_1r_2}}\prod_{u_2=1,u_2\neq u_1}^{q_1}(a_{j_1j_2u_1}-a_{j_1j_2u_2})^{n_{j_1u_2}}>0$ and $a_{j_2j_20}\prod_{s_2=1,s_2\neq s_1}^{q_2}(a_{j_2j_2s_1}-a_{j_2j_2s_2})^{n_{j_2s_2}}>0$,

$$(a_{j_1j_1r_1}, a_{j_2j_2s_1}) = \underbrace{([2(n_{j_1u_1}1 - n_{j_2s_1}1)]^{\text{th}}\text{II:II},[(2m_{j_1r_1}1),(2n_{j_2s_1}1+1)]^{\text{th}}\text{SO})}_{([2(n_{j_1u_1}1-n_{j_2s_1}1)],[(2m_{j_1r_1}1),(2n_{j_2s_1}1+1)])\text{-increasing-inflection source}}.$$

The equilibrium is a $([2(n_{j_1u_1}1 - n_{j_2s_1}1)]^{\text{th}}\text{II:II},[(2m_{j_1r_1}1),(2n_{j_2s_1}1+1)]^{\text{th}}\text{SO})$-increasing-inflection source at $(x_{j_1}^*, x_{j_2}^*) = (a_{j_1j_1r_1}, a_{j_2j_2s_1})$.

- For $a_{j_1j_10}\prod_{r_2=1,r_2\neq r_1}^{p_1}(a_{j_1j_1r_1}-a_{j_1j_1r_2})^{m_{j_1r_2}}\prod_{u_2=1,u_2\neq u_1}^{q_1}(a_{j_1j_2u_1}-a_{j_1j_2u_2})^{n_{j_1u_2}}<0$ and $a_{j_2j_20}\prod_{s_2=1,s_2\neq s_1}^{q_2}(a_{j_2j_2s_1}-a_{j_2j_2s_2})^{n_{j_2s_2}}>0$,

$$(a_{j_1j_1r_1}, a_{j_2j_2s_1}) = \underbrace{([2(n_{j_1u_1}1 - n_{j_2s_1}1)]^{\text{th}}\text{DI:DI},[(2m_{j_1r_1}1),(2n_{j_2s_1}1+1)]^{\text{th}}\text{SO})}_{([2(n_{j_1u_1}1-n_{j_2s_1}1)],[(2m_{j_1r_1}1),(2n_{j_2s_1}1+1)])\text{-decreasing - inflection source}}.$$

The equilibrium is a $([2(n_{j_1u_1}1 - n_{j_2s_1}1)]^{\text{th}}\text{DI:DI},[(2m_{j_1r_1}1),(2n_{j_2s_1}1+1)]^{\text{th}}\text{SO})$-decreasing-inflection source at $(x_{j_1}^*, x_{j_2}^*) = (a_{j_1j_1r_1}, a_{j_2j_2s_1})$.

- For $a_{j_1j_10}\prod_{r_2=1,r_2\neq r_1}^{p_1}(a_{j_1j_1r_1}-a_{j_1j_1r_2})^{m_{j_1r_2}}\prod_{u_2=1,u_2\neq u_1}^{q_1}(a_{j_1j_2u_1}-a_{j_1j_2u_2})^{n_{j_1u_2}}>0$ and $a_{j_2j_20}\prod_{s_2=1,s_2\neq s_1}^{q_2}(a_{j_2j_2s_1}-a_{j_2j_2s_2})^{n_{j_2s_2}}<0$,

$$(a_{j_1j_1r_1}, a_{j_2j_2s_1}) = \underbrace{([2(n_{j_1u_1}1 - n_{j_2s_1}1)]^{\text{th}}\text{DI:DI},[(2m_{j_1r_1}1),(2n_{j_2s_1}1+1)]^{\text{th}}\text{SI})}_{([2(n_{j_1u_1}1-n_{j_2s_1}1)],[(2m_{j_1r_1}1),(2n_{j_2s_1}1+1)])\text{-decreasing - inflection sink}}.$$

The equilibrium is a $([2(n_{j_1u_1 1}-n_{j_2s_1 1})]^{\text{th}}\text{DI:DI},[(2m_{j_1r_1 1}),(2n_{j_2s_1 1}+1)]^{\text{th}}\text{SI})$-decreasing-inflection sink at $(x^*_{j_1},x^*_{j_2})=(a_{j_1j_1r_1},a_{j_2j_2s_1})$.

- For $a_{j_1j_1 0}\prod_{r_2=1,r_2\neq r_1}^{p_1}(a_{j_1j_1r_1}-a_{j_1j_1r_2})^{m_{j_1r_2}}\prod_{u_2=1,u_2\neq u_1}^{q_1}(a_{j_1j_2u_1}-a_{j_1j_2u_2})^{n_{j_1u_2}}<0$ and $a_{j_2j_2 0}\prod_{s_2=1,s_2\neq s_1}^{q_2}(a_{j_2j_2s_1}-a_{j_2j_2s_2})^{n_{j_2s_2}}<0$,

$$(a_{j_1j_1r_1},a_{j_2j_2s_1})=\underbrace{([2(n_{j_1u_1 1}-n_{j_2s_1 1})]^{\text{th}}\text{II:II},[(2m_{j_1r_1 1}),(2n_{j_2s_1 1}+1)]^{\text{th}}\text{SI}])}_{([2(n_{j_1u_1 1}-n_{j_2s_1 1})],[(2m_{j_1r_1 1}),(2n_{j_2s_1 1}+1)])\text{-increasing-inflection sink}}.$$

The equilibrium is a $([2(n_{j_1u_1 1}-n_{j_2s_1 1})]^{\text{th}}\text{II:II},[(2m_{j_1r_1 1}),(2n_{j_2s_1 1}+1)]^{\text{th}}\text{SI})$-increasing-inflection sink at $(x^*_{j_1},x^*_{j_2})=(a_{j_1j_1r_1},a_{j_2j_2s_1})$.

($\text{ii}_{2\text{a}2\text{f}}$) For $m_{j_1r_1}=2m_{j_1r_1 1}$, $n_{j_1u_1}=2n_{j_1u_1 1}$, and $n_{j_2s_1}=2n_{j_2s_1 1}+1$, the equilibrium of $(x^*_{j_1},x^*_{j_2})=(a_{j_1j_1r_1},a_{j_2j_2s_1})$ on the infinite-equilibrium $x^*_{j_2}=a_{j_1j_2u_1}=a_{j_2j_2s_1}$ has the following properties as in Eqs. (1.219)–(1.222).

- For $a_{j_1j_1 0}\prod_{r_2=1,r_2\neq r_1}^{p_1}(a_{j_1j_1r_1}-a_{j_1j_1r_2})^{m_{j_1r_2}}\prod_{u_2=1,u_2\neq u_1}^{q_1}(a_{j_1j_2u_1}-a_{j_1j_2u_2})^{n_{j_1u_2}}>0$ and $a_{j_2j_2 0}\prod_{s_2=1,s_2\neq s_1}^{q_2}(a_{j_2j_2s_1}-a_{j_2j_2s_2})^{n_{j_2s_2}}>0$,

$$(a_{j_1j_1r_1},a_{j_2j_2s_1})=\underbrace{([2(n_{j_1u_1 1}-n_{j_2s_1 1})-1]^{\text{th}}\text{UP:UP},[(2m_{j_1r_1 1}),(2n_{j_2s_1 1}+1)]^{\text{th}}\text{SO})}_{([2(n_{j_1u_1 1}-n_{j_2s_1 1})-1],[(2m_{j_1r_1 1}),(2n_{j_2s_1 1}+1)])\text{-up-parabola source}}.$$

The equilibrium is a $([2(n_{j_1u_1 1}-n_{j_2s_1 1})-1]^{\text{th}}\text{UP:UP},[(2m_{j_1r_1 1}),(2n_{j_2s_1 1}+1)]^{\text{th}}\text{SO})$-up-parabola source at $(x^*_{j_1},x^*_{j_2})=(a_{j_1j_1r_1},a_{j_2j_2s_1})$.

- For $a_{j_1j_1 0}\prod_{r_2=1,r_2\neq r_1}^{p_1}(a_{j_1j_1r_1}-a_{j_1j_1r_2})^{m_{j_1r_2}}\prod_{u_2=1,u_2\neq u_1}^{q_1}(a_{j_1j_2u_1}-a_{j_1j_2u_2})^{n_{j_1u_2}}<0$ and $a_{j_2j_2 0}\prod_{s_2=1,s_2\neq s_1}^{q_2}(a_{j_2j_2s_1}-a_{j_2j_2s_2})^{n_{j_2s_2}}>0$,

$$(a_{j_1j_1r_1},a_{j_2j_2s_1})=\underbrace{([2(n_{j_1u_1 1}-n_{j_2s_1 1})-1]^{\text{th}}\text{DP:DP},[(2m_{j_1r_1 1}),(2n_{j_2s_1 1}+1)]^{\text{th}}\text{SO})}_{([2(n_{j_1u_1 1}-n_{j_2s_1 1})-1],[(2m_{j_1r_1 1}),(2n_{j_2s_1 1}+1)])\text{-down-parabola source}}.$$

The equilibrium is a $([2(n_{j_1u_1 1}-n_{j_2s_1 1})-1]^{\text{th}}\text{DP:DP},[(2m_{j_1r_1 1}),(2n_{j_2s_1 1}+1)]^{\text{th}}\text{SO})$-down-parabola source at $(x^*_{j_1},x^*_{j_2})=(a_{j_1j_1r_1},a_{j_2j_2s_1})$.

- For $a_{j_1j_1 0}\prod_{r_2=1,r_2\neq r_1}^{p_1}(a_{j_1j_1r_1}-a_{j_1j_1r_2})^{m_{j_1r_2}}\prod_{u_2=1,u_2\neq u_1}^{q_1}(a_{j_1j_2u_1}-a_{j_1j_2u_2})^{n_{j_1u_2}}>0$ and $a_{j_2j_2 0}\prod_{s_2=1,s_2\neq s_1}^{q_2}(a_{j_2j_2s_1}-a_{j_2j_2s_2})^{n_{j_2s_2}}<0$,

$$(a_{j_1j_1r_1},a_{j_2j_2s_1})=\underbrace{([2(n_{j_1u_1 1}-n_{j_2s_1 1})-1]^{\text{th}}\text{DP:DP},[(2m_{j_1r_1 1}),(2n_{j_2s_1 1}+1)]^{\text{th}}\text{SI})}_{([2(n_{j_1u_1 1}-n_{j_2s_1 1})-1],[(2m_{j_1r_1 1}),(2n_{j_2s_1 1}+1)])\text{-down-parabola sink}}.$$

The equilibrium is a $([2(n_{j_1u_1 1}-n_{j_2s_1 1})-1]^{\text{th}}\text{DP:DP},[(2m_{j_1r_1 1}),(2n_{j_2s_1 1}+1)]^{\text{th}}\text{SI})$-down-parabola sink at $(x^*_{j_1},x^*_{j_2})=(a_{j_1j_1r_1},a_{j_2j_2s_1})$.

- For $a_{j_1j_1 0}\prod_{r_2=1,r_2\neq r_1}^{p_1}(a_{j_1j_1r_1}-a_{j_1j_1r_2})^{m_{j_1r_2}}\prod_{u_2=1,u_2\neq u_1}^{q_1}(a_{j_1j_2u_1}-a_{j_1j_2u_2})^{n_{j_1u_2}}<0$ and $a_{j_2j_2 0}\prod_{s_2=1,s_2\neq s_1}^{q_2}(a_{j_2j_2s_1}-a_{j_2j_2s_2})^{n_{j_2s_2}}<0$,

$$(a_{j_1j_1r_1}, a_{j_2j_2s_1}) = \underbrace{([2(n_{j_1u_1 1} - n_{j_2s_1 1}) - 1]^{\text{th}}\text{UP:UP},[(2m_{j_1r_1 1}), (2n_{j_2s_1 1} + 1)]^{\text{th}}\text{SI})}_{([2(n_{j_1u_1 1}-n_{j_2s_1 1})-1],[(2m_{j_1r_1 1}),(2n_{j_2s_1 1}+1)])\text{-up-parabola sink}}.$$

The equilibrium is a $([2(n_{j_1u_1 1} - n_{j_2s_1 1}) - 1]^{\text{th}}\text{UP:UP},[(2m_{j_1r_1 1}), (2n_{j_2s_1 1} + 1)]^{\text{th}}$ SI)-up-parabola sink at $(x_{j_1}^*, x_{j_2}^*) = (a_{j_1j_1r_1}, a_{j_2j_2s_1})$.

(ii_{2a2g}) For $m_{j_1r_1} = 2m_{j_1r_1 1} + 1$, $n_{j_1u_1} = 2n_{j_1u_1 1}$, and $n_{j_2s_1} = 2n_{j_2s_1 1}$, the equilibrium of $(x_{j_1}^*, x_{j_2}^*) = (a_{j_1j_1r_1}, a_{j_2j_2s_1})$ on the infinite-equilibrium $x_{j_2}^* = a_{j_1j_2u_1} = a_{j_2j_2s_1}$ has the following properties as in Eqs. (1.223)–(1.226).

- For $a_{j_1j_10}\prod_{r_2=1,r_2\neq r_1}^{p_1}(a_{j_1j_1r_1} - a_{j_1j_1r_2})^{m_{j_1r_2}}\prod_{u_2=1,u_2\neq u_1}^{q_1}(a_{j_1j_2u_1} - a_{j_1j_2u_2})^{n_{j_1u_2}} > 0$ and $a_{j_2j_20}\prod_{s_2=1,s_2\neq s_1}^{q_2}(a_{j_2j_2s_1} - a_{j_2j_2s_2})^{n_{j_2s_2}} > 0$,

$$(a_{j_1j_1r_1}, a_{j_2j_2s_1}) = \underbrace{([2(n_{j_1u_1 1} - n_{j_2s_1 1}) + 1]^{\text{th}}\text{UP:UP},[(2m_{j_1r_1 1}), (2n_{j_2s_1 1})]^{\text{th}}\text{US})}_{([2(n_{j_1u_1 1}-n_{j_2s_1 1})+1],[(2m_{j_1r_1 1}),(2n_{j_2s_1 1})])\text{-up-parabola upper-saddle}}.$$

The equilibrium is a $([2(n_{j_1u_1 1} - n_{j_2s_1 1}) + 1]^{\text{th}}\text{UP:UP},[(2m_{j_1r_1 1}), (2n_{j_2s_1 1})]^{\text{th}}\text{US})$-up-parabola upper-saddle at $(x_{j_1}^*, x_{j_2}^*) = (a_{j_1j_1r_1}, a_{j_2j_2s_1})$.

- For $a_{j_1j_10}\prod_{r_2=1,r_2\neq r_1}^{p_1}(a_{j_1j_1r_1} - a_{j_1j_1r_2})^{m_{j_1r_2}}\prod_{u_2=1,u_2\neq u_1}^{q_1}(a_{j_1j_2u_1} - a_{j_1j_2u_2})^{n_{j_1u_2}} < 0$ and $a_{j_2j_20}\prod_{s_2=1,s_2\neq s_1}^{q_2}(a_{j_2j_2s_1} - a_{j_2j_2s_2})^{n_{j_2s_2}} > 0$,

$$(a_{j_1j_1r_1}, a_{j_2j_2s_1}) = \underbrace{([2(n_{j_1u_1 1} - n_{j_2s_1 1}) + 1]^{\text{th}}\text{DP:DP},[(2m_{j_1r_1 1}), (2n_{j_2s_1 1})]^{\text{th}}\text{US})}_{([2(n_{j_1u_1 1}-n_{j_2s_1 1})+1],[(2m_{j_1r_1 1}),(2n_{j_2s_1 1})])\text{-down-parabola upper-saddle}}.$$

The equilibrium is a $([2(n_{j_1u_1 1} - n_{j_2s_1 1}) + 1]^{\text{th}}\text{DP:DP},[(2m_{j_1r_1 1}), (2n_{j_2s_1 1})]^{\text{th}}\text{US})$-down-parabola upper-saddle at $(x_{j_1}^*, x_{j_2}^*) = (a_{j_1j_1r_1}, a_{j_2j_2s_1})$.

- For $a_{j_1j_10}\prod_{r_2=1,r_2\neq r_1}^{p_1}(a_{j_1j_1r_1} - a_{j_1j_1r_2})^{m_{j_1r_2}}\prod_{u_2=1,u_2\neq u_1}^{q_1}(a_{j_1j_2u_1} - a_{j_1j_2u_2})^{n_{j_1u_2}} > 0$ and $a_{j_2j_20}\prod_{s_2=1,s_2\neq s_1}^{q_2}(a_{j_2j_2s_1} - a_{j_2j_2s_2})^{n_{j_2s_2}} < 0$,

$$(a_{j_1j_1r_1}, a_{j_2j_2s_1}) = \underbrace{([2(n_{j_1u_1 1} - n_{j_2s_1 1}) + 1]^{\text{th}}\text{DP:DP},[(2m_{j_1r_1 1}), (2n_{j_2s_1 1})]^{\text{th}}\text{LS})}_{([2(n_{j_1u_1 1}-n_{j_2s_1 1})+1],[(2m_{j_1r_1 1}),(2n_{j_2s_1 1})])\text{-down-parabola lower-saddle}}.$$

The equilibrium is a $([2(n_{j_1u_1 1} - n_{j_2s_1 1}) + 1]^{\text{th}}\text{DP:DP},[(2m_{j_1r_1 1}), (2n_{j_2s_1 1})]^{\text{th}}\text{LS})$-down-parabola lower-saddle at $(x_{j_1}^*, x_{j_2}^*) = (a_{j_1j_1r_1}, a_{j_2j_2s_1})$.

- For $a_{j_1j_10}\prod_{r_2=1,r_2\neq r_1}^{p_1}(a_{j_1j_1r_1} - a_{j_1j_1r_2})^{m_{j_1r_2}}\prod_{u_2=1,u_2\neq u_1}^{q_1}(a_{j_1j_2u_1} - a_{j_1j_2u_2})^{n_{j_1u_2}} < 0$ and $a_{j_2j_20}\prod_{s_2=1,s_2\neq s_1}^{q_2}(a_{j_2j_2s_1} - a_{j_2j_2s_2})^{n_{j_2s_2}} < 0$,

$$(a_{j_1j_1r_1}, a_{j_2j_2s_1}) = \underbrace{([2(n_{j_1u_1 1} - n_{j_2s_1 1}) + 1]^{\text{th}}\text{UP:UP},[(2m_{j_1r_1 1}), (2n_{j_2s_1 1})]^{\text{th}}\text{LS})}_{([2(n_{j_1u_1 1}-n_{j_2s_1 1})+1],[(2m_{j_1r_1 1}),(2n_{j_2s_1 1})])\text{-up-parabola lower-saddle}}.$$

The equilibrium is a $([2(n_{j_1u_1 1} - n_{j_2s_1 1}) + 1]^{\text{th}}\text{UP:UP},[(2m_{j_1r_1 1}), (2n_{j_2s_1 1})]^{\text{th}}\text{LS})$-up-parabola lower-saddle at $(x^*_{j_1}, x^*_{j_2}) = (a_{j_1j_1r_1}, a_{j_2j_2s_1})$.

(ii_{2a2h}) For $m_{j_1r_1} = 2m_{j_1r_1 1}$, $n_{j_1u_1} = 2n_{j_1u_1 1}$, and $n_{j_2s_1} = 2n_{j_2s_1 1}$, the equilibrium of $(x^*_{j_1}, x^*_{j_2}) = (a_{j_1j_1r_1}, a_{j_2j_2s_1})$ on the infinite-equilibrium $x^*_{j_2} = a_{j_1j_2u_1} = a_{j_2j_2s_1}$ has the following properties as in Eqs. (1.227)–(1.230).

- For $a_{j_1j_1 0}\prod_{r_2=1,r_2\neq r_1}^{p_1}(a_{j_1j_1r_1} - a_{j_1j_1r_2})^{m_{j_1r_2}}\prod_{u_2=1,u_2\neq u_1}^{q_1}(a_{j_1j_2u_1} - a_{j_1j_2u_2})^{n_{j_1u_2}} > 0$ and $a_{j_2j_2 0}\prod_{s_2=1,s_2\neq s_1}^{q_2}(a_{j_2j_2s_1} - a_{j_2j_2s_2})^{n_{j_2s_2}} > 0$,

$$(a_{j_1j_1r_1}, a_{j_2j_2s_1}) = \underbrace{([2(n_{j_1u_1 1} - n_{j_2s_1 1})]^{\text{th}}\text{II:II},[(2m_{j_1r_1 1}), (2n_{j_2s_1 1})]^{\text{th}}\text{US})}_{([2(n_{j_1u_1 1}-n_{j_2s_1 1})],[(2m_{j_1r_1 1}),(2n_{j_2s_1 1})])\text{-increasing inflection upper-saddle}}.$$

The equilibrium is a $([2(n_{j_1u_1 1} - n_{j_2s_1 1})]^{\text{th}}\text{II:II},[(2m_{j_1r_1 1}), (2n_{j_2s_1 1})]^{\text{th}}\text{US})$-increasing inflection upper-saddle at $(x^*_{j_1}, x^*_{j_2}) = (a_{j_1j_1r_1}, a_{j_2j_2s_1})$.
- For $a_{j_1j_1 0}\prod_{r_2=1,r_2\neq r_1}^{p_1}(a_{j_1j_1r_1} - a_{j_1j_1r_2})^{m_{j_1r_2}}\prod_{u_2=1,u_2\neq u_1}^{q_1}(a_{j_1j_2u_1} - a_{j_1j_2u_2})^{n_{j_1u_2}} < 0$ and $a_{j_2j_2 0}\prod_{s_2=1,s_2\neq s_1}^{q_2}(a_{j_2j_2s_1} - a_{j_2j_2s_2})^{n_{j_2s_2}} > 0$,

$$(a_{j_1j_1r_1}, a_{j_2j_2s_1}) = \underbrace{([2(n_{j_1u_1 1} - n_{j_2s_1 1})]^{\text{th}}\text{DI:DI},[(2m_{j_1r_1 1}), (2n_{j_2s_1 1})]^{\text{th}}\text{US})}_{([2(n_{j_1u_1 1}-n_{j_2s_1 1})],[(2m_{j_1r_1 1}),(2n_{j_2s_1 1})])\text{-decreasing inflection upper-saddle}}.$$

The equilibrium is a $([2(n_{j_1u_1 1} - n_{j_2s_1 1})]^{\text{th}}\text{DI:DI},[(2m_{j_1r_1 1}), (2n_{j_2s_1 1})]^{\text{th}}\text{US})$-decreasing inflection upper-saddle at $(x^*_{j_1}, x^*_{j_2}) = (a_{j_1j_1r_1}, a_{j_2j_2s_1})$.
- For $a_{j_1j_1 0}\prod_{r_2=1,r_2\neq r_1}^{p_1}(a_{j_1j_1r_1} - a_{j_1j_1r_2})^{m_{j_1r_2}}\prod_{u_2=1,u_2\neq u_1}^{q_1}(a_{j_1j_2u_1} - a_{j_1j_2u_2})^{n_{j_1u_2}} > 0$ and $a_{j_2j_2 0}\prod_{s_2=1,s_2\neq s_1}^{q_2}(a_{j_2j_2s_1} - a_{j_2j_2s_2})^{n_{j_2s_2}} < 0$,

$$(a_{j_1j_1r_1}, a_{j_2j_2s_1}) = \underbrace{([2(n_{j_1u_1 1} - n_{j_2s_1 1})]^{\text{th}}\text{DI:DI},[(2m_{j_1r_1 1}), (2n_{j_2s_1 1})]^{\text{th}}\text{LS})}_{([2(n_{j_1u_1 1}-n_{j_2s_1 1})],[(2m_{j_1r_1 1}),(2n_{j_2s_1 1})])\text{-decreasing inflection lower-saddle}}.$$

The equilibrium is a $([2(n_{j_1u_1 1} - n_{j_2s_1 1})]^{\text{th}}\text{DI:DI},[(2m_{j_1r_1 1}), (2n_{j_2s_1 1})]^{\text{th}}\text{LS})$-decreasing inflection lower-saddle at $(x^*_{j_1}, x^*_{j_2}) = (a_{j_1j_1r_1}, a_{j_2j_2s_1})$
- For $a_{j_1j_1 0}\prod_{r_2=1,r_2\neq r_1}^{p_1}(a_{j_1j_1r_1} - a_{j_1j_1r_2})^{m_{j_1r_2}}\prod_{u_2=1,u_2\neq u_1}^{q_1}(a_{j_1j_2u_1} - a_{j_1j_2u_2})^{n_{j_1u_2}} < 0$ and $a_{j_2j_2 0}\prod_{s_2=1,s_2\neq s_1}^{q_2}(a_{j_2j_2s_1} - a_{j_2j_2s_2})^{n_{j_2s_2}} < 0$,

$$(a_{j_1j_1r_1}, a_{j_2j_2s_1}) = \underbrace{([2(n_{j_1u_1 1} - n_{j_2s_1 1})]^{\text{th}}\text{II:II},[(2m_{j_1r_1 1}), (2n_{j_2s_1 1})]^{\text{th}}\text{LS})}_{([2(n_{j_1u_1 1}-n_{j_2s_1 1})],[(2m_{j_1r_1 1}),(2n_{j_2s_1 1})])\text{-increasing inflection lower-saddle}}.$$

The equilibrium is a $([2(n_{j_1u_1 1} - n_{j_2s_1 1})]^{\text{th}}\text{II:II},[(2m_{j_1r_1 1}), (2n_{j_2s_1 1})]^{\text{th}}\text{LS})$-increasing inflection lower-saddle at $(x^*_{j_1}, x^*_{j_2}) = (a_{j_1j_1r_1}, a_{j_2j_2s_1})$.

(ii$_{2b}$) For, at and $\bar{x}_{j_1} \neq a_{j_1j_1r}$, if

$$\begin{aligned}
&\frac{dx_{j_1}}{dx_{j_2}}\bigg|_{x_{j2}^*=a_{j_2j_2s_1}\pm\varepsilon}\\
&=\lim_{\varepsilon\to 0}\frac{a_{j_1j_10}\prod_{r=1}^{p_1}(\bar{x}_{j_1}-a_{j_1j_1r})^{m_{j_1r}}\prod_{u_2=1,u_2\neq u_1}^{q_1}(a_{j_1j_2u_1}-a_{j_1j_2u_2})^{n_{j_1u_2}}}{a_{j_2j_20}(\pm\varepsilon)^{n_{j_2s_1}-n_{j_1u_1}}\prod_{s_2=1,s_2\neq s_1}^{q_2}(a_{j_2j_2s_1}-a_{j_sj_2s_2})^{n_{j_2s_2}}}\\
&=+\infty,
\end{aligned}$$

the flow at are a $[2(n_{j_2s_1 1}-n_{j_1u_1 1})]^{\text{th}}$-order increasing-concave flows for $n_{j_2s_1} = 2n_{j_2s_1 1}+1, n_{j_1u_1} = 2n_{j_1u_1 1}+1$ and $n_{j_2s_1} = 2n_{j_2s_1 1}, n_{j_1u_1} = 2n_{j_1u_1 1}$), if

$$\begin{aligned}
&\frac{dx_{j_1}}{dx_{j_2}}\bigg|_{x_{j2}^*=a_{j_2j_2s_1}\pm\varepsilon}\\
&=\lim_{\varepsilon\to 0}\frac{a_{j_1j_10}\prod_{r=1}^{p_1}(\bar{x}_{j_1}-a_{j_1j_1r})^{m_{j_1r}}\prod_{u_2=1,u_2\neq u_1}^{q_1}(a_{j_1j_2u_1}-a_{j_1j_2u_2})^{n_{j_1u_2}}}{a_{j_2j_20}(\pm\varepsilon)^{n_{j_2s_1}-n_{j_1u_1}}\prod_{s_2=1,s_2\neq s_1}^{q_2}(a_{j_2j_2s_1}-a_{j_sj_2s_2})^{n_{j_2s_2}}}\\
&=-\infty,
\end{aligned}$$

the flow at is a $[2(n_{j_2s_1 1}-n_{j_1u_1 1})]^{\text{th}}$-order decreasing-concave flow for $n_1 = 2n_{11}+1, n_2 = 2n_{21}+1$ and $n_1 = 2n_{11}, n_2 = 2n_{21}$).

For $x_{j2}^* = a_{j_2j_21} = a_{j_1j_21}$ and $\bar{x}_{j_1} \neq a_{j_1j_11}$, if

$$\begin{aligned}
&\frac{dx_{j_1}}{dx_{j_2}}\bigg|_{x_{j2}^*=a_{j_2j_2s_1}\pm\varepsilon}\\
&=\lim_{\varepsilon\to 0}\frac{a_{j_1j_10}\prod_{r=1}^{p_1}(\bar{x}_{j_1}-a_{j_1j_1r})^{m_{j_1r}}\prod_{u_2=1,u_2\neq u_1}^{q_1}(a_{j_1j_2u_1}-a_{j_1j_2u_2})^{n_{j_1u_2}}}{a_{j_2j_20}(\pm\varepsilon)^{n_{j_2s_1}-n_{j_1u_1}}\prod_{s_2=1,s_2\neq s_1}^{q_2}(a_{j_2j_2s_1}-a_{j_sj_2s_2})^{n_{j_2s_2}}}\\
&=\pm\infty,
\end{aligned}$$

the flow at is a $[2(n_{j_2s_1 1}-n_{j_1u_1 1})\pm 1]^{\text{th}}$-order up-asymptote flow for $n_{j_2s_1} = 2n_{j_2s_1 1}, n_{j_1u_1} = 2n_{j_1u_1 1}+1$ and $n_{j_2s_1} = 2n_{j_2s_1 1}+1, n_{j_1u_1} = 2n_{j_1u_1 1}$), if

$$\begin{aligned}
&\frac{dx_{j_1}}{dx_{j_2}}\bigg|_{x_{j2}^*=a_{j_2j_2s_1}\pm\varepsilon}\\
&=\lim_{\varepsilon\to 0}\frac{a_{j_1j_10}\prod_{r=1}^{p_1}(\bar{x}_{j_1}-a_{j_1j_1r})^{m_{j_1r}}\prod_{u_2=1,u_2\neq u_1}^{q_1}(a_{j_1j_2u_1}-a_{j_1j_2u_2})^{n_{j_1u_2}}}{a_{j_2j_20}(\pm\varepsilon)^{n_{j_2s_1}-n_{j_1u_1}}\prod_{s_2=1,s_2\neq s_1}^{q_2}(a_{j_2j_2s_1}-a_{j_sj_2s_2})^{n_{j_2s_2}}}\\
&=\mp\infty,
\end{aligned}$$

the flow at is a $[2(n_{j_2s_1 1}-n_{j_1u_1 1})\pm 1]^{\text{th}}$-order down-asymptote flow for $n_{j_2s_1} = 2n_{j_2s_1 1}, n_{j_1u_1} = 2n_{j_1u_1 1}+1$ and $n_{j_2s_1} = 2n_{j_2s_1 1}+1, n_{j_1u_1} = 2n_{j_1u_1 1}$).

The variational equation at is

$$\Delta \dot{x}_{j_2} = a_{j_2 j_2 0} \prod_{s_2=1, s_2 \neq s_1}^{q_2} (a_{j_2 j_2 s_1} - a_{j_s j_2 s_2})^{n_{j_2 s_2}} (\Delta x_{j_2})^{n_{j_2 s_1}}.$$

For

$$a_{j_2 j_2 0} \prod_{s_2=1, s_2 \neq s_1}^{q_2} (a_{j_2 j_2 s_1} - a_{j_s j_2 s_2})^{n_{j_2 s_2}} > 0, \text{ and}$$
$$a_{j_2 j_2 0} \prod_{s_2=1, s_2 \neq s_1}^{q_2} (a_{j_2 j_2 s_1} - a_{j_s j_2 s_2})^{n_{j_2 s_2}} < 0,$$

the flows at are $(2n_{j_2 s_1 1}+1)^{\text{th}}$ source and sink flows for $n_{j_2 s_1} = 2n_{j_2 s_1 1}+1$, and are $(2n_{j_2 s_1 1})^{\text{th}}$-upper-saddle and lower-saddle flows for $n_{j_2 s_1} = 2n_{j_2 s_1 1}$.

Therefore, the infinite-equilibrium of has the following properties.

$(\text{ii}_{2\text{b}1})$ For $n_{j_2 s_1} > n_{j_1 u_1}$, 16 types of infinite-equilibriums exist at $x_{j_2}^* = a_{j_1 j_1 r_1} = a_{j_2 j_2 s_1}$.

$(\text{ii}_{2\text{b}1\text{a}})$ For $n_{j_1 u_1} = 2n_{j_1 u_1 1}+1$ and $n_{j_2 s_1} = 2n_{j_2 s_1 1}+1$, the infinite-equilibrium of $x_{j_2}^* = a_{j_1 j_2 u_1} = a_{j_2 j_2 s_1}$ with $\overline{x}_{j_1} \neq a_{j_1 j_1 r_1}$ has the following properties as in Eqs. (1.231)–(1.234).

- For $a_{j_1 j_1 0} \prod_{r=1}^{p_1} (\overline{x}_{j_1} - a_{j_1 j_1 r})^{m_{j_1 r}} \prod_{u_2=1, u_2 \neq u_1}^{q_1} (a_{j_1 j_2 u_1} - a_{j_1 j_2 u_2})^{n_{j_1 u_2}} > 0$ and $a_{j_2 j_2 0} \prod_{s_2=1, s_2 \neq s_1}^{q_2} (a_{j_2 j_2 s_1} - a_{j_2 j_2 s_2})^{n_{j_2 s_2}} > 0$,

$$(\overline{x}_{j_1}, a_{j_2 j_2 s_1}) = \underbrace{([2(n_{j_2 s_1 1} - n_{j_1 u_1 1})]^{\text{th}}\text{IC}, (2n_{j_2 s_1 1}+1)^{\text{th}}\text{SO})}_{([2(n_{j_2 s_1 1} - n_{j_1 u_1 1})], (2n_{j_2 s_1 1}+1))\text{-increasing-concve source}}.$$

  The infinite-equilibrium is a $([2(n_{j_2 s_1 1} - n_{j_1 u_1 1})]^{\text{th}}\text{IC}, (2n_{j_2 s_1 1}+1)^{\text{th}}\text{SO})$-increasing-concave source at $x_{j_2}^* = a_{j_1 j_2 u_1} = a_{j_2 j_2 s_1}$.
- For $a_{j_1 j_1 0} \prod_{r=1}^{p_1} (\overline{x}_{j_1} - a_{j_1 j_1 r})^{m_{j_1 r}} \prod_{u_2=1, u_2 \neq u_1}^{q_1} (a_{j_1 j_2 u_1} - a_{j_1 j_2 u_2})^{n_{j_1 u_2}} < 0$ and $a_{j_2 j_2 0} \prod_{s_2=1, s_2 \neq s_1}^{q_2} (a_{j_2 j_2 s_1} - a_{j_2 j_2 s_2})^{n_{j_2 s_2}} > 0$,

$$(\overline{x}_{j_1}, a_{j_2 j_2 s_1}) = \underbrace{([[2(n_{j_2 s_1 1} - n_{j_1 u_1 1})]^{\text{th}}\text{DC}, (2n_{j_2 s_1 1}+1)^{\text{th}}\text{SO})}_{([2(n_{j_2 s_1 1} - n_{j_1 u_1 1})], (2n_{j_2 s_1 1}+1))\text{-decreasing - concave source}}.$$

  The infinite-equilibrium is a $([[2(n_{j_2 s_1 1} - n_{j_1 u_1 1})]^{\text{th}}\text{DC}, (2n_{j_2 s_1 1}+1)^{\text{th}}\text{SO})$-decreasing-concave source at $x_{j_2}^* = a_{j_1 j_2 u_1} = a_{j_2 j_2 s_1}$.
- For $a_{j_1 j_1 0} \prod_{r=1}^{p_1} (\overline{x}_{j_1} - a_{j_1 j_1 r})^{m_{j_1 r}} \prod_{u_2=1, u_2 \neq u_1}^{q_1} (a_{j_1 j_2 u_1} - a_{j_1 j_2 u_2})^{n_{j_1 u_2}} > 0$ and $a_{j_2 j_2 0} \prod_{s_2=1, s_2 \neq s_1}^{q_2} (a_{j_2 j_2 s_1} - a_{j_2 j_2 s_2})^{n_{j_2 s_2}} < 0$,

$$(\overline{x}_{j_1}, a_{j_2 j_2 s_1}) = \underbrace{([[2(n_{j_2 s_1 1} - n_{j_1 u_1 1})]^{\text{th}}\text{DC}, (2n_{j_2 s_1 1}+1)^{\text{th}}\text{SI})}_{([2(n_{j_2 s_1 1} - n_{j_1 u_1 1})], (2n_{j_2 s_1 1}+1))\text{-decreasing-concave sink}}.$$

The infinite-equilibrium is a $([2(n_{j_2 s_1 1} - n_{j_1 u_1 1})]^{\text{th}}\text{DC}, (2n_{j_2 s_1 1}+1)]^{\text{th}}\text{SI})$-decreasing-concave sink at $x^*_{j_2} = a_{j_1 j_2 u_1} = a_{j_2 j_2 s_1}$.

- For $a_{j_1 j_1 0}\prod_{r=1}^{p_1}(\overline{x}_{j_1} - a_{j_1 j_1 r})^{m_{j_1 r}}\prod_{u_2=1,u_2\neq u_1}^{q_1}(a_{j_1 j_2 u_1} - a_{j_1 j_2 u_2})^{n_{j_1 u_2}} < 0$ and $a_{j_2 j_2 0}\prod_{s_2=1,s_2\neq s_1}^{q_2}(a_{j_2 j_2 s_1} - a_{j_2 j_2 s_2})^{n_{j_2 s_2}} < 0$,

$$(\overline{x}_{j_1}, a_{j_2 j_2 s_1}) = \underbrace{([2(n_{j_2 s_1 1} - n_{j_1 u_1 1})]^{\text{th}}\text{IC}, (2n_{j_2 s_1 1} + 1)^{\text{th}}\text{SI})}_{([2(n_{j_2 s_1 1} - n_{j_1 u_1 1})], (2n_{j_2 s_1 1}+1))\text{-increasing-concave sink}}.$$

The infinite-equilibrium is a $([2(n_{j_1 u_1 1} - n_{j_2 s_1 1})]^{\text{th}}\text{II}, (2n_{j_2 s_1 1}+1)^{\text{th}}\text{SI})$-increasing-concave sink at $x^*_{j_2} = a_{j_1 j_2 u_1} = a_{j_2 j_2 s_1}$.

($\text{ii}_{2\text{b}1\text{b}}$) For $n_{j_1 u_1} = 2n_{j_1 u_1 1}$ and $n_{j_2 s_1} = 2n_{j_2 s_1 1} + 1$, the infinite-equilibrium of $x^*_{j_2} = a_{j_1 j_2 u_1} = a_{j_2 j_2 s_1}$ with $\overline{x}_{j_1} \neq a_{j_1 j_1 r_1}$ has the following properties as in Eqs. (1.235)–(1.238).

- For $a_{j_1 j_1 0}\prod_{r=1}^{p_1}(\overline{x}_{j_1} - a_{j_1 j_1 r})^{m_{j_1 r}}\prod_{u_2=1,u_2\neq u_1}^{q_1}(a_{j_1 j_2 u_1} - a_{j_1 j_2 u_2})^{n_{j_1 u_2}} > 0$ and $a_{j_2 j_2 0}\prod_{s_2=1,s_2\neq s_1}^{q_2}(a_{j_2 j_2 s_1} - a_{j_2 j_2 s_2})^{n_{j_2 s_2}} > 0$,

$$(\overline{x}_{j_1}, a_{j_2 j_2 s_1}) = \underbrace{([2(n_{j_2 s_1 1} - n_{j_1 u_1 1}) + 1]^{\text{th}}\text{DA}, (2n_{j_2 s_1 1} + 1)^{\text{th}}\text{SO})}_{([2(n_{j_2 s_1 1} - n_{j_1 u_1 1})+1], (2n_{j_2 s_1 1}+1))\text{-down-asymptote source}}.$$

The infinite-equilibrium is a $([2(n_{j_2 s_1 1} - n_{j_1 u_1 1})+1]^{\text{th}}\text{UP}, (2n_{j_2 s_1 1}+1)^{\text{th}}\text{SO})$-down-asymptote source at $x^*_{j_2} = a_{j_1 j_2 u_1} = a_{j_2 j_2 s_1}$.

- For $a_{j_1 j_1 0}\prod_{r=1}^{p_1}(\overline{x}_{j_1} - a_{j_1 j_1 r})^{m_{j_1 r}}\prod_{u_2=1,u_2\neq u_1}^{q_1}(a_{j_1 j_2 u_1} - a_{j_1 j_2 u_2})^{n_{j_1 u_2}} < 0$ and $a_{j_2 j_2 0}\prod_{s_2=1,s_2\neq s_1}^{q_2}(a_{j_2 j_2 s_1} - a_{j_2 j_2 s_2})^{n_{j_2 s_2}} > 0$,

$$(\overline{x}_{j_1}, a_{j_2 j_2 s_1}) = \underbrace{([2(n_{j_2 s_1 1} - n_{j_1 u_1 1}) + 1]^{\text{th}}\text{UA}, (2n_{j_2 s_1 1} + 1)^{\text{th}}\text{SO})}_{([2(n_{j_2 s_1 1} - n_{j_1 u_1 1})+1], (2n_{j_2 s_1 1}+1))\text{-up-asymptote source}}$$

The infinite-equilibrium is a $([2(n_{j_2 s_1 1} - n_{j_1 u_1 1})+1]^{\text{th}}\text{UA}, (2n_{j_2 s_1 1}+1)^{\text{th}}\text{SO})$-up asymptote source at $x^*_{j_2} = a_{j_1 j_2 u_1} = a_{j_2 j_2 s_1}$.

- For $a_{j_1 j_1 0}\prod_{r=1}^{p_1}(\overline{x}_{j_1} - a_{j_1 j_1 r})^{m_{j_1 r}}\prod_{u_2=1,u_2\neq u_1}^{q_1}(a_{j_1 j_2 u_1} - a_{j_1 j_2 u_2})^{n_{j_1 u_2}} > 0$ and $a_{j_2 j_2 0}\prod_{s_2=1,s_2\neq s_1}^{q_2}(a_{j_2 j_2 s_1} - a_{j_2 j_2 s_2})^{n_{j_2 s_2}} < 0$,

$$(\overline{x}_{j_1}, a_{j_2 j_2 s_1}) = \underbrace{([2(n_{j_2 s_1 1} - n_{j_1 u_1 1}) + 1]^{\text{th}}\text{UA}, (2n_{j_2 s_1 1} + 1)^{\text{th}}\text{SI})}_{([2(n_{j_2 s_1 1} - n_{j_1 u_1 1})+1], (2n_{j_2 s_1 1}+1))\text{-up-asymtote sink}}.$$

The infinite-equilibrium is a $([2(n_{j_2 s_1 1} - n_{j_1 u_1 1})+1]^{\text{th}}\text{UA}, (2n_{j_2 s_1 1}+1)^{\text{th}}\text{SI})$-up asymptote sink at $x^*_{j_2} = a_{j_1 j_2 u_1} = a_{j_2 j_2 s_1}$.

- For $a_{j_1 j_1 0}\prod_{r=1}^{p_1}(\overline{x}_{j_1} - a_{j_1 j_1 r})^{m_{j_1 r}}\prod_{u_2=1,u_2\neq u_1}^{q_1}(a_{j_1 j_2 u_1} - a_{j_1 j_2 u_2})^{n_{j_1 u_2}} < 0$ and $a_{j_2 j_2 0}\prod_{s_2=1,s_2\neq s_1}^{q_2}(a_{j_2 j_2 s_1} - a_{j_2 j_2 s_2})^{n_{j_2 s_2}} < 0$,

$$(\overline{x}_{j_1}, a_{j_2j_2s_1}) = \underbrace{([2(n_{j_2s_1}1 - n_{j_1u_1}1) + 1]^{\text{th}}\text{DA}, (2n_{j_2s_1}1 + 1)^{\text{th}}\text{SI})}_{([2(n_{j_2s_1}1 - n_{j_1u_1}1)+1],(2n_{j_2s_1}1+1))\text{-down-asymptote sink}}.$$

The infinite-equilibrium is a $([2(n_{j_2s_1}1 - n_{j_1u_1}1)+1]^{\text{th}}\text{DA}, (2n_{j_2s_1}1+1)^{\text{th}}\text{SI})$-down-asymptote sink at $x_{j_2}^* = a_{j_1j_2u_1} = a_{j_2j_2s_1}$.

($\text{ii}_{2\text{b}1\text{c}}$) For $n_{j_1u_1} = 2n_{j_1u_1}1 + 1$ and $n_{j_2s_1} = 2n_{j_2s_1}1$, the infinite-equilibrium of $x_{j_2}^* = a_{j_1j_2u_1} = a_{j_2j_2s_1}$ with $\overline{x}_{j_1} \neq a_{j_1j_1r_1}$ has the following properties as in Eqs. (1.239)–(1.242).

- For $a_{j_1j_10}\prod_{r=1}^{p_1}(\overline{x}_{j_1} - a_{j_1j_1r})^{m_{j_1r}}\prod_{u_2=1,u_2\neq u_1}^{q_1}(a_{j_1j_2u_1} - a_{j_1j_2u_2})^{n_{j_1u_2}} > 0$ and $a_{j_2j_20}\prod_{s_2=1,s_2\neq s_1}^{q_2}(a_{j_2j_2s_1} - a_{j_2j_2s_2})^{n_{j_2s_2}} > 0$,

$$(\overline{x}_{j_1}, a_{j_2j_2s_1}) = \underbrace{([2(n_{j_2s_1}1 - n_{j_1u_1}1) - 1]^{\text{th}}\text{DA}, (2n_{j_2s_1}1)^{\text{th}}\text{US})}_{([2(n_{j_2s_1}1 - n_{j_1u_1}1)-1],(2n_{j_2s_1}1))\text{-down-asymptote upper-saddle}}.$$

The infinite-equilibrium is a $([2(n_{j_2s_1}1 - n_{j_1u_1}1) - 1]^{\text{th}}\text{DA}, (2n_{j_2s_1}1)^{\text{th}}\text{US})$-down-asymptote upper-saddle at $x_{j_2}^* = a_{j_1j_2u_1} = a_{j_2j_2s_1}$.
- For $a_{j_1j_10}\prod_{r=1}^{p_1}(\overline{x}_{j_1} - a_{j_1j_1r})^{m_{j_1r}}\prod_{u_2=1,u_2\neq u_1}^{q_1}(a_{j_1j_2u_1} - a_{j_1j_2u_2})^{n_{j_1u_2}} < 0$ and $a_{j_2j_20}\prod_{s_2=1,s_2\neq s_1}^{q_2}(a_{j_2j_2s_1} - a_{j_2j_2s_2})^{n_{j_2s_2}} > 0$,

$$(\overline{x}_{j_1}, a_{j_2j_2s_1}) = \underbrace{([2(n_{j_2s_1}1 - n_{j_1u_1}1) - 1]^{\text{th}}\text{UA}, (2n_{j_2s_1}1)^{\text{th}}\text{US})}_{([2(n_{j_2s_1}1 - n_{j_1u_1}1)-1],(2n_{j_2s_1}1))\text{-up-asymptote upper-saddle}}.$$

The infinite-equilibrium is a $([2(n_{j_2s_1}1 - n_{j_1u_1}1) - 1]^{\text{th}}\text{UA}, (2n_{j_2s_1}1)^{\text{th}}\text{US})$-up-asymptote upper-saddle at $x_{j_2}^* = a_{j_1j_2u_1} = a_{j_2j_2s_1}$.
- For $a_{j_1j_10}\prod_{r=1}^{p_1}(\overline{x}_{j_1} - a_{j_1j_1r})^{m_{j_1r}}\prod_{u_2=1,u_2\neq u_1}^{q_1}(a_{j_1j_2u_1} - a_{j_1j_2u_2})^{n_{j_1u_2}} > 0$ and $a_{j_2j_20}\prod_{s_2=1,s_2\neq s_1}^{q_2}(a_{j_2j_2s_1} - a_{j_2j_2s_2})^{n_{j_2s_2}} < 0$,

$$(\overline{x}_{j_1}, a_{j_2j_2s_1}) = \underbrace{([2(n_{j_2s_1}1 - n_{j_1u_1}1) - 1]^{\text{th}}\text{UA}, (2n_{j_2s_1}1)^{\text{th}}\text{LS})}_{([2(n_{j_2s_1}1 - n_{j_1u_1}1)-1],(2n_{j_2s_1}1))\text{-up-asymptote lower-saddle}}.$$

The infinite-equilibrium is a $([2(n_{j_2s_1}1 - n_{j_1u_1}1) - 1]^{\text{th}}\text{UA}, (2n_{j_2s_1}1)^{\text{th}}\text{LS})$-up-asymptote lower-saddle at $x_{j_2}^* = a_{j_1j_2u_1} = a_{j_2j_2s_1}$.
- For $a_{j_1j_10}\prod_{r=1}^{p_1}(\overline{x}_{j_1} - a_{j_1j_1r})^{m_{j_1r}}\prod_{u_2=1,u_2\neq u_1}^{q_1}(a_{j_1j_2u_1} - a_{j_1j_2u_2})^{n_{j_1u_2}} < 0$ and $a_{j_2j_20}\prod_{s_2=1,s_2\neq s_1}^{q_2}(a_{j_2j_2s_1} - a_{j_2j_2s_2})^{n_{j_2s_2}} < 0$,

$$(\overline{x}_{j_1}, a_{j_2j_2s_1}) = \underbrace{([2(n_{j_2s_1}1 - n_{j_1u_1}1) - 1]^{\text{th}}\text{DA}, (2n_{j_2s_1}1)^{\text{th}}\text{LS})}_{([2(n_{j_2s_1}1 - n_{j_1u_1}1)-1],(2n_{j_2s_1}1))\text{-down-asymprote lower-saddle}}.$$

The infinite-equilibrium is a $([2(n_{j_2s_1}1 - n_{j_1u_1}1) - 1]^{\text{th}}\text{DA}, (2n_{j_2s_1}1)^{\text{th}}\text{LS})$-down-asymptote lower-saddle at $x_{j_2}^* = a_{j_1j_2u_1} = a_{j_2j_2s_1}$.

($\text{ii}_{2\text{b1d}}$) For $n_{j_1u_1} = 2n_{j_1u_1}1$ and $n_{j_2s_1} = 2n_{j_2s_1}1$, the infinite-equilibrium of $x^*_{j_2} = a_{j_1j_2u_1} = a_{j_2j_2s_1}$ with $\overline{x}_{j_1} \neq a_{j_1j_1r_1}$ has the following properties as in Eqs. (1.243)–(1.246).

- For $a_{j_1j_10}\prod_{r=1}^{p_1}(\overline{x}_{j_1} - a_{j_1j_1r})^{m_{j_1r}}\prod_{u_2=1,u_2\neq u_1}^{q_1}(a_{j_1j_2u_1} - a_{j_1j_2u_2})^{n_{j_1u_2}} > 0$ and $a_{j_2j_20}\prod_{s_2=1,s_2\neq s_1}^{q_2}(a_{j_2j_2s_1} - a_{j_2j_2s_2})^{n_{j_2s_2}} > 0$,

$$(\overline{x}_{j_1}, a_{j_2j_2s_1}) = \underbrace{([2(n_{j_2s_1}1 - n_{j_1u_1}1)]^{\text{th}}\text{IC}, (2n_{j_2s_1}1)]^{\text{th}}\text{US})}_{([2(n_{j_2s_1}1-n_{j_1u_1}1)],(2n_{j_2s_1}1)])\text{-increasing-concave upper-saddle}}.$$

The infinite-equilibrium is a $([2(n_{j_2s_1}1-n_{j_1u_1}1)]^{\text{th}}\text{IC},(2n_{j_2s_1}1)^{\text{th}}\text{US})$-increasing concave upper-saddle at $x^*_{j_2} = a_{j_1j_2u_1} = a_{j_2j_2s_1}$.
- For $a_{j_1j_10}\prod_{r=1}^{p_1}(\overline{x}_{j_1} - a_{j_1j_1r})^{m_{j_1r}}\prod_{u_2=1,u_2\neq u_1}^{q_1}(a_{j_1j_2u_1} - a_{j_1j_2u_2})^{n_{j_1u_2}} < 0$ and $a_{j_2j_20}\prod_{s_2=1,s_2\neq s_1}^{q_2}(a_{j_2j_2s_1} - a_{j_2j_2s_2})^{n_{j_2s_2}} > 0$,

$$(\overline{x}_{j_1}, a_{j_2j_2s_1}) = \underbrace{([2(n_{j_2s_1}1 - n_{j_1u_1}1)]^{\text{th}}\text{DC}, (2n_{j_2s_1}1)]^{\text{th}}\text{US})}_{([2(n_{j_2s_1}1-n_{j_1u_1}1)],(2n_{j_2s_1}1)])\text{-decreasing concave upper-saddle}}.$$

The infinite-equilibrium is a $([2(n_{j_2s_1}1-n_{j_1u_1}1)]^{\text{th}}\text{DC},(2n_{j_2s_1}1)^{\text{th}}\text{US})$-decreasing concave upper-saddle at $x^*_{j_2} = a_{j_1j_2u_1} = a_{j_2j_2s_1}$.
- For $a_{j_1j_10}\prod_{r=1}^{p_1}(\overline{x}_{j_1} - a_{j_1j_1r})^{m_{j_1r}}\prod_{u_2=1,u_2\neq u_1}^{q_1}(a_{j_1j_2u_1} - a_{j_1j_2u_2})^{n_{j_1u_2}} > 0$ and $a_{j_2j_20}\prod_{s_2=1,s_2\neq s_1}^{q_2}(a_{j_2j_2s_1} - a_{j_2j_2s_2})^{n_{j_2s_2}} < 0$,

$$(\overline{x}_{j_1}, a_{j_2j_2s_1}) = \underbrace{([2(n_{j_2s_1}1 - n_{j_1u_1}1)]^{\text{th}}\text{DC}, (2n_{j_2s_1}1)]^{\text{th}}\text{US})}_{([2(n_{j_2s_1}1-n_{j_1u_1}1)],(2n_{j_2s_1}1)])\text{-decreasing concave upper-saddle}}.$$

The infinite-equilibrium is a $([2(n_{j_2s_1}1-n_{j_1u_1}1)]^{\text{th}}\text{DC},(2n_{j_2s_1}1)^{\text{th}}\text{LS})$-decreasing concave lower-saddle at $x^*_{j_2} = a_{j_1j_2u_1} = a_{j_2j_2s_1}$.
- For $a_{j_1j_10}\prod_{r=1}^{p_1}(\overline{x}_{j_1} - a_{j_1j_1r})^{m_{j_1r}}\prod_{u_2=1,u_2\neq u_1}^{q_1}(a_{j_1j_2u_1} - a_{j_1j_2u_2})^{n_{j_1u_2}} < 0$ and $a_{j_2j_20}\prod_{s_2=1,s_2\neq s_1}^{q_2}(a_{j_2j_2s_1} - a_{j_2j_2s_2})^{n_{j_2s_2}} < 0$,

$$(\overline{x}_{j_1}, a_{j_2j_2s_1}) = \underbrace{([2(n_{j_2s_1}1 - n_{j_1u_1}1)]^{\text{th}}\text{IC}, (2n_{j_2s_1}1)]^{\text{th}}\text{LS})}_{([2(n_{j_2s_1}1-n_{j_1u_1}1)],(2n_{j_2s_1}1)])\text{-increasing inflection lower-saddle}}.$$

The infinite-equilibrium is a $([2(n_{j_2s_1}1-n_{j_1u_1}1)]^{\text{th}}\text{IC},(2n_{j_2s_1}1)^{\text{th}}\text{LS})$-increasing concave lower-saddle at $x^*_{j_2} = a_{j_1j_2u_1} = a_{j_2j_2s_1}$.

($\text{ii}_{2\text{b2}}$) At $x^*_{j_1} = a_{j_1j_1r_1}$, the variational equation for $\overline{x}_{j_2} \neq a_{j_1j_2u} = a_{j_2j_2s}$ is

$$\Delta\dot{x}_{j_1} = a_{j_1j_10}\prod_{r_2=1,r_2\neq r_1}^{p_1}(a_{j_1j_1r_1} - a_{j_1j_1r_2})^{m_{j_1r_2}}\prod_{u=1}^{q_1}(\overline{x}_{j_2} - a_{j_1j_2u})^{n_{j_1u}}(\Delta x_{j_1})^{m_{j_1r_1}}.$$

For

$$a_{j_1j_10}\prod\nolimits_{r_2=1,r_2\neq r_1}^{p_1}(a_{j_1j_1r_1}-a_{j_1j_1r_2})^{m_{j_1r_2}}\prod\nolimits_{u=1}^{q_1}(\overline{x}_{j_2}-a_{j_1j_2u})^{n_{j_1u}}>0,\ \text{and}$$
$$a_{j_1j_10}\prod\nolimits_{r_2=1,r_2\neq r_1}^{p_1}(a_{j_1j_1r_1}-a_{j_1j_1r_2})^{m_{j_1r_2}}\prod\nolimits_{u=1}^{q_1}(\overline{x}_{j_2}-a_{j_1j_2u})^{n_{j_1u}}<0,$$

the flows at $x^*_{j_1}=a_{j_1j_1r_1}$ and $\overline{x}_{j_2}\neq a_{j_2j_2s_1}=a_{j_1j_2u_1}$ are $(2m_{j_1r_1 1}+1)^{\text{th}}$-source and sink flows, and $(2m_{j_1r_11})^{\text{th}}$-upper saddle and lower-saddle flows.

On both sides of $\overline{x}_{j_1}=a_{j_1j_1r_1}$, the concave or asymptote flows in case (ii$_{2b1}$) will be changed from increasing to decreasing or up to down, and vice verso for $m_1=(2m_{11}+1)$, and the concave and asymptote flows will not be changed between increasing and decreasing or between up and down for $m_1=2m_{11}$.

Therefore, from case (ii$_{2b1}$) at $\overline{x}_{j_1}=a_{j_1j_1r_1}$ with the above analysis, there are 32 types of equilibrium on the infinite-equilibrium of $x^*_{j_2}=a_{j_1j_2u_1}=a_{j_2j_2s_1}$.

(ii$_{2b2a}$) For $m_{j_1r_1}=2m_{j_1r_11}+1$,and $n_{j_2s_1}=2n_{j_2s_11}+1$, the equilibrium of $(x^*_{j_1},x^*_{j_2})=(a_{j_1j_1r_1},a_{j_2j_2s_1})$ on the infinite-equilibrium $x^*_{j_2}=a_{j_1j_2u_1}=a_{j_2j_2s_1}$ has the following properties as in Eqs. (1.247)–(1.250)

- For $a_{j_1j_10}\prod_{r_2=1,r_2\neq r_1}^{p_1}(a_{j_1j_1r_1}-a_{j_1j_1r_2})^{m_{j_1r_2}}\prod_{u_2=1,u_2\neq u_1}^{q_1}(a_{j_1j_2u_1}-a_{j_1j_2u_2})^{n_{j_1u_2}}>0$ and $a_{j_2j_20}\prod_{s_2=1,s_2\neq s_1}^{q_2}(a_{j_2j_2s_1}-a_{j_2j_2s_2})^{n_{j_2s_2}}>0$,

$$(a_{j_1j_1r_1},a_{j_2j_2s_1})$$
$$=\underbrace{([2(n_{j_2s_11}-n_{j_1u_11})]^{\text{th}}\text{DC:IC},[(2m_{j_1r_11}+1),(2n_{j_2s_11}+1)]^{\text{th}}\text{SO})}_{([2(n_{j_2s_11}-n_{j_1u_11})],[(2m_{j_1r_11}+1),(2n_{j_2s_11}+1)])\text{-decreasing-to-increasing concave source}}.$$

  The equilibrium is a $([2(n_{j_2s_11}-n_{j_1u_11})]^{\text{th}}$DC:IC,$[(2m_{j_1r_11}+1),(2n_{j_2s_11}+1)]^{\text{th}}$ SO)-decreasing-to-increasing-concave source at $(x^*_{j_1},x^*_{j_2})=(a_{j_1j_1r_1},a_{j_2j_2s_1})$.
- For $a_{j_1j_10}\prod_{r_2=1,r_2\neq r_1}^{p_1}(a_{j_1j_1r_1}-a_{j_1j_1r_2})^{m_{j_1r_2}}\prod_{u_2=1,u_2\neq u_1}^{q_1}(a_{j_1j_2u_1}-a_{j_1j_2u_2})^{n_{j_1u_2}}<0$ and $a_{j_2j_20}\prod_{s_2=1,s_2\neq s_1}^{q_2}(a_{j_2j_2s_1}-a_{j_2j_2s_2})^{n_{j_2s_2}}>0$,

$$(a_{j_1j_1r_1},a_{j_2j_2s_1})$$
$$=\underbrace{([2(n_{j_2s_11}-n_{j_1u_11})]^{\text{th}}\text{IC:DC},\ [(2m_{j_1r_11}+1),(2n_{j_2s_11}+1)]^{\text{th}}\text{SO})}_{([2(n_{j_2s_11}-n_{j_1u_11})],[(2m_{j_1r_11}+1),(2n_{j_2s_11}+1)])\text{-increasing-to-decreasing concave source}}.$$

  The equilibrium is a $[2(n_{j_2s_11}-n_{j_1u_11})]^{\text{th}}$IC:DC, $[(2m_{j_1r_11}+1),(2n_{j_2s_11}+1)]^{\text{th}}$ SO)-increasing-to-decreasing-concave source at $(x^*_{j_1},x^*_{j_2})=(a_{j_1j_1r_1},a_{j_2j_2s_1})$.
- For $a_{j_1j_10}\prod_{r_2=1,r_2\neq r_1}^{p_1}(a_{j_1j_1r_1}-a_{j_1j_1r_2})^{m_{j_1r_2}}\prod_{u_2=1,u_2\neq u_1}^{q_1}(a_{j_1j_2u_1}-a_{j_1j_2u_2})^{n_{j_1u_2}}>0$ and $a_{j_2j_20}\prod_{s_2=1,s_2\neq s_1}^{q_2}(a_{j_2j_2s_1}-a_{j_2j_2s_2})^{n_{j_2s_2}}<0$,

$$
\begin{aligned}
&(a_{j_1j_1r_1}, a_{j_2j_2s_1}) \\
&= \underbrace{([2(n_{j_2s_1}1 - n_{j_1u_1}1)]^{\text{th}}\text{IC:DC}, [(2m_{j_1r_1}1+1), (2n_{j_2s_1}1+1)]^{\text{th}}\text{SI})}_{([2(n_{j_2s_1}1-n_{j_1u_1}1)],[(2m_{j_1r_1}1+1),(2n_{j_2s_1}1+1)])\text{-increasing-to-decreasing concave sink}}.
\end{aligned}
$$

The equilibrium is a ($[2(n_{j_2s_1}1 - n_{j_1u_1}1)]^{\text{th}}$IC:DC, $[(2m_{j_1r_1}1+1), (2n_{j_2s_1}1+1)]^{\text{th}}$ SI)-increasing-to-decreasing-concave sink at $(x^*_{j_1}, x^*_{j_2}) = (a_{j_1j_1r_1}, a_{j_2j_2s_1})$.

- For $a_{j_1j_10}\prod_{r_2=1,r_2\neq r_1}^{p_1}(a_{j_1j_1r_2} - a_{j_1j_1r_1})^{m_{j_1r_2}}\prod_{u_2=1,u_2\neq u_1}^{q_1}(a_{j_1j_2u_1} - a_{j_1j_2u_2})^{n_{j_1u_2}} < 0$ and $a_{j_2j_20}\prod_{s_2=1,s_2\neq s_1}^{q_2}(a_{j_2j_2s_1} - a_{j_2j_2s_2})^{n_{j_2s_2}} < 0$,

$$
\begin{aligned}
&(a_{j_1j_1r_1}, a_{j_2j_2s_1}) \\
&= \underbrace{([2(n_{j_2s_1}1 - n_{j_1u_1}1)]^{\text{th}}\text{DC:IC}, [(2m_{j_1r_1}1+1), (2n_{j_2s_1}1+1)]^{\text{th}}\text{SI})}_{([2(n_{j_2s_1}1-n_{j_1u_1}1)],[(2m_{j_1r_1}1+1),(2n_{j_2s_1}1+1)])\text{-decreasing-to-increasing concave sink}}.
\end{aligned}
$$

The equilibrium is a ($[2(n_{j_2s_1}1 - n_{j_1u_1}1)]^{\text{th}}$DC:IC, $[(2m_{j_1r_1}1+1), (2n_{j_2s_1}1+1)]^{\text{th}}$ SI)-decreasing-to-increasing-concave sink at $(x^*_{j_1}, x^*_{j_2}) = (a_{j_1j_1r_1}, a_{j_2j_2s_1})$.

(ii_{2b2b}) For $m_{j_1r_1} = 2m_{j_1r_1}1+1$,$n_{j_1u_1} = 2n_{j_1u_1}1$ and $n_{j_2s_1} = 2n_{j_2s_1}1+1$, the equilibrium of $(x^*_{j_1}, x^*_{j_2}) = (a_{j_1j_1r_1}, a_{j_2j_2s_1})$ on the infinite-equilibrium $x^*_{j_2} = a_{j_1j_2u_1} = a_{j_2j_2s_1}$ has the following properties as in Eqs. (1.251)–(1.254)

- For $a_{j_1j_10}\prod_{r_2=1,r_2\neq r_1}^{p_1}(a_{j_1j_1r_1} - a_{j_1j_1r_2})^{m_{j_1r_2}}\prod_{u_2=1,u_2\neq u_1}^{q_1}(a_{j_1j_2u_1} - a_{j_1j_2u_2})^{n_{j_1u_2}} > 0$ and $a_{j_2j_20}\prod_{s_2=1,s_2\neq s_1}^{q_2}(a_{j_2j_2s_1} - a_{j_2j_2s_2})^{n_{j_2s_2}} > 0$,

$$
\begin{aligned}
&(a_{j_1j_1r_1}, a_{j_2j_2s_1}) \\
&= \underbrace{([2(n_{j_2s_1}1 - n_{j_1u_1}1)+1]^{\text{th}}\text{UA:DA}, [(2m_{j_1r_1}1+1), (2n_{j_2s_1}1+1)]^{\text{th}}\text{SO})}_{([2(n_{j_2s_1}1-n_{j_1u_1}1)+1],[(2m_{j_1r_1}1+1),(2n_{j_2s_1}1+1)])\text{-up-to-down asymptote source}}.
\end{aligned}
$$

The equilibrium is a ($[2(n_{j_2s_1}1 - n_{j_1u_1}1)+1]^{\text{th}}$UA:DA, $[(2m_{j_1r_1}1+1), (2n_{j_2s_1}1+1)]^{\text{th}}$ SO)-up-to-down asymptote source at $(x^*_{j_1}, x^*_{j_2}) = (a_{j_1j_1r_1}, a_{j_2j_2s_1})$.

- For $a_{j_1j_10}\prod_{r_2=1,r_2\neq r_1}^{p_1}(a_{j_1j_1r_1} - a_{j_1j_1r_2})^{m_{j_1r_2}}\prod_{u_2=1,u_2\neq u_1}^{q_1}(a_{j_1j_2u_1} - a_{j_1j_2u_2})^{n_{j_1u_2}} < 0$ and $a_{j_2j_20}\prod_{s_2=1,s_2\neq s_1}^{q_2}(a_{j_2j_2s_1} - a_{j_2j_2s_2})^{n_{j_2s_2}} > 0$,

$$
\begin{aligned}
&(a_{j_1j_1r_1}, a_{j_2j_2s_1}) \\
&= \underbrace{([2(n_{j_2s_1}1 - n_{j_1u_1}1)+1]^{\text{th}}\text{DA:UA}, [(2m_{j_1r_1}1+1), (2n_{j_2s_1}1+1)]^{\text{th}}\text{SO})}_{([2(n_{j_2s_1}1-n_{j_1u_1}1)+1],[(2m_{j_1r_1}1+1),(2n_{j_2s_1}1+1)])\text{-down-to-up asymptote source}}.
\end{aligned}
$$

The equilibrium is a ($[2(n_{j_2s_1}1 - n_{j_1u_1}1)+1]^{\text{th}}$DA:UA,$[(2m_{j_1r_1}1+1), (2n_{j_2s_1}1+1)]^{\text{th}}$ SO)SO)-down-to-up asymptote source at $(x^*_{j_1}, x^*_{j_2}) = (a_{j_1j_1r_1}, a_{j_2j_2s_1})$.

- For $a_{j_1j_10}\prod_{r_2=1,r_2\neq r_1}^{p_1}(a_{j_1j_1r_1} - a_{j_1j_1r_2})^{m_{j_1r_2}}\prod_{u_2=1,u_2\neq u_1}^{q_1}(a_{j_1j_2u_1} - a_{j_1j_2u_2})^{n_{j_1u_2}} > 0$ and $a_{j_2j_20}\prod_{s_2=1,s_2\neq s_1}^{q_2}(a_{j_2j_2s_1} - a_{j_2j_2s_2})^{n_{j_2s_2}} < 0$,

$$(a_{j_1j_1r_1}, a_{j_2j_2s_1}) = \underbrace{([2(n_{j_2s_1 1} - n_{j_1u_1 1}) + 1]^{\text{th}}\text{DA:UA}, [(2m_{j_1r_1 1} + 1), (2n_{j_2s_1 1} + 1)]^{\text{th}}\text{SI})}_{([2(n_{j_2s_1 1}-n_{j_1u_1 1})+1],[(2m_{j_1r_1 1}+1),(2n_{j_2s_1 1}+1)])\text{-down-to-up asymptote sink}}.$$

The equilibrium is a $([2(n_{j_2s_1 1}-n_{j_1u_1 1})+1]^{\text{th}}$DA:UA,$[(2m_{j_1r_1 1}+1),(2n_{j_2s_1 1}+1)]^{\text{th}}$ SI)-down-to-up asymptote sink at $(x_{j_1}^*, x_{j_2}^*) = (a_{j_1j_1r_1}, a_{j_2j_2s_1})$.

- For $a_{j_1j_10}\prod_{r_2=1,r_2\neq r_1}^{p_1}(a_{j_1j_1r_1} - a_{j_1j_1r_2})^{m_{j_1r_2}}\prod_{u_2=1,u_2\neq u_1}^{q_1}(a_{j_1j_2u_1} - a_{j_1j_2u_2})^{n_{j_1u_2}} < 0$ and $a_{j_2j_20}\prod_{s_2=1,s_2\neq s_1}^{q_2}(a_{j_2j_2s_1} - a_{j_2j_2s_2})^{n_{j_2s_2}} < 0$,

$$(a_{j_1j_1r_1}, a_{j_2j_2s_1}) = \underbrace{([2(n_{j_2s_1 1} - n_{j_1u_1 1}) + 1]^{\text{th}}\text{UA:DA}, [(2m_{j_1r_1 1} + 1), (2n_{j_2s_1 1} + 1)]^{\text{th}}\text{SI})}_{([2(n_{j_2s_1 1}-n_{j_1u_1 1})+1],[(2m_{j_1r_1 1}+1),(2n_{j_2s_1 1}+1)])\text{-up-to-down asymptote sink}}.$$

The equilibrium is a $([2(n_{j_2s_1 1}-n_{j_1u_1 1})+1]^{\text{th}}$UA:DA,$[(2m_{j_1r_1 1}+1),(2n_{j_2s_1 1}+1)]^{\text{th}}$ SI)-up-to-down asymptote sink at $(x_{j_1}^*, x_{j_2}^*) = (a_{j_1j_1r_1}, a_{j_2j_2s_1})$.

(ii$_{2b2c}$) For $m_{j_1r_1} = 2m_{j_1r_1 1} + 1$,$n_{j_1u_1} = 2n_{j_1u_1 1} + 1$ and $n_{j_2s_1} = 2n_{j_2s_1 1}$, the equilibrium of $(x_{j_1}^*, x_{j_2}^*) = (a_{j_1j_1r_1}, a_{j_2j_2s_1})$ on the infinite-equilibrium $x_{j_2}^* = a_{j_1j_2u_1} = a_{j_2j_2s_1}$ has the following properties as in Eqs. (1.255)–(1.258).

- For $a_{j_1j_10}\prod_{r_2=1,r_2\neq r_1}^{p_1}(a_{j_1j_1r_1} - a_{j_1j_1r_2})^{m_{j_1r_2}}\prod_{u_2=1,u_2\neq u_1}^{q_1}(a_{j_1j_2u_1} - a_{j_1j_2u_2})^{n_{j_1u_2}} > 0$ and $a_{j_2j_20}\prod_{s_2=1,s_2\neq s_1}^{q_2}(a_{j_2j_2s_1} - a_{j_2j_2s_2})^{n_{j_2s_2}} > 0$,

$$(a_{j_1j_1r_1}, a_{j_2j_2s_1}) = \underbrace{([2(n_{j_2s_1 1} - n_{j_1u_1 1}) - 1]^{\text{th}}\text{UA:DA}, [(2m_{j_1r_1 1} + 1), (2n_{j_2s_1 1})]^{\text{th}}\text{US})}_{([2(n_{j_2s_1 1}-n_{j_1u_1 1})-1],[(2m_{j_1r_1 1}+1),(2n_{j_2s_1 1})])\text{-up-to-down asymptote upper-saddle}}.$$

The equilibrium is a $([2(n_{j_2s_1 1}-n_{j_1u_1 1})-1]^{\text{th}}$UA:DA, $[(2m_{j_1r_1 1}+1), (2n_{j_2s_1 1})]^{\text{th}}$ US)-up-to-down asymptote upper-saddle at $(x_{j_1}^*, x_{j_2}^*) = (a_{j_1j_1r_1}, a_{j_2j_2s_1})$.

- For $a_{j_1j_10}\prod_{r_2=1,r_2\neq r_1}^{p_1}(a_{j_1j_1r_1} - a_{j_1j_1r_2})^{m_{j_1r_2}}\prod_{u_2=1,u_2\neq u_1}^{q_1}(a_{j_1j_2u_1} - a_{j_1j_2u_2})^{n_{j_1u_2}} < 0$ and $a_{j_2j_20}\prod_{s_2=1,s_2\neq s_1}^{q_2}(a_{j_2j_2s_1} - a_{j_2j_2s_2})^{n_{j_2s_2}} > 0$,

$$(a_{j_1j_1r_1}, a_{j_2j_2s_1}) = \underbrace{([2(n_{j_2s_1 1} - n_{j_1u_1 1}) - 1]^{\text{th}}\text{DA:UA}, [(2m_{j_1r_1 1} + 1), (2n_{j_2s_1 1})]^{\text{th}}\text{US})}_{([2(n_{j_2s_1 1}-n_{j_1u_1 1})-1],[(2m_{j_1r_1 1}+1),(2n_{j_2s_1 1})])\text{-down-to-up asymptote upper-saddle}}.$$

The equilibrium is a $([2(n_{j_2s_1 1}-n_{j_1u_1 1})-1]^{\text{th}}$DA:UA, $[(2m_{j_1r_1 1}+1), (2n_{j_2s_1 1})]^{\text{th}}$ US)-down-to-up asymptote upper-saddle at $(x_{j_1}^*, x_{j_2}^*) = (a_{j_1j_1r_1}, a_{j_2j_2s_1})$.

- For $a_{j_1j_10}\prod_{r_2=1,r_2\neq r_1}^{p_1}(a_{j_1j_1r_1} - a_{j_1j_1r_2})^{m_{j_1r_2}}\prod_{u_2=1,u_2\neq u_1}^{q_1}(a_{j_1j_2u_1} - a_{j_1j_2u_2})^{n_{j_1u_2}} > 0$ and $a_{j_2j_20}\prod_{s_2=1,s_2\neq s_1}^{q_2}(a_{j_2j_2s_1} - a_{j_2j_2s_2})^{n_{j_2s_2}} < 0$,

$$(a_{j_1j_1r_1}, a_{j_2j_2s_1}) = \underbrace{([2(n_{j_2s_1 1} - n_{j_1u_1 1}) - 1]^{\text{th}}\text{DA:UA}, [(2m_{j_1r_1 1} + 1), (2n_{j_2s_1 1})]^{\text{th}}\text{LS})}_{([2(n_{j_2s_1 1}-n_{j_1u_1 1})-1],[(2m_{j_1r_1 1}+1),(2n_{j_2s_1 1})])\text{-down-to-up asymptote lower-saddle}}.$$

The equilibrium is a $([2(n_{j_2s_1 1}-n_{j_1u_1 1})-1]^{\text{th}}\text{DA:UA}, [(2m_{j_1r_1 1}+1), (2n_{j_2s_1 1})]^{\text{th}}$ LS)-down-to-up asymptote lower-saddle at $(x_{j_1}^*, x_{j_2}^*) = (a_{j_1j_1r_1}, a_{j_2j_2s_1})$.

- For $a_{j_1j_10}\prod_{r_2=1,r_2\neq r_1}^{p_1}(a_{j_1j_1r_1} - a_{j_1j_1r_2})^{m_{j_1r_2}}\prod_{u_2=1,u_2\neq u_1}^{q_1}(a_{j_1j_2u_1} - a_{j_1j_2u_2})^{n_{j_1u_2}} < 0$ and $a_{j_2j_20}\prod_{s_2=1,s_2\neq s_1}^{q_2}(a_{j_2j_2s_1} - a_{j_2j_2s_2})^{n_{j_2s_2}} < 0$,

$$(a_{j_1j_1r_1}, a_{j_2j_2s_1}) = \underbrace{([2(n_{j_2s_1 1} - n_{j_1u_1 1}) - 1]^{\text{th}}\text{UA:DA}, [(2m_{j_1r_1 1} + 1), (2n_{j_2s_1 1})]^{\text{th}}\text{LS})}_{([2(n_{j_2s_1 1}-n_{j_1u_1 1})-1],[(2m_{j_1r_1 1}+1),(2n_{j_2s_1 1})])\text{-up-to-down asymptote lower-saddle}}.$$

The equilibrium is a $([2(n_{j_2s_1 1}-n_{j_1u_1 1})-1]^{\text{th}}\text{DA:UA}, [(2m_{j_1r_1 1}+1), (2n_{j_2s_1 1})]^{\text{th}}$ LS)-up-to-down asymptote lower-saddle at $(x_{j_1}^*, x_{j_2}^*) = (a_{j_1j_1r_1}, a_{j_2j_2s_1})$.

($\text{ii}_{2\text{b}2\text{d}}$) For $m_{j_1r_1} = 2m_{j_1r_1 1} + 1, n_{j_1u_1} = 2n_{j_1u_1 1}$ and $n_{j_2s_1} = 2n_{j_2s_1 1}$, the equilibrium of $(x_{j_1}^*, x_{j_2}^*) = (a_{j_1j_1r_1}, a_{j_2j_2s_1})$ on the infinite-equilibrium $x_{j_2}^* = a_{j_1j_2u_1} = a_{j_2j_2s_1}$ has the following properties as in Eqs. (1.259)–(1.162).

- For $a_{j_1j_10}\prod_{r_2=1,r_2\neq r_1}^{p_1}(a_{j_1j_1r_1} - a_{j_1j_1r_2})^{m_{j_1r_2}}\prod_{u_2=1,u_2\neq u_1}^{q_1}(a_{j_1j_2u_1} - a_{j_1j_2u_2})^{n_{j_1u_2}} > 0$ and $a_{j_2j_20}\prod_{s_2=1,s_2\neq s_1}^{q_2}(a_{j_2j_2s_1} - a_{j_2j_2s_2})^{n_{j_2s_2}} > 0$,

$$(a_{j_1j_1r_1}, a_{j_2j_2s_1}) = \underbrace{([2(n_{j_2s_1 1} - n_{j_1u_1 1})]^{\text{th}}\text{DC:IC},[(2m_{j_1r_1 1} + 1), (2n_{j_2s_1 1})]^{\text{th}}\text{US})}_{([2(n_{j_2s_1 1}-n_{j_1u_1 1})],[(2m_{j_1r_1 1}+1),(2n_{j_2s_1 1})])\text{-decreasing-to-increasing concave upper-saddle}}.$$

The equilibrium is a $([2(n_{j_2s_1 1}-n_{j_1u_1 1})]^{\text{th}}\text{DC:IC},[(2m_{j_1r_1 1}+1), (2n_{j_2s_1 1})]^{\text{th}}\text{US})$-decreasing-to-increasing concave upper-saddle at $(x_{j_1}^*, x_{j_2}^*) = (a_{j_1j_1r_1}, a_{j_2j_2s_1})$.

- For $a_{j_1j_10}\prod_{r_2=1,r_2\neq r_1}^{p_1}(a_{j_1j_1r_1} - a_{j_1j_1r_2})^{m_{j_1r_2}}\prod_{u_2=1,u_2\neq u_1}^{q_1}(a_{j_1j_2u_1} - a_{j_1j_2u_2})^{n_{j_1u_2}} < 0$ and $a_{j_2j_20}\prod_{s_2=1,s_2\neq s_1}^{q_2}(a_{j_2j_2s_1} - a_{j_2j_2s_2})^{n_{j_2s_2}} > 0$,

$$(a_{j_1j_1r_1}, a_{j_2j_2s_1}) = \underbrace{([2(n_{j_2s_1 1} - n_{j_1u_1 1})]^{\text{th}}\text{IC:DC}, [(2m_{j_1r_1 1} + 1), (2n_{j_2s_1 1})]^{\text{th}}\text{US})}_{([2(n_{j_2s_1 1}-n_{j_1u_1 1})],[(2m_{j_1r_1 1}+1),(2n_{j_2s_1 1})])\text{-decreasing-to-increasing concave upper-saddle}}.$$

The equilibrium is a $([2(n_{j_1u_1 1}-n_{j_2s_1 1})]^{\text{th}}\text{IC:DC}, [(2m_{j_1r_1 1}+1), (2n_{j_2s_1 1})]^{\text{th}}\text{US})$-increasing-to-decreasing concave upper-saddle at $(x_{j_1}^*, x_{j_2}^*) = (a_{j_1j_1r_1}, a_{j_2j_2s_1})$.

- For $a_{j_1j_10}\prod_{r_2=1,r_2\neq r_1}^{p_1}(a_{j_1j_1r_1} - a_{j_1j_1r_2})^{m_{j_1r_2}}\prod_{u_2=1,u_2\neq u_1}^{q_1}(a_{j_1j_2u_1} - a_{j_1j_2u_2})^{n_{j_1u_2}} > 0$ and $a_{j_2j_20}\prod_{s_2=1,s_2\neq s_1}^{q_2}(a_{j_2j_2s_1} - a_{j_2j_2s_2})^{n_{j_2s_2}} < 0$,

$$(a_{j_1j_1r_1}, a_{j_2j_2s_1}) = \underbrace{([2(n_{j_2s_11} - n_{j_1u_11})]^{\text{th}}\text{IC:DC}, [(2m_{j_1r_11}+1), (2n_{j_2s_11})]^{\text{th}}\text{LS})}_{([2(n_{j_2s_11}-n_{j_1u_11})],[(2m_{j_1r_11}+1),(2n_{j_2s_11})])\text{-decreasing-to-increasing concave lower-saddle}}.$$

The equilibrium is a $([2(n_{j_2s_11} - n_{j_1u_11})]^{\text{th}}\text{IC:DC}, [(2m_{j_1r_11}+1), (2n_{j_2s_11})]^{\text{th}}\text{LS})$-increasing-to-decreasing concave lower-saddle at $(x^*_{j_1}, x^*_{j_2}) = (a_{j_1j_1r_1}, a_{j_2j_2s_1})$.

- For $a_{j_1j_10}\prod_{r_2=1,r_2\neq r_1}^{p_2}(a_{j_1j_1r_1} - a_{j_1j_1r_2})^{m_{j_1r_2}}\prod_{u_2=1,u_2\neq u_1}^{q_2}(a_{j_1j_2u_1} - a_{j_1j_2u_2})^{n_{j_1u_2}} < 0$ and $a_{j_2j_20}\prod_{s_2=1,s_2\neq s_1}^{q_2}(a_{j_2j_2s_1} - a_{j_2j_2s_2})^{n_{j_2s_2}} < 0$,

$$(a_{j_1j_1r_1}, a_{j_2j_2s_1}) = \underbrace{([2(n_{j_2s_11} - n_{j_1u_11})]^{\text{th}}\text{DC:IC},[(2m_{j_1r_11}+1), (2n_{j_2s_11})]^{\text{th}}\text{LS})}_{([2(n_{j_2s_11}-n_{j_1u_11})],[(2m_{j_1r_11}+1),(2n_{j_2s_11})])\text{-decreasing-to-increasing concave lower-saddle}}.$$

The equilibrium is a $([2(n_{j_1u_11} - n_{j_2s_11})]^{\text{th}}\text{DC:IC},[(2m_{j_1r_11}+1), (2n_{j_2s_11})]^{\text{th}}\text{LS})$-decreasing-to-increasing concave lower-saddle at $(x^*_{j_1}, x^*_{j_2}) = (a_{j_1j_1r_1}, a_{j_2j_2s_1})$.

(ii_{2b2e}) For $m_{j_1r_1} = 2m_{j_1r_11}$, $n_{j_1u_1} = 2n_{j_1u_11}+1$ and $n_{j_2s_1} = 2n_{j_2s_11}+1$, the equilibrium of $(x^*_{j_1}, x^*_{j_2}) = (a_{j_1j_1r_1}, a_{j_2j_2s_1})$ on the infinite-equilibrium $x^*_{j_2} = a_{j_1j_2u_1} = a_{j_2j_2s_1}$ has the following properties as in Eqs. (1.263)–(1.266).

- For $a_{j_1j_10}\prod_{r_2=1,r_2\neq r_1}^{p_1}(a_{j_1j_1r_1} - a_{j_1j_1r_2})^{m_{j_1r_2}}\prod_{u_2=1,u_2\neq u_1}^{q_1}(a_{j_1j_2u_1} - a_{j_1j_2u_2})^{n_{j_1u_2}} > 0$ and $a_{j_2j_20}\prod_{s_2=1,s_2\neq s_1}^{q_2}(a_{j_2j_2s_1} - a_{j_2j_2s_2})^{n_{j_2s_2}} > 0$,

$$(a_{j_1j_1r_1}, a_{j_2j_2s_1}) = \underbrace{([2(n_{j_2s_11} - n_{j_1u_11})]^{\text{th}}\text{IC:IC}, [(2m_{j_1r_11}), (2n_{j_2s_11}+1)]^{\text{th}}\text{SO})}_{([2(n_{j_2s_11}-n_{j_1u_11})],[(2m_{j_1r_11}),(2n_{j_2s_11}+1)])\text{-increasing concave source}}.$$

The equilibrium is a $([2(n_{j_2s_11} - n_{j_1u_11})]^{\text{th}}\text{IC:IC}, [(2m_{j_1r_11}), (2n_{j_2s_11}+1)]^{\text{th}}\text{SO})$-increasing-concave source at $(x^*_{j_1}, x^*_{j_2}) = (a_{j_1j_1r_1}, a_{j_2j_2s_1})$.

- For $a_{j_1j_10}\prod_{r_2=1,r_2\neq r_1}^{p_1}(a_{j_1j_1r_1} - a_{j_1j_1r_2})^{m_{j_1r_2}}\prod_{u_2=1,u_2\neq u_1}^{q_1}(a_{j_1j_2u_1} - a_{j_1j_2u_2})^{n_{j_1u_2}} < 0$ and $a_{j_2j_20}\prod_{s_2=1,s_2\neq s_1}^{q_2}(a_{j_2j_2s_1} - a_{j_2j_2s_2})^{n_{j_2s_2}} > 0$,

$$(a_{j_1j_1r_1}, a_{j_2j_2s_1}) = \underbrace{([2(n_{j_2s_11} - n_{j_1u_11})]^{\text{th}}\text{DC:DC}, [(2m_{j_1r_11}), (2n_{j_2s_11}+1)]^{\text{th}}\text{SO})}_{([2(n_{j_2s_11}-n_{j_1u_11})],[(2m_{j_1r_11}),(2n_{j_2s_11}+1)])\text{-decreasing concave source}}.$$

The equilibrium is a $([2(n_{j_2s_11} - n_{j_1u_11})]^{\text{th}}\text{DC:DC}, [(2m_{j_1r_11}), (2n_{j_2s_11}+1)]^{\text{th}}\text{SO})$-decreasing-concave source at $(x^*_{j_1}, x^*_{j_2}) = (a_{j_1j_1r_1}, a_{j_2j_2s_1})$.

- For $a_{j_1j_10}\prod_{r_2=1,r_2\neq r_1}^{p_1}(a_{j_1j_1r_1} - a_{j_1j_1r_2})^{m_{j_1r_2}}\prod_{u_2=1,u_2\neq u_1}^{q_1}(a_{j_1j_2u_1} - a_{j_1j_2u_2})^{n_{j_1u_2}} > 0$ and $a_{j_2j_20}\prod_{s_2=1,s_2\neq s_1}^{q_2}(a_{j_2j_2s_1} - a_{j_2j_2s_2})^{n_{j_2s_2}} < 0$,

$$(a_{j_1j_1r_1}, a_{j_2j_2s_1}) = \underbrace{([2(n_{j_2s_11} - n_{j_1u_11})]^{\text{th}}\text{DC:DC}, [(2m_{j_1r_11}), (2n_{j_2s_11}+1)]^{\text{th}}\text{SI})}_{([2(n_{j_2s_11}-n_{j_1u_11})],[(2m_{j_1r_11}),(2n_{j_2s_11}+1)])\text{-decreasing concave sink}}.$$

The equilibrium is a $([2(n_{j_2s_1}1-n_{j_1u_1}1)]^{\text{th}}\text{DC:DC}, [(2m_{j_1r_1}1), (2n_{j_2s_1}1+1)]^{\text{th}}\text{SI})$-decreasing-concave sink at $(x^*_{j_1}, x^*_{j_2}) = (a_{j_1j_1r_1}, a_{j_2j_2s_1})$.

- For $a_{j_1j_10}\prod_{r_2=1,r_2\neq r_1}^{p_1}(a_{j_1j_1r_1}-a_{j_1j_1r_2})^{m_{j_1r_2}}\prod_{u_2=1,u_2\neq u_1}^{q_1}(a_{j_1j_2u_1}-a_{j_1j_2u_2})^{n_{j_1u_2}}<0$ and $a_{j_2j_20}\prod_{s_2=1,s_2\neq s_1}^{q_2}(a_{j_2j_2s_1}-a_{j_2j_2s_2})^{n_{j_2s_2}}<0$,

$$(a_{j_1j_1r_1}, a_{j_2j_2s_1}) = \underbrace{([2(n_{j_2s_1}1-n_{j_1u_1}1)]^{\text{th}}\text{IC:IC}, [(2m_{j_1r_1}1), (2n_{j_2s_1}1+1)]^{\text{th}}\text{SI})}_{([2(n_{j_2s_1}1-n_{j_1u_1}1)],[(2m_{j_1r_1}1),(2n_{j_2s_1}1+1)])\text{-increasing concave sink}}.$$

The equilibrium is a $([2(n_{j_1u_1}1-n_{j_2s_1}1)]^{\text{th}}\text{IC:IC}, [(2m_{j_1r_1}1), (2n_{j_2s_1}1+1)]^{\text{th}}\text{SI})$-increasing-concave sink at $(x^*_{j_1}, x^*_{j_2}) = (a_{j_1j_1r_1}, a_{j_2j_2s_1})$.

($\text{ii}_{2\text{b}2\text{f}}$) For $m_{j_1r_1}=2m_{j_1r_1}1$, $n_{j_1u_1}=2n_{j_1u_1}1$ and $n_{j_2s_1}=2n_{j_2s_1}1+1$, the equilibrium of $(x^*_{j_1}, x^*_{j_2}) = (a_{j_1j_1r_1}, a_{j_2j_2s_1})$ on the infinite-equilibrium $x^*_{j_2}=a_{j_1j_2u_1}=a_{j_2j_2s_1}$ has the following properties as in Eqs. (1.267)–(1.270).

- For $a_{j_1j_10}\prod_{r_2=1,r_2\neq r_1}^{p_1}(a_{j_1j_1r_1}-a_{j_1j_1r_2})^{m_{j_1r_2}}\prod_{u_2=1,u_2\neq u_1}^{q_1}(a_{j_1j_2u_1}-a_{j_1j_2u_2})^{n_{j_1u_2}}>0$ and $a_{j_2j_20}\prod_{s_2=1,s_2\neq s_1}^{q_2}(a_{j_2j_2s_1}-a_{j_2j_2s_2})^{n_{j_2s_2}}>0$,

$$\begin{aligned}&(a_{j_1j_1r_1}, a_{j_2j_2s_1})\\&=\underbrace{([2(n_{j_2s_1}1-n_{j_1u_1}1)+1]^{\text{th}}\text{DA:DA}, [(2m_{j_1r_1}1), (2n_{j_2s_1}1+1)]^{\text{th}}\text{SO})}_{([2(n_{j_2s_1}1-n_{j_1u_1}1)+1],[(2m_{j_1r_1}1),(2n_{j_2s_1}1+1)])\text{-down-asymptote source}}.\end{aligned}$$

The equilibrium is a $([2(n_{j_2s_1}1-n_{j_1u_1}1)+1]^{\text{th}}\text{DA:DA}, [(2m_{j_1r_1}1), (2n_{j_2s_1}1+1)]^{\text{th}}\text{SO})$-down-asymptote source at $(x^*_{j_1}, x^*_{j_2}) = (a_{j_1j_1r_1}, a_{j_2j_2s_1})$.

- For $a_{j_1j_10}\prod_{r_2=1,r_2\neq r_1}^{p_1}(a_{j_1j_1r_1}-a_{j_1j_1r_2})^{m_{j_1r_2}}\prod_{u_2=1,u_2\neq u_1}^{q_1}(a_{j_1j_2u_1}-a_{j_1j_2u_2})^{n_{j_1u_2}}<0$ and $a_{j_2j_20}\prod_{s_2=1,s_2\neq s_1}^{q_2}(a_{j_2j_2s_1}-a_{j_2j_2s_2})^{n_{j_2s_2}}>0$,

$$\begin{aligned}&(a_{j_1j_1r_1}, a_{j_2j_2s_1})\\&=\underbrace{([2(n_{j_2s_1}1-n_{j_1u_1}1)+1]^{\text{th}}\text{UA:UA}, [(2m_{j_1r_1}1), (2n_{j_2s_1}1+1)]^{\text{th}}\text{SO})}_{([2(n_{j_2s_1}1-n_{j_1u_1}1)+1],[(2m_{j_1r_1}1),(2n_{j_2s_1}1+1)])\text{-up-asymptote source}}.\end{aligned}$$

The equilibrium is a $([2(n_{j_2s_1}1-n_{j_1u_1}1)+1]^{\text{th}}\text{UA:UA}, [(2m_{j_1r_1}1), (2n_{j_2s_1}1+1)]^{\text{th}}\text{SO})$-up-asymptote source at $(x^*_{j_1}, x^*_{j_2}) = (a_{j_1j_1r_1}, a_{j_2j_2s_1})$.

- For $a_{j_1j_10}\prod_{r_2=1,r_2\neq r_1}^{p_1}(a_{j_1j_1r_1}-a_{j_1j_1r_2})^{m_{j_1r_2}}\prod_{u_2=1,u_2\neq u_1}^{q_1}(a_{j_1j_2u_1}-a_{j_1j_2u_2})^{n_{j_1u_2}}>0$ and $a_{j_2j_20}\prod_{s_2=1,s_2\neq s_1}^{q_2}(a_{j_2j_2s_1}-a_{j_2j_2s_2})^{n_{j_2s_2}}<0$,

$$\begin{aligned}&(a_{j_1j_1r_1}, a_{j_2j_2s_1})\\&=\underbrace{([2(n_{j_2s_1}1-n_{j_1u_1}1)+1]^{\text{th}}\text{UA:UA}, [(2m_{j_1r_1}1), (2n_{j_2s_1}1+1)]^{\text{th}}\text{SI})}_{([2(n_{j_2s_1}1-n_{j_1u_1}1)+1],[(2m_{j_1r_1}1),(2n_{j_2s_1}1+1)])\text{-up-asymptote sink}}.\end{aligned}$$

The equilibrium is a $([2(n_{j_2s_1 1}-n_{j_1u_1 1})+1]^{\text{th}}\text{UA:UA}, [(2m_{j_1r_1 1}), (2n_{j_2s_1 1}+1)]^{\text{th}}$ SI)-up-asymptote sink at $(x_{j_1}^*, x_{j_2}^*) = (a_{j_1j_1r_1}, a_{j_2j_2s_1})$.

- For $a_{j_1j_1 0}\prod_{r_2=1,r_2\neq r_1}^{p_1}(a_{j_1j_1r_1}-a_{j_1j_1r_2})^{m_{j_1r_2}}\prod_{u_2=1,u_2\neq u_1}^{q_1}(a_{j_1j_2u_1}-a_{j_1j_2u_2})^{n_{j_1u_2}}<0$ and $a_{j_2j_2 0}\prod_{s_2=1,s_2\neq s_1}^{q_2}(a_{j_2j_2s_1}-a_{j_2j_2s_2})^{n_{j_2s_2}}<0$,

$$(a_{j_1j_1r_1}, a_{j_2j_2s_1}) = \underbrace{([2(n_{j_2s_1 1}-n_{j_1u_1 1})+1]^{\text{th}}\text{DA:DA}, [(2m_{j_1r_1 1}), (2n_{j_2s_1 1}+1)]^{\text{th}}\text{SI})}_{([2(n_{j_2s_1 1}-n_{j_1u_1 1})+1],[(2m_{j_1r_1 1}),(2n_{j_2s_1 1}+1)])\text{-down-asymptote sink}}.$$

The equilibrium is a $([2(n_{j_2s_1 1}-n_{j_1u_1 1})+1]^{\text{th}}\text{DA:DA}, [(2m_{j_1r_1 1}), (2n_{j_2s_1 1}+1)]^{\text{th}}$ SI)-down-asymptote sink at $(x_{j_1}^*, x_{j_2}^*) = (a_{j_1j_1r_1}, a_{j_2j_2s_1})$.

(ii_{2b2g}) For $m_{j_1r_1}=2m_{j_1r_1 1}$, $n_{j_1u_1}=2n_{j_1u_1 1}+1$ and $n_{j_2s_1}=2n_{j_2s_1 1}$, the equilibrium of $(x_{j_1}^*, x_{j_2}^*) = (a_{j_1j_1r_1}, a_{j_2j_2s_1})$ on the infinite-equilibrium $x_{j_2}^*=a_{j_1j_2u_1}=a_{j_2j_2s_1}$ has the following properties as in Eqs. (1.271)–(1.274)

- For $a_{j_1j_1 0}\prod_{r_2=1,r_2\neq r_1}^{p_1}(a_{j_1j_1r_1}-a_{j_1j_1r_2})^{m_{j_1r_2}}\prod_{u_2=1,u_2\neq u_1}^{q_1}(a_{j_1j_2u_1}-a_{j_1j_2u_2})^{n_{j_1u_2}}>0$ and $a_{j_2j_2 0}\prod_{s_2=1,s_2\neq s_1}^{q_2}(a_{j_2j_2s_1}-a_{j_2j_2s_2})^{n_{j_2s_2}}>0$,

$$(a_{j_1j_1r_1}, a_{j_2j_2s_1}) = \underbrace{([2(n_{j_2s_1 1}-n_{j_1u_1 1})-1]^{\text{th}}\text{DA:DA}, [(2m_{j_1r_1 1}), (2n_{j_2s_1 1})]^{\text{th}}\text{US})}_{([2(n_{j_2s_1 1}-n_{j_1u_1 1})-1],[(2m_{j_1r_1 1}),(2n_{j_2s_1 1})])\text{-down-asymptote upper-saddle}}.$$

The equilibrium is a $([2(n_{j_2s_1 1}-n_{j_1u_1 1})-1]^{\text{th}}\text{DA:DA}, [(2m_{j_1r_1 1}), (2n_{j_2s_1 1})]^{\text{th}}\text{US})$-down-asymptote upper-saddle at $(x_{j_1}^*, x_{j_2}^*) = (a_{j_1j_1r_1}, a_{j_2j_2s_1})$.

- For $a_{j_1j_1 0}\prod_{r_2=1,r_2\neq r_1}^{p_1}(a_{j_1j_1r_1}-a_{j_1j_1r_2})^{m_{j_1r_2}}\prod_{u_2=1,u_2\neq u_1}^{q_1}(a_{j_1j_2u_1}-a_{j_1j_2u_2})^{n_{j_1u_2}}<0$ and $a_{j_2j_2 0}\prod_{s_2=1,s_2\neq s_1}^{q_2}(a_{j_2j_2s_1}-a_{j_2j_2s_2})^{n_{j_2s_2}}>0$,

$$(a_{j_1j_1r_1}, a_{j_2j_2s_1}) = \underbrace{([2(n_{j_2s_1 1}-n_{j_1u_1 1})-1]^{\text{th}}\text{UA:UA}, [(2m_{j_1r_1 1}), (2n_{j_2s_1 1})]^{\text{th}}\text{US})}_{([2(n_{j_2s_1 1}-n_{j_1u_1 1})-1],[(2m_{j_1r_1 1}),(2n_{j_2s_1 1})])\text{-up-asymptote upper-saddle}}.$$

The equilibrium is a $([2(n_{j_2s_1 1}-n_{j_1u_1 1})-1]^{\text{th}}\text{UA:UA}, [(2m_{j_1r_1 1}), (2n_{j_2s_1 1})]^{\text{th}}\text{US})$-up-asymptote upper-saddle at $(x_{j_1}^*, x_{j_2}^*) = (a_{j_1j_1r_1}, a_{j_2j_2s_1})$.

- For $a_{j_1j_1 0}\prod_{r_2=1,r_2\neq r_1}^{p_1}(a_{j_1j_1r_1}-a_{j_1j_1r_2})^{m_{j_1r_2}}\prod_{u_2=1,u_2\neq u_1}^{q_1}(a_{j_1j_2u_1}-a_{j_1j_2u_2})^{n_{j_1u_2}}>0$ and $a_{j_2j_2 0}\prod_{s_2=1,s_2\neq s_1}^{q_2}(a_{j_2j_2s_1}-a_{j_2j_2s_2})^{n_{j_2s_2}}<0$,

$$(a_{j_1j_1r_1}, a_{j_2j_2s_1}) = \underbrace{([2(n_{j_2s_1 1}-n_{j_1u_1 1})-1]^{\text{th}}\text{UA:UA}, [(2m_{j_1r_1 1}), (2n_{j_2s_1 1})]^{\text{th}}\text{LS})}_{([2(n_{j_2s_1 1}-n_{j_1u_1 1})-1],[(2m_{j_1r_1 1}),(2n_{j_2s_1 1})])\text{-up-asymptote lower-saddle}}.$$

The equilibrium is a $([2(n_{j_2s_1 1}-n_{j_1u_1 1})-1]^{\text{th}}\text{UA:UA}, [(2m_{j_1r_1 1}), (2n_{j_2s_1 1})]^{\text{th}}\text{LS})$-up-asymptote lower-saddle at $(x_{j_1}^*, x_{j_2}^*) = (a_{j_1j_1r_1}, a_{j_2j_2s_1})$.

- For $a_{j_1j_10}\prod_{r_2=1,r_2\neq r_1}^{p_1}(a_{j_1j_1r_1}-a_{j_1j_1r_2})^{m_{j_1r_2}}\prod_{u_2=1,u_2\neq u_1}^{q_1}(a_{j_1j_2u_1}-a_{j_1j_2u_2})^{n_{j_1u_2}}<0$ and $a_{j_2j_20}\prod_{s_2=1,s_2\neq s_1}^{q_2}(a_{j_2j_2s_1}-a_{j_2j_2s_2})^{n_{j_2s_2}}<0$,

$$(a_{j_1j_1r_1},a_{j_2j_2s_1})=\underbrace{([2(n_{j_2s_1}1-n_{j_1u_1}1)-1]^{\text{th}}\text{DA:DA},[(2m_{j_1r_1}1),(2n_{j_2s_1}1)]^{\text{th}}\text{LS})}_{([2(n_{j_2s_1}1-n_{j_1u_1}1)-1],[(2m_{j_1r_1}1),(2n_{j_2s_1}1)])\text{-down-asymptote lower-saddle}}.$$

The equilibrium is a $([2(n_{j_2s_1}1-n_{j_1u_1}1)-1]^{\text{th}}\text{DA:DA},[(2m_{j_1r_1}1),(2n_{j_2s_1}1)]^{\text{th}}\text{LS})$-down-asymptote lower-saddle at $(x_{j_1}^*,x_{j_2}^*)=(a_{j_1j_1r_1},a_{j_2j_2s_1})$.

($\text{ii}_{2\text{b}2\text{h}}$) For $m_{j_1r_1}=2m_{j_1r_1}1$, $n_{j_1u_1}=2n_{j_1u_1}1$ and $n_{j_2s_1}=2n_{j_2s_1}1$, the equilibrium of $(x_{j_1}^*,x_{j_2}^*)=(a_{j_1j_1r_1},a_{j_2j_2s_1})$ on the infinite-equilibrium $x_{j_2}^*=a_{j_1j_2u_1}=a_{j_2j_2s_1}$ has the following properties as in Eqs. (1.275)–(1.278)

- For $a_{j_1j_10}\prod_{r_2=1,r_2\neq r_1}^{p_1}(a_{j_1j_1r_1}-a_{j_1j_1r_2})^{m_{j_1r_2}}\prod_{u_2=1,u_2\neq u_1}^{q_1}(a_{j_1j_2u_1}-a_{j_1j_2u_2})^{n_{j_1u_2}}>0$ and $a_{j_2j_20}\prod_{s_2=1,s_2\neq s_1}^{q_2}(a_{j_2j_2s_1}-a_{j_2j_2s_2})^{n_{j_2s_2}}>0$,

$$(a_{j_1j_1r_1},a_{j_2j_2s_1})=\underbrace{([2(n_{j_2s_1}1-n_{j_1u_1}1)]^{\text{th}}\text{IC:IC},[(2m_{j_1r_1}1),(2n_{j_2s_1}1)]^{\text{th}}\text{US})}_{([2(n_{j_2s_1}1-n_{j_1u_1}1)],[(2m_{j_1r_1}1),(2n_{j_2s_1}1)])\text{-increasing-concave upper-saddle}}.$$

The equilibrium is a $([2(n_{j_2s_1}1-n_{j_1u_1}1)]^{\text{th}}\text{IC:IC},[(2m_{j_1r_1}1),(2n_{j_2s_1}1)]^{\text{th}}\text{US})$-increasing concave upper-saddle at $(x_{j_1}^*,x_{j_2}^*)=(a_{j_1j_1r_1},a_{j_2j_2s_1})$.

- For $a_{j_1j_10}\prod_{r_2=1,r_2\neq r_1}^{p_1}(a_{j_1j_1r_1}-a_{j_1j_1r_2})^{m_{j_1r_2}}\prod_{u_2=1,u_2\neq u_1}^{q_1}(a_{j_1j_2u_1}-a_{j_1j_2u_2})^{n_{j_1u_2}}<0$ and $a_{j_2j_20}\prod_{s_2=1,s_2\neq s_1}^{q_2}(a_{j_2j_2s_1}-a_{j_2j_2s_2})^{n_{j_2s_2}}>0$,

$$(a_{j_1j_1r_1},a_{j_2j_2s_1})=\underbrace{([2(n_{j_2s_1}1-n_{j_1u_1}1)]^{\text{th}}\text{DC:DC},[(2m_{j_1r_1}1),(2n_{j_2s_1}1)]^{\text{th}}\text{US})}_{([2(n_{j_2s_1}1-n_{j_1u_1}1)],[(2m_{j_1r_1}1),(2n_{j_2s_1}1)])\text{-decreasing - concave upper-saddle}}.$$

The equilibrium is a $([2(n_{j_2s_1}1-n_{j_1u_1}1)]^{\text{th}}\text{DC:DC},[(2m_{j_1r_1}1),(2n_{j_2s_1}1)]^{\text{th}}\text{US})$-decreasing concave upper-saddle at $(x_{j_1}^*,x_{j_2}^*)=(a_{j_1j_1r_1},a_{j_2j_2s_1})$.

- For $a_{j_1j_10}\prod_{r_2=1,r_2\neq r_1}^{p_1}(a_{j_1j_1r_1}-a_{j_1j_1r_2})^{m_{j_1r_2}}\prod_{u_2=1,u_2\neq u_1}^{q_1}(a_{j_1j_2u_1}-a_{j_1j_2u_2})^{n_{j_1u_2}}>0$ and $a_{j_2j_20}\prod_{s_2=1,s_2\neq s_1}^{q_2}(a_{j_2j_2s_1}-a_{j_2j_2s_2})^{n_{j_2s_2}}<0$,

$$(a_{j_1j_1r_1},a_{j_2j_2s_1})=\underbrace{([2(n_{j_2s_1}1-n_{j_1u_1}1)]^{\text{th}}\text{DC:DC},[(2m_{j_1r_1}1),(2n_{j_2s_1}1)]^{\text{th}}\text{LS})}_{([2(n_{j_2s_1}1-n_{j_1u_1}1)],[(2m_{j_1r_1}1),(2n_{j_2s_1}1)])\text{-decreasing - concave lower-saddle}}.$$

The equilibrium is a $([2(n_{j_2s_1}1-n_{j_1u_1}1)]^{\text{th}}\text{DC:DC},[(2m_{j_1r_1}1),(2n_{j_2s_1}1)]^{\text{th}}\text{LS})$-decreasing concave lower-saddle at $(x_{j_1}^*,x_{j_2}^*)=(a_{j_1j_1r_1},a_{j_2j_2s_1})$.

- For $a_{j_1j_10}\prod_{r_2=1,r_2\neq r_1}^{p_1}(a_{j_1j_1r_1}-a_{j_1j_1r_2})^{m_{j_1r_2}}\prod_{u_2=1,u_2\neq u_1}^{q_1}(a_{j_1j_2u_1}-a_{j_1j_2u_2})^{n_{j_1u_2}}<0$ and $a_{j_2j_20}\prod_{s_2=1,s_2\neq s_1}^{q_2}(a_{j_2j_2s_1}-a_{j_2j_2s_2})^{n_{j_2s_2}}<0$,

$$(a_{j_1 j_1 r_1}, a_{j_2 j_2 s_1}) = \underbrace{([2(n_{j_2 s_1 1} - n_{j_1 u_1 1})]^{\text{th}}\text{IC:IC}, [(2m_{j_1 r_1 1}), (2n_{j_2 s_1 1})]^{\text{th}}\text{LS})}_{([2(n_{j_2 s_1 1} - n_{j_1 u_1 1})],[(2m_{j_1 r_1 1}),(2n_{j_2 s_1 1})])\text{-increasing-concave lower-saddle}}.$$

The equilibrium is a $([2(n_{j_1 u_1 1} - n_{j_2 s_1 1})]^{\text{th}}\text{IC:IC}, [(2m_{j_1 r_1 1}), (2n_{j_2 s_1 1})]^{\text{th}}\text{LS})$-increasing concave lower-saddle at $(x_{j_1}^*, x_{j_2}^*) = (a_{j_1 j_1 r_1}, a_{j_2 j_2 s_1})$.

2.2.3 *Hybrid Networks of Simple Equilibriums and Flows*

In this section, the theorem about the hybrid networks of simple equilibriums and 1-dimensional flows will be proved, and the infinite-equilibriums will be discussed.

(iii) Consider

$$\begin{aligned}
\dot{x}_{j_2} &= a_{j_2 j_2 0} \prod_{s=1}^{n_2} (x_{j_2} - a_{j_2 j_2 s}) \\
\dot{x}_{j_1} &= a_{j_1 j_1 0} \prod_{r=1}^{m_1} (x_{j_1} - a_{j_1 j_1 r}) \prod_{u=1}^{n_1} (x_{j_2} - a_{j_1 j_2 u}).
\end{aligned}$$

For $a_{j_1 j_2 u} \neq a_{j_2 j_2 s}$ $(u = 1, 2, \cdots, n_1; s = 1, 2, \cdots, n_2)$, using Eqs. (1.136)–(1.140),

$$\begin{aligned}
q_1 &= n_1, q_2 = n_2, p_1 = m_1; \\
m_{j_1 r_i} &= 1 (i = 1, 2, \cdots, m_1), \\
n_{j_2 u_j} &= 1 (j = l, 2, \cdots, n_1), \\
n_{j_2 s_l} &= 1 (l = l, 2, \cdots, n_2);
\end{aligned}$$

the first integral manifold is obtained as

$$\begin{aligned}
&\sum_{r_1=1}^{m_1} A_{r_1} \ln \frac{|x_{j_1} - a_{j_1 j_1 r_1}|}{|x_{j_1 0} - a_{j_1 j_1 r_1}|} = \frac{a_{j_1 j_1 0}}{a_{j_2 j_2 0}} \left\{ \sum_{\beta_0 = n_2}^{n_1} b_{j_1 j_2 \beta_0} \sum_{\substack{\alpha_1 + \beta_1 = \beta_0 - 1 \\ \alpha_1 \in [0, \beta_0 - n_2], \\ \beta_1 \in [n_2 - 1, \beta_0 - 1]}} \sum_{\substack{\alpha_2 + \beta_2 = \beta_1 - 2 \\ \alpha_2 \in [0, \beta_1 - n_2], \\ \beta_2 \in [n_2 - 2, \beta_1 - 2]}} \cdots \right. \\
&\sum_{\substack{\alpha_{n_2 - 1} + \beta_{n_2 - 1} = \beta_{n_2 - 2} - n_2 + 1 \\ \alpha_{n_2 - 1} \in [0, \beta_{n_2 - 2} - n_2] \\ \beta_{n_2 - 1} \in [1, \beta_{n_2 - 2} - n_2 + 1]}} \prod_{j=1}^{n_2 - 1} \frac{(\beta_{j-2} - j)!}{\alpha_j ! \beta_j !} (a_{j_2 j_2 s_{j+1}} - a_{j_2 j_2 s_j})^{\alpha_j} \\
&\times \frac{1}{\beta_{n_2 - 1}} [(x_{j_2} - a_{j_2 j_2 n_2})^{\beta_{n_2 - 1}} - (x_{j_2 0} - a_{j_2 j_2 n_2})^{\beta_{n_2 - 1}}]
\end{aligned}$$

$$
\begin{aligned}
&-\sum_{\beta_0=n_2}^{n_1} b_{j_1j_2\beta_0}\sum_{k=2}^{n_2-1}\Bigg[\sum_{\substack{\alpha_1\in[0,\beta_0-n_2),\\ \beta_1\in(n_2-1,\beta_0-1]}}^{\alpha_1+\beta_1=\beta_0-1}\ \sum_{\substack{\alpha_2\in[0,\beta_1-n_2]\\ \beta_2\in(n_2-2,\beta_1-2)}}^{\alpha_2+\beta_2=\beta_1-2}\cdots\\
&\times\Bigg(\sum_{\substack{\alpha_{k-1}\in(\beta_{k-2}-n_2+1,\beta_{k-2}-k+1]\\ \beta_{k-1}\in[0,n_2-k)}}^{\alpha_{k-1}+\beta_{k-1}=\beta_{k-2}-k+1}\prod_{j=1}^{k-1}\frac{(\beta_{j-2}-j)!}{\alpha_j!\beta_j!}(a_{j_2j_2s_{j+1}}-a_{j_2j_2s_j})^{\alpha_j}\\
&\times\sum_{\rho=k}^{n_2}A_{\rho 1}^{(s_k)}\ln\frac{|x_{j_2}-a_{j_2j_2s_\rho}|}{|x_{j_20}-a_{j_2j_2s_\rho}|}\\
&+\sum_{\substack{\alpha_{k-1}\in(\beta_{k-2}-n_2,\beta_{k-2}-n_2+1]\\ \beta_{k-1}\in[n_2-k,n_2-k+1)}}^{\alpha_{k-1}+\beta_{k-1}=\beta_{k-2}-k+1}\prod_{j=1}^{k-1}\frac{(\beta_{j-2}-j)!}{\alpha_j!\beta_j!}(a_{j_2j_2s_{j+1}}-a_{j_2j_2s_j})^{\alpha_j}\\
&\times\sum_{\rho=k+1}^{n_2}B_{\rho 1}^{(s_k)}\ln\frac{|x_{j_2}-a_{j_2j_2s_\rho}|}{|x_{j_20}-a_{j_2j_2s_\rho}|}\Bigg)\Bigg]+b_{j_1j_20}\sum_{\rho=1}^{n_2}A_{\rho 1}^{(s_1)}\ln\frac{|x_{j_2}-a_{j_2j_2s_\rho}|}{|x_{j_20}-a_{j_2j_2s_\rho}|}\\
&+\sum_{\beta_0=1}^{n_2-1} b_{j_1j_2\beta_0}\sum_{\rho=2}^{n_2}B_{\rho 1}^{(s_1)}\frac{|x_{j_2}-a_{j_2j_2s_\rho}|}{|x_{j_20}-a_{j_2j_2s_\rho}|}\Bigg\};
\end{aligned}
$$

where

$$
\begin{aligned}
A_{r_1}&=\frac{1}{\prod_{r_2=1,r_2\neq r_1}^{m_1}(a_{j_1j_1r_1}-a_{j_1j_1r_\sigma})};\\
b_{j_1j_2\beta_0}&=\sum_{t_1,l_1=0}^{t_1+l_1=1}\cdots\sum_{t_q,l_q=0}^{t_q+l_q=1}\prod_{k=1}^{n_1}(a_{j_2j_2s_1}-a_{j_1j_2u_k})^{t_k}\delta_{\beta_0}^{\Sigma_{r=1}^{n_1}l_r};\\
&\lim_{\beta_{n_2-1}\to 0}\frac{1}{\beta_{n_2-1}}[(x_{j_2}-a_{j_2j_2s_{n_2}})^{\beta_{n_2-1}}-(x_{j_20}-a_{j_2j_2s_{n_2}})^{\beta_{n_2-1}}]=\ln\frac{|x_{j_2}-a_{j_2j_2s_{n_2}}|}{|x_{j_20}-a_{j_2j_2s_{n_2}}|};\\
A_{\rho 1}^{(s_k)}&=\frac{1}{\prod_{\sigma=k,\sigma\neq\rho}^{n_2}(a_{j_2j_2s_\rho}-a_{j_2j_2s_\sigma})},\ \text{for } \rho=\{k,k+1,\cdots,n_2\},\\
B_{\rho 1}^{(s_k)}&=\frac{(a_{j_2j_2s_\rho}-a_{j_2j_2s_k})^{\beta_{k-1}-k}}{\prod_{\sigma=k+1,\sigma\neq\rho}^{n_2}(a_{j_2j_2s_\rho}-a_{j_2j_2s_\sigma})}, for \rho=\{k+1,k+2,\cdots,n_2\};\\
k&\in\{1,2,\cdots,n_2-1\}.
\end{aligned}
$$

(iii_1) For $a_{j_1j_2u_1}\neq a_{j_2j_2s_1}(s=1,2,\cdots,n_2;u=1,2,\cdots,n_1)$, the simple 1-dimensional flows and simple equilibriums exist.

(iii_{1a}) At the equilibrium of $x_{j_2}^*=a_{j_1j_2u_1}$ and $\overline{x}_{j_1}\neq a_{j_1j_1r}$, in phase space,

$$\begin{aligned}&\frac{dx_{j_1}}{dx_{j_2}}\Big|_{x_{j_2}^*=a_{j_1j_2s_1}}\\&=\frac{a_{j_1j_10}(x_{j_2}-a_{j_1j_2u_1})\prod_{r=1}^{m_1}(\overline{x}_{j_1}-a_{j_1j_1r})\prod_{u_2=1,u_2\neq u_1}^{n_1}(x_{j_2}-a_{j_1j_2u_2})}{a_{j_2j_20}\prod_{s=1}^{n_1}(x_{j_2}-a_{j_2j_2s})}\Big|_{x_{j_2}^*=a_{j_1j_2u_1}}\\&=0.\end{aligned}$$

If

$$\frac{d^2x_{j_1}}{dx_{j_2}^2}\Big|_{x_{j_2}^*=a_{j_1j_2u_1}}=\frac{a_{j_1j_10}\prod_{r=1}^{m_1}(\overline{x}_{j_1}-a_{j_1j_1r})\prod_{u_2=1,u_2\neq u_1}^{n_1}(a_{j_1j_2u_1}-a_{j_1j_2u_2})}{a_{j_2j_20}\prod_{s=1}^{n_2}(a_{j_1j_2u_1}-a_{j_2j_2s})}>0,$$

the up-parabola flow at $x_{j_2}^*=a_{j_1j_2u_1}$ exists. If

$$\frac{d^2x_{j_1}}{dx_{j_2}^2}\Big|_{x_{j_2}^*=a_{j_1j_2u_1}}=\frac{a_{j_1j_10}\prod_{r=1}^{m_1}(\overline{x}_{j_1}-a_{j_1j_1r})\prod_{u_2=1,u_2\neq u_1}^{n_1}(a_{j_1j_2u_1}-a_{j_1j_2u_2})}{a_{j_2j_20}\prod_{s=1}^{n_2}(a_{j_1j_2u_1}-a_{j_2j_2s})}<0,$$

the down-parabola flow at $x_{j_2}^*=a_{j_1j_2u_2}$ exists. Let

$$\dot{x}_{j_2}=a_{j_2j_20}\prod_{s=1}^{n_2}(a_{j_1j_2u_1}-a_{j_2j_2s}).$$

The parabola flows at $x_{j_2}^*=a_{j_1j_2u_2}$ are positive and negative for $\dot{x}_{j_2}>0$ and $\dot{x}_{j_2}<0$, respectively.

At the equilibrium of $x_{j_1}^*=a_{j_1j_1r_1}$ and $\overline{x}_{j_2}\neq a_{j_1j_2u},a_{j_2j_2s}$, the variational equation is

$$\begin{aligned}\dot{x}_{j_2}&=a_{j_2j_20}\prod_{s=1}^{n_2}(\overline{x}_{j_2}-a_{j_2j_2s}),\\\Delta\dot{x}_{j_1}&=a_{j_1j_10}\prod_{r_1=1,r_2\neq r_1}^{m_1}(a_{j_1j_1r_1}-a_{j_1j_1r_2})\prod_{u=1}^{n_1}(\overline{x}_{j_2}-a_{j_1j_2u})(\Delta x_{j_1}).\end{aligned}$$

For

$$\begin{aligned}&a_{j_1j_10}\prod_{r_1=1,r_2\neq r_1}^{m_1}(a_{j_1j_1r_1}-a_{j_1j_1r_2})\prod_{u=1}^{n_1}(\overline{x}_{j_2}-a_{j_1j_2u})>0,\ \text{and}\\&a_{j_1j_10}\prod_{r_1=1,r_2\neq r_1}^{m_1}(a_{j_1j_1r_1}-a_{j_1j_1r_2})\prod_{u=1}^{n_1}(\overline{x}_{j_2}-a_{j_1j_2u})<0,\end{aligned}$$

the flows at $x_{j_1}^*=a_{j_1j_1r_1}$ and $\overline{x}_{j_2}\neq a_{j_1j_2u}$ are source and sink, respectively.

Let $\overline{x}_{j_1}=a_{j_1j_1r_1}$ and $\overline{x}_{j_2}=a_{j_1j_2u_1}\neq a_{j_2j_2s}$. Therefore, the equilibrium of $(x_{j_1}^*,x_{j_2}^*)=(a_{j_1j_1r_1},a_{j_1j_2u_1})(r_1=1,2,\cdots,m_1;u_1=1,2,\cdots,n_1)$ for the -direction has the following properties as in Eqs. (1.282)–(1.285).

- For $a_{j_1j_10}\prod_{r_2=1,r_2\neq r_1}^{m_1}(a_{j_1j_1r_1}-a_{j_1j_1r_2})\prod_{u_2=1,u_2\neq u_1}^{n_1}(a_{j_1j_2u_2}-a_{j_1j_2u_2})>0$ and $a_{j_2j_20}\prod_{s=1}^{n_2}(a_{j_1j_2u_1}-a_{j_2j_2s})>0$,

$$(a_{j_1j_1r_1}, a_{j_1j_2u_1}) = \underbrace{\text{(DP:UP, pF)}}_{\text{hyperbolic flow }(+)}.$$

- For $a_{j_1j_10}\prod_{r_2=1,r_2\neq r_1}^{m_1}(a_{j_1j_1r_1}-a_{j_1j_1r_2})\prod_{u_2=1,u_2\neq u_1}^{n_1}(a_{j_1j_2u_2}-a_{j_1j_2u_2}) < 0$ and $a_{j_2j_20}\prod_{s=1}^{n_2}(a_{j_1j_2u_1}-a_{j_2j_2s}) > 0$,

$$(a_{j_1j_1r_1}, a_{j_1j_2u_1}) = \underbrace{\text{(UP:DP, pF)}}_{\text{hyperbolic - secant flow }(+)}.$$

- For $a_{j_1j_10}\prod_{r_2=1,r_2\neq r_1}^{m_1}(a_{j_1j_1r_1}-a_{j_1j_1r_2})\prod_{u_2=1,u_2\neq u_1}^{n_1}(a_{j_1j_2u_2}-a_{j_1j_2u_2}) > 0$ and $a_{j_2j_20}\prod_{s=1}^{n_2}(a_{j_1j_2u_1}-a_{j_2j_2s}) < 0$,

$$(a_{j_1j_1r_1}, a_{j_1j_2u_1}) = \underbrace{\text{(UP:DP, nF)}}_{\text{hyperbolic - secant flow }(-)}.$$

- For $a_{j_1j_10}\prod_{r_2=1,r_2\neq r_1}^{m_1}(a_{j_1j_1r_1}-a_{j_1j_1r_2})\prod_{u_2=1,u_2\neq u_1}^{n_1}(a_{j_1j_2u_2}-a_{j_1j_2u_2}) < 0$ and $a_{j_2j_20}\prod_{s=1}^{n_2}(a_{j_1j_2u_1}-a_{j_2j_2s}) < 0$,

$$(a_{j_1j_1r_1}, a_{j_1j_2u_1}) = \underbrace{\text{(DP:UP,nF)}}_{\text{hyperbolic flow }(-)}.$$

(iii$_{1b}$) At $(x_{j_1}^*, x_{j_2}^*) = (a_{j_1j_1r_1}, a_{j_2j_2s_1})(r_1 = 1, 2, \cdots, m_1; s_1 = 1, 2, \cdots, n_2)$,

$$\begin{aligned}
\Delta\dot{x}_{j_2} &= a_{j_2j_20}\prod_{s_2=1,s_2\neq s_1}^{n_2}(a_{j_2j_2s_1}-a_{j_2j_2s_2})(\Delta x_{j_2}),\\
\Delta\dot{x}_{j_1} &= a_{j_1j_10}\prod_{r_1=1,r_2\neq r_1}^{m_1}(a_{j_1j_1r_1}-a_{j_1j_1r_2})\prod_{u=1}^{n_1}(\overline{x}_{j_2}-a_{j_1j_2u})(\Delta x_{j_1}).
\end{aligned}$$

For

$$\begin{aligned}
&a_{j_2j_20}\prod_{s_2=1,s_2\neq s_1}^{n_2}(a_{j_2j_2s_1}-a_{j_2j_2s_2}) > 0, \text{ and}\\
&a_{j_2j_20}\prod_{s_2=1,s_2\neq s_1}^{n_2}(a_{j_2j_2s_1}-a_{j_2j_2s_2}) < 0,
\end{aligned}$$

the flows at $x_{j_2}^* = a_{j_2j_2s_1}$ are source and sink, respectively. For

$$\begin{aligned}
&a_{j_1j_10}\prod_{r_2=1,r_2\neq r_1}^{m_1}(a_{j_1j_1r_1}-a_{j_1j_1r_2})\prod_{u=1}^{n_1}(\overline{x}_{j_2}-a_{j_1j_2u}) > 0 \text{ and}\\
&a_{j_1j_10}\prod_{r_2=1,r_2\neq r_1}^{m_1}(a_{j_1j_1r_1}-a_{j_1j_1r_2})\prod_{u=1}^{n_1}(\overline{x}_{j_2}-a_{j_1j_2u}) < 0,
\end{aligned}$$

the flow at $x_{j_1}^* = a_{j_1j_1r_1}$ are source and sink, respectively.

Therefore, the equilibrium of $(x_{j_1}^*, x_{j_2}^*) = (a_{j_1 j_1 r_1}, a_{j_2 j_2 s_1})$ $(r_1 = 1, 2, \cdots, m_1; s_1 = 1, 2, \cdots, n)$ has the following properties as in Eqs. (1.286)–(1.289).

- For $a_{j_1 j_1 0} \prod_{r_2=1, r_2 \neq r_1}^{m_1} (a_{j_1 j_1 r_1} - a_{j_1 j_1 r_2}) \prod_{u=1}^{n_1} (a_{j_2 j_2 s_1} - a_{j_1 j_2 u}) < 0$ and $a_{j_2 j_2 0} \prod_{s_2=1, s_2 \neq s_1}^{n_2} (a_{j_2 j_2 s_1} - a_{j_2 j_2 s_2}) > 0$,

$$(a_{j_1 j_1 r_1}, a_{j_2 j_2 s_1}) = \underbrace{(\text{SO,SO})}_{\text{source}}.$$

- For $a_{j_1 j_1 0} \prod_{r_2=1, r_2 \neq r_1}^{m_1} (a_{j_1 j_1 r_1} - a_{j_1 j_1 r_2}) \prod_{u=1}^{n_1} (a_{j_2 j_2 s_1} - a_{j_1 j_2 u}) < 0$ and $a_{j_2 j_2 0} \prod_{s_2=1, s_2 \neq s_1}^{n_2} (a_{j_2 j_2 s_1} - a_{j_2 j_2 s_2}) > 0$,

$$(a_{j_1 j_1 r_1}, a_{j_2 j_2 s_1}) = \underbrace{(\text{SI,SO})}_{saddle}.$$

- For $a_{j_1 j_1 0} \prod_{r_2=1, r_2 \neq r_1}^{m_1} (a_{j_1 j_1 r_1} - a_{j_1 j_1 r_2}) \prod_{u=1}^{n_1} (a_{j_2 j_2 s_1} - a_{j_1 j_2 u}) > 0$ and $a_{j_2 j_2 0} \prod_{s_2=1, s_2 \neq s_1}^{n_2} (a_{j_2 j_2 s_1} - a_{j_2 j_2 s_2}) < 0$,

$$(a_{j_1 j_1 r_1}, a_{j_2 j_2 s_1}) = \underbrace{(\text{SO,SI})}_{\text{saddle}}.$$

- For $a_{j_1 j_1 0} \prod_{r_2=1, r_2 \neq r_1}^{m_1} (a_{j_1 j_1 r_1} - a_{j_1 j_1 r_2}) \prod_{u=1}^{n_1} (a_{j_2 j_2 s_1} - a_{j_1 j_2 u}) < 0$ and $a_{j_2 j_2 0} \prod_{s_2=1, s_2 \neq s_1}^{n_2} (a_{j_2 j_2 s_1} - a_{j_2 j_2 s_2}) < 0$,

$$(a_{j_1 j_1 r_1}, a_{j_2 j_2 s_1}) = \underbrace{(\text{SI,SI})}_{\text{sink}}.$$

Define

$$\begin{aligned}
&\bigcup_{r=1}^{m_1} \bigcup_{l=1}^{n_1+n_2} (a_{j_1 j_1 r}, c_{j_2 j_2 l}) \\
&= \bigcup_{r=1}^{m_1} \bigcup_{u=1}^{n_1} \underbrace{(a_{j_1 j_1 r}, a_{j_1 j_2 u})}_{\text{(XX,xF)}} \oplus \bigcup_{r=1}^{m_1} \bigcup_{s=1}^{n_2} \underbrace{(a_{j_1 j_1 r}, a_{j_2 j_2 s})}_{\text{XX}}, \\
&\{c_{j_2 j_2 l} | l = 1, 2, \cdots, n_1 + n_2\} \\
&= \text{sort}\{a_{j_1 j_2 u}, a_{j_2 j_2 s} | u = 1, 2, \cdots, n_1; s = 1, 2, \cdots, n_2\}, \\
&\text{for } c_{j_2 j_2 l_1} < c_{j_2 j_2 l_2} \text{ for } l_1 < l_2;
\end{aligned}$$

where

$$\bigcup\nolimits_{r=1}^{m_1}\bigcup\nolimits_{u=1}^{n_1}\underbrace{(a_{j_1j_1r},a_{j_1j_2u})}_{\text{XX}}$$

$$=\begin{bmatrix}(a_{j_1j_1 1},a_{j_1j_2n_1}) & \cdots & (a_{j_1j_1(m_1-1)},a_{j_1j_2n_1}) & (a_{j_1j_1m_1},a_{j_1j_2n_1})\\(a_{j_1j_1 1},a_{j_1j_2(n_1-1)}) & \cdots & (a_{j_1j_1(m_1-1)},a_{j_1j_2(n_1-1)}) & (a_{j_1j_1m_1},a_{j_1j_2(n_1-1)})\\ \vdots & \vdots & \vdots & \vdots\\(a_{j_1j_1 1},a_{j_1j_2 1}) & \cdots & (a_{j_1j_1(m_1-1)},a_{j_1j_2 1}) & (a_{j_1j_1m_1},a_{j_1j_2 1})\end{bmatrix}_{n_1\times m_1}.$$

and

$$\bigcup\nolimits_{r=1}^{m_1}\bigcup\nolimits_{s=1}^{n_2}\underbrace{(a_{j_1j_1r},a_{j_2j_2s})}_{\text{XX}}$$

$$=\begin{bmatrix}(a_{j_1j_1 1},a_{j_2j_2n_2}) & \cdots & (a_{j_1j_1(m_1-1)},a_{j_2j_2n_2}) & (a_{j_1j_1m_1},a_{j_2j_2n_2})\\(a_{j_1j_1 1},a_{j_2j_2(n_2-1)}) & \cdots & (a_{j_1j_1(m_1-1)},a_{j_2j_2(n_2-1)}) & (a_{j_1j_1m_1},a_{j_2j_2(n_2-1)})\\ \vdots & \vdots & \vdots & \vdots\\(a_{j_1j_1 1},a_{j_2j_2 1}) & \cdots & (a_{j_1j_1(m_1-1)},a_{j_2j_2 1}) & (a_{j_1j_1m_1},a_{j_2j_2 1})\end{bmatrix}_{n_2\times m_1}.$$

(iii$_2$) For $a_{j_1j_2u_1}=a_{j_2j_2s_1}$, the dynamical system becomes

$$\begin{aligned}\dot{x}_{j_1}&=a_{j_1j_10}(x_{j_2}-a_{j_1j_2u_1})\prod\nolimits_{r=1}^{m}(x_{j_1}-a_{j_1j_1r})\prod\nolimits_{u_2=1,u_2\neq u_1}^{n_1}(x_{j_2}-a_{j_1j_2u_2}),\\ \dot{x}_{j_2}&=a_{j_2j_20}(x_{j_2}-a_{j_2j_2s_1})\prod\nolimits_{s_2=1,s_2\neq s_1}^{n_2}(x_{j_2}-a_{j_2j_2s_2}).\end{aligned}$$

The first integral manifolds can be determined from Eqs. (1.280) through the equivalent dynamical system as

$$\begin{aligned}\dot{x}_{j_1}&=a_{j_1j_10}\prod\nolimits_{r=1}^{m}(x_{j_1}-a_{j_1j_1r})\prod\nolimits_{u_2=1,u_2\neq u_1}^{n_1}(x_{j_2}-a_{j_1j_2u_2}),\\ \dot{x}_{j_2}&=a_{j_2j_20}\prod\nolimits_{s_2=1,s_2\neq s_1}^{n_1}(x_{j_2}-a_{j_2j_2s_2}).\\ &n_1\Rightarrow n_1-1,\,u_{j+1}\Rightarrow u_j\ (j=1,2,\cdots,n_1-1);\\ &n_2\Rightarrow n_2-1,\,s_{i+1}\Rightarrow s_i\ (i=1,2,\cdots,n_2-1).\end{aligned}$$

Define a network matrix of 1-dimensional flows and equilibriums with an infinite-equilibrium as

$$\begin{aligned}
\mathrm{NM} &= \underbrace{(a_{j_1j_1r_1}, a_{j_2j_2s_1})}_{\text{(XX, YY)}} \oplus \bigcup\nolimits_{r_2=1,r_2\neq r_1}^{m_1} \underbrace{(a_{j_1j_1r_2}, a_{j_2j_2s_1})}_{\text{(XX,YY)}} \\
&\oplus \bigcup\nolimits_{r=1}^{m_1} \bigcup\nolimits_{u_2=1,u_2\neq u_1}^{n_1} \underbrace{(a_{j_1j_1r}, a_{j_1j_2u_2})}_{\text{(XX,xF)}} \\
&\oplus \bigcup\nolimits_{r=1}^{m_1} \bigcup\nolimits_{s_2=1,s_2\neq s_1}^{n_2} \underbrace{(a_{j_1j_1r}, a_{j_2j_2s_2})}_{\text{ZZ}},
\end{aligned}$$

For $x_{j_2}^* = a_{j_2j_2s_1} = a_{j_1j_2u_1}$ and $\overline{x}_{j_1} \neq a_{j_1j_1r}$, if

$$\frac{dx_{j_1}}{dx_{j_2}}\bigg|_{x_{j_2}^*=a_{j_1j_2u_1}} = \frac{a_{j_1j_10}\prod_{r=1}^{m_1}(\overline{x}_{j_1} - a_{j_1j_1r})\prod_{u_2=1,u_2\neq u_1}^{n_1}(a_{j_1j_2u_1} - a_{j_1j_2u_2})}{a_{j_2j_20}\prod_{s_2=1,s_2\neq s_1}^{n_2}(a_{j_1j_2u_1} - a_{j_2j_2s_2})} > 0,$$

the flow at $x_{j_2}^* = a_{j_2j_2s_1} = a_{j_1j_2u_1}$ is an increasing-inflection flow, and if

$$\frac{dx_{j_1}}{dx_{j_2}}\bigg|_{x_{j_2}^*=a_{j_1j_2u_1}} = \frac{a_{j_1j_10}\prod_{r=1}^{m_1}(\overline{x}_{j_1} - a_{j_1j_1r})\prod_{u_2=1,u_2\neq u_1}^{n_1}(a_{j_1j_2u_1} - a_{j_1j_2u_2})}{a_{j_2j_20}\prod_{s_2=1,s_2\neq s_1}^{n_2}(a_{j_1j_2u_1} - a_{j_2j_2s_2})} < 0,$$

the flow at $x_{j_2}^* = a_{j_2j_2s_1} = a_{j_1j_2u_1}$ is a decreasing-inflection flow.

The variational equation at $x_{j_2}^* = a_{j_2j_2s_1} = a_{j_1j_2u_1}$ is

$$\Delta\dot{x}_{j_2} = a_{j_2j_20}\prod\nolimits_{s_2=1,s_2\neq s_1}^{n_2}(a_{j_2j_2s_1} - a_{j_2j_2s_2})(\Delta x_{j_2}).$$

For

$$\begin{aligned}
&a_{j_2j_20}\prod\nolimits_{s_2=1,s_2\neq s_1}^{n_2}(a_{j_2j_2s_1} - a_{j_2j_2s_2}) > 0 \text{ and} \\
&a_{j_2j_20}\prod\nolimits_{s_2=1,s_2\neq s_1}^{n_2}(a_{j_2j_2s_1} - a_{j_2j_2s_2}) < 0
\end{aligned}$$

the flows at $x_{j_2}^* = a_{j_2j_2s_1} = a_{j_1j_2u_1}$ are source and sink, respectively.

At $x_{j_1}^* = a_{j_1j_1r_1}$, the variational equation for $\overline{x}_{j_2} \neq a_{j_1j_2u}$ is

$$\Delta\dot{x}_{j_1} = a_{j_1j_10}\prod\nolimits_{r_2=1,r_2\neq r_1}^{m_1}(a_{j_1j_1r_1} - a_{j_1j_1r_2})\prod\nolimits_{u_2=1}^{n_1}(\overline{x}_{j_1} - a_{j_1j_2u})(\Delta x_{j_1}).$$

For

$$a_{j_1j_10}\prod\nolimits_{r_2=1,r_2\neq r_1}^{m_1}(a_{j_1j_1r_1}-a_{j_1j_1r_2})\prod\nolimits_{u_2=1}^{n_1}(\overline{x}_{j_1}-a_{j_1j_2u})>0\text{ and}$$
$$a_{j_1j_10}\prod\nolimits_{r_2=1,r_2\neq r_1}^{m_1}(a_{j_1j_1r_1}-a_{j_1j_1r_2})\prod\nolimits_{u_2=1}^{n_1}(\overline{x}_{j_1}-a_{j_1j_2u})<0,$$

the flows at $x_{j_1}^*=a_{j_1j_1r_1}$ and $\overline{x}_{j_2}\neq a_{j_1j_2u}$ are source and sink, respectively.

Thus, based on the above analysis, for

$$a_{j_2j_20}\prod\nolimits_{s_2=1,s_2\neq s_1}^{n_2}(a_{j_2j_2s_1}-a_{j_2j_2s_2})>0,\text{ and}$$
$$a_{j_2j_20}\prod\nolimits_{s_2=1,s_2\neq s_1}^{n_2}(a_{j_2j_2s_1}-a_{j_2j_2s_2})<0$$

the flows at $x_{j_2}^*=a_{j_2j_2s_1}=a_{j_1j_2u_1}$ and on both sides of $x_{j_1}^*=a_{j_1j_1r_1}$ are source and sink flows, respectively.

(iii$_{2a}$) The infinite-equilibrium of $x_{j_2}^*=a_{j_1j_2u_1}=a_{j_2j_2s_1}$ with $\overline{x}_{j_1}\neq a_{j_1j_1r_1}$ has the following properties as in Eqs. (1.296)–(1.299).

- For $a_{j_1j_10}\prod_{r=1}^{m_1}(\overline{x}_{j_1}-a_{j_1j_1r})\prod_{u_2=1,u_2\neq u_1}^{n_1}(a_{j_1j_2u_1}-a_{j_1j_2u_2})>0$ and $a_{j_2j_20}\prod_{s_2=1,s_2\neq s_1}^{n_2}(a_{j_2j_2s_1}-a_{j_2j_2s_2})>0$,

$$(\overline{x}_{j_1},a_{j_2j_2s_1})=\underbrace{(\text{II,SO})}_{\text{increasing-inflection source}}.$$

  The infinite-equilibrium is a (II,SO)-increasing-inflection source at $x_{j_2}^*=a_{j_1j_2u_1}=a_{j_2j_2s_1}$.

- For $a_{j_1j_10}\prod_{r=1}^{m_1}(\overline{x}_{j_1}-a_{j_1j_1r})\prod_{u_2=1,u_2\neq u_1}^{n_1}(a_{j_1j_2u_1}-a_{j_1j_2u_2})<0$ and $a_{j_2j_20}\prod_{s_2=1,s_2\neq s_1}^{n_2}(a_{j_2j_2s_1}-a_{j_2j_2s_2})>0$,

$$(\overline{x}_{j_1},a_{j_2j_2s_1})=\underbrace{(\text{DI,SO})}_{\text{decreasing-inflection source}}.$$

  The infinite-equilibrium is a (DI,SO)-decreasing-inflection source at $x_{j_2}^*=a_{j_1j_2u_1}=a_{j_2j_2s_1}$.

- For $a_{j_1j_10}\prod_{r=1}^{m_1}(\overline{x}_{j_1}-a_{j_1j_1r})\prod_{u_2=1,u_2\neq u_1}^{n_1}(a_{j_1j_2u_1}-a_{j_1j_2u_2})>0$ and $a_{j_2j_20}\prod_{s_2=1,s_2\neq s_1}^{n_2}(a_{j_2j_2s_1}-a_{j_2j_2s_2})<0$,

$$(\overline{x}_{j_1},a_{j_2j_2s_1})=\underbrace{(\text{DI,SI})}_{\text{decreasing-inflection sink}}.$$

The infinite-equilibrium is a (DI,SI)- decreasing-inflection sink at $x_{j_2}^* = a_{j_1j_2u_1} = a_{j_2j_2s_1}$.

- For $a_{j_1j_10}\prod_{r=1}^{m_1}(\overline{x}_{j_1} - a_{j_1j_1r})\prod_{u_2=1,u_2\neq u_1}^{n_1}(a_{j_1j_2u_1} - a_{j_1j_2u_2}) < 0$ and $a_{j_2j_20}\prod_{s_2=1,s_2\neq s_1}^{n_2}(a_{j_2j_2s_1} - a_{j_2j_2s_2}) < 0$,

$$(\overline{x}_{j_1}, a_{j_2j_2s_1}) = \underbrace{\text{(II,SI)}}_{\text{increasing-inflection sink}}.$$

The infinite-equilibrium is a (II,SI)-increasing-inflection sink at $x_{j_2}^* = a_{j_1j_2u_1} = a_{j_2j_2s_1}$.

(iii$_{2b}$) On both sides of $\overline{x}_{j_2} = a_{j_1j_1r_1}$, the inflection flows will be switched. Therefore, the equilibrium of $(x_{j_1}^*, x_{j_2}^*) = (a_{j_1j_1r_1}, a_{j_2j_2s_1})$ on the infinite-equilibrium of $x_{j_2}^* = a_{j_1j_2u_1} = a_{j_2j_2s_1}$ has the following properties as in Eqs. (1.300)–(1.303).

- For $a_{j_1j_10}\prod_{r_2=1,r_2\neq r_1}^{m_1}(a_{j_1j_1r_1} - a_{j_1j_1r_2})\prod_{u_2=1,u_2\neq u_1}^{n_2}(a_{j_1j_2u_1} - a_{j_1j_2u_2}) > 0$ and $a_{j_2j_20}\prod_{s_2=1,s_2\neq s_1}^{n_2}(a_{j_2j_2s_1} - a_{j_2j_2s_2}) > 0$,

$$(a_{j_1j_1r_1}, a_{j_2j_2s_1}) = \underbrace{\text{(DI:II,SO)}}_{\text{decreasing-to-increasing-inflection source}}.$$

The equilibrium is a (DI:II,SO)-decreasing-to-increasing-inflection source at $(x_{j_1}^*, x_{j_2}^*) = (a_{j_1j_1r_1}, a_{j_2j_2s_1})$.

- For $a_{j_1j_10}\prod_{r_2=1,r_2\neq r_1}^{m_1}(a_{j_1j_1r_1} - a_{j_1j_1r_2})\prod_{u_2=1,u_2\neq u_1}^{n_2}(a_{j_1j_2u_1} - a_{j_1j_2u_2}) < 0$ and $a_{j_2j_20}\prod_{s_2=1,s_2\neq s_1}^{n_2}(a_{j_2j_2s_1} - a_{j_2j_2s_2}) > 0$,

$$(a_{j_1j_1r_1}, a_{j_2j_2s_1}) = \underbrace{\text{(II:DI,SO)}}_{\text{increasing-to-decreasing-inflection source}}.$$

The equilibrium is an (II:DI,SO)-increasing-to-decreasing-inflection source at $(x_{j_1}^*, x_{j_2}^*) = (a_{j_1j_1r_1}, a_{j_2j_2s_1})$.

- For $a_{j_1j_10}\prod_{r_2=1,r_2\neq r_1}^{m_1}(a_{j_1j_1r_1} - a_{j_1j_1r_2})\prod_{u_2=1,u_2\neq u_1}^{n_2}(a_{j_1j_2u_1} - a_{j_1j_2u_2}) > 0$ and $a_{j_2j_20}\prod_{s_2=1,s_2\neq s_1}^{n_2}(a_{j_2j_2s_1} - a_{j_2j_2s_2}) < 0$,

$$(a_{j_1j_1r_1}, a_{j_2j_2s_1}) = \underbrace{\text{(II:DI,SI)}}_{\text{increasing-to-decreasing-inflection sink}}.$$

The equilibrium is an (II:DI,SO)-increasing-to-decreasing-inflection sink at $(x_{j_1}^*, x_{j_2}^*) = (a_{j_1j_1r_1}, a_{j_2j_2s_1})$.

- For $a_{j_1j_10}\prod_{r_2=1,r_2\neq r_1}^{m_1}(a_{j_1j_1r_1} - a_{j_1j_1r_2})\prod_{u_2=1,u_2\neq u_1}^{n_2}(a_{j_1j_2u_1} - a_{j_1j_2u_2}) < 0$ and $a_{j_2j_20}\prod_{s_2=1,s_2\neq s_1}^{n_2}(a_{j_2j_2s_1} - a_{j_2j_2s_2}) < 0$,

$$(a_{j_1j_1r_1}, a_{j_2j_2s_1}) = \underbrace{\text{(DI:II,SI)}}_{\text{decreasing-to-increasing-inflection source}}.$$

The equilibrium is a (DI:II,SI)-decreasing-to-increasing-inflection sink at $(x^*_{j_1}, x^*_{j_2}) = (a_{j_1j_1r_1}, a_{j_2j_2s_1})$.

2.3 Appearing and Switching for Equilibriums and Flows

In this section, the theorem on appearing and switching bifurcations of singular equilibriums and 1-dimensional flows in the product polynomial systems will be proved.

2.3.1 Appearing and Switching for 1-Dimensional Flows

(iv) The 1-dimensional flow at $(x^*_{j_1}, x^*_{j_2}) = (a_{j_1j_11}, a_{j_1j_21})$ with $a_{j_2j_21} \neq a_{j_1j_21}$ has four classes of 1-diemsnional flow bifurcations with $\dot{x}^*_{j_2} = a_{j_2j_20}(a_{j_1j_21} - a_{j_2j_21})^{n_2}$.

(iv$_1$) For $m_1 = 2m_{11}+1$ and $n_1 = 2n_{11}+1$ the equilibrium of $(x^*_{j_1}, x^*_{j_2}) = (a_{j_1j_11}, a_{j_1j_21})$ for the x_{j_1}-direction has the following bifurcation properties with

$$\begin{aligned}
&2m_{11} + 1 - \sum\nolimits_{r_1=1}^{p_1} m_{j_1r_1} = 2\sum\nolimits_{\rho_1=1}^{l_1} m_{\rho_1},\\
&2n_{11} + 1 - \sum\nolimits_{u_1=1}^{q_1} n_{j_1u_1} = 2\sum\nolimits_{\sigma_1=1}^{l_2} n_{\sigma_1};\\
&\sum\nolimits_{r_2=1}^{\bar{p}_1} m_{j_1r_2} = 2m_{11} + 1, \sum\nolimits_{u_2=1}^{\bar{q}_1} n_{j_1u_2} = 2n_{11} + 1;\\
&\sum\nolimits_{r_i=1}^{p_1} m_{j_1r_i} = 2m_{11} + 1, \sum\nolimits_{u_i=1}^{q_1} n_{j_1u_i} = 2n_{11} + 1, \textit{ for } i = 1, 2;
\end{aligned}$$

l_1, l_2-quadratic polynomials are without real roots and, ρ_1^{th} and σ_1^{th}-quadratic polynomials are with powers m_{ρ_1} and n_{σ_1}, respectively.

The bifurcation process is discussed through differential equations as follows. There are four bifurcations (I$_1$)–(I$_3$).

(I$_1$) Consider a dynamical system having a single equilibrium (A$_1$) as

$$\begin{aligned}
\dot{x}_{j_1} &= a_{j_1j_10}(x_{j_1} - b_{j_1j_11})\prod\nolimits_{r=1}^{m_{11}} [(x_{j_1} - a^{(r)}_{j_1j_1})^2 + \Delta^{(r)}_{j_1j_1}]\\
&\quad \times (x_{j_2} - b_{j_1j_21})\prod\nolimits_{u=1}^{n_{11}} [(x_{j_2} - a^{(u)}_{j_1j_2})^2 + \Delta^{(u)}_{j_1j_2}],\\
\dot{x}_{j_2} &= a_{j_2j_20}(x_{j_2} - a_{j_2j_21})^{n_2};
\end{aligned}$$

where $a_{j_1j_11} = b_{j_1j_11}$ and $a_{j_1j_21} = b_{j_1j_21} \neq a_{j_2j_21}$. Once

$$\Delta_{j_1j_1}^{(r)} = \Delta_{j_1j_1} = 0\,(r = 1, 2, \cdots, m_{11}),$$
$$\Delta_{j_1j_2}^{(u)} = \Delta_{j_1j_2} = 0\,(u = 1, 2, \cdots, n_{11}),$$

if

$$a_{j_1j_11} = b_{j_1j_11} = a_{j_1j_1}^{(r)}\,(r = 1, 2, \cdots, m_{11}),$$
$$a_{j_1j_21} = b_{j_1j_21} = a_{j_1j_2}^{(u)}\,(u = 1, 2, \cdots, n_{11}),$$

the foregoing differential equations becomes a dynamical system with a singular equilibrium (B_1) as

$$\dot{x}_{j_1} = a_{j_1j_10}(x_{j_1} - a_{j_1j_11})^{2m_{11}+1}(x_{j_2} - a_{j_1j_21})^{2n_{11}+1},$$
$$\dot{x}_{j_2} = a_{j_2j_20}(x_{j_2} - a_{j_2j_21})^{n_2}.$$

If

$$\Delta_{j_1j_1}^{(r)} = \Delta_{j_1j_1} + \delta_r\,(r = 1, 2, \cdots, m_{11}),\ \delta_r > 0;$$
$$\Delta_{j_1j_2}^{(u)} = \Delta_{j_1j_2} + \delta_u\,(u = 1, 2, \cdots, n_{11}),\ \delta_u > 0;$$

then

$$a_{j_1j_11}^{(r)}, a_{j_1j_12}^{(r)} = a_{j_1j_11} \pm \varepsilon_r,\ (r = 1, 2, \cdots, m_{11}),$$
$$\{a_{j_1j_11}, a_{j_1j_12}, \cdots, a_{j_1j_1(2m_{11}+1)}\}$$
$$= \text{sort}\{b_{j_1j_11}, a_{j_1j_11}^{(r)}, a_{j_1j_12}^{(r)} | r = 1, 2, \cdots, m_{11}\};$$
$$a_{j_1j_21}^{(u)}, a_{j_1j_22}^{(u)} = a_{j_1j_21} \pm \varepsilon_u,\ (u = 1, 2, \cdots, n_1),$$
$$\{a_{j_1j_21}, a_{j_1j_22}, \cdots, a_{j_1j_2(2n_{11}+1)}\}$$
$$= \text{sort}\{b_{j_1j_21}, a_{j_1j_21}^{(u)}, a_{j_1j_22}^{(u)} | u = 1, 2, \cdots, n_{11}\}.$$

Thus, the differential equation becomes a dynamical system with a non-singular equilibrium network (C_1) as

$$\dot{x}_{j_1} = a_{j_1j_10}\prod\nolimits_{r=1}^{2m_{11}+1}(x_{j_1} - a_{j_1j_1r})\prod\nolimits_{u=1}^{2n_{11}+1}(x_{j_2} - a_{j_1j_2u}),$$
$$\dot{x}_{j_2} = a_{j_2j_20}(x_{j_2} - a_{j_2j_21})^{n_2}.$$

- From the above bifurcation process analysis, at least, only $(m_{11}+1)$-parameter variations in the x_{j_1}-direction are engaged in such a bifurcation from a 1-dimensional flow to the 1-dimensional flow array of $1 \times (2m_{11} + 1)$ through the $(2m_{11} + 1)^{\text{th}}$-order source and sink bifurcations in the x_{j_1}-direction. The higher-order source and sink flow bifurcations are developed.

Thus, the 1-dimensional flow appearing or vanishing bifurcation route is as follows.

$$\underbrace{(a_{j_1j_1 1},\dot{x}_{j_2})}_{\text{XX}} \rightleftharpoons \underbrace{(a_{j_1j_1 1},\dot{x}_{j_2})}_{(2m_{11}+1)^{\text{th}}\text{order XX}} \rightleftharpoons \bigcup\nolimits_{r=1}^{2m_{11}+1} \underbrace{(a_{j_1j_1 r},\dot{x}_{j_2})}_{\text{XX}}.$$

- From the above bifurcation process analysis, at least, only $(n_{11}+1)$-parameter variations in the x_{j_1}-direction are engaged in such a bifurcation from a 1-dimensional flow to the 1-dimensional flow array of $(2n_{11}+1)\times 1$ through the $(2n_{11}+1)^{\text{th}}$-order up-parabola and down-parabola bifurcations in the x_{j_1}-direction. The higher-order up-parabola and down-parabola flow bifurcations are developed.

Thus, the 1-dimensional flow appearing or vanishing bifurcation route is as follows.

$$\underbrace{(a_{j_1j_2 1},\dot{x}_{j_2})}_{\text{YY}} \rightleftharpoons \underbrace{(a_{j_1j_2 1},\dot{x}_{j_2})}_{(2n_{11}+1)^{\text{th}}\text{order YY}} \rightleftharpoons \bigcup\nolimits_{u=1}^{2n_{11}+1} \underbrace{(a_{j_1j_2 u},\dot{x}_{j_2})}_{\text{YY}}.$$

- From the above bifurcation process analysis, $(m_{11}+n_{11}+2)$-parameter variations in the two directions are engaged in such a bifurcation from a parabola-flow to a $(2m_{11}+1)\times(2n_{11}+1)$ matrix of hyperbolic and hyperbolic-secant flows through the $(2m_{11}+1)$-source and sink bifurcations in the-direction, and the $(2n_{11}+1)^{\text{th}}$-order up-parabola and down-parabola flows bifurcations in the-direction. The higher-order hyperbolic and hyperbolic-recant flow bifurcations are developed.

Thus, the 1-dimensional flow appearing or vanishing bifurcation route is as follows.

$$\begin{aligned}\underbrace{(a_{j_1j_1 1},a_{j_1j_2 1})}_{\text{(XX,xF)-flow}} &\rightleftharpoons \underbrace{([(2m_{11}+1),(2n_{11}+1)]\text{XX,xF})}_{[(2m_{11}+1),(2n_{11}+1)]\text{-XX}}\\ &\rightleftharpoons \bigcup\nolimits_{r=1}^{2m_{11}+1}\bigcup\nolimits_{u=1}^{2n_{11}+1} \underbrace{(a_{j_1j_1 r},a_{j_1j_2 u})}_{\text{(YY,xF)-flow}},\end{aligned}$$

which are the higher-order hyperbolic and hyperbolic-secant flows bifurcations.

(I_2) From Case (I_1), consider a dynamical system having singular equilibriums (A_2) as

$$\begin{aligned}\dot{x}_{j_1} &= a_{j_1j_1 0}\prod\nolimits_{r_1=1}^{p_1}(x_{j_1}-b_{j_1j_1r_1})^{m_{j_1r_1}}\prod\nolimits_{r=1}^{l_1}[(x_{j_1}-a_{j_1j_1}^{(r)})^2+\Delta_{j_1j_1}^{(r)}]^{m_{j_1r}}\\ &\times\prod\nolimits_{u_1=1}^{q_1}(x_{j_2}-b_{j_1j_2u_1})^{n_{j_1u_1}}\prod\nolimits_{u=1}^{l_2}[(x_{j_2}-a_{j_1j_2}^{(u)})^2+\Delta_{j_1j_2}^{(u)}]^{n_{j_1u}},\\ \dot{x}_{j_2} &= a_{j_2j_2 0}(x_{j_2}-a_{j_2j_2 1})^{n_2};\end{aligned}$$

where

$$
\begin{aligned}
&2m_{11}+1-\sum\nolimits_{r_1=1}^{p_1} m_{j_1r_1}=2\sum\nolimits_{\rho_1=1}^{l_1} m_{\rho_1},\\
&2n_{11}+1-\sum\nolimits_{u_1=1}^{q_1} n_{j_1u_1}=2\sum\nolimits_{\sigma_1=1}^{l_2} n_{\sigma_1};\\
&\sum\nolimits_{r_2=1}^{\bar{p}_1} m_{j_1r_2}=2m_{11}+1,\ \sum\nolimits_{u_2=1}^{\bar{q}_1} n_{j_1u_2}=2n_{11}+1;
\end{aligned}
$$

l_1, l_2-quadratic polynomials are without real roots and ρ_1^{th} and σ_1^{th}-quadratic polynomials are with powers m_{ρ_1} and n_{σ_1}, respectively.

Once

$$
\begin{aligned}
\Delta_{j_1j_1}^{(r)}&=0\,(r=1,2,\cdots,l_1),\\
\Delta_{j_1j_2}^{(u)}&=0\,(u=1,2,\cdots,l_2);
\end{aligned}
$$

if

$$
\begin{aligned}
a_{j_1j_1 1}&\equiv a_{j_1j_1r_1}=a_{j_1j_1}^{(r)}\ (r_1=1,2,\cdots,p_1;\,r=1,2,\cdots,l_1),\\
a_{j_1j_2 1}&\equiv a_{j_1j_2u_1}=a_{j_1j_2}^{(u)}\ (u_1=1,2,\cdots,q_1;\,u=1,2,\cdots,l_2),
\end{aligned}
$$

the foregoing differential equations become a dynamical system with a 1-dimensional flow (B_2) as

$$
\begin{aligned}
\dot{x}_{j_1}&=a_{j_1j_1 0}(x_{j_1}-a_{j_1j_1 1})^{2m_{11}+1}(x_{j_2}-a_{j_1j_2 1})^{2n_{11}+1},\\
\dot{x}_{j_2}&=a_{j_2j_2 0}(x_{j_2}-a_{j_2j_2 1})^{n_2}.
\end{aligned}
$$

If

$$
\begin{aligned}
\Delta_{j_1j_1}^{(r)}&=\Delta_{j_1j_1}+\delta_r\ (r=1,2,\cdots,l_1),\ \delta_r>0;\\
\Delta_{j_1j_2}^{(u)}&=\Delta_{j_1j_2}+\delta_u\ (u=1,2,\cdots,l_2),\ \delta_s>0;
\end{aligned}
$$

then

$$
\begin{aligned}
&a_{j_1j_1 1}^{(r)},a_{j_1j_1 2}^{(r)}=a_{j_1j_1 1}\pm\varepsilon_r,\ (r=1,2,\cdots,l_1),\\
&\{a_{j_1j_1 1},a_{j_1j_1 2},\cdots,a_{j_1j_1\bar{p}_1}\}\\
&=\text{sort}\{a_{j_1j_1 1},a_{j_1j_1 2},\cdots,a_{j_1j_1p_1};a_{j_1j_1 1}^{(r)},a_{j_1j_1 2}^{(r)}|r=1,2,\cdots,l_1\};\\
&a_{j_1j_2 1}^{(u)},a_{j_1j_2 2}^{(u)}=a_{j_1j_2 1}\pm\varepsilon_u,\ (u=1,2,\cdots,l_2),\\
&\{a_{j_1j_2 1},a_{j_1j_2 2},\cdots,a_{j_1j_2\bar{q}_1}\}\\
&=\text{sort}\{a_{j_1j_2 1},a_{j_1j_2 2},\cdots,a_{j_1j_2q_1};a_{j_2j_1 1}^{(u)},a_{j_2j_1 2}^{(u)}|u=1,2,\cdots,l_2\}.
\end{aligned}
$$

Thus, the differential equation becomes a dynamical system with singular 1-dimensional flows (C_2) as

$$\dot{x}_{j_1} = a_{j_1 j_1 0} \prod\nolimits_{r_2=1}^{\overline{p}_1} (x_{j_1} - a_{j_1 j_1 r_2})^{m_{j_1 r_2}} \prod\nolimits_{u_2=1}^{\overline{q}_1} (x_{j_2} - a_{j_1 j_1 u_2})^{n_{j_1 u_2}},$$
$$\dot{x}_{j_2} = a_{j_2 j_2 0}(x_{j_2} - a_{j_2 j_2 1})^{n_2}$$

where

$$\Sigma_{r_2=1}^{\overline{p}_1} m_{j_1 r_2} = 2m_{11} + 1,\ \Sigma_{u_2=1}^{\overline{q}_1} n_{j_1 u_2} = 2n_{11} + 1.$$

- From the above bifurcation process analysis, only $\overline{p}_1$-parameter variations in the x_{j_1}-direction from a p_1 to $\overline{p}_1$ 1-dimensional flow array through the $(2m_{11} + 1)^{\text{th}}$-order source and sink flow bifurcations in the x_{j_1}-direction.

  Thus, the 1-dimensional flow appearing or vanishing bifurcation route from a p_1 to $\overline{p}_1$ 1-dimensional flow array is expressed as

$$\bigcup\nolimits_{r_1=1}^{p_1} \underbrace{(a_{j_1 j_1 r_1}, \dot{x}_{j_2})}_{(m_{j_1 r_1})^{\text{th}}\text{order XX}} \rightleftharpoons \underbrace{(a_{j_1 j_1 1}, \dot{x}_{j_2})}_{(2m_{11}+1)^{\text{th}}\text{ order XX}} \rightleftharpoons \bigcup\nolimits_{r_2=1}^{\overline{p}_1} \underbrace{(a_{j_1 j_1 r_2}, \dot{x}_{j_2})}_{(m_{j_1 r_2})^{\text{th}}\text{order XX}}.$$

- From the above bifurcation process analysis, only $\overline{q}_1$-parameter variations in the x_{j_1}-direction from a q_1 to $\overline{q}_1$ 1-dimensional flow array through the $(2n_{11} + 1)^{\text{th}}$-order up-parabola and down-parabola flow bifurcations in the x_{j_1}-direction.

  Thus, the 1-dimensional flow appearing or vanishing bifurcation route from a q_1 to $\overline{q}_1$ 1-dimensional flow array is expressed as

$$\bigcup\nolimits_{u_1=1}^{q_1} \underbrace{(a_{j_2 j_1 u_1}, \dot{x}_{j_2})}_{(n_{j_1 u_1})^{\text{th}}\text{order YY}} \rightleftharpoons \underbrace{(a_{j_1 j_2 1}, \dot{x}_{j_2})}_{(2n_{11}+1)^{\text{th}}\text{ order YY}} \rightleftharpoons \bigcup\nolimits_{u_2=1}^{\overline{q}_1} \underbrace{(a_{j_1 j_2 u_2}, \dot{x}_{j_2})}_{(n_{j_1 u_2})^{\text{th}}\text{order YY}}.$$

- From the above bifurcation process analysis, only $\overline{q}_1 + \overline{p}_1$-parameter variations in the x_{j_1}-direction from a $q_1 \times p_1$ to $\overline{q}_1 \times \overline{p}_1$ matrix of 1-dimensional flows through the $(2m_{11} + 1)^{\text{th}}$-order source and sink flows bifurcations in the x_{j_1}-direction, and the $(2n_{11} + 1)^{\text{th}}$-order up-parabola and down-parabola flows bifurcations in the x_{j_1}-direction.

  Thus, the 1-dimensional flow appearing or vanishing bifurcation route is as follows:

$$\bigcup\nolimits_{r_1=1}^{p_1} \bigcup\nolimits_{u_1=1}^{q_1} \underbrace{(a_{j_1 j_1 r_1}, a_{j_1 j_2 u_1})}_{([(m_{j_1 r_1}),(n_{j_1 u_1})]\text{XX,xF})\text{-flow}} \rightleftharpoons \underbrace{[(2m_{11}+1), (2n_{11} + 1)]\text{XX, xF})}_{[(2m_{11}+1),(2n_{11}+1)]\text{-XX}}$$
$$\rightleftharpoons \bigcup\nolimits_{r_2=1}^{\overline{p}_1} \bigcup\nolimits_{u_2=1}^{\overline{q}_1} \underbrace{(a_{j_1 j_1 r_2}, a_{j_1 j_2 u_2})}_{([(m_{j_1 r_2}),(n_{j_1 u_1})]\text{XX,xF})\text{-flow}},$$

  which are the higher-order hyperbolic and hyperbolic-secant flows bifurcations.

(I_3) Consider two dynamical systems with the same 1-dimensional flows with locations switched (A_3,C_3) for $i = 1, 2$ as

$$\dot{x}_{j_1} = a_{j_1 j_1 0} \prod_{r_i=1}^{p_1} (x_{j_1} - a_{j_1 j_1 r_i})^{m_{j_1 r_i}} \prod_{u_i=1}^{q_1} (x_{j_2} - a_{j_1 j_2 u_i})^{n_{j_1 u_i}},$$
$$\dot{x}_{j_2} = a_{j_2 j_2 0} (x_{j_2} - a_{j_2 j_2 1})^{n_2};$$

where

$$\sum_{r_i=1}^{p_1} m_{j_1 r_i} = 2m_{11} + 1, \sum_{u_i=1}^{q_1} n_{j_1 u_i} = 2n_{11} + 1, \text{ for } i = 1, 2.$$

Consider a dynamical system as a singular 1-dimensional flow (B_3) as

$$\dot{x}_{j_1} = a_{j_1 j_1 0}(x_{j_1} - a_{j_1 j_1 1})^{2m_{11}+1}(x_{j_2} - a_{j_1 j_2 1})^{2n_{11}+1},$$
$$\dot{x}_{j_2} = a_{j_2 j_2 0}(x_{j_2} - a_{j_2 j_2 1})^{n_2}.$$

For $r_i = 1, 2, \cdots, p_1$ and $u_i = 1, 2, \cdots, q_1$ $(i = 1, 2)$, define a function as

$$\Delta_{j_1 j_1}^{(r_1 r_2)} = (a_{j_1 j_1 r_1} - a_{j_1 j_1 r_2})^2 \; (r_1, r_2 \in \{1, 2, \cdots, p_1\}, r_1 \neq r_2),$$
$$\Delta_{j_1 j_2}^{(u_1 u_2)} = (a_{j_1 j_2 u_1} - a_{j_1 j_2 u_2})^2 \; (u_1, u_2 \in \{1, 2, \cdots, q_1\}, u_1 \neq u_2).$$

For $r_i = 1, 2, \cdots, p$ and $s_i = 1, 2, \cdots, q_1$ $(i = 1, 2)$, if

$$\Delta_{j_1 j_1}^{(r_1 r_2)} = (a_{j_1 j_1 r_1} - a_{j_1 j_1 r_2})^2 = 0 \text{ and } \Delta_{j_1 j_2}^{(u_1 u_2)} = (a_{j_1 j_2 u_1} - a_{j_1 j_2 u_2})^2 = 0,$$

there are three bifurcations as follows.

- Two 1-dimensional flows of and switching at point with the same order singularity are given through

$$\underbrace{\bigcup_{r_1=1}^{p_1} (a_{j_1 j_1 r_1}, \dot{x}_{j_2})}_{(m_{j_1 r_1})^{\text{th}} \text{order XX}} \rightleftharpoons \underbrace{(a_{j_1 j_1 1}, \dot{x}_{j_2})}_{(2m_{11}+1)^{\text{th}} \text{ order XX}} \rightleftharpoons \underbrace{\bigcup_{r_2=1}^{p_1} (a_{j_1 j_1 r_2}, \dot{x}_{j_2})}_{(m_{j_1 r_2})^{\text{th}} \text{order XX}}$$
$$\left\{a_{j_1 j_1 r_1} \rightleftharpoons a_{j_1 j_1 r_2} \middle| m_{j_1 r_1} = m_{j_1 r_2}; r_1, r_2 \in \{1, 2, \cdots, p_1\}, r_1 \neq r_2\right\};$$

  and

$$a_{j_1 j_1 r_1} = a_{j_1 j_1 1} - \varepsilon_{r_1}, a_{j_1 j_1 r_2} = a_{j_1 j_1 1} + \varepsilon_{r_2},$$
$$(a_{j_1 j_1 r_1} = a_{j_1 j_1 1} + \varepsilon_{r_1}, a_{j_1 j_1 r_2} = a_{j_1 j_1 1} - \varepsilon_{r_2});$$
$$\varepsilon_{r_1}, \varepsilon_{r_2} > 0; \; r_i = 1, 2, \cdots, p_1 \; (i = 1, 2).$$

- Two 1-dimensional flows of $(a_{j_1 j_2 u_1}, \dot{x}_{j_2})$ and $(a_{j_1 j_2 u_2}, \dot{x}_{j_2})$ switching at point $(a_{j_1 j_2 1}, \dot{x}_{j_2})$ with the same order singularity are given through

$$\bigcup_{u_1=1}^{q_1} \underbrace{(a_{j_1j_2u_1}, \dot{x}_{j_2})}_{(n_{j_1u_1})^{\text{th}}\text{order YY}} \rightleftharpoons \underbrace{(a_{j_1j_21}, \dot{x}_{j_2})}_{(2n_{11}+1)^{\text{th}}\text{ order YY}} \rightleftharpoons \bigcup_{u_2=1}^{q_1} \underbrace{(a_{j_1j_2u_2}, \dot{x}_{j_2})}_{(n_{j_1u_2})^{\text{th}}\text{order YY}}$$

$$\left\{a_{j_1j_2u_1} \rightleftharpoons a_{j_1j_2u_2} \middle| n_{j_1u_1} = n_{j_1u_2}; u_1, u_2 \in \{1, 2, \cdots, q\}, u_1 \neq u_2\right\};$$

and

$$\begin{aligned}
&a_{j_1j_2u_1} = a_{j_1j_21} - \varepsilon_{u_1}, a_{j_1j_2u_2} = a_{j_1j_21} + \varepsilon_{u_2},\\
&(a_{j_1j_2u_1} = a_{j_1j_21} + \varepsilon_{u_1}, a_{j_1j_2u_2} = a_{j_1j_21} - \varepsilon_{u_2});\\
&\varepsilon_{u_1}, \varepsilon_{u_2} > 0;\ u_i = 1, 2, \cdots, q_1\ (i = 1, 2).
\end{aligned}$$

- Two 1-dimensional flows of $(a_{j_1j_1r_1}, a_{j_1j_2u_1})$ and $(a_{j_1j_1r_2}, a_{j_1j_2u_2})$ switching at point $(a_{j_1j_11}, a_{j_1j_21})$ with the same order singularity are given through

$$\begin{aligned}
\bigcup_{r_1=1}^{p_1}\bigcup_{u_1=1}^{q_1} \underbrace{(a_{j_1j_1r_1}, a_{j_1j_2u_1})}_{([(m_{j_1r_1}),(n_{j_1u_1})]\text{XX,xF})\text{-flow}} &\rightleftharpoons \underbrace{([(2m_{11}+1),(2\text{n}_{11}+1)]\text{XX, xF})}_{[(2m_{11}+1),(2\text{n}_{11}+1)]\text{-XX}}\\
&\rightleftharpoons \bigcup_{r_2=1}^{p_1}\bigcup_{u_2=1}^{q_1} \underbrace{(a_{j_1j_1r_2}, a_{j_1j_2u_2})}_{([(m_{j_1r_2}),(n_{j_1r_2})]\text{XX,xF})\text{-flow}}
\end{aligned}$$

with

$$\begin{aligned}
&a_{j_1j_1r_1} = a_{j_1j_11} - \varepsilon_{r_1}, a_{j_1j_1r_2} = a_{j_1j_11} + \varepsilon_{r_2},\\
&(a_{j_1j_1r_1} = a_{j_1j_11} + \varepsilon_{r_1}, a_{j_1j_1r_2} = a_{j_1j_11} - \varepsilon_{r_2}),\\
&\varepsilon_{r_1}, \varepsilon_{r_2} > 0,\ r_i = 1, 2, \cdots, p_1\ (i = 1, 2);\\
&a_{j_1j_2u_1} = a_{j_1j_21} - \varepsilon_{u_1}, a_{j_1j_2u_2} = a_{j_1j_21} + \varepsilon_{u_2},\\
&(a_{j_1j_2u_1} = a_{j_1j_21} + \varepsilon_{u_1}, a_{j_1j_2u_2} = a_{j_1j_21} - \varepsilon_{u_2}),\\
&\varepsilon_{u_1}, \varepsilon_{u_2} > 0,\ u_i = 1, 2, \cdots, q_1\ (i = 1, 2).
\end{aligned}$$

The switching bifurcations are the higher-order hyperbolic and hyperbolic-secant flows bifurcations.

In summary, from the analysis in case (I_1)–(I_3), the 1-dimensional flow of $(x_{j_1}^*, x_{j_2}^*) = (a_{j_1j_11}, a_{j_1j_21})$ for $m_1 = 2m_{11} + 1$ and $n_1 = 2n_{11} + 1$ has the bifurcation properties as stated in $(\text{iv}_{1\text{a}})$-$(\text{iv}_{1\text{d}})$ through Eqs. (1.305)–(1.324).

(iv_2) For $m_1 = 2m_{11}$ and $n_1 = 2n_{11} + 1$ the equilibrium of $(x_{j_1}^*, x_{j_2}^*) = (a_{j_1j_11}, a_{j_1j_21})$ for the x_{j_1}-direction has the following bifurcation properties with

$$\begin{aligned}
&2m_{11} - \sum\nolimits_{r_1=1}^{p_1} m_{j_1r_1} = 2\sum\nolimits_{\rho_1=1}^{l_1} m_{\rho_1},\\
&2n_{11} + 1 - \sum\nolimits_{u_1=1}^{q_1} n_{j_1u_1} = 2\sum\nolimits_{\sigma_1=1}^{l_2} n_{\sigma_1};
\end{aligned}$$

$$\sum_{r_2=1}^{\bar{p}_1} m_{j_1 r_2} = 2m_{11}, \sum_{u_2=1}^{\bar{q}_1} n_{j_1 u_2} = 2n_{11} + 1;$$
$$\sum_{r_i=1}^{p_1} m_{j_1 r_i} = 2m_{11}, \sum_{u_i=1}^{q_1} n_{j_1 u_i} = 2n_{11} + 1, for\, i = 1, 2;$$

l_1, l_2-quadratic polynomials are without real roots and ρ_1^{th} and σ_1^{th} -quadratic polynomials are with powers m_{ρ_1} and n_{σ_1}, respectively.

The bifurcation process is discussed through differential equations as follows. There are four bifurcations (I_1)–(I_3).

(I_1) Consider a dynamical system having a single equilibrium (A_1) as

$$\begin{aligned}\dot{x}_{j_1} =& a_{j_1 j_1 0} \prod_{r=1}^{m_{11}} [(x_{j_1} - a_{j_1 j_1}^{(r)})^2 + \Delta_{j_1 j_1}^{(r)}] \\ & \times (x_{j_2} - b_{j_1 j_2 1}) \prod_{u=1}^{n_{11}} [(x_{j_2} - a_{j_1 j_2}^{(u)})^2 + \Delta_{j_1 j_2}^{(u)}], \\ \dot{x}_{j_2} =& a_{j_2 j_2 0}(x_{j_2} - a_{j_2 j_2 1})^{n_2};\end{aligned}$$

where $a_{j_1 j_2 1} = b_{j_1 j_2 1} \neq a_{j_2 j_2 1}$. Once

$$\begin{aligned}\Delta_{j_1 j_1}^{(r)} &= \Delta_{j_1 j_1} = 0\,(r = 1, 2, \cdots, m_{11}), \\ \Delta_{j_1 j_2}^{(u)} &= \Delta_{j_1 j_2} = 0\,(u = 1, 2, \cdots, n_{11}),\end{aligned}$$

if

$$\begin{aligned}a_{j_1 j_1 1} = b_{j_1 j_1 1} &= a_{j_1 j_1}^{(r)}\,(r = 1, 2, \cdots, m_{11}), \\ a_{j_1 j_2 1} = b_{j_1 j_2 1} &= a_{j_1 j_2}^{(u)}\,(u = 1, 2, \cdots, n_{11}),\end{aligned}$$

the foregoing differential equations become a dynamical system with a singular equilibrium (B_1) as

$$\begin{aligned}\dot{x}_{j_1} &= a_{j_1 j_1 0}(x_{j_1} - a_{j_1 j_1 1})^{2m_{11}}(x_{j_2} - a_{j_1 j_2 1})^{2n_{11}+1}, \\ \dot{x}_{j_2} &= a_{j_2 j_2 0}(x_{j_2} - a_{j_2 j_2 1})^{n_2}.\end{aligned}$$

If

$$\begin{aligned}\Delta_{j_1 j_1}^{(r)} &= \Delta_{j_1 j_1} + \delta_r\,(r = 1, 2, \cdots, m_{11}),\ \delta_r > 0; \\ \Delta_{j_1 j_2}^{(u)} &= \Delta_{j_1 j_2} + \delta_u\,(u = 1, 2, \cdots, n_{11}),\ \delta_s > 0;\end{aligned}$$

then

$$
\begin{aligned}
&a^{(r)}_{j_1j_11}, a^{(r)}_{j_1j_12} = a_{j_1j_11} \pm \varepsilon_r, \ (r = 1, 2, \cdots, m_{11}),\\
&\{a_{j_1j_11}, a_{j_1j_12}, \cdots, a_{j_1j_1(2m_{11})}\} = \text{sort}\{a^{(r)}_{j_1j_11}, a^{(r)}_{j_1j_12} | r = 1, 2, \cdots, m_{11}\};\\
&a^{(u)}_{j_1j_21}, a^{(u)}_{j_1j_22} = a_{j_1j_21} \pm \varepsilon_u, \ (u = 1, 2, \cdots, n_1),\\
&\{a_{j_1j_21}, a_{j_1j_22}, \cdots, a_{j_1j_2(2n_{11}+1)}\} = \text{sort}\{b_{j_1j_21}, a^{(u)}_{j_1j_21}, a^{(u)}_{j_1j_22} | u = 1, 2, \cdots, n_{11}\}.
\end{aligned}
$$

Thus, the differential equation becomes a dynamical system with a non-singular equilibrium network (C_1) as

$$
\begin{aligned}
\dot{x}_{j_1} &= a_{j_1j_10} \prod_{r=1}^{2m_{11}} (x_{j_1} - a_{j_1j_1r}) \prod_{u=1}^{2n_{11}+1} (x_{j_2} - a_{j_1j_2u}),\\
\dot{x}_{j_2} &= a_{j_2j_20}(x_{j_2} - a_{j_2j_21})^{n_2}.
\end{aligned}
$$

- From the above bifurcation process analysis, at least, only m_{11}-parameter variations in the x_{j_1}-direction are engaged in such a bifurcation from a 1-dimensional flow to the 1-dimensional flow array of $1 \times (2m_{11})$ through the $(2m_{11})^{\text{th}}$-order source and sink bifurcations in the x_{j_1}-direction. The higher-order source and sink flow bifurcations are developed.

  Thus, the 1-dimensional flow appearing or vanishing bifurcation route is as follows.

$$
\underbrace{(a_{j_1j_11}, \dot{x}_{j_2})}_{\text{XX}} \rightleftharpoons \underbrace{(a_{j_1j_11}, \dot{x}_{j_2})}_{(2m_{11})^{\text{th}}\text{order XX}} \rightleftharpoons \bigcup_{r=1}^{2m_{11}} \underbrace{(a_{j_1j_1r}, \dot{x}_{j_2})}_{\text{XX}}.
$$

- From the above bifurcation process analysis, at least, only $(n_{11} + 1)$-parameter variations in the x_{j_1}-direction are engaged in such a bifurcation from a 1-dimensional flow to the 1-dimensional flow array of $(2n_{11} + 1) \times 1$ through the $(2n_{11} + 1)^{\text{th}}$-order up-parabola and down-parabola bifurcations in the x_{j_1}-direction. The higher-order up-parabola and down-parabola flow bifurcations are developed.

  Thus, the 1-dimensional flow appearing or vanishing bifurcation route is as follows.

$$
\underbrace{(a_{j_1j_21}, \dot{x}_{j_2})}_{\text{YY}} \rightleftharpoons \underbrace{(a_{j_1j_21}, \dot{x}_{j_2})}_{(2n_{11}+1)^{\text{th}}\text{order YY}} \rightleftharpoons \bigcup_{u=1}^{2n_{11}+1} \underbrace{(a_{j_1j_2u}, \dot{x}_{j_2})}_{\text{YY}}.
$$

- From the above bifurcation process analysis, $(m_{11} + n_{11} + 1)$-parameter variations in the two directions are engaged in such a bifurcation from a parabola-flow to a $(2m_{11}) \times (2n_{11} + 1)$ matrix of hyperbolic and hyperbolic-secant flows through the $(2m_{11})^{\text{th}}$-order upper-saddle and lower-saddle bifurcations in the-direction, and the $(2n_{11}+1)^{\text{th}}$-order up-parabola and down-parabola flows bifurcations in the-direction. The higher-order hyperbolic-to-hyperbolic-recant flow bifurcations are developed.

Thus, the 1-dimensional flow appearing or vanishing bifurcation route is as follows.

$$\underbrace{(a_{j_1j_21},\dot{x}_{j_2})}_{\text{(XX,xF)-flow}} \rightleftharpoons \underbrace{[(2m_{11}),(2n_{11}+1)]\text{XX},xF)}_{[(2m_{11}),(2n_{11}+1)]\text{-XX}}$$
$$\rightleftharpoons \bigcup\nolimits_{r=1}^{2m_{11}}\bigcup\nolimits_{u=1}^{2n_{11}+1}\underbrace{(a_{j_1j_1r},a_{j_1j_2u})}_{\text{(YY,xF)-flow}}.$$

which are the higher-order hyperbolic-to-hyperbolic-secant flows bifurcations.

(I_2) From Case (I_1), consider a dynamical system having singular equilibriums (A_2) as

$$\begin{aligned}
\dot{x}_{j_1} &= a_{j_1j_10}\prod\nolimits_{r_1=1}^{p_1}(x_{j_1}-b_{j_1j_1r_1})^{m_{j_1r_1}}\prod\nolimits_{r=1}^{l_1}[(x_{j_1}-a_{j_1j_1}^{(r)})^2+\Delta_{j_1j_1}^{(r)}]^{m_{j_1r}}\\
&\quad\times\prod\nolimits_{u_1=1}^{q_1}(x_{j_2}-b_{j_1j_2u_1})^{n_{j_1u_1}}\prod\nolimits_{u=1}^{l_2}[(x_{j_2}-a_{j_1j_2}^{(u)})^2+\Delta_{j_1j_2}^{(u)}]^{n_{j_1u}},\\
\dot{x}_{j_2} &= a_{j_2j_20}(x_{j_2}-a_{j_2j_21})^{n_2};
\end{aligned}$$

where

$$\begin{aligned}
&2m_{11}-\sum\nolimits_{r_1=1}^{p_1}m_{j_1r_1}=2\sum\nolimits_{\rho_1=1}^{l_1}m_{\rho_1},\\
&2n_{11}+1-\sum\nolimits_{u_1=1}^{q_1}n_{j_1u_1}=2\sum\nolimits_{\sigma_1=1}^{l_2}n_{\sigma_1};\\
&\sum\nolimits_{r_2=1}^{\bar{p}_1}m_{j_1r_2}=2m_{11},\sum\nolimits_{u_2=1}^{\bar{q}_1}n_{j_1u_2}=2n_{11}+1;
\end{aligned}$$

l_1,l_2-quadratic polynomials are without real roots and ρ_1^{th} and σ_1^{th}-quadratic polynomials are with powers m_{ρ_1} and n_{σ_1}, respectively.

Once

$$\begin{aligned}
\Delta_{j_1j_1}^{(r)} &= 0\,(r=1,2,\cdots,l_1),\\
\Delta_{j_1j_2}^{(u)} &= 0\,(u=1,2,\cdots,l_2);
\end{aligned}$$

if

$$a_{j_1j_11}\equiv a_{j_1j_1r_1}=a_{j_1j_1}^{(r)}\,(r_1=1,2,\cdots,p_1;r=1,2,\cdots,l_1),$$

$$a_{j_1j_21} \equiv a_{j_1j_2u_1} = a^{(u)}_{j_1j_2}\ (u_1 = 1, 2, \cdots, q_1; u = 1, 2, \cdots, l_2),$$

the foregoing differential equations becomes a dynamical system with a 1-dimensional flow (B_2) as

$$\begin{aligned}\dot{x}_{j_1} &= a_{j_1j_10}(x_{j_1} - a_{j_1j_11})^{2m_{11}}(x_{j_2} - a_{j_1j_21})^{2n_{11}+1},\\ \dot{x}_{j_2} &= a_{j_2j_20}(x_{j_2} - a_{j_2j_21})^{n_2}.\end{aligned}$$

If

$$\begin{aligned}\Delta^{(r)}_{j_1j_1} &= \Delta_{j_1j_1} + \delta_r\ (r = 1, 2, \cdots, l_1),\ \delta_r > 0;\\ \Delta^{(u)}_{j_1j_2} &= \Delta_{j_1j_2} + \delta_u\ (u = 1, 2, \cdots, l_2),\ \delta_u > 0;\end{aligned}$$

then

$$\begin{aligned}&a^{(r)}_{j_1j_11}, a^{(r)}_{j_1j_12} = a_{j_1j_11} \pm \varepsilon_r,\ (r = 1, 2, \cdots, l_1),\\ &\{a_{j_1j_11}, a_{j_1j_12}, \cdots, a_{j_1j_1\overline{p}_1}\}\\ &= \text{sort}\{a_{j_1j_11}, a_{j_1j_12}, \cdots, a_{j_1j_1p_1}; a^{(r)}_{j_1j_11}, a^{(r)}_{j_1j_12} | r = 1, 2, \cdots, l_1\};\\ &a^{(u)}_{j_1j_21}, a^{(u)}_{j_1j_22} = a_{j_1j_21} \pm \varepsilon_u,\ (u = 1, 2, \cdots, l_2),\\ &\{a_{j_1j_21}, a_{j_1j_22}, \cdots, a_{j_1j_2\overline{q}_1}\}\\ &= \text{sort}\{a_{j_1j_21}, a_{j_1j_22}, \cdots, a_{j_1j_2q_1}; a^{(u)}_{j_2j_11}, a^{(u)}_{j_2j_12} | u = 1, 2, \cdots, l_2\}.\end{aligned}$$

Thus, the differential equation becomes a dynamical system with singular 1-dimensional flows (C_2) as

$$\begin{aligned}\dot{x}_{j_1} &= a_{j_1j_10}\prod\nolimits_{r_2=1}^{\overline{p}_1}(x_{j_1} - a_{j_1j_1r_2})^{m_{j_1r_2}}\prod\nolimits_{u_2=1}^{\overline{q}_1}(x_{j_2} - a_{j_1j_1u_2})^{n_{j_1u_2}},\\ \dot{x}_{j_2} &= a_{j_2j_20}(x_{j_2} - a_{j_2j_21})^{n_2}\end{aligned}$$

where

$$\Sigma^{\overline{p}_1}_{r_2=1}m_{j_1r_2} = 2m_{11},\ \Sigma^{\overline{q}_1}_{u_2=1}n_{j_1u_2} = 2n_{11} + 1.$$

- From the above bifurcation process analysis, only $\overline{p}_1$-parameter variations in the x_{j_1}-direction from a p_1 to $\overline{p}_1$ 1-dimensional flow array through the $(2m_{11})^{\text{th}}$-order source and sink flow bifurcations in the x_{j_1}-direction.

  Thus, the 1-dimensional flow appearing or vanishing bifurcation route from a p_1 to p_2 1-dimensional flow array is expressed as

$$\bigcup_{r_1=1}^{p_1} \underbrace{(a_{j_1j_1r_1}, \dot{x}_{j_2})}_{(m_{j_1r_1})^{\text{th}}\text{order XX}} \rightleftharpoons \underbrace{(a_{j_1j_11}, \dot{x}_{j_2})}_{(2m_{11})^{\text{th}}\text{ order XX}} \rightleftharpoons \bigcup_{r_2=1}^{\overline{p}_1} \underbrace{(a_{j_1j_1r_2}, \dot{x}_{j_2})}_{(m_{j_1r_2})^{\text{th}}\text{order XX}}.$$

- From the above bifurcation process analysis, only $\overline{q}_1$-parameter variations in the x_{j_1}-direction from a q_1 to $\overline{q}_1$ 1-dimensional flow array through the $(2n_{11}+1)^{\text{th}}$ order up-parabola and down-parabola flow bifurcations in the x_{j_1}-direction.

  Thus, the 1-dimensional flow appearing or vanishing bifurcation route from a q_1 to $\overline{q}_1$ 1-dimensional flow array is expressed as

$$\bigcup_{u_1=1}^{q_1} \underbrace{(a_{j_2j_1u_1}, \dot{x}_{j_2})}_{(n_{j_1u_1})^{\text{th}}\text{order YY}} \rightleftharpoons \underbrace{(a_{j_1j_21}, \dot{x}_{j_2})}_{(2n_1+1)^{\text{th}}\text{ order YY}} \rightleftharpoons \bigcup_{u_2=1}^{\overline{q}_1} \underbrace{(a_{j_1j_2u_2}, \dot{x}_{j_2})}_{(n_{j_1u_2})^{\text{th}}\text{order YY}}.$$

- From the above bifurcation process analysis, only $\overline{q}_1+\overline{p}_1$-parameter variations in the x_{j_1}-direction from a $q_1 \times p_1$ to $\overline{q}_1 \times \overline{p}_1$ matrix of 1-dimensional flows through the $(2m_{11})^{\text{th}}$-order upper and lower-saddle flows bifurcations in the x_{j_1}-direction, and the $(2n_{11}+1)^{\text{th}}$-order up-parabola and down-parabola flows bifurcations in the x_{j_1}-direction.

  Thus, the 1-dimensional flow appearing or vanishing bifurcation route is as follows:

$$\begin{aligned}\bigcup_{r_1=1}^{p_1}\bigcup_{u_1=1}^{q_1} \underbrace{(a_{j_1j_1r_1}, a_{j_1j_2u_1})}_{([(m_{j_1r_1}),(n_{j_1u_1})]\text{XX,xF})\text{-flow}} &\rightleftharpoons \underbrace{([(2m_{11}), (2n_{11}+1)]\text{XX, xF})}_{[(2m_{11}),(2n_{11}+1)]\text{-XX}} \\ &\rightleftharpoons \bigcup_{r_2=1}^{\overline{p}_1}\bigcup_{u_2=1}^{\overline{q}_1} \underbrace{(a_{j_1j_1r_2}, a_{j_1j_2u_2})}_{([(m_{j_1r_2}),(n_{j_1u_1})]\text{XX,xF})\text{-flow}}.\end{aligned}$$

  which are the higher-order hyperbolic-to-hyperbolic-secant flows bifurcations.

(I_3) Consider two dynamical systems with the same 1-dimensional flows with locations switched (A_3,C_3) for $i=1,2$ as

$$\begin{aligned}\dot{x}_{j_1} &= a_{j_1j_10}\prod_{r_i=1}^{p_1}(x_{j_1}-a_{j_1j_1r_i})^{m_{j_1r_i}}\prod_{u_i=1}^{q_1}(x_{j_1}-a_{j_2j_1u_i})^{n_{j_1u_i}},\\ \dot{x}_{j_2} &= a_{j_2j_20}(x_{j_2}-a_{j_2j_21})^{n_2};\end{aligned}$$

where

$$\sum_{r_i=1}^{p_1} m_{j_1r_i} = 2m_{11}, \sum_{u_i=1}^{q_1} n_{j_1u_i} = 2n_{11}+1, \text{ for } i=1,2.$$

Consider a dynamical system as a singular 1-dimensional flow (B_3) as

$$\begin{aligned}
\dot{x}_{j_1} &= a_{j_1 j_1 0}(x_{j_1} - a_{j_1 j_1 1})^{2m_{11}}(x_{j_2} - a_{j_1 j_2 1})^{2n_{11}+1},\\
\dot{x}_{j_2} &= a_{j_2 j_1 0}(x_{j_2} - a_{j_2 j_2 1})^{n_2}.
\end{aligned}$$

For $r_i = 1, 2, \cdots, p_1$ and $s_i = 1, 2, \cdots, q_1$ $(i = 1, 2)$, define a function as

$$\begin{aligned}
\Delta_{j_1 j_1}^{(r_1 r_2)} &= (a_{j_1 j_1 r_1} - a_{j_1 j_1 r_2})^2 \; (r_1, r_2 \in \{1, 2, \cdots, p_1\}, r_1 \neq r_2),\\
\Delta_{j_1 j_2}^{(u_1 u_2)} &= (a_{j_1 j_2 u_1} - a_{j_1 j_2 u_2})^2 \; (u_1, u_2 \in \{1, 2, \cdots, q_1\}, u_1 \neq u_2).
\end{aligned}$$

For $r_i = 1, 2, \cdots, p_1$ and $s_i = 1, 2, \cdots, q_1$ $(i = 1, 2)$, if

$$\Delta_{j_1 j_1}^{(r_1 r_2)} = (a_{j_1 j_1 r_1} - a_{j_1 j_1 r_2})^2 = 0 \text{ and } \Delta_{j_1 j_2}^{(u_1 u_2)} = (a_{j_1 j_2 u_1} - a_{j_1 j_2 u_2})^2 = 0,$$

there are three bifurcations as follows.

- Two 1-dimensional flows of and switching at point with the same order singularity are given through

$$\bigcup_{r_1=1}^{p_1} \underbrace{(a_{j_1 j_1 r_1}, \dot{x}_{j_2})}_{(m_{j_1 r_1})^{\text{th}} \text{order XX}} \rightleftharpoons \underbrace{(a_{j_1 j_1 1}, \dot{x}_{j_2})}_{(2m_{11})^{\text{th}} \text{ order XX}} \rightleftharpoons \bigcup_{r_2=1}^{p_1} \underbrace{(a_{j_1 j_1 r_2}, \dot{x}_{j_2})}_{(m_{j_1 r_2})^{\text{th}} \text{order XX}}$$
$$\left\{a_{j_1 j_1 r_1} \rightleftharpoons a_{j_1 j_1 r_2} \middle| m_{j_1 r_1} = m_{j_1 r_2};\, r_1, r_2 \in \{1, 2, \cdots, p_1\}, r_1 \neq r_2\right\};$$

  and

$$\begin{aligned}
&a_{j_1 j_1 r_1} = a_{j_1 j_1 1} - \varepsilon_{r_1}, a_{j_1 j_1 r_2} = a_{j_1 j_1 1} + \varepsilon_{r_2},\\
&(a_{j_1 j_1 r_1} = a_{j_1 j_1 1} + \varepsilon_{r_1}, a_{j_1 j_1 r_2} = a_{j_1 j_1 1} - \varepsilon_{r_2});\\
&\varepsilon_{r_1}, \varepsilon_{r_2} > 0;\; r_i = 1, 2, \cdots, p_1 \; (i = 1, 2).
\end{aligned}$$

- Two 1-dimensional flows of $(a_{j_1 j_2 u_1}, \dot{x}_{j_2})$ and $(a_{j_1 j_2 u_1}, \dot{x}_{j_2})$ switching at point $(a_{j_1 j_2 1}, \dot{x}_{j_2})$ with the same order singularity are given through

$$\bigcup_{u_1=1}^{q_1} \underbrace{(a_{j_1 j_2 u_1}, \dot{x}_{j_2})}_{(n_{j_1 u_1})^{\text{th}} \text{order YY}} \rightleftharpoons \underbrace{(a_{j_1 j_2 1}, \dot{x}_{j_2})}_{(2n_{11}+1)^{\text{th}} \text{ order YY}} \rightleftharpoons \bigcup_{u_2=1}^{q_1} \underbrace{(a_{j_1 j_2 u_2}, \dot{x}_{j_2})}_{(n_{j_1 u_2})^{\text{th}} \text{order YY}}$$
$$\left\{a_{j_1 j_2 u_1} \rightleftharpoons a_{j_1 j_2 u_2} \middle| n_{j_1 u_1} = n_{j_1 u_2};\, u_1, u_2 \in \{1, 2, \cdots, q\}, u_1 \neq u_2\right\};$$

  and

$$\begin{aligned}
&a_{j_1 j_2 u_1} = a_{j_1 j_2 1} - \varepsilon_{u_1}, a_{j_1 j_2 u_2} = a_{j_1 j_2 1} + \varepsilon_{u_2},\\
&(a_{j_1 j_2 u_1} = a_{j_1 j_2 1} + \varepsilon_{u_1}, a_{j_1 j_2 u_2} = a_{j_1 j_2 1} - \varepsilon_{u_2});\\
&\varepsilon_{u_1}, \varepsilon_{u_2} > 0;\; u_i = 1, 2, \cdots, q_1 \; (i = 1, 2).
\end{aligned}$$

- Two 1-dimensional flows of $(a_{j_1j_1r_1}, a_{j_1j_2u_1})$ and $(a_{j_1j_1r_2}, a_{j_1j_2u_2})$ switching at point $(a_{j_1j_1 1}, a_{j_1j_2 1})$ with the same order singularity are given through

$$\bigcup_{r_1=1}^{p_1}\bigcup_{u_1=1}^{q_1}\underbrace{(a_{j_1j_1r_1}, a_{j_1j_2u_1})}_{([(m_{j_1r_1}),(n_{j_1u_1})]\text{XX,xF})\text{-flow}} \rightleftharpoons \underbrace{([(2m_{11}),(2\text{n}_{11}+1)]\text{XX, xF})}_{[(2m_{11}),(2\text{n}_{11}+1)]\text{-XX}}$$
$$\rightleftharpoons \bigcup_{r_2=1}^{\bar{p}_1}\bigcup_{u_2=1}^{\bar{q}_1}\underbrace{(a_{j_1j_1r_2}, a_{j_1j_2u_2})}_{([(m_{j_1r_2}),(n_{j_1u_2})]\text{XX,xF})\text{-flow}}$$

with

$$\begin{aligned}
&a_{j_1j_1r_1} = a_{j_1j_1 1} - \varepsilon_{r_1}, a_{j_1j_1r_2} = a_{j_1j_1 1} + \varepsilon_{r_2},\\
&(a_{j_1j_1r_1} = a_{j_1j_1 1} + \varepsilon_{r_1}, a_{j_1j_1r_2} = a_{j_1j_1 1} - \varepsilon_{r_2}),\\
&\varepsilon_{r_1}, \varepsilon_{r_2} > 0,\ r_i = 1, 2, \cdots, p_1\ (i = 1, 2);\\
&a_{j_1j_2u_1} = a_{j_1j_2 1} - \varepsilon_{u_1}, a_{j_1j_2u_2} = a_{j_1j_2 1} + \varepsilon_{u_2},\\
&(a_{j_1j_2u_1} = a_{j_1j_2 1} + \varepsilon_{u_1}, a_{j_1j_2u_2} = a_{j_1j_2 1} - \varepsilon_{u_2}),\\
&\varepsilon_{u_1}, \varepsilon_{u_2} > 0,\ u_i = 1, 2, \cdots, q_1\ (i = 1, 2).
\end{aligned}$$

The switching bifurcations are the higher-order hyperbolic-to-hyperbolic-secant flows bifurcations.

In summary, from the analysis in case (I_1)-(I_3), the 1-dimensional flow of $(x^*_{j_1}, x^*_{j_2}) = (a_{j_1j_1 1}, a_{j_1j_2 1})$ for $m_1 = 2m_{11}$ and $n_1 = 2n_{11} + 1$ has the bifurcation properties as stated in (iv_{2a})-(iv_{2d}) through Eqs. (1.326)–(1.345).

(iv_3) For $m_1 = 2m_{11} + 1$ and $n_1 = 2n_{11}$ the equilibrium of $(x^*_{j_1}, x^*_{j_2}) = (a_{j_1j_1 1}, a_{j_1j_2 1})$ for the x_{j_1}-direction has the following bifurcation properties with

$$\begin{aligned}
&2m_{11} + 1 - \sum\nolimits_{r_1=1}^{p_1} m_{j_1r_1} = 2\sum\nolimits_{\rho_1=1}^{l_1} m_{\rho_1},\\
&2n_{11} - \sum\nolimits_{u_1=1}^{q_1} n_{j_1u_1} = 2\sum\nolimits_{\sigma_1=1}^{l_2} n_{\sigma_1};\\
&\sum\nolimits_{r_2=1}^{\bar{p}_1} m_{j_1r_2} = 2m_{11} + 1, \sum\nolimits_{u_2=1}^{\bar{q}_1} n_{j_1u_2} = 2n_{11};\\
&\sum\nolimits_{r_i=1}^{p_1} m_{j_1r_i} 2m_{11} + 1, \sum\nolimits_{u_i=1}^{q_1} n_{j_1u_i} = 2n_{11}, for\, i = 1, 2;
\end{aligned}$$

l_1, l_2 -quadratic polynomials are without real roots and ρ_1^{th} and σ_1^{th} -quadratic polynomials are with powers m_{ρ_1} and n_{σ_1}, respectively.

The bifurcation process is discussed through differential equations as follows. There are four bifurcations (I_1)-(I_3).

(I_1) Consider a dynamical system having a single equilibrium (A_1) as

$$\begin{aligned}\dot{x}_{j_1} =& a_{j_1j_10}(x_{j_1} - b_{j_1j_11}) \prod\nolimits_{r=1}^{m_{11}} [(x_{j_1} - a_{j_1j_1}^{(r)})^2 + \Delta_{j_1j_1}^{(r)}] \\ &\times \prod\nolimits_{u=1}^{n_{11}} [(x_{j_2} - a_{j_1j_2}^{(u)})^2 + \Delta_{j_1j_2}^{(u)}], \\ \dot{x}_{j_2} =& a_{j_2j_20}(x_{j_2} - a_{j_2j_21})^{n_2};\end{aligned}$$

where $a_{j_1j_11} = b_{j_1j_11}$. Once

$$\begin{aligned}\Delta_{j_1j_1}^{(r)} &= \Delta_{j_1j_1} = 0\,(r = 1, 2, \cdots, m_{11}), \\ \Delta_{j_1j_2}^{(u)} &= \Delta_{j_1j_2} = 0\,(u = 1, 2, \cdots, n_{11}),\end{aligned}$$

if

$$\begin{aligned}a_{j_1j_11} &= b_{j_1j_11} = a_{j_1j_1}^{(r)}\ (r = 1, 2, \cdots, m_{11}), \\ a_{j_1j_21} &= b_{j_1j_21} = a_{j_1j_2}^{(u)}\ (u = 1, 2, \cdots, n_{11}),\end{aligned}$$

the foregoing differential equation becomes a dynamical system with a singular equilibrium (B_1) as

$$\begin{aligned}\dot{x}_{j_1} &= a_{j_1j_10}(x_{j_1} - a_{j_1j_11})^{2m_{11}+1}(x_{j_2} - a_{j_1j_21})^{2n_{11}}, \\ \dot{x}_{j_2} &= a_{j_2j_20}(x_{j_2} - a_{j_2j_21})^{n_2}.\end{aligned}$$

If

$$\begin{aligned}\Delta_{j_1j_1}^{(r)} &= \Delta_{j_1j_1} + \delta_r\,(r = 1, 2, \cdots, m_{11}),\ \delta_r > 0; \\ \Delta_{j_1j_2}^{(u)} &= \Delta_{j_1j_2} + \delta_s\,(u = 1, 2, \cdots, n_{11}),\ \delta_u > 0;\end{aligned}$$

then

$$\begin{aligned}& a_{j_1j_11}^{(r)}, a_{j_1j_12}^{(r)} = a_{j_1j_11} \pm \varepsilon_r,\ (r = 1, 2, \cdots, m_{11}), \\ & \{a_{j_1j_11}, a_{j_1j_12}, \cdots, a_{j_1j_1(2m_{11})}\} = \text{sort}\{a_{j_1j_11}^{(r)}, a_{j_1j_12}^{(r)} | r = 1, 2, \cdots, m_{11}\}; \\ & a_{j_1j_21}^{(u)}, a_{j_1j_22}^{(u)} = a_{j_1j_21} \pm \varepsilon_u,\ (u = 1, 2, \cdots, n_1), \\ & \{a_{j_1j_21}, a_{j_1j_22}, \cdots, a_{j_1j_2(2n_{11}+1)}\} = \text{sort}\{b_{j_1j_21}, a_{j_1j_21}^{(u)}, a_{j_1j_22}^{(u)} | u = 1, 2, \cdots, n_{11}\}.\end{aligned}$$

Thus, the differential equation becomes a dynamical system with a non-singular equilibrium network (C_1) as

$$\dot{x}_{j_1} = a_{j_1j_10} \prod\nolimits_{r=1}^{2m_{11}+1} (x_{j_1} - a_{j_1j_1r}) \prod\nolimits_{u=1}^{2n_{11}} (x_{j_2} - a_{j_1j_2u}),$$

$$\dot{x}_{j_2} = a_{j_2j_20}(x_{j_2} - a_{j_2j_21})^{n_2}.$$

- From the above bifurcation process analysis, at least, only $(m_{11} + 1)$-parameter variations in the x_{j_1}-direction are engaged in such a bifurcation from a 1-dimensional flow to the 1-dimensional flow array of $1 \times (2m_{11} + 1)$ through the $(2m_{11} + 1)^{\text{th}}$-order source and sink bifurcations in the x_{j_1}-direction. The higher-order source and sink flow bifurcations are developed.

  Thus, the 1-dimensional flow appearing or vanishing bifurcation route is as follows.

$$\underbrace{(a_{j_1j_1 1}, \dot{x}_{j_2})}_{\text{XX}} \rightleftharpoons \underbrace{(a_{j_1j_1 1}, \dot{x}_{j_2})}_{(2m_{11}+1)^{\text{th}}\text{order XX}} \rightleftharpoons \bigcup\nolimits_{r=1}^{2m_{11}+1} \underbrace{(a_{j_1j_1 r}, \dot{x}_{j_2})}_{\text{XX}}.$$

- From the above bifurcation process analysis, at least, only (n_{11})-parameter variations in the x_{j_1}-direction are engaged in such a bifurcation from a 1-dimensional flow to the 1-dimensional flow array of $(2n_{11}) \times 1$ through the $(2n_{11})^{\text{th}}$-order increasing and decreasing-inflection flows bifurcations in the x_{j_1}-direction. The higher-order increasing-inflection and decreasing-inflection flow bifurcations are developed.

  Thus, the 1-dimensional flow appearing or vanishing bifurcation route is as follows.

$$\underbrace{(a_{j_1j_2 1}, \dot{x}_{j_2})}_{\text{YY}} \rightleftharpoons \underbrace{(a_{j_1j_2 1}, \dot{x}_{j_2})}_{(2n_{11})^{\text{th}}\text{order YY}} \rightleftharpoons \bigcup\nolimits_{r=1}^{2n_{11}} \underbrace{(a_{j_1j_2 r}, \dot{x}_{j_2})}_{\text{YY}}.$$

- From the above bifurcation process analysis, $(m_{11} + n_{11} + 1)$-parameter variations in the two directions are engaged in such a bifurcation from a parabola-flow to a $(2m_{11} + 1) \times (2n_{11})$ matrix of inflection flows through the $(2m_{11}+1)^{\text{th}}$-order source and sink bifurcations in the-direction, and the $(2n_{11})^{\text{th}}$-order increasing and decreasing-inflection flows bifurcations in the-direction. The higher-order increasing and decreasing-inflection flow bifurcations are developed.

  Thus, the 1-dimensional flow appearing or vanishing bifurcation route is as follows.

$$\begin{aligned}\underbrace{(a_{j_1j_1 1}, \dot{x}_{j_2})}_{\text{(XX,xF)-flow}} &\rightleftharpoons \underbrace{([(2m_{11}+1), (2n_{11})]\text{XX, xF})}_{[(2m_{11}+1),(2n_1)]\text{-XX}} \\ &\rightleftharpoons \bigcup\nolimits_{r=1}^{2m_{11}+1} \bigcup\nolimits_{u=1}^{2n_{11}} \underbrace{(a_{j_1j_1 r}, a_{j_1j_2 u})}_{\text{(YY,xF)-flow}},\end{aligned}$$

which are the higher-order inflection source and sink flows bifurcations.

(I_2) From Case (I_1), consider a dynamical system having singular equilibriums (A_2) as

$$\begin{aligned}\dot{x}_{j_1} &= a_{j_1j_10}\prod_{r_1=1}^{p_1}(x_{j_1}-b_{j_1j_1r_1})^{m_{j_1r_1}}\prod_{r=1}^{l_1}[(x_{j_1}-a_{j_1j_1}^{(r)})^2+\Delta_{j_1j_1}^{(r)}]^{m_{j_1r}}\\&\quad\times\prod_{u_1=1}^{q_1}(x_{j_2}-b_{j_1j_2u_1})^{m_{j_1u_1}}\prod_{u=1}^{l_2}[(x_{j_2}-a_{j_1j_2}^{(u)})^2+\Delta_{j_1j_2}^{(u)}]^{m_{j_1u}},\\ \dot{x}_{j_2} &= a_{j_2j_20}(x_{j_2}-a_{j_2j_21})^{n_2};\end{aligned}$$

where

$$\begin{aligned}&2m_{11}+1-\sum_{r_1=1}^{p_1}m_{j_1r_1}=2\sum_{\rho_1=1}^{l_1}m_{\rho_1},\\&2n_{11}-\sum_{u_1=1}^{q_1}n_{j_1u_1}=2\sum_{\sigma_1=1}^{l_2}n_{\sigma_1};\\&\sum_{r_2=1}^{\bar{p}_1}m_{j_1r_2}=2m_{11}+1,\sum_{u_2=1}^{\bar{q}_1}n_{j_1u_2}=2n_{11};\end{aligned}$$

l_1, l_2 -quadratic polynomials are without real roots and ρ_1^{th} and σ_1^{th} -quadratic polynomials are with powers m_{ρ_1} and n_{σ_1}, respectively.

Once

$$\begin{aligned}\Delta_{j_1j_1}^{(r)}&=0\,(r=1,2,\cdots,l_1),\\ \Delta_{j_1j_2}^{(u)}&=0\,(u=1,2,\cdots,l_2);\end{aligned}$$

if

$$\begin{aligned}a_{j_1j_11}\equiv a_{j_1j_1r_1}&=a_{j_1j_1}^{(r)}\,(r_1=1,2,\cdots,p_1;\,r=1,2,\cdots,l_1),\\ a_{j_1j_21}\equiv a_{j_1j_2u_1}&=a_{j_1j_2}^{(u)}\,(u_1=1,2,\cdots,q_1;\,u=1,2,\cdots,l_2),\end{aligned}$$

the foregoing differential equations become a dynamical system with a 1-dimensional flow (B_2) as

$$\begin{aligned}\dot{x}_{j_1}&=a_{j_1j_10}(x_{j_1}-a_{j_1j_11})^{2m_{11}+1}(x_{j_2}-a_{j_1j_21})^{2n_{11}},\\ \dot{x}_{j_2}&=a_{j_2j_20}(x_{j_2}-a_{j_2j_21})^{n_2}.\end{aligned}$$

If

$$\Delta_{j_1j_1}^{(r)}=\Delta_{j_1j_1}+\delta_r\,(r=1,2,\cdots,l_1),\ \delta_r>0;$$

$$\Delta_{j_1j_2}^{(u)} = \Delta_{j_1j_2} + \delta_u \, (u = 1, 2, \cdots, l_2), \ \delta_u > 0;$$

then

$$\begin{aligned}
&a_{j_1j_11}^{(r)}, a_{j_1j_12}^{(r)} = a_{j_1j_11} \pm \varepsilon_r, \ (r = 1, 2, \cdots, l_1), \\
&\{a_{j_1j_11}, a_{j_1j_12}, \cdots, a_{j_1j_1\overline{p}_1}\} \\
&= \mathrm{sort}\{a_{j_1j_11}, a_{j_1j_12}, \cdots, a_{j_1j_1p_1}; a_{j_1j_11}^{(r)}, a_{j_1j_11}^{(r)} | r = 1, 2, \cdots, l_1\}; \\
&a_{j_1j_21}^{(u)}, a_{j_1j_22}^{(u)} = a_{j_1j_21} \pm \varepsilon_u, \ (u = 1, 2, \cdots, l_2), \\
&\{a_{j_1j_21}, a_{j_1j_22}, \cdots, a_{j_1j_2\overline{q}_1}\} \\
&= \mathrm{sort}\{a_{j_1j_21}, a_{j_1j_22}, \cdots, a_{j_1j_2q_1}; a_{j_2j_11}^{(u)}, a_{j_2j_12}^{(u)} | u = 1, 2, \cdots, l_2\}.
\end{aligned}$$

Thus, the differential equation becomes a dynamical system with singular 1-dimensional flows (C_2) as

$$\begin{aligned}
\dot{x}_{j_1} &= a_{j_1j_10} \prod\nolimits_{r_2=1}^{\overline{p}_1} (x_{j_1} - a_{j_1j_1r_2})^{m_{j_1r_2}} \prod\nolimits_{u_2=1}^{\overline{q}_1} (x_{j_2} - a_{j_1j_2u_2})^{n_{j_1u_2}}, \\
\dot{x}_{j_2} &= a_{j_2j_20}(x_{j_2} - a_{j_2j_21})^{n_2}
\end{aligned}$$

where

$$\Sigma_{r_2=1}^{\overline{p}_1} m_{j_1r_2} = 2m_{11} + 1, \ \Sigma_{u_2=1}^{\overline{q}_1} n_{j_1u_2} = 2n_{11}.$$

- From the above bifurcation process analysis, only $\overline{p}_1$-parameter variations in the x_{j_1}-direction from a p_1 to $\overline{p}_1$ 1-dimensional flow array through the $(2m_{11} + 1)^{\text{th}}$-order source and sink flow bifurcations in the x_{j_1}-direction.

  Thus, the 1-dimensional flow appearing or vanishing bifurcation route from a p_1 to p_2 1-dimensional flow array is expressed as

$$\bigcup\nolimits_{r_1=1}^{p_1} \underbrace{(a_{j_1j_1r_1}, \dot{x}_{j_2})}_{(m_{j_1r_1})^{\text{th}}\text{order XX}} \rightleftharpoons \underbrace{(a_{j_1j_11}, \dot{x}_{j_2})}_{(2m_{11}+1)^{\text{th}}\text{ order XX}} \rightleftharpoons \bigcup\nolimits_{r_2=1}^{\overline{p}_1} \underbrace{(a_{j_1j_1r_2}, \dot{x}_{j_2})}_{(m_{j_1r_2})^{\text{th}}\text{order XX}}.$$

- From the above bifurcation process analysis, only $\overline{q}_1$-parameter variations in the x_{j_1}-direction from a q_1 to $\overline{q}_1$ 1-dimensional flow array through the $(2n_{11})^{\text{th}}$-order increasing and decreasing inflection flow bifurcations in the x_{j_1}-direction.

  Thus, the 1-dimensional flow appearing or vanishing bifurcation route from a q_1 to $\overline{q}_1$ 1-dimensional flow array is expressed as

$$\bigcup_{u_1=1}^{q_1} \underbrace{(a_{j_2j_1u_1}, \dot{x}_{j_2})}_{(n_{j_1u_1})^{\text{th}}\text{order YY}} \rightleftharpoons \underbrace{(a_{j_1j_21}, \dot{x}_{j_2})}_{(2n_{11})^{\text{th}}\text{ order YY}} \rightleftharpoons \bigcup_{u_2=1}^{\bar{q}_1} \underbrace{(a_{j_1j_2u_2}, \dot{x}_{j_2})}_{(n_{j_1u_2})^{\text{th}}\text{order YY}}.$$

- From the above bifurcation process analysis, only $\bar{q}_1 + \bar{p}_1$-parameter variations in the x_{j_1}-direction from a $q_1 \times p_1$ to $\bar{q}_1 \times \bar{p}_1$ matrix of 1-dimensional flows through the $(2m_{11}+1)^{\text{th}}$-order source and sink flows bifurcations in the x_{j_1}-direction, and the $(2n_{11})^{\text{th}}$-order increasing and decreasing-inflection flows bifurcations in the x_{j_1}-direction.

  Thus, the 1-dimensional flow appearing or vanishing bifurcation route is as follows:

$$\bigcup_{r_1=1}^{p_1} \bigcup_{u_1=1}^{q_1} \underbrace{(a_{j_1j_1r_1}, a_{j_1j_2u_1})}_{([(m_{j_1r_1}),(n_{j_1u_1})]\text{XX,xF})\text{-flow}} \rightleftharpoons \underbrace{([(2m_{11}+1), (2n_{11})]\text{XX, xF})}_{[(2m_{11}+1),(2n_{11})]\text{-XX}}$$
$$\rightleftharpoons \bigcup_{r_2=1}^{\bar{p}_1} \bigcup_{u_2=1}^{\bar{q}_1} \underbrace{(a_{j_1j_1r_2}, a_{j_1j_2u_2})}_{([(m_{j_1r_2}),(n_{j_1u_1})]\text{XX,xF})\text{-flow}},$$

  which are the higher-order inflection source and sink flows bifurcations.

(I_3) Consider two dynamical systems with the same 1-dimensional flows with locations switched (A_3,C_3) for $i = 1, 2$ as

$$\dot{x}_{j_1} = a_{j_1j_10} \prod_{r_i=1}^{p_1} (x_{j_1} - a_{j_1j_1r_i})^{m_{j_1r_i}} \prod_{u_i=1}^{q_1} (x_{j_2} - a_{j_1j_2u_i})^{n_{j_1u_i}},$$
$$\dot{x}_{j_2} = a_{j_2j_20}(x_{j_2} - a_{j_2j_21})^{n_2};$$

where

$$\sum_{r_i=1}^{p_1} m_{j_1r_i} = 2m_{11} + 1, \sum_{u_i=1}^{q_1} n_{j_1u_i} = 2n_{11}, \text{ for } i = 1, 2.$$

Consider a dynamical system as a singular 1-dimensional flow (B_3) as

$$\dot{x}_{j_1} = a_{j_1j_10}(x_{j_1} - a_{j_1j_11})^{2m_{11}+1}(x_{j_2} - a_{j_1j_21})^{2n_{11}},$$
$$\dot{x}_{j_2} = a_{j_2j_10}(x_{j_2} - a_{j_2j_21})^{n_2}.$$

For $r_i = 1, 2, \cdots, p$ and $s_i = 1, 2, \cdots, q$ (), define a function as

$$\Delta_{j_1j_1}^{(r_1r_2)} = (a_{j_1j_1r_1} - a_{j_1j_1r_2})^2 \ (r_1, r_2 \in \{1, 2, \cdots, p_1\}, r_1 \neq r_2),$$
$$\Delta_{j_1j_2}^{(u_1u_2)} = (a_{j_1j_2u_1} - a_{j_1j_2u_2})^2 \ (u_1, u_2 \in \{1, 2, \cdots, q_1\}, u_1 \neq u_2).$$

For $r_i = 1, 2, \cdots, p$ and $s_i = 1, 2, \cdots, q$ $(i = 1, 2)$, if

$$\Delta_{j_1j_1}^{(r_1r_2)} = (a_{j_1j_1r_1} - a_{j_1j_1r_2})^2 = 0 \text{ and } \Delta_{j_1j_2}^{(u_1u_2)} = (a_{j_1j_2u_1} - a_{j_1j_2u_2})^2 = 0,$$

there are three bifurcations as follows.

- Two 1-dimensional flows of and switching at point with the same order singularity are given through

$$\bigcup_{r_1=1}^{p_1} \underbrace{(a_{j_1j_1r_1}, \dot{x}_{j_2})}_{(m_{j_1r_1})^{\text{th}}\text{order XX}} \rightleftharpoons \underbrace{(a_{j_1j_1 1}, \dot{x}_{j_2})}_{(2m_{11}+1)^{\text{th}}\text{ order XX}} \rightleftharpoons \bigcup_{r_2=1}^{p_1} \underbrace{(a_{j_1j_1r_2}, \dot{x}_{j_2})}_{(m_{j_1r_2})^{\text{th}}\text{order XX}}$$
$$\{a_{j_1j_1r_1} \rightleftharpoons a_{j_1j_1r_2} | m_{j_1r_1} = m_{j_1r_2}; r_1, r_2 \in \{1, 2, \cdots, p_1\}, r_1 \neq r_2\};$$

and

$$\begin{aligned} &a_{j_1j_1r_1} = a_{j_1j_11} - \varepsilon_{r_1}, a_{j_1j_1r_2} = a_{j_1j_11} + \varepsilon_{r_2}, \\ &(a_{j_1j_1r_1} = a_{j_1j_11} + \varepsilon_{r_1}, a_{j_1j_1r_2} = a_{j_1j_11} - \varepsilon_{r_2}); \\ &\varepsilon_{r_1}, \varepsilon_{r_2} > 0;\ r_i = 1, 2, \cdots, p_1\ (i = 1, 2). \end{aligned}$$

- Two 1-dimensional flows of $(a_{j_1j_2u_1}, \dot{x}_{j_2})$ and $(a_{j_1j_2u_1}, \dot{x}_{j_2})$ switching at point $(a_{j_1j_21}, \dot{x}_{j_2})$ with the same order singularity are given through

$$\bigcup_{u_1=1}^{q_1} \underbrace{(a_{j_1j_2u_1}, \dot{x}_{j_2})}_{(n_{j_1u_1})^{\text{th}}\text{order YY}} \rightleftharpoons \underbrace{(a_{j_1j_21}, \dot{x}_{j_2})}_{(2n_1)^{\text{th}}\text{ order YY}} \rightleftharpoons \bigcup_{u_2=1}^{q_1} \underbrace{(a_{j_1j_2u_2}, \dot{x}_{j_2})}_{(n_{j_1u_2})^{\text{th}}\text{order YY}}$$
$$\{a_{j_1j_2u_1} \rightleftharpoons a_{j_1j_2u_2} | n_{j_1u_1} = n_{j_1u_2}; u_1, u_2 \in \{1, 2, \cdots, q\}, u_1 \neq u_2\};$$

and

$$\begin{aligned} &a_{j_1j_2u_1} = a_{j_1j_21} - \varepsilon_{u_1}, a_{j_1j_2u_2} = a_{j_1j_21} + \varepsilon_{u_2}, \\ &(a_{j_1j_2u_1} = a_{j_1j_21} + \varepsilon_{u_1}, a_{j_1j_2u_2} = a_{j_1j_21} - \varepsilon_{u_2}); \\ &\varepsilon_{u_1}, \varepsilon_{u_2} > 0;\ u_i = 1, 2, \cdots, q_1\ (i = 1, 2). \end{aligned}$$

- Two 1-dimensional flows of $(a_{j_1j_1r_1}, a_{j_1j_2u_1})$ and $(a_{j_1j_1r_2}, a_{j_1j_2u_2})$ switching at point $(a_{j_1j_11}, a_{j_1j_21})$ with the same order singularity are given through

$$\begin{aligned} \bigcup_{r_1=1}^{p_1} \bigcup_{u_1=1}^{q_1} \underbrace{(a_{j_1j_1r_1}, a_{j_1j_2u_1})}_{([(m_{j_1r_1}),(n_{j_1u_1})]\text{XX,xF})\text{-flow}} &\rightleftharpoons \underbrace{([(2m_{11}+1),(2\mathrm{n}_{11})]\text{XX, xF})}_{[(2m_{11}+1),(2\mathrm{n}_{11})]\text{-XX}} \\ &\rightleftharpoons \bigcup_{r_2=1}^{p_1} \bigcup_{u_2=1}^{q_1} \underbrace{(a_{j_1j_1r_2}, a_{j_1j_2u_2})}_{([(m_{j_1r_2}),(n_{j_1u_2})]\text{XX,xF})\text{-flow}} \end{aligned}$$

with

$$\begin{aligned}
&a_{j_1j_1r_1}=a_{j_1j_1 1}-\varepsilon_{r_1}, a_{j_1j_1r_2}=a_{j_1j_1 1}+\varepsilon_{r_2},\\
&(a_{j_1j_1r_1}=a_{j_1j_1 1}+\varepsilon_{r_1}, a_{j_1j_1r_2}=a_{j_1j_1 1}-\varepsilon_{r_2}),\\
&\varepsilon_{r_1},\varepsilon_{r_2}>0,\ r_i=1,2,\cdots,p_1\ (i=1,2);\\
&a_{j_1j_2u_1}=a_{j_1j_2 1}-\varepsilon_{u_1}, a_{j_1j_2u_2}=a_{j_1j_2 1}+\varepsilon_{u_2},\\
&(a_{j_1j_2u_1}=a_{j_1j_2 1}+\varepsilon_{u_1}, a_{j_1j_2u_2}=a_{j_1j_2 1}-\varepsilon_{u_2}),\\
&\varepsilon_{u_1},\varepsilon_{u_2}>0,\ u_i=1,2,\cdots,q_1\ (i=1,2).
\end{aligned}$$

In summary, from the analysis in case (I_1)-(I_3), the 1-dimensional flow of $(x_{j_1}^*, x_{j_2}^*)=(a_{j_1j_1 1}, a_{j_1j_2 1})$ for $m_1=2m_{11}+1$ and $n_1=2n_{11}$ has the bifurcation properties as stated in (iv_{3a})–(iv_{3d}) through Eqs. (1.347)–(1.366).

(iv_4) For $m_1=2m_{11}$ and $n_1=2n_{11}$ the equilibrium of $(x_{j_1}^*, x_{j_2}^*)=(a_{j_1j_1 1}, a_{j_1j_2 1})$ for the x_{j_1}-direction has the following bifurcation properties with

$$\begin{aligned}
&2m_{11}-\sum\nolimits_{r_1=1}^{p_1} m_{j_1r_1}=2\sum\nolimits_{\rho_1=1}^{l_1} m_{\rho_1},\\
&2n_{11}-\sum\nolimits_{u_1=1}^{q_1} n_{j_1u_1}=2\sum\nolimits_{\sigma_1=1}^{l_2} n_{\sigma_1};\\
&\sum\nolimits_{r_2=1}^{\bar{p}_1} m_{j_1r_2}=2m_{11}, \sum\nolimits_{u_2=1}^{\bar{q}_1} n_{j_1u_2}=2n_{11};\\
&\sum\nolimits_{r_i=1}^{p_1} m_{j_1r_i}=2m_{11}, \sum\nolimits_{u_i=1}^{q_1} n_{j_1u_i}=2n_{11},\ \text{for}\ i=1,2;
\end{aligned}$$

l_1, l_2 -quadratic polynomials are without real roots and ρ_1^{th} and σ_1^{th} -quadratic polynomials are with powers m_{ρ_1} and n_{σ_1},respectively.

The bifurcation process is discussed through differential equations as follows. There are four bifurcations (I_1)-(I_3).

(I_1) Consider a dynamical system having a single equilibrium (A_1) as

$$\begin{aligned}
&\dot{x}_{j_1}=a_{j_1j_1 0}\prod\nolimits_{r=1}^{m_{11}}[(x_{j_1}-a_{j_1j_1}^{(r)})^2+\Delta_{j_1j_1}^{(r)}]\prod\nolimits_{u=1}^{n_{11}}[(x_{j_2}-a_{j_1j_2}^{(u)})^2+\Delta_{j_1j_2}^{(u)}],\\
&\dot{x}_{j_2}=a_{j_2j_2 0}(x_{j_2}-a_{j_2j_2 1})^{n_2}.
\end{aligned}$$

Once

$$\begin{aligned}
&\Delta_{j_1j_1}^{(r)}=\Delta_{j_1j_1}=0\,(r=1,2,\cdots,m_{11}),\\
&\Delta_{j_1j_2}^{(u)}=\Delta_{j_1j_2}=0\,(u=1,2,\cdots,n_{11}),
\end{aligned}$$

if

$$a_{j_1j_11} = b_{j_1j_11} = a^{(r)}_{j_1j_1}\ (r = 1, 2, \cdots, m_{11}),$$
$$a_{j_1j_21} = b_{j_1j_21} = a^{(u)}_{j_1j_2}\ (u = 1, 2, \cdots, n_{11}),$$

the foregoing differential equations becomes a dynamical system with a singular equilibrium (B_1) as

$$\dot{x}_{j_1} = a_{j_1j_10}(x_{j_1} - a_{j_1j_11})^{2m_{11}}(x_{j_2} - a_{j_1j_21})^{2n_{11}},$$
$$\dot{x}_{j_2} = a_{j_2j_20}(x_{j_2} - a_{j_2j_21})^{n_2}.$$

If

$$\Delta^{(r)}_{j_1j_1} = \Delta_{j_1j_1} + \delta_r\ (r = 1, 2, \cdots, m_{11}),\ \delta_r > 0;$$
$$\Delta^{(u)}_{j_1j_2} = \Delta_{j_1j_2} + \delta_u\ (u = 1, 2, \cdots, n_{11}),\ \delta_u > 0;$$

then

$$a^{(r)}_{j_1j_11}, a^{(r)}_{j_1j_12} = a_{j_1j_11} \pm \varepsilon_r,\ (r = 1, 2, \cdots, m_{11}),$$
$$\{a_{j_1j_11}, a_{j_1j_12}, \cdots, a_{j_1j_1(2m_{11})}\} = \mathrm{sort}\{a^{(r)}_{j_1j_11}, a^{(r)}_{j_1j_12} | r = 1, 2, \cdots, m_{11}\};$$
$$a^{(u)}_{j_1j_21}, a^{(u)}_{j_1j_22} = a_{j_1j_21} \pm \varepsilon_u,\ (u = 1, 2, \cdots, n_1),$$
$$\{a_{j_1j_21}, a_{j_1j_22}, \cdots, a_{j_1j_2(2n_{11})}\} = \mathrm{sort}\{a^{(u)}_{j_1j_21}, a^{(u)}_{j_1j_22} | u = 1, 2, \cdots, n_{11}\}.$$

Thus, the differential equation becomes a dynamical system with a non-singular equilibrium network (C_1) as

$$\dot{x}_{j_1} = a_{j_1j_10}\prod_{r=1}^{2m_{11}}(x_{j_1} - a_{j_1j_1r})\prod_{u=1}^{2n_{11}}(x_{j_2} - a_{j_1j_2u}),$$
$$\dot{x}_{j_2} = a_{j_2j_20}(x_{j_2} - a_{j_2j_21})^{n_2}.$$

- From the above bifurcation process analysis, at least, only (m_{11})-parameter variations in the x_{j_1}-direction are engaged in such a bifurcation from a 1-dimensional flow to the 1-dimensional flow array of $1 \times (2m_{11})$ through the $(2m_{11})^{\text{th}}$-order upper-saddle and lower-saddle bifurcations in the x_{j_1}-direction. The higher-order upper and lower-saddle flow bifurcations are developed.

  Thus, the 1-dimensional flow appearing or vanishing bifurcation route is as follows.

$$\underbrace{(a_{j_1j_11}, \dot{x}_{j_2})}_{\text{XX}} \rightleftharpoons \underbrace{(a_{j_1j_11}, \dot{x}_{j_2})}_{(2m_{11})^{\text{th}}\text{order XX}} \rightleftharpoons \bigcup_{r=1}^{2m_{11}} \underbrace{(a_{j_1j_1r}, \dot{x}_{j_2})}_{\text{XX}}.$$

- From the above bifurcation process analysis, at least, only n_{11}-parameter variations in the x_{j_1}-direction are engaged in such a bifurcation from a 1-dimensional flow to the 1-dimensional flow array of $(2n_{11}) \times 1$ through the $(2n_{11})^{\text{th}}$-order increasing and decreasing-inflection bifurcations in the x_{j_1}-direction. The higher-order increasing-inflection and decreasing-inflection flow bifurcations are developed.

  Thus, the 1-dimensional flow appearing or vanishing bifurcation route is as follows.

$$\underbrace{(a_{j_1j_21}, \dot{x}_{j_2})}_{\text{YY}} \rightleftharpoons \underbrace{(a_{j_1j_21}, \dot{x}_{j_2})}_{(2n_{11})^{\text{th}}\text{order YY}} \rightleftharpoons \bigcup\nolimits_{u=1}^{2n_{11}} \underbrace{(a_{j_1j_2u}, \dot{x}_{j_2})}_{\text{YY}}.$$

- From the above bifurcation process analysis, $(m_{11} + n_{11})$-parameter variations in the two directions are engaged in such a bifurcation from a parabola-flow to a $(2m_{11}) \times (2n_{11})$ matrix of inflection flows through the $(2m_{11})^{\text{th}}$-order source and sink bifurcations in the-direction, and the $(2n_{11})^{\text{th}}$ order increasing and decreasing-inflection flows bifurcations in the-direction. The higher-order increasing and decreasing-inflection flow bifurcations are developed.

  Thus, the 1-dimensional flow appearing or vanishing bifurcation route is as follows.

$$\underbrace{(a_{j_1j_11}, \dot{x}_{j_1})}_{(\text{XX,xF})\text{-flow}} \rightleftharpoons \underbrace{[(2m_{11}), (2n_{11})]\text{XX, xF})}_{[(2m_{11}),(2n_{11})]\text{-XX}} \rightleftharpoons \bigcup\nolimits_{r=1}^{2m_{11}} \bigcup\nolimits_{u=1}^{2n_{11}} \underbrace{(a_{j_1j_1r}, a_{j_1j_2u})}_{(\text{YY,xF})\text{-flow}},$$

  which are the higher-order inflection upper-saddle and lower-saddle flows bifurcations.

(I_2) From Case (I_1), consider a dynamical system having singular equilibriums (A_2) as

$$\begin{aligned}
\dot{x}_{j_1} =& a_{j_1j_10} \prod\nolimits_{r_1=1}^{p_1} (x_{j_1} - b_{j_1j_1r_1})^{m_{j_1r_1}} \prod\nolimits_{r=1}^{l_1} [(x_{j_1} - a_{j_1j_1}^{(r)})^2 + \Delta_{j_1j_1}^{(r)}]^{m_{j_1r}} \\
& \times \prod\nolimits_{u_1=1}^{q_1} (x_{j_2} - b_{j_1j_2u_1})^{m_{j_1u_1}} \prod\nolimits_{u=1}^{l_2} [(x_{j_2} - a_{j_1j_2}^{(u)})^2 + \Delta_{j_1j_2}^{(u)}]^{m_{j_1u}}, \\
\dot{x}_{j_2} =& a_{j_2j_20}(x_{j_2} - a_{j_2j_21})^{n_2};
\end{aligned}$$

where

$$\begin{aligned}
2m_{11} - \sum\nolimits_{r_1=1}^{p_1} m_{j_1r_1} &= 2\sum\nolimits_{\rho_1=1}^{l_1} m_{\rho_1}, \\
2n_{11} - \sum\nolimits_{u_1=1}^{q_1} n_{j_1u_1} &= 2\sum\nolimits_{\sigma_1=1}^{l_2} n_{\sigma_1};
\end{aligned}$$

$$\sum_{r_2=1}^{\overline{p}_1} m_{j_1 r_2} = 2m_{11}, \sum_{u_2=1}^{\overline{q}_1} n_{j_1 u_2} = 2n_{11};$$

l_1, l_2 -quadratic polynomials are without real roots and ρ_1^{th} and σ_1^{th} -quadratic polynomials are with powers m_{ρ_1} and n_{σ_1},respectively.

Once

$$\Delta_{j_1 j_1}^{(r)} = 0\,(r = 1, 2, \cdots, l_1),$$
$$\Delta_{j_1 j_2}^{(u)} = 0\,(u = 1, 2, \cdots, l_2);$$

if

$$a_{j_1 j_1 1} \equiv a_{j_1 j_1 r_1} = a_{j_1 j_1}^{(r)}\ (r_1 = 1, 2, \cdots, p_1; r = 1, 2, \cdots, l_1),$$
$$a_{j_1 j_2 1} \equiv a_{j_1 j_2 u_1} = a_{j_1 j_2}^{(u)}\ (u_1 = 1, 2, \cdots, q_1; u = 1, 2, \cdots, l_2),$$

the foregoing differential equations become a dynamical system with a 1-dimensional flow (B_2) as

$$\dot{x}_{j_1} = a_{j_1 j_1 0}(x_{j_1} - a_{j_1 j_1 1})^{2m_{11}}(x_{j_2} - a_{j_1 j_2 1})^{2n_{11}},$$
$$\dot{x}_{j_2} = a_{j_2 j_2 0}(x_{j_2} - a_{j_2 j_2 1})^{n_2}.$$

If

$$\Delta_{j_1 j_1}^{(r)} = \Delta_{j_1 j_1} + \delta_r\,(r = 1, 2, \cdots, l_1),\ \delta_r > 0;$$
$$\Delta_{j_1 j_2}^{(u)} = \Delta_{j_1 j_2} + \delta_u\,(u = 1, 2, \cdots, l_2),\ \delta_u > 0;$$

then

$$a_{j_1 j_1 1}^{(r)}, a_{j_1 j_1 2}^{(r)} = a_{j_1 j_1 1} \pm \varepsilon_r,\ (r = 1, 2, \cdots, l_1),$$
$$\{a_{j_1 j_1 1}, a_{j_1 j_1 2}, \cdots, a_{j_1 j_1 \overline{p}_1}\}$$
$$= \text{sort}\{a_{j_1 j_1 1}, a_{j_1 j_1 2}, \cdots, a_{j_1 j_1 p_1}; a_{j_1 j_1 1}^{(r)}, a_{j_1 j_1 2}^{(r)} | r = 1, 2, \cdots, l_1\};$$
$$a_{j_1 j_2 1}^{(u)}, a_{j_1 j_2 2}^{(u)} = a_{j_1 j_2 1} \pm \varepsilon_u,\ (u = 1, 2, \cdots, l_2),$$
$$\{a_{j_1 j_2 1}, a_{j_1 j_2 2}, \cdots, a_{j_1 j_2 \overline{q}_1}\}$$
$$= \text{sort}\{a_{j_1 j_2 1}, a_{j_1 j_2 2}, \cdots, a_{j_1 j_2 q_1}; a_{j_1 j_2 1}^{(u)}, a_{j_1 j_2 2}^{(u)} | u = 1, 2, \cdots, l_2\}.$$

Thus, the differential equation becomes a dynamical system with singular 1-dimensional flows (C_2) as

$$\dot{x}_{j_1} = a_{j_1j_10}\prod_{r_2=1}^{\overline{p}_1}(x_{j_1} - a_{j_1j_1r_2})^{m_{j_1r_2}}\prod_{u_2=1}^{\overline{q}_1}(x_{j_2} - a_{j_1j_2u_2})^{n_{j_1u_2}},$$
$$\dot{x}_{j_2} = a_{j_2j_20}(x_{j_2} - a_{j_2j_21})^{n_2}$$

where

$$\Sigma_{r_2=1}^{\overline{p}_1}m_{j_1r_2} = 2m_{11},\ \Sigma_{u_2=1}^{\overline{q}_1}n_{j_1u_2} = 2n_{11}.$$

- From the above bifurcation process analysis, only $\overline{p}_1$-parameter variations in the x_{j_1}-direction from a p_1 to $\overline{p}_1$ 1-dimensional flow array through the $(2m_{11})^{\text{th}}$-order source and sink flow bifurcations in the x_{j_1}-direction.

  Thus, the 1-dimensional flow appearing or vanishing bifurcation route from a p_1 to $\overline{p}_1$ 1-dimensional flow array is expressed as

$$\bigcup_{r_1=1}^{p_1}\underbrace{(a_{j_1j_1r_1},\dot{x}_{j_2})}_{(m_{j_1r_1})^{\text{th}}\text{order XX}} \rightleftharpoons \underbrace{(a_{j_1j_11},\dot{x}_{j_2})}_{(2m_{11}+1)^{\text{th}}\text{ order XX}} \rightleftharpoons \bigcup_{r_2=1}^{\bar{p}_1}\underbrace{(a_{j_1j_1r_2},\dot{x}_{j_2})}_{(m_{j_1r_2})^{\text{th}}\text{order XX}}.$$

- From the above bifurcation process analysis, only $\overline{q}_1$-parameter variations in the x_{j_1}-direction from a q_1 to $\overline{q}_1$ 1-dimensional flow array through the $(2n_{11})^{\text{th}}$-order increasing and decreasing-inflection flow bifurcations in the x_{j_1}-direction.

  Thus, the 1-dimensional flow appearing or vanishing bifurcation route from a q_1 to $\overline{q}_1$ 1-dimensional flow array is expressed as

$$\bigcup_{u_1=1}^{q_1}\underbrace{(a_{j_2j_1u_1},\dot{x}_{j_2})}_{(n_{j_1u_1})^{\text{th}}\text{order YY}} \rightleftharpoons \underbrace{(a_{j_1j_21},\dot{x}_{j_2})}_{(2n_{11})^{\text{th}}\text{ order YY}} \rightleftharpoons \bigcup_{u_2=1}^{\overline{q}_1}\underbrace{(a_{j_1j_2u_2},\dot{x}_{j_2})}_{(n_{j_1u_2})^{\text{th}}\text{order YY}}.$$

- From the above bifurcation process analysis, only $\overline{q}_1+\overline{p}_1$-parameter variations in the x_{j_1}-direction from a $q_1 \times p_1$ to $\overline{q}_1 \times \overline{p}_1$ matrix of 1-dimensional flows through the $(2m_{11})^{\text{th}}$-order upper-saddle and lower-saddle flows bifurcations in the x_{j_1}-direction, and the $(2n_{11})^{\text{th}}$-order up-parabola and down-parabola flows bifurcations in the x_{j_1}-direction.

  Thus, the 1-dimensional flow appearing or vanishing bifurcation route is as follows:

$$\begin{aligned}\bigcup_{r_1=1}^{p_1}\bigcup_{u_1=1}^{q_1}\underbrace{(a_{j_1j_1r_1},a_{j_1j_2u_1})}_{([(m_{j_1r_1}),(n_{j_1u_1})]\text{XX,xF})\text{-flow}} &\rightleftharpoons \underbrace{([(2m_{11}),(2n_{11})]\text{XX, xF})}_{[(2m_{11}),(2n_{11})]\text{-XX}}\\ &\rightleftharpoons \bigcup_{r_2=1}^{\overline{p}_1}\bigcup_{u_2=1}^{\overline{q}_1}\underbrace{(a_{j_1j_1r_2},a_{j_1j_2u_2})}_{([(m_{j_1r_2}),(n_{j_1u_1})]\text{XX,xF})\text{-flow}},\end{aligned}$$

which are the higher-order inflection upper-saddle and lower-saddle flows bifurcations.

(I_3) Consider two dynamical systems with the same 1-dimensional flows with locations switched (A,C_3) for $i = 1, 2$ as$_3$

$$\begin{aligned}\dot{x}_{j_1} &= a_{j_1j_10}\prod\nolimits_{r_i=1}^{p_1}(x_{j_1} - a_{j_1j_1r_i})^{m_{j_1r_i}}\prod\nolimits_{u_i=1}^{q_1}(x_{j_2} - a_{j_1j_2u_i})^{n_{j_1u_i}},\\ \dot{x}_{j_2} &= a_{j_2j_20}(x_{j_2} - a_{j_2j_21})^{n_2};\end{aligned}$$

where

$$\sum\nolimits_{r_i=1}^{p_1} m_{j_1r_i} = 2m_{11} + 1, \sum\nolimits_{u_i=1}^{q_1} n_{j_1u_i} = 2n_{11}, \text{ for } i = 1, 2.$$

Consider a dynamical system as a singular 1-dimensional flow (B_3) as

$$\begin{aligned}\dot{x}_{j_1} &= a_{j_1j_10}(x_{j_1} - a_{j_1j_11})^{2m_{11}}(x_{j_2} - a_{j_1j_21})^{2n_{11}},\\ \dot{x}_{j_2} &= a_{j_2j_10}(x_{j_2} - a_{j_2j_21})^{n_2}.\end{aligned}$$

For $r_i = 1, 2, \cdots, p_1$ and $s_i = 1, 2, \cdots, q_1$ $(i = 1, 2)$, define a function as

$$\begin{aligned}\Delta_{j_1j_1}^{(r_1r_2)} &= (a_{j_1j_1r_1} - a_{j_1j_1r_2})^2\ (r_1, r_2 \in \{1, 2, \cdots, p_1\}, r_1 \neq r_2),\\ \Delta_{j_1j_2}^{(u_1u_2)} &= (a_{j_1j_2u_1} - a_{j_1j_2u_2})^2\ (u_1, u_2 \in \{1, 2, \cdots, q_1\}, u_1 \neq u_2).\end{aligned}$$

For $r_i = 1, 2, \cdots, p_1$ and $s_i = 1, 2, \cdots, q_1$ $(i = 1, 2)$, if

$$\Delta_{j_1j_1}^{(r_1r_2)} = (a_{j_1j_1r_1} - a_{j_1j_1r_2})^2 = 0 \text{ and } \Delta_{j_1j_2}^{(u_1u_2)} = (a_{j_1j_2u_1} - a_{j_1j_2u_2})^2 = 0,$$

there are three bifurcations as follows.

- Two 1-dimensional flows of and switching at point with the same order singularity are given through

$$\bigcup\nolimits_{r_1=1}^{p_1}\underbrace{(a_{j_1j_1r_1}, \dot{x}_{j_2})}_{(m_{j_1r_1})^{\text{th}}\text{order XX}} \rightleftharpoons \underbrace{(a_{j_1j_11}, \dot{x}_{j_2})}_{(2m_{11})^{\text{th}}\text{ order XX}} \rightleftharpoons \bigcup\nolimits_{r_2=1}^{p_1}\underbrace{(a_{j_1j_1r_2}, \dot{x}_{j_2})}_{(m_{j_1r_2})^{\text{th}}\text{order XX}}$$
$$\left\{a_{j_1j_1r_1} \rightleftharpoons a_{j_1j_1r_2} \middle| m_{j_1r_1} = m_{j_1r_2}; r_1, r_2 \in \{1, 2, \cdots, p_1\}, r_1 \neq r_2\right\};$$

  and

$$\begin{aligned}&a_{j_1j_1r_1} = a_{j_1j_11} - \varepsilon_{r_1}, a_{j_1j_1r_2} = a_{j_1j_11} + \varepsilon_{r_2},\\ &(a_{j_1j_1r_1} = a_{j_1j_11} + \varepsilon_{r_1}, a_{j_1j_1r_2} = a_{j_1j_11} - \varepsilon_{r_2});\\ &\varepsilon_{r_1}, \varepsilon_{r_2} > 0;\ r_i = 1, 2, \cdots, p_1\ (i = 1, 2).\end{aligned}$$

- Two 1-dimensional flows of $(a_{j_1j_2s_1}, \dot{x}_{j_2})$ and $(a_{j_1j_2s_1}, \dot{x}_{j_2})$ switching at point $(a_{j_1j_2 1}, \dot{x}_{j_2})$ with the same order singularity are given through

$$\bigcup_{u_1=1}^{q_1} \underbrace{(a_{j_1j_2u_1}, \dot{x}_{j_2})}_{(n_{j_1u_1})^{\text{th}}\text{order YY}} \rightleftharpoons \underbrace{(a_{j_1j_2 1}, \dot{x}_{j_2})}_{(2n_{11})^{\text{th}}\text{ order YY}} \rightleftharpoons \bigcup_{u_2=1}^{q_1} \underbrace{(a_{j_1j_2u_2}, \dot{x}_{j_2})}_{(n_{j_1u_2})^{\text{th}}\text{order YY}}$$

$$\{a_{j_1j_2u_1} \rightleftharpoons a_{j_1j_2u_2} | n_{j_1u_1} = n_{j_1u_2}; u_1, u_2 \in \{1, 2, \cdots, q\}, u_1 \neq u_2\};$$

and

$$\begin{aligned} &a_{j_1j_2u_1} = a_{j_1j_2 1} - \varepsilon_{u_1}, a_{j_1j_2u_2} = a_{j_1j_2 1} + \varepsilon_{u_2}, \\ &(a_{j_1j_2u_1} = a_{j_1j_2 1} + \varepsilon_{u_1}, a_{j_1j_2u_2} = a_{j_1j_2 1} - \varepsilon_{u_2}); \\ &\varepsilon_{u_1}, \varepsilon_{u_2} > 0;\ u_i = 1, 2, \cdots, q_1\ (i = 1, 2). \end{aligned}$$

- Two 1-dimensional flows of $(a_{j_1j_1r_1}, a_{j_1j_2u_1})$ and $(a_{j_1j_1r_2}, a_{j_1j_2u_2})$ switching at point $(a_{j_1j_1 1}, a_{j_1j_2 1})$ with the same order singularity are given through

$$\begin{aligned} \bigcup_{r_1=1}^{p} \bigcup_{u_1=1}^{q} \underbrace{(a_{j_1j_1r_1}, a_{j_1j_2u_1})}_{([(m_{j_1r_1}),(n_{j_1u_1})]\text{XX,xF})\text{-flow}} &\rightleftharpoons \underbrace{([(2m_{11}), (2n_{11})]\text{XX, xF})}_{[(2m_{11}),(2n_{11})]\text{-XX}} \\ &\rightleftharpoons \bigcup_{r_2=1}^{p} \bigcup_{u_2=1}^{q} \underbrace{(a_{j_1j_1r_2}, a_{j_1j_2u_2})}_{([(m_{j_1r_2}),(n_{j_1u_2})]\text{XX,xF})\text{-flow}} \end{aligned}$$

with

$$\begin{aligned} &a_{j_1j_1r_1} = a_{j_1j_1 1} - \varepsilon_{r_1}, a_{j_1j_1r_2} = a_{j_1j_1 1} + \varepsilon_{r_2}, \\ &(a_{j_1j_1r_1} = a_{j_1j_1 1} + \varepsilon_{r_1}, a_{j_1j_1r_2} = a_{j_1j_1 1} - \varepsilon_{r_2}), \\ &\varepsilon_{r_1}, \varepsilon_{r_2} > 0,\ r_i = 1, 2, \cdots, p_1\ (i = 1, 2); \\ &a_{j_1j_2u_1} = a_{j_1j_2 1} - \varepsilon_{u_1}, a_{j_1j_2u_2} = a_{j_1j_2 1} + \varepsilon_{u_2}, \\ &(a_{j_1j_2u_1} = a_{j_1j_2 1} + \varepsilon_{u_1}, a_{j_1j_2u_2} = a_{j_1j_2 1} - \varepsilon_{u_2}), \\ &\varepsilon_{u_1}, \varepsilon_{u_2} > 0,\ u_i = 1, 2, \cdots, q_1\ (i = 1, 2). \end{aligned}$$

The switching bifurcations are the higher-order inflection upper-saddle and lower-saddle flows bifurcations.

In summary, from the analysis in case (I_1)-(I_3), the 1-dimensional flow of $(x_{j_1}^*, x_{j_2}^*) = (a_{j_1j_1 1}, a_{j_1j_2 1})$ for $m_1 = 2m_{11}$ and $n_1 = 2n_{11}$ has the bifurcation properties as stated in (iv_{4a})-(iv_{4d}) through Eqs. (1.368)–(1.387).

Note that the equation in the x_{j_2}-direction can be considered as follows.

$$\begin{aligned} \dot{x}_{j_2} &= a_{j_2j_2 0} \prod_{s=1}^{n_2} (x_{j_2} - a_{j_2j_2s});\ \text{or} \\ \dot{x}_{j_2} &= a_{j_2j_2 0} \prod_{s=1}^{q_2} (x_{j_2} - a_{j_2j_2s})^{n_{j_2s}}. \end{aligned}$$

2.3.2 Appearing and Switching Bifurcations for Equilibriums

(v) The equilibrium of $(x^*_{j_1}, x^*_{j_2}) = (a_{j_1j_11}, a_{j_2j_21})$ with $a_{j_2j_21} \neq a_{j_1j_21}$ has four classes of equilibrium bifurcations.

(v_1) For $m_1 = 2m_{11}+1$ and $n_2 = 2n_{21}+1$ the equilibrium of $(x^*_{j_1}, x^*_{j_2}) = (a_{j_1j_11}, a_{j_2j_21})$ has the following bifurcation properties with

$$
\begin{aligned}
&2m_{11} + 1 - \sum\nolimits_{r_1=1}^{p_1} m_{j_1r_1} = 2\sum\nolimits_{\rho_1=1}^{l_1} m_{\rho_1},\\
&2n_{21} + 1 - \sum\nolimits_{s_1=1}^{q_2} n_{j_2s_1} = 2\sum\nolimits_{\sigma_1=1}^{l_2} n_{\sigma_1};\\
&\sum\nolimits_{r_2=1}^{\bar{p}_1} m_{j_1r_2} = 2m_{11} + 1, \sum\nolimits_{s_2=1}^{\bar{q}_2} n_{j_2s_2} = 2n_{21} + 1;\\
&\sum\nolimits_{r_i=1}^{p_1} m_{j_1r_i} = 2m_{11} + 1, \sum\nolimits_{s_i=1}^{q_1} n_{j_2s_i} = 2n_{21} + 1, \mathit{for}\, i = 1, 2;
\end{aligned}
$$

l_1, l_2-quadratic polynomials are without real roots and ρ_1^{th} and σ_1^{th}-quadratic polynomials are with powers m_{ρ_1} and n_{σ_1}, respectively.

The bifurcation process is discussed through differential equations as follows. There are four bifurcations (I_1)-(I_3).

(I_1) Consider a dynamical system having a single equilibrium (A_1) as

$$
\begin{aligned}
\dot{x}_{j_1} &= a_{j_1j_10}(x_{j_1} - b_{j_1j_11})\prod\nolimits_{r=1}^{m_{11}} [(x_{j_1} - a^{(r)}_{j_1j_1})^2 + \Delta^{(r)}_{j_1j_1}](x_{j_2} - a_{j_1j_21})^{n_1},\\
\dot{x}_{j_2} &= a_{j_2j_20}(x_{j_2} - b_{j_2j_21})\prod\nolimits_{s=1}^{n_{21}} [(x_{j_2} - a^{(s)}_{j_2j_2})^2 + \Delta^{(s)}_{j_2j_2}];
\end{aligned}
$$

where $a_{j_1j_11} = b_{j_1j_11}$ and $a_{j_2j_21} = b_{j_2j_21} \neq a_{j_1j_21}$. Once

$$
\begin{aligned}
\Delta^{(r)}_{j_1j_1} &= \Delta_{j_1j_1} = 0\,(r = 1, 2, \cdots, m_{11}),\\
\Delta^{(s)}_{j_2j_2} &= \Delta_{j_2j_2} = 0\,(s = 1, 2, \cdots, n_{21}),
\end{aligned}
$$

if

$$
\begin{aligned}
a_{j_1j_11} &= b_{j_1j_11} = a^{(r)}_{j_1j_1}\ (r = 1, 2, \cdots, m_{11}),\\
a_{j_2j_21} &= b_{j_2j_21} = a^{(s)}_{j_2j_2}\ (s = 1, 2, \cdots, n_{21}),
\end{aligned}
$$

the foregoing differential equations become a dynamical system with a singular equilibrium (B_1) as

$$
\begin{aligned}
\dot{x}_{j_1} &= a_{j_1j_10}(x_{j_1} - a_{j_1j_11})^{2m_{11}+1}(x_{j_2} - a_{j_1j_21})^{n_1},\\
\dot{x}_{j_2} &= a_{j_2j_20}(x_{j_2} - a_{j_2j_21})^{2n_{21}+1}.
\end{aligned}
$$

If

$$\Delta_{j_1j_1}^{(r)} = \Delta_{j_1j_1} + \delta_r\, (r = 1, 2, \cdots, m_{11}),\ \delta_r > 0;$$
$$\Delta_{j_2j_2}^{(s)} = \Delta_{j_2j_2} + \delta_s\, (s = 1, 2, \cdots, n_{21}),\ \delta_s > 0;$$

then

$$a_{j_1j_1 1}^{(r)}, a_{j_1j_1 2}^{(r)} = a_{j_1j_1 1} \pm \varepsilon_r,\ (r = 1, 2, \cdots, m_{11}),$$
$$\{a_{j_1j_1 1}, a_{j_1j_1 2}, \cdots, a_{j_1j_1(2m_{11}+1)}\} = \text{sort}\{b_{j_1j_1 1}, a_{j_1j_1 1}^{(r)}, a_{j_1j_1 2}^{(r)} | r = 1, 2, \cdots, m_{11}\};$$
$$a_{j_2j_2 1}^{(s)}, a_{j_2j_2 2}^{(s)} = a_{j_2j_2 1} \pm \varepsilon_s,\ (s = 1, 2, \cdots, n_{21}),$$
$$\{a_{j_2j_2 1}, a_{j_2j_2 2}, \cdots, a_{j_2j_2(2n_{21}+1)}\} = \text{sort}\{b_{j_2j_2 1}, a_{j_2j_2 1}^{(s)}, a_{j_2j_2 2}^{(s)} | s = 1, 2, \cdots, n_{21}\}.$$

Thus, the differential equation becomes a dynamical system with a non-singular equilibrium network (C_1) as

$$\dot{x}_{j_1} = a_{j_1j_1 0} \prod\nolimits_{r=1}^{2m_{11}+1} (x_{j_1} - a_{j_1j_1 r})(x_{j_2} - a_{j_1j_2 1})^{n_1},$$
$$\dot{x}_{j_2} = a_{j_2j_2 0} \prod\nolimits_{s=1}^{2n_{21}+1} (x_{j_2} - a_{j_2j_2 s}).$$

- From the above bifurcation process analysis, at least, only $(m_{11}+1)$-parameter variations in the x_{j_1}-direction are engaged in such a bifurcation from a 1-dimensional flow to the 1-dimensional flow array of $1 \times (2m_{11}+1)$ through the $(2m_{11}+1)^{\text{th}}$-order source and sink flows bifurcations in the x_{j_1}-direction. The higher-order source and sink flow bifurcations are developed.

  Thus, the 1-dimensional flow appearing or vanishing bifurcation route is as follows.

$$\underbrace{(a_{j_1j_1 1}, \dot{x}_{j_2})}_{\text{XX}} \rightleftharpoons \underbrace{(a_{j_1j_1 1}, \dot{x}_{j_2})}_{(2m_{11}+1)^{\text{th}}\text{order XX}} \rightleftharpoons \bigcup\nolimits_{r=1}^{2m_{11}+1} \underbrace{(a_{j_1j_1 r}, \dot{x}_{j_2})}_{\text{XX}}.$$

- From the above bifurcation process analysis, at least, only $(n_{21}+1)$-parameter variations in the x_{j_1}-direction are engaged in such a bifurcation from a 1-dimensional flow to the 1-dimensional flow array of $(2n_{21}+1) \times 1$ through the $(2n_{21}+1)^{\text{th}}$-order source and sink flows bifurcations in the x_{j_2}-direction. The higher-order source and sink flow bifurcations are developed.

  Thus, the 1-dimensional flow appearing or vanishing bifurcation route is as follows.

$$\underbrace{(\dot{x}_{j_1}, a_{j_2j_21})}_{\text{YY}} \rightleftharpoons \underbrace{(\dot{x}_{j_1}, a_{j_2j_21})}_{(2n_{21}+1)^{\text{th}}\text{order YY}} \rightleftharpoons \bigcup\nolimits_{s=1}^{2n_{21}+1} \underbrace{(\dot{x}_{j_1}, a_{j_2j_2s})}_{\text{YY}}.$$

- From the above bifurcation process analysis, $(m_{11}+n_{21}+2)$-parameter variations in the two directions are engaged in such a bifurcation from one of saddle, source and sink to a $(2m_{11}+1)\times(2n_{21}+1)$ matrix of saddle, source and sink through the $(2m_{11}+1)^{\text{th}}$-order source and sink bifurcations in the x_{j_1}-direction, and the $(2n_{11}+1)^{\text{th}}$-order source and sink flows bifurcations in the x_{j_2}-direction. The higher-order source, sink and saddle bifurcations are developed.

  Thus, the equilibrium network appearing or vanishing bifurcation route is as follows.

$$\underbrace{(a_{j_1j_11}, a_{j_2j_21})}_{\text{XX}} \rightleftharpoons \underbrace{((2m_{11}+1)\text{YY},(2n_{21}+1)\text{ZZ})}_{((2m_{11}+1),(2n_{21}+1))\text{-XX}} \rightleftharpoons \bigcup\nolimits_{r=1}^{2m_{11}+1}\bigcup\nolimits_{s=1}^{2n_{21}+1} \underbrace{(a_{j_1j_1r}, a_{j_2j_2s})}_{\text{XX}},$$

  which are the higher-order source, sink and saddle bifurcations.

(I_2) From Case (I_1), consider a dynamical system having singular equilibriums (A_2) as

$$\begin{aligned}
\dot{x}_{j_1} &= a_{j_1j_10}\prod\nolimits_{r_1=1}^{p_1}(x_{j_1}-b_{j_1j_1r_1})^{m_{j_1r_1}}\prod\nolimits_{r=1}^{l_1}[(x_{j_1}-a_{j_1j_1}^{(r)})^2+\Delta_{j_1j_1}^{(r)}]^{m_{j_1r}}(x_{j_2}-a_{j_1j_21})^{n_2},\\
\dot{x}_{j_2} &= a_{j_2j_20}\prod\nolimits_{s_1=1}^{q_2}(x_{j_2}-b_{j_2j_2s_1})^{n_{j_2s_1}}\prod\nolimits_{s=1}^{l_2}[(x_{j_2}-a_{j_2j_2}^{(s)})^2+\Delta_{j_2j_2}^{(s)}]^{n_{j_2s}};
\end{aligned}$$

where

$$\begin{aligned}
&2m_{11}+1-\sum\nolimits_{r_1=1}^{p_1} m_{j_1r_1} = 2\sum\nolimits_{\rho_1=1}^{l_1} m_{\rho_1},\\
&2n_{21}+1-\sum\nolimits_{s_1=1}^{q_2} n_{j_2s_1} = 2\sum\nolimits_{\sigma_1=1}^{l_2} n_{\sigma_1};\\
&\sum\nolimits_{r_2=1}^{\bar{p}_1} m_{j_1r_2} = 2m_{11}+1, \sum\nolimits_{s_2=1}^{\bar{q}_2} n_{j_2s_2} = 2n_{21}+1;
\end{aligned}$$

l_1, l_2-quadratic polynomials are without real roots and ρ_1^{th} and σ_1^{th}-quadratic polynomials are with powers m_{ρ_1} and n_{σ_1},respectively.

Once

$$\begin{aligned}
\Delta_{j_1j_1}^{(r)} &= 0\,(r=1,2,\cdots,l_1),\\
\Delta_{j_2j_2}^{(s)} &= 0\,(s=1,2,\cdots,l_2);
\end{aligned}$$

if

$$a_{j_1j_11}\equiv a_{j_1j_1r_1}=a_{j_1j_1}^{(r)}\,(r_1=1,2,\cdots,p_1;r=1,2,\cdots,l_1),$$

$$a_{j_2j_21} \equiv a_{j_2j_2s_1} = a^{(s)}_{j_2j_2}\ (s_1 = 1, 2, \cdots, q_2;\ s = 1, 2, \cdots, l_2),$$

the foregoing differential equations become a dynamical system with a 1-dimensional flow (B_2) as

$$\begin{aligned}
\dot{x}_{j_1} &= a_{j_1j_10}(x_{j_1} - a_{j_1j_11})^{2m_{11}+1}(x_{j_2} - a_{j_1j_21})^{n_1},\\
\dot{x}_{j_2} &= a_{j_2j_20}(x_{j_2} - a_{j_2j_21})^{2n_{21}+1}.
\end{aligned}$$

If

$$\begin{aligned}
\Delta^{(r)}_{j_1j_1} &= \Delta_{j_1j_1} + \delta_r\ (r = 1, 2, \cdots, l_1),\ \delta_r > 0;\\
\Delta^{(s)}_{j_2j_2} &= \Delta_{j_2j_2} + \delta_s\ (s = 1, 2, \cdots, l_2),\ \delta_s > 0;
\end{aligned}$$

then

$$\begin{aligned}
&a^{(r)}_{j_1j_11}, a^{(r)}_{j_1j_12} = a_{j_1j_11} \pm \varepsilon_r,\ (r = 1, 2, \cdots, l_1),\\
&\{a_{j_1j_11}, a_{j_1j_12}, \cdots, a_{j_1j_1\bar{p}_1}\}\\
&= \text{sort}\{a_{j_1j_11}, a_{j_1j_12}, \cdots, a_{j_1j_1p_1};\ a^{(r)}_{j_1j_11}, a^{(r)}_{j_1j_12} | r = 1, 2, \cdots, l_1\};\\
&a^{(s)}_{j_2j_21}, a^{(s)}_{j_2j_22} = a_{j_2j_21} \pm \varepsilon_s,\ (s = 1, 2, \cdots, l_2),\\
&\{a_{j_2j_21}, a_{j_2j_22}, \cdots, a_{j_2j_2\bar{q}_2}\}\\
&= \text{sort}\{a_{j_2j_21}, a_{j_2j_22}, \cdots, a_{j_2j_2q_2};\ a^{(s)}_{j_2j_21}, a^{(s)}_{j_2j_22} | s = 1, 2, \cdots, l_2\}.
\end{aligned}$$

Thus, the differential equation becomes a dynamical system with singular 1-dimensional flows (C_2) as

$$\begin{aligned}
\dot{x}_{j_1} &= a_{j_1j_10}\prod\nolimits_{r_2=1}^{\bar{p}_1}(x_{j_1} - a_{j_1j_1r_2})^{m_{j_1r_2}}(x_{j_2} - a_{j_1j_21})^{n_1},\\
\dot{x}_{j_2} &= a_{j_2j_20}\prod\nolimits_{s_2=1}^{\bar{q}_2}(x_{j_2} - a_{j_2j_2s_2})^{n_{j_2s_2}}
\end{aligned}$$

where

$$\Sigma^{\bar{p}_1}_{r_2=1} m_{j_1r_2} = 2m_{11} + 1,\ \Sigma^{\bar{q}_2}_{u_2=1} n_{j_2s_2} = 2n_{21} + 1.$$

- From the above bifurcation process analysis, only $\bar{p}_1$-parameter variations in the x_{j_1}-direction from a p_1 to $\bar{p}_1$ 1-dimensional flow array through the $(2m_1 + 1)^{\text{th}}$-order source and sink flow bifurcations in the x_{j_1}-direction.

Thus, the 1-dimensional flow appearing or vanishing bifurcation route from a p_1 to $\overline{p}_1$ 1-dimensional flow array is expressed as

$$\bigcup_{r_1=1}^{p_1} \underbrace{(a_{j_1j_1r_1}, \dot{x}_{j_2})}_{(\mathrm{m}_{j_1r_1})^{\text{th}}\text{order XX}} \rightleftharpoons \underbrace{(a_{j_1j_11}, \dot{x}_{j_2})}_{(2m_{11}+1)^{\text{th}}\text{ order XX}} \rightleftharpoons \bigcup_{r_2=1}^{\overline{p}_1} \underbrace{(a_{j_1j_1r_2}, \dot{x}_{j_2})}_{(\mathrm{m}_{j_1r_2})^{\text{th}}\text{order XX}}.$$

- From the above bifurcation process analysis, only $\overline{q}_2$-parameter variations in the x_{j_2}-direction from a q_2 to $\overline{q}_2$ 1-dimensional flow array through the $(2n_{21}+1)^{\text{th}}$-order source and sink flow bifurcations in the x_{j_2}-direction.

  Thus, the 1-dimensional flow appearing or vanishing bifurcation route from a q_2 to $\overline{q}_2$ 1-dimensional flow array is expressed as

$$\bigcup_{s_1=1}^{q_2} \underbrace{(\dot{x}_{j_1}, a_{j_2j_2s_1})}_{(n_{j_2s_1})^{\text{th}}\text{order XX}} \rightleftharpoons \underbrace{(\dot{x}_{j_1}, a_{j_2j_21})}_{(2n_{21}+1)^{\text{th}}\text{ order YY}} \rightleftharpoons \bigcup_{s_2=1}^{\overline{q}_2} \underbrace{(\dot{x}_{j_1}, a_{j_2j_2s_2})}_{(n_{j_2s_2})^{\text{th}}\text{order YY}}.$$

- From the above bifurcation process analysis, only $\overline{q}_2+\overline{p}_1$-parameter variations from a $q_2\times p_1$ to $\overline{q}_2\times\overline{p}_1$ matrix of 1-dimensional flows through the $(2m_{11}+1)^{\text{th}}$-order source and sink flows bifurcations in the x_{j_1}-direction, and the $(2n_{21}+1)^{\text{th}}$-order source and sink flows bifurcations in the x_{j_2}-direction.

  Thus, the singular-equilibrium network appearing or vanishing bifurcation route is as follows:

$$\begin{aligned}\bigcup_{r_1=1}^{p_1}\bigcup_{s_1=1}^{q_2} \underbrace{(a_{j_1j_1r_1}, a_{j_2j_2s_1})}_{((m_{j_1r_1}),(n_{j_2s_1}))\text{-XX}} &\rightleftharpoons \underbrace{((2m_{11}+1)\text{YY},(2n_{21}+1)\text{ZZ})}_{((2m_{11}+1),(2n_{21}+1))\text{-XX}}\\ &\rightleftharpoons \bigcup_{r_2=1}^{\overline{p}_1}\bigcup_{s_2=1}^{\overline{q}_2} \underbrace{(a_{j_1j_1r_2}, a_{j_2j_2s_2})}_{((m_{j_1r_2}),(n_{j_2s_2}))\text{-XX}},\end{aligned}$$

  which are the higher-order source, sink and saddle bifurcations.

(I_3) Consider two dynamical systems with the same 1-dimensional flows with locations switched (A_3,C_3) for $i = 1, 2$ as

$$\begin{aligned}\dot{x}_{j_1} &= a_{j_1j_10}\prod_{r_i=1}^{p_1}(x_{j_1}-a_{j_1j_1r_i})^{m_{j_1r_i}}(x_{j_2}-a_{j_1j_21})^{n_1},\\ \dot{x}_{j_2} &= a_{j_2j_20}\prod_{s_i=1}^{q_1}(x_{j_2}-a_{j_2j_2s_i})^{n_{j_2s_i}};\end{aligned}$$

where

$$\sum_{r_i=1}^{p_1} m_{j_1r_i} = 2m_{11}+1, \sum_{s_i=1}^{q_2} n_{j_1s_i} = 2n_{21}+1, \text{ for } i = 1, 2.$$

Consider a dynamical system as a singular 1-dimensional flow (B_3) as

$$\dot{x}_{j_1} = a_{j_1 j_1 0}(x_{j_1} - a_{j_1 j_1 1})^{2m_{11}+1}(x_{j_2} - a_{j_1 j_2 1})^{n_1},$$
$$\dot{x}_{j_2} = a_{j_2 j_2 0}(x_{j_2} - a_{j_2 j_2 1})^{2n_{21}+1}.$$

For $r_i = 1, 2, \cdots, p_1$ and $s_i = 1, 2, \cdots, q_2$ $(i = 1, 2)$, define a function as

$$\Delta_{j_1 j_1}^{(r_1 r_2)} = (a_{j_1 j_1 r_1} - a_{j_1 j_1 r_2})^2 \; (r_1, r_2 \in \{1, 2, \cdots, p_1\}, r_1 \neq r_2),$$
$$\Delta_{j_2 j_2}^{(s_1 s_2)} = (a_{j_2 j_2 s_1} - a_{j_2 j_2 s_2})^2 \; (s_1, s_2 \in \{1, 2, \cdots, q_2\}, s_1 \neq s_2).$$

For $r_i = 1, 2, \cdots, p_1$ and $s_i = 1, 2, \cdots, q_2$ $(i = 1, 2)$, if

$$\Delta_{j_1 j_1}^{(r_1 r_2)} = (a_{j_1 j_1 r_1} - a_{j_1 j_1 r_2})^2 = 0 \text{ and } \Delta_{j_2 j_2}^{(s_1 s_2)} = (a_{j_2 j_2 s_1} - a_{j_2 j_2 s_2})^2 = 0,$$

there are three bifurcations as follows.

- Two 1-dimensional flows of $(a_{j_1 j_1 r_1}, \dot{x}_{j_2})$ and $(a_{j_1 j_1 r_2}, \dot{x}_{j_2})$ switching at point $(a_{j_1 j_1 1}, \dot{x}_{j_2})$ with the same order singularity are given through

$$\bigcup\nolimits_{r_1=1}^{p_1} \underbrace{(a_{j_1 j_1 r_1}, \dot{x}_{j_2})}_{(m_{j_1 r_1})^{\text{th}}\text{order } XX} \rightleftharpoons \underbrace{(a_{j_1 j_1 1}, \dot{x}_{j_2})}_{(2m_{11}+1)^{\text{th}} \text{ order XX}} \rightleftharpoons \bigcup\nolimits_{r_2=1}^{p_1} \underbrace{(a_{j_1 j_1 r_2}, \dot{x}_{j_2})}_{(m_{j_1 r_2})^{\text{th}}\text{order XX}}$$
$$\left\{a_{j_1 j_1 r_1} \rightleftharpoons a_{j_1 j_1 r_2} \middle| m_{j_1 r_1} = m_{j_1 r_2}; r_1, r_2 \in \{1, 2, \cdots, p_1\}, r_1 \neq r_2\right\};$$

  and

$$a_{j_1 j_1 r_1} = a_{j_1 j_1 1} - \varepsilon_{r_1}, a_{j_1 j_1 r_2} = a_{j_1 j_1 1} + \varepsilon_{r_2},$$
$$(a_{j_1 j_1 r_1} = a_{j_1 j_1 1} + \varepsilon_{r_1}, a_{j_1 j_1 r_2} = a_{j_1 j_1 1} - \varepsilon_{r_2});$$
$$\varepsilon_{r_1}, \varepsilon_{r_2} > 0; \; r_i = 1, 2, \cdots, p_1 \; (i = 1, 2).$$

- Two 1-dimensional flows of $(\dot{x}_{j_1}, a_{j_2 j_2 s_1})$ and $(\dot{x}_{j_1}, a_{j_2 j_2 s_2})$ switching at point $(\dot{x}_{j_1}, a_{j_2 j_2 1})$ with the same order singularity are given through

$$\bigcup\nolimits_{s_1=1}^{q_2} \underbrace{(\dot{x}_{j_1}, a_{j_2 j_2 s_1})}_{(n_{j_2 s_1})^{\text{th}}\text{order YY}} \rightleftharpoons \underbrace{(\dot{x}_{j_1}, a_{j_2 j_2 1})}_{(2n_{21}+1)^{\text{th}} \text{ order YY}} \rightleftharpoons \bigcup\nolimits_{u_2=1}^{q_2} \underbrace{(\dot{x}_{j_1}, a_{j_2 j_2 s_2})}_{(n_{j_2 s_2})^{\text{th}}\text{order YY}}$$
$$\left\{a_{j_2 j_2 s_1} \rightleftharpoons a_{j_2 j_2 s_2} \middle| n_{j_2 s_1} = n_{j_2 s_2}; s_1, s_2 \in \{1, 2, \cdots, q_2\}, s_1 \neq s_2\right\};$$

  and

$$a_{j_2 j_2 s_1} = a_{j_2 j_2 1} - \varepsilon_{s_1}, a_{j_2 j_2 u_2} = a_{j_2 j_2 1} + \varepsilon_{s_2},$$
$$(a_{j_2 j_2 s_1} = a_{j_2 j_2 1} + \varepsilon_{s_1}, a_{j_2 j_2 s_2} = a_{j_2 j_2 1} - \varepsilon_{s_2});$$
$$\varepsilon_{s_1}, \varepsilon_{s_2} > 0; \; s_i = 1, 2, \cdots, q_2 \; (i = 1, 2).$$

- Two singular-equilibrium networks of $(a_{j_1j_1r_1}, a_{j_2j_2s_1})$ and $(a_{j_1j_1r_2}, a_{j_2j_2s_2})$ switching at point $(a_{j_1j_1 1}, a_{j_2j_2 1})$ with the same order singularity are given through

$$\bigcup\nolimits_{r_1=1}^{p_1}\bigcup\nolimits_{s_1=1}^{q_2}\underbrace{(a_{j_1j_1r_1}, a_{j_2j_2s_1})}_{((m_{j_1r_1}),(n_{j_2s_1}))\text{-}XX} \rightleftharpoons \underbrace{([(2m_{11}+1)\text{YY},(2n_{21}+1)\text{ZZ}])}_{((2m_{11}+1),(2n_{21}+1))\text{-XX}}$$
$$\rightleftharpoons \bigcup\nolimits_{r_2=1}^{p_1}\bigcup\nolimits_{s_2=1}^{q_2}\underbrace{(a_{j_1j_1r_2}, a_{j_2j_2s_2})}_{((m_{j_1r_2}),(n_{j_2s_2}))\text{-}XX}$$

with

$$\begin{aligned}
a_{j_1j_1r_1} &= a_{j_1j_1 1} - \varepsilon_{r_1}, a_{j_1j_1r_2} = a_{j_1j_1 1} + \varepsilon_{r_2},\\
(a_{j_1j_1r_1} &= a_{j_1j_1 1} + \varepsilon_{r_1}, a_{j_1j_1r_2} = a_{j_1j_1 1} - \varepsilon_{r_2}),\\
\varepsilon_{r_1}, \varepsilon_{r_2} &> 0,\ r_i = 1, 2, \cdots, p_1\ (i = 1, 2);\\
a_{j_2j_2s_1} &= a_{j_2j_2 1} - \varepsilon_{s_1}, a_{j_2j_2s_2} = a_{j_2j_2 1} + \varepsilon_{s_2},\\
(a_{j_2j_2s_1} &= a_{j_2j_2 1} + \varepsilon_{s_1}, a_{j_2j_2s_2} = a_{j_2j_2 1} - \varepsilon_{s_2}),\\
\varepsilon_{s_1}, \varepsilon_{s_2} &> 0,\ s_i = 1, 2, \cdots, q_2\ (i = 1, 2).
\end{aligned}$$

The switching bifurcations of two equilibrium networks are the higher-order source, sink and saddle bifurcations.

In summary, from the analysis in case (I_1)–(I_3), the 1-dimensional flow of $(x^*_{j_1}, x^*_{j_2}) = (a_{j_1j_1 1}, a_{j_2j_2 1})$ for $m_1 = 2m_{11} + 1$ and $n_2 = 2n_{21} + 1$ has the bifurcation properties as stated in cases (v_{1a})–(v_{1d}) through Eqs. (1.389)–(1.414).

(v_2) For $m_1 = 2m_{11} + 1$ and $n_2 = 2n_{21}$ the equilibrium of $(x^*_{j_1}, x^*_{j_2}) = (a_{j_1j_1 1}, a_{j_2j_2 1})$ has the following bifurcation properties with

$$\begin{aligned}
&2m_{11} + 1 - \sum\nolimits_{r_1=1}^{p_1} m_{j_1r_1} = 2\sum\nolimits_{\rho_1=1}^{l_1} m_{\rho_1},\\
&2n_{21} - \sum\nolimits_{s_1=1}^{q_2} n_{j_2s_1} = 2\sum\nolimits_{\sigma_1=1}^{l_2} n_{\sigma_1};\\
&\sum\nolimits_{r_2=1}^{\bar{p}_1} m_{j_1r_2} = 2m_{11} + 1, \sum\nolimits_{s_2=1}^{\bar{q}_2} n_{j_2s_2} = 2n_{21};\\
&\sum\nolimits_{r_i=1}^{p_1} m_{j_1r_i} = 2m_{11} + 1, \sum\nolimits_{s_i=1}^{q_1} n_{j_2s_i} = 2n_{21}, \mathit{for}\, i = 1, 2;
\end{aligned}$$

l_1, l_2-quadratic polynomials are without real roots and ρ_1^{th} and σ_1^{th}-quadratic polynomials are with powers m_{ρ_1} and n_{σ_1}, respectively.

The bifurcation process is discussed through differential equations as follows. There are four bifurcations (I_1)-(I_3).

(I_1) Consider a dynamical system having a single equilibrium (A_1) as

$$\dot{x}_{j_1} = a_{j_1j_1 0}(x_{j_1} - b_{j_1j_1 1})\prod_{r=1}^{m_{11}} [(x_{j_1} - a_{j_1j_1}^{(r)})^2 + \Delta_{j_1j_1}^{(r)}](x_{j_2} - a_{j_1j_2 1})^{n_1},$$
$$\dot{x}_{j_2} = a_{j_2j_2 0}\prod_{s=1}^{n_{21}} [(x_{j_2} - a_{j_2j_2}^{(s)})^2 + \Delta_{j_2j_2}^{(s)}];$$

where $a_{j_1j_1 1} = b_{j_1j_1 1}$. Once

$$\Delta_{j_1j_1}^{(r)} = \Delta_{j_1j_1} = 0\,(r = 1, 2, \cdots, m_{11}),$$
$$\Delta_{j_2j_2}^{(s)} = \Delta_{j_2j_2} = 0\,(s = 1, 2, \cdots, n_{21}),$$

if

$$a_{j_1j_1 1} = b_{j_1j_1 1} = a_{j_1j_1}^{(r)}\ (r = 1, 2, \cdots, m_{11}),$$
$$a_{j_2j_2 1} = b_{j_2j_2 1} = a_{j_2j_2}^{(s)}\ (s = 1, 2, \cdots, n_{21}),$$

the foregoing differential equations become a dynamical system with a singular equilibrium (B_1) as

$$\dot{x}_{j_1} = a_{j_1j_1 0}(x_{j_1} - a_{j_1j_1 1})^{2m_{11}+1}(x_{j_2} - a_{j_1j_2 1})^{n_1},$$
$$\dot{x}_{j_2} = a_{j_2j_2 0}(x_{j_2} - a_{j_2j_2 1})^{2n_{21}}.$$

If

$$\Delta_{j_1j_1}^{(r)} = \Delta_{j_1j_1} + \delta_r\,(r = 1, 2, \cdots, m_{11}),\ \delta_r > 0;$$
$$\Delta_{j_2j_2}^{(s)} = \Delta_{j_2j_2} + \delta_s\,(s = 1, 2, \cdots, n_{21}),\ \delta_s > 0;$$

then

$$a_{j_1j_1 1}^{(r)}, a_{j_1j_1 2}^{(r)} = a_{j_1j_1 1} \pm \varepsilon_r,\ (r = 1, 2, \cdots, m_{11}),$$
$$\{a_{j_1j_1 1}, a_{j_1j_1 2}, \cdots, a_{j_1j_1 (2m_{11}+1)}\} = \text{sort}\{b_{j_1j_1 1}, a_{j_1j_1 1}^{(r)} a_{j_1j_1 2}^{(r)} | r = 1, 2, \cdots, m_{11}\};$$
$$a_{j_2j_2 1}^{(s)}, a_{j_2j_2 2}^{(s)} = a_{j_2j_2 1} \pm \varepsilon_s,\ (s = 1, 2, \cdots, n_1),$$
$$\{a_{j_2j_2 1}, a_{j_2j_2 2}, \cdots, a_{j_2j_2 (2n_{21})}\} = \text{sort}\{a_{j_2j_2 1}^{(s)}, a_{j_2j_2 2}^{(s)} | s = 1, 2, \cdots, n_{21}\}.$$

Thus, the differential equation becomes a dynamical system with a non-singular equilibrium network (C_1) as

$$\dot{x}_{j_1} = a_{j_1j_1 0}\prod_{r=1}^{2m_{11}+1} (x_{j_1} - a_{j_1j_1 r})(x_{j_2} - a_{j_1j_2 1})^{n_1},$$
$$\dot{x}_{j_2} = a_{j_2j_2 0}\prod_{s=1}^{2n_{21}} (x_{j_2} - a_{j_2j_2 s}).$$

- From the above bifurcation process analysis, at least, only $(m_{11}+1)$-parameter variations in the x_{j_1}-direction are engaged in such a bifurcation from a 1-dimensional flow to the 1-dimensional flow array of $1 \times (2m_{11}+1)$ through the $(2m_{11}+1)^{\text{th}}$-order source and sink bifurcations in the x_{j_1}-direction. The higher-order source and sink flow bifurcations are developed.

  Thus, the 1-dimensional flow appearing or vanishing bifurcation route is as follows.

$$\underbrace{(a_{j_1j_1}1, \dot{x}_{j_2})}_{\text{XX}} \rightleftharpoons \underbrace{(a_{j_1j_1}1, \dot{x}_{j_2})}_{(2m_{11}+1)^{\text{th}}\text{order XX}} \rightleftharpoons \bigcup\nolimits_{r=1}^{2m_{11}+1} \underbrace{(a_{j_1j_1r}, \dot{x}_{j_2})}_{\text{XX}}.$$

- From the above bifurcation process analysis, at least, only n_{21}-parameter variations in the x_{j_2}-direction are engaged in such a bifurcation from a 2-dimensional flows to the 1-dimensional flow array of $(2n_{21}) \times 1$ through the $(2n_{21})^{\text{th}}$-order source and sink bifurcations in the x_{j_2}-direction. The higher-order upper and lower-saddle flow bifurcations are developed.

  Thus, the 1-dimensional flow appearing or vanishing bifurcation route is as follows.

$$\underbrace{(\dot{x}_{j_1}, \dot{x}_{j_2})}_{\text{YY}} \rightleftharpoons \underbrace{(\dot{x}_{j_1}, a_{j_2j_2}1)}_{(2n_{21})^{\text{th}}\text{order YY}} \rightleftharpoons \bigcup\nolimits_{s=1}^{2n_{21}} \underbrace{(\dot{x}_{j_1}, a_{j_2j_2s},)}_{\text{YY}}.$$

- From the above bifurcation process analysis, $(m_{11}+n_{21}+1)$-parameter variations in the two directions are engaged in such a bifurcation from one of saddle, source and sink to a $(2m_{11}+1) \times (2n_{21})$ matrix of saddle, source and sink through the $(2m_{11}+1)^{\text{th}}$-order source and sink flows bifurcations in the x_{j_1}-direction, and the $(2n_{21})^{\text{th}}$-order upper and lower-saddle flows bifurcations in the x_{j_2}-direction. The higher-order saddle-source and saddle-sink bifurcations are developed.

  Thus, the equilibrium appearing or vanishing bifurcation route is as follows.

$$\underbrace{(a_{j_1j_1}1, \dot{x}_{j_2})}_{\text{(XX,xF)-flow}} \rightleftharpoons \underbrace{((2m_{11}+1)\text{YY}, (2n_{21})\text{ZZ})}_{((2m_{11}+1),(2n_{21}))\text{-XX}} \rightleftharpoons \bigcup\nolimits_{r=1}^{2m_{11}+1} \bigcup\nolimits_{s=1}^{2n_{21}+1} \underbrace{(a_{j_1j_1r}, a_{j_2j_2s})}_{\text{XX}},$$

  which are the higher-order saddle-source and saddle-sink bifurcations.

(I_2) From Case (I_1), consider a dynamical system having singular equilibriums (A_2) as

$$\dot{x}_{j_1} = a_{j_1j_1 0} \prod\nolimits_{r_1=1}^{p_1} (x_{j_1} - b_{j_1j_1r_1})^{m_{j_1r_1}} \prod\nolimits_{r=1}^{l_1} [(x_{j_1} - a_{j_1j_1}^{(r)})^2 + \Delta_{j_1j_1}^{(r)}]^{m_{j_1r}} (x_{j_2} - a_{j_1j_2}1)^{n_1},$$

$$\dot{x}_{j_2} = a_{j_2 j_2 0} \prod_{s_1=1}^{q_2} (x_{j_2} - b_{j_2 j_2 s_1})^{n_{j_2 s_1}} \prod_{s=1}^{l_2} [(x_{j_2} - a_{j_2 j_2}^{(s)})^2 + \Delta_{j_2 j_2}^{(s)}]^{n_{j_2 s}};$$

where

$$\begin{aligned} &2m_{11} + 1 - \sum_{r_1=1}^{p_1} m_{j_1 r_1} = 2 \sum_{\rho_1=1}^{l_1} m_{\rho_1}, \\ &2n_{21} - \sum_{s_1=1}^{q_2} n_{j_2 s_1} = 2 \sum_{\sigma_1=1}^{l_2} n_{\sigma_1}; \\ &\sum_{r_2=1}^{\bar{p}_1} m_{j_1 r_2} = 2m_{11} + 1, \sum_{s_2=1}^{\bar{q}_2} n_{j_2 s_2} = 2n_{21}; \end{aligned}$$

l_1, l_2-quadratic polynomials are without real roots and ρ_1^{th} and σ_1^{th}-quadratic polynomials are with powers m_{ρ_1} and n_{σ_1},respectively.

Once

$$\begin{aligned} \Delta_{j_1 j_1}^{(r)} &= 0 \, (r = 1, 2, \cdots, l_1), \\ \Delta_{j_2 j_2}^{(s)} &= 0 \, (s = 1, 2, \cdots, l_2); \end{aligned}$$

if

$$\begin{aligned} a_{j_1 j_1 1} &\equiv a_{j_1 j_1 r_1} = a_{j_1 j_1}^{(r)} \, (r_1 = 1, 2, \cdots, p_1; r = 1, 2, \cdots, l_1), \\ a_{j_2 j_2 1} &\equiv a_{j_2 j_2 s_1} = a_{j_2 j_2}^{(s)} \, (s_1 = 1, 2, \cdots, q_2; s = 1, 2, \cdots, l_2), \end{aligned}$$

the foregoing differential equations become a dynamical system with a 1-dimensional flow (B_2) as

$$\begin{aligned} \dot{x}_{j_1} &= a_{j_1 j_1 0} (x_{j_1} - a_{j_1 j_1 1})^{2m_{11}+1} (x_{j_2} - a_{j_1 j_2 1})^{n_1}, \\ \dot{x}_{j_2} &= a_{j_2 j_2 0} (x_{j_2} - a_{j_2 j_2 1})^{2n_{21}}. \end{aligned}$$

If

$$\begin{aligned} \Delta_{j_1 j_1}^{(r)} &= \Delta_{j_1 j_1} + \delta_r \, (r = 1, 2, \cdots, l_1), \ \delta_r > 0; \\ \Delta_{j_2 j_2}^{(s)} &= \Delta_{j_2 j_2} + \delta_s \, (s = 1, 2, \cdots, l_2), \ \delta_s > 0; \end{aligned}$$

then

$$\begin{aligned} &a_{j_1 j_1 1}^{(r)}, a_{j_1 j_1 2}^{(r)} = a_{j_1 j_1 1} \pm \varepsilon_r, \ (r = 1, 2, \cdots, l_1), \\ &\{a_{j_1 j_1 1}, a_{j_1 j_1 2}, \cdots, a_{j_1 j_1 \bar{p}_1}\} \\ &= \text{sort}\{a_{j_1 j_1 1}, a_{j_1 j_1 2}, \cdots, a_{j_1 j_1 p_1}; a_{j_1 j_1 1}^{(r)}, a_{j_1 j_1 2}^{(r)} | r = 1, 2, \cdots, l_1\}; \end{aligned}$$

$$\begin{aligned}
&a_{j_2j_21}^{(s)}, a_{j_2j_22}^{(s)} = a_{j_2j_21} \pm \varepsilon_s,\ (s = 1, 2, \cdots, l_2),\\
&\{a_{j_2j_21}, a_{j_2j_22}, \cdots, a_{j_2j_2\overline{q}_2}\}\\
&= \mathrm{sort}\{a_{j_1j_21}, a_{j_1j_22}, \cdots, a_{j_1j_2q_2};\, a_{j_2j_11}^{(s)}, a_{j_2j_12}^{(s)} | s = 1, 2, \cdots, l_2\}.
\end{aligned}$$

Thus, the differential equation becomes a dynamical system with singular 1-dimensional flows (C_2) as

$$\begin{aligned}
\dot{x}_{j_1} &= a_{j_1j_10} \prod\nolimits_{r_2=1}^{\overline{p}_1} (x_{j_1} - a_{j_1j_1r_2})^{m_{j_1r_2}} (x_{j_2} - a_{j_1j_21})^{n_1},\\
\dot{x}_{j_2} &= a_{j_2j_20} \prod\nolimits_{s_2=1}^{\overline{q}_2} (x_{j_2} - a_{j_2j_1s_2})^{n_{j_2s_2}}
\end{aligned}$$

where

$$\Sigma_{r_2=1}^{\overline{p}_1} m_{j_1r_2} = 2m_{11} + 1,\ \Sigma_{u_2=1}^{\overline{q}_2} n_{j_2s_2} = 2n_{21}.$$

- From the above bifurcation process analysis, only $\overline{p}_1$-parameter variations in the x_{j_1}-direction from a p_1 to $\overline{p}_1$ 1-dimensional flow array through the $(2m_1 + 1)^{\text{th}}$-order source and sink flows bifurcations in the x_{j_1}-direction.

  Thus, the 1-dimensional flow appearing or vanishing bifurcation route from a p_1 to $\overline{p}_1$ 1-dimensional flow array is expressed as

$$\bigcup\nolimits_{r_1=1}^{p_1} \underbrace{(a_{j_1j_1r_1}, \dot{x}_{j_2})}_{(m_{j_1r_1})^{\text{th}}\text{order XX}} \rightleftharpoons \underbrace{(a_{j_1j_11}, \dot{x}_{j_2})}_{(2m_{11}+1)^{\text{th}}\text{ order XX}} \rightleftharpoons \bigcup\nolimits_{r_2=1}^{\overline{p}_1} \underbrace{(a_{j_1j_1r_2}, \dot{x}_{j_2})}_{(m_{j_1r_2})^{\text{th}}\text{order XX}}.$$

- From the above bifurcation process analysis, only $\overline{q}_2$-parameter variations in the x_{j_2}-direction from a q_2 to $\overline{q}_2$ 1-dimensional flow array through the $(2n_{21})^{\text{th}}$-order upper and lower-saddle flow bifurcations in the x_{j_2}-direction.

  Thus, the 1-dimensional flow appearing or vanishing bifurcation route from a q_2 to $\overline{q}_2$ 1-dimensional flow array is expressed as

$$\bigcup\nolimits_{s_1=1}^{q_2} \underbrace{(\dot{x}_{j_1}, a_{j_2j_2s_1})}_{(n_{j_2s_1})^{\text{th}}\text{order YY}} \rightleftharpoons \underbrace{(\dot{x}_{j_1}, a_{j_2j_21})}_{(2n_{21})^{\text{th}}\text{ order YY}} \rightleftharpoons \bigcup\nolimits_{s_2=1}^{\overline{q}_2} \underbrace{(\dot{x}_{j_2}, a_{j_2j_2s_2})}_{(n_{j_2s_2})^{\text{th}}\text{order YY}}.$$

- From the above bifurcation process analysis, only $\overline{q}_2 + \overline{p}_1$-parameter variations in the x_{j_1}-direction from a $q_2 \times p_1$ to $\overline{q}_1 \times \overline{p}_1$ matrix of 1-dimensional flows through the $(2m_{11} + 1)^{\text{th}}$-order source and sink flows bifurcations in the x_{j_1}-direction, and the $(2n_{21})^{\text{th}}$-order upper and lower-saddle flows bifurcations in the x_{j_1}-direction.

Thus, the equilibriums appearing or vanishing bifurcation route is as follows:

$$\bigcup_{r_1=1}^{p_1}\bigcup_{s_1=1}^{q_2}\underbrace{(a_{j_1j_1r_1},a_{j_2j_2s_1})}_{((m_{j_1r_1}),(n_{j_2s_1}))\text{-XX}} \rightleftharpoons \underbrace{((2m_{11}+1)\text{YY},(2n_{21})\text{ZZ})}_{((2m_{11}+1),(2n_{21}))\text{-XX}}$$
$$\rightleftharpoons \bigcup_{r_2=1}^{\bar{p}_1}\bigcup_{s_2=1}^{\bar{q}_2}\underbrace{(a_{j_1j_1r_2},a_{j_2j_2s_2})}_{((m_{j_1r_2}),(n_{j_2s_2}))\text{-XX}},$$

which are the higher-order saddle-source and saddle-sink bifurcations.

(I_3) Consider two dynamical systems with the same 1-dimensional flows with locations switched (A_3,C_3) for $i=1,2$ as

$$\dot{x}_{j_1}=a_{j_1j_10}\prod_{r_i=1}^{p_1}(x_{j_1}-a_{j_1j_1r_i})^{m_{j_1r_i}}(x_{j_2}-a_{j_1j_21})^{n_1},$$
$$\dot{x}_{j_2}=a_{j_2j_20}\prod_{s_i=1}^{q_1}(x_{j_2}-a_{j_2j_2s_i})^{n_{j_2s_i}};$$

where

$$\sum_{r_i=1}^{p_1}m_{j_1r_i}=2m_{11}+1,\sum_{s_i=1}^{q_2}n_{j_2s_i}=2n_{21},\ \text{for } i=1,2.$$

Consider a dynamical system as a singular 1-dimensional flow (B_3) as

$$\dot{x}_{j_1}=a_{j_1j_10}(x_{j_1}-a_{j_1j_11})^{2m_{11}+1}(x_{j_2}-a_{j_1j_21})^{n_1},$$
$$\dot{x}_{j_2}=a_{j_2j_10}(x_{j_2}-a_{j_2j_21})^{2n_{21}}.$$

For $r_i=1,2,\cdots,p_1$ and $s_i=1,2,\cdots,q_2$ $(i=1,2)$, define a function as

$$\Delta_{j_1j_1}^{(r_1r_2)}=(a_{j_1j_1r_1}-a_{j_1j_1r_2})^2\ (r_1,r_2\in\{1,2,\cdots,p_1\},r_1\neq r_2),$$
$$\Delta_{j_2j_2}^{(s_1s_2)}=(a_{j_2j_2s_1}-a_{j_2j_2s_2})^2\ (s_1,s_2\in\{1,2,\cdots,q_2\},s_1\neq s_2).$$

For $r_i=1,2,\cdots,p_1$ and $s_i=1,2,\cdots,q_2$ $(i=1,2)$, if

$$\Delta_{j_1j_1}^{(r_1r_2)}=(a_{j_1j_1r_1}-a_{j_1j_1r_2})^2=0 \text{ and } \Delta_{j_2j_2}^{(s_1s_2)}=(a_{j_2j_2s_1}-a_{j_2j_2s_2})^2=0,$$

there are three bifurcations as follows.

- Two 1-dimensional flows of $(a_{j_1j_1r_1},\dot{x}_{j_2})$ and $(a_{j_1j_1r_2},\dot{x}_{j_2})$ switching at point $(a_{j_1j_11},\dot{x}_{j_2})$ with the same order singularity are given through

$$\bigcup_{r_1=1}^{p_1} \underbrace{(a_{j_1j_1r_1}, \dot{x}_{j_2})}_{(m_{j_1r_1})^{\text{th}}\text{order XX}} \rightleftharpoons \underbrace{(a_{j_1j_11}, \dot{x}_{j_2})}_{(2m_{11}+1)^{\text{th}}\text{ order XX}} \rightleftharpoons \bigcup_{r_2=1}^{p_1} \underbrace{(a_{j_1j_1r_2}, \dot{x}_{j_2})}_{(m_{j_1r_2})^{\text{th}}\text{order XX}}$$

$$\{a_{j_1j_1r_1} \rightleftharpoons a_{j_1j_1r_2} | m_{j_1r_1} = m_{j_1r_2}; r_1, r_2 \in \{1, 2, \cdots, p_1\}, r_1 \neq r_2\};$$

and

$$\begin{aligned} a_{j_1j_1r_1} &= a_{j_1j_11} - \varepsilon_{r_1}, a_{j_1j_1r_2} = a_{j_1j_11} + \varepsilon_{r_2}, \\ (a_{j_1j_1r_1} &= a_{j_1j_11} + \varepsilon_{r_1}, a_{j_1j_1r_2} = a_{j_1j_11} - \varepsilon_{r_2}); \\ \varepsilon_{r_1}, \varepsilon_{r_2} &> 0; \ r_i = 1, 2, \cdots, p_1 \ (i = 1, 2). \end{aligned}$$

- Two 1-dimensional flows of $(\dot{x}_{j_1}, a_{j_2j_2s_1})$ and $(\dot{x}_{j_1}, a_{j_2j_2s_2})$ switching at point $(\dot{x}_{j_1}, a_{j_2j_21})$ with the same order singularity are given through

$$\bigcup_{s_1=1}^{q_2} \underbrace{(\dot{x}_{j_1}, a_{j_2j_2s_1})}_{(n_{j_2s_1})^{\text{th}}\text{order YY}} \rightleftharpoons \underbrace{(\dot{x}_{j_1}, a_{j_2j_21})}_{(2n_{21})^{\text{th}}\text{ order YY}} \rightleftharpoons \bigcup_{u_2=1}^{q_2} \underbrace{(\dot{x}_{j_1}, a_{j_2j_2s_2})}_{(n_{j_2s_2})^{\text{th}}\text{order YY}}$$

$$\{a_{j_2j_2s_1} \rightleftharpoons a_{j_2j_2s_2} | n_{j_2s_1} = n_{j_2s_2}; s_1, s_2 \in \{1, 2, \cdots, q_1\}, s_1 \neq s_2\};$$

and

$$\begin{aligned} a_{j_2j_2s_1} &= a_{j_2j_21} - \varepsilon_{s_1}, a_{j_2j_2u_2} = a_{j_2j_21} + \varepsilon_{s_2}, \\ (a_{j_2j_2s_1} &= a_{j_2j_21} + \varepsilon_{s_1}, a_{j_2j_2s_2} = a_{j_2j_21} - \varepsilon_{s_2}); \\ \varepsilon_{s_1}, \varepsilon_{s_2} &> 0; \ s_i = 1, 2, \cdots, q_1 \ (i = 1, 2). \end{aligned}$$

- Two singular-equilibrium networks of $(a_{j_1j_1r_1}, a_{j_2j_2s_1})$ and $(a_{j_1j_1r_2}, a_{j_2j_2s_2})$ switching at point $(a_{j_1j_11}, a_{j_1j_21})$ with the same order singularity are given through

$$\begin{aligned} \bigcup_{r_1=1}^{p_1} \bigcup_{s_1=1}^{q_2} \underbrace{(a_{j_1j_1r_1}, a_{j_2j_2s_1})}_{((m_{j_1r_1}),(n_{j_2s_1}))\text{-XX}} &\rightleftharpoons \underbrace{((2m_{11}+1)\text{YY},(2n_{21})\text{ZZ})}_{((2m_{11}+1),(2n_{21}))\text{-XX}} \\ &\rightleftharpoons \bigcup_{r_2=1}^{p_1} \bigcup_{s_2=1}^{q_2} \underbrace{(a_{j_1j_1r_2}, a_{j_2j_2s_2})}_{((m_{j_1r_2}),(n_{j_2s_2}))\text{-XX}} \end{aligned}$$

with

$$\begin{aligned} a_{j_1j_1r_1} &= a_{j_1j_11} - \varepsilon_{r_1}, a_{j_1j_1r_2} = a_{j_1j_11} + \varepsilon_{r_2}, \\ (a_{j_1j_1r_1} &= a_{j_1j_11} + \varepsilon_{r_1}, a_{j_1j_1r_2} = a_{j_1j_11} - \varepsilon_{r_2}), \\ \varepsilon_{r_1}, \varepsilon_{r_2} &> 0, \ r_i = 1, 2, \cdots, p_1 \ (i = 1, 2); \\ a_{j_2j_2s_1} &= a_{j_1j_21} - \varepsilon_{s_1}, a_{j_2j_2s_2} = a_{j_2j_21} + \varepsilon_{s_2}, \\ (a_{j_2j_2s_1} &= a_{j_1j_21} + \varepsilon_{s_1}, a_{j_2j_2s_2} = a_{j_2j_21} - \varepsilon_{s_2}), \end{aligned}$$

$$\varepsilon_{s_1}, \varepsilon_{s_2} > 0,\ s_i = 1, 2, \cdots, q_2\ (i = 1, 2).$$

The switching bifurcations are the higher-order saddle-source and saddle-sink bifurcations.

In summary, from the analysis in cases (I_1)-(I_3), the equilibrium of $(x^*_{j_1}, x^*_{j_2}) = (a_{j_1j_1 1}, a_{j_2j_2 1})$ for $m_1 = 2m_{11} + 1$ and $n_2 = 2n_{21}$ has the bifurcation properties as stated in (v_{2a})–(v_{2d}) through Eqs. (1.415)–(1.429).

(v_3) For $m_1 = 2m_{11}$ and $n_2 = 2n_{21} + 1$ the equilibrium of $(x^*_{j_1}, x^*_{j_2}) = (a_{j_1j_1 1}, a_{j_2j_2 1})$ has the following bifurcation properties with

$$\begin{aligned}
&2m_{11} - \sum\nolimits_{r_1=1}^{p_1} m_{j_1 r_1} = 2\sum\nolimits_{\rho_1=1}^{l_1} m_{\rho_1},\\
&2n_{21} + 1 - \sum\nolimits_{s_1=1}^{q_2} n_{j_2 s_1} = 2\sum\nolimits_{\sigma_1=1}^{l_2} n_{\sigma_1};\\
&\sum\nolimits_{r_2=1}^{\bar{p}_1} m_{j_1 r_2} = 2m_{11}, \sum\nolimits_{s_2=1}^{\bar{q}_2} n_{j_2 s_2} = 2n_{21} + 1;\\
&\sum\nolimits_{r_i=1}^{p_1} m_{j_1 r_i} = 2m_{11}, \sum\nolimits_{s_i=1}^{q_1} n_{j_2 s_i} = 2n_{21} + 1,\ for\ i = 1, 2;
\end{aligned}$$

l_1, l_2-quadratic polynomials are without real roots and ρ_1^{th} and σ_1^{th}-quadratic polynomials are with powers m_{ρ_1} and n_{σ_1}, respectively.

The bifurcation process is discussed through differential equations as follows. There are four bifurcations (I_1)-(I_3).

(I_1) Consider a dynamical system having a single equilibrium (A_1) as

$$\begin{aligned}
\dot{x}_{j_1} &= a_{j_1j_1 0}\prod\nolimits_{r=1}^{m_{11}} [(x_{j_1} - a^{(r)}_{j_1j_1})^2 + \Delta^{(r)}_{j_1j_1}](x_{j_2} - a_{j_1j_2 1})^{n_1},\\
\dot{x}_{j_2} &= a_{j_2j_2 0}(x_{j_2} - b_{j_2j_2 1})\prod\nolimits_{s=1}^{n_{21}} [(x_{j_2} - a^{(s)}_{j_2j_2})^2 + \Delta^{(s)}_{j_2j_2}];
\end{aligned}$$

where $a_{j_2j_2 1} = b_{j_2j_2 1} \neq a_{j_1j_2 1}$. Once

$$\begin{aligned}
\Delta^{(r)}_{j_1j_1} &= \Delta_{j_1j_1} = 0\,(r = 1, 2, \cdots, m_{11}),\\
\Delta^{(s)}_{j_2j_2} &= \Delta_{j_2j_2} = 0\,(s = 1, 2, \cdots, n_{21}),
\end{aligned}$$

if

$$\begin{aligned}
a_{j_1j_1 1} &= b_{j_1j_1 1} = a^{(r)}_{j_1j_1}\ (r = 1, 2, \cdots, m_{11}),\\
a_{j_2j_2 1} &= b_{j_2j_2 1} = a^{(s)}_{j_2j_2}\ (s = 1, 2, \cdots, n_{21}),
\end{aligned}$$

the foregoing differential equations become a dynamical system with a singular equilibrium (B_1) as

$$\dot{x}_{j_1} = a_{j_1 j_1 0}(x_{j_1} - a_{j_1 j_1 1})^{2m_{11}}(x_{j_2} - a_{j_1 j_2 1})^{n_1},$$
$$\dot{x}_{j_2} = a_{j_2 j_2 0}(x_{j_2} - a_{j_2 j_2 1})^{2n_{21}+1}.$$

If

$$\Delta_{j_1 j_1}^{(r)} = \Delta_{j_1 j_1} + \delta_r \ (r = 1, 2, \cdots, m_{11}), \ \delta_r > 0;$$
$$\Delta_{j_2 j_2}^{(s)} = \Delta_{j_2 j_2} + \delta_s \ (s = 1, 2, \cdots, n_{21}), \ \delta_s > 0;$$

then

$$a_{j_1 j_1 1}^{(r)}, a_{j_1 j_1 2}^{(r)} = a_{j_1 j_1 1} \pm \varepsilon_r, \ (r = 1, 2, \cdots, m_{11}),$$
$$\{a_{j_1 j_1 1}, a_{j_1 j_1 2}, \cdots, a_{j_1 j_1 (2m_{11})}\} = \mathrm{sort}\{a_{j_1 j_1 1}^{(r)}, a_{j_1 j_1 2}^{(r)} | r = 1, 2, \cdots, m_{11}\};$$
$$a_{j_2 j_2 1}^{(s)}, a_{j_2 j_2 2}^{(s)} = a_{j_2 j_2 1} \pm \varepsilon_s, \ (s = 1, 2, \cdots, n_1),$$
$$\{a_{j_2 j_2 1}, a_{j_2 j_2 2}, \cdots, a_{j_2 j_2 (2n_{21}+1)}\} = \mathrm{sort}\{b_{j_2 j_2 1}, a_{j_2 j_2 1}^{(s)}, a_{j_2 j_2 2}^{(s)} | s = 1, 2, \cdots, n_{21}\}.$$

Thus, the differential equation becomes a dynamical system with a non-singular equilibrium network (C_1) as

$$\dot{x}_{j_1} = a_{j_1 j_1 0} \prod_{r=1}^{2m_{11}} (x_{j_1} - a_{j_1 j_1 r})(x_{j_2} - a_{j_1 j_2 1})^{n_1},$$
$$\dot{x}_{j_2} = a_{j_2 j_2 0} \prod_{s=1}^{2n_{21}+1} (x_{j_2} - a_{j_2 j_2 s}).$$

- From the above bifurcation process analysis, at least, only m_{11}-parameter variations in the x_{j_1}-direction are engaged in such a bifurcation from a 1-dimensional flow to the 1-dimensional flow array of $1 \times (2m_{11})$ through the $(2m_{11})^{\text{th}}$-order upper-saddle and lower-saddle flows bifurcations in the x_{j_1}-direction. The higher-order upper-saddle and lower-saddle flow bifurcations are developed.

  Thus, the 1-dimensional flow appearing or vanishing bifurcation route is as follows.

$$\underbrace{(\dot{x}_{j_1}, \dot{x}_{j_2})}_{\text{(xF, xF)-flow}} \rightleftharpoons \underbrace{(a_{j_1 j_1 1}, \dot{x}_{j_2})}_{(2m_{11})^{\text{th}}\text{order XX}} \rightleftharpoons \bigcup_{r=1}^{2m_{11}} \underbrace{(a_{j_1 j_1 r}, \dot{x}_{j_2})}_{\text{YY}}.$$

- From the above bifurcation process analysis, at least, only $(n_{21}+1)$-parameter variations in the x_{j_1}-direction are engaged in such a bifurcation from a 1-dimensional flow to the 1-dimensional flow array of $(2n_{21}+1)\times 1$ through the $(2n_{21}+1)^{\text{th}}$-order source and sink bifurcations in the x_{j_2}-direction. The higher-order source and sink flow bifurcations are developed.

  Thus, the 1-dimensional flow appearing or vanishing bifurcation route is as follows.

$$\underbrace{(\dot{x}_{j_1}, a_{j_2j_21})}_{\text{YY}} \rightleftharpoons \underbrace{(\dot{x}_{j_1}, a_{j_2j_21})}_{(2n_{21}+1)^{\text{th}}\text{order YY}} \rightleftharpoons \bigcup\nolimits_{s=1}^{2n_{21}+1} \underbrace{(\dot{x}_{j_1}, a_{j_2j_2s})}_{\text{YY}}.$$

- From the above bifurcation process analysis, $(m_{11}+n_{21}+1)$-parameter variations in the two directions are engaged in such a bifurcation from one of saddle, source and sink to a $(2m_{11})\times(2n_{21}+1)$ matrix of saddle, source and sink through the $(2m_{11})^{\text{th}}$-order upper-saddle and lower-saddle bifurcations in the x_{j_1}-direction, and the $(2n_{11}+1)^{\text{th}}$-order source and sink flows bifurcations in the x_{j_2}-direction. The higher-order saddle-source and saddle-sink bifurcations are developed.

  Thus, the equilibrium network appearing or vanishing bifurcation route is as follows.

$$\underbrace{(\dot{x}_{j_1}, a_{j_2j_21})}_{\text{(xF,XX)-flow}} \rightleftharpoons \underbrace{((2m_{11})\text{YY},(2n_{21}+1)\text{ZZ})}_{((2m_1),(2n_{21}+1))\text{-XX}} \rightleftharpoons \bigcup\nolimits_{r=1}^{2m_{11}} \bigcup\nolimits_{s=1}^{2n_{21}+1} \underbrace{(a_{j_1j_1r}, a_{j_2j_2s})}_{\text{XX}},$$

  which are the higher-order saddle-source and saddle-sink bifurcations.

(I_2) From Case (I_1), consider a dynamical system having singular equilibriums (A_2) as

$$\begin{aligned}
\dot{x}_{j_1} &= a_{j_1j_10} \prod\nolimits_{r_1=1}^{p_1} (x_{j_1} - b_{j_1j_1r_1})^{m_{j_1r_1}} \prod\nolimits_{r=1}^{l_1} [(x_{j_1} - a_{j_1j_1}^{(r)})^2 + \Delta_{j_1j_1}^{(r)}]^{m_{j_1r}} (x_{j_2} - a_{j_1j_21})^{n_1},\\
\dot{x}_{j_2} &= a_{j_2j_20} \prod\nolimits_{s_1=1}^{q_2} (x_{j_2} - b_{j_2j_2s_1})^{n_{j_2s_1}} \prod\nolimits_{s=1}^{l_2} [(x_{j_2} - a_{j_2j_2}^{(s)})^2 + \Delta_{j_2j_2}^{(s)}]^{n_{j_2s}};
\end{aligned}$$

where

$$\begin{aligned}
&2m_{11} - \sum\nolimits_{r_1=1}^{p_1} m_{j_1r_1} = 2\sum\nolimits_{\rho_1=1}^{l_1} m_{\rho_1},\\
&2n_{21} + 1 - \sum\nolimits_{s_1=1}^{q_2} n_{j_2s_1} = 2\sum\nolimits_{\sigma_1=1}^{l_2} n_{\sigma_1};\\
&\sum\nolimits_{r_2=1}^{\bar{p}_1} m_{j_1r_2} = 2m_{11}, \sum\nolimits_{s_2=1}^{\bar{q}_2} n_{j_2s_2} = 2n_{21} + 1;
\end{aligned}$$

l_1, l_2-quadratic polynomials are without real roots and ρ_1^{th} and σ_1^{th}-quadratic polynomials are with powers m_{ρ_1} and n_{σ_1}, respectively.

Once

$$\Delta_{j_1 j_1}^{(r)} = 0\,(r = 1, 2, \cdots, l_1),$$
$$\Delta_{j_2 j_2}^{(s)} = 0\,(s = 1, 2, \cdots, l_2);$$

if

$$a_{j_1 j_1 1} \equiv a_{j_1 j_1 r_1} = a_{j_1 j_1}^{(r)}\ (r_1 = 1, 2, \cdots, p_1;\, r = 1, 2, \cdots, l_1),$$
$$a_{j_2 j_2 1} \equiv a_{j_2 j_2 s_1} = a_{j_2 j_2}^{(s)}\ (s_1 = 1, 2, \cdots, q_2;\, s = 1, 2, \cdots, l_2),$$

the foregoing differential equations become a dynamical system with a 1-dimensional flow (B_2) as

$$\dot{x}_{j_1} = a_{j_1 j_1 0}(x_{j_1} - a_{j_1 j_1 1})^{2m_{11}}(x_{j_2} - a_{j_1 j_2 1})^{n_1},$$
$$\dot{x}_{j_2} = a_{j_2 j_2 0}(x_{j_2} - a_{j_2 j_2 1})^{2n_{21}+1}.$$

If

$$\Delta_{j_1 j_1}^{(r)} = \Delta_{j_1 j_1} + \delta_r\ (r = 1, 2, \cdots, l_1),\ \delta_r > 0;$$
$$\Delta_{j_2 j_2}^{(s)} = \Delta_{j_2 j_2} + \delta_s\ (s = 1, 2, \cdots, l_2),\ \delta_s > 0;$$

then

$$a_{j_1 j_1 1}^{(r)}, a_{j_1 j_1 2}^{(r)} = a_{j_1 j_1 1} \pm \varepsilon_r,\ (r = 1, 2, \cdots, l_1),$$
$$\{a_{j_1 j_1 1}, a_{j_1 j_1 2}, \cdots, a_{j_1 j_1 \bar{p}_1}\}$$
$$= \text{sort}\{a_{j_1 j_1 1}, a_{j_1 j_1 2}, \cdots, a_{j_1 j_1 p_1};\, a_{j_1 j_1 1}^{(r)}, a_{j_1 j_1 2}^{(r)} | r = 1, 2, \cdots, l_1\};$$
$$a_{j_2 j_2 1}^{(s)}, a_{j_2 j_2 2}^{(s)} = a_{j_2 j_2 1} \pm \varepsilon_s,\ (s = 1, 2, \cdots, l_2),$$
$$\{a_{j_2 j_2 1}, a_{j_2 j_2 2}, \cdots, a_{j_2 j_2 \bar{q}_2}\}$$
$$= \text{sort}\{a_{j_1 j_2 1}, a_{j_1 j_2 2}, \cdots, a_{j_1 j_2 q_2};\, a_{j_2 j_1 1}^{(s)}, a_{j_2 j_1 2}^{(s)} | s = 1, 2, \cdots, l_2\}.$$

Thus, the differential equation becomes a dynamical system with singular 1-dimensional flows (C_2) as

$$\dot{x}_{j_1} = a_{j_1 j_1 0} \prod_{r_2=1}^{\bar{p}_1} (x_{j_1} - a_{j_1 j_1 r_2})^{m_{j_1 r_2}} (x_{j_2} - a_{j_1 j_2 1})^{n_1},$$
$$\dot{x}_{j_2} = a_{j_2 j_2 0} \prod_{s_2=1}^{\bar{q}_2} (x_{j_2} - a_{j_2 j_1 s_2})^{n_{j_2 s_2}}$$

where

$$\Sigma_{r_2=1}^{\overline{p}_1} m_{j_1 r_2} = 2m_{11}, \; \Sigma_{u_2=1}^{\overline{q}_2} n_{j_2 s_2} = 2n_{21} + 1.$$

- From the above bifurcation process analysis, only $\overline{p}_1$-parameter variations in the x_{j_1}-direction from a p_1 to $\overline{p}_1$ 1-dimensional flow array through the $(2m_1)^{\text{th}}$-order upper-saddle and lower-saddle flow bifurcations in the x_{j_1}-direction.

  Thus, the 1-dimensional flow appearing or vanishing bifurcation route from a p_1 to $\overline{p}_1$ 1-dimensional flow array is expressed as

$$\bigcup_{r_1=1}^{p_1} \underbrace{(a_{j_1 j_1 r_1}, \dot{x}_{j_2})}_{(\mathrm{m}_{j_1 r_1})^{\text{th}}\text{order XX}} \rightleftharpoons \underbrace{(a_{j_1 j_1 1}, \dot{x}_{j_2})}_{(2m_{11})^{\text{th}}\text{ order XX}} \rightleftharpoons \bigcup_{r_2=1}^{\overline{p}_1} \underbrace{(a_{j_1 j_1 r_2}, \dot{x}_{j_2})}_{(\mathrm{m}_{j_1 r_2})^{\text{th}}\text{order XX}} .$$

- From the above bifurcation process analysis, only $\overline{q}_2$-parameter variations in the x_{j_2}-direction from a q_2 to $\overline{q}_2$ 1-dimensional flow array through the $(2n_{21} + 1)^{\text{th}}$-order source and sink flows bifurcations in the x_{j_2}-direction.

  Thus, the 1-dimensional flow appearing or vanishing bifurcation route from a q_2 to $\overline{q}_2$ 1-dimensional flow array is expressed as

$$\bigcup_{s_1=1}^{q_2} \underbrace{(\dot{x}_{j_1}, a_{j_2 j_2 s_1})}_{(n_{j_2 s_1})^{\text{th}}\text{order YY}} \rightleftharpoons \underbrace{(\dot{x}_{j_1}, a_{j_2 j_2 1})}_{(2n_{21}+1)^{\text{th}}\text{ order YY}} \rightleftharpoons \bigcup_{s_2=1}^{\overline{q}_2} \underbrace{(\dot{x}_{j_2}, a_{j_2 j_2 s_2})}_{(n_{j_2 s_2})^{\text{th}}\text{order YY}} .$$

- From the above bifurcation process analysis, only $\overline{q}_2 + \overline{p}_1$-parameter variations in the x_{j_1}-direction from a $q_2 \times p_1$ to $\overline{q}_2 \times \overline{p}_1$ matrix of 1-dimensional flows through the $(2m_{11})^{\text{th}}$-upper and lower-saddle flows bifurcations in the x_{j_1}-direction, and the $(2n_{21} + 1)^{\text{th}}$-source and sink flows bifurcations in the x_{j_2}-direction.

  Thus, the singular-equilibrium network appearing or vanishing bifurcation route is as follows:

$$\bigcup_{r_1=1}^{p_1} \bigcup_{s_1=1}^{q_2} \underbrace{(a_{j_1 j_1 r_1}, a_{j_2 j_2 s_1})}_{((m_{j_1 r_1}),(n_{j_2 s_1}))\text{-XX}} \rightleftharpoons \underbrace{((2m_{11})\text{YY},(2n_{21} + 1)\text{ZZ})}_{((2m_{11}),(2n_{21}+1))\text{-XX}}$$
$$\rightleftharpoons \bigcup_{r_2=1}^{\overline{p}_1} \bigcup_{s_2=1}^{\overline{q}_2} \underbrace{(a_{j_1 j_1 r_2}, a_{j_2 j_2 s_2})}_{((m_{j_1 r_2}),(n_{j_2 s_2}))\text{-XX}},$$

  which are the higher-order saddle-source and saddle-sink bifurcations.

(I_3) Consider two dynamical systems with the same 1-dimensional flows with locations switched (A_3,C_3) for $i = 1, 2$ as

$$\begin{aligned}\dot{x}_{j_1} &= a_{j_1j_10}\prod\nolimits_{r_i=1}^{p_1}(x_{j_1} - a_{j_1j_1r_i})^{m_{j_1r_i}}(x_{j_2} - a_{j_1j_21})^{n_1},\\ \dot{x}_{j_2} &= a_{j_2j_20}\prod\nolimits_{s_i=1}^{q_1}(x_{j_2} - a_{j_2j_2s_i})^{n_{j_2s_i}};\end{aligned}$$

where

$$\sum\nolimits_{r_i=1}^{p_1} m_{j_1r_i} = 2m_{11}, \sum\nolimits_{s_i=1}^{q_2} n_{j_1s_i} = 2n_{21} + 1, \text{ for } i = 1, 2.$$

Consider a dynamical system as a singular 1-dimensional flow (B_3) as

$$\begin{aligned}\dot{x}_{j_1} &= a_{j_1j_10}(x_{j_1} - a_{j_1j_11})^{2m_{11}}(x_{j_2} - a_{j_1j_21})^{n_1},\\ \dot{x}_{j_2} &= a_{j_2j_10}(x_{j_2} - a_{j_2j_21})^{2n_{21}+1}.\end{aligned}$$

For $r_i = 1, 2, \cdots, p_1$ and $s_i = 1, 2, \cdots, q_2$ $(i = 1, 2)$, define a function as

$$\begin{aligned}\Delta_{j_1j_1}^{(r_1r_2)} &= (a_{j_1j_1r_1} - a_{j_1j_1r_2})^2\ (r_1, r_2 \in \{1, 2, \cdots, p_1\}, r_1 \neq r_2),\\ \Delta_{j_2j_2}^{(s_1s_2)} &= (a_{j_2j_2s_1} - a_{j_2j_2s_2})^2\ (s_1, s_2 \in \{1, 2, \cdots, q_2\}, s_1 \neq s_2).\end{aligned}$$

For $r_i = 1, 2, \cdots, p_1$ and $s_i = 1, 2, \cdots, q_2$ $(i = 1, 2)$, if

$$\Delta_{j_1j_1}^{(r_1r_2)} = (a_{j_1j_1r_1} - a_{j_1j_1r_2})^2 = 0 \text{ and } \Delta_{j_2j_2}^{(s_1s_2)} = (a_{j_2j_2s_1} - a_{j_2j_2s_2})^2 = 0,$$

there are three bifurcations as follows.

- Two 1-dimensional flows of $(a_{j_1j_1r_1}, \dot{x}_{j_2})$ and $(a_{j_1j_1r_2}, \dot{x}_{j_2})$ switching at point $(a_{j_1j_11}, \dot{x}_{j_2})$ with the same order singularity are given through

$$\bigcup\nolimits_{r_1=1}^{p_1}\underbrace{(a_{j_1j_1r_1}, \dot{x}_{j_2})}_{(m_{j_1r_1})^{\text{th}}\text{order XX}} \rightleftharpoons \underbrace{(a_{j_1j_11}, \dot{x}_{j_2})}_{(2m_{11})^{\text{th}}\text{ order XX}} \rightleftharpoons \bigcup\nolimits_{r_2=1}^{p_1}\underbrace{(a_{j_1j_1r_2}, \dot{x}_{j_2})}_{(m_{j_1r_2})^{\text{th}}\text{order XX}}$$
$$\left\{a_{j_1j_1r_1} \rightleftharpoons a_{j_1j_1r_2}\middle| m_{j_1r_1} = m_{j_1r_2}; r_1, r_2 \in \{1, 2, \cdots, p_1\}, r_1 \neq r_2\right\};$$

  and

$$\begin{aligned}&a_{j_1j_1r_1} = a_{j_1j_11} - \varepsilon_{r_1}, a_{j_1j_1r_2} = a_{j_1j_11} + \varepsilon_{r_2},\\ &(a_{j_1j_1r_1} = a_{j_1j_11} + \varepsilon_{r_1}, a_{j_1j_1r_2} = a_{j_1j_11} - \varepsilon_{r_2});\\ &\varepsilon_{r_1}, \varepsilon_{r_2} > 0;\ r_i = 1, 2, \cdots, p_1\ (i = 1, 2).\end{aligned}$$

- Two 1-dimensional flows of $(\dot{x}_{j_1}, a_{j_2j_2s_1})$ and $(\dot{x}_{j_1}, a_{j_2j_2s_2})$ switching at point $(\dot{x}_{j_1}, a_{j_2j_2 1})$ with the same order singularity are given through

$$\bigcup_{s_1=1}^{q_2} \underbrace{(\dot{x}_{j_1}, a_{j_2j_2s_1})}_{(n_{j_1s_1})^{\text{th}}\text{order YY}} \rightleftharpoons \underbrace{(\dot{x}_{j_1}, a_{j_2j_2 1})}_{(2n_{21}+1)^{\text{th}}\text{ order YY}} \rightleftharpoons \bigcup_{u_2=1}^{q_2} \underbrace{(\dot{x}_{j_1}, a_{j_2j_2s_2})}_{(n_{j_1s_2})^{\text{th}}\text{order YY}}$$

$$\{a_{j_2j_2s_1} \rightleftharpoons a_{j_2j_2s_2} | n_{j_2s_1} = n_{j_2s_2}; s_1, s_2 \in \{1, 2, \cdots, q_2\}, s_1 \neq s_2\};$$

and

$$\begin{aligned}
a_{j_2j_2s_1} &= a_{j_2j_2 1} - \varepsilon_{s_1}, a_{j_2j_2u_2} = a_{j_2j_2 1} + \varepsilon_{s_2},\\
(a_{j_2j_2s_1} &= a_{j_2j_2 1} + \varepsilon_{s_1}, a_{j_2j_2s_2} = a_{j_2j_2 1} - \varepsilon_{s_2});\\
\varepsilon_{s_1}, \varepsilon_{s_2} &> 0;\ s_i = 1, 2, \cdots, q_2\ (i = 1, 2).
\end{aligned}$$

- Two singular-equilibrium networks of $(a_{j_1j_1r_1}, a_{j_2j_2s_1})$ and $(a_{j_1j_1r_2}, a_{j_2j_2s_2})$ switching at point $(a_{j_1j_1 1}, a_{j_1j_2 1})$ with the same order singularity are given through

$$\begin{aligned}
&\bigcup_{r_1=1}^{p_1} \bigcup_{s_1=1}^{q_2} \underbrace{(a_{j_1j_1r_1}, a_{j_2j_2s_1})}_{((m_{j_1r_1}),(n_{j_2s_1}))\text{-XX}} \rightleftharpoons \underbrace{((2m_{11})YY, (2n_{21}+1)ZZ)}_{((2m_{11}),(2n_{21}+1))\text{-XX}}\\
&\qquad \rightleftharpoons \bigcup_{r_2=1}^{p_1} \bigcup_{s_2=1}^{q_2} \underbrace{(a_{j_1j_1r_2}, a_{j_2j_2s_2})}_{((m_{j_1r_2}),(n_{j_2s_2}))\text{-XX}}
\end{aligned}$$

with

$$\begin{aligned}
a_{j_1j_1r_1} &= a_{j_1j_1 1} - \varepsilon_{r_1}, a_{j_1j_1r_2} = a_{j_1j_1 1} + \varepsilon_{r_2},\\
(a_{j_1j_1r_1} &= a_{j_1j_1 1} + \varepsilon_{r_1}, a_{j_1j_1r_2} = a_{j_1j_1 1} - \varepsilon_{r_2}),\\
\varepsilon_{r_1}, \varepsilon_{r_2} &> 0,\ r_i = 1, 2, \cdots, p_1\ (i = 1, 2);\\
a_{j_2j_2s_1} &= a_{j_1j_2 1} - \varepsilon_{s_1}, a_{j_2j_2s_2} = a_{j_2j_2 1} + \varepsilon_{s_2},\\
(a_{j_2j_2s_1} &= a_{j_1j_2 1} + \varepsilon_{s_1}, a_{j_2j_2s_2} = a_{j_2j_2 1} - \varepsilon_{s_2}),\\
\varepsilon_{s_1}, \varepsilon_{s_2} &> 0,\ s_i = 1, 2, \cdots, q_2\ (i = 1, 2).
\end{aligned}$$

The switching bifurcations of two equilibrium networks are the higher-order saddle-source and saddle-sink bifurcations.

In summary, from the analysis in case (I_1)-(I_3), the 1-dimensional flow of $(x_{j_1}^*, x_{j_2}^*) = (a_{j_1j_1 1}, a_{j_2j_2 1})$ for $m_1 = 2m_{11}$ and $n_2 = 2n_{21} + 1$ has the bifurcation properties as stated in cases (v_{3a})-(v_{3d}) through Eqs. (1.430)–(1.449).

(v_4) For $m_1 = 2m_{11}$ and $n_2 = 2n_{21}$ the equilibrium of $(x_{j_1}^*, x_{j_2}^*) = (a_{j_1j_1 1}, a_{j_2j_2 1})$ has the following bifurcation properties with

$$
\begin{aligned}
&2m_{11} - \sum\nolimits_{r_1=1}^{p_1} m_{j_1 r_1} = 2\sum\nolimits_{\rho_1=1}^{l_1} m_{\rho_1},\\
&2n_{21} - \sum\nolimits_{s_1=1}^{q_2} n_{j_2 s_1} = 2\sum\nolimits_{\sigma_1=1}^{l_2} n_{\sigma_1};\\
&\sum\nolimits_{r_2=1}^{\bar{p}_1} m_{j_1 r_2} = 2m_{11}, \sum\nolimits_{s_2=1}^{\bar{q}_2} n_{j_2 s_2} = 2n_{21};\\
&\sum\nolimits_{r_i=1}^{p_1} m_{j_1 r_i} = 2m_{11}, \sum\nolimits_{s_i=1}^{q_1} n_{j_2 s_i} = 2n_{21}, \textit{for } i = 1, 2;
\end{aligned}
$$

l_1, l_2-quadratic polynomials are without real roots and ρ_1^{th} and σ_1^{th}-quadratic polynomials are with powers m_{ρ_1} and n_{σ_1}, respectively.

The bifurcation process is discussed through differential equations as follows. There are four bifurcations (I_1)-(I_3).

(I_1) Consider a dynamical system having a single equilibrium (A_1) as

$$
\begin{aligned}
\dot{x}_{j_1} &= a_{j_1 j_1 0} \prod\nolimits_{r=1}^{m_{11}} [(x_{j_1} - a_{j_1 j_1}^{(r)})^2 + \Delta_{j_1 j_1}^{(r)}](x_{j_2} - a_{j_1 j_2 1})^{n_1},\\
\dot{x}_{j_2} &= a_{j_2 j_2 0} \prod\nolimits_{s=1}^{n_{21}} [(x_{j_2} - a_{j_2 j_2}^{(s)})^2 + \Delta_{j_2 j_2}^{(s)}].
\end{aligned}
$$

Once

$$
\begin{aligned}
\Delta_{j_1 j_1}^{(r)} &= \Delta_{j_1 j_1} = 0\,(r = 1, 2, \cdots, m_{11}),\\
\Delta_{j_2 j_2}^{(s)} &= \Delta_{j_2 j_2} = 0\,(s = 1, 2, \cdots, n_{21}),
\end{aligned}
$$

if

$$
\begin{aligned}
a_{j_1 j_1 1} &= b_{j_1 j_1 1} = a_{j_1 j_1}^{(r)}\ (r = 1, 2, \cdots, m_{11}),\\
a_{j_2 j_2 1} &= b_{j_2 j_2 1} = a_{j_2 j_2}^{(s)}\ (s = 1, 2, \cdots, n_{21}),
\end{aligned}
$$

the foregoing differential equations become a dynamical system with a singular equilibrium (B_1) as

$$
\begin{aligned}
\dot{x}_{j_1} &= a_{j_1 j_1 0}(x_{j_1} - a_{j_1 j_1 1})^{2m_{11}}(x_{j_2} - a_{j_1 j_2 1})^{n_1},\\
\dot{x}_{j_2} &= a_{j_2 j_2 0}(x_{j_2} - a_{j_2 j_2 1})^{2n_{21}}.
\end{aligned}
$$

If

$$
\begin{aligned}
\Delta_{j_1 j_1}^{(r)} &= \Delta_{j_1 j_1} + \delta_r\,(r = 1, 2, \cdots, m_{11}),\ \delta_r > 0;\\
\Delta_{j_2 j_2}^{(s)} &= \Delta_{j_2 j_2} + \delta_s\,(s = 1, 2, \cdots, n_{21}),\ \delta_s > 0;
\end{aligned}
$$

then

$$\begin{aligned}
&a^{(r)}_{j_1j_11}, a^{(r)}_{j_1j_12} = a_{j_1j_11} \pm \varepsilon_r,\ (r = 1, 2, \cdots, m_{11}),\\
&\{a_{j_1j_11}, a_{j_1j_12}, \cdots, a_{j_1j_1(2m_{11})}\} = \mathrm{sort}\{a^{(r)}_{j_1j_11}, a^{(r)}_{j_1j_11} | r = 1, 2, \cdots, m_{11}\};\\
&a^{(s)}_{j_2j_21}, a^{(s)}_{j_2j_22} = a_{j_2j_21} \pm \varepsilon_s,\ (s = 1, 2, \cdots, n_1),\\
&\{a_{j_2j_21}, a_{j_2j_22}, \cdots, a_{j_2j_2(2n_{21})}\} = \mathrm{sort}\{a^{(s)}_{j_2j_21}, a^{(s)}_{j_2j_22} | s = 1, 2, \cdots, n_{21}\}.
\end{aligned}$$

Thus, the differential equation becomes a dynamical system with a non-singular equilibrium network (C_1) as

$$\begin{aligned}
\dot{x}_{j_1} &= a_{j_1j_10} \prod\nolimits_{r=1}^{2m_{11}} (x_{j_1} - a_{j_1j_1r})(x_{j_2} - a_{j_1j_21})^{n_1},\\
\dot{x}_{j_2} &= a_{j_2j_20} \prod\nolimits_{s=1}^{2n_{21}} (x_{j_2} - a_{j_2j_2s}).
\end{aligned}$$

- From the above bifurcation process analysis, at least, only m_{11}-parameter variations in the x_{j_1}-direction are engaged in such a bifurcation from a 2-dimensional flow to the 1-dimensional flow array of $1 \times (2m_{11})$ through the $(2m_{11})^{\text{th}}$-upper and lower-saddle bifurcations in the x_{j_1}-direction. The higher-order upper-saddle and lower-saddle flow bifurcations are developed.

  Thus, the 1-dimensional flow appearing or vanishing bifurcation route is as follows.

$$\underbrace{(\dot{x}_{j_1}, \dot{x}_{j_2})}_{\text{(xF,xF)-flow}} \rightleftharpoons \underbrace{(a_{j_1j_11}, \dot{x}_{j_2})}_{(2m_{11})^{\text{th}}\text{order XX}} \rightleftharpoons \bigcup\nolimits_{r=1}^{2m_{11}} \underbrace{(a_{j_1j_1r}, \dot{x}_{j_2})}_{\text{YY}}.$$

- From the above bifurcation process analysis, at least, only $(n_{21}+1)$-parameter variations in the x_{j_1}-direction are engaged in such a bifurcation from a 2-dimensional flow to the 1-dimensional flow array of $(2n_{21}) \times 1$ through the $(2n_{21})^{\text{th}}$-upper and lower-saddle flow bifurcations in the x_{j_2}-direction. The higher-order double-saddle bifurcations are developed.

  Thus, the 1-dimensional flow appearing or vanishing bifurcation route is as follows.

$$\underbrace{(\dot{x}_{j_1}, \dot{x}_{j_2})}_{\text{(xF,xF)-flow}} \rightleftharpoons \underbrace{(\dot{x}_{j_1}, a_{j_2j_21})}_{(2n_{21})^{\text{th}}\text{order YY}} \rightleftharpoons \bigcup\nolimits_{s=1}^{2n_{21}} \underbrace{(\dot{x}_{j_1}, a_{j_2j_2s})}_{\text{YY}}.$$

- From the above bifurcation process analysis, $(m_{11}+n_{21})$-parameter variations in the two directions are engaged in such a bifurcation from one of saddle, source and sink to a $(2m_{11}) \times (2n_{21})$ matrix of saddle, source and sink through the $(2m_{11})^{\text{th}}$-upper and lower-saddle flow bifurcations in the x_{j_1}-direction, and the $(2n_{11})^{\text{th}}$-upper and lower-saddle flows bifurcations in the x_{j_2}-direction. The higher-order double-saddle bifurcations are developed.

Thus, the equilibrium network appearing or vanishing bifurcation route is as follows.

$$\underbrace{(\dot{x}_{j_1}, \dot{x}_{j_2})}_{\text{(xF,xF)-flow}} \rightleftharpoons \underbrace{((2m_{11})\text{YY},(2n_{21})\text{ZZ})}_{((2m_1),(2n_{21}))\text{-XX}} \rightleftharpoons \bigcup_{r=1}^{2m_{11}} \bigcup_{s=1}^{2n_{21}} \underbrace{(a_{j_1j_1r}, a_{j_2j_2s})}_{\text{XX}},$$

which are the higher-order double-saddle bifurcations.

(I_2) From Case (I_1), consider a dynamical system having singular equilibriums (A_2) as

$$\begin{aligned}
\dot{x}_{j_1} &= a_{j_1j_10} \prod_{r_1=1}^{p_1} (x_{j_1} - b_{j_1j_1r_1})^{m_{j_1r_1}} \prod_{r=1}^{l_1} [(x_{j_1} - a_{j_1j_1}^{(r)})^2 + \Delta_{j_1j_1}^{(r)}]^{m_{j_1r}} (x_{j_2} - a_{j_1j_21})^{n_1},\\
\dot{x}_{j_2} &= a_{j_2j_20} \prod_{s_1=1}^{q_2} (x_{j_2} - b_{j_2j_2s_1})^{n_{j_2s_1}} \prod_{s=1}^{l_2} [(x_{j_2} - a_{j_2j_2}^{(s)})^2 + \Delta_{j_2j_2}^{(s)}]^{n_{j_2s}};
\end{aligned}$$

where

$$\begin{aligned}
&2m_{11} - \sum_{r_1=1}^{p_1} m_{j_1r_1} = 2\sum_{\rho_1=1}^{l_1} m_{\rho_1},\\
&2n_{21} - \sum_{s_1=1}^{q_2} n_{j_2s_1} = 2\sum_{\sigma_1=1}^{l_2} n_{\sigma_1};\\
&\sum_{r_2=1}^{\bar{p}_1} m_{j_1r_2} = 2m_{11}, \sum_{s_2=1}^{\bar{q}_2} n_{j_2s_2} = 2n_{21};
\end{aligned}$$

l_1, l_2-quadratic polynomials are without real roots and ρ_1^{th} and σ_1^{th}-quadratic polynomials are with powers m_{ρ_1} and n_{σ_1}, respectively.

Once

$$\begin{aligned}
\Delta_{j_1j_1}^{(r)} &= 0\,(r = 1, 2, \cdots, l_1),\\
\Delta_{j_2j_2}^{(s)} &= 0\,(s = 1, 2, \cdots, l_2);
\end{aligned}$$

if

$$\begin{aligned}
a_{j_1j_11} &\equiv a_{j_1j_1r_1} = a_{j_1j_1}^{(r)}\ (r_1 = 1, 2, \cdots, p_1; r = 1, 2, \cdots, l_1),\\
a_{j_2j_21} &\equiv a_{j_2j_2s_1} = a_{j_2j_2}^{(s)}\ (s_1 = 1, 2, \cdots, q_2; s = 1, 2, \cdots, l_2),
\end{aligned}$$

the foregoing differential equations become a dynamical system with a 1-dimensional flow (B_2) as

$$\begin{aligned}\dot{x}_{j_1} &= a_{j_1j_10}(x_{j_1} - a_{j_1j_11})^{2m_{11}}(x_{j_2} - a_{j_1j_21})^{n_1},\\ \dot{x}_{j_2} &= a_{j_2j_20}(x_{j_2} - a_{j_2j_21})^{2n_{21}}.\end{aligned}$$

If

$$\begin{aligned}\Delta_{j_1j_1}^{(r)} &= \Delta_{j_1j_1} + \delta_r\ (r = 1, 2, \cdots, l_1),\ \delta_r > 0;\\ \Delta_{j_2j_2}^{(s)} &= \Delta_{j_2j_2} + \delta_s\ (s = 1, 2, \cdots, l_2),\ \delta_s > 0;\end{aligned}$$

then

$$\begin{aligned}&a_{j_1j_11}^{(r)}, a_{j_1j_12}^{(r)} = a_{j_1j_11} \pm \varepsilon_r,\ (r = 1, 2, \cdots, l_1),\\ &\{a_{j_1j_11}, a_{j_1j_12}, \cdots, a_{j_1j_1\overline{p}_1}\}\\ &= \text{sort}\{a_{j_1j_11}, a_{j_1j_12}, \cdots, a_{j_1j_1p_1}; a_{j_1j_11}^{(r)}, a_{j_1j_12}^{(r)} | r = 1, 2, \cdots, l_1\};\\ &a_{j_2j_21}^{(s)}, a_{j_2j_22}^{(s)} = a_{j_2j_21} \pm \varepsilon_s,\ (s = 1, 2, \cdots, l_2),\\ &\{a_{j_2j_21}, a_{j_2j_22}, \cdots, a_{j_2j_2\overline{q}_2}\}\\ &= \text{sort}\{a_{j_1j_21}, a_{j_1j_22}, \cdots, a_{j_1j_2q_2}; a_{j_2j_11}^{(s)}, a_{j_2j_12}^{(s)} | s = 1, 2, \cdots, l_2\}.\end{aligned}$$

Thus, the differential equation becomes a dynamical system with singular 1-dimensional flows (C_2) as

$$\begin{aligned}\dot{x}_{j_1} &= a_{j_1j_10}\prod\nolimits_{r_2=1}^{\overline{p}_1}(x_{j_1} - a_{j_1j_1r_2})^{m_{j_1r_2}}(x_{j_2} - a_{j_1j_21})^{n_1},\\ \dot{x}_{j_2} &= a_{j_2j_20}\prod\nolimits_{s_2=1}^{\overline{q}_2}(x_{j_2} - a_{j_2j_1s_2})^{n_{j_2s_2}}\end{aligned}$$

where

$$\Sigma_{r_2=1}^{\overline{p}_1} m_{j_1r_2} = 2m_{11},\ \Sigma_{u_2=1}^{\overline{q}_2} n_{j_2s_2} = 2n_{21}.$$

- From the above bifurcation process analysis, only $\overline{p}_1$-parameter variations in the x_{j_1}-direction from a p_1 to $\overline{p}_1$ 1-dimensional flow array through the $(2m_1)^{\text{th}}$-upper and lower-saddle flow bifurcations in the x_{j_1}-direction.

  Thus, the 1-dimensional flow appearing or vanishing bifurcation route from a p_1 to $\overline{p}_1$ 1-dimensional flow array is expressed as

$$\bigcup\nolimits_{r_1=1}^{p_1} \underbrace{(a_{j_1 j_1 r_1}, \dot{x}_{j_2})}_{(\mathrm{m}_{j_1 r_1})^{\mathrm{th}}\text{order XX}} \rightleftharpoons \underbrace{(a_{j_1 j_1 1}, \dot{x}_{j_2})}_{(2m_{11})^{\mathrm{th}}\text{ order XX}} \rightleftharpoons \bigcup\nolimits_{r_2=1}^{\overline{p}_1} \underbrace{(a_{j_1 j_1 r_2}, \dot{x}_{j_2})}_{(\mathrm{m}_{j_1 r_2})^{\mathrm{th}}\text{order XX}}.$$

- From the above bifurcation process analysis, only $\overline{q}_2$-parameter variations in the x_{j_2}-direction from a q_2 to $\overline{q}_2$ 1-dimensional flow array through the $(2n_{21})^{\text{th}}$-upper and lower-saddle flow bifurcations in the x_{j_2}-direction.

  Thus, the 1-dimensional flow appearing or vanishing bifurcation route from a q_2 to $\overline{q}_2$ 1-dimensional flow array is expressed as

$$\bigcup\nolimits_{s_1=1}^{q_2} \underbrace{(\dot{x}_{j_1}, a_{j_2 j_2 s_1})}_{(n_{j_2 s_1})^{\mathrm{th}}\text{order YY}} \rightleftharpoons \underbrace{(\dot{x}_{j_1}, a_{j_2 j_2 1})}_{(2n_{21})^{\mathrm{th}}\text{ order YY}} \rightleftharpoons \bigcup\nolimits_{s_2=1}^{\overline{q}_2} \underbrace{(\dot{x}_{j_2}, a_{j_2 j_2 s_2})}_{(n_{j_2 s_2})^{\mathrm{th}}\text{order YY}}.$$

- From the above bifurcation process analysis, only $\overline{q}_2 + \overline{p}_1$-parameter variations in the x_{j_1}-direction from a $q_2 \times p_1$ to $\overline{q}_2 \times \overline{p}_1$ matrix of 1-dimensional flows through the $(2m_{11})^{\text{th}}$-upper and lower-saddle flows bifurcations in the x_{j_1}-direction, and the $(2n_{21})^{\text{th}}$-upper and lower-saddle flows bifurcations in the x_{j_2}-direction.

  Thus, the singular-equilibrium network appearing or vanishing bifurcation route is as follows:

$$\begin{aligned}\bigcup\nolimits_{r_1=1}^{p_1} \bigcup\nolimits_{s_1=1}^{q_2} \underbrace{(a_{j_1 j_1 r_1}, a_{j_2 j_2 s_1})}_{((m_{j_1 r_1}),(n_{j_2 s_1}))\text{-XX}} &\rightleftharpoons \underbrace{(2m_{11})\text{YY},(2n_{11})\text{ZZ})}_{((2m_{11}),(2n_{11}))\text{-XX}}\\ &\rightleftharpoons \bigcup\nolimits_{r_2=1}^{\overline{p}_1} \bigcup\nolimits_{s_2=1}^{\overline{q}_2} \underbrace{(a_{j_1 j_1 r_2}, a_{j_2 j_2 s_2})}_{((m_{j_1 r_2}),(n_{j_2 s_2}))\text{-XX}},\end{aligned}$$

  which are the higher-order double-saddle bifurcations.

(I_3) Consider two dynamical systems with the same 1-dimensional flows with locations switched (A_3,C_3) for $i = 1, 2$ as

$$\begin{aligned}\dot{x}_{j_1} &= a_{j_1 j_1 0} \prod\nolimits_{r_i=1}^{p_1} (x_{j_1} - a_{j_1 j_1 r_i})^{m_{j_1 r_i}} (x_{j_2} - a_{j_1 j_2 1})^{n_1},\\ \dot{x}_{j_2} &= a_{j_2 j_2 0} \prod\nolimits_{s_i=1}^{q_1} (x_{j_2} - a_{j_2 j_2 s_i})^{n_{j_2 s_i}};\end{aligned}$$

where

$$\sum\nolimits_{r_i=1}^{p_1} m_{j_1 r_i} = 2m_{11}, \sum\nolimits_{s_i=1}^{q_2} n_{j_1 s_i} = 2n_{21}, \text{ for } i = 1, 2.$$

Consider a dynamical system as a singular 1-dimensional flow (B_3) as

$$\dot{x}_{j_1} = a_{j_1 j_1 0}(x_{j_1} - a_{j_1 j_1 1})^{2m_{11}}(x_{j_2} - a_{j_1 j_2 1})^{n_1},$$
$$\dot{x}_{j_2} = a_{j_2 j_1 0}(x_{j_2} - a_{j_2 j_2 1})^{2n_{21}}.$$

For $r_i = 1, 2, \cdots, p_1$ and $s_i = 1, 2, \cdots, q_2$ $(i = 1, 2)$, define a function as

$$\Delta_{j_1 j_1}^{(r_1 r_2)} = (a_{j_1 j_1 r_1} - a_{j_1 j_1 r_2})^2 \; (r_1, r_2 \in \{1, 2, \cdots, p_1\}, r_1 \neq r_2),$$
$$\Delta_{j_2 j_2}^{(s_1 s_2)} = (a_{j_2 j_2 s_1} - a_{j_2 j_2 s_2})^2 \; (s_1, s_2 \in \{1, 2, \cdots, q_2\}, s_1 \neq s_2).$$

For $r_i = 1, 2, \cdots, p_1$ and $s_i = 1, 2, \cdots, q_2$ $(i = 1, 2)$, if

$$\Delta_{j_1 j_1}^{(r_1 r_2)} = (a_{j_1 j_1 r_1} - a_{j_1 j_1 r_2})^2 = 0 \text{ and } \Delta_{j_2 j_2}^{(s_1 s_2)} = (a_{j_2 j_2 s_1} - a_{j_2 j_2 s_2})^2 = 0,$$

there are three bifurcations as follows.

- Two 1-dimensional flows of $(a_{j_1 j_1 r_1}, \dot{x}_{j_2})$ and $(a_{j_1 j_1 r_2}, \dot{x}_{j_2})$ switching at point $(a_{j_1 j_1 1}, \dot{x}_{j_2})$ with the same order singularity are given through

$$\bigcup\nolimits_{r_1=1}^{p_1} \underbrace{(a_{j_1 j_1 r_1}, \dot{x}_{j_2})}_{(m_{j_1 r_1})^{\text{th}} \text{order } XX} \rightleftharpoons \underbrace{(a_{j_1 j_1 1}, \dot{x}_{j_2})}_{(2m_{11})^{\text{th}} \text{ order XX}} \rightleftharpoons \bigcup\nolimits_{r_2=1}^{p_1} \underbrace{(a_{j_1 j_1 r_2}, \dot{x}_{j_2})}_{(m_{j_1 r_2})^{\text{th}} \text{order } XX}$$
$$\{a_{j_1 j_1 r_1} \rightleftharpoons a_{j_1 j_1 r_2} | m_{j_1 r_1} = m_{j_1 r_2};\, r_1, r_2 \in \{1, 2, \cdots, p_1\}, r_1 \neq r_2\};$$

  and

$$a_{j_1 j_1 r_1} = a_{j_1 j_1 1} - \varepsilon_{r_1}, a_{j_1 j_1 r_2} = a_{j_1 j_1 1} + \varepsilon_{r_2},$$
$$(a_{j_1 j_1 r_1} = a_{j_1 j_1 1} + \varepsilon_{r_1}, a_{j_1 j_1 r_2} = a_{j_1 j_1 1} - \varepsilon_{r_2});$$
$$\varepsilon_{r_1}, \varepsilon_{r_2} > 0;\; r_i = 1, 2, \cdots, p_1 \; (i = 1, 2).$$

- Two 1-dimensional flows of $(\dot{x}_{j_1}, a_{j_2 j_2 s_1})$ and $(\dot{x}_{j_1}, a_{j_2 j_2 s_2})$ switching at point $(\dot{x}_{j_1}, a_{j_2 j_2 1})$ with the same order singularity are given through

$$\bigcup\nolimits_{r_1=1}^{p_1} \bigcup\nolimits_{s_1=1}^{q_2} \underbrace{(a_{j_1 j_1 r_1}, a_{j_2 j_2 s_1})}_{((m_{j_1 r_1}),(n_{j_2 s_1}))\text{-XX}} \rightleftharpoons \underbrace{((2m_{11})\text{YY}, (2n_{21})\text{ZZ})}_{((2m_{11}),(2n_{21}))\text{-XX}}$$
$$\rightleftharpoons \bigcup\nolimits_{r_2=1}^{p_1} \bigcup\nolimits_{s_2=1}^{q_2} \underbrace{(a_{j_1 j_1 r_2}, a_{j_2 j_2 s_2})}_{((m_{j_1 r_2}),(n_{j_2 s_2}))\text{-XX}}$$

  and

$$a_{j_2 j_2 s_1} = a_{j_2 j_2 1} - \varepsilon_{s_1}, a_{j_2 j_2 u_2} = a_{j_2 j_2 1} + \varepsilon_{s_2},$$
$$(a_{j_2 j_2 s_1} = a_{j_2 j_2 1} + \varepsilon_{s_1}, a_{j_2 j_2 s_2} = a_{j_2 j_2 1} - \varepsilon_{s_2});$$

$$\varepsilon_{s_1}, \varepsilon_{s_2} > 0;\ s_i = 1, 2, \cdots, q_2\ (i = 1, 2).$$

- Two singular-equilibrium networks of $(a_{j_1j_1r_1}, a_{j_2j_2s_1})$ and $(a_{j_1j_1r_2}, a_{j_2j_2s_2})$ switching at point $(a_{j_1j_1 1}, a_{j_1j_2 1})$ with the same order singularity are given through

$$\bigcup_{r_1=1}^{p_1} \bigcup_{s_1=1}^{q_2} \underbrace{(a_{j_1j_1r_1}, a_{j_2j_2s_1})}_{((m_{j_1r_1}),(n_{j_2s_1}))\text{-XX}} \rightleftharpoons \underbrace{((2m_{11})\text{YY}, (2n_{21})\text{ZZ})}_{((2m_{11}),(2n_{21}))\text{-XX}}$$
$$\rightleftharpoons \bigcup_{r_2=1}^{p_1} \bigcup_{s_2=1}^{q_2} \underbrace{(a_{j_1j_1r_2}, a_{j_2j_2s_2})}_{((m_{j_1r_2}),(n_{j_2s_2}))\text{-XX}}$$

with

$$\begin{aligned}
a_{j_1j_1r_1} &= a_{j_1j_1 1} - \varepsilon_{r_1}, a_{j_1j_1r_2} = a_{j_1j_1 1} + \varepsilon_{r_2},\\
(a_{j_1j_1r_1} &= a_{j_1j_1 1} + \varepsilon_{r_1}, a_{j_1j_1r_2} = a_{j_1j_1 1} - \varepsilon_{r_2}),\\
\varepsilon_{r_1}, \varepsilon_{r_2} &> 0,\ r_i = 1, 2, \cdots, p_1\ (i = 1, 2);\\
a_{j_2j_2s_1} &= a_{j_1j_2 1} - \varepsilon_{s_1}, a_{j_2j_2s_2} = a_{j_2j_2 1} + \varepsilon_{s_2},\\
(a_{j_2j_2s_1} &= a_{j_1j_2 1} + \varepsilon_{s_1}, a_{j_2j_2s_2} = a_{j_2j_2 1} - \varepsilon_{s_2}),\\
\varepsilon_{s_1}, \varepsilon_{s_2} &> 0,\ \text{s}_i = 1, 2, \cdots, q_2\ (i = 1, 2).
\end{aligned}$$

The switching bifurcations of two equilibrium networks are the higher-order double-saddle bifurcations.

In summary, from the analysis in case (I_1)-(I_3), the 1-dimensional flow of $(x_{j_1}^*, x_{j_2}^*) = (a_{j_1j_1 1}, a_{j_2j_2 1})$ for $m_1 = 2m_{11}$ and $n_2 = 2n_{21}$ has the bifurcation properties as stated in cases $(\text{v}_{3\text{a}})$-$(\text{v}_{3\text{d}})$ through Eqs. (1.432)–(1.470).

Note that the equation in the x_{j_1}-direction can be considered as follows.

$$\begin{aligned}
\dot{x}_{j_1} &= a_{j_1j_1 0} \prod_{r=1}^{p_1} (x_{j_1} - a_{j_1j_1r})^{m_{j_1r}} \prod_{u_i=1}^{q_1} (x_{j_2} - a_{j_1j_2u_i})^{n_{j_1u}},\\
\dot{x}_{j_1} &= a_{j_1j_1 0} \prod_{r_i=1}^{p_1} (x_{j_1} - a_{j_1j_1r_i})^{m_{j_1r_i}} \prod_{u=1}^{n_1} (x_{j_2} - a_{j_1j_2u});
\end{aligned}$$

which can be used for discussion for case (v).

2.4 Network Switching for Equilibriums and Flows

In this section, the theorem about the switching bifurcations of hybrid networks of simple and singular equilibriums and 1-dimensional flows in product polynomial systems will be proved.

(vi) Consider

$$\begin{aligned}\dot{x}_{j_1} &= a_{j_1 j_1 0}(x_{j_1} - a_{j_1 j_1 1})^{m_1}(x_{j_2} - a_{j_1 j_2 1})^{n_1},\\ \dot{x}_{j_2} &= a_{j_2 j_2 0}(x_{j_2} - a_{j_2 j_2 1})^{n_2}.\end{aligned}$$

For $a_{j_2 j_2 1} = a_{j_1 j_2 1}$, there are 64 types of singular infinite-equilibriums at $x_{j_2}^{*} = a_{j_2 j_2 1} = a_{j_1 j_2 1}$. For $n_1 \geq n_2$, 32 types of singular infinite-equilibriums are

$$(a_{j_1 j_1 1}, a_{j_2 j_2 1}) = \underbrace{([n_1 - n_2]^{\text{th}}\text{XX},[m_1, n_2]^{\text{th}}\text{YY})}_{([n_1 - n_2],[m_1, n_2])\text{-XXYY}},$$

where

$$\begin{aligned}&\text{XX} \in \{\text{DI:II, II:DI, II:DI, DI:II}\},\\ &\text{YY} \in \{\text{SO, SO, SI, SI}\};\end{aligned}$$

$$\begin{aligned}&\text{XX} \in \{\text{DP:UP, UP:DP, UP:DP, DP:UP}\},\\ &\text{YY} \in \{\text{SO, SO, SI, SI}\};\end{aligned}$$

$$\begin{aligned}&\text{XX} \in \{\text{DP:UP, UP:DP, UP:DP, DP:UP}\},\\ &\text{YY} \in \{\text{US, US, LS, LS}\};\end{aligned}$$

$$\begin{aligned}&\text{XX} \in \{\text{DI:II, II:DI, II:DI, DI:II}\},\\ &\text{YY} \in \{\text{US, US, LS, LS}\};\end{aligned}$$

$$\begin{aligned}&\text{XX} \in \{\text{II:II, DI:DI, DI:DI, II:II}\},\\ &\text{YY} \in \{\text{SO, SO, SI, SI}\};\end{aligned}$$

$$\begin{aligned}&\text{XX} \in \{\text{UP:UP, DP:DP, DP:DP, UP:UP}\},\\ &\text{YY} \in \{\text{SO, SO, SI, SI}\};\end{aligned}$$

$$\begin{aligned}&\text{XX} \in \{\text{UP:UP, DP:DP, DP:DP, UP:UP}\},\\ &\text{YY} \in \{\text{US, US, LS, LS}\};\end{aligned}$$

$$\begin{aligned}&\text{XX} \in \{\text{II:II, DI:DI, DI:DI, II:II}\},\\ &\text{YY} \in \{\text{US, US, LS, LS}\}.\end{aligned}$$

For $n_1 < n_2$, 32 types of singular infinite-equilibriums are

$$(a_{j_1 j_1 1}, a_{j_2 j_2 1}) = \underbrace{([n_2 - n_1]^{\text{th}}\text{ZZ},[m_1, n_2]^{\text{th}}\text{YY})}_{([n_2 - n_1],[m_1, n_2])\text{-ZZYY}},$$

where

$$\begin{aligned}&\text{ZZ} \in \{\text{DC:IC, IC:DC, IC:DC, DC:IC}\},\\ &\text{YY} \in \{\text{SO, SO, SI, SI}\};\end{aligned}$$

$$\begin{aligned}
&\text{XX} \in \{\text{UA:DA, DA:UA, DA:UA, UA:DA}\},\\
&\text{YY} \in \{\text{SO, SO, SI, SI}\};\\
&\text{XX} \in \{\text{UA:DA, DA:UA, DA:UA, UA:DA}\},\\
&\text{YY} \in \{\text{US, US, LS, LS}\};\\
&\quad\text{ZZ} \in \{\text{DC:IC, IC:DC, IC:DC, DC:IC}\},\\
&\quad\text{YY} \in \{\text{US, US, LS, LS}\};\\
&\quad\text{ZZ} \in \{\text{IC:IC, DC:DC, DC:DC, IC:IC}\},\\
&\quad\text{YY} \in \{\text{SO, SO, SI, SI}\};\\
&\text{XX} \in \{\text{DA:DA, UA:UA, UA:UA, DA:DA}\},\\
&\text{YY} \in \{\text{SO, SO, SI, SI}\};\\
&\text{XX} \in \{\text{DA:DA, UA:UA, UA:UA, DA:DA}\},\\
&\text{YY} \in \{\text{US, US, LS, LS}\};\\
&\quad\text{ZZ} \in \{\text{IC:IC, DC:DC, DC:DC, IC:IC}\},\\
&\quad\text{YY} \in \{\text{US, US, LS, LS}\}.
\end{aligned}$$

(vi$_1$) For $n_2 < n_1$, the singularity in the x_{j_1}-direction more than singularity in the x_{j_2}-direction, and the following three ($[n_1 - n_2]^{\text{th}}$XX,$[m_1, n_2]^{\text{th}}$YY)-XXYY switching bifurcations are given as follows.

(vi$_{1a}$) For

$$\begin{aligned}
\dot{x}_{j_1} &= a_{j_1j_10}(x_{j_1} - a_{j_1j_11})^{m_1}(x_{j_2} - a_{j_1j_21})^{n_1},\\
\dot{x}_{j_2} &= a_{j_2j_20}(x_{j_2} - a_{j_2j_21})^{n_2}.
\end{aligned}$$

with $a_{j_2j_21} = a_{j_1j_21}$, once

$$a^-_{j_2j_21} \rightleftharpoons a^+_{j_1j_21},\, a^-_{j_1j_21} \rightleftharpoons a^+_{j_2j_21}$$

the ($[n_1 - n_2]^{\text{th}}$XX,$[m_1, n_2]^{\text{th}}$YY)-XXYY switching bifurcation is from a singular equilibrium with a singular 1-dimensional flow switching to a singular 1-dimensional flow with a singular equilibrium.

$$\left\{\begin{array}{c}\underbrace{(a_{j_1j_11}, a_{j_2j_21})}_{(m_1,n_2)-YY_1}\\ \underbrace{(a_{j_1j_11}, a_{j_1j_21})}_{((m_1,n_1)^{\text{th}}XX_1,xF)}\end{array}\right\} \rightleftharpoons \underbrace{([n_1 - n_2]^{\text{th}}\text{XX},[m_1, n_2]^{\text{th}}\text{YY})}_{([n_1-n_2],[m_1,n_2])\text{-XXYY}} \rightleftharpoons \left\{\begin{array}{c}\underbrace{(a_{j_1j_11}, a_{j_1j_21})}_{((m_1,n_1)^{\text{th}}XX_2,xF)}\\ \underbrace{(a_{j_1j_11}, a_{j_2j_21})}_{(m_1,n_2)-YY_2}\end{array}\right\}$$

with

$$
\begin{aligned}
&XX_1, XX_2 \in \{\text{DP:UP, UP:DP; UP:UP, DP:DP; }{}_{\text{II:DI}}\text{SO}, {}_{\text{DI:II}}\text{SO},\\
&{}_{\text{II:DI}}\text{SI}, {}_{\text{DI:II}}\text{SI}; {}_{\text{II:II}}\text{US}, {}_{\text{II:II}}\text{LS}, {}_{\text{DI:DI}}\text{US}, {}_{\text{DI:DI}}\text{LS}\};\\
&xF_1, xF_2 \in \{\text{pF, nF}\};\\
&YY_1, YY_2 \in \{\text{sink, source, saddles; upper (lower)-saddle-source,}\\
&\text{upper (lower)-saddle-sink; double-saddles}\}.
\end{aligned}
$$

(vi_{1b}) For

$$
\begin{aligned}
\dot{x}_{j_1} &= a_{j_1j_10}\prod_{r=1}^{m_1}(x_{j_1} - a_{j_1j_1r})\prod_{u=1}^{n_1}(x_{j_2} - a_{j_1j_2u}),\\
\dot{x}_{j_2} &= a_{j_2j_20}\prod_{s=1}^{n_2}(x_{j_2} - a_{j_2j_21}),
\end{aligned}
$$

with

$$
\begin{aligned}
\dot{x}_{j_1} &= a_{j_1j_10}(x_{j_1} - a_{j_1j_11})^{m_1}(x_{j_2} - a_{j_1j_21})^{n_1},\\
\dot{x}_{j_2} &= a_{j_2j_20}(x_{j_2} - a_{j_2j_21})^{n_2},
\end{aligned}
$$

with $a_{j_2j_21} = a_{j_1j_21}$.

Once

$$
a_{j_1j_2u}(u = 1, 2, \cdots, n_1) = a_{j_1j_21} = a_{j_2j_21} = a_{j_2j_2s}(s = 1, 2, \cdots, n_2),
$$

and two of $\sum_{k=1}^{n_2} C_{n_1+1}^{k} C_{n_2-1}^{k-1}$ -sequences ($n_2 \leq n_1$) are switched as

$$
\text{sort}(\bigcup_{u_1=1}^{n_1} a^{-}_{j_1j_2u_1}, \bigcup_{s_1=1}^{n_2} a^{-}_{j_2j_2s_1}) \rightleftharpoons \text{sort}(\bigcup_{u_2=1}^{n_1} a^{+}_{j_1j_2u_2}, \bigcup_{s_2=1}^{n_2} a^{+}_{j_2j_2s_2}),
$$

the $([n_1 - n_2]^{\text{th}}\text{XX},[m_1, n_2]^{\text{th}}\text{YY})$-XXYY switching bifurcation is for two $m_1 \times (n_1 + n_2)$ networks of non-singular equilibrium and hyperbolic and hyperbolic-secant flows.

$$
\begin{aligned}
&\bigcup_{r=1}^{m_1}\bigcup_{u_1=1}^{n_1}\underbrace{(a_{j_1j_1r}, a_{j_1j_2u_1})}_{(XX_1,xF_1)} \oplus \bigcup_{r=1}^{m_1}\bigcup_{s_1=1}^{n_2}\underbrace{(a_{j_1j_1r}, a_{j_2j_2s_1})}_{YY_1}\\
&\rightleftharpoons \underbrace{([n_1 - n_2]^{\text{th}}\text{XX},[m_1, n_2]^{\text{th}}\text{YY})}_{([n_1-n_2],[m_1,n_2])\text{-XXYY}}\\
&\rightleftharpoons \bigcup_{r=1}^{m_1}\bigcup_{u_2=1}^{n_1}\underbrace{(a_{j_1j_1r}, a_{j_1j_2u_1})}_{(XX_2,xF_2)} \oplus \bigcup_{r=1}^{m_1}\bigcup_{s_2=1}^{n_1}\underbrace{(a_{j_1j_1r}, a_{j_2j_2s_2})}_{YY_2},
\end{aligned}
$$

with

$$\begin{aligned}
&\mathrm{YY}_1, \mathrm{YY}_2 \in \{\text{sink, source, saddle}\};\\
&\mathrm{xF}_1, \mathrm{xF}_2 \in \{\mathrm{pF}, \mathrm{nF}\};\\
&\mathrm{XX}_1, \mathrm{XX}_2 \in \{\text{hyperbolic flow, hypersecant flow}\}.
\end{aligned}$$

(vi_{1c}) For

$$\begin{aligned}
\dot{x}_{j_1} &= a_{j_1j_10}\prod\nolimits_{r=1}^{p_1}(x_{j_1}-a_{j_1j_1r})^{m_{j_1r}}\prod\nolimits_{u=1}^{q_1}(x_{j_2}-a_{j_1j_2u})^{n_{j_1u}},\\
\dot{x}_{j_2} &= a_{j_2j_20}\prod\nolimits_{s=1}^{q_2}(x_{j_2}-a_{j_2j_2s})^{n_{j_2s}};\\
\sum\nolimits_{r=1}^{p_1}& m_{j_1r}=m_1, \sum\nolimits_{u=1}^{q_1} n_{j_1u}=n_1, \sum\nolimits_{s=1}^{q_2} n_{j_1s}=n_2;
\end{aligned}$$

with

$$\begin{aligned}
\dot{x}_{j_1} &= a_{j_1j_10}(x_{j_1}-a_{j_1j_11})^{m_1}(x_{j_2}-a_{j_1j_21})^{n_1},\\
\dot{x}_{j_2} &= a_{j_2j_20}(x_{j_2}-a_{j_2j_21})^{n_2};
\end{aligned}$$

with $a_{j_2j_21}=a_{j_1j_21}$, once

$$a_{j_1j_2u}(u=1,2,\cdots,q_1)=a_{j_1j_21}=a_{j_2j_21}=a_{j_2j_2s}(s=1,2,\cdots,q_2),$$

and two of $\sum_{k=1}^{q_2}C_{q_1+1}^{k}C_{q_2-1}^{k-1}$ -sequences ($q_2\le q_1$) are switched as

$$\mathrm{sort}(\bigcup\nolimits_{u_1=1}^{q_1} a_{j_1j_2u_1}^{-}, \bigcup\nolimits_{s_1=1}^{q_2} a_{j_2j_2s_1}^{-}) \rightleftharpoons \mathrm{sort}(\bigcup\nolimits_{u_2=1}^{q_1} a_{j_1j_2u_2}^{+}, \bigcup\nolimits_{s_2=1}^{q_2} a_{j_2j_2s_2}^{+}),$$

The ($[n_1-n_2]^{\mathrm{th}}$XX,$[m_1,n_2]^{\mathrm{th}}$YY)-XXYY switching bifurcation is for two $p_1\times(q_1+q_2)$ networks of singular equilibriums and singular 1-dimensional flows.

$$\begin{aligned}
&\bigcup\nolimits_{r=1}^{p_1}\bigcup\nolimits_{u_1=1}^{q_1}\underbrace{(a_{j_1j_1r}, a_{j_1j_2u_1})}_{([m_r,n_{u_1}]\mathrm{XX}_1,\mathrm{xF}_1)}\oplus\bigcup\nolimits_{r=1}^{p_1}\bigcup\nolimits_{s_1=1}^{q_2}\underbrace{(a_{j_1j_1r}, a_{j_2j_2s_1})}_{(\mathrm{m}_r,\mathrm{n}_{s_1})\text{-}\mathrm{YY}_1}\\
&\rightleftharpoons\underbrace{([n_1-n_2]^{\mathrm{th}}\mathrm{XX},[m_1,n_2]^{\mathrm{th}}\mathrm{YY})}_{([n_1-n_2],[m_1,n_2])\text{-XXYY}}\\
&\rightleftharpoons\bigcup\nolimits_{r=1}^{p_1}\bigcup\nolimits_{u_2=1}^{q_1}\underbrace{(a_{j_1j_1r}, a_{j_1j_2u_2})}_{([m_r,n_{u_2}]XX_1,\mathrm{xF}_2)}\oplus\bigcup\nolimits_{r=1}^{p_1}\bigcup\nolimits_{s_2=1}^{q_2}\underbrace{(a_{j_1j_1r}, a_{j_2j_2s_2})}_{(\mathrm{m}_r,\mathrm{n}_{s_2})\text{-}\mathrm{YY}_1}
\end{aligned}$$

with

$$\begin{aligned}&XX_1, XX_2 \in \{DP{:}UP,\ UP{:}DP;\ UP{:}UP,\ DP{:}DP;\ {}_{II:DI}SO,\ {}_{DI:II}SO,\\&{}_{II:DI}SI,\ {}_{DI:II}SI;\ {}_{II:II}US,\ {}_{II:II}LS,\ {}_{DI:DI}US,\ {}_{DI:DI}LS\};\\&xF_1, xF_2 \in \{pF,\ nF\};\\&YY_1, YY_2 \in \{\text{sink, source, saddles; upper (lower)-saddle-source,}\\&\text{upper (lower)-saddle-sink; double-saddles}\}.\end{aligned}$$

and

$$m_1 = \sum\nolimits_{r_i=1}^{p_1} m_{1r_i},\ n_1 = \sum\nolimits_{u_i=1}^{q_1} n_{1u_i},\ n_2 = \sum\nolimits_{s_i=1}^{q_1} n_{2s_i};\ i = 1, 2.$$

(vi$_2$) For $n_1 < n_2$, the singularity in the x_{j_2}-direction more than singularity in x_{j_1}-direction, and the following three ($[n_2 - n_1]^{\text{th}}$ZZ,$[m_1, n_2]^{\text{th}}$YY)-ZZYY switching bifurcations are given as follows.

(vi$_{2a}$) For

$$\begin{aligned}\dot{x}_{j_1} &= a_{j_1j_10}(x_{j_1} - a_{j_1j_11})^{m_1}(x_{j_2} - a_{j_1j_21})^{n_1},\\ \dot{x}_{j_2} &= a_{j_2j_20}(x_{j_2} - a_{j_2j_21})^{n_2}.\end{aligned}$$

with $a_{j_2j_21} = a_{j_1j_21}$, once

$$a^{-}_{j_2j_21} \rightleftharpoons a^{+}_{j_1j_21},\ a^{-}_{j_1j_21} \rightleftharpoons a^{+}_{j_2j_21},$$

the ($[n_2 - n_1]^{\text{th}}$ZZ,$[m_1, n_2]^{\text{th}}$YY)-ZZYY switching bifurcation is from a singular equilibrium with a singular 1-dimensional flow switching to a singular 1-dimensional flow with another singular equilibrium.

$$\left\{\begin{array}{c}\underbrace{(a_{j_1j_11}, a_{j_2j_21})}_{(m_1,n_2)-YY_1}\\ \underbrace{(a_{j_1j_11}, a_{j_1j_21})}_{((m_1,n_1)^{\text{th}}XX_1,xF)}\end{array}\right\} \rightleftharpoons \underbrace{([n_2 - n_1]^{\text{th}}\text{ZZ},[m_1, n_2]^{\text{th}}\text{YY})}_{([n_2-n_1],[m_1,n_2])\text{-ZZYY}} \rightleftharpoons \left\{\begin{array}{c}\underbrace{(a_{j_1j_11}, a_{j_1j_21})}_{((m_1,n_1)^{\text{th}}XX_2,xF)}\\ \underbrace{(a_{j_1j_11}, a_{j_2j_21})}_{(m_1,n_2)-YY_2}\end{array}\right\}$$

with

$$\begin{aligned}&XX_1, XX_2 \in \{DP{:}UP,\ UP{:}DP;\ UP{:}UP,\ DP{:}DP;\ {}_{II:DI}SO,\ {}_{DI:II}SO,\\&{}_{II:DI}SI,\ {}_{DI:II}SI;\ {}_{II:II}US,\ {}_{II:II}LS,\ {}_{DI:DI}US,\ {}_{DI:DI}LS\};\\&xF_1, xF_2 \in \{pF,\ nF\};\\&YY_1, YY_2 \in \{\text{sink, source, saddles; upper (lower)-saddle-source,}\\&\text{upper (lower)-saddle-sink; double-saddles}\}.\end{aligned}$$

(vi_{2b}) For

$$\begin{aligned}\dot{x}_{j_1} &= a_{j_1j_10}\prod_{r=1}^{m_1}(x_{j_1}-a_{j_1j_1r})\prod_{u=1}^{n_1}(x_{j_2}-a_{j_1j_2u}),\\ \dot{x}_{j_2} &= a_{j_2j_20}\prod_{s=1}^{n_2}(x_{j_2}-a_{j_2j_21}),\end{aligned}$$

with

$$\begin{aligned}\dot{x}_{j_1} &= a_{j_1j_10}(x_{j_1}-a_{j_1j_11})^{m_1}(x_{j_2}-a_{j_1j_21})^{n_1},\\ \dot{x}_{j_2} &= a_{j_2j_20}(x_{j_2}-a_{j_2j_21})^{n_2},\end{aligned}$$

with $a_{j_2j_21}=a_{j_1j_21}$.

Once

$$a_{j_1j_2u}(u=1,2,\cdots,n_1)=a_{j_1j_21}=a_{j_2j_21}=a_{j_2j_2s}(s=1,2,\cdots,n_2),$$

and two of $\sum_{k=1}^{n_1}C_{n_2+1}^{k}C_{n_1-1}^{k-1}$-sequences for $n_1<n_2$ are switched as

$$\text{sort}(\bigcup_{u_1=1}^{n_1}a_{j_1j_2u_1}^{-},\bigcup_{s_1=1}^{n_2}a_{j_2j_2s_1}^{-})\rightleftharpoons\text{sort}(\bigcup_{u_2=1}^{n_1}a_{j_1j_2u_2}^{+},\bigcup_{s_2=1}^{n_2}a_{j_2j_2s_2}^{+}),$$

the $([n_2-n_1]^{\text{th}}\text{ZZ},[m_1,n_2]^{\text{th}}\text{YY})$-ZZYY switching bifurcation is for two $m_1\times(n_1+n_2)$ networks of non-singular equilibrium and hyperbolic and hyperbolic-secant flows.

$$\begin{aligned}&\bigcup_{r=1}^{m_1}\bigcup_{u_1=1}^{n_1}\underbrace{(a_{j_1j_1r},a_{j_1j_2u_1})}_{(\text{XX}_1,\text{xF}_1)}\oplus\bigcup_{r=1}^{m_1}\bigcup_{s_1=1}^{n_2}\underbrace{(a_{j_1j_1r},a_{j_2j_2s_1})}_{\text{YY}_1}\\ &\rightleftharpoons\underbrace{([n_2-n_1]^{\text{th}}\text{ZZ},[m_1,n_2]^{\text{th}}\text{YY})}_{([n_2-n_1],[m_1,n_2])\text{-ZZYY}}\\ &\rightleftharpoons\bigcup_{r=1}^{m_1}\bigcup_{u_2=1}^{n_1}\underbrace{(a_{j_1j_1r},a_{j_1j_2u_1})}_{(XX_2,\text{xF}_2)}\oplus\bigcup_{r=1}^{m_1}\bigcup_{s_2=1}^{n_1}\underbrace{(a_{j_1j_1r},a_{j_2j_2s_2})}_{\text{YY}_2},\end{aligned}$$

with

$$\begin{aligned}&\text{YY}_1,\text{YY}_2\in\{\text{sink, source, saddle}\};\\ &\text{xF}_1,\text{xF}_2\in\{\text{pF, nF}\};\\ &\text{XX}_1,\text{XX}_2\in\{\text{hyperbolic flow, hyperbolic-secant flow}\}.\end{aligned}$$

(vi_{2c}) For

$$\dot{x}_{j_1}=a_{j_1j_10}\prod_{r=1}^{p_1}(x_{j_1}-a_{j_1j_1r})^{m_{j_1r}}\prod_{u=1}^{q_1}(x_{j_2}-a_{j_1j_2u})^{n_{j_1u}},$$

$$\dot{x}_{j_2} = a_{j_2 j_2 0} \prod_{s=1}^{q_2} (x_{j_2} - a_{j_2 j_2 s})^{n_{j_2 s}};$$
$$\sum_{r=1}^{p_1} m_{j_1 r} = m_1, \sum_{u=1}^{q_1} n_{j_1 u} = n_1, \sum_{s=1}^{q_2} n_{j_1 s} = n_2;$$

with

$$\dot{x}_{j_1} = a_{j_1 j_1 0}(x_{j_1} - a_{j_1 j_1 1})^{m_1}(x_{j_2} - a_{j_1 j_2 1})^{n_1},$$
$$\dot{x}_{j_2} = a_{j_2 j_2 0}(x_{j_2} - a_{j_2 j_2 1})^{n_2};$$

with $a_{j_2 j_2 1} = a_{j_1 j_2 1}$.

Once

$$a_{j_1 j_2 u}(u = 1, 2, \cdots, q_1) = a_{j_1 j_2 1} = a_{j_2 j_2 1} = a_{j_2 j_2 s}(s = 1, 2, \cdots, q_2),$$

and two of $\sum_{k=1}^{q_1} C_{q_2+1}^{k} C_{q_1-1}^{k-1}$ -sequences for $q_1 < q_2$ are switched as

$$\text{sort}(\bigcup_{u_1=1}^{q_1} a_{j_1 j_2 u_1}^{-}, \bigcup_{s_1=1}^{q_2} a_{j_2 j_2 s_1}^{-}) \rightleftharpoons \text{sort}(\bigcup_{u_2=1}^{q_1} a_{j_1 j_2 u_2}^{+}, \bigcup_{s_2=1}^{q_2} a_{j_2 j_2 s_2}^{+}),$$

the $([n_2 - n_1]^{\text{th}}\text{ZZ},[m_1, n_2]^{\text{th}}\text{YY})$-ZZYY switching bifurcation is for two $p_1 \times (q_1 + q_2)$ networks of singular equilibriums and singular 1-dimensional flows.

$$\begin{aligned}
&\bigcup_{r=1}^{p_1} \bigcup_{u_1=1}^{q_1} \underbrace{(a_{j_1 j_1 r}, a_{j_1 j_2 u_1})}_{([m_r, n_{u_1}]\text{XX}_1, \text{xF}_1)} \oplus \bigcup_{r=1}^{p_1} \bigcup_{s_1=1}^{q_2} \underbrace{(a_{j_1 j_1 r}, a_{j_2 j_2 s_1})}_{(\text{m}_r, \text{n}_{s_1})\text{ - YY}_1} \\
&\rightleftharpoons \underbrace{([n_2 - n_1]^{\text{th}}\text{ZZ},[m_1, n_2]^{\text{th}}\text{YY})}_{([n_2 - n_1],[m_1, n_2])\text{-ZZYY}} \\
&\rightleftharpoons \bigcup_{r=1}^{p_1} \bigcup_{u_2=1}^{q_1} \underbrace{(a_{j_1 j_1 r}, a_{j_1 j_2 u_2})}_{([m_r, n_{u_2}]XX_1, \text{xF}_2)} \oplus \bigcup_{r=1}^{p_1} \bigcup_{s_2=1}^{q_2} \underbrace{(a_{j_1 j_1 r}, a_{j_2 j_2 s_2})}_{(\text{m}_r, \text{n}_{s_2})\text{ - YY}_1}
\end{aligned}$$

with

$$\begin{aligned}
&\text{XX}_1, \text{XX}_2 \in \{\text{DP:UP, UP:DP; UP:UP, DP:DP; }_{\text{II:DI}}\text{SO, }_{\text{DI:II}}\text{SO,} \\
&_{\text{II:DI}}\text{SI, }_{\text{DI:II}}\text{SI; }_{\text{II:II}}\text{US, }_{\text{II:II}}\text{LS, }_{\text{DI:DI}}\text{US, }_{\text{DI:DI}}\text{LS}\}; \\
&\text{xF}_1, \text{xF}_2 \in \{\text{pF, nF}\}; \\
&\text{YY}_1, \text{YY}_2 \in \{\text{sink, source, saddles; upper (lower)-saddle-source,} \\
&\text{upper (lower)-saddle-sink; double-saddles}\}.
\end{aligned}$$

and

$$m_1 = \sum_{r_i=1}^{p_1} m_{1r_i}, n_1 = \sum_{u_i=1}^{q_1} n_{1u_i}, n_2 = \sum_{s_i=1}^{q_1} n_{2s_i};\ i = 1, 2.$$

In the end, the theorem is proved. ■

Chapter 3
Singular Equilibria and Flows, and Simple Networks

In this chapter, the singular 1-dimensional flows and equilibriums in 2-dimensional polynomial systems are illustrated to help one understand the global bifurcations for networks of the hyperbolic and hyperbolic-secant flows with source, sink and saddles. The singular saddles, sources and sinks, saddle-sources, and saddle-sinks, double-saddles are presented, and the singular hyperbolic-flows, hyperbolic-to-hyperbolic-secant flows, inflection-source and sink flows, inflection-saddle flows are also presented. The infinite-equilibriums in the self and product polynomial systems are the appearing and switching bifurcations for hybrid networks of simple and singular equilibriums and flows.

3.1 $[(2m_{11}+1), (2n_{11}+1)]$ and $((2m_{11}+1), (2n_{21}+1))$-Singular Systems

In this section, there are four global bifurcations of the 1-diemsnional flows through singular hyperbolic flows, and four global bifurcations of equilibriums, through singular source, sink and saddle bifurcations. The corresponding global bifurcations generate networks of hyperbolic and hyperbolic-secant flows with sources, sinks and saddles. The infinite-equilibriums are the appearing and switching bifurcations for the singular flows and singular equilibriums and for hybrid networks of hyperbolic flows and source, sink and saddles.

A. C. J. Luo, *Two-Dimensional Self and Product Polynomial Systems*,
https://doi.org/10.1007/978-981-96-5483-3_3

3.1.1 $[(2m_{11}+1),(2n_{11}+1)]$-Flows and $((2m_{11}+1),(2n_{21}+1))$-Equilibriums

Consider a polynomial dynamical system for $a_{221} \neq a_{121}$ as

$$\begin{aligned}\dot{x}_2 &= a_{220}(x_2 - a_{221})^{2n_{21}+1},\\ \dot{x}_1 &= a_{110}(x_1 - a_{111})^{2m_{11}+1}(x_2 - a_{121})^{2n_{11}+1}.\end{aligned} \tag{3.1}$$

The first integral manifold is

$$\begin{aligned}&-\frac{1}{2m_{11}}\left[\frac{1}{(x_1 - a_{111})^{2m_{11}}} - \frac{1}{(x_{10} - a_{111})^{2m_{11}}}\right]\\ &= \frac{a_{110}}{a_{220}}\sum_{\alpha,\beta=0}^{\alpha+\beta=2n_{11}+1}\frac{(2n_{11}+1)!}{\alpha!\beta!}\frac{(a_{221}-a_{121})^{\alpha}}{\beta - 2n_{21}}\\ &\quad\times\left[(x_2 - a_{221})^{\beta-2n_{21}} - (x_2 - a_{221})^{\beta-2n_{21}}\right]\\ &\text{for } 2m_{11}+1 \neq 1,\ \beta - 2n_{21} \neq 0;\\ &-\frac{1}{2m_{11}}\left[\frac{1}{(x_1 - a_{111})^{2m_{11}}} - \frac{1}{(x_{10} - a_{111})^{2m_{11}}}\right] = \ln\frac{|x_1 - a_{111}|}{|x_{10} - a_{111}|}\\ &\text{for } 2m_{11}+1 = 1,\\ &\frac{1}{\beta - 2n_{21}}\left[(x_2 - a_{221})^{\beta-2n_{21}} - (x_2 - a_{221})^{\beta-2n_{21}}\right] = \ln\frac{|x_2 - a_{221}|}{|x_{20} - a_{221}|}\\ &\text{for } \beta - 2n_{21} = 0.\end{aligned} \tag{3.2}$$

For $(x_1^*, x_2^*) = (a_{111}, a_{121})$ with $\dot{x}_1 = 0$, the $[(2m_{11}+1),(2n_{11}+1)]$-singular flows exist with $\dot{x}_2 = a_{220}(a_{121} - a_{221})^{2n_{21}+1}$. For $(x_1^*, x_2^*) = (a_{111}, a_{121})$, the 1-dimensional flow direction is determined by $\dot{x}_2 \neq 0s$. For $(x_1^*, x_2^*) = (a_{111}, a_{221})$ with $\dot{x}_1 = \dot{x}_2 = 0$, the $((2m_{11}+1),(2n_{21}+1))$- singular equilibriums exist.

(A) For $a_{121} > a_{221}$,

$$\left\{\begin{array}{l}(a_{111}, a_{121})\\(a_{111}, a_{221})\end{array}\right\} = \left\{\begin{array}{c}\underbrace{([(2m_{11}+1),(2n_{11}+1)]\text{DP:UP, pF})}_{[(2m_{11}+1),(2n_{11}+1)]\text{-hyperbolic flow }(+)}\\ \underbrace{((2m_{11}+1)\text{SI}, (2n_{21}+1)\text{SO})}_{[(2m_{11}+1),(2n_{21}+1)]\text{-saddle}}\end{array}\right\},$$

for $a_{110} > 0$, $a_{220} > 0$;

$$\left\{\begin{array}{l}(a_{111}, a_{121})\\(a_{111}, a_{221})\end{array}\right\} = \left\{\begin{array}{c}\underbrace{([(2m_{11}+1),(2n_{11}+1)]\text{UP:DP, pF})}_{[(2m_{11}+1),(2n_{11}+1)]\text{-hyperbolic-secant flow }(+)}\\ \underbrace{((2m_{11}+1)\text{SO}, (2n_{21}+1)\text{SO})}_{[(2m_{11}+1),(2n_{21}+1)]\text{-source}}\end{array}\right\},$$

for $a_{110} < 0, a_{220} > 0$;

$$\left\{\begin{array}{l}(a_{111}, a_{121})\\(a_{111}, a_{221})\end{array}\right\} = \left\{\begin{array}{c}\underbrace{([(2m_{11}+1),(2n_{11}+1)]\text{UP:DP, nF})}_{[(2m_{11}+1),(2n_{11}+1)]\text{-hyperbolic-secant flow }(-)}\\ \underbrace{((2m_{11}+1)\text{SI},(2n_{21}+1)\text{SI})}_{[(2m_{11}+1),(2n_{21}+1)]\text{-sink}}\end{array}\right\},$$

for $a_{110} > 0, a_{220} < 0$;

$$\left\{\begin{array}{l}(a_{111}, a_{121})\\(a_{111}, a_{221})\end{array}\right\} = \left\{\begin{array}{c}\underbrace{([(2m_{11}+1),(2n_{11}+1)]\text{DP:UP, nF})}_{[(2m_{11}+1),(2n_{11}+1)]\text{-hyperbolic flow }(-)}\\ \underbrace{((2m_{11}+1)\text{SO},(2n_{21}+1)\text{SI})}_{[(2m_{11}+1),(2n_{21}+1)]\text{-saddle}}\end{array}\right\},$$

for $a_{110} < 0, a_{220} < 0$. (3.3)

(B) For $a_{121} < a_{221}$,

$$\left\{\begin{array}{l}(a_{111}, a_{221})\\(a_{111}, a_{121})\end{array}\right\} = \left\{\begin{array}{c}\underbrace{((2m_{11}+1)\text{SO},(2n_{21}+1)\text{SO})}_{[(2m_{11}+1),(2n_{21}+1)]\text{-source}}\\ \underbrace{([(2m_{11}+1),(2n_{11}+1)]\text{UP:DP, nF})}_{[(2m_{11}+1),(2n_{11}+1)]\text{-hyperbolic-secant flow }(-)}\end{array}\right\},$$

for $a_{110} > 0, a_{220} > 0$;

$$\left\{\begin{array}{l}(a_{111}, a_{221})\\(a_{111}, a_{121})\end{array}\right\} = \left\{\begin{array}{c}\underbrace{((2m_{11}+1)\text{SI},(2n_{21}+1)\text{SO})}_{[(2m_{11}+1),(2n_{21}+1)]\text{-saddle}}\\ \underbrace{([(2m_{11}+1),(2n_{11}+1)]\text{DP:UP, nF})}_{[(2m_{11}+1),(2n_{11}+1)]\text{-hyperbolic flow }(-)}\end{array}\right\},$$

for $a_{110} < 0, a_{220} > 0$;

$$\left\{\begin{array}{l}(a_{111}, a_{221})\\(a_{111}, a_{121})\end{array}\right\} = \left\{\begin{array}{c}\underbrace{((2m_{11}+1)\text{SO},(2n_{21}+1)\text{SI})}_{[(2m_{11}+1),(2n_{21}+1)]\text{-saddle}}\\ \underbrace{([(2m_{11}+1),(2n_{11}+1)]\text{DP:UP, pF})}_{[(2m_{11}+1),(2n_{11}+1)]\text{-hyperbolic flow }(+)}\end{array}\right\},$$

for $a_{110} > 0, a_{220} < 0$;

$$\left\{\begin{array}{l}(a_{111}, a_{221})\\(a_{111}, a_{121})\end{array}\right\} = \left\{\begin{array}{c}\underbrace{((2m_{11}+1)\text{SI},(2n_{21}+1)\text{SI})}_{[(2m_{11}+1),(2n_{21}+1)]\text{-sink}}\\ \underbrace{([(2m_{11}+1),(2n_{11}+1)]\text{UP:DP, pF})}_{[(2m_{11}+1),(2n_{11}+1)]\text{-hyperbolic-secant flow }(+)}\end{array}\right\},$$

for $a_{110} < 0, a_{220} < 0$. (3.4)

– The $([(2m_{11}+1),(2n_{11}+1)]\text{DP:UP, pF})$ is for a positive singular hyperbolic flow.

- The $([(2m_{11}+1), (2n_{11}+1)]\mathrm{UP{:}DP, pF})$ is for a positive singular hyperbolic-secant flow.
- The $([(2m_{11}+1), (2n_{11}+1)]\mathrm{UP{:}DP, nF})$ is for a negative singular hyperbolic-secant flow.
- The $([(2m_{11}+1), (2n_{11}+1)]\mathrm{DP{:}UP, nF})$ is for a negative singular hyperbolic flow.

The acronyms "UP" and "DP" are for up-parabola and down-parabola flows, respectively. "pF" and "nF" are for positive and negative flows, respectively. DP:UP is for a hyperbolic flow, and UP:DP is for a hyperbolic-secant flow.

- The $((2m_{11}+1)\mathrm{SO}, (2n_{21}+1)\mathrm{SO}]$ is for a $((2m_{11}+1), (2n_{21}+1)]$-singular source.
- The $((2m_{11}+1)\mathrm{SI}, (2n_{21}+1)\mathrm{SO}]$ is for a $((2m_{11}+1), (2n_{21}+1)]$-singular saddle.
- The $((2m_{11}+1)\mathrm{SO}, (2n_{21}+1)\mathrm{SI}]$ is for a $((2m_{11}+1), (2n_{21}+1)]$-singular saddle.
- The $((2m_{11}+1)\mathrm{SI}, (2n_{21}+1)\mathrm{SI}]$ is for a $((2m_{11}+1), (2n_{21}+1)]$-singular sink.

The singular hyperbolic and hyperbolic-secant flows with singular source, sink and saddles are illustrated in Figs. 3.1 and 3.2. Through such singular flows, the corresponding global bifurcations generate from a hyperbolic (hyperbolic-secant) flow to a network of singular hyperbolic and hyperbolic-secant flows. Through such singular equilibriums, the corresponding global bifurcations generate from one of source, sink and saddles to a network of singular source, sink and saddle equilibriums.

3.1.2 $[(2m_{11}+1), (2n_{11}+1)]$ and $((2m_{11}+1), (2n_{21}+1))$-*Infinite-Equilibriums*

Consider a polynomial dynamical system for $a_{221} = a_{121}$ as

$$\begin{aligned}
\dot{x}_2 &= a_{220}(x_2 - a_{221})^{2n_{21}+1},\\
\dot{x}_1 &= a_{110}(x_1 - a_{111})^{2m_{11}+1}(x_2 - a_{121})^{2n_{11}+1}.
\end{aligned} \tag{3.5}$$

The first integral manifold is

$$\begin{aligned}
&-\frac{1}{2m_{11}}\left[\frac{1}{(x_1 - a_{111})^{2m_{11}}} - \frac{1}{(x_{10} - a_{111})^{2m_{11}}}\right]\\
&= \frac{a_{110}}{a_{220}}\frac{1}{2(n_{11} - n_{21}) + 1}[(x_2 - a_{121})^{2(n_{11}-n_{21})+1} - (x_{20} - a_{121})^{2(n_{11}-n_{21})+1}]\\
&\text{for } 2m_{11} + 1 \neq 1;
\end{aligned}$$

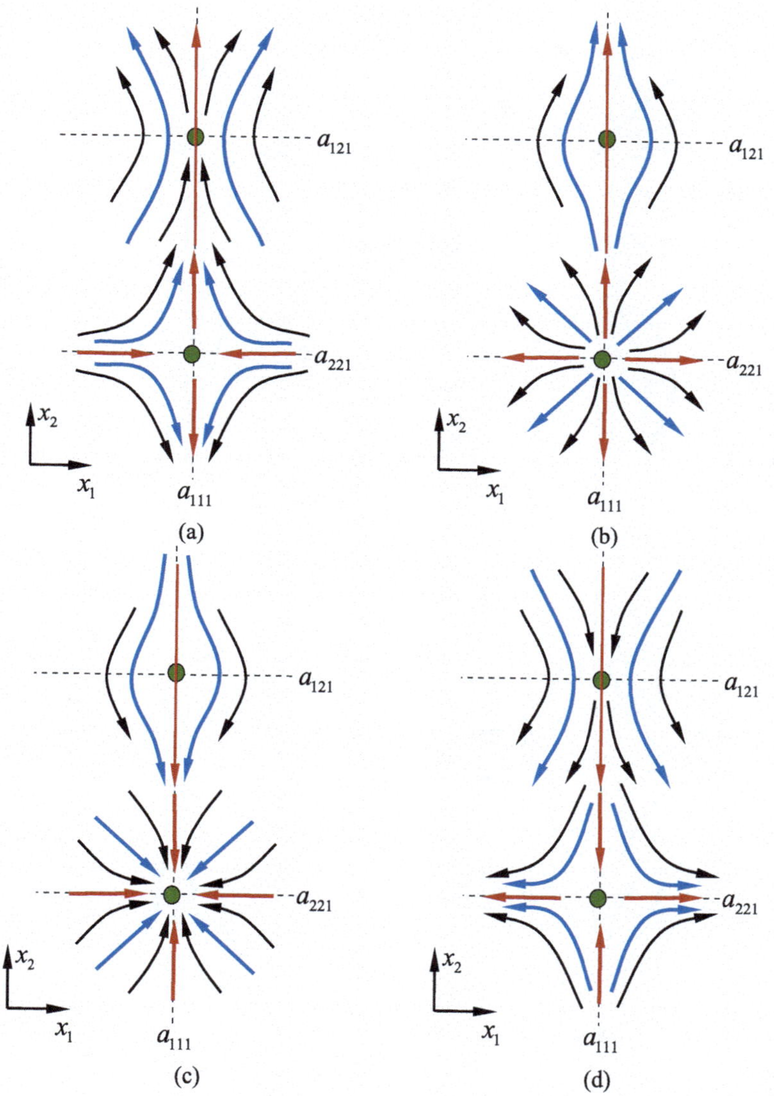

Fig. 3.1 Phase portraits ($a_{121} > a_{221}$) of the ($[(2m_{11}+1),(2n_{11}+1)]$, xF)-hyperbolic and hyperbolic-secant singular flows at $(x_1^*, x_2^*) = (a_{111}, a_{121})$ and $((2m_{11}+1),(2n_{21}+1))$-saddle, source and sink at $(x_1^*, x_2^*) = (a_{111}, a_{221})$ for 2-dimensional polynomial systems: **a** ($a_{110} > 0$, $a_{220} > 0$), **b** ($a_{110} < 0$, ${}_{110}a_{220} > 0$), **c** ($a > 0$, $a_{220} < 0$), **d** ($a_{110} < 0$, $a_{220} < 0$)

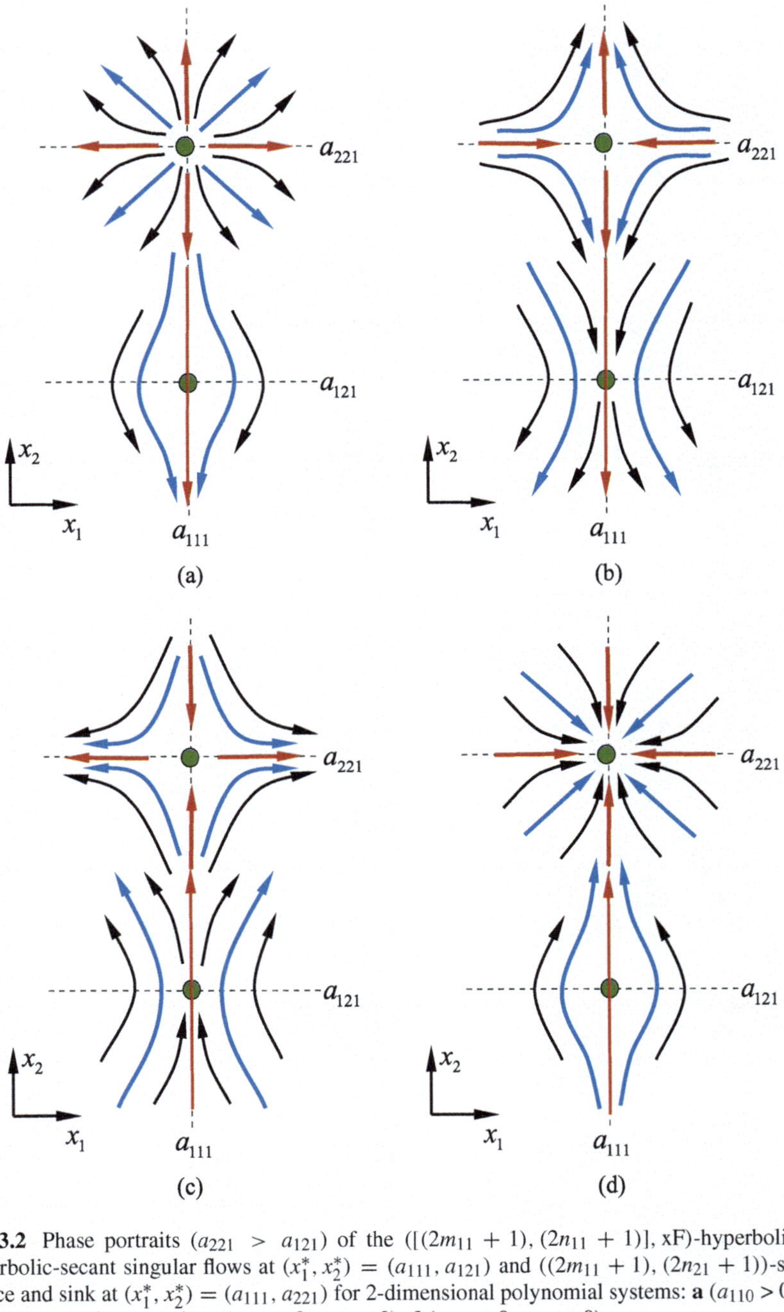

Fig. 3.2 Phase portraits ($a_{221} > a_{121}$) of the ([$(2m_{11}+1)$, $(2n_{11}+1)$], xF)-hyperbolic and hyperbolic-secant singular flows at $(x_1^*, x_2^*) = (a_{111}, a_{121})$ and ($(2m_{11}+1)$, $(2n_{21}+1)$)-saddle, source and sink at $(x_1^*, x_2^*) = (a_{111}, a_{221})$ for 2-dimensional polynomial systems: **a** ($a_{110} > 0$, $a_{220} > 0$), **b** ($a_{110} < 0$, $a_{220} > 0$), **c** ($a_{110} > 0$, $a_{220} < 0$), **d** ($a_{110} < 0$, $a_{220} < 0$)

$$-\frac{1}{2m_{11}}\Big[\frac{1}{(x_1-a_{111})^{2m_{11}}}-\frac{1}{(x_{10}-a_{111})^{2m_{11}}}\Big]=\ln\frac{|x_1-a_{111}|}{|x_{10}-a_{111}|}$$
for $2m_{11}+1=1$. (3.6)

(A) For $2n_{11}+1\geq 2n_{21}+1$, the corresponding infinite-equilibriums are given.

$$\begin{aligned}
(a_{111},a_{221})=&\underbrace{([2(n_{11}-n_{21})]^{\text{th}}\text{DI:II},[(2m_{11}+1),(2n_{21}+1)]^{\text{th}}\text{SO})}_{([2(n_{11}-n_{21})],[(2m_{11}+1),(2n_{21}+1)])\text{-decreasing-to-increasing inflection-source}},\\
&\text{for } a_{110}>0,a_{220}>0;\\
(a_{111},a_{221})=&\underbrace{([2(n_{11}-n_{21})]^{\text{th}}\text{II:DI},[(2m_{11}+1),(2n_{21}+1)]^{\text{th}}\text{SO})}_{([2(n_{11}-n_{21})],[(2m_{11}+1),(2n_{21}+1)])\text{-increasing-to-decreasing inflection-source}},\\
&\text{for } a_{110}<0,a_{220}>0;\\
(a_{111},a_{221})=&\underbrace{([2(n_{11}-n_{21})]^{\text{th}}\text{II:DI},[(2m_{11}+1),(2n_{21}+1)]^{\text{th}}\text{SI})}_{([2(n_{11}-n_{21})],[(2m_{11}+1),(2n_{21}+1)])\text{-increasing-to-decreasing inflection-sink}},\\
&\text{for } a_{110}>0,a_{220}<0;\\
(a_{111},a_{221})=&\underbrace{([2(n_{11}-n_{21})]^{\text{th}}\text{DI:II},[(2m_{11}+1),(2n_{21}+1)]^{\text{th}}\text{SI})}_{([2(n_{11}-n_{21})],[(2m_{11}+1),(2n_{21}+1)])\text{-decreasing-to-increasing inflection-sink}},\\
&\text{for } a_{110}<0,a_{220}<0.
\end{aligned}\tag{3.7}$$

(B) For $2n_{11}+1<2n_{21}+1$, the infinite-equilibriums are given as follows.

$$\begin{aligned}
(a_{111},a_{221})=&\underbrace{([2(n_{21}-n_{11})]^{\text{th}}\text{DC:IC},[(2m_{11}+1),(2n_{21}+1)]^{\text{th}}\text{SO})}_{([2(n_{21}-n_{11})],[(2m_{11}+1),(2n_{21}+1)])\text{-decreasing-to-increasing concave-source}},\\
&\text{for } a_{110}>0,a_{220}>0;\\
(a_{111},a_{221})=&\underbrace{([2(n_{21}-n_{11})]^{\text{th}}\text{IC:DC},[(2m_{11}+1),(2n_{21}+1)]^{\text{th}}\text{SO})}_{([2(n_{21}-n_{11})],[(2m_{11}+1),(2n_{21}+1)])\text{-increasing-to-decreasing concave-source}},\\
&\text{for } a_{110}<0,a_{220}>0;\\
(a_{111},a_{221})=&\underbrace{([2(n_{21}-n_{11})]^{\text{th}}\text{IC:DC},[(2m_{11}+1),(2n_{21}+1)]^{\text{th}}\text{SI})}_{([2(n_{21}-n_{11})],[(2m_{11}+1),(2n_{21}+1)])\text{-increasing-to-decreasing concave-sink}},\\
&\text{for } a_{110}>0,a_{220}<0;\\
(a_{111},a_{221})=&\underbrace{([2(n_{21}-n_{11})]^{\text{th}}\text{DC:IC},[(2m_{11}+1),(2n_{21}+1)]^{\text{th}}\text{SI})}_{([2(n_{21}-n_{11})],[(2m_{11}+1),(2n_{21}+1)])\text{-decreasing-to-increasing concave-sink}},\\
&\text{for } a_{110}<0,a_{220}<0.
\end{aligned}\tag{3.8}$$

For $x_2^*=a_{121}=a_{221}$, the infinite-equilibrium exists. The infinite-equilibriums are illustrated in Figs. 3.3 and 3.4 for $n_{11}\geq n_{21}$ and $n_{11}<n_{21}$, respectively. For $n_{11}\geq$

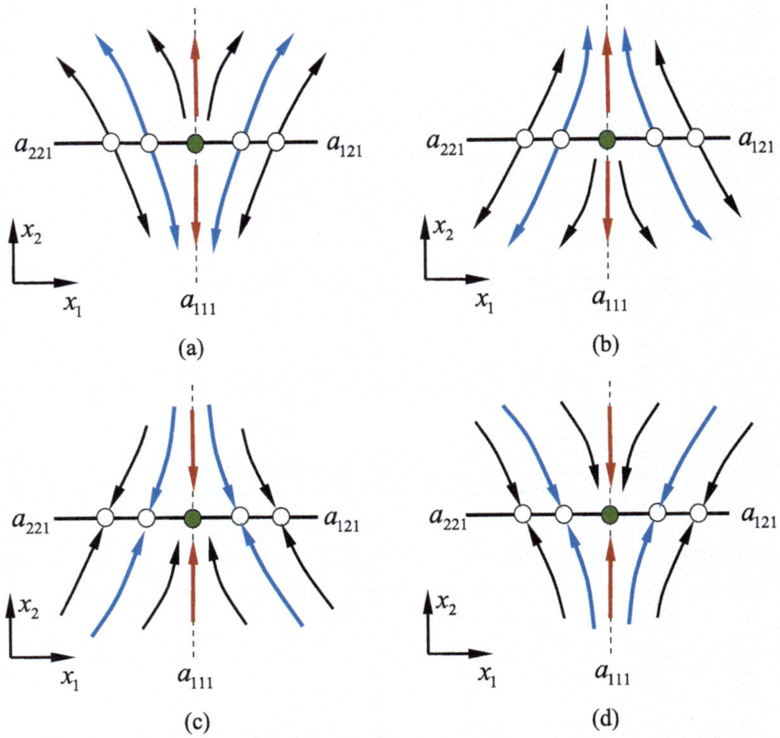

Fig. 3.3 The infinite-equilibriums ($a_{121} = a_{221}, 2n_{11} + 1 \geq 2n_{21} + 1, m_1 = 2m_{11} + 1$) for 2-dimensional polynomial systems: **a** ($a_{110} > 0, a_{220} > 0$), **b** ($a_{110} < 0, a_{220} > 0$), **c** ($a_{110} > 0, a_{220} < 0$), **d** ($a_{110} < 0, a_{220} < 0$)

n_{21}, the singular decreasing-to-increasing and increasing-to-decreasing inflection-source and sink are presented. For $n_{11} < n_{21}$, the singular decreasing-to-increasing and increasing-to-decreasing concave-source and sink are presented.

3.1.3 $[(2m_{11} + 1), (2n_{11} + 1)]$ *and* $((2m_{11} + 1), (2n_{21} + 1))$*-Hybrid Networks*

Consider

$$\begin{aligned}\dot{x}_2 &= a_{220} \prod_{s=1}^{2n_{21}+1} (x_2 - a_{22s}),\\ \dot{x}_1 &= a_{110} \prod_{r=1}^{2m_{11}+1} (x_1 - a_{11r}) \prod_{u=1}^{2n_{11}+1} (x_2 - a_{12u}).\end{aligned} \tag{3.9}$$

The first integral manifold is

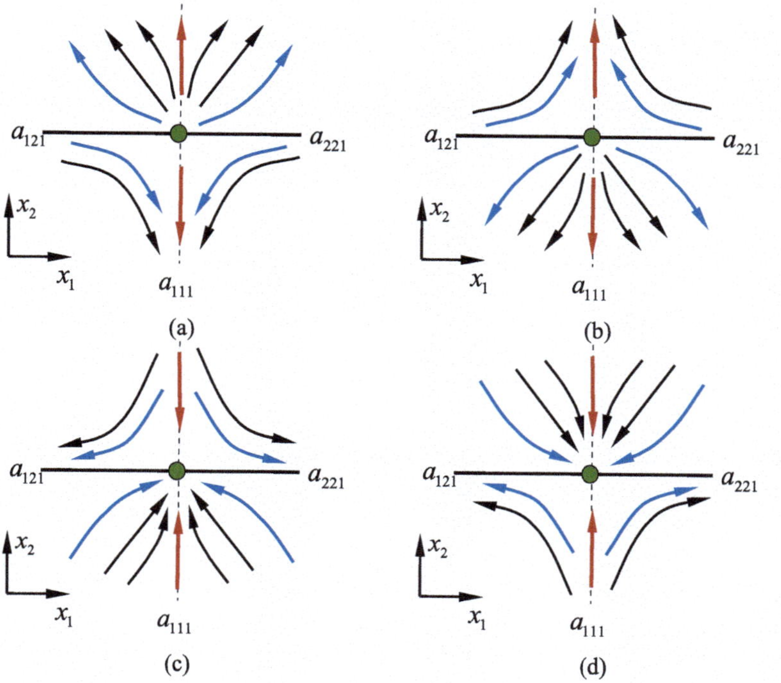

Fig. 3.4 The infinite-equilibriums ($a_{121} = a_{221}, 2n_{11}+1 < 2n_{21}+1, m_1 = 2m_{11}+1$) for 2-dimensional polynomial systems: **a** ($a_{110} > 0$, $a_{220} > 0$), **b** ($a_{110} < 0$, $a_{220} > 0$), **c** ($a_{110} > 0$, $a_{220} < 0$), **d** ($a_{110} < 0$, $a_{220} < 0$)

$$\sum_{r_1=1}^{2m_{11}+1} A_{r_1} \ln \frac{|x_1 - a_{11r_1}|}{|x_{10} - a_{11r_1}|} = \frac{a_{110}}{a_{220}} \Bigg\{ \sum_{\beta_0=2n_{21}+1}^{2n_{11}+1} b_{12\beta_0} \sum_{\substack{\alpha_1\in[0,\beta_0-2n_{21}-1],\\ \beta_1\in[2n_{21},\beta_0-1]}}^{\alpha_1+\beta_1=\beta_0-1} \sum_{\substack{\alpha_2\in[0,\beta_1-2n_{21}-1]\\ \beta_2\in[2n_{21}-1,\beta_1-2]}}^{\alpha_2+\beta_2=\beta_1-2} \cdots$$

$$\sum_{\substack{\alpha_{2n_{21}}\in[0,\beta_{2n_{21}-1}-2n_{21}-1]\\ \beta_{2n_{21}}\in[1,\beta_{2n_{21}-1}-2n_{21}]}}^{\alpha_{2n_{21}}+\beta_{2n_{21}}=\beta_{2n_{21}-1}-2n_{21}} \prod_{j=1}^{2n_{21}} \frac{(\beta_{j-1}-j)!}{\alpha_j!\beta_j!}(a_{22s_{j+1}} - a_{22s_j})^{\alpha_j}$$

$$\times \frac{1}{\beta_{2n_{21}}}[(x_2 - a_{22s_{2n_{21}+1}})^{\beta_{2n_{21}}} - (x_{20} - a_{22s_{2n_{21}+1}})^{\beta_{2n_{21}}}]$$

$$- \sum_{\beta_0=2n_2+1}^{2n_{21}+1} b_{12\beta_0} \sum_{k=2}^{2n_{21}} \Bigg[\sum_{\substack{\alpha_1\in[0,\beta_0-2n_{21}-1),\\ \beta_1\in(2n_{21},\beta_0-1]}}^{\alpha_1+\beta_1=\beta_0-1} \sum_{\substack{\alpha_2\in[0,\beta_1-2n_{21}-1]\\ \beta_2\in(2n_{21}-1,\beta_1-2)}}^{\alpha_2+\beta_2=\beta_1-2} \cdots$$

$$\times \Bigg(\sum_{\substack{\alpha_{k-1}\in(\beta_{k-2}-2n_{21},\beta_{k-2}-k+1]\\ \beta_{k-1}\in[0,2n_{21}+1-k)}}^{\alpha_{k-1}+\beta_{k-1}=\beta_{k-2}-k+1} \prod_{j=1}^{k-1} \frac{(\beta_{j-1}-j)!}{\alpha_j!\beta_j!}(a_{22s_{j+1}} - a_{22s_j})^{\alpha_j}$$

$$\times \sum_{\rho=k}^{2n_{21}+1} A_{\rho 1}^{(s_k)} \ln \frac{|x_{22} - a_{22s_\rho}|}{|x_{20} - a_{22r_\rho}|}$$

$$+ \sum_{\substack{\alpha_{k-1} \in (\beta_{k-2}-(2n_{21}-1, \beta_{k-2}-2n_{21}] \\ \beta_{k-1} \in [2n_{21}+1-k, 2n_{21}-k+2)}}^{\alpha_{k-1}+\beta_{k-1}=\beta_{k-2}-k+1} \prod_{j=1}^{k-1} \frac{(\beta_{j-1}-j)!}{\alpha_j! \beta_j!} (a_{22s_{j+1}} - a_{22s_j})^{\alpha_j}$$

$$\times \sum_{\rho=k+1}^{2n_{21}+1} B_{\rho 1}^{(s_k)} \ln \frac{|x_2 - a_{22s_\rho}|}{|x_{20} - a_{22s_\rho}|} \Bigg) \Bigg] + b_{j_1 j_2 0} \sum_{\rho=1}^{2n_{21}+1} A_{\rho 1}^{(s_1)} \ln \frac{|x_2 - a_{22s_\rho}|}{|x_{20} - a_{22s_\rho}|}$$

$$+ \sum_{\beta_0=1}^{2n_{21}} b_{j_1 j_2 \beta_0} \sum_{\rho=2}^{2n_{21}+1} B_{\rho 1}^{(s_1)} \frac{|x_2 - a_{22s_\rho}|}{|x_{20} - a_{22s_\rho}|} \Bigg\}; \tag{3.10}$$

where

$$A_{r_1} = \frac{1}{\prod_{r_2=1, r_2 \neq r_1}^{2m_{11}+1} (a_{11r_1} - a_{11r_\sigma})};$$

$$b_{j_1 j_2 \beta_0} = \sum_{t_1, l_1=0}^{t_1+l_1=1} \cdots \sum_{t_q, l_q=0}^{t_q+l_q=1} \prod_{k=1}^{2n_{11}+1} (a_{22s_1} - a_{12u_k})^{t_k} \delta_{\beta_0}^{\Sigma_{r=1}^{2n_{21}+1} l_r};$$

$$\lim_{\beta_{2n_{21}} \to 0} \frac{1}{\beta_{2n_{21}}} [(x_2 - a_{22s_{2n_{21}+1}})^{\beta_{2n_{21}}} - (x_{20} - a_{22s_{2n_{21}+1}})^{\beta_{2n_{21}}}] = \ln \frac{|x_2 - a_{22s_{2n_{21}+1}}|}{|x_{20} - a_{22s_{2n_{21}+1}}|}; \tag{3.11}$$

$$A_{\rho 1}^{(s_k)} = \frac{1}{\prod_{\sigma=k, \sigma \neq \rho}^{2n_{21}+1} (a_{22s_\rho} - a_{22s_\sigma})}, \text{ for } \rho \in \{k, k+1, \cdots, 2n_{21}+1\},$$

$$B_{\rho 1}^{(s_k)} = \frac{(a_{22s_\rho} - a_{22s_k})^{\beta_{k-1}-k}}{\prod_{\sigma=k+1, \sigma \neq \rho}^{2n_{21}+1} (a_{22s_\rho} - a_{22s_\sigma})}, \text{ for} \rho \in \{k+1, k+2, \cdots, 2n_{21}+1\};$$

$$k \in \{1, 2, \cdots, 2n_{21}\}. \tag{3.12}$$

For all cases, the $(2m_{11}+1) \times (2n_{21}+1)$-simple equilibriums plus $(2m_{11}+1) \times (2n_{11}+1)$-hyperbolic flows are based on the following inflection-source and sink bifurcations for $2n_{11}+1 \geq 2n_{21}+1$:

- $([2(n_{11}-n_{21})]^{\text{th}}$ DI:II, $[(2m_{11}+1), (2n_{21}+1)]^{\text{th}}$SO)-decreasing-to-increasing inflection-source bifurcation for ($a_{110} > 0, a_{220} > 0$),
- $([2(n_{11}-n_{21})]^{\text{th}}$ II:DI, $[(2m_{11}+1), (2n_{21}+1)]^{\text{th}}$SO)-increasing-to-decreasing inflection-source bifurcation for ($a_{110} < 0, a_{220} > 0$),
- $([2(n_{11}-n_{21})]^{\text{th}}$ II:DI, $[(2m_{11}+1), (2n_{21}+1)]^{\text{th}}$SI)-increasing-to-decreasing inflection-sink bifurcation for ($a_{110} > 0, a_{220} < 0$),
- $([2(n_{11}-n_{21})]^{\text{th}}$ DI:II, $[(2m_{11}+1), (2n_{21}+1)]^{\text{th}}$SI)-decreasing-to-increasing inflection-sink bifurcation for ($a_{110} < 0, a_{220} < 0$);

or based on the following concave-source and sink bifurcations for $2n_{11}+1 < 2n_{21}+1$:

- $([2(n_{21}-n_{11})]^{\text{th}}$ DC:IC, $[(2m_{11}+1), (2n_{21}+1)]^{\text{th}}$SO)-decreasing-to-increasing concave-source bifurcation for ($a_{110} > 0, a_{220} > 0$),
- $([2(n_{11}-n_{21})]^{\text{th}}$ IC:DC, $[(2m_{11}+1), (2n_{21}+1)]^{\text{th}}$SO)-increasing-to-decreasing concave-source bifurcation for ($a_{110} < 0, a_{220} > 0$),
- $([2(n_{11}-n_{21})]^{\text{th}}$ IC:DC, $[(2m_{11}+1), (2n_{21}+1)]^{\text{th}}$SI)-increasing-to-decreasing concave-sink bifurcation for ($a_{110} > 0, a_{220} < 0$),
- $([2(n_{11}-n_{21})]^{\text{th}}$ DC:IC, $[(2m_{11}+1), (2n_{21}+1)]^{\text{th}}$SI)-decreasing-to-increasing concave-sink bifurcation for ($a_{110} < 0, a_{220} < 0$).

Consider $2n_{11}+1 = 2n_{21}+1$ for illustrations, phase portraits for the 2-dimensional systems for the hybrid networks of hyperbolic and hyperbolic-secant flows with the simple equilibriums of the saddles, sinks and sources with $x_1^* = a_{111}, a_{112}, \ldots, a_{11(2m_1+1)}$ and $x_2^* = a_{121}, a_{221}, \ldots, a_{12(2n_{11}+1)}, a_{22(2n_{21}+1)}$ are presented in Fig. 3.5 (a–d) for ($a_{110} > 0$, $a_{220} > 0$), ($a_{110} < 0$, $a_{220} > 0$) , ($a_{110} > 0$, $a_{220} > 0$) and ($a_{110} < 0$, $a_{220} < 0$).

$$
\begin{Bmatrix}
(a_{111}, a_{22(2n_{21}+1)}) & \cdots & (a_{11(2m_{11})}, a_{22(2n_{21}+1)}) & (a_{11(2m_{11}+1)}, a_{22(2n_{21}+1)}) \\
(a_{111}, a_{12(2n_{11}+1)}) & \cdots & (a_{11(2m_{11})}, a_{12(2n_{11}+1)}) & (a_{11(2m_{11}+1)}, a_{12(2n_{11}+1)}) \\
(a_{111}, a_{22(2n_{21})}) & \cdots & (a_{11(2m_{11})}, a_{22(2n_{21})}) & (a_{11(2m_{11}+1)}, a_{22(2n_{21})}) \\
\vdots & \vdots & \vdots & \vdots \\
(a_{111}, a_{221}) & \cdots & (a_{11(2m_{11})}, a_{221}) & (a_{11(2m_{11}+1)}, a_{221}) \\
(a_{111}, a_{121}) & \cdots & (a_{11(2m_{11})}, a_{121}) & (a_{11(2m_{11}+1)}, a_{121})
\end{Bmatrix}
$$

$$
= \begin{Bmatrix}
\underbrace{(\text{SO, SO})}_{\text{source}} & \cdots & \underbrace{(\text{SI, SO})}_{\text{saddle}} & \underbrace{(\text{SO, SO})}_{\text{source}} \\
\underbrace{(\text{UP:DP, nF})}_{\text{hyperbolic-secant flow }(-)} & \cdots & \underbrace{(\text{DP:UP, nF})}_{\text{hyperbolic flow }(-)} & \underbrace{(\text{UP:DP, nF})}_{\text{hyperbolic-secant flow }(-)} \\
\underbrace{(\text{SI, SI})}_{\text{sink}} & \cdots & \underbrace{(\text{SO, SI})}_{\text{saddle}} & \underbrace{(\text{SI, SI})}_{\text{sink}} \\
\vdots & \vdots & \vdots & \vdots \\
\underbrace{(\text{SO, SO})}_{\text{source}} & \cdots & \underbrace{(\text{SI, SO})}_{\text{saddle}} & \underbrace{(\text{SO, SO})}_{\text{source}} \\
\underbrace{(\text{UP:DP, nF})}_{\text{hyperbolic-secant flow }(-)} & \cdots & \underbrace{(\text{DP:UP, nF})}_{\text{hyperbolic flow }(-)} & \underbrace{(\text{UP:DP, nF})}_{\text{hyperbolic-secant flow }(-)}
\end{Bmatrix}_{(2n_{11}+2n_{21}+2)\times(2m_{11}+1)}
$$

for $a_{110} > 0, a_{220} > 0$; (3.13)

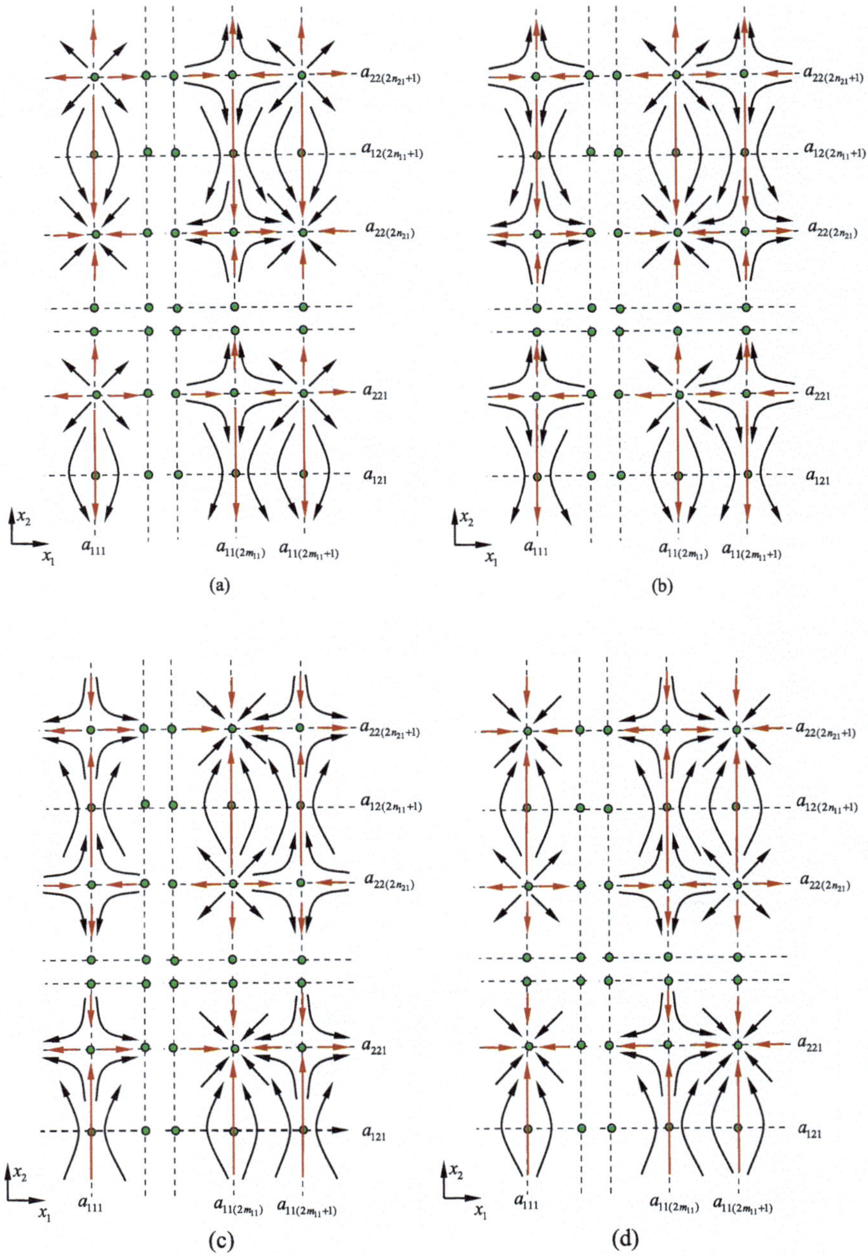

Fig. 3.5 Phase portraits for 2-dimensional systems with $x_1^* = a_{111}, a_{112}, \ldots, a_{11(2m_{11}+1)}$ and $x_2^* = a_{121}, a_{221}, \ldots, a_{12(2n_{11}+1)}, a_{22(2n_{21}+1)}$. The networks of flows with sink, source and saddle: **a** ($a_{110} > 0$, $a_{220} > 0$), **b** ($a_{110} < 0$, $a_{220} > 0$), **c** ($a_{110} > 0$, $a_{220} < 0$), **d** ($a_{110} < 0$, $a_{220} < 0$)

$$\left\{\begin{array}{cccc} (a_{111},a_{22(2n_{21}+1)}) & \cdots & (a_{11(2m_{11})},a_{22(2n_{21}+1)}) & (a_{11(2m_{11}+1)},a_{22(2n_{21}+1)}) \\ (a_{111},a_{12(2n_{11}+1)}) & \cdots & (a_{11(2m_{11})},a_{12(2n_{11}+1)}) & (a_{11(2m_{11}+1)},a_{12(2n_{11}+1)}) \\ (a_{111},a_{22(2n_{21})}) & \cdots & (a_{11(2m_{11})},a_{22(2n_{21})}) & (a_{11(2m_{11}+1)},a_{22(2n_{21})}) \\ \vdots & \vdots & \vdots & \vdots \\ (a_{111},a_{221}) & \cdots & (a_{11(2m_{11})},a_{221}) & (a_{11(2m_{11}+1)},a_{221}) \\ (a_{111},a_{121}) & \cdots & (a_{11(2m_{11})},a_{121}) & (a_{11(2m_{11}+1)},a_{121}) \end{array}\right\}$$
$$= \left\{\begin{array}{cccc} \underbrace{(\text{SI, SO})}_{\text{saddle}} & \cdots & \underbrace{(\text{SO, SO})}_{\text{source}} & \underbrace{(\text{SI, SO})}_{\text{saddle}} \\ \underbrace{(\text{DP:UP, nF})}_{\text{hyperbolic flow }(-)} & \cdots & \underbrace{(\text{UP:DP, nF})}_{\text{hyperbolic-secant flow }(-)} & \underbrace{(\text{DP:UP, nF})}_{\text{hyperbolic flow }(-)} \\ \underbrace{(\text{SO, SI})}_{\text{saddle}} & \cdots & \underbrace{(\text{SI, SI})}_{\text{sink}} & \underbrace{(\text{SO, SI})}_{\text{saddle}} \\ \vdots & \vdots & \vdots & \vdots \\ \underbrace{(\text{SI, SO})}_{\text{saddle}} & \cdots & \underbrace{(\text{SO, SO})}_{\text{source}} & \underbrace{(\text{SI, SO})}_{\text{saddle}} \\ \underbrace{(\text{DP:UP, nF})}_{\text{hyperbolic flow }(-)} & \cdots & \underbrace{(\text{UP:DP, nF})}_{\text{hyperbolic-secant flow }(-)} & \underbrace{(\text{DP:UP, nF})}_{\text{hyperbolic flow }(-)} \end{array}\right\}_{(2n_{11}+2n_{21}+2)\times(2m_{11}+1)}$$

for $a_{110} < 0, a_{220} > 0$; (3.14)

$$\left\{\begin{array}{cccc} (a_{111},a_{22(2n_{21}+1)}) & \cdots & (a_{11(2m_{11})},a_{22(2n_{21}+1)}) & (a_{11(2m_{11}+1)},a_{22(2n_{21}+1)}) \\ (a_{111},a_{12(2n_{11}+1)}) & \cdots & (a_{11(2m_{11})},a_{12(2n_{11}+1)}) & (a_{11(2m_{11}+1)},a_{12(2n_{11}+1)}) \\ (a_{111},a_{22(2n_{21})}) & \cdots & (a_{11(2m_{11})},a_{22(2n_{21})}) & (a_{11(2m_{11}+1)},a_{22(2n_{21})}) \\ \vdots & \vdots & \vdots & \vdots \\ (a_{111},a_{221}) & \cdots & (a_{11(2m_{11})},a_{221}) & (a_{11(2m_{11}+1)},a_{221}) \\ (a_{111},a_{121}) & \cdots & (a_{11(2m_{11})},a_{121}) & (a_{11(2m_{11}+1)},a_{121}) \end{array}\right\}$$
$$= \left\{\begin{array}{cccc} \underbrace{(\text{SO, SI})}_{\text{saddle}} & \cdots & \underbrace{(\text{SI, SI})}_{\text{sink}} & \underbrace{(\text{SO, SI})}_{\text{saddle}} \\ \underbrace{(\text{DP:UP, pF})}_{\text{hyperbolic flow }(+)} & \cdots & \underbrace{(\text{UP:DP, pF})}_{\text{hyperbolic-secant flow }(+)} & \underbrace{(\text{DP:UP, pF})}_{\text{hyperbolic flow }(+)} \\ \underbrace{(\text{SI, SO})}_{\text{saddle}} & \cdots & \underbrace{(\text{SO, SO})}_{\text{source}} & \underbrace{(\text{SI, SO})}_{\text{saddle}} \\ \vdots & \vdots & \vdots & \vdots \\ \underbrace{(\text{SO, SI})}_{\text{saddle}} & \cdots & \underbrace{(\text{SI, SI})}_{\text{sink}} & \underbrace{(\text{SO, SI})}_{\text{saddle}} \\ \underbrace{(\text{DP:UP, pF})}_{\text{hyperbolic flow }(+)} & \cdots & \underbrace{(\text{UP:DP, pF})}_{\text{hyperbolic-secant flow }(+)} & \underbrace{(\text{DP:UP, pF})}_{\text{hyperbolic flow }(+)} \end{array}\right\}_{(2n_{11}+2n_{21}+2)\times(2m_{11}+1)}$$

for $a_{110} > 0, a_{220} < 0$; (3.15)

$$\left\{\begin{array}{cccc} (a_{111}, a_{22(2n_{21}+1)}) & \cdots & (a_{11(2m_{11})}, a_{22(2n_{21}+1)}) & (a_{11(2m_{11}+1)}, a_{22(2n_{21}+1)}) \\ (a_{111}, a_{12(2n_{11}+1)}) & \cdots & (a_{11(2m_{11})}, a_{12(2n_{11}+1)}) & (a_{11(2m_{11}+1)}, a_{12(2n_{11}+1)}) \\ (a_{111}, a_{22(2n_{21})}) & \cdots & (a_{11(2m_{11})}, a_{22(2n_{21})}) & (a_{11(2m_{11}+1)}, a_{22(2n_{21})}) \\ \vdots & \vdots & \vdots & \vdots \\ (a_{111}, a_{221}) & \cdots & (a_{11(2m_{11})}, a_{221}) & (a_{11(2m_{11}+1)}, a_{221}) \\ (a_{111}, a_{121}) & \cdots & (a_{11(2m_{11})}, a_{121}) & (a_{11(2m_{11}+1)}, a_{121}) \end{array}\right\}$$

$$= \left\{\begin{array}{cccc} \underbrace{(\mathrm{SI},\mathrm{SI})}_{\text{sink}} & \cdots & \underbrace{(\mathrm{SO},\mathrm{SI})}_{\text{saddle}} & \underbrace{(\mathrm{SI},\mathrm{SI})}_{\text{sink}} \\ \underbrace{(\mathrm{UP{:}DP},\mathrm{pF})}_{\text{hyperbolic-secant flow }(+)} & \cdots & \underbrace{(\mathrm{DP{:}UP},\mathrm{pF})}_{\text{hyperbolic flow }(+)} & \underbrace{(\mathrm{UP{:}DP},\mathrm{pF})}_{\text{hyperbolic-secant flow }(+)} \\ \underbrace{(\mathrm{SO},\mathrm{SO})}_{\text{source}} & \cdots & \underbrace{(\mathrm{SI},\mathrm{SO})}_{\text{saddle}} & \underbrace{(\mathrm{SO},\mathrm{SO})}_{\text{source}} \\ \vdots & \vdots & \vdots & \vdots \\ \underbrace{(\mathrm{SI},\mathrm{SI})}_{\text{sink}} & \cdots & \underbrace{(\mathrm{SO},\mathrm{SI})}_{\text{saddle}} & \underbrace{(\mathrm{SI},\mathrm{SI})}_{\text{sink}} \\ \underbrace{(\mathrm{UP{:}DP},\mathrm{pF})}_{\text{hyperbolic-secant flow }(-)} & \cdots & \underbrace{(\mathrm{DP{:}UP},\mathrm{pF})}_{\text{hyperbolic flow }(+)} & \underbrace{(\mathrm{UP{:}DP},\mathrm{pF})}_{\text{hyperbolic-secant flow }(+)} \end{array}\right\}_{(2n_{11}+2n_{21}+2)\times(2m_{11}+1)}$$

for $a_{110} < 0, a_{220} < 0$. (3.16)

Consider $2n_{11} + 1 = 2n_{21} + 1$ for illustrations. Phase portraits for the 2-dimensional systems for the hybrid networks of hyperbolic and hyperbolic-secant flows with the simple equilibriums of the saddles, sinks and sources with $x_1^* = a_{111}, a_{112}, \ldots, a_{11(2m_{11}+1)}$ and $x_2^* = a_{221}, a_{121}, \ldots, a_{22(2n_{21}+1)}, a_{12(2n_{11}+1)}$ are presented in Fig. 3.6 (a–d) for ($a_{110} > 0, a_{220} > 0$), ($a_{110} < 0, a_{220} > 0$), ($a_{110} > 0, a_{220} < 0$) and ($a_{110} < 0, a_{220} < 0$).

$$\left\{\begin{array}{cccc} (a_{111}, a_{12(2n_{11}+1)}) & \cdots & (a_{11(2m_{11})}, a_{12(2n_{11}+1)}) & (a_{11(2m_{11}+1)}, a_{12(2n_{11}+1)}) \\ (a_{111}, a_{22(2n_{21}+1)}) & \cdots & (a_{11(2m_{11})}, a_{22(2n_{21}+1)}) & (a_{11(2m_{11}+1)}, a_{22(2n_{21}+1)}) \\ (a_{111}, a_{12(2n_{11})}) & \cdots & (a_{11(2m_{11})}, a_{12(2n_{11})}) & (a_{11(2m_{11}+1)}, a_{12(2n_{11})}) \\ \vdots & \vdots & \vdots & \vdots \\ (a_{111}, a_{121}) & \cdots & (a_{11(2m_{11})}, a_{121}) & (a_{11(2m_{11}+1)}, a_{121}) \\ (a_{111}, a_{221}) & \cdots & (a_{11(2m_{11})}, a_{221}) & (a_{11(2m_{11}+1)}, a_{221}) \end{array}\right\}$$

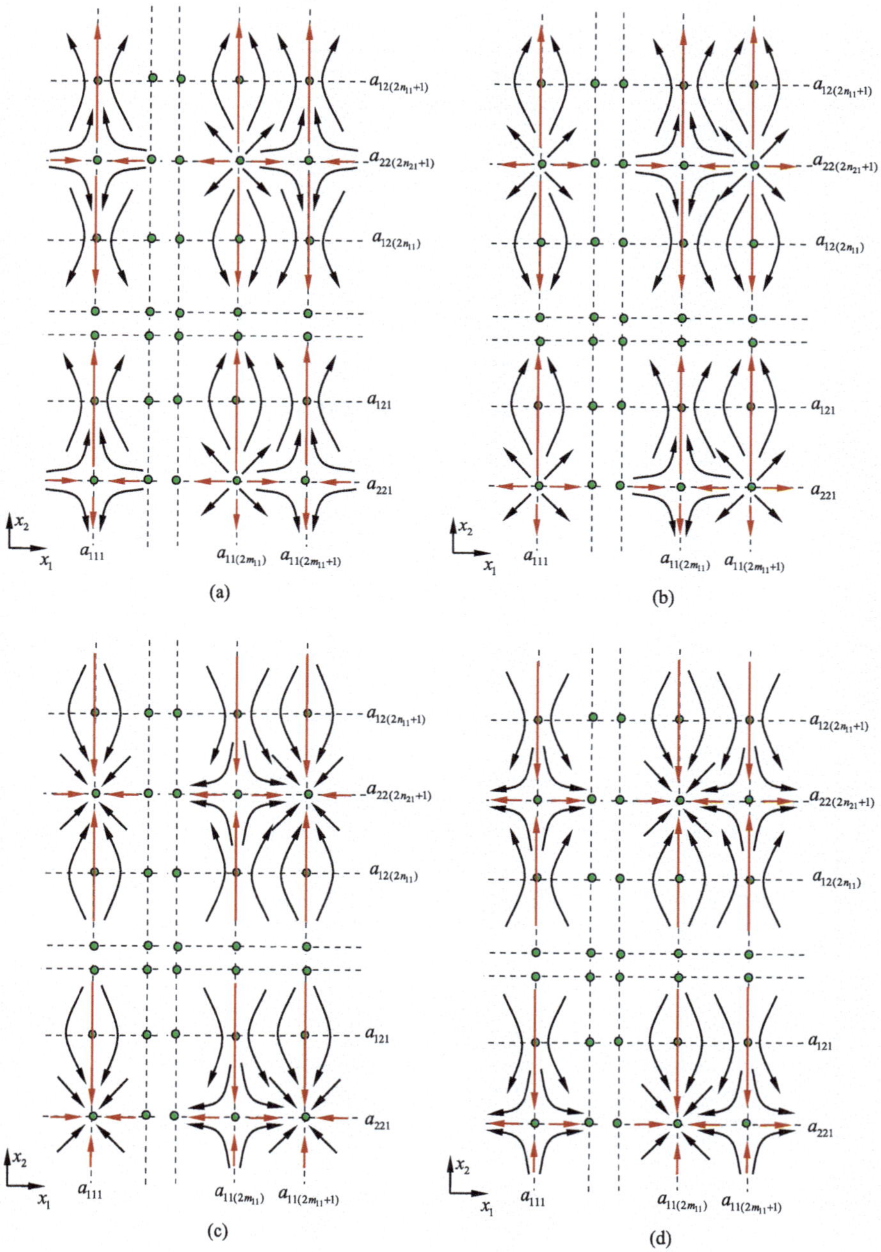

Fig. 3.6 Phase portraits for 2-dimensional systems with $x_1^* = a_{111}, a_{112}, \ldots, a_{11(2m_{11}+1)}$ and $x_2^* = a_{221}, a_{121}, \ldots, a_{22(2n_{21}+1)}, a_{12(2n_{11}+1)}$. The networks of flows with sink, source and saddle: **a** ($a_{110} > 0$, $a_{220} > 0$), **b** ($a_{110} < 0$, $a_{220} > 0$), **c** ($a_{110} > 0$, $a_{220} < 0$), **d** ($a_{110} < 0$, $a_{220} < 0$)

$$
= \left\{ \begin{array}{cccc}
\underbrace{(\mathrm{DP{:}UP,\ pF})}_{\text{hyperbolic flow }(+)} & \cdots & \underbrace{(\mathrm{UP{:}DP,\ pF})}_{\text{hyperbolic-secant flow }(+)} & \underbrace{(\mathrm{DP{:}UP,\ pF})}_{\text{hyperbolic flow }(+)} \\
\underbrace{(\mathrm{SI,\ SO})}_{\text{saddle}} & \cdots & \underbrace{(\mathrm{SO,\ SO})}_{\text{source}} & \underbrace{(\mathrm{SI,\ SO})}_{\text{saddle}} \\
\underbrace{(\mathrm{DP{:}UP,\ nF})}_{\text{hyperbolic flow }(-)} & \cdots & \underbrace{(\mathrm{UP{:}DP,\ pF})}_{\text{hyperbolic-secant flow }(-)} & \underbrace{(\mathrm{DP{:}UP,\ nF})}_{\text{hyperbolic flow }(-)} \\
\vdots & \vdots & \vdots & \vdots \\
\underbrace{(\mathrm{DP{:}UP,\ pF})}_{\text{hyperbolic flow }(+)} & \cdots & \underbrace{(\mathrm{UP{:}DP,\ pF})}_{\text{hyperbolic-secant flow }(+)} & \underbrace{(\mathrm{DP{:}UP,\ pF})}_{\text{hyperbolic flow }(+)} \\
\underbrace{(\mathrm{SI,\ SO})}_{\text{saddle}} & \cdots & \underbrace{(\mathrm{SO,\ SO})}_{\text{source}} & \underbrace{(\mathrm{SI,\ SO})}_{\text{saddle}}
\end{array} \right\}_{(2n_{11}+2n_{21}+2)\times(2m_{11}+1)}
$$

for $a_{110} > 0, a_{220} > 0$; (3.17)

$$
\left\{ \begin{array}{cccc}
(a_{111}, a_{12(2n_{11}+1)}) & \cdots & (a_{11(2m_{11})}, a_{12(2n_{11}+1)}) & (a_{11(2m_{11}+1)}, a_{12(2n_{11}+1)}) \\
(a_{111}, a_{22(2n_{21}+1)}) & \cdots & (a_{11(2m_{11})}, a_{22(2n_{21}+1)}) & (a_{11(2m_{11}+1)}, a_{22(2n_{21}+1)}) \\
(a_{111}, a_{12(2n_{11})}) & \cdots & (a_{11(2m_{11})}, a_{12(2n_{11})}) & (a_{11(2m_{11}+1)}, a_{12(2n_{11})}) \\
\vdots & \vdots & \vdots & \vdots \\
(a_{111}, a_{121}) & \cdots & (a_{11(2m_{11})}, a_{121}) & (a_{11(2m_{11}+1)}, a_{121}) \\
(a_{111}, a_{221}) & \cdots & (a_{11(2m_{11})}, a_{221}) & (a_{11(2m_{11}+1)}, a_{221})
\end{array} \right\}
$$

$$
= \left\{ \begin{array}{cccc}
\underbrace{(\mathrm{UP{:}DP,\ pF})}_{\text{hyperbolic-secant flow }(+)} & \cdots & \underbrace{(\mathrm{DP{:}UP,\ pF})}_{\text{hyperbolic flow }(+)} & \underbrace{(\mathrm{UP{:}DP,\ pF})}_{\text{hyperbolic-secant flow }(+)} \\
\underbrace{(\mathrm{SO,\ SO})}_{\text{source}} & \cdots & \underbrace{(\mathrm{SI,\ SO})}_{\text{saddle}} & \underbrace{(\mathrm{SO,\ SO})}_{\text{source}} \\
\underbrace{(\mathrm{UP{:}DP,\ nF})}_{\text{hyperbolic-secant flow }(-)} & \cdots & \underbrace{(\mathrm{DP{:}UP,\ nF})}_{\text{hyperbolic flow }(-)} & \underbrace{(\mathrm{UP{:}DP,\ nF})}_{\text{hyperbolic-secant flow }(-)} \\
\vdots & \vdots & \vdots & \vdots \\
\underbrace{(\mathrm{UP{:}DP,\ pF})}_{\text{hyperbolic-secant flow }(+)} & \cdots & \underbrace{(\mathrm{DP{:}UP,\ pF})}_{\text{hyperbolic flow }(+)} & \underbrace{(\mathrm{UP{:}DP,\ pF})}_{\text{hyperbolic-secant flow }(+)} \\
\underbrace{(\mathrm{SO,\ SO})}_{\text{source}} & \cdots & \underbrace{(\mathrm{SI,\ SO})}_{\text{saddle}} & \underbrace{(\mathrm{SO,\ SO})}_{\text{source}}
\end{array} \right\}_{(2n_{11}+2n_{21}+2)\times(2m_{11}+1)}
$$

for $a_{110} < 0, a_{220} > 0$; (3.18)

$$\left\{\begin{array}{cccc}
(a_{111},a_{12(2n_{11}+1)}) & \cdots & (a_{11(2m_{11})},a_{12(2n_{11}+1)}) & (a_{11(2m_{11}+1)},a_{12(2n_{11}+1)})\\
(a_{111},a_{22(2n_{21}+1)}) & \cdots & (a_{11(2m_{11})},a_{22(2n_{21}+1)}) & (a_{11(2m_{11}+1)},a_{22(2n_{21}+1)})\\
(a_{111},a_{12(2n_{11})}) & \cdots & (a_{11(2m_{11})},a_{12(2n_{11})}) & (a_{11(2m_{11}+1)},a_{12(2n_{11})})\\
\vdots & \vdots & \vdots & \vdots\\
(a_{111},a_{121}) & \cdots & (a_{11(2m_{11})},a_{121}) & (a_{11(2m_{11}+1)},a_{121})\\
(a_{111},a_{221}) & \cdots & (a_{11(2m_{11})},a_{221}) & (a_{11(2m_{11}+1)},a_{221})
\end{array}\right\}$$

$$=\left\{\begin{array}{cccc}
\underbrace{\text{(UP:DP, nF)}}_{\text{hyperbolic-secant flow }(-)} & \cdots & \underbrace{\text{(DP:UP, nF)}}_{\text{hyperbolic flow }(-)} & \underbrace{\text{(UP:DP, nF)}}_{\text{hyperbolic-secant flow }(-)}\\
\underbrace{\text{(SI, SI)}}_{\text{sink}} & \cdots & \underbrace{\text{(SO, SI)}}_{\text{saddle}} & \underbrace{\text{(SI, SI)}}_{\text{sink}}\\
\underbrace{\text{(UP:DP, pF)}}_{\text{hyperbolic-secant flow }(+)} & \cdots & \underbrace{\text{(DP:UP, pF)}}_{\text{hyperbolic flow }(+)} & \underbrace{\text{(UP:DP, pF)}}_{\text{hyperbolic-secant flow }(+)}\\
\vdots & \vdots & \vdots & \vdots\\
\underbrace{\text{(UP:DP, nF)}}_{\text{hyperbolic-secant flow }(-)} & \cdots & \underbrace{\text{(DP:UP, nF)}}_{\text{hyperbolic flow }(-)} & \underbrace{\text{(UP:DP, nF)}}_{\text{hyperbolic-secant flow }(-)}\\
\underbrace{\text{(SI, SI)}}_{\text{sink}} & \cdots & \underbrace{\text{(SO, SI)}}_{\text{saddle}} & \underbrace{\text{(SI, SI)}}_{\text{sink}}
\end{array}\right\}_{(2n_{11}+2n_{21}+2)\times(2m_{11}+1)}$$

for $a_{110}>0, a_{220}<0$; (3.19)

$$\left\{\begin{array}{cccc}
(a_{111},a_{12(2n_{11}+1)}) & \cdots & (a_{11(2m_{11})},a_{12(2n_{11}+1)}) & (a_{11(2m_{11}+1)},a_{12(2n_{11}+1)})\\
(a_{111},a_{22(2n_{21}+1)}) & \cdots & (a_{11(2m_{11})},a_{22(2n_{21}+1)}) & (a_{11(2m_{11}+1)},a_{22(2n_{21}+1)})\\
(a_{111},a_{12(2n_{11})}) & \cdots & (a_{11(2m_{11})},a_{12(2n_{11})}) & (a_{11(2m_{11}+1)},a_{12(2n_{11})})\\
\vdots & \vdots & \vdots & \vdots\\
(a_{111},a_{121}) & \cdots & (a_{11(2m_{11})},a_{121}) & (a_{11(2m_{11}+1)},a_{121})\\
(a_{111},a_{221}) & \cdots & (a_{11(2m_{11})},a_{221}) & (a_{11(2m_{11}+1)},a_{221})
\end{array}\right\}$$

$$=\left\{\begin{array}{cccc}
\underbrace{\text{(DP:UP, nF)}}_{\text{hyperbolic flow }(-)} & \cdots & \underbrace{\text{(UP:DP, nF)}}_{\text{hyperbolic-secant flow }(-)} & \underbrace{\text{(DP:UP, nF)}}_{\text{hyperbolic flow }(-)}\\
\underbrace{\text{(SO, SI)}}_{\text{saddle}} & \cdots & \underbrace{\text{(SI, SI)}}_{\text{sink}} & \underbrace{\text{(SO, SI)}}_{\text{saddle}}\\
\underbrace{\text{(DP:UP, pF)}}_{\text{hyperbolic flow }(+)} & \cdots & \underbrace{\text{(UP:DP, pF)}}_{\text{hyperbolic-secant flow }(+)} & \underbrace{\text{(DP:UP, pF)}}_{\text{hyperbolic flow }(+)}\\
\vdots & \vdots & \vdots & \vdots\\
\underbrace{\text{(DP:UP, nF)}}_{\text{hyperbolic flow }(-)} & \cdots & \underbrace{\text{(UP:DP, nF)}}_{\text{hyperbolic-secant flow }(-)} & \underbrace{\text{(DP:UP, nF)}}_{\text{hyperbolic flow }(-)}\\
\underbrace{\text{(SO, SI)}}_{\text{saddle}} & \cdots & \underbrace{\text{(SI, SI)}}_{\text{sink}} & \underbrace{\text{(SO, SI)}}_{\text{saddle}}
\end{array}\right\}_{(2n_{11}+2n_{21}+2)\times(2m_{11}+1)}$$

for $a_{110}<0, a_{220}<0$. (3.20)

3.2 $[(2m_{11}+1),(2n_{11})]$ and $((2m_{11}+1),(2n_{21}+1))$-Singular Systems

In this section, there are four global bifurcations of the 1-diemsnional flows through *singular inflection* flows, and four global bifurcations of equilibriums, through singular source, sink and saddle bifurcations. The corresponding global bifurcations generate networks of hyperbolic and hyperbolic-secant flows with sources, sinks and saddles. The infinite-equilibriums are the appearing and switching bifurcations for the singular flows and singular equilibriums and for hybrid networks of hyperbolic flows and source, sink and saddles.

3.2.1 $[(2m_{11}+1),(2n_{11})]$-Flows and $((2m_{11}+1),(2n_{21}+1))$-Equilibriums

Consider a polynomial dynamical system for $a_{221}\neq a_{121}$ as

$$\begin{aligned}\dot{x}_2 &= a_{220}(x_2-a_{221})^{2n_{21}+1},\\ \dot{x}_1 &= a_{110}(x_1-a_{111})^{2m_{11}+1}(x_2-a_{121})^{2n_{11}}.\end{aligned}\tag{3.21}$$

The first integral manifold is

$$\begin{aligned}&-\frac{1}{2m_{11}}\left[\frac{1}{(x_1-a_{111})^{2m_{11}}}-\frac{1}{(x_{10}-a_{111})^{2m_{11}}}\right]\\ &=\frac{a_{110}}{a_{220}}\sum_{\alpha,\beta=0}^{\alpha+\beta=2n_{11}}\frac{(2n_{11})!}{\alpha!\beta!}\frac{(a_{221}-a_{121})^{\alpha}}{\beta-2n_{21}}[(x_2-a_{221})^{\beta-2n_{21}}-(x_2-a_{221})^{\beta-2n_{21}}]\\ &\text{for } 2m_{11}+1\neq 1,\ \beta-2n_{21}\neq 0;\\ &-\frac{1}{2m_{11}}\left[\frac{1}{(x_1-a_{111})^{2m_{11}}}-\frac{1}{(x_{10}-a_{111})^{2m_{11}}}\right]=\ln\frac{|x_1-a_{111}|}{|x_{10}-a_{111}|}\\ &\text{for } 2m_{11}+1=1;\\ &\frac{1}{\beta-2n_{21}}[(x_2-a_{221})^{\beta-2n_{21}}-(x_2-a_{221})^{\beta-2n_{21}}]=\ln\frac{|x_2-a_{221}|}{|x_{20}-a_{221}|}\\ &\text{for } \beta-2n_{21}=0.\end{aligned}\tag{3.22}$$

For $(x_1^*,x_2^*)=(a_{111},a_{121})$ with $\dot{x}_1=0$, the $[(2m_{11}+1),(2n_{11})]$-singular flows exist with $\dot{x}_2=a_{220}(a_{121}-a_{221})^{2n_{21}+1}$. For $(x_1^*,x_2^*)=(a_{111},a_{121})$, the 1-dimensional flow direction is determined by $\dot{x}_2\neq 0s$. For $(x_1^*,x_2^*)=(a_{111},a_{221})$ with $\dot{x}_1=\dot{x}_2=0$, the $((2m_{11}+1),(2n_{21}+1))$- singular equilibriums exist.

(A) For $a_{121}>a_{221}$,

$$\left\{\begin{array}{l}(a_{111},a_{121})\\(a_{111},a_{221})\end{array}\right\}=\left\{\begin{array}{c}\underbrace{([(2m_{11}+1),(2n_{11})]_{\mathrm{DI:II}}\mathrm{SO},\mathrm{pF})}_{[(2m_{11}+1),(2n_{11})]\text{-decreasing-to-increasing inflection source flow }(+)}\\\underbrace{((2m_{11}+1)\mathrm{SO},(2n_{21}+1)\mathrm{SO})}_{[(2m_{11}+1),(2n_{21}+1)]\text{-source}}\end{array}\right\},$$

for $a_{110}>0, a_{220}>0$;

$$\left\{\begin{array}{l}(a_{111},a_{121})\\(a_{111},a_{221})\end{array}\right\}=\left\{\begin{array}{c}\underbrace{([(2m_{11}+1),(2n_{11})]_{\mathrm{II:DI}}\mathrm{SI},\mathrm{pF})}_{[(2m_{11}+1),(2n_{11})]\text{-increasing-to-decreasing inflection sink flow }(+)}\\\underbrace{((2m_{11}+1)\mathrm{SI},(2n_{21}+1)\mathrm{SO})}_{[(2m_{11}+1),(2n_{21}+1)]\text{-saddle}}\end{array}\right\},$$

for $a_{110}<0, a_{220}>0$;

$$\left\{\begin{array}{l}(a_{111},a_{121})\\(a_{111},a_{221})\end{array}\right\}=\left\{\begin{array}{c}\underbrace{([(2m_{11}+1),(2n_{11})]_{\mathrm{II:DI}}\mathrm{SO},\mathrm{nF})}_{[(2m_{11}+1),(2n_{11})]\text{-increasing-to-decreasing inflection source flow }(-)}\\\underbrace{((2m_{11}+1)\mathrm{SO},(2n_{21}+1)\mathrm{SI})}_{[(2m_{11}+1),(2n_{21}+1)]\text{-saddle}}\end{array}\right\},$$

for $a_{110}>0, a_{220}<0$;

$$\left\{\begin{array}{l}(a_{111},a_{121})\\(a_{111},a_{221})\end{array}\right\}=\left\{\begin{array}{c}\underbrace{([(2m_{11}+1),(2n_{11})]_{\mathrm{DI:II}}\mathrm{SI},\mathrm{nF})}_{[(2m_{11}+1),(2n_{11})]\text{-decreasing-to-increasing inflection sink flow }(-)}\\\underbrace{((2m_{11}+1)\mathrm{SI},(2n_{21}+1)\mathrm{SI})}_{[(2m_{11}+1),(2n_{21}+1)]\text{ - sink}}\end{array}\right\},$$

for $a_{110}<0, a_{220}<0$. (3.23)

(B) For $a_{121}<a_{221}$,

$$\left\{\begin{array}{l}(a_{111},a_{221})\\(a_{111},a_{121})\end{array}\right\}=\left\{\begin{array}{c}\underbrace{((2m_{11}+1)\mathrm{SO},(2n_{21}+1)\mathrm{SO})}_{[(2m_{11}+1),(2n_{21}+1)]\text{-source}}\\\underbrace{([(2m_{11}+1),(2n_{11})]_{\mathrm{DI:II}}\mathrm{SO},\mathrm{nF})}_{[(2m_{11}+1),(2n_{11})]\text{-decreasing-to-increasing inflection source flow }(-)}\end{array}\right\},$$

for $a_{110}>0, a_{220}>0$;

$$\left\{\begin{array}{l}(a_{111},a_{221})\\(a_{111},a_{121})\end{array}\right\}=\left\{\begin{array}{c}\underbrace{((2m_{11}+1)\mathrm{SI},(2n_{21}+1)\mathrm{SO})}_{[(2m_{11}+1),(2n_{21}+1)]\text{-saddle}}\\\underbrace{([(2m_{11}+1),(2n_{11})]_{\mathrm{II:DI}}\mathrm{SI},\mathrm{nF})}_{[(2m_{11}+1),(2n_{11})]\text{-increasing-to-decreasing inflection sink flow }(-)}\end{array}\right\},$$

for $a_{110}<0, a_{220}>0$;

$$\left\{\begin{array}{l}(a_{111},a_{221})\\(a_{111},a_{121})\end{array}\right\}=\left\{\begin{array}{c}\underbrace{((2m_{11}+1)\mathrm{SO},(2n_{21}+1)\mathrm{SI})}_{[(2m_{11}+1),(2n_{21}+1)]\text{-saddle}}\\\underbrace{([(2m_{11}+1),(2n_{11})]_{\mathrm{II:DI}}\mathrm{SO},\mathrm{pF})}_{[(2m_{11}+1),(2n_{11})]\text{-increasing-to-decreasing inflection source flow }(+)}\end{array}\right\},$$

for $a_{110} > 0, a_{220} < 0$;

$$
\left\{ \begin{array}{l} (a_{111}, a_{221}) \\ (a_{111}, a_{121}) \end{array} \right\} = \left\{ \begin{array}{c} \underbrace{((2m_{11}+1)\text{SI}, (2n_{21}+1)\text{SI})}_{[(2m_{11}+1),(2n_{21}+1)]\text{-sink}} \\ \underbrace{([(2m_{11}+1), (2n_{11})]_{\text{DI:II}}\text{SI,pF})}_{[(2m_{11}+1),(2n_{11})]\text{-decreasing-to-increasing inflection sink flow }(+)} \end{array} \right\},
$$

for $a_{110} < 0, a_{220} < 0$. (3.24)

- The $([(2m_{11}+1), (2n_{11})]_{\text{DI:II}}\,\text{SO, pF})$ is for a positive singular decreasing-to-increasing inflection source flow.
- The $([(2m_{11}+1), (2n_{11})]_{\text{II:DI}}\,\text{SI, pF})$ is for a positive singular increasing-to-decreasing sink flow.
- The $([(2m_{11}+1), (2n_{11})]_{\text{II:DI}}\,\text{SO, nF})$ is for a negative singular increasing-to-decreasing inflection source flow.
- The $([(2m_{11}+1), (2n_{11})]_{\text{DI:II}}\,\text{SI, nF})$ is for a negative singular decreasing-to-increasing inflection source flow.

The acronyms "SO" and "SI" are for source and sink flows, respectively. "pF" and "nF" are for positive and negative flows, respectively.

- The $((2m_{11}+1)\text{SO}, (2n_{21}+1)\text{SO}]$ is for a $((2m_{11}+1), (2n_{21}+1)]$-singular source.
- The $((2m_{11}+1)\text{SI}, (2n_{21}+1)\text{SO}]$ is for a $((2m_{11}+1), (2n_{21}+1)]$-singular saddle.
- The $((2m_{11}+1)\text{SO}, (2n_{21}+1)\text{SI}]$ is for a $((2m_{11}+1), (2n_{21}+1)]$-singular saddle.
- The $((2m_{11}+1)\text{SI}, (2n_{21}+1)\text{SI}]$ is for a $((2m_{11}+1), (2n_{21}+1)]$-singular sink.

The singular hyperbolic and hyperbolic-secant flows with singular source, sink and saddles are illustrated in Figs. 3.7 and 3.8. Through such singular flows, the corresponding global bifurcations generate from a hyperbolic (hyperbolic-secant) flow to a network of singular hyperbolic and hyperbolic-secant flows. Through such singular equilibriums, the corresponding global bifurcations generate from one of source, sink and saddles to a network of singular source, sink and saddle equilibriums.

3.2.2 $[(2m_{11}+1), (2n_{11})]$ and $((2m_{11}+1), (2n_{21}+1))$-*Infinite-Equilibriums*

Consider a polynomial dynamical system for $a_{221} = a_{121}$ as

$$
\begin{aligned}
\dot{x}_2 &= a_{220}(x_2 - a_{221})^{2n_{21}+1}, \\
\dot{x}_1 &= a_{110}(x_1 - a_{111})^{2m_{11}+1}(x_2 - a_{121})^{2n_{11}}.
\end{aligned} \tag{3.25}
$$

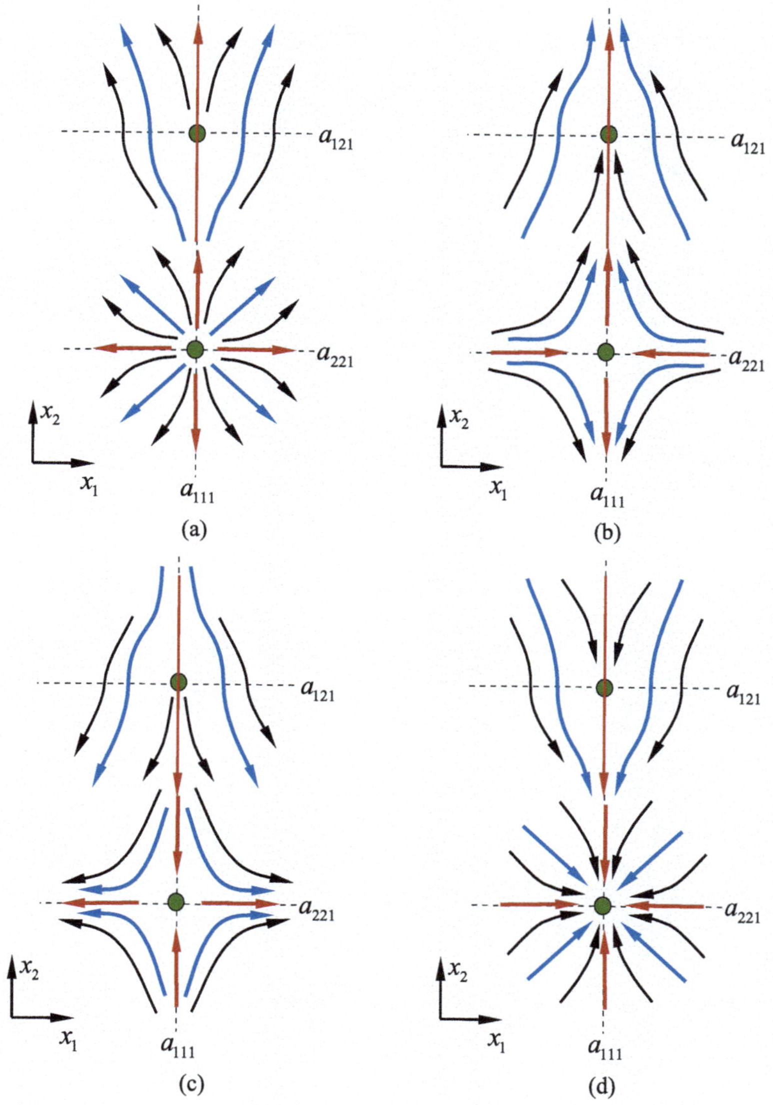

Fig. 3.7 Phase portraits $(a_{121} > a_{221})$ of the $([(2m_{11}+1), (2n_{11})]$, xF)-decreasing-to-increasing increasing-to-decreasing inflection-source and sink singular flows at $(x_1^*, x_2^*) = (a_{111}, a_{121})$ and $((2m_{11}+1), (2n_{21}+1))$-saddle, source and sink at $(x_1^*, x_2^*) = (a_{111}, a_{221})$ for 2-dimensional polynomial systems: **a** $(a_{110} > 0, a_{220} > 0)$, **b** $(a_{110} < 0, a_{220} > 0)$, **c** $(a_{110} > 0, a_{220} < 0)$, **d** $(a_{110} < 0, a_{220} < 0)$

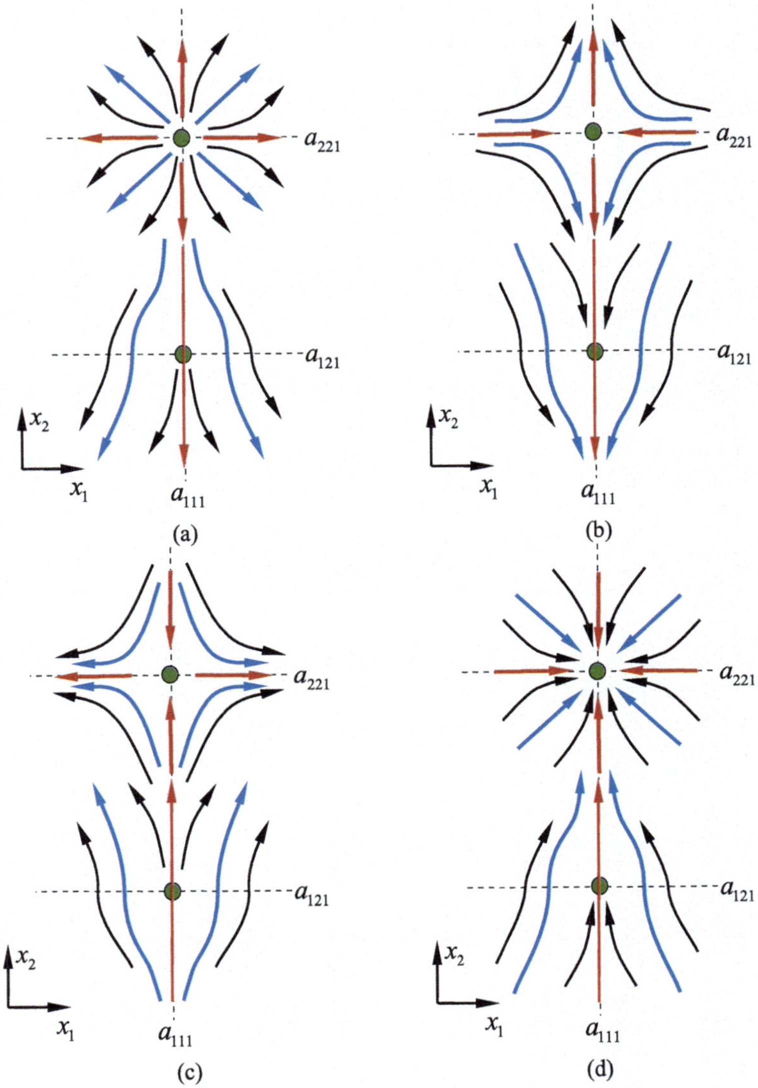

Fig. 3.8 Phase portraits ($a_{221} > a_{121}$) of the ($[(2m_{11}+1), (2n_{11})]$, xF)-decreasing-to-increasing increasing-to-decreasing inflection-source and sink singular flows at $(x_1^*, x_2^*) = (a_{111}, a_{121})$ and $((2m_{11}+1), (2n_{21}+1))$-saddle, source and sink at $(x_1^*, x_2^*) = (a_{111}, a_{221})$ for 2-dimensional polynomial systems: **a** ($a_{110} > 0, a_{220} > 0$), **b** ($a_{110} < 0, a_{220} > 0$), **c** ($a_{110} > 0, a_{220} < 0$), **d** ($a_{110} < 0, a_{220} < 0$)

The first integral manifold is

$$
\begin{aligned}
&-\frac{1}{2m_{11}}\left[\frac{1}{(x_1-a_{111})^{2m_{11}}}-\frac{1}{(x_{10}-a_{111})^{2m_{11}}}\right]\\
&=\frac{a_{110}}{a_{220}}\frac{1}{2(n_{11}-n_{21})}[(x_2-a_{121})^{2(n_{11}-n_{21})}-(x_2-a_{121})^{2(n_{11}-n_{21})}]\\
&\text{for } 2m_{11}+1\neq 1,\ 2(n_{11}-n_{21})\neq 0;\\
&-\frac{1}{2m_{11}}\left[\frac{1}{(x_1-a_{111})^{2m_{11}}}-\frac{1}{(x_{10}-a_{111})^{2m_{11}}}\right]=\ln\frac{|x_1-a_{111}|}{|x_{10}-a_{111}|}\\
&\text{for } 2m_{11}+1=1,\\
&\frac{1}{2(n_{11}-n_{21})}[(x_2-a_{121})^{2(n_{11}-n_{21})}-(x_2-a_{121})^{2(n_{11}-n_{21})}]=\ln\frac{|x_2-a_{121}|}{|x_{20}-a_{121}|}\\
&\text{for } 2(n_{11}-n_{21})=0.
\end{aligned}
\tag{3.26}
$$

(A) For $2n_{11}\geq 2n_{21}+1$, the corresponding infinite-equilibriums are given.

$$
\begin{aligned}
&(a_{111},a_{221})=\underbrace{([2(n_{11}-n_{21})-1]^{\text{th}}\text{DP:UP},\ [(2m_{11}+1),(2n_{21}+1)]^{\text{th}}\text{SO})}_{([2(n_{11}-n_{21})-1],[(2m_{11}+1),(2n_{21}+1)])\text{-hyperbolic source}},\\
&\text{for } a_{110}>0,\ a_{220}>0;\\
&(a_{111},a_{221})=\underbrace{([2(n_{11}-n_{21})-1]^{\text{th}}\text{UP:DP},\ [(2m_{11}+1),(2n_{21}+1)]^{\text{th}}\text{SO})}_{([2(n_{11}-n_{21})-1],[(2m_{11}+1),(2n_{21}+1)])\text{-hyperbolic-secant source}},\\
&\text{for } a_{110}<0,\ a_{220}>0;\\
&(a_{111},a_{221})=\underbrace{([2(n_{11}-n_{21})-1]^{\text{th}}\text{UP:DP},\ [(2m_{11}+1),(2n_{21}+1)]^{\text{th}}\text{SI})}_{([2(n_{11}-n_{21})-1],[(2m_{11}+1),(2n_{21}+1)])\text{-hyperbolic-secant sink}},\\
&\text{for } a_{110}>0,\ a_{220}<0;\\
&(a_{111},a_{221})=\underbrace{([2(n_{11}-n_{21})-1]^{\text{th}}\text{DP:UP},\ [(2m_{11}+1),(2n_{21}+1)]^{\text{th}}\text{SI})}_{([2(n_{11}-n_{21})-1],[(2m_{11}+1),(2n_{21}+1)])\text{-hyperbolic sink}},\\
&\text{for } a_{110}<0,\ a_{220}<0.
\end{aligned}
\tag{3.27}
$$

(B) For $2n_{11}<2n_{21}+1$, the infinite-equilibriums are given as follows.

$$
\begin{aligned}
&(a_{111},a_{221})=\underbrace{([2(n_{21}-n_{11})+1]^{\text{th}}\text{UA:DA},\ [(2m_{11}+1),(2n_{21}+1)]^{\text{th}}\text{SO})}_{([2(n_{21}-n_{11})+1],[(2m_{11}+1),(2n_{21}+1)])\text{-up-to-down asymptote-source}},\\
&\text{for } a_{110}>0,\ a_{220}>0;\\
&(a_{111},a_{221})=\underbrace{([2(n_{21}-n_{11})]^{\text{th}}\text{DA:UA},\ [(2m_{11}+1),(2n_{21}+1)]^{\text{th}}\text{SO})}_{([2(n_{21}-n_{11})],[(2m_{11}+1),(2n_{21}+1)])\text{-down-to-up asymptote-source}},\\
&\text{for } a_{110}<0,\ a_{220}>0;
\end{aligned}
$$

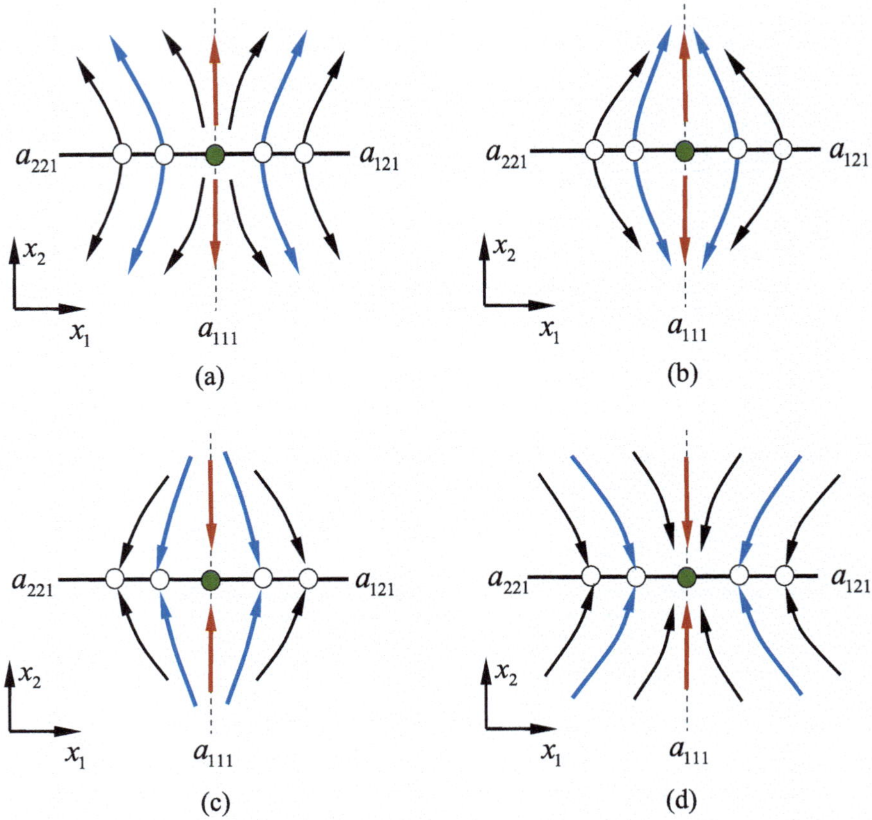

Fig. 3.9 The infinite-equilibriums ($a_{121} = a_{221}, 2n_{11} \geq 2n_{21} + 1, m_1 = 2m_{11} + 1$) for 2-dimensional polynomial systems: **a** ($a_{110} > 0$, $a_{220} > 0$), **b** ($a_{110} < 0$, $a_{220} > 0$), **c** ($a_{110} > 0$, $a_{220} < 0$), **d** ($a_{110} < 0$, $a_{220} < 0$)

$$\begin{aligned}
&(a_{111}, a_{221}) = \underbrace{([2(n_{21} - n_{11}){+}1]^{\text{th}}\text{DA:UA}, [(2m_{11}{+}1), (2n_{21}{+}1)]^{\text{th}}\text{SI})}_{([2(n_{21}-n_{11})+1],[(2m_{11}+1),(2n_{21}+1)])\text{-down-to-up asymptote-sink}},\\
&\text{for } a_{110} > 0, a_{220} < 0;\\
&(a_{111}, a_{221}) = \underbrace{([2(n_{21} - n_{11}){+}1]^{\text{th}}\text{UA:DA}, [(2m_{11}{+}1), (2n_{21}{+}1)]^{\text{th}}\text{SI})}_{([2(n_{21}-n_{11})+1],[(2m_{11}+1),(2n_{21}+1)])\text{-up-to-down asymptote-sink}},\\
&\text{for } a_{110} < 0, a_{220} < 0.
\end{aligned} \tag{3.28}$$

For $x_2^* = a_{121} = a_{221}$, the infinite-equilibrium exists. The infinite-equilibriums are illustrated in Figs. 3.9 and 3.10 for $2n_{11} \geq 2n_{21} + 1$ and $2n_{11} < 2n_{21} + 1$, respectively. For $2n_{11} \geq 2n_{21} + 1$, the singular decreasing-to-increasing and increasing-to-decreasing inflection-source and sink are presented. For $2n_{11} < 2n_{21} + 1$, the singular up-to-down and down-to-up asymptote-source and sink are presented.

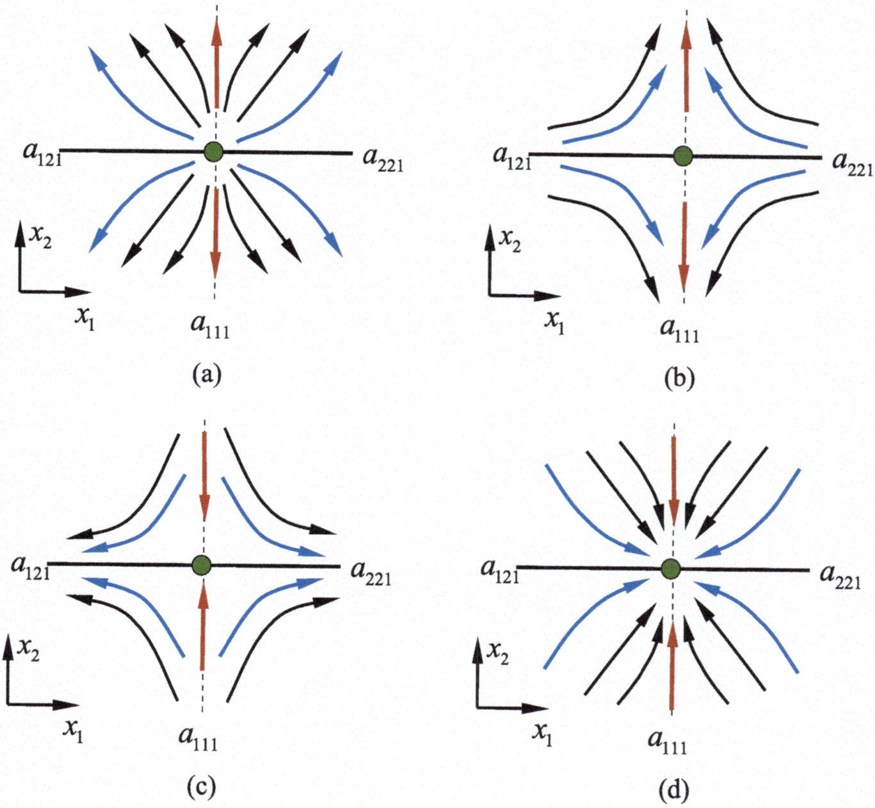

Fig. 3.10 The infinite-equilibriums ($a_{121} = a_{221}, 2n_{11} < 2n_{21}+1, m_1 = 2m_{11}+1$) for 2-dimensional polynomial systems: **a** ($a_{110} > 0, a_{220} > 0$), **b** ($a_{110} < 0, a_{220} > 0$), **c** ($a_{110} > 0, a_{220} < 0$), **d** ($a_{110} < 0, a_{220} < 0$)

3.2.3 $[(2m_{11}+1),(2n_{11})]$ and $((2m_{11}+1),(2n_{21}+1))$-Hybrid Networks

Consider

$$\begin{aligned}\dot{x}_2 &= a_{220}\prod_{s=1}^{2n_{21}+1}(x_2 - a_{22s}),\\ \dot{x}_1 &= a_{110}\prod_{r=1}^{2m_{11}}(x_1 - a_{11r})\prod_{u=1}^{2n_{11}}(x_2 - a_{12u}).\end{aligned} \tag{3.29}$$

The first integral manifold is

$$\sum_{r_1=1}^{2m_{11}} A_{r_1}\ln\frac{|x_1 - a_{11r_1}|}{|x_{10} - a_{11r_1}|} = \frac{a_{110}}{a_{220}}\Bigg\{\sum_{\beta_0=2n_{21}+1}^{2n_{11}} b_{12\beta_0}\sum_{\substack{\alpha_1\in[0,\beta_0-(2n_{21}+1)],\\ \beta_1\in[2n_{21},\beta_0-1]}}^{\alpha_1+\beta_1=\beta_0-1}\ \sum_{\substack{\alpha_2\in[0,\beta_1-(2n_{21}+1)]\\ \beta_2\in[2n_{21}-1,\beta_1-2]}}^{\alpha_2+\beta_2=\beta_1-2}\cdots$$

$$\sum_{\substack{\alpha_{2n_{21}}\in[0,\beta_{2n_{21}-1}-2n_{21}]\\ \beta_{2n_{21}}\in[1,\beta_{2n_{21}-1}-2n_{21}]}}^{\alpha_{2n_{21}}+\beta_{2n_{21}}=\beta_{2n_{21}-1}-2n_{21}} \prod_{j=1}^{2n_{21}} \frac{(\beta_{j-1}-j)!}{\alpha_j!\beta_j!}(a_{22s_{j+1}}-a_{22s_j})^{\alpha_j}$$

$$\times \frac{1}{\beta_{2n_{21}}}[(x_2-a_{22s_{2n_{21}+1}})^{\beta_{2n_{21}}}-(x_{20}-a_{22s_{2n_{21}+1}})^{\beta_{2n_{21}}}]$$

$$-\sum_{\beta_0=n_{21}+1}^{2n_{21}} b_{12\beta_0}\sum_{k=2}^{2n_{21}}\Bigg[\sum_{\substack{\alpha_1\in[0,\beta_0-2n_{21}-1),\\ \beta_1\in(2n_{21},\beta_0-1]}}^{\alpha_1+\beta_1=\beta_0-1}\ \sum_{\substack{\alpha_2\in[0,\beta_1-2n_{21}]\\ \beta_2\in(2n_{21}-1,\beta_1-2)}}^{\alpha_2+\beta_2=\beta_1-2}\cdots$$

$$\times\Bigg(\sum_{\substack{\alpha_{k-1}\in(\beta_{k-2}-2n_{21},\beta_{k-2}-k+1]\\ \beta_{k-1}\in[0,2n_{21}+1-k)}}^{\alpha_{k-1}+\beta_{k-1}=\beta_{k-2}-k+1}\prod_{j=1}^{k-1}\frac{(\beta_{j-1}-j)!}{\alpha_j!\beta_j!}(a_{22s_{j+1}}-a_{22s_j})^{\alpha_j}\sum_{\rho=k}^{2n_{21}+1}A_{\rho 1}^{(s_k)}\ln\frac{|x_{22}-a_{22s_\rho}|}{|x_{20}-a_{22r_\rho}|}$$

$$+\sum_{\substack{\alpha_{k-1}\in(\beta_{k-2}-2n_{21}-1,\beta_{k-2}-2n_{21}]\\ \beta_{k-1}\in[2n_{21}+1-k,2n_{21}-k+2)}}^{\alpha_{k-1}+\beta_{k-1}=\beta_{k-2}-k+1}\prod_{j=1}^{k-1}\frac{(\beta_{j-1}-j)!}{\alpha_j!\beta_j!}(a_{22s_{j+1}}-a_{22s_j})^{\alpha_j}$$

$$\times\sum_{\rho=k+1}^{2n_{21}+1}B_{\rho 1}^{(s_k)}\ln\frac{|x_2-a_{22s_\rho}|}{|x_{20}-a_{22s_\rho}|}\Bigg)\Bigg]+b_{j_1j_20}\sum_{\rho=1}^{2n_{21}+1}A_{\rho 1}^{(s_1)}\ln\frac{|x_2-a_{22s_\rho}|}{|x_{20}-a_{22s_\rho}|}$$

$$+\sum_{\beta_0=1}^{2n_{21}}b_{j_1j_2\beta_0}\sum_{\rho=2}^{2n_{21}+1}B_{\rho 1}^{(s_1)}\frac{|x_2-a_{22s_\rho}|}{|x_{20}-a_{22s_\rho}|}\Bigg\}; \tag{3.30}$$

where

$$A_{r_1}=\frac{1}{\prod_{r_2=1,r_2\neq r_1}^{2m_{11}+1}(a_{11r_1}-a_{11r_\sigma})};$$

$$b_{j_1j_2\beta_0}=\sum_{t_1,l_1=0}^{t_1+l_1=1}\cdots\sum_{t_q,l_q=0}^{t_q+l_q=1}\prod_{k=1}^{2n_{11}}(a_{22s_1}-a_{12u_k})^{t_k}\delta_{\beta_0}^{\Sigma_{r=1}^{2n_{21}}l_r};$$

$$\lim_{\beta_{2n_{21}}\to 0}\frac{1}{\beta_{2n_{21}}}[(x_2-a_{22s_{2n_{21}+1}})^{\beta_{2n_{21}}}-(x_{20}-a_{22s_{2n_{21}+1}})^{\beta_{2n_{21}}}]=\ln\frac{|x_2-a_{22s_{2n_{21}+1}}|}{|x_{20}-a_{22s_{2n_{21}+1}}|}; \tag{3.31}$$

$$A_{\rho 1}^{(s_k)}=\frac{1}{\prod_{\sigma=k,\sigma\neq\rho}^{2n_{21}+1}(a_{22s_\rho}-a_{22s_\sigma})},\ \text{for } \rho\in\{k,k+1,\cdots,2n_{21}+1\},$$

$$B_{\rho 1}^{(s_k)}=\frac{(a_{22s_\rho}-a_{22s_k})^{\beta_{k-1}-k}}{\prod_{\sigma=k+1,\sigma\neq\rho}^{2n_{21}+1}(a_{22s_\rho}-a_{22s_\sigma})},\ \text{for} \rho\in\{k+1,k+2,\cdots,2n_{21}+1\};$$

$$k\in\{1,2,\cdots,2n_{21}\}. \tag{3.32}$$

For all cases, the $(2m_{11}+1)\times(2n_{21}+1)$-simple equilibriums plus $(2m_{11}+1)\times(2n_{11})$-hyperbolic flows are based on the following hyperbolic and hyperbolic-secant-source and sink bifurcations for $2n_{11}\geq 2n_{21}+1$:

- $[2(n_{11}-n_{21})-1]^{\text{th}}$ DP:UP, $[(2m_{11}+1),(2n_{11}+1)]^{\text{th}}$SO)-hyperbolic source bifurcation for ($a_{110}>0, a_{220}>0$),
- $[2(n_{11}-n_{21})-1]^{\text{th}}$ UP:DP, $[(2m_{11}+1),(2n_{11}+1)]^{\text{th}}$SO)-hyperbolic-secant source bifurcation for ($a_{110}<0,\ _{220}>0$),
- $[2(n_{11}-n_{21})+1]^{\text{th}}$ UP:DP, $[(2m_{11}+1),(2n_{11}+1)]^{\text{th}}$SI)-hyperbolic-secant sink bifurcation for ($a_{110}>0, a_{220}<0$),
- $[2(n_{11}-n_{21})+1]^{\text{th}}$ DP:UP, $[(2m_{11}+1),(2n_{21}+1)]^{\text{th}}$SI)-hyperbolic sink bifurcation for ($a_{110}<0, a_{220}<0$);

or based on the following up-to-down/down-to-up asymptote-source and sink bifurcations for $2n_{11}<2n_{21}+1$:

- $[2(n_{21}-n_{11})-1]^{\text{th}}$ UA:DA, $[(2m_{11}+1),(2n_{21}+1)]^{\text{th}}$ SO)-up-to-down asymptote-source bifurcation for ($a_{110}>0, a_{220}>0$),
- $[2(n_{21}-n_{11})-1]^{\text{th}}$ DA:UA, $[(2m_{11}+1),(2n_{21}+1)]^{\text{th}}$ SO)-down-to-up asymptote-source bifurcation for ($a_{110}<0, a_{220}>0$),
- $[2(n_{21}-n_{11})-1]^{\text{th}}$ DA:UA, $[(2m_{11}+1),,2n_{21}+1)]^{\text{th}}$ SI)-down-to-up asymptote-sink bifurcation for ($a_{110}>0, a_{220}<0$),
- $[2(n_{21}-n_{11})-1]^{\text{th}}$ UA:DA, $[(2m_{11}+1),(2n_{21}+1)]^{\text{th}}$ SO)-up-to-down asymptote-sink bifurcation for ($a_{110}<0, a_{220}<0$).

Consider $2n_{11}=2n_{21}$ for illustrations, phase portraits for the 2-dimensional systems for the hybrid networks of hyperbolic and hyperbolic-secant flows with the simple equilibriums of the saddles, sinks and sources with $x_1^*=a_{111},a_{112},\ldots,a_{11(2m_{11}+1)}$ and $x_2^*=a_{121},a_{221},\ldots,a_{12(2n_{11}+1)},a_{22(2n_{21}+1)}$ are presented in Fig. 3.11 (a–d) for ($a_{110}>0, a_{220}>0$), ($a_{110}<0, a_{220}>0$), ($a_{110}>0, a_{220}<0$) and ($a_{110}<0, a_{220}<0$).

$$\left\{\begin{array}{cccc}(a_{111},a_{22(2n_{21}+1)}) & \cdots\ (a_{11(2m_{11})},a_{22(2n_{21}+1)}) & (a_{11(2m_{11}+1)},a_{22(2n_{21}+1)})\\ (a_{111},a_{22(2n_{21})}) & \cdots\ (a_{11(2m_{11})},a_{22(2n_{21})}) & (a_{11(2m_{11}+1)},a_{22(2n_{21})})\\ (a_{111},a_{12(2n_{11})}) & \cdots\ (a_{11(2m_{11})},a_{12(2n_{11}+1)}) & (a_{11(2m_{11}+1)},a_{12(2n_{11})})\\ \vdots & \vdots\quad\quad\vdots & \vdots\\ (a_{111},a_{221}) & \cdots\ (a_{11(2m_{11})},a_{221}) & (a_{11(2m_{11}+1)},a_{221})\\ (a_{111},a_{121}) & \cdots\ (a_{11(2m_{11})},a_{121}) & (a_{11(2m_{11}+1)},a_{121})\end{array}\right\}$$

$$
= \left\{ \begin{array}{cccc}
\underbrace{(\text{SO, SO})}_{\text{source}} & \cdots & \underbrace{(\text{SI, SO})}_{\text{saddle}} & \underbrace{(\text{SO, SO})}_{\text{source}} \\
\underbrace{(\text{SO, SI})}_{\text{saddle}} & \cdots & \underbrace{(\text{SI, SI})}_{\text{sink}} & \underbrace{(\text{SO, SI})}_{\text{saddle}} \\
\underbrace{(\text{DP:UP, pF})}_{\text{hyperbolic flow }(+)} & \cdots & \underbrace{(\text{UP:DP, pF})}_{\text{hyperbolic-secant flow }(+)} & \underbrace{(\text{DP:UP, pF})}_{\text{hyperbolic flow }(+)} \\
\vdots & \vdots & \vdots & \vdots \\
\underbrace{(\text{SI, SO})}_{\text{saddle}} & \cdots & \underbrace{(\text{SO, SO})}_{\text{source}} & \underbrace{(\text{SI, SO})}_{\text{saddle}} \\
\underbrace{(\text{DP:UP, nF})}_{\text{hyperbolic flow }(-)} & \cdots & \underbrace{(\text{UP:DP, nF})}_{\text{hyperbolic-secant flow }(-)} & \underbrace{(\text{DP:UP, nF})}_{\text{hyperbolic flow }(-)}
\end{array} \right\}_{(2n_{11}+2n_{21}+1)\times(2m_{11}+1)}
$$

for $a_{110} > 0, a_{220} > 0;$ (3.33)

$$
\left\{ \begin{array}{cccc}
(a_{111}, a_{22(2n_{21}+1)}) & \cdots & (a_{11(2m_{11})}, a_{22(2n_{21}+1)}) & (a_{11(2m_{11}+1)}, a_{22(2n_{21}+1)}) \\
(a_{111}, a_{22(2n_{21})}) & \cdots & (a_{11(2m_{11})}, a_{22(2n_{21})}) & (a_{11(2m_{11}+1)}, a_{22(2n_{21})}) \\
(a_{111}, a_{12(2n_{11})}) & \cdots & (a_{11(2m_{11})}, a_{12(2n_{11}+1)}) & (a_{11(2m_{11}+1)}, a_{12(2n_{11})}) \\
\vdots & \vdots & \vdots & \vdots \\
(a_{111}, a_{221}) & \cdots & (a_{11(2m_{11})}, a_{221}) & (a_{11(2m_{11}+1)}, a_{221}) \\
(a_{111}, a_{121}) & \cdots & (a_{11(2m_{11})}, a_{121}) & (a_{11(2m_{11}+1)}, a_{121})
\end{array} \right\}
$$

$$
= \left\{ \begin{array}{cccc}
\underbrace{(\text{SI, SO})}_{\text{saddle}} & \cdots & \underbrace{(\text{SO, SO})}_{\text{source}} & \underbrace{(\text{SI, SO})}_{\text{saddle}} \\
\underbrace{(\text{SI, SI})}_{\text{sink}} & \cdots & \underbrace{(\text{SO, SI})}_{\text{saddle}} & \underbrace{(\text{SI, SI})}_{\text{sink}} \\
\underbrace{(\text{UP:DP, pF})}_{\text{hyperbolic-secant flow }(+)} & \cdots & \underbrace{(\text{DP:UP, pF})}_{\text{hyperbolic flow }(+)} & \underbrace{(\text{UP:DP, pF})}_{\text{hyperbolic-secant flow }(+)} \\
\vdots & \vdots & \vdots & \vdots \\
\underbrace{(\text{SO, SO})}_{\text{source}} & \cdots & \underbrace{(\text{SI, SO})}_{\text{saddle}} & \underbrace{(\text{SO, SO})}_{\text{source}} \\
\underbrace{(\text{UP:DP, nF})}_{\text{hyperbolic-secant flow }(-)} & \cdots & \underbrace{(\text{DP:UP, nF})}_{\text{hyperbolic flow }(-)} & \underbrace{(\text{UP:DP, nF})}_{\text{hyperbolic-secant flow }(-)}
\end{array} \right\}_{(2n_{11}+2n_{21}+1)\times(2m_{11}+1)}
$$

for $a_{110} < 0, a_{220} > 0;$ (3.34)

$$
\left\{ \begin{array}{cccc}
(a_{111}, a_{22(2n_{21}+1)}) & \cdots & (a_{11(2m_{11})}, a_{22(2n_{21}+1)}) & (a_{11(2m_{11}+1)}, a_{22(2n_{21}+1)}) \\
(a_{111}, a_{22(2n_{21})}) & \cdots & (a_{11(2m_{11})}, a_{22(2n_{21})}) & (a_{11(2m_{11}+1)}, a_{22(2n_{21})}) \\
(a_{111}, a_{12(2n_{11})}) & \cdots & (a_{11(2m_{11})}, a_{12(2n_{11}+1)}) & (a_{11(2m_{11}+1)}, a_{12(2n_{11})}) \\
\vdots & \vdots & \vdots & \vdots \\
(a_{111}, a_{221}) & \cdots & (a_{11(2m_{11})}, a_{221}) & (a_{11(2m_{11}+1)}, a_{221}) \\
(a_{111}, a_{121}) & \cdots & (a_{11(2m_{11})}, a_{121}) & (a_{11(2m_{11}+1)}, a_{121})
\end{array} \right\}
$$

$$
= \left\{ \begin{array}{cccc}
\underbrace{(\text{SO, SI})}_{\text{saddle}} & \cdots & \underbrace{(\text{SI, SI})}_{\text{sink}} & \underbrace{(\text{SO, SI})}_{\text{saddle}} \\
\underbrace{(\text{SO, SO})}_{\text{source}} & \cdots & \underbrace{(\text{SI, SO})}_{\text{saddle}} & \underbrace{(\text{SO, SO})}_{\text{source}} \\
\underbrace{(\text{UP:DP, nF})}_{\text{hyperbolic-secant flow }(-)} & \cdots & \underbrace{(\text{DP:UP, nF})}_{\text{hyperbolic flow }(-)} & \underbrace{(\text{UP:DP, nF})}_{\text{hyperbolic-secant flow }(-)} \\
\vdots & \vdots & \vdots & \vdots \\
\underbrace{(\text{SI, SI})}_{\text{sink}} & \cdots & \underbrace{(\text{SO, SI})}_{\text{saddle}} & \underbrace{(\text{SI, SI})}_{\text{sink}} \\
\underbrace{(\text{UP:DP, pF})}_{\text{hyperbolic-secant flow }(+)} & \cdots & \underbrace{(\text{DP:UP, pF})}_{\text{hyperbolic flow }(+)} & \underbrace{(\text{UP:DP, pF})}_{\text{hyperbolic-secant flow }(+)}
\end{array} \right\}_{(2n_{11}+2n_{21}+1)\times(2m_{11}+1)}
$$

for $a_{110} > 0, a_{220} < 0$; (3.35)

$$
\left\{ \begin{array}{cccc}
(a_{111}, a_{22(2n_{21}+1)}) & \cdots & (a_{11(2m_{11})}, a_{22(2n_{21}+1)}) & (a_{11(2m_{11}+1)}, a_{22(2n_{21}+1)}) \\
(a_{111}, a_{22(2n_{21})}) & \cdots & (a_{11(2m_{11})}, a_{22(2n_{21})}) & (a_{11(2m_{11}+1)}, a_{22(2n_{21})}) \\
(a_{111}, a_{12(2n_{11})}) & \cdots & (a_{11(2m_{11})}, a_{12(2n_{11}+1)}) & (a_{11(2m_{11}+1)}, a_{12(2n_{11})}) \\
\vdots & \vdots & \vdots & \vdots \\
(a_{111}, a_{221}) & \cdots & (a_{11(2m_{11})}, a_{221}) & (a_{11(2m_{11}+1)}, a_{221}) \\
(a_{111}, a_{121}) & \cdots & (a_{11(2m_{11})}, a_{121}) & (a_{11(2m_{11}+1)}, a_{121})
\end{array} \right\}
$$

$$
= \left\{ \begin{array}{cccc}
\underbrace{(\text{SO, SO})}_{\text{source}} & \cdots & \underbrace{(\text{SO, SI})}_{\text{saddle}} & \underbrace{(\text{SI, SI})}_{\text{sink}} \\
\underbrace{(\text{SI, SO})}_{\text{saddle}} & \cdots & \underbrace{(\text{SO, SO})}_{\text{source}} & \underbrace{(\text{SI, SO})}_{\text{saddle}} \\
\underbrace{(\text{DP:UP, nF})}_{\text{hyperbolic flow }(-)} & \cdots & \underbrace{(\text{UP:DP, nF})}_{\text{hyperbolic-secant flow }(-)} & \underbrace{(\text{DP:UP, nF})}_{\text{hyperbolic flow }(-)} \\
\vdots & \vdots & \vdots & \vdots \\
\underbrace{(\text{SO, SI})}_{\text{saddle}} & \cdots & \underbrace{(\text{SI, SI})}_{\text{sink}} & \underbrace{(\text{SO, SI})}_{\text{saddle}} \\
\underbrace{(\text{DP:UP, pF})}_{\text{hyperbolic flow }(+)} & \cdots & \underbrace{(\text{UP:DP, pF})}_{\text{hyperbolic-secant flow }(+)} & \underbrace{(\text{DP:UP, pF})}_{\text{hyperbolic flow }(+)}
\end{array} \right\}_{(2n_{11}+2n_{21}+1)\times(2m_{11}+1)}
$$

for $a_{110} < 0, a_{220} < 0$. (3.36)

Consider $2n_{11} = 2n_{21}$ for illustrations, phase portraits for the 2-dimensional systems for the hybrid networks of hyperbolic and hyperbolic-secant flows with the simple equilibriums of the saddles, sinks and sources with $x_1^* = a_{111}, a_{112}, \ldots, a_{11(2m_1+1)}$ and $x_2^* = a_{221}, a_{121}, \ldots, a_{12(2n_{11})}, a_{22(2n_{21}+1)}$ are presented in Fig. 3.12 (a–d) for ($a_{110} > 0, a_{220} > 0$), ($a_{110} < 0, a_{220} > 0$), ($a_{110} > 0, a_{220} < 0$) and ($a_{110} < 0, a_{220} < 0$).

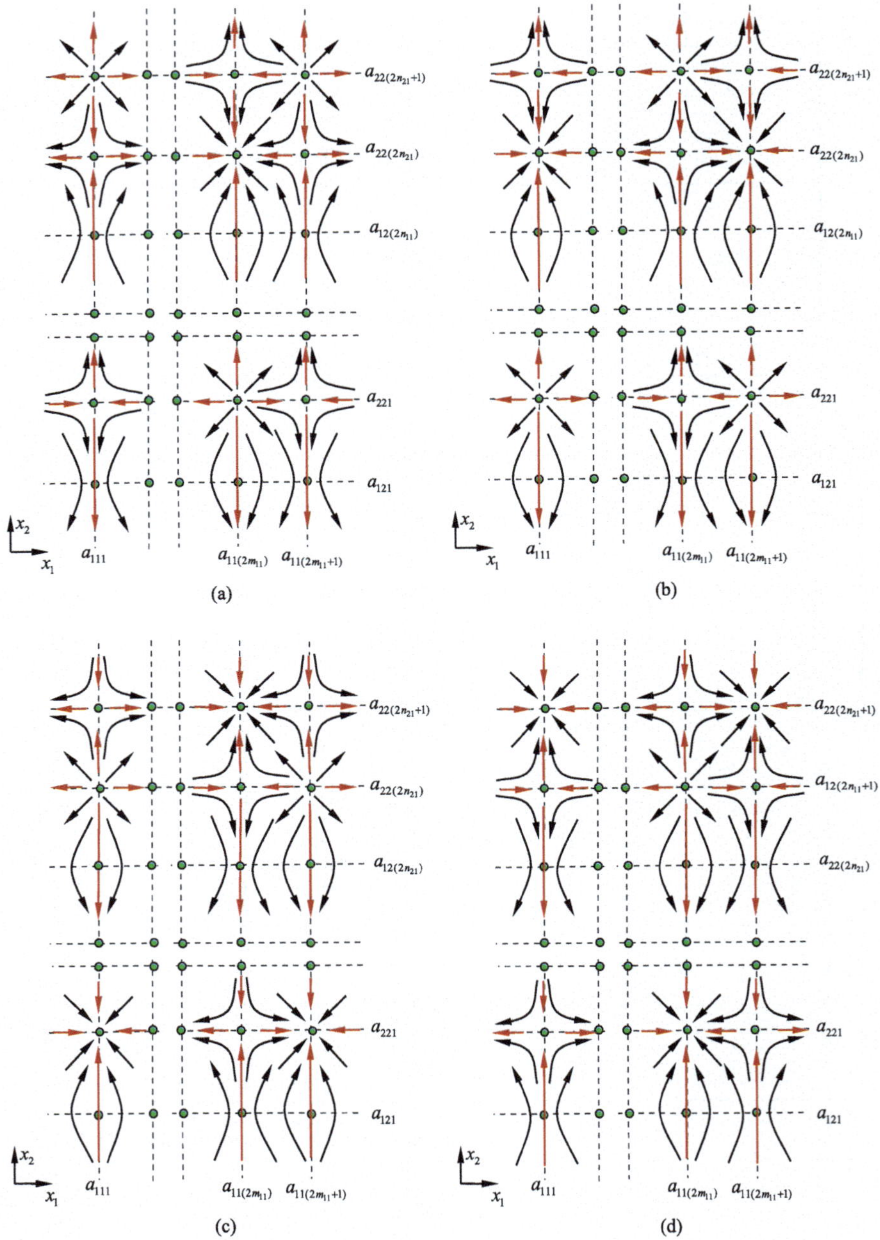

Fig. 3.11 Phase portraits for 2-dimensional systems with $x_1^* = a_{111}, a_{112}, \ldots, a_{11(2m_1+1)}$ and $x_2^* = a_{121}, a_{221}, \ldots, a_{12(2n_{11}+1)}, a_{22(2n_{21}+1)}$. The networks of flows with sink, source and saddle: **a** ($a_{110} > 0, a_{220} > 0$), **b** ($a_{110} < 0, a_{220} > 0$), **c** ($a_{110} > 0, a_{220} < 0$), **d** ($a_{110} < 0, a_{220} < 0$)

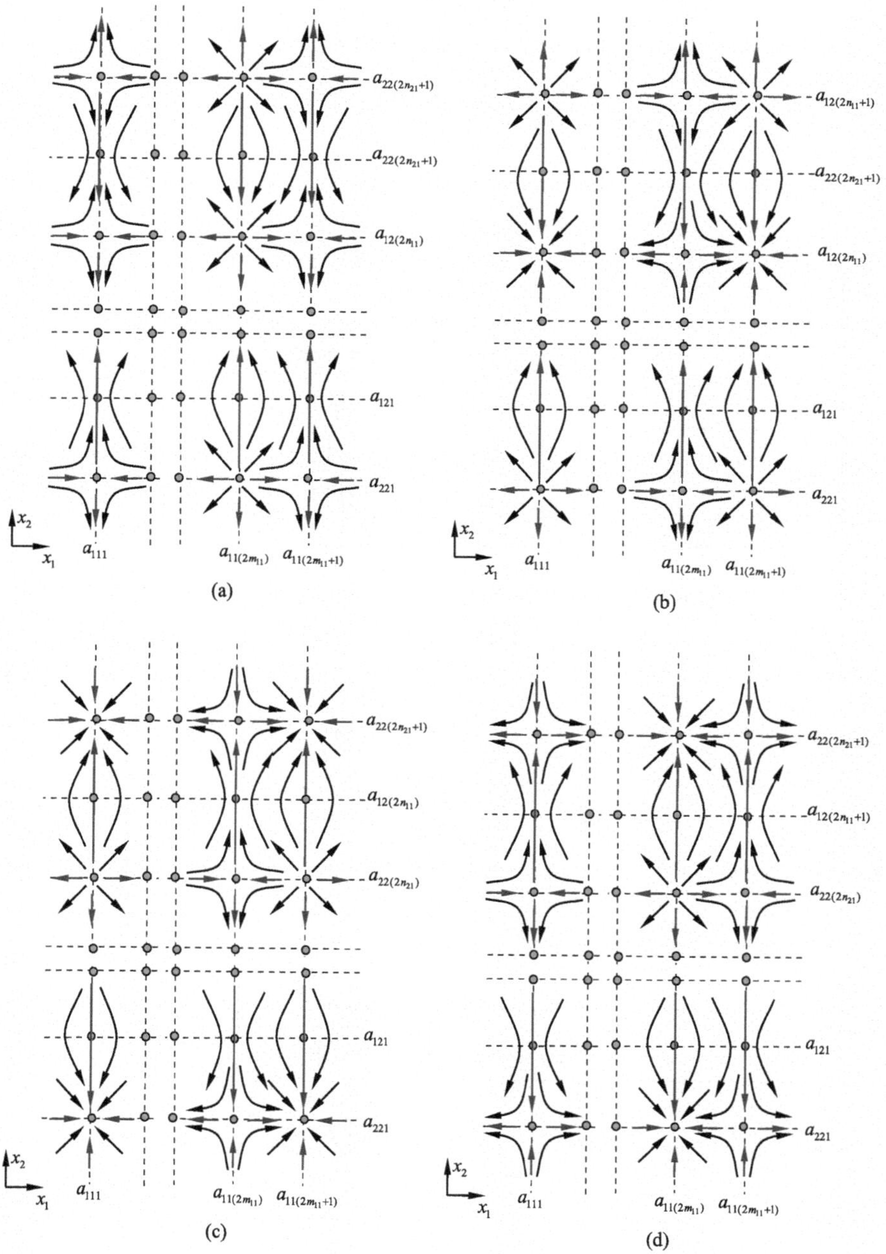

Fig. 3.12 Phase portraits for 2-dimensional systems with $x_1^* = a_{111}, a_{112}, \ldots, a_{11(2m_{11}+1)}$ and $x_2^* = a_{221}, a_{121}, \ldots, a_{12(2n_{11})}, a_{22(2n_{21}+1)}$. The networks of flows with sink, source and saddle: **a** $(a_{110} > 0, a_{220} > 0)$, **b** $(a_{110} < 0, a_{220} > 0)$, **c** $(a_{110} > 0, a_{220} < 0)$, **d** $(a_{110} < 0, a_{220} < 0)$

$$
\begin{Bmatrix}
(a_{111}, a_{22(2n_{21}+1)}) & \cdots & (a_{11(2m_{11})}, a_{22(2n_{21}+1)}) & (a_{11(2m_{11}+1)}, a_{22(2n_{21}+1)}) \\
(a_{111}, a_{12(2n_{11})}) & \cdots & (a_{11(2m_{11})}, a_{12(2n_{11})}) & (a_{11(2m_{11}+1)}, a_{12(2n_{11})}) \\
(a_{111}, a_{22(2n_{21})}) & \cdots & (a_{11(2m_{11})}, a_{22(2n_{21})}) & (a_{11(2m_{11}+1)}, a_{22(2n_{21})}) \\
\vdots & \vdots & \vdots & \vdots \\
(a_{111}, a_{121}) & \cdots & (a_{11(2m_{11})}, a_{121}) & (a_{11(2m_{11}+1)}, a_{121}) \\
(a_{111}, a_{221}) & \cdots & (a_{11(2m_{11})}, a_{221}) & (a_{11(2m_{11}+1)}, a_{221})
\end{Bmatrix}
$$

$$
= \begin{Bmatrix}
\underbrace{(\text{SO, SO})}_{\text{source}} & \cdots & \underbrace{(\text{SI, SO})}_{\text{saddle}} & \underbrace{(\text{SO, SO})}_{\text{source}} \\
\underbrace{(\text{UP:DP, nF})}_{\text{hyperbolic-secant flow }(-)} & \cdots & \underbrace{(\text{DP:UP, nF})}_{\text{hyperbolic flow }(-)} & \underbrace{(\text{UP:DP, nF})}_{\text{hyperbolic-secant flow }(-)} \\
\underbrace{(\text{SI, SI})}_{\text{sink}} & \cdots & \underbrace{(\text{SO, SI})}_{\text{saddle}} & \underbrace{(\text{SI,SI})}_{\text{sink}} \\
\vdots & \vdots & \vdots & \vdots \\
\underbrace{(\text{UP:DP, pF})}_{\text{hyperbolic-secant flow }(+)} & \cdots & \underbrace{(\text{DP:UP, pF})}_{\text{hyperbolic flow }(+)} & \underbrace{(\text{UP:DP, pF})}_{\text{hyperbolic-secant flow }(+)} \\
\underbrace{(\text{SO ,SO})}_{\text{source}} & \cdots & \underbrace{(\text{SI, SO})}_{\text{saddle}} & \underbrace{(\text{SO, SO})}_{\text{source}}
\end{Bmatrix}_{(2n_{11}+2n_{21}+1)\times(2m_{11}+1)}
$$

for $a_{110} > 0, a_{220} > 0$; (3.37)

$$
\begin{Bmatrix}
(a_{111}, a_{22(2n_{21}+1)}) & \cdots & (a_{11(2m_{11})}, a_{22(2n_{21}+1)}) & (a_{11(2m_{11}+1)}, a_{22(2n_{21}+1)}) \\
(a_{111}, a_{12(2n_{11})}) & \cdots & (a_{11(2m_{11})}, a_{12(2n_{11})}) & (a_{11(2m_{11}+1)}, a_{12(2n_{11})}) \\
(a_{111}, a_{22(2n_{21})}) & \cdots & (a_{11(2m_{11})}, a_{22(2n_{21})}) & (a_{11(2m_{11}+1)}, a_{22(2n_{21})}) \\
\vdots & \vdots & \vdots & \vdots \\
(a_{111}, a_{121}) & \cdots & (a_{11(2m_{11})}, a_{121}) & (a_{11(2m_{11}+1)}, a_{121}) \\
(a_{111}, a_{221}) & \cdots & (a_{11(2m_{11})}, a_{221}) & (a_{11(2m_{11}+1)}, a_{221})
\end{Bmatrix}
$$

$$
= \begin{Bmatrix}
\underbrace{(\text{SI, SO})}_{\text{saddle}} & \cdots & \underbrace{(\text{SO, SO})}_{\text{source}} & \underbrace{(\text{SI, SO})}_{\text{saddle}} \\
\underbrace{(\text{DP:UP, pF})}_{\text{hyperbolic flow }(-)} & \cdots & \underbrace{(\text{UP:DP, nF})}_{\text{hyperbolic-secant flow }(-)} & \underbrace{(\text{DP:UP, nF})}_{\text{hyperbolic flow }(-)} \\
\underbrace{(\text{SO, SI})}_{\text{saddle}} & \cdots & \underbrace{(\text{SI, SI})}_{\text{sink}} & \underbrace{(\text{SO, SI})}_{\text{saddle}} \\
\vdots & \vdots & \vdots & \vdots \\
\underbrace{(\text{DP:UP, pF})}_{\text{hyperbolic flow }(+)} & \cdots & \underbrace{(\text{UP:DP, pF})}_{\text{hyperbolic-secant flow }(+)} & \underbrace{(\text{DP:UP, pF})}_{\text{hyperbolic flow }(+)} \\
\underbrace{(\text{SI, SO})}_{\text{saddle}} & \cdots & \underbrace{(\text{SO, SO})}_{\text{source}} & \underbrace{(\text{SI, SO})}_{\text{saddle}}
\end{Bmatrix}_{(2n_{11}+2n_{21}+1)\times(2m_{11}+1)}
$$

for $a_{110} < 0, a_{220} > 0$; (3.38)

$$\left\{\begin{array}{cccc} (a_{111},a_{22(2n_{21}+1)}) & \cdots & (a_{11(2m_{11})},a_{22(2n_{21}+1)}) & (a_{11(2m_{11}+1)},a_{22(2n_{21}+1)}) \\ (a_{111},a_{12(2n_{11})}) & \cdots & (a_{11(2m_{11})},a_{12(2n_{11})}) & (a_{11(2m_{11}+1)},a_{12(2n_{11})}) \\ (a_{111},a_{22(2n_{21})}) & \cdots & (a_{11(2m_{11})},a_{22(2n_{21})}) & (a_{11(2m_{11}+1)},a_{22(2n_{21})}) \\ \vdots & \vdots & \vdots & \vdots \\ (a_{111},a_{121}) & \cdots & (a_{11(2m_{11})},a_{121}) & (a_{11(2m_{11}+1)},a_{121}) \\ (a_{111},a_{221}) & \cdots & (a_{11(2m_{11})},a_{221}) & (a_{11(2m_{11}+1)},a_{221}) \end{array}\right\}$$

$$= \left\{\begin{array}{cccc} \underbrace{(\mathrm{SO},\mathrm{SI})}_{\text{saddle}} & \cdots & \underbrace{(\mathrm{SI},\mathrm{SI})}_{\text{sink}} & \underbrace{(\mathrm{SO},\mathrm{SI})}_{\text{saddle}} \\ \underbrace{(\mathrm{DP{:}UP},\mathrm{pF})}_{\text{hyperbolic flow }(+)} & \cdots & \underbrace{(\mathrm{UP{:}DP},\mathrm{pF})}_{\text{hyperbolic-secant flow }(+)} & \underbrace{(\mathrm{DP{:}UP},\mathrm{pF})}_{\text{hyperbolic flow }(+)} \\ \underbrace{(\mathrm{SI},\mathrm{SO})}_{\text{saddle}} & \cdots & \underbrace{(\mathrm{SO},\mathrm{SO})}_{\text{source}} & \underbrace{(\mathrm{SI},\mathrm{SO})}_{\text{saddle}} \\ \vdots & \vdots & \vdots & \vdots \\ \underbrace{(\mathrm{DP{:}UP},\mathrm{nF})}_{\text{hyperbolic flow }(-)} & \cdots & \underbrace{(\mathrm{UP{:}DP},\mathrm{nF})}_{\text{hyperbolic-secant flow }(-)} & \underbrace{(\mathrm{DP{:}UP},\mathrm{nF})}_{\text{hyperbolic flow }(-)} \\ \underbrace{(\mathrm{SO},\mathrm{SI})}_{\text{saddle}} & \cdots & \underbrace{(\mathrm{SI},\mathrm{SI})}_{\text{sink}} & \underbrace{(\mathrm{SO},\mathrm{SI})}_{\text{saddle}} \end{array}\right\}_{(2n_{11}+2n_{21}+1)\times(2m_{11}+1)}$$

for $a_{110}>0,\ a_{220}<0;$ (3.39)

$$\left\{\begin{array}{cccc} (a_{111},a_{22(2n_{21}+1)}) & \cdots & (a_{11(2m_{11})},a_{22(2n_{21}+1)}) & (a_{11(2m_{11}+1)},a_{22(2n_{21}+1)}) \\ (a_{111},a_{12(2n_{11})}) & \cdots & (a_{11(2m_{11})},a_{12(2n_{11})}) & (a_{11(2m_{11}+1)},a_{12(2n_{11})}) \\ (a_{111},a_{22(2n_{21})}) & \cdots & (a_{11(2m_{11})},a_{22(2n_{21})}) & (a_{11(2m_{11}+1)},a_{22(2n_{21})}) \\ \vdots & \vdots & \vdots & \vdots \\ (a_{111},a_{121}) & \cdots & (a_{11(2m_{11})},a_{121}) & (a_{11(2m_{11}+1)},a_{121}) \\ (a_{111},a_{221}) & \cdots & (a_{11(2m_{11})},a_{221}) & (a_{11(2m_{11}+1)},a_{221}) \end{array}\right\}$$

$$= \left\{\begin{array}{cccc} \underbrace{(\mathrm{SI},\mathrm{SI})}_{\text{sink}} & \cdots & \underbrace{(\mathrm{SO},\mathrm{SI})}_{\text{saddle}} & \underbrace{(\mathrm{SI},\mathrm{SI})}_{\text{sink}} \\ \underbrace{(\mathrm{DP{:}UP},\mathrm{pF})}_{\text{hyperbolic flow }(+)} & \cdots & \underbrace{(\mathrm{DP{:}UP},\mathrm{pF})}_{\text{hyperbolic flow }(+)} & \underbrace{(\mathrm{UP{:}DP},\mathrm{pF})}_{\text{hyperbolic-secant flow }(+)} \\ \underbrace{(\mathrm{SO},\mathrm{SO})}_{\text{source}} & \cdots & \underbrace{(\mathrm{SI},\mathrm{SO})}_{\text{saddle}} & \underbrace{(\mathrm{SO},\mathrm{SO})}_{\text{source}} \\ \vdots & \vdots & \vdots & \vdots \\ \underbrace{(\mathrm{UP{:}DP},\mathrm{nF})}_{\text{hyperbolic-secant flow }(-)} & \cdots & \underbrace{(\mathrm{DP{:}UP},\mathrm{nF})}_{\text{hyperbolic flow }(-)} & \underbrace{(\mathrm{UP{:}DP},\mathrm{nF})}_{\text{hyperbolic-secant flow }(-)} \\ \underbrace{(\mathrm{SI},\mathrm{SI})}_{\text{sink}} & \cdots & \underbrace{(\mathrm{SO},\mathrm{SI})}_{\text{saddle}} & \underbrace{(\mathrm{SI},\mathrm{SI})}_{\text{sink}} \end{array}\right\}_{(2n_{11}+2n_{21}+1)\times(2m_{11}+1)}$$

for $a_{110}<0,\ a_{220}<0.$ (3.40)

3.3 $[(2m_{11}+1),(2n_{11}+1)]$ and $((2m_{11}+1),(2n_{21}))$-Singular Systems

In this section, there are four global bifurcations of the 1-diemsnional flows through singular hyperbolic flows, and four global bifurcations of equilibriums, through singular saddle-source and saddle-sink bifurcations. The corresponding global bifurcations generate networks of hyperbolic and hyperbolic-secant flows with source, sink, and saddles. The infinite-equilibriums are the appearing and switching bifurcations for the singular flows and singular equilibriums and for hybrid networks of hyperbolic flows and source, sink and saddles.

3.3.1 *$[(2m_{11}+1),(2n_{11}+1)]$-Flows and $((2m_{11}+1),(2n_{21}))$-Equilibriums*

Consider a polynomial dynamical system for $a_{221}\neq a_{121}$ as

$$\begin{aligned}\dot{x}_2&=a_{220}(x_2-a_{221})^{2n_{21}},\\ \dot{x}_1&=a_{110}(x_1-a_{111})^{2m_{11}+1}(x_2-a_{121})^{2n_{11}+1}.\end{aligned}\tag{3.41}$$

The first integral manifold is

$$\begin{aligned}&-\frac{1}{2m_{11}}\left[\frac{1}{(x_1-a_{111})^{2m_{11}}}-\frac{1}{(x_{10}-a_{111})^{2m_{11}}}\right]\\ &=\frac{a_{110}}{a_{220}}\sum_{\alpha,\beta=0}^{\alpha+\beta=2n_{11}+1}\frac{(2n_{11}+1)!}{\alpha!\beta!}\frac{(a_{221}-a_{121})^{\alpha}}{\beta-2n_{21}+1}\\ &\quad\times[(x_2-a_{221})^{\beta-2n_{21}+1}-(x_2-a_{221})^{\beta-2n_{21}+1}]\\ &\text{for } 2m_{11}+1\neq 1,\beta-2n_{21}+1\neq 0;\\ &-\frac{1}{2m_{11}}\left[\frac{1}{(x_1-a_{111})^{2m_{11}}}-\frac{1}{(x_{10}-a_{111})^{2m_{11}}}\right]=\ln\frac{|x_1-a_{111}|}{|x_{10}-a_{111}|}\\ &\text{for } 2m_{11}+1=1,\\ &\frac{1}{\beta-2n_{21}+1}[(x_2-a_{221})^{\beta-2n_{21}+1}-(x_2-a_{221})^{\beta-2n_{21}+1}]=\ln\frac{|x_2-a_{221}|}{|x_{20}-a_{221}|}\\ &\text{for } \beta-2n_{21}+1=0.\end{aligned}\tag{3.42}$$

For $(x_1^*,x_2^*)=(a_{111},a_{121})$ with $\dot{x}_1=0$, the $[(2m_{11}+1),(2n_{11}+1)]$-singular flows exist with $\dot{x}_2=a_{220}(a_{121}-a_{221})^{2n_{21}}$. For $(x_1^*,x_2^*)=(a_{111},a_{121})$, the 1-dimensional flow direction is determined by $\dot{x}_2\neq 0s$. For $(x_1^*,x_2^*)=(a_{111},a_{221})$ with $\dot{x}_1=\dot{x}_2=0$, the $((2m_{11}+1),(2n_{21}))$- singular equilibriums exist.

(A) For $a_{121} > a_{221}$,

$$\begin{Bmatrix} (a_{111}, a_{121}) \\ (a_{111}, a_{221}) \end{Bmatrix} = \left\{ \begin{array}{c} \underbrace{([(2m_{11}+1), (2n_{11}+1)]\text{DP:UP, pF})}_{[(2m_{11}+1),(2n_{11}+1)]\text{-hyperbolic flow }(+)} \\ \underbrace{((2m_{11}+1)\text{SI}, (2n_{21})\text{US})}_{[(2m_{11}+1),(2n_{21})]\text{-upper-saddle-sink}} \end{array} \right\},$$

for $a_{110} > 0, a_{220} > 0$;

$$\begin{Bmatrix} (a_{111}, a_{121}) \\ (a_{111}, a_{221}) \end{Bmatrix} = \left\{ \begin{array}{c} \underbrace{([(2m_{11}+1), (2n_{11}+1)]\text{UP:DP, pF})}_{[(2m_{11}+1),(2n_{11}+1)]\text{-hyperbolic-secant flow }(+)} \\ \underbrace{((2m_{11}+1)\text{SO}, (2n_{21})\text{US})}_{[(2m_{11}+1),(2n_{21})]\text{-upper-saddle-source}} \end{array} \right\},$$

for $a_{110} < 0, a_{220} > 0$;

$$\begin{Bmatrix} (a_{111}, a_{121}) \\ (a_{111}, a_{221}) \end{Bmatrix} = \left\{ \begin{array}{c} \underbrace{([(2m_{11}+1), (2n_{11}+1)]\text{UP:DP, nF})}_{[(2m_{11}+1),(2n_{11}+1)]\text{-hyperbolic-secant flow }(-)} \\ \underbrace{((2m_{11}+1)\text{SI},(2n_{21})\text{LS})}_{[(2m_{11}+1),(2n_{21})]\text{-lower-saddle-sink}} \end{array} \right\},$$

for $a_{110} > 0, a_{220} < 0$;

$$\begin{Bmatrix} (a_{111}, a_{121}) \\ (a_{111}, a_{221}) \end{Bmatrix} = \left\{ \begin{array}{c} \underbrace{([(2m_{11}+1), (2n_{11}+1)]\text{DP:UP, nF})}_{[(2m_{11}+1),(2n_{11}+1)]\text{-hyperbolic flow }(-)} \\ \underbrace{((2m_{11}+1)\text{SO},(2n_{21})\text{LS})}_{[(2m_{11}+1),(2n_{21})]\text{-lower-saddle-sink}} \end{array} \right\},$$

for $a_{110} < 0, a_{220} < 0$. (3.43)

(B) For $a_{121} < a_{221}$,

$$\begin{Bmatrix} (a_{111}, a_{221}) \\ (a_{111}, a_{121}) \end{Bmatrix} = \left\{ \begin{array}{c} \underbrace{((2m_{11}+1)\text{SO}, (2n_{21})\text{US})}_{[(2m_{11}+1),(2n_{21})]\text{-upper-saddle-source}} \\ \underbrace{([(2m_{11}+1), (2n_{11}+1)]\text{UP:DP, pF})}_{[(2m_{11}+1),(2n_{11}+1)]\text{-hyperbolic-secant flow }(+)} \end{array} \right\},$$

for $a_{110} > 0, a_{220} > 0$;

$$\begin{Bmatrix} (a_{111}, a_{221}) \\ (a_{111}, a_{121}) \end{Bmatrix} = \left\{ \begin{array}{c} \underbrace{((2m_{11}+1)\text{SI}, (2n_{21})\text{US})}_{[(2m_{11}+1),(2n_{21})]\text{-upper-saddle-sink}} \\ \underbrace{([(2m_{11}+1), (2n_{11}+1)]\text{DP:UP, pF})}_{[(2m_{11}+1),(2n_{11}+1)]\text{-hyperbolic flow }(+)} \end{array} \right\},$$

for $a_{110} < 0, a_{220} > 0$;

$$\left\{\begin{matrix}(a_{111}, a_{221})\\(a_{111}, a_{121})\end{matrix}\right\} = \left\{\begin{matrix}\underbrace{((2m_{11}+1)\text{SO}, (2n_{21})\text{LS})}_{[(2m_{11}+1),(2n_{21})]\text{-lower-saddle-source}}\\ \underbrace{([(2m_{11}+1), (2n_{11}+1)]\text{DP:UP, pF})}_{[(2m_{11}+1),(2n_{11}+1)]\text{-hyperbolic flow }(+)}\end{matrix}\right\},$$

for $a_{110} > 0, a_{220} < 0$;

$$\left\{\begin{matrix}(a_{111}, a_{221})\\(a_{111}, a_{121})\end{matrix}\right\} = \left\{\begin{matrix}\underbrace{((2m_{11}+1)\text{SI}, (2n_{21})\text{LS})}_{[(2m_{11}+1),(2n_{21})]\text{-lower-saddle-sink}}\\ \underbrace{([(2m_{11}+1), (2n_{11}+1)]\text{UP:DP, pF})}_{[(2m_{11}+1),(2n_{11}+1)]\text{-hyperbolic-secant flow }(+)}\end{matrix}\right\},$$

for $a_{110} < 0, a_{220} < 0.$ (3.44)

The $([(2m_{11}+1), (2n_{11}+1)]\text{DP:UP, pF})$ is for a positive singular hyperbolic flow. The $([(2m_{11}+1), (2n_{11}+1)]\text{UP:DP, pF})$ is for a positive singular hyperbolic-secant flow. The $([(2m_{11}+1), (2n_{11}+1)]\text{UP:DP, nF})$ is for a negative singular hyperbolic-secant flow. The $([(2m_{11}+1), (2n_{11}+1)]\text{DP:UP, nF})$ is for a negative singular hyperbolic flow. The acronyms "UP" and "DP" are for up-parabola and down-parabola flows, respectively. "pF" and "nF" are for positive and negative flows, respectively. DP:UP is for a hyperbolic flow, and UP:DP is for a hyperbolic-secant flow. The $((2m_{11}+1)\text{SO}, (2n_{21})\text{US}]$ is for a $((2m_{11}+1), (2n_{21})]$-upper-saddle-source. The $((2m_{11}+1)\text{SI}, (2n_{21})\text{US}]$ is for a $((2m_{11}+1), (2n_{21})]$-upper-saddle-sink. The $((2m_{11}+1)\text{SO}, (2n_{21})\text{LS}]$ is for a $((2m_{11}+1), (2n_{21})]$-lower-saddle-source. The $((2m_{11}+1)\text{SI}, (2n_{21})\text{LS}]$ is for a $((2m_{11}+1), (2n_{21})]$-lower-saddle-sink.

The singular hyperbolic and hyperbolic-secant flows with saddle-sources and saddle-sinks are illustrated in Figs. 3.13 and 3.14. Through such singular flows, the corresponding global bifurcations generate from a hyperbolic (hyperbolic-secant) flow to a network of singular hyperbolic and hyperbolic-secant flows. Through such singular equilibriums, the corresponding global bifurcations generate from one of source, sink and saddles to a network of singular source, sink and saddle equilibriums.

3.3.2 $[(2m_{11}+1), (2n_{11}+1)]$ and $((2m_{11}+1), (2n_{21}))$-*Infinite-Equilibriums*

Consider a polynomial dynamical system for $a_{221} = a_{121}$ as

$$\begin{aligned}\dot{x}_2 &= a_{220}(x_2 - a_{221})^{2n_{21}},\\ \dot{x}_1 &= a_{110}(x_1 - a_{111})^{2m_{11}+1}(x_2 - a_{121})^{2n_{11}+1}.\end{aligned} \tag{3.45}$$

The first integral manifold is

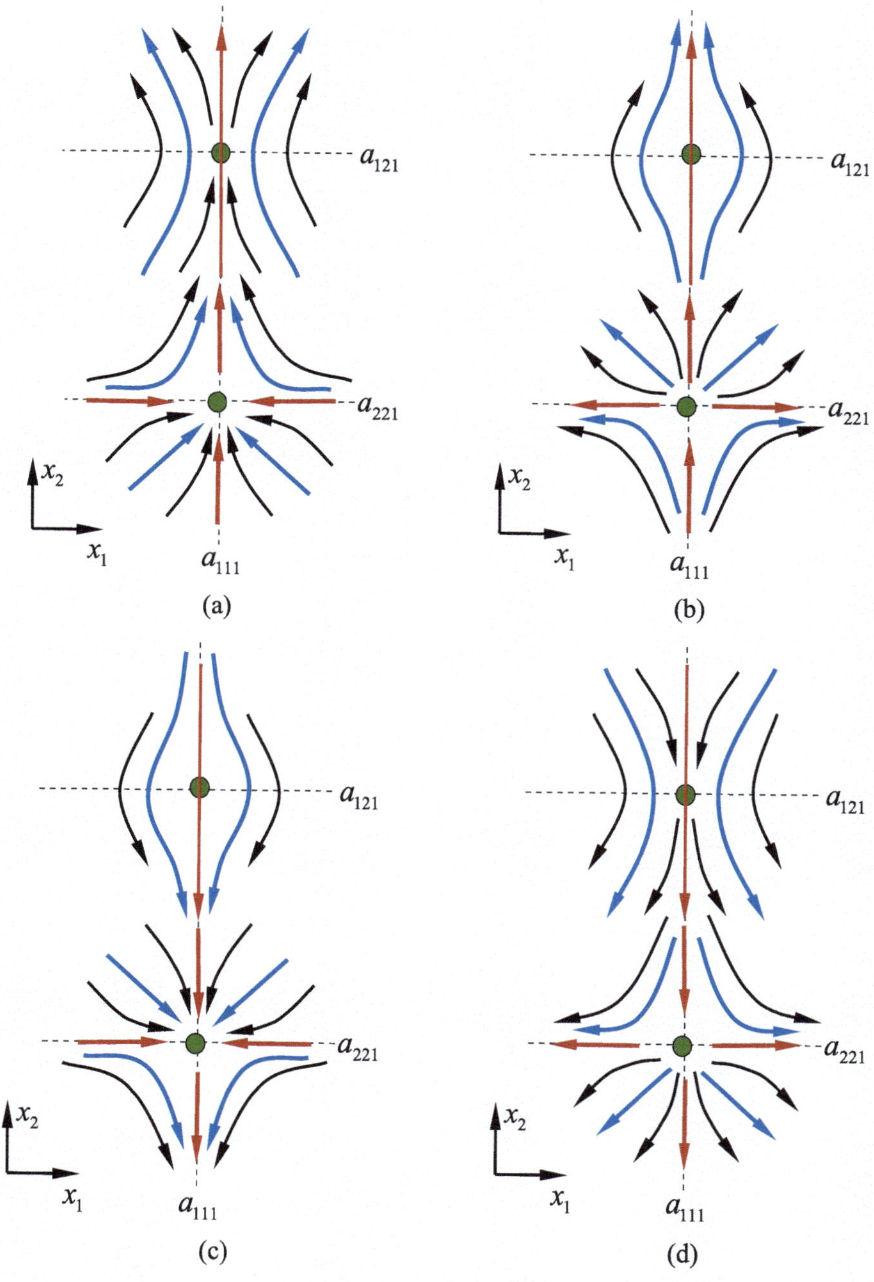

Fig. 3.13 Phase portraits ($a_{121} > a_{221}$) of the ([$(2m_{11}+1)$, $(2n_{11}+1)$], xF)-hyperbolic and hyperbolic-secant singular flows at $(x_1^*, x_2^*) = (a_{111}, a_{121})$ and ($(2m_{11}+1)$, $(2n_{21})$)-saddle-source and saddle-sink at $(x_1^*, x_2^*) = (a_{111}, a_{221})$ for 2-dimensional polynomial systems: **a** ($a_{110} > 0$, $a_{220} > 0$), **b** ($a_{110} < 0$, $a_{220} > 0$), **c** ($a_{110} > 0$, $a_{220} < 0$), **d** ($a_{110} < 0$, $a_{220} < 0$)

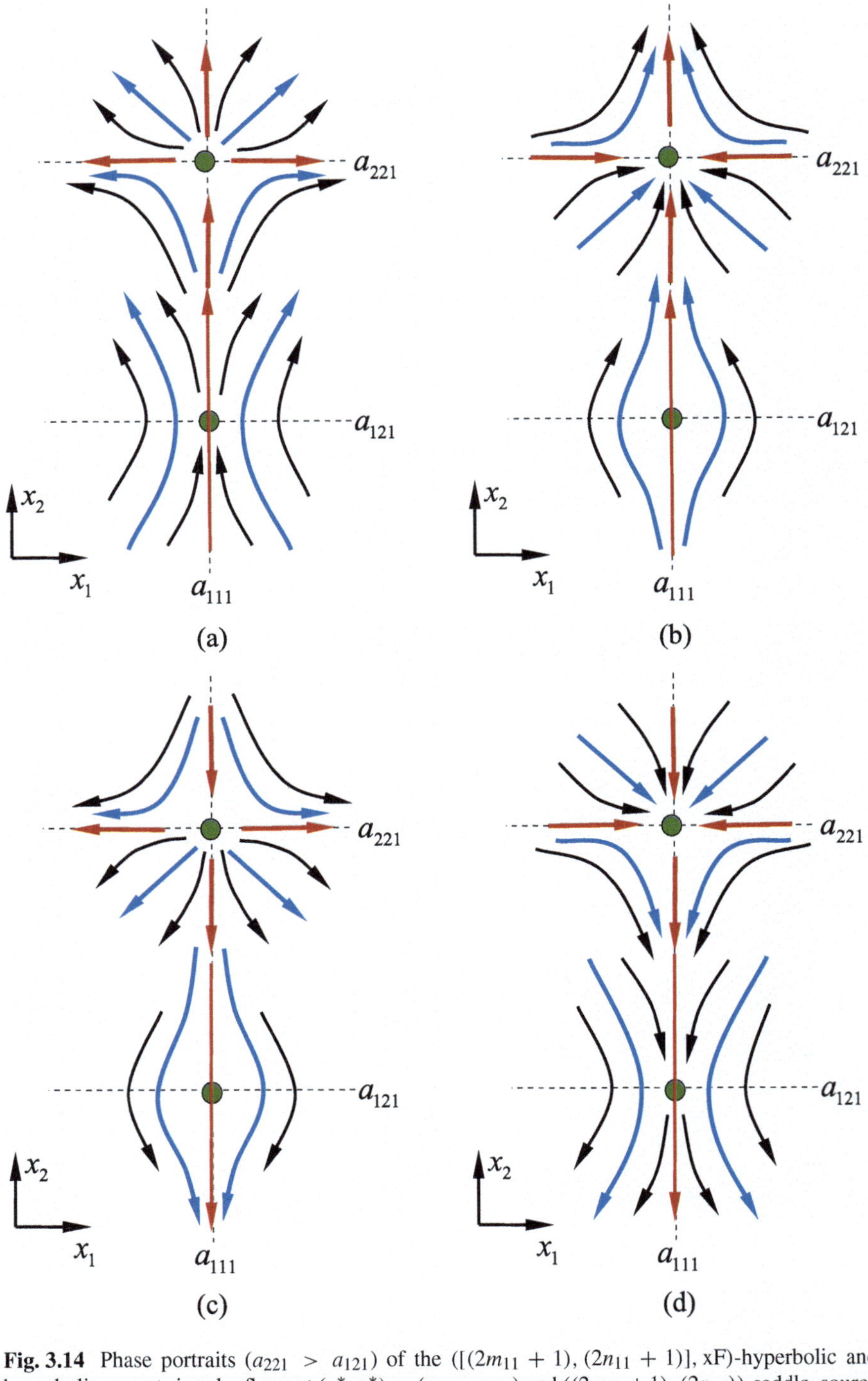

Fig. 3.14 Phase portraits ($a_{221} > a_{121}$) of the ([$(2m_{11}+1)$, $(2n_{11}+1)$], xF)-hyperbolic and hyperbolic-secant singular flows at $(x_1^*, x_2^*) = (a_{111}, a_{121})$ and $((2m_{11}+1), (2n_{21}))$-saddle, source and sink at $(x_1^*, x_2^*) = (a_{111}, a_{221})$ for 2-dimensional polynomial systems: **a** ($a_{110} > 0$, $a_{220} > 0$), **b** ($a_{110} < 0$, $a_{220} > 0$), **c** ($a_{110} > 0$, $a_{220} < 0$), **d** ($a_{110} < 0$, $a_{220} < 0$)

$$
\begin{aligned}
&-\frac{1}{2m_{11}}[\frac{1}{(x_1-a_{111})^{2m_{11}}}-\frac{1}{(x_{10}-a_{111})^{2m_{11}}}]\\
&=\frac{a_{110}}{a_{220}}\frac{1}{2(n_{11}-n_{21})+2}[(x_2-a_{121})^{2(n_{11}-n_{21})+2}-(x_2-a_{121})^{2(n_{11}-n_{21})+2}]\\
&\text{for } 2m_{11}+1\neq 1, 2(n_{11}-n_{21})+2\neq 0;\\
&-\frac{1}{2m_{11}}[\frac{1}{(x_1-a_{111})^{2m_{11}}}-\frac{1}{(x_{10}-a_{111})^{2m_{11}}}]=\ln\frac{|x_1-a_{111}|}{|x_{10}-a_{111}|}\\
&\text{for } 2m_{11}+1=1;\\
&\frac{1}{2(n_{11}-n_{21})+2}[(x_2-a_{221})^{2(n_{11}-n_{21})+2}-(x_2-a_{221})^{2(n_{11}-n_{21})+2}]=\ln\frac{|x_2-a_{221}|}{|x_{10}-a_{221}|}\\
&\text{for } 2(n_{11}-n_{21})+2=0.
\end{aligned}
\tag{3.46}
$$

(A) For $2n_{11}+1\geq 2n_{21}$, the corresponding infinite-equilibriums are given.

$$
\begin{aligned}
&(a_{111},a_{221})=\underbrace{([2(n_{11}-n_{21})+1]^{\text{th}}\text{DP:UP},[(2m_{11}+1),(2n_{21})]^{\text{th}}\text{US})}_{([2(n_{11}-n_{21})+1],[(2m_{11}+1),(2n_{21})])\text{-hyperbolic upper-saddle}},\\
&\text{for } a_{110}>0, a_{220}>0;\\
&(a_{111},a_{221})=\underbrace{([2(n_{11}-n_{21})+1]^{\text{th}}\text{UP:DP},[(2m_{11}+1),(2n_{21})]^{\text{th}}\text{US})}_{([2(n_{11}-n_{21})+1],[(2m_{11}+1),(2n_{21}+1)])\text{-hyperbolic-secant upper-saddle}},\\
&\text{for } a_{110}<0, a_{220}>0;\\
&(a_{111},a_{221})=\underbrace{([2(n_{11}-n_{21})+1]^{\text{th}}\text{UP:DP},[(2m_{11}+1),(2n_{21})]^{\text{th}}\text{LS})}_{([2(n_{11}-n_{21})+1],[(2m_{11}+1),(2n_{21}+1)])\text{-hyperbolic-secant lower-saddle}},\\
&\text{for } a_{110}>0, a_{220}<0;\\
&(a_{111},a_{221})=\underbrace{([2(n_{11}-n_{21})+1]^{\text{th}}\text{DP:UP},[(2m_{11}+1),(2n_{21})]^{\text{th}}\text{LS})}_{([2(n_{11}-n_{21})+1],[(2m_{11}+1),(2n_{21}+1)])\text{-hyperbolic lower-saddle}},\\
&\text{for } a_{110}<0, a_{220}<0.
\end{aligned}
\tag{3.47}
$$

(B) For $2n_{11}+1<2n_{21}+1$, the infinite-equilibriums are given as follows.

$$
\begin{aligned}
&(a_{111},a_{221})=\underbrace{([2(n_{21}-n_{11})-1]^{\text{th}}\text{UA:DA},[(2m_{11}+1),(2n_{21})]^{\text{th}}\text{US})}_{([2(n_{21}-n_{11})-1],[(2m_{11}+1),(2n_{21}+1)])\text{-up-to-down asymptote upper-saddle}},\\
&\text{for } a_{110}>0, a_{220}>0;\\
&(a_{111},a_{221})=\underbrace{([2(n_{21}-n_{11})-1]^{\text{th}}\text{DA:UA},[(2m_{11}+1),(2n_{21})]^{\text{th}}\text{US})}_{([2(n_{21}-n_{11})-1],[(2m_{11}+1),(2n_{21}+1)])\text{-down-to-up asymptote upper-saddle}},\\
&\text{for } a_{110}<0, a_{220}>0;\\
&(a_{111},a_{221})=\underbrace{([2(n_{21}-n_{11})-1]^{\text{th}}\text{DA:UA},[(2m_{11}+1),(2n_{21})]^{\text{th}}\text{LS})}_{([2(n_{21}-n_{11})-1],[(2m_{11}+1),(2n_{21}+1)])\text{-down-to-up asymptote lower-saddle}},
\end{aligned}
$$

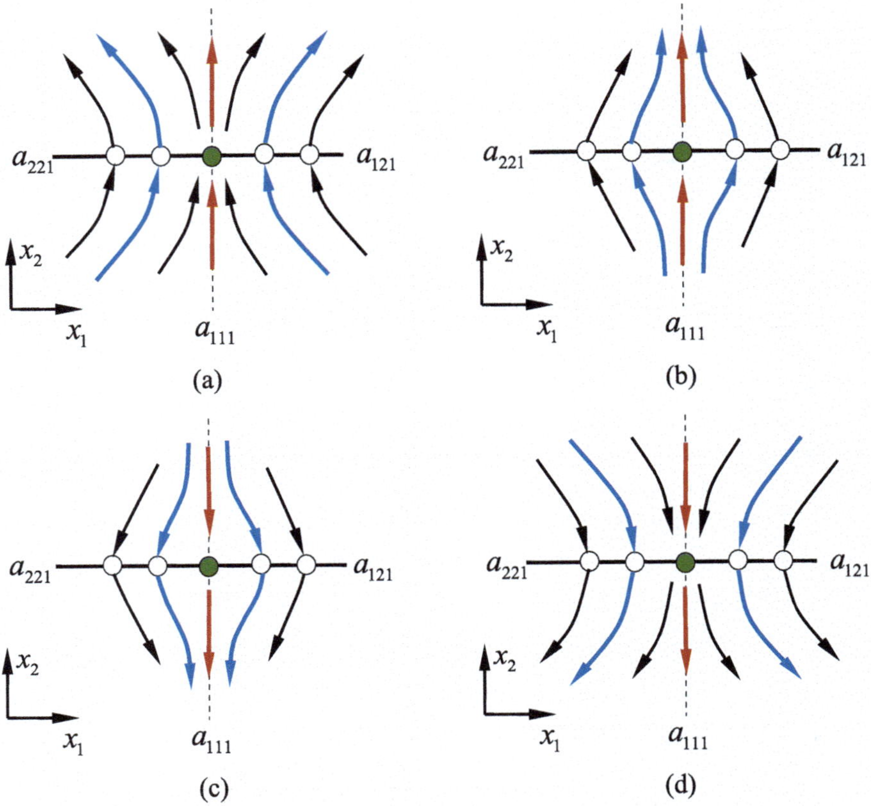

Fig. 3.15 The infinite-equilibriums ($a_{121} = a_{221}, 2n_{11} + 1 \geq 2n_{21}, m_1 = 2m_{11} + 1$) for 2-dimensional polynomial systems: **a** ($a_{110} > 0, a_{220} > 0$), **b** ($a_{110} < 0, a_{220} > 0$), **c** ($a_{110} > 0, a_{220} < 0$), **d** ($a_{110} < 0, a_{220} < 0$)

$$\begin{aligned}
&\text{for } a_{110} > 0, a_{220} < 0; \\
&(a_{111}, a_{221}) = \underbrace{([2(n_{21} - n_{11}) - 1]^{\text{th}}\text{UA:DA},[(2m_{11}+1), (2n_{21})]^{\text{th}}\text{LS})}_{([2(n_{21}-n_{11})-1],[(2m_{11}+1),(2n_{21}+1)])\text{-up-to-down asymptote lower-saddle}}, \\
&\text{for } a_{110} < 0, a_{220} < 0.
\end{aligned} \tag{3.48}$$

For $x_2^* = a_{121} = a_{221}$, the infinite-equilibrium exists. The infinite-equilibriums are illustrated in Fig. 3.15 and 3.16 for $2n_{11} + 1 \geq 2n_{21}$ and $2n_{11} + 1 < 2n_{21}$, respectively. For $2n_{11} + 1 \geq 2n_{21}$, the singular decreasing-to-increasing and increasing-to-decreasing inflection-source and sink are presented. For $2n_{11} + 1 < 2n_{21}$, the singular decreasing-to-increasing and increasing-to-decreasing concave-source and sink are presented.

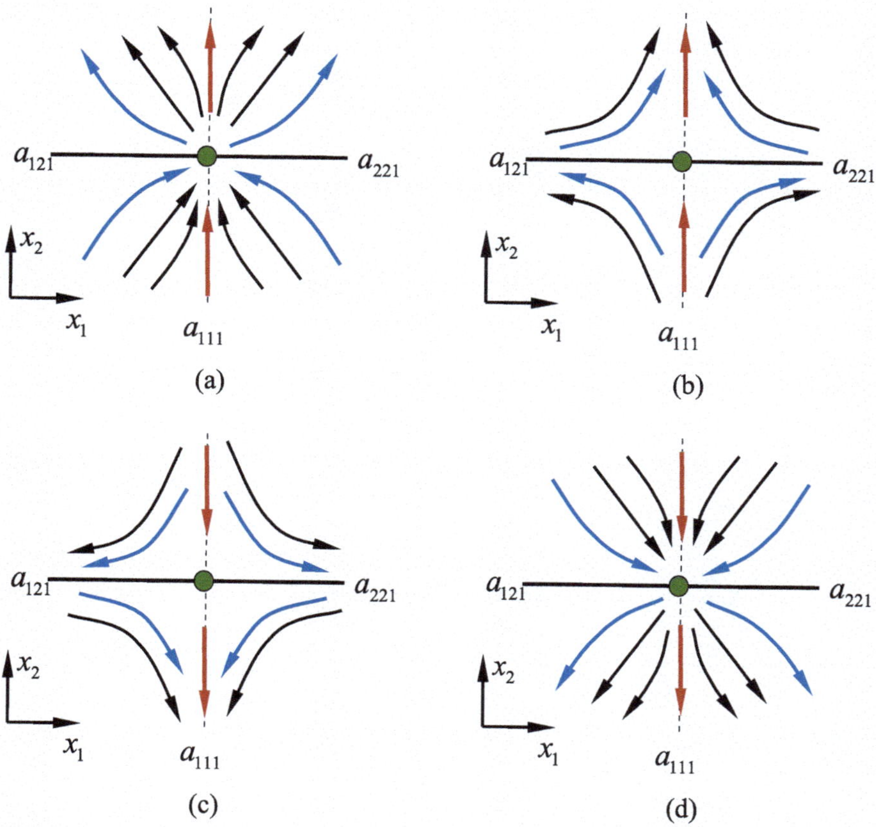

Fig. 3.16 The infinite-equilibriums ($a_{121} = a_{221}, 2n_{11} + 1 < 2n_{21}, m_1 = 2m_{11} + 1$) for 2-dimensional polynomial systems: **a** ($a_{110} > 0, a_{220} > 0$), **b** ($a_{110} < 0, a_{220} > 0$), **c** ($a_{110} > 0, a_{220} < 0$), **d** ($a_{110} < 0, a_{220} < 0$)

3.3.3 [(2m_{11} + 1), (2n_{11} + 1)] and ((2m_{11} + 1), (2n_{21}))-Hybrid Networks

Consider

$$\dot{x}_2 = a_{220} \prod_{s=1}^{2n_{21}} (x_2 - a_{22s}),$$
$$\dot{x}_1 = a_{110} \prod_{r=1}^{2m_{11}+1} (x_1 - a_{11r}) \prod_{u=1}^{2n_{11}+1} (x_2 - a_{12u}). \quad (3.49)$$

The first integral manifold is

$$
\begin{aligned}
&\sum_{r_1=1}^{2m_{11}+1} A_{r_1} \ln \frac{|x_1 - a_{11r_1}|}{|x_{10} - a_{11r_1}|} = \frac{a_{110}}{a_{220}} \Bigg\{ \sum_{\beta_0=2n_{21}}^{2n_{11}+1} b_{12\beta_0} \sum_{\substack{\alpha_1 \in [0,\beta_0-2n_{21}],\\ \beta_1\in[2n_{21}-1,\beta_0-1]}}^{\alpha_1+\beta_1=\beta_0-1} \sum_{\substack{\alpha_2\in[0,\beta_1-2n_{21}]\\ \beta_2\in[2n_{21}-2,\beta_1-2]}}^{\alpha_2+\beta_2=\beta_1-2} \cdots \\
&\sum_{\substack{\alpha_{2n_{21}-1}\in[0,\beta_{2n_{21}-2}-2n_{21}]\\ \beta_{2n_{21}-1}\in[1,\beta_{2n_{21}-2}-2n_{21}+1]}}^{\alpha_{2n_{21}-1}+\beta_{2n_{21}-1}=\beta_{2n_{21}-1}-2n_{21}+1} \prod_{j=1}^{2n_{21}-1} \frac{(\beta_{j-1}-j)!}{\alpha_j!\beta_j!}(a_{22s_{j+1}} - a_{22s_j})^{\alpha_j} \\
&\times \frac{1}{\beta_{2n_{21}-1}}[(x_2 - a_{22s_{2n_{21}}})^{\beta_{2n_{21}-1}} - (x_{20} - a_{22s_{2n_{21}}})^{\beta_{2n_{21}-1}}] \\
&- \sum_{\beta_0=2n_{21}}^{2n_{11}+1} b_{12\beta_0} \sum_{k=2}^{2n_{21}-1} \Bigg[\sum_{\substack{\alpha_1\in[0,\beta_0-2n_{21}),\\ \beta_1\in(2n_{21}-1,\beta_0-1]}}^{\alpha_1+\beta_1=\beta_0-1} \sum_{\substack{\alpha_2\in[0,\beta_1-2n_{21}]\\ \beta_2\in(2n_{21}-2,\beta_1-2)}}^{\alpha_2+\beta_2=\beta_1-2} \cdots \\
&\times \Bigg(\sum_{\substack{\alpha_{k-1}\in(\beta_{k-2}-2n_{21},\beta_{k-2}-k+1]\\ \beta_{k-1}\in[0,2n_{21}+1-k)}}^{\alpha_{k-1}+\beta_{k-1}=\beta_{k-2}-k+1} \prod_{j=1}^{k-1} \frac{(\beta_{j-1}-j)!}{\alpha_j!\beta_j!}(a_{22s_{j+1}} - a_{22s_j})^{\alpha_j} \sum_{\rho=k}^{2n_{21}} A_{\rho 1}^{(s_k)} \ln \frac{|x_{22} - a_{22s_\rho}|}{|x_{20} - a_{22r_\rho}|} \\
&+ \sum_{\substack{\alpha_{k-1}\in(\beta_{k-2}-2n_{21},\beta_{k-2}-2n_{21}+1]\\ \beta_{k-1}\in[(2n_{21}-k),2n_{21}-k+1)}}^{\alpha_{k-1}+\beta_{k-1}=\beta_{k-2}-k+1} \prod_{j=1}^{k-1} \frac{(\beta_{j-1}-j)!}{\alpha_j!\beta_j!}(a_{22s_{j+1}} - a_{22s_j})^{\alpha_j} \\
&\times \sum_{\rho=k+1}^{2n_{21}} B_{\rho 1}^{(s_k)} \ln \frac{|x_2 - a_{22s_\rho}|}{|x_{20} - a_{22s_\rho}|} \Bigg) \Bigg] + b_{j_1 j_2 0} \sum_{\rho=1}^{2n_{21}} A_{\rho 1}^{(s_1)} \ln \frac{|x_2 - a_{22s_\rho}|}{|x_{20} - a_{22s_\rho}|} \\
&+ \sum_{\beta_0=1}^{2n_{21}-1} b_{j_1 j_2 \beta_0} \sum_{\rho=2}^{2n_{21}} B_{\rho 1}^{(s_1)} \frac{|x_2 - a_{22s_\rho}|}{|x_{20} - a_{22s_\rho}|} \Bigg\};
\end{aligned}
\tag{3.50}
$$

where

$$
\begin{aligned}
A_{r_1} &= \frac{1}{\prod_{r_2=1, r_2\neq r_1}^{2m_{11}+1} (a_{11r_1} - a_{11r_\sigma})}; \\
b_{j_1 j_2 \beta_0} &= \sum_{t_1, l_1=0}^{t_1+l_1=1} \cdots \sum_{t_q, l_q=0}^{t_q+l_q=1} \prod_{k=1}^{2n_{11}+1} (a_{22s_1} - a_{12u_k})^{t_k} \delta_{\beta_0}^{\Sigma_{r=1}^{2n_{21}} l_r}; \\
&\lim_{\beta_{2n_{21}-1}\to 0} \frac{1}{\beta_{2n_{21}-1}}[(x_2 - a_{22s_{2n_{21}}})^{\beta_{2n_{21}-1}} - (x_{20} - a_{22s_{2n_{21}}})^{\beta_{2n_{21}-1}}] = \ln \frac{|x_2 - a_{22s_{2n_{21}}}|}{|x_{20} - a_{22s_{2n_{21}}}|};
\end{aligned}
\tag{3.51}
$$

$$
\begin{aligned}
A_{\rho 1}^{(s_k)} &= \frac{1}{\prod_{\sigma=k,\sigma\neq\rho}^{2n_{21}+1}(a_{22s_\rho}-a_{22s_\sigma})}, \text{ for } \rho\in\{k,k+1,\cdots,2n_{21}\},\\
B_{\rho 1}^{(s_k)} &= \frac{(a_{22s_\rho}-a_{22s_k})^{\beta_{k-1}-k}}{\prod_{\sigma=k+1,\sigma\neq\rho}^{2n_{21}+1}(a_{22s_\rho}-a_{22s_\sigma})}, \text{for} \rho\in\{k+1,k+2,\cdots,2n_{21}\};\\
k &\in \{1,2,\cdots,2n_{21}-1\}.
\end{aligned}
\tag{3.52}
$$

For all cases, the $(2m_{11}+1)\times(2n_{21})$-simple equilibriums plus $(2m_{11}+1)\times(2n_{11}+1)$-hyperbolic flows are based on the following hyperbolic and hyperbolic-secant upper/lower-saddle bifurcations for $2n_{11}+1\geq 2n_{21}$:

- $[2(n_{11}-n_{21})+1]^{\text{th}}$DP:UP, $[(2m_{11}+1),(2n_{21})]^{\text{th}}$US)-hyperbolic upper-saddle bifurcation for ($a_{110}>0, a_{220}>0$),
- $[2(n_{11}-n_{21})+1]^{\text{th}}$UP:DP, $[(2m_{11}+1),(2n_{21})]^{\text{th}}$US)-hyperbolic-secant upper-saddle bifurcation for ($a_{110}<0, a_{220}>0$),
- $[2(n_{11}-n_{21})+1]^{\text{th}}$UP:DP, $[(2m_{11}+1),(2n_{21})]^{\text{th}}$LS) hyperbolic-secant lower-saddle bifurcation for ($a_{110}>0, a_{220}<0$),
- $[2(n_{11}-n_{21})+1]^{\text{th}}$DP:UP, $[(2m_{11}+1),(2n_{21})]^{\text{th}}$LS)-hyperbolic lower-saddle bifurcation for ($a_{110}<0, a_{220}<0$);

or based on the following up-to-down and down-to-up asymptote bifurcations for $2n_{11}+1<2n_{21}$:

- $[2(n_{21}-n_{11})-1]^{\text{th}}$UA:DA, $[(2m_{11}+1),(2n_{21})]^{\text{th}}$US)-up-to-down asymptote upper-saddle bifurcation for ($a_{110}>0, a_{220}>0$),
- $[2(n_{21}-n_{11})-1]^{\text{th}}$DA:UA, $[(2m_{11}+1),(2n_{21})]^{\text{th}}$US)-down-to-up asymptote upper-saddle bifurcation for ($a_{110}<0, a_{220}>0$),
- $[2(n_{21}-n_{11})-1]^{\text{th}}$DA:UA, $[(2m_{11}+1),(2n_{21})]^{\text{th}}$LS)-down-to-up asymptote lower-saddle bifurcation for ($a_{110}>0, a_{220}<0$),
- $[2(n_{21}-n_{11})-1]^{\text{th}}$UA:DA, $[(2m_{11}+1),(2n_{21})]^{\text{th}}$LS)-up-to-down asymptote lower-saddle bifurcation for ($a_{110}<0, a_{220}<0$).

Consider $2n_{11}+1=2n_{21}+1$ for illustrations, phase portraits for the 2-dimensional systems for the hybrid networks of hyperbolic and hyperbolic-secant flows with the simple equilibriums of the saddles, sinks and sources with $x_1^*=a_{111},a_{112},\ldots,a_{11(2m_1+1)}$ and $x_2^*=a_{121},a_{221},\ldots,a_{22(2n_{21})},a_{12(2n_{11}+1)}$ are presented in Fig. 3.17 (a–d) for $(a_{110}>0, a_{220}>0)$, $(a_{110}<0, a_{220}>0)$, $(a_{110}>0, a_{220}<0)$ and $(a_{110}<0, a_{220}<0)$.

$$
\left\{\begin{array}{cccc}
(a_{111},a_{12(2n_{11}+1)}) & \cdots & (a_{11(2m_{11})},a_{12(2n_{11}+1)}) & (a_{11(2m_{11}+1)},a_{12(2n_{11}+1)})\\
(a_{111},a_{22(2n_{21})}) & \cdots & (a_{11(2m_{11})},a_{22(2n_{21})}) & (a_{11(2m_{11}+1)},a_{22(2n_{21})})\\
(a_{111},a_{12(2n_{11})}) & \cdots & (a_{11(2m_{11})},a_{12(2n_{11})}) & (a_{11(2m_{11}+1)},a_{12(2n_{11})})\\
\vdots & \vdots & \vdots & \vdots\\
(a_{111},a_{221}) & \cdots & (a_{11(2m_{11})},a_{221}) & (a_{11(2m_{11}+1)},a_{221})\\
(a_{111},a_{121}) & \cdots & (a_{11(2m_{11})},a_{121}) & (a_{11(2m_{11}+1)},a_{121})
\end{array}\right\}
$$

$$
= \left\{ \begin{array}{cccc}
\underbrace{(\text{UP:DP, pF})}_{\text{hyperbolic-secant flow }(+)} & \cdots & \underbrace{(\text{UP:DP, pF})}_{\text{hyperbolic-secant flow }(+)} & \underbrace{(\text{DP:DP, pF})}_{\text{hyperbolic flow }(+)} \\
\underbrace{(\text{SI, SO})}_{\text{saddle}} & \cdots & \underbrace{(\text{SO, SO})}_{\text{source}} & \underbrace{(\text{SI, SO})}_{\text{saddle}} \\
\underbrace{(\text{UP:DP, nF})}_{\text{hyperbolic-secant flow }(-)} & \cdots & \underbrace{(\text{UP:DP, nF})}_{\text{hyperbolic-secant flow }(-)} & \underbrace{(\text{UP:DP, nF})}_{\text{hyperbolic-secant flow }(-)} \\
\vdots & \vdots & \vdots & \vdots \\
\underbrace{(\text{SO, SI})}_{\text{saddle}} & \cdots & \underbrace{(\text{SI, SI})}_{\text{saddle}} & \underbrace{(\text{SO, SI})}_{\text{source}} \\
\underbrace{(\text{UP:DP, pF})}_{\text{hyperbolic-secant flow }(+)} & \cdots & \underbrace{(\text{DP:UP, pF})}_{\text{hyperbolic flow }(+)} & \underbrace{(\text{UP:DP, pF})}_{\text{hyperbolic-secant flow }(+)}
\end{array} \right\}_{(2n_{11}+2n_{21}+1)\times(2m_{11}+1)}
$$

for $a_{110} > 0, a_{220} > 0$; (3.53)

$$
\left\{ \begin{array}{cccc}
(a_{111}, a_{12(2n_{11}+1)}) & \cdots & (a_{11(2m_{11})}, a_{12(2n_{11}+1)}) & (a_{11(2m_{11}+1)}, a_{12(2n_{11}+1)}) \\
(a_{111}, a_{22(2n_{21})}) & \cdots & (a_{11(2m_{11})}, a_{22(2n_{21})}) & (a_{11(2m_{11}+1)}, a_{22(2n_{21})}) \\
(a_{111}, a_{12(2n_{11})}) & \cdots & (a_{11(2m_{11})}, a_{12(2n_{11})}) & (a_{11(2m_{11}+1)}, a_{12(2n_{11})}) \\
\vdots & \vdots & \vdots & \vdots \\
(a_{111}, a_{221}) & \cdots & (a_{11(2m_{11})}, a_{221}) & (a_{11(2m_{11}+1)}, a_{221}) \\
(a_{111}, a_{121}) & \cdots & (a_{11(2m_{11})}, a_{121}) & (a_{11(2m_{11}+1)}, a_{121})
\end{array} \right\}
$$

$$
= \left\{ \begin{array}{cccc}
\underbrace{(\text{UP:DP, pF})}_{\text{hyperbolic-secant flow }(+)} & \cdots & \underbrace{(\text{DP:UP, pF})}_{\text{hyperbolic flow }(+)} & \underbrace{(\text{UP:DP, pF})}_{\text{hyperbolic-secant flow }(+)} \\
\underbrace{(\text{SO, SO})}_{\text{source}} & \cdots & \underbrace{(\text{SI, SO})}_{\text{saddle}} & \underbrace{(\text{SO, SO})}_{\text{source}} \\
\underbrace{(\text{UP:DP, nF})}_{\text{hyperbolic-secant flow }(-)} & \cdots & \underbrace{(\text{DP:UP, nF})}_{\text{hyperbolic flow }(-)} & \underbrace{(\text{UP:DP, nF})}_{\text{hyperbolic-secant flow }(-)} \\
\vdots & \vdots & \vdots & \vdots \\
\underbrace{(\text{SO, SI})}_{\text{sink}} & \cdots & \underbrace{(\text{SO, SI})}_{\text{saddle}} & \underbrace{(\text{SI, SI})}_{\text{sink}} \\
\underbrace{(\text{UP:DP, pF})}_{\text{hyperbolic-secant flow }(+)} & \cdots & \underbrace{(\text{DP:UP, pF})}_{\text{hyperbolic flow }(+)} & \underbrace{(\text{UP:DP, pF})}_{\text{hyperbolic-secant flow }(+)}
\end{array} \right\}_{(2n_{11}+2n_{21}+1)\times(2m_{11}+1)}
$$

for $a_{110} < 0, a_{220} > 0$; (3.54)

$$
\left\{\begin{matrix}
(a_{111},a_{12(2n_{11}+1)}) & \cdots & (a_{11(2m_{11})},a_{12(2n_{11}+1)}) & (a_{11(2m_{11}+1)},a_{12(2n_{11}+1)})\\
(a_{111},a_{22(2n_{21})}) & \cdots & (a_{11(2m_{11})},a_{22(2n_{21})}) & (a_{11(2m_{11}+1)},a_{22(2n_{21})})\\
(a_{111},a_{12(2n_{11})}) & \cdots & (a_{11(2m_{11})},a_{12(2n_{11})}) & (a_{11(2m_{11}+1)},a_{12(2n_{11})})\\
\vdots & \vdots & \vdots & \vdots\\
(a_{111},a_{221}) & \cdots & (a_{11(2m_{11})},a_{221}) & (a_{11(2m_{11}+1)},a_{221})\\
(a_{111},a_{121}) & \cdots & (a_{11(2m_{11})},a_{121}) & (a_{11(2m_{11}+1)},a_{121})
\end{matrix}\right\}
$$

$$
=\left\{\begin{matrix}
\underbrace{(\text{UP:DP, nF})}_{\text{hyperbolic-secant flow }(-)} & \cdots & \underbrace{(\text{DP:UP, nF})}_{\text{hyperbolic flow }(-)} & \underbrace{(\text{UP:DP, nF})}_{\text{hyperbolic-secant flow }(-)}\\
\underbrace{(\text{SI, SI})}_{\text{sink}} & \cdots & \underbrace{(\text{SO, SI})}_{\text{saddle}} & \underbrace{(\text{SI, SI})}_{\text{sink}}\\
\underbrace{(\text{UP:DP, pF})}_{\text{hyperbolic-secant flow }(+)} & \cdots & \underbrace{(\text{DP:UP, pF})}_{\text{hyperbolic flow }(+)} & \underbrace{(\text{UP:DP, pF})}_{\text{hyperbolic-secant flow }(+)}\\
\vdots & \vdots & \vdots & \vdots\\
\underbrace{(\text{SO, SO})}_{\text{source}} & \cdots & \underbrace{(\text{SI, SO})}_{\text{saddle}} & \underbrace{(\text{SO, SO})}_{\text{source}}\\
\underbrace{(\text{UP:DP, nF})}_{\text{hyperbolic-secant flow }(-)} & \cdots & \underbrace{(\text{DP:UP, nF})}_{\text{hyperbolic flow }(-)} & \underbrace{(\text{UP:DP, nF})}_{\text{hyperbolic-secant flow }(-)}
\end{matrix}\right\}_{(2n_{11}+2n_{21}+1)\times(2m_{11}+1)}
$$

for $a_{110}>0,\ a_{220}<0;$ (3.55)

$$
\left\{\begin{matrix}
(a_{111},a_{12(2n_{11}+1)}) & \cdots & (a_{11(2m_{11})},a_{12(2n_{11}+1)}) & (a_{11(2m_{11}+1)},a_{12(2n_{11}+1)})\\
(a_{111},a_{22(2n_{21})}) & \cdots & (a_{11(2m_{11})},a_{22(2n_{21})}) & (a_{11(2m_{11}+1)},a_{22(2n_{21})})\\
(a_{111},a_{12(2n_{11})}) & \cdots & (a_{11(2m_{11})},a_{12(2n_{11})}) & (a_{11(2m_{11}+1)},a_{12(2n_{11})})\\
\vdots & \vdots & \vdots & \vdots\\
(a_{111},a_{221}) & \cdots & (a_{11(2m_{11})},a_{221}) & (a_{11(2m_{11}+1)},a_{221})\\
(a_{111},a_{121}) & \cdots & (a_{11(2m_{11})},a_{121}) & (a_{11(2m_{11}+1)},a_{121})
\end{matrix}\right\}
$$

$$
=\left\{\begin{matrix}
\underbrace{(\text{DP:UP, nF})}_{\text{hyperbolic flow }(-)} & \cdots & \underbrace{(\text{UP:DP, nF})}_{\text{hyperbolic-secant flow }(-)} & \underbrace{(\text{DP:UP, nF})}_{\text{hyperbolic flow }(-)}\\
\underbrace{(\text{SO, SI})}_{\text{saddle}} & \cdots & \underbrace{(\text{SI, SI})}_{\text{sink}} & \underbrace{(\text{SO, SI})}_{\text{saddle}}\\
\underbrace{(\text{UP:DP,pF})}_{\text{hyperbolic-secant flow }(+)} & \cdots & \underbrace{(\text{UP:DP, pF})}_{\text{hyperbolic-secant flow }(+)} & \underbrace{(\text{DP:UP, pF})}_{\text{hyperbolic flow }(+)}\\
\vdots & \vdots & \vdots & \vdots\\
\underbrace{(\text{SI, SO})}_{\text{saddle}} & \cdots & \underbrace{(\text{SO, SO})}_{\text{source}} & \underbrace{(\text{SI, SO})}_{\text{saddle}}\\
\underbrace{(\text{DP:UP, nF})}_{\text{hyperbolic flow }(-)} & \cdots & \underbrace{(\text{UP:DP, nF})}_{\text{hyperbolic-secant flow }(-)} & \underbrace{(\text{DP:UP, nF})}_{\text{hyperbolic flow }(-)}
\end{matrix}\right\}_{(2n_{11}+2n_{21}+1)\times(2m_{11}+1)}
$$

for $a_{110}<0,\ a_{220}<0.$ (3.56)

Consider $n_{11} = n_{21}$ for illustrations. Phase portraits for the 2-dimensional systems for the hybrid networks of hyperbolic and hyperbolic-secant flows with the simple equilibriums of the saddles, sinks and sources with $x_1^* = a_{111}, a_{112}, \ldots, a_{11(2m_{11}+1)}$ and $x_2^* = a_{221}, a_{121}, \ldots, a_{12(2n_{11})}, a_{12(2n_{11}+1)}$ are presented in Fig. 3.18 (a–d) for ($a_{110} > 0$, $a_{220} > 0$), ($a_{110} < 0$, $a_{220} > 0$), ($a_{110} > 0$, $a_{220} < 0$) and ($a_{110} < 0$, $a_{220} < 0$).

$$\begin{Bmatrix} (a_{111}, a_{12(2n_{11}+1)}) & \cdots & (a_{11(2m_{11})}, a_{12(2n_{11}+1)}) & (a_{11(2m_{11}+1)}, a_{12(2n_{11}+1)}) \\ (a_{111}, a_{12(2n_{11})}) & \cdots & (a_{11(2m_{11})}, a_{12(2n_{11})}) & (a_{11(2m_{11}+1)}, a_{12(2n_{11})}) \\ (a_{111}, a_{22(2n_{21})}) & \cdots & (a_{11(2m_{11})}, a_{22(2n_{21})}) & (a_{11(2m_{11}+1)}, a_{22(2n_{21})}) \\ \vdots & \vdots & \vdots & \vdots \\ (a_{111}, a_{121}) & \cdots & (a_{11(2m_{11})}, a_{121}) & (a_{11(2m_{11}+1)}, a_{121}) \\ (a_{111}, a_{221}) & \cdots & (a_{11(2m_{11})}, a_{221}) & (a_{11(2m_{11}+1)}, a_{221}) \end{Bmatrix}$$
$$= \begin{Bmatrix} \underbrace{(\text{DP:UP, pF})}_{\text{hyperbolic-secant flow }(+)} & \cdots & \underbrace{(\text{UP:DP, pF})}_{\text{hyperbolic-secant flow }(+)} & \underbrace{(\text{DP:UP, pF})}_{\text{hyperbolic flow }(+)} \\ \underbrace{(\text{UP:DP, pF})}_{\text{hyperbolic-secant flow }(+)} & \cdots & \underbrace{(\text{DP:UP, pF})}_{\text{hyperbolic flow }(+)} & \underbrace{(\text{UP:DP, pF})}_{\text{hyperbolic-secant flow }(+)} \\ \underbrace{(\text{SO, SO})}_{\text{source}} & \cdots & \underbrace{(\text{SI, SO})}_{\text{saddle}} & \underbrace{(\text{SO, SO})}_{\text{secant}} \\ \vdots & \vdots & \vdots & \vdots \\ \underbrace{(\text{UP:DP, nF})}_{\text{hyperbolic-secant flow }(-)} & \cdots & \underbrace{(\text{DP:UP, nF})}_{\text{hyperbolic flow }(-)} & \underbrace{(\text{UP:DP, nF})}_{\text{hyperbolic-secant flow }(-)} \\ \underbrace{(\text{SI, SI})}_{\text{sink}} & \cdots & \underbrace{(\text{SO, SI})}_{\text{saddle}} & \underbrace{(\text{SI, SI})}_{\text{sink}} \end{Bmatrix}_{(2n_{11}+2n_{21}+1)\times(2m_{11}+1)}$$

for $a_{110} > 0, a_{220} > 0$; (3.57)

$$\begin{Bmatrix} (a_{111}, a_{12(2n_{11}+1)}) & \cdots & (a_{11(2m_{11})}, a_{12(2n_{11}+1)}) & (a_{11(2m_{11}+1)}, a_{12(2n_{11}+1)}) \\ (a_{111}, a_{12(2n_{11})}) & \cdots & (a_{11(2m_{11})}, a_{12(2n_{11})}) & (a_{11(2m_{11}+1)}, a_{12(2n_{11})}) \\ (a_{111}, a_{22(2n_{21})}) & \cdots & (a_{11(2m_{11})}, a_{22(2n_{21})}) & (a_{11(2m_{11}+1)}, a_{22(2n_{21})}) \\ \vdots & \vdots & \vdots & \vdots \\ (a_{111}, a_{121}) & \cdots & (a_{11(2m_{11})}, a_{121}) & (a_{11(2m_{11}+1)}, a_{121}) \\ (a_{111}, a_{221}) & \cdots & (a_{11(2m_{11})}, a_{221}) & (a_{11(2m_{11}+1)}, a_{221}) \end{Bmatrix}$$

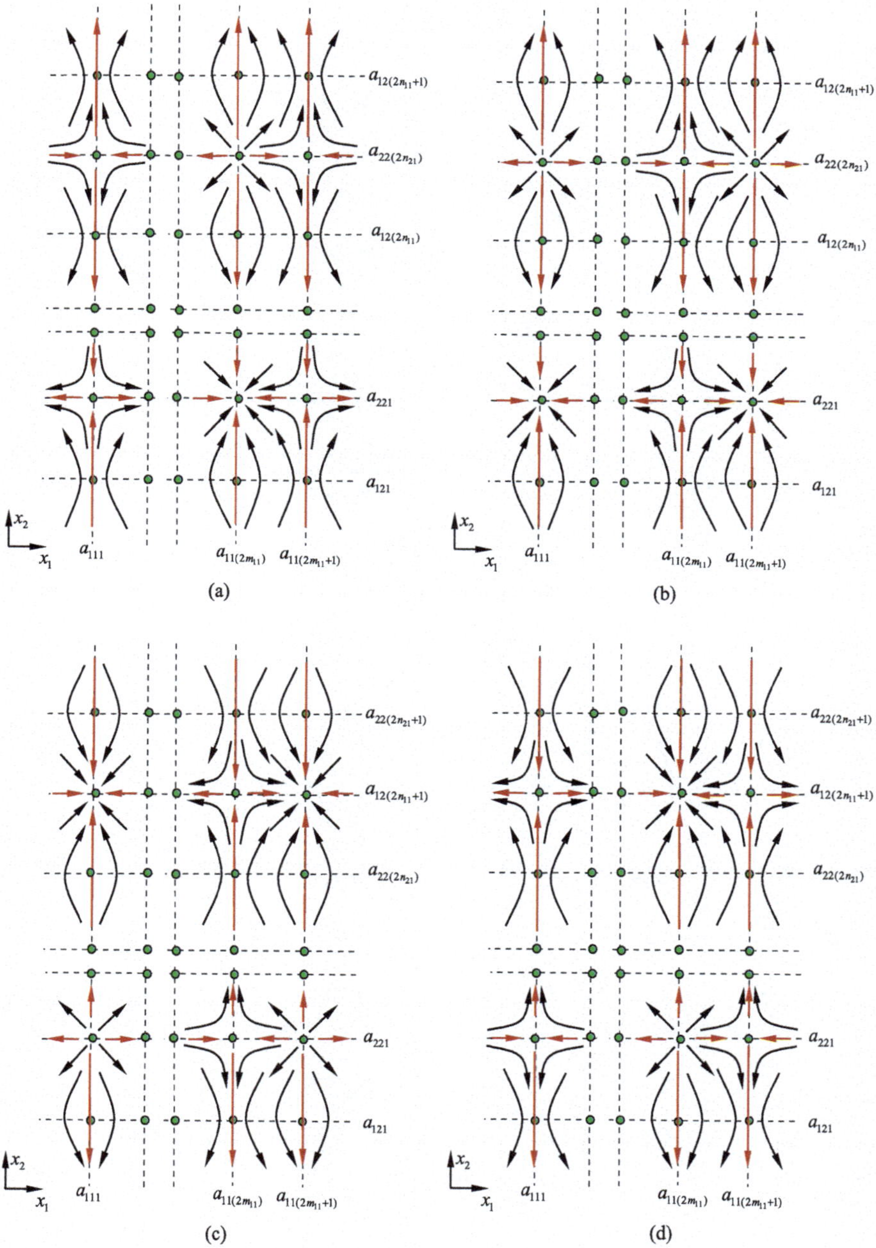

Fig. 3.17 Phase portraits for 2-dimensional systems with $x_1^* = a_{111}, a_{112}, \ldots, a_{11(2m_1+1)}$ and $x_2^* = a_{121}, a_{221}, \ldots, a_{22(2n_{21})}, a_{12(2n_{11}+1)}$. The networks of flows with sink, source and saddle: **a** ($a_{110} > 0$, $a_{220} > 0$), **b** ($a_{110} < 0$, $a_{220} > 0$), **c** ($a_{110} > 0$, $a_{220} < 0$), **d** ($a_{110} < 0$, $a_{220} < 0$)

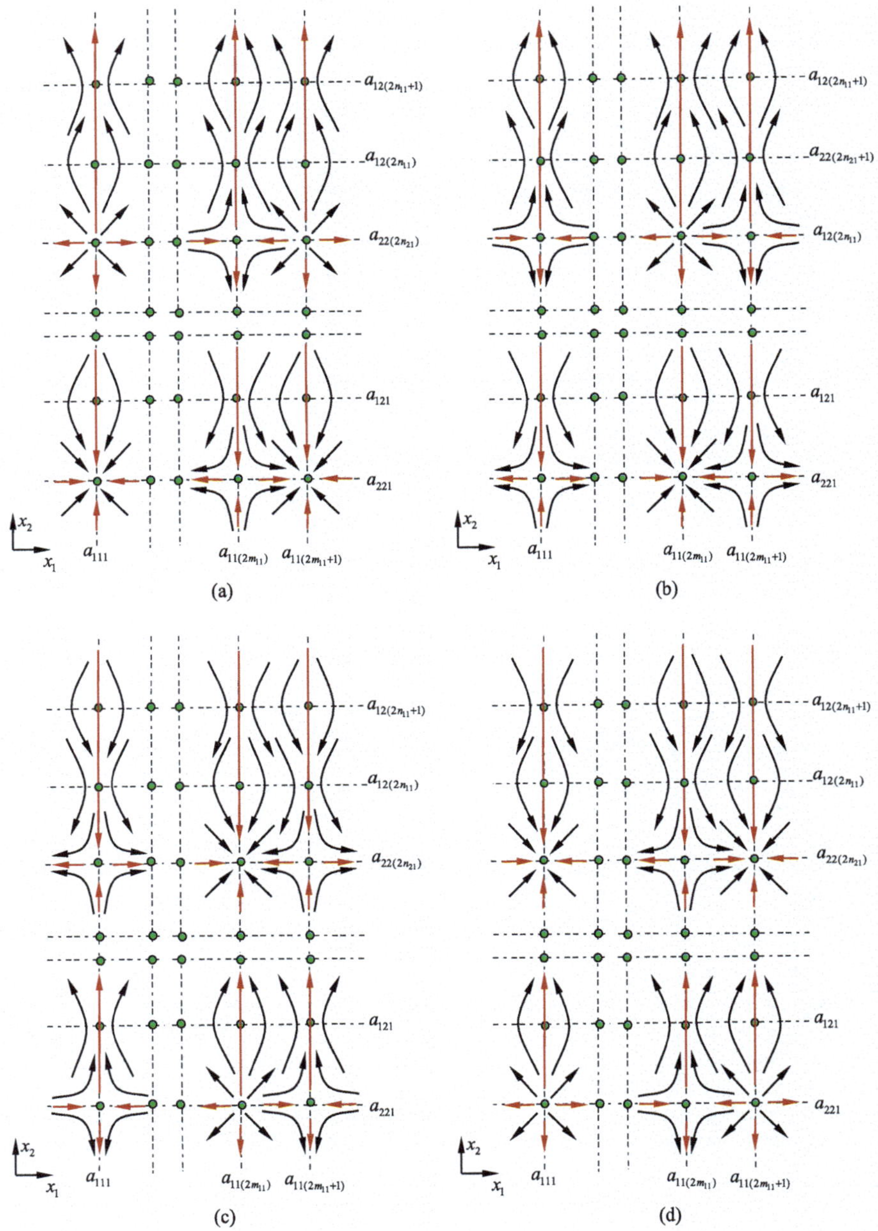

Fig. 3.18 Phase portraits for 2-dimensional systems with $x_1^* = a_{111}, a_{112}, \ldots, a_{11(2m_{11}+1)}$ and $x_2^* = a_{221}, a_{121}, \ldots, a_{12(2n_{11})}, a_{12(2n_{11}+1)}$. The networks of flows with sink, source and saddle: **a** ($a_{110} > 0$, $a_{220} > 0$), **b** ($a_{110} < 0$, $a_{220} > 0$), **c** ($a_{110} > 0$, $a_{220} < 0$), **d** ($a_{110} < 0$, $a_{220} < 0$)

$$
= \left\{ \begin{array}{cccc}
\underbrace{(\text{UP:DP, pF})}_{\text{hyperbolic-secant flow (+)}} & \cdots & \underbrace{(\text{DP:UP, pF})}_{\text{hyperbolic flow (+)}} & \underbrace{(\text{UP:DP, pF})}_{\text{hyperbolic-secant flow (+)}} \\
\underbrace{(\text{DP:UP, pF})}_{\text{hyperbolic flow (+)}} & \cdots & \underbrace{(\text{UP:DP, pF})}_{\text{hyperbolic-secant flow (+)}} & \underbrace{(\text{DP:UP, pF})}_{\text{hyperbolic flow (+)}} \\
\underbrace{(\text{SI, SO})}_{\text{saddle}} & \cdots & \underbrace{(\text{SO, SO})}_{\text{source}} & \underbrace{(\text{SI, SO})}_{\text{saddle}} \\
\vdots & \vdots & \vdots & \vdots \\
\underbrace{(\text{DP:UP, nF})}_{\text{hyperbolic flow }(-)} & \cdots & \underbrace{(\text{UP:DP, nF})}_{\text{hyperbolic-secant flow }(-)} & \underbrace{(\text{DP:UP, nF})}_{\text{hyperbolic flow }(-)} \\
\underbrace{(\text{SO, SI})}_{\text{saddle}} & \cdots & \underbrace{(\text{SI, SI})}_{\text{sink}} & \underbrace{(\text{SO, SI})}_{\text{saddle}}
\end{array} \right\}_{(2n_{11}+2n_{21}+1)\times(2m_{11}+1)}
$$

for $a_{110} < 0, a_{220} > 0$; (3.58)

$$
\left\{ \begin{array}{cccc}
(a_{111}, a_{12(2n_{11}+1)}) & \cdots & (a_{11(2m_{11})}, a_{12(2n_{11}+1)}) & (a_{11(2m_{11}+1)}, a_{12(2n_{11}+1)}) \\
(a_{111}, a_{12(2n_{11})}) & \cdots & (a_{11(2m_{11})}, a_{12(2n_{11})}) & (a_{11(2m_{11}+1)}, a_{12(2n_{11})}) \\
(a_{111}, a_{22(2n_{21})}) & \cdots & (a_{11(2m_{11})}, a_{22(2n_{21})}) & (a_{11(2m_{11}+1)}, a_{22(2n_{21})}) \\
\vdots & \vdots & \vdots & \vdots \\
(a_{111}, a_{121}) & \cdots & (a_{11(2m_{11})}, a_{121}) & (a_{11(2m_{11}+1)}, a_{121}) \\
(a_{111}, a_{221}) & \cdots & (a_{11(2m_{11})}, a_{221}) & (a_{11(2m_{11}+1)}, a_{221})
\end{array} \right\}
$$

$$
= \left\{ \begin{array}{cccc}
\underbrace{(\text{UP:DP, nF})}_{\text{hyperbolic-secant flow }(-)} & \cdots & \underbrace{(\text{DP:UP, nF})}_{\text{hyperbolic flow }(-)} & \underbrace{(\text{UP:DP, nF})}_{\text{hyperbolic-secant flow }(-)} \\
\underbrace{(\text{DP:UP, nF})}_{\text{hyperbolic flow }(-)} & \cdots & \underbrace{(\text{UP:DP, nF})}_{\text{hyperbolic-secant flow }(-)} & \underbrace{(\text{DP:UP, nF})}_{\text{hyperbolic flow }(-)} \\
\underbrace{(\text{SO, SI})}_{\text{saddle}} & \cdots & \underbrace{(\text{SI, SI})}_{\text{sink}} & \underbrace{(\text{SO, SI})}_{\text{saddle}} \\
\vdots & \vdots & \vdots & \vdots \\
\underbrace{(\text{DP:UP, pF})}_{\text{hyperbolic flow (+)}} & \cdots & \underbrace{(\text{UP:DP, pF})}_{\text{hyperbolic-secant flow (+)}} & \underbrace{(\text{DP:UP, pF})}_{\text{hyperbolic flow (+)}} \\
\underbrace{(\text{SI, SO})}_{\text{saddle}} & \cdots & \underbrace{(\text{SO, SO})}_{\text{source}} & \underbrace{(\text{SI, SO})}_{\text{saddle}}
\end{array} \right\}_{(2n_{11}+2n_{21}+1)\times(2m_{11}+1)}
$$

for $a_{110} > 0, a_{220} < 0$; (3.59)

$$\left\{\begin{array}{cccc}(a_{111},a_{12(2n_{11}+1)}) & \cdots & (a_{11(2m_{11})},a_{12(2n_{11}+1)}) & (a_{11(2m_{11}+1)},a_{12(2n_{11}+1)})\\(a_{111},a_{12(2n_{11})}) & \cdots & (a_{11(2m_{11})},a_{12(2n_{11})}) & (a_{11(2m_{11}+1)},a_{12(2n_{11})})\\(a_{111},a_{22(2n_{21})}) & \cdots & (a_{11(2m_{11})},a_{22(2n_{21})}) & (a_{11(2m_{11}+1)},a_{22(2n_{21})})\\\vdots & \vdots & \vdots & \vdots\\(a_{111},a_{121}) & \cdots & (a_{11(2m_{11})},a_{121}) & (a_{11(2m_{11}+1)},a_{121})\\(a_{111},a_{221}) & \cdots & (a_{11(2m_{11})},a_{221}) & (a_{11(2m_{11}+1)},a_{221})\end{array}\right\}$$

$$=\left\{\begin{array}{cccc}\underbrace{(\text{DP:UP, nF})}_{\text{hyperbolic flow }(-)} & \cdots & \underbrace{(\text{UP:DP, nF})}_{\text{hyperbolic-secant flow }(-)} & \underbrace{(\text{DP:UP, nF})}_{\text{hyperbolic flow }(-)}\\\underbrace{(\text{UP:DP, nF})}_{\text{hyperbolic-secant flow }(-)} & \cdots & \underbrace{(\text{DP:UP, nF})}_{\text{hyperbolic flow }(-)} & \underbrace{(\text{UP:DP, nF})}_{\text{hyperbolic-secant flow }(-)}\\\underbrace{(\text{SI, SI})}_{\text{sink}} & \cdots & \underbrace{(\text{SO, SI})}_{\text{saddle}} & \underbrace{(\text{SI,SI})}_{\text{sink}}\\\vdots & \vdots & \vdots & \vdots\\\underbrace{(\text{UP:DP, pF})}_{\text{hyperbolic-secant flow }(+)} & \cdots & \underbrace{(\text{DP:UP, pF})}_{\text{hyperbolic flow }(+)} & \underbrace{(\text{UP:DP, pF})}_{\text{hyperbolic-secant flow }(+)}\\\underbrace{(\text{SO, SO})}_{\text{source}} & \cdots & \underbrace{(\text{SI, SO})}_{\text{saddle}} & \underbrace{(\text{SO, SO})}_{\text{source}}\end{array}\right\}_{(2n_{11}+2n_{21}+1)\times(2m_{11}+1}$$

for $a_{110} < 0, a_{220} < 0$. (3.60)

3.4 $[(2m_{11}+1),(2n_{11})]$ and $((2m_{11}+1),(2n_{21}))$-Singular Systems

In this section, there are four global bifurcations of the 1-diemsnional flows through singular inflection-source and sink flows, and four global bifurcations of equilibriums, through singular saddle-source and saddle-sink bifurcations. The corresponding global bifurcations generate networks of hyperbolic and hyperbolic-secant flows with sources, sinks and saddles. The infinite-equilibriums are the appearing and switching bifurcations for the singular flows and singular equilibriums and for hybrid networks of hyperbolic flows and source, sink and saddles.

3.4.1 $[(2m_{11}+1),(2n_{11})]$-*Flows and* $((2m_{11}+1),(2n_{21}))$-*Equilibriums*

Consider a polynomial dynamical system for $a_{221} \neq a_{121}$ as

$$\dot{x}_2 = a_{220}(x_2 - a_{221})^{2n_{21}},$$

$$\dot{x}_1 = a_{110}(x_1 - a_{111})^{2m_{11}+1}(x_2 - a_{121})^{2n_{11}}. \tag{3.61}$$

The first integral manifold is

$$\begin{aligned}
&-\frac{1}{2m_{11}}\left[\frac{1}{(x_1 - a_{111})^{2m_{11}}} - \frac{1}{(x_{10} - a_{111})^{2m_{11}}}\right] \\
&= \frac{a_{110}}{a_{220}} \sum_{\alpha,\beta=0}^{\alpha+\beta=2n_{11}} \frac{(2n_{11})!}{\alpha!\beta!} \frac{(a_{221} - a_{121})^{\alpha}}{\beta - 2n_{21} + 1}\left[(x_2 - a_{221})^{\beta-2n_{21}+1} - (x_2 - a_{221})^{\beta-2n_{21}+1}\right] \\
&\text{for } 2m_{11} + 1 \neq 1, \beta - 2n_{21} + 1 \neq 0; \\
&-\frac{1}{2m_{11}}\left[\frac{1}{(x_1 - a_{111})^{2m_{11}}} - \frac{1}{(x_{10} - a_{111})^{2m_{11}}}\right] = \ln\frac{|x_1 - a_{111}|}{|x_{10} - a_{111}|} \\
&\text{for } 2m_{11} + 1 = 1, \\
&\frac{1}{\beta - 2n_{21} + 1}\left[(x_2 - a_{221})^{\beta-2n_{21}+1} - (x_2 - a_{221})^{\beta-2n_{21}+1}\right] = \ln\frac{|x_2 - a_{221}|}{|x_{20} - a_{221}|} \\
&\text{for } \beta - 2n_{21} + 1 = 0.
\end{aligned} \tag{3.62}$$

For $(x_1^*, x_2^*) = (a_{111}, a_{121})$ with $\dot{x}_1 = 0$, the $[(2m_{11} + 1), (2n_{11})]$-singular flows exist with $\dot{x}_2 = a_{220}(a_{121} - a_{221})^{2n_{21}}$. For $(x_1^*, x_2^*) = (a_{111}, a_{121})$, the 1-dimensional flow direction is determined by $\dot{x}_2 \neq 0s$. For $(x_1^*, x_2^*) = (a_{111}, a_{221})$ with $\dot{x}_1 = \dot{x}_2 = 0$, the $((2m_{11} + 1), (2n_{21}))$- singular equilibriums exist.

(A) For $a_{121} > a_{221}$,

$$\begin{Bmatrix} (a_{111}, a_{121}) \\ (a_{111}, a_{221}) \end{Bmatrix} = \begin{Bmatrix} \underbrace{([(2m_{11}+1), (2n_{11})]_{\text{DI:II}}\text{SO, pF})}_{[(2m_{11}+1),(2n_{11})]\text{-decreasing-to-increasing inflection source flow }(+)} \\ \underbrace{((2m_{11} + 1)\text{SI},(2n_{21})\text{US})}_{[(2m_{11}+1),(2n_{21})]\text{-upper-saddle sink}} \end{Bmatrix},$$

for $a_{110} > 0, a_{220} > 0$;

$$\begin{Bmatrix} (a_{111}, a_{121}) \\ (a_{111}, a_{221}) \end{Bmatrix} = \begin{Bmatrix} \underbrace{([(2m_{11}+1), (2n_{11})]_{\text{II:DI}}\text{SI, pF})}_{[(2m_{11}+1),(2n_{11})]\text{-increasing-to-decreasing inflection sink flow }(+)} \\ \underbrace{((2m_{11} + 1)\text{SO}, (2n_{21})\text{US})}_{[(2m_{11}+1),(2n_{21})]\text{-upper-saddle source}} \end{Bmatrix},$$

for $a_{110} < 0, a_{220} > 0$;

$$\begin{Bmatrix} (a_{111}, a_{121}) \\ (a_{111}, a_{221}) \end{Bmatrix} = \begin{Bmatrix} \underbrace{([(2m_{11}+1), (2n_{11})]_{\text{II:DI}}\text{SO, nF})}_{[(2m_{11}+1),(2n_{11})]\text{-hincreasing-to-decreasing inflection source flow }(-)} \\ \underbrace{((2m_{11} + 1)\text{SI}, (2n_{21})\text{LS})}_{[(2m_{11}+1),(2n_{21})]\text{-lower-saddle-sink}} \end{Bmatrix},$$

for $a_{110} > 0, a_{220} < 0$;

$$\left\{\begin{array}{l}(a_{111},a_{121})\\(a_{111},a_{221})\end{array}\right\}=\left\{\begin{array}{c}\underbrace{([(2m_{11}+1),(2n_{11})]_{\mathrm{DI:II}}\mathrm{SI},\,\mathrm{nF})}_{[(2m_{11}+1),(2n_{11})]\text{-decreasing-to-increasing inflection sink flow }(-)}\\\underbrace{((2m_{11}+1)\mathrm{SO},(2n_{21})\mathrm{LS})}_{[(2m_{11}+1),(2n_{21})]\text{-lower-saddle source}}\end{array}\right\},$$

for $a_{110}<0,a_{220}<0$. (3.63)

(B) For $a_{121}<a_{221}$,

$$\left\{\begin{array}{l}(a_{111},a_{221})\\(a_{111},a_{121})\end{array}\right\}=\left\{\begin{array}{c}\underbrace{((2m_{11}+1)\mathrm{SO},(2n_{21})\mathrm{LS})}_{[(2m_{11}+1),(2n_{21})]\text{-lower-saddle source}}\\\underbrace{([(2m_{11}+1),(2n_{11})]_{\mathrm{II:DI}}\mathrm{SO},\mathrm{pF})}_{[(2m_{11}+1),(2n_{11})]\text{-increasing-to-decreasing inflection source flow }(+)}\end{array}\right\},$$

for $a_{110}>0,a_{220}>0$;

$$\left\{\begin{array}{l}(a_{111},a_{221})\\(a_{111},a_{121})\end{array}\right\}=\left\{\begin{array}{c}\underbrace{((2m_{11}+1)\mathrm{SI},(2n_{21})\mathrm{US})}_{[(2m_{11}+1),(2n_{21})]\text{-upper-saddle sink}}\\\underbrace{([(2m_{11}+1),(2n_{11})]_{\mathrm{DI:II}}\mathrm{SI},\,\mathrm{pF})}_{[(2m_{11}+1),(2n_{11})]\text{-decreasing-to-increasing inflection sink flow }(+)}\end{array}\right\},$$

for $a_{110}<0,a_{220}>0$;

$$\left\{\begin{array}{l}(a_{111},a_{221})\\(a_{111},a_{121})\end{array}\right\}=\left\{\begin{array}{c}\underbrace{((2m_{11}+1)\mathrm{SO},(2n_{21})\mathrm{LS})}_{[(2m_{11}+1),(2n_{21})]\text{-lower-saddle source}}\\\underbrace{([(2m_{11}+1),(2n_{11})]_{\mathrm{DI:II}}\mathrm{SO},\,\mathrm{nF})}_{[(2m_{11}+1),(2n_{11})]\text{-decreasing-to-increasing inflection source flow }(-)}\end{array}\right\},$$

for $a_{110}>0,a_{220}<0$;

$$\left\{\begin{array}{l}(a_{111},a_{221})\\(a_{111},a_{121})\end{array}\right\}=\left\{\begin{array}{c}\underbrace{((2m_{11}+1)\mathrm{SI},(2n_{21})\mathrm{LS})}_{[(2m_{11}+1),(2n_{21})]\text{-lower-saddle sink}}\\\underbrace{([(2m_{11}+1),(2n_{11})]_{\mathrm{II:DI}}\mathrm{SI},\,\mathrm{nF})}_{[(2m_{11}+1),(2n_{11})]\text{-increasing-to-decreasing inflection sink flow }(-)}\end{array}\right\},$$

for $a_{110}<0,a_{220}<0$. (3.64)

The $([(2m_{11}+1),(2n_{11})]_{\mathrm{DI:II}}\mathrm{SO},\mathrm{pF})$ is for a positive singular decreasing-to-increasing inflection-source flow. The $([(2m_{11}+1),(2n_{11})]_{\mathrm{II:DI}}\mathrm{SI},\mathrm{pF})$ is for a positive singular increasing-to-decreasing inflection sink flow. The $([(2m_{11}+1),(2n_{11})]_{\mathrm{II:DI}}\mathrm{SO},\mathrm{nF})$ is for a negative singular increasing-to-decreasing inflection source flow. The $([(2m_{11}+1),(2n_{11})]_{\mathrm{DI:II}}\mathrm{SO},\mathrm{nF})$ is for a negative singular decreasing-to-increasing inflection source flow. The acronyms "DI" and "II" are for decreasing-inflection and increasing-inflection flows, respectively. The $((2m_{11}+1)\text{SO},(2n_{21})\text{US})$ is for a $((2m_{11}+1),(2n_{21}))$-saddle-source. The $((2m_{11}+1)\text{SI},(2n_{21})\text{US})$ is for a $((2m_{11}+1),(2n_{21}))$-upper-saddle sink. The $((2m_{11}+1)\text{SO},(2n_{21})\text{LS})$ is for a $((2m_{11}+1),(2n_{21}))$-lower-saddle source. The $((2m_{11}+1)\text{SI},(2n_{21})\text{LS})$ is for a $((2m_{11}+1),(2n_{21}))$-singular lower-saddle sink.

The singular inflection flows with singular saddle-sources and saddle-sinks are illustrated in Figs. 3.19 and 3.20. Through such singular flows, the corresponding global bifurcations generate from an inflection flow to a network of singular hyperbolic and hyperbolic-secant flows. Through such singular equilibriums, the corresponding global bifurcations generate from a flow to a network of singular source, sink and saddle equilibriums.

3.4.2 $[(2m_{11}+1),(2n_{11})]$ and $((2m_{11}+1),(2n_{21}))$-*Infinite-Equilibriums*

Consider a polynomial dynamical system for $a_{221}=a_{121}$ as

$$\begin{aligned}\dot{x}_2&=a_{220}(x_2-a_{221})^{2n_{21}},\\ \dot{x}_1&=a_{110}(x_1-a_{111})^{2m_{11}+1}(x_2-a_{121})^{2n_{11}}.\end{aligned}\tag{3.65}$$

The first integral manifold is

$$\begin{aligned}&-\frac{1}{2m_{11}}\left[\frac{1}{(x_1-a_{11})^{2m_{11}}}-\frac{1}{(x_{10}-a_{11})^{2m_{11}}}\right]\\ &\quad=\frac{a_{10}}{a_{220}}\frac{1}{2(n_{11}-n_{21})+1}\left[(x_2-a_{121})^{2(n_{11}-n_{21})-1}-(x_2-a_{121})^{2(n_{11}-n_{21})-1}\right]\\ &\text{for } 2m_{11}+1\neq 1;\\ &-\frac{1}{2m_{11}}\left[\frac{1}{(x_1-a_{11})^{2m_{11}}}-\frac{1}{(x_{10}-a_{11})^{2m_{11}}}\right]=\ln\frac{|x_1-a_{11}|}{|x_{10}-a_{11}|}\\ &\text{for } 2m_{11}+1=1.\end{aligned}\tag{3.66}$$

(A) For $2n_{11}\geq 2n_{21}$, the corresponding infinite-equilibriums are given.

$$(a_{111},a_{221})=\underbrace{([2(n_{11}-n_{21})]^{\text{th}}\text{DI:II},[(2m_{11}+1),(2n_{21})]^{\text{th}}\text{US})}_{([2(n_{11}-n_{21})],[(2m_{11}+1),(2n_{21})])\text{-decreasing-to-increasing inflection-upper-saddle}},$$

for $a_{110}>0$, $a_{220}>0$;

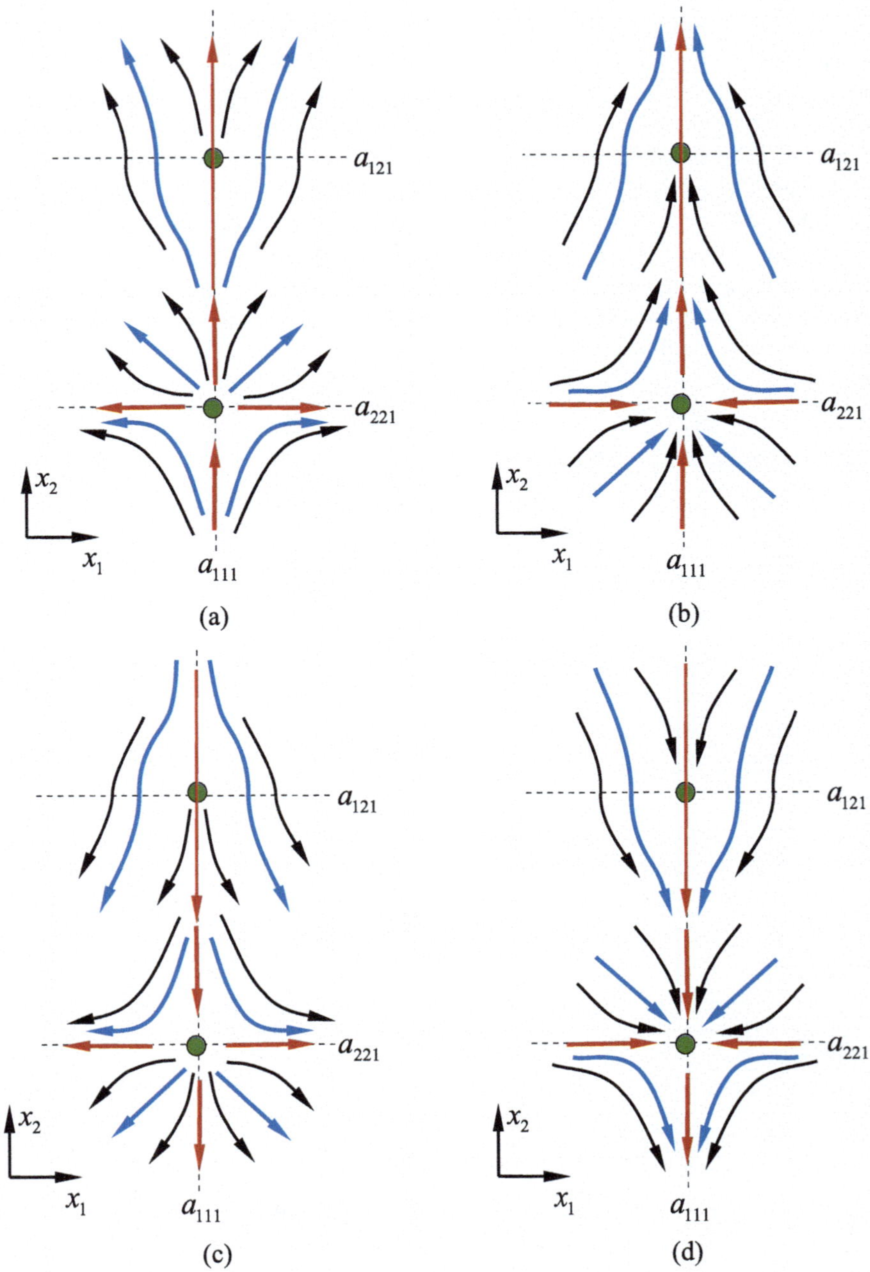

Fig. 3.19 Phase portraits ($a_{121} > a_{221}$) of the ([$(2m_{11}+1)$, $(2n_{11})$], xF)-inflection-source and inflection-sink flows at $(x_1^*, x_2^*) = (a_{111}, a_{121})$ and ($(2m_{11}+1)$, $(2n_{21})$)-saddle-sources, and saddle-sinks at $(x_1^*, x_2^*) = (a_{111}, a_{221})$ for 2-dimensional polynomial systems: **a** ($a_{110} > 0, a_{220} > 0$), **b** ($a_{110} < 0, a_{220} > 0$), **c**($a_{110} > 0, a_{220} < 0$), **d** ($a_{110} < 0, a_{220} < 0$)

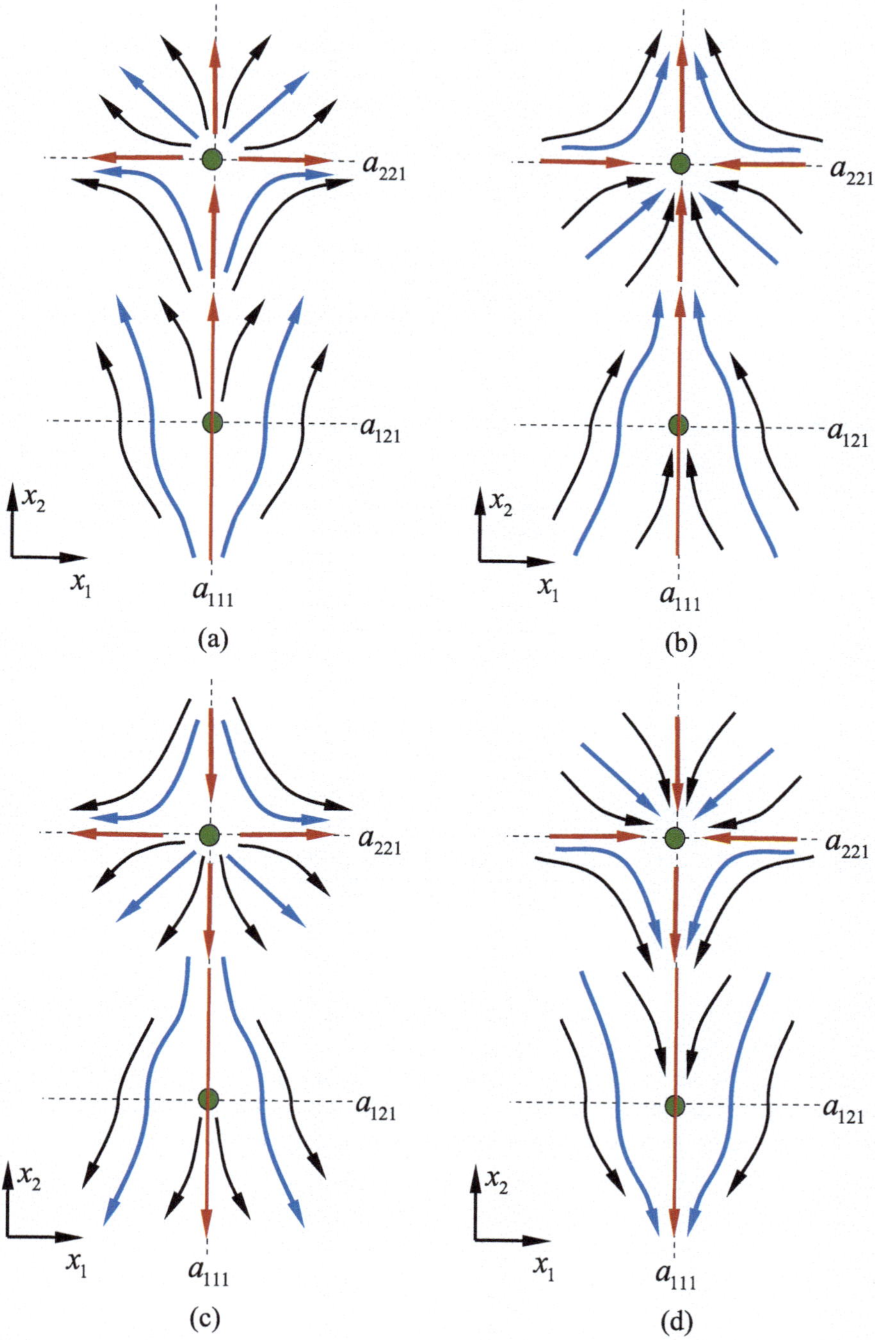

Fig. 3.20 Phase portraits ($a_{221} > a_{121}$) of the ($[(2m_{11}+1),(2n_{11})]$, xF)-inflection-source and inflection-sink flows at $(x_1^*, x_2^*) = (a_{111}, a_{121})$ and $((2m_{11}+1),(2n_{21}))$-saddle-sources and saddle-sinks at $(x_1^*, x_2^*) = (a_{111}, a_{221})$ for 2-dimensional polynomial systems: **a** ($a_{110} > 0$, $a_{220} > 0$), **b** ($a_{110} < 0$, $a_{220} > 0$), **c** ($a_{110} > 0$, $a_{220} < 0$), **d** ($a_{110} < 0$, $a_{220} < 0$)

$$(a_{111}, a_{221}) = \underbrace{([2(n_{11} - n_{21})]^{\text{th}}\text{II:DI},[(2m_{11}+1), (2n_{21})]^{\text{th}}\text{US})}_{([2(n_{11}-n_{21})],[(2m_{11}+1),(2n_{21})])\text{-increasing-to-decreasing inflection-upper-saddle}},$$
for $a_{110} < 0, a_{220} > 0$;
$$(a_{111}, a_{221}) = \underbrace{([2(n_{11} - n_{21})]^{\text{th}}\text{II:DI},[(2m_{11}+1), (2n_{21})]^{\text{th}}\text{LS})}_{([2(n_{11}-n_{21})],[(2m_{11}+1),(2n_{21})])\text{-increasing-to-decreasing inflection-lower-saddle}},$$
for $a_{110} > 0, a_{220} < 0$;
$$(a_{111}, a_{221}) = \underbrace{([2(n_{11} - n_{21})]^{\text{th}}\text{DI:II},[(2m_{11}+1), (2n_{21})]^{\text{th}}\text{LS})}_{([2(n_{11}-n_{21})],[(2m_{11}+1),(2n_{21})])\text{-decreasing-to-increasing inflection-lower-saddle}},$$
for $a_{110} < 0, a_{220} < 0$. (3.67)

(B) For $2n_{11} < 2n_{21}$, the infinite-equilibriums are given as follows.

$$(a_{111}, a_{221}) = \underbrace{([2(n_{21} - n_{11})]^{\text{th}}\text{DC:IC}, [(2m_{11}+1), (2n_{21})]^{\text{th}}\text{US})}_{([2(n_{21}-n_{11})],[(2m_{11}+1),(2n_{21})])\text{-decreasing-to-increasing concave-upper-saddle}},$$
for $a_{110} > 0, a_{220} > 0$;
$$(a_{111}, a_{221}) = \underbrace{([2(n_{21} - n_{11})]^{\text{th}}\text{IC:DC}, [(2m_{11}+1), (2n_{21})]^{\text{th}}\text{US})}_{([2(n_{21}-n_{11})],[(2m_{11}+1),(2n_{21})])\text{-increasing-to-decreasing concave-upper-saddle}},$$
for $a_{110} < 0, a_{220} > 0$;
$$(a_{111}, a_{221}) = \underbrace{([2(n_{21} - n_{11})]^{\text{th}}\text{IC:DC},[(2m_{11}+1), (2n_{21})]^{\text{th}}\text{LS})}_{([2(n_{21}-n_{11})],[(2m_{11}+1),(2n_{21})])\text{-increasing-to-decreasing concave-lower-saddle}},$$
for $a_{110} > 0, a_{220} < 0$;
$$(a_{111}, a_{221}) = \underbrace{([2(n_{21} - n_{11})]^{\text{th}}\text{DC:IC},[(2m_{11}+1), (2n_{21})]^{\text{th}}\text{LS})}_{([2(n_{21}-n_{11})],[(2m_{11}+1),(2n_{21})])\text{-decreasing-to-increasing concave-lower-saddle}},$$
for $a_{110} < 0, a_{220} < 0$. (3.68)

For $x_2^* = a_{121} = a_{221}$, the infinite-equilibrium exists. The infinite-equilibriums are illustrated in Figs. 3.21 and 3.22 for $n_{11} \geq n_{21}$ and $n_{11} < n_{21}$, respectively. For $n_{11} \geq n_{21}$, the singular decreasing-to-increasing and increasing-to-decreasing inflection-source and sink are presented. For $n_{11} < n_{21}$, the singular decreasing-to-increasing and increasing-to-decreasing concave-source and sink are presented.

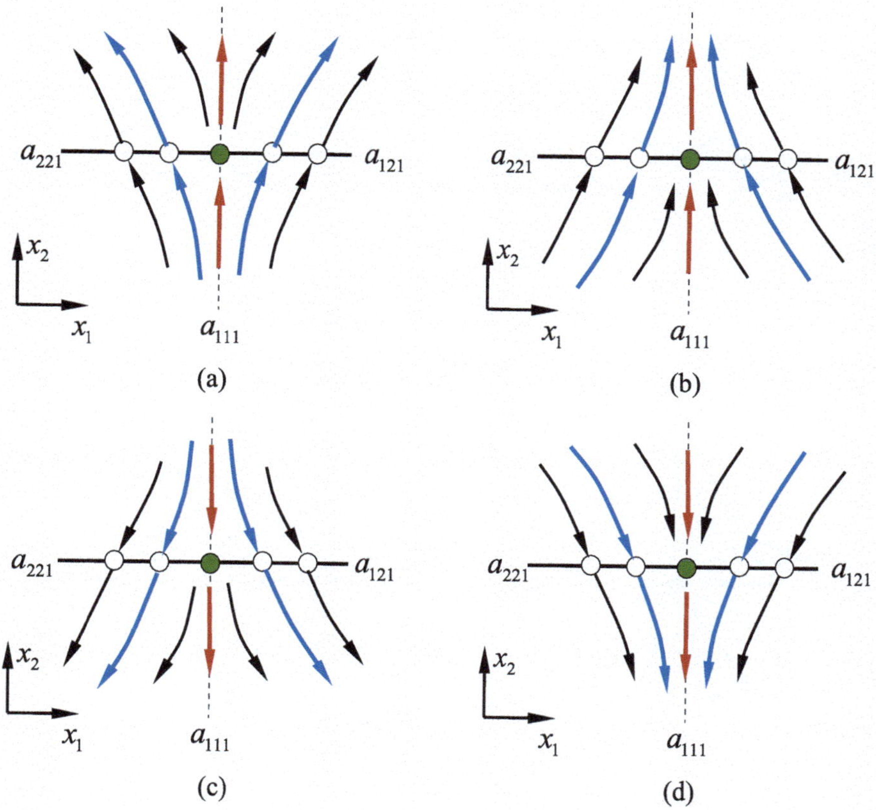

Fig. 3.21 The infinite-equilibriums ($a_{121}=a_{221}$, $2n_{11}\geq 2n_{21}$, $m_1=2m_{11}+1$) for 2-dimensional polynomial systems: **a** ($a_{110}>0$, $a_{220}>0$), **b** ($a_{110}<0$, $a_{220}>0$), **c** ($a_{110}>0$, $a_{220}<0$), **d** ($a_{110}<0$, $a_{220}<0$)

3.4.3 $[(2m_{11}+1),(2n_{11})]$ *and* $((2m_{11}+1),(2n_{21}))$*-Hybrid Networks*

Consider

$$\begin{aligned}\dot{x}_2 &= a_{220}\prod_{s=1}^{2n_{21}}(x_2-a_{22s}),\\ \dot{x}_1 &= a_{110}\prod_{r=1}^{2m_{11}+1}(x_1-a_{11r})\prod_{u=1}^{2n_{11}}(x_2-a_{12u}).\end{aligned}\tag{3.69}$$

The first integral manifold is

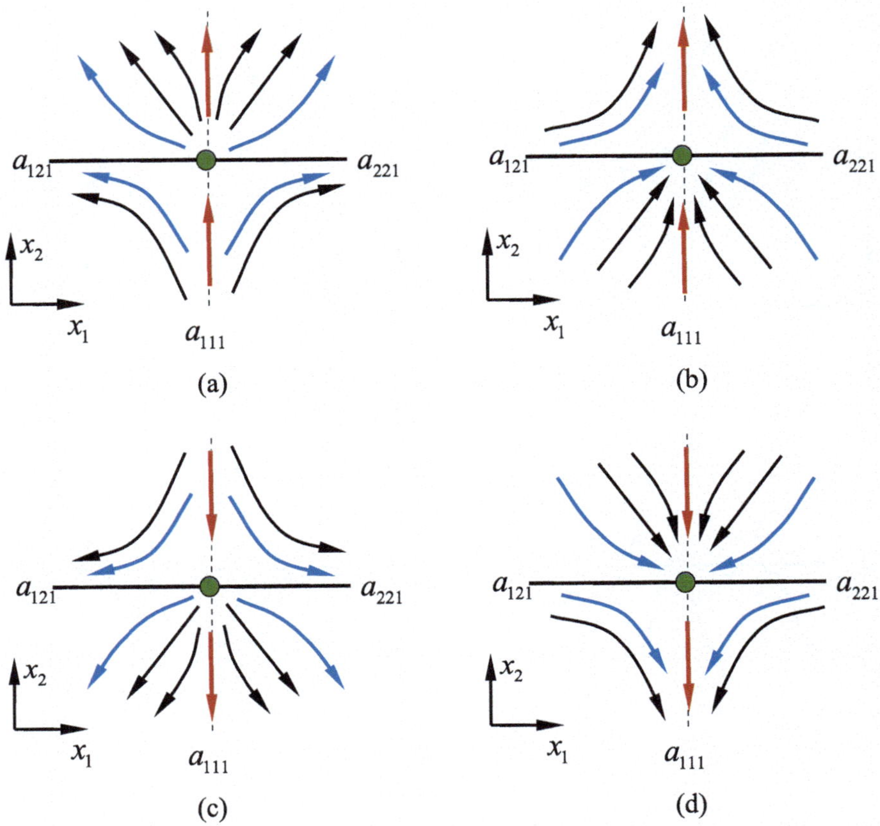

Fig. 3.22 The infinite-equilibriums ($a_{121} = a_{221}$, $2n_{11} < 2n_{21}$, $m_1 = 2m_{11} + 1$) for 2-dimensional polynomial systems: **a** ($a_{110} > 0$, $a_{220} > 0$), **b** ($a_{110} < 0$, $a_{220} > 0$), **c** ($a_{110} > 0$, $a_{220} < 0$), **d** ($a_{110} < 0$, $a_{220} < 0$)

$$
\begin{aligned}
\sum_{r_1=1}^{2m_{11}+1} A_{r_1} \ln \frac{|x_1 - a_{11r_1}|}{|x_{10} - a_{11r_1}|} = \frac{a_{110}}{a_{220}} \Bigg\{ & \sum_{\beta_0=2n_{21}+1}^{2n_{11}} b_{12\beta_0} \sum_{\substack{\alpha_1 \in [0, \beta_0 - 2n_{21}], \\ \beta_1 \in [2n_{21}, \beta_0 - 1]}}^{\alpha_1+\beta_1=\beta_0-1} \sum_{\substack{\alpha_2 \in [0, \beta_1 - 2n_{21} - 1)] \\ \beta_2 \in [2n_{21}-1, \beta_1 - 2]}}^{\alpha_2+\beta_2=\beta_1-2} \cdots \\
& \sum_{\substack{\alpha_{2n_{21}-1} \in [0, \beta_{2n_{21}-2} - 2n_{21}] \\ \beta_{2n_{21}-1} \in [1, \beta_{2n_{21}-2} - 2n_{21} + 1)]}}^{\alpha_{2n_{21}-1}+\beta_{2n_{21}-1}=\beta_{2n_{21}-2}-2n_{21}+1} \prod_{j=1}^{2n_{21}-1} \frac{(\beta_{j-1} - j)!}{\alpha_j! \beta_j!} (a_{22s_{j+1}} - a_{22s_j})^{\alpha_j} \\
& \times \frac{1}{\beta_{2n_{21}-1}} [(x_2 - a_{22s_{2n_{21}}})^{\beta_{2n_{21}-1}} - (x_{20} - a_{22s_{2n_{21}}})^{\beta_{2n_{21}-1}}] \\
& - \sum_{\beta_0=2n_{21}}^{2n_{11}} b_{12\beta_0} \sum_{k=2}^{2n_{21}-1} \Bigg[\sum_{\substack{\alpha_1 \in [0, \beta_0 - 2n_{21}), \\ \beta_1 \in (2n_{21}-1, \beta_0 - 1]}}^{\alpha_1+\beta_1=\beta_0-1} \sum_{\substack{\alpha_2 \in [0, \beta_1 - 2n_{21}] \\ \beta_2 \in (2n_{21}-2, \beta_1 - 2)}}^{\alpha_2+\beta_2=\beta_1-2} \cdots
\end{aligned}
$$

$$
\begin{aligned}
&\times\Bigg(\sum_{\substack{\alpha_{k-1}\in(\beta_{k-2}-2n_{21},\beta_{k-2}-k+1]\\ \beta_{k-1}\in[0,2n_{21}-k)}}^{\alpha_{k-1}+\beta_{k-1}=\beta_{k-2}-k+1}\prod_{j=1}^{k-1}\frac{(\beta_{j-1}-j)!}{\alpha_j!\beta_j!}(a_{22s_{j+1}}-a_{22s_j})^{\alpha_j}\sum_{\rho=k}^{2n_{21}}A_{\rho 1}^{(s_k)}\ln\frac{|x_{22}-a_{22s_\rho}|}{|x_{20}-a_{22r_\rho}|}\\
&+\sum_{\substack{\alpha_{k-1}\in(\beta_{k-2}-2n_{21},\beta_{k-2}-2n_{21}+1]\\ \beta_{k-1}\in[2n_{21}-k,2n_{21}-k+1)}}^{\alpha_{k-1}+\beta_{k-1}=\beta_{k-2}-k+1}\prod_{j=1}^{k-1}\frac{(\beta_{j-1}-j)!}{\alpha_j!\beta_j!}(a_{22s_{j+1}}-a_{22s_j})^{\alpha_j}\\
&\times\sum_{\rho=k+1}^{2n_{21}}B_{\rho 1}^{(s_k)}\ln\frac{|x_2-a_{22s_\rho}|}{|x_{20}-a_{22s_\rho}|}\Bigg)\Bigg]+b_{j_1j_20}\sum_{\rho=1}^{2n_{21}+1}A_{\rho 1}^{(s_1)}\ln\frac{|x_2-a_{22s_\rho}|}{|x_{20}-a_{22s_\rho}|}\\
&+\sum_{\beta_0=1}^{2n_{21}-1}b_{j_1j_2\beta_0}\sum_{\rho=2}^{2n_{21}}B_{\rho 1}^{(s_1)}\frac{|x_2-a_{22s_\rho}|}{|x_{20}-a_{22s_\rho}|}\Bigg\};
\end{aligned}
\tag{3.70}
$$

where

$$
\begin{aligned}
A_{r_1}&=\frac{1}{\prod_{r_2=1,r_2\neq r_1}^{2m_{11}+1}(a_{11r_1}-a_{11r_\sigma})};\\
b_{j_1j_2\beta_0}&=\sum_{t_1,l_1=0}^{t_1+l_1=1}\cdots\sum_{t_q,l_q=0}^{t_q+l_q=1}\prod_{k=1}^{2n_{11}}(a_{22s_1}-a_{12u_k})^{t_k}\delta_{\beta_0}^{\Sigma_{r=1}^{2n_{11}}l_r};\\
\lim_{\beta_{2n_{21}-1}\to 0}&\frac{1}{\beta_{2n_{21}-1}}[(x_2-a_{22s_{2n_{21}}})^{\beta_{2n_{21}-1}}-(x_{20}-a_{22s_{2n_{21}}})^{\beta_{2n_{21}-1}}]=\ln\frac{|x_2-a_{22s_{2n_{21}}}|}{|x_{20}-a_{22s_{2n_{21}}}|};
\end{aligned}
\tag{3.71}
$$

$$
\begin{aligned}
A_{\rho 1}^{(s_k)}&=\frac{1}{\prod_{\sigma=k,\sigma\neq\rho}^{2n_{21}}(a_{22s_\rho}-a_{22s_\sigma})},\ \text{for }\rho\in\{k,k+1,\cdots,2n_{21}\},\\
B_{\rho 1}^{(s_k)}&=\frac{(a_{22s_\rho}-a_{22s_k})^{\beta_{k-1}-k}}{\prod_{\sigma=k+1,\sigma\neq\rho}^{2n_{21}}(a_{22s_\rho}-a_{22s_\sigma})},\ \text{for}\rho\in\{k+1,k+2,\cdots,2n_{21}\};\\
k&\in\{1,2,\cdots,2n_{21}-1\}.
\end{aligned}
\tag{3.72}
$$

For all cases, the $(2m_{11}+1)\times(2n_{21})$-simple equilibriums plus $(2m_{11}+1)\times(2n_{11})$-hyperbolic flows are based on the following inflection-upper-saddle and lower-saddle bifurcations for $2n_{11}\geq 2n_{21}$:

- $[2(n_{11}-n_{21})]^{\text{th}}$ DI:II, $[(2m_{11}+1),(2n_{21})]^{\text{th}}$US)-decreasing-to-increasing inflection-upper-saddle bifurcation for ($a_{110}>0, a_{220}>0$),
- $[2(n_{11}-n_{21})]^{\text{th}}$ II:DI, $[(2m_{11}+1),(2n_{21})]^{\text{th}}$US)-increasing-to-decreasing inflection-upper-saddle bifurcation for ($a_{110}<0, a_{220}>0$),
- $[2(n_{11}-n_{21})]^{\text{th}}$ II:DI, $[(2m_{11}+1),(2n_{21})]^{\text{th}}$LS)-increasing-to-decreasing inflection-lower-saddle bifurcation for ($a_{110}>0, a_{220}<0$),

- $[2(n_{11} - n_{21})]^{\text{th}}$ DI:II, $[(2m_{11}+1), (2n_{21})]^{\text{th}}$LS)-decreasing-to-increasing inflection-lower-saddle bifurcation for ($a_{110} < 0, a_{220} < 0$);

or based on the following concave-upper-saddle and lower-saddle bifurcations for $2n_{11} < 2n_{21}$:

- $[2(n_{21} - n_{11})]^{\text{th}}$ DC:IC, $[(2m_{11}+1), (2n_{21})]^{\text{th}}$US)-decreasing-to-increasing concave-upper-saddle bifurcation for ($a_{110} > 0, a_{220} > 0$),
- $[2(n_{11} - n_{21})]^{\text{th}}$ IC:DC, $[(2m_{11}+1), (2n_{21})]^{\text{th}}$US)-increasing-to-decreasing concave-upper-saddle bifurcation for ($a_{110} < 0, a_{220} > 0$),
- $[2(n_{11} - n_{21})]^{\text{th}}$ IC:DC, $[(2m_{11}+1), (2n_{21})]^{\text{th}}$LS)-increasing-to-decreasing concave-lower-saddle bifurcation for ($a_{110} > 0, a_{220} < 0$),
- $[2(n_{11} - n_{21})]^{\text{th}}$ DC:IC, $[(2m_{11}+1), (2n_{21})]^{\text{th}}$LS)-decreasing-to-increasing concave-lower-saddle bifurcation for ($a_{110} < 0, a_{220} < 0$).

Consider $2n_{11} = 2n_{21}$ for illustrations, phase portraits for the 2-dimensional systems for the hybrid networks of hyperbolic and hyperbolic-secant flows with the simple equilibriums of the saddles, sinks and sources with $x_1^* = a_{111}, a_{112}, \ldots, a_{11(2m_{11}+1)}$ and $x_2^* = a_{121}, a_{221}, \ldots, a_{12(2n_{11})}, a_{22(2n_{21})}$ are presented in Fig. 3.23 (a–d) for ($a_{110} > 0, a_{220} > 0$), ($a_{110} < 0, a_{220} > 0$), ($a_{110} > 0, a_{220} < 0$) and ($a_{110} < 0, a_{220} < 0$).

$$
\left\{\begin{array}{cccc}
(a_{111}, a_{22(2n_{21})}) & \cdots & (a_{11(2m_{11})}, a_{22(2n_{21})}) & (a_{11(2m_{11}+1)}, a_{22(2n_{21})}) \\
(a_{111}, a_{12(2n_{11})}) & \cdots & (a_{11(2m_{11})}, a_{12(2n_{11})}) & (a_{11(2m_{11}+1)}, a_{12(2n_{11})}) \\
(a_{111}, a_{22(2n_{21}-1)}) & \cdots & (a_{11(2m_{11})}, a_{22(2n_{21}-1)}) & (a_{11(2m_{11}+1)}, a_{22(2n_{21}-1)}) \\
\vdots & \vdots & \vdots & \vdots \\
(a_{111}, a_{221}) & \cdots & (a_{11(2m_{11})}, a_{221}) & (a_{11(2m_{11}+1)}, a_{221}) \\
(a_{111}, a_{121}) & \cdots & (a_{11(2m_{11})}, a_{121}) & (a_{11(2m_{11}+1)}, a_{121})
\end{array}\right\}
$$

$$
= \left\{\begin{array}{cccc}
\underbrace{(\text{SO, SO})}_{\text{source}} & \cdots & \underbrace{(\text{SI, SO})}_{\text{saddle}} & \underbrace{(\text{SO, SO})}_{\text{source}} \\
\underbrace{(\text{UP:DP,nF})}_{\text{hyperbolic-secant flow }(-)} & \cdots & \underbrace{(\text{DP:UP, nF})}_{\text{hyperbolic flow }(-)} & \underbrace{(\text{UP:DP, nF})}_{\text{hyperbolic-secant flow }(-)} \\
\underbrace{(\text{SI, SI})}_{\text{sink}} & \cdots & \underbrace{(\text{SO, SI})}_{\text{saddle}} & \underbrace{(\text{SI, SI})}_{\text{sink}} \\
\vdots & \vdots & \vdots & \vdots \\
\underbrace{(\text{SI, SI})}_{\text{sink}} & \cdots & \underbrace{(\text{SO, SI})}_{\text{saddle}} & \underbrace{(\text{SI, SI})}_{\text{source}} \\
\underbrace{(\text{UP:DP, pF})}_{\text{hyperbolic-secant flow }(+)} & \cdots & \underbrace{(\text{DP:UP, pF})}_{\text{hyperbolic flow }(+)} & \underbrace{(\text{UP:DP, pF})}_{\text{hyperbolic-secant flow }(+)}
\end{array}\right\}_{(2n_{11}+2n_{21})\times(2m_{11}+1)}
$$

for $a_{110} > 0, a_{220} > 0$; (3.73)

$$
\left\{\begin{array}{cccc}
(a_{111},a_{22(2n_{21})}) & \cdots & (a_{11(2m_{11})},a_{22(2n_{21})}) & (a_{11(2m_{11}+1)},a_{22(2n_{21})})\\
(a_{111},a_{12(2n_{11})}) & \cdots & (a_{11(2m_{11})},a_{12(2n_{11})}) & (a_{11(2m_{11}+1)},a_{12(2n_{11})})\\
(a_{111},a_{22(2n_{21}-1)}) & \cdots & (a_{11(2m_{11})},a_{22(2n_{21}-1)}) & (a_{11(2m_{11}+1)},a_{22(2n_{21}-1)})\\
\vdots & \vdots & \vdots & \vdots\\
(a_{111},a_{221}) & \cdots & (a_{11(2m_{11})},a_{221}) & (a_{11(2m_{11}+1)},a_{221})\\
(a_{111},a_{121}) & \cdots & (a_{11(2m_{11})},a_{121}) & (a_{11(2m_{11}+1)},a_{121})
\end{array}\right\}
$$

$$
=\left\{\begin{array}{cccc}
\underbrace{(\mathrm{SI},\mathrm{SO})}_{\text{saddle}} & \cdots & \underbrace{(\mathrm{SO},\mathrm{SO})}_{\text{source}} & \underbrace{(\mathrm{SI},\mathrm{SO})}_{\text{saddle}}\\
\underbrace{(\mathrm{DP{:}UP},\mathrm{nF})}_{\text{hyperbolic flow }(-)} & \cdots & \underbrace{(\mathrm{UP{:}DP},\mathrm{nF})}_{\text{hyperbolic-secant flow }(-)} & \underbrace{(\mathrm{DP{:}UP},\mathrm{nF})}_{\text{hyperbolic flow }(-)}\\
\underbrace{(\mathrm{SO},\mathrm{SI})}_{\text{saddle}} & \cdots & \underbrace{(\mathrm{SI},\mathrm{SI})}_{\text{sink}} & \underbrace{(\mathrm{SO},\mathrm{SI})}_{\text{saddle}}\\
\vdots & \vdots & \vdots & \vdots\\
\underbrace{(\mathrm{SO},\mathrm{SI})}_{\text{saddle}} & \cdots & \underbrace{(\mathrm{SI},\mathrm{SI})}_{\text{sink}} & \underbrace{(\mathrm{SO},\mathrm{SI})}_{\text{saddle}}\\
\underbrace{(\mathrm{DP{:}UP},\mathrm{pF})}_{\text{hyperbolic flow }(+)} & \cdots & \underbrace{(\mathrm{UP{:}DP},\mathrm{pF})}_{\text{hyperbolic-secant flow }(+)} & \underbrace{(\mathrm{DP{:}UP},\mathrm{pF})}_{\text{hyperbolic flow }(+)}
\end{array}\right\}_{(2n_{11}+2n_{21})\times(2m_{11}+1)}
$$

for $a_{110}<0,\ a_{220}>0$; (3.74)

$$
\left\{\begin{array}{cccc}
(a_{111},a_{22(2n_{21}+1)}) & \cdots & (a_{11(2m_{11})},a_{22(2n_{21}+1)}) & (a_{11(2m_{11}+1)},a_{22(2n_{21}+1)})\\
(a_{111},a_{12(2n_{11}+1)}) & \cdots & (a_{11(2m_{11})},a_{12(2n_{11}+1)}) & (a_{11(2m_{11}+1)},a_{12(2n_{11}+1)})\\
(a_{111},a_{22(2n_{21})}) & \cdots & (a_{11(2m_{11})},a_{22(2n_{21})}) & (a_{11(2m_{11}+1)},a_{22(2n_{21})})\\
\vdots & \vdots & \vdots & \vdots\\
(a_{111},a_{221}) & \cdots & (a_{11(2m_{11})},a_{221}) & (a_{11(2m_{11}+1)},a_{221})\\
(a_{111},a_{121}) & \cdots & (a_{11(2m_{11})},a_{121}) & (a_{11(2m_{11}+1)},a_{121})
\end{array}\right\}
$$

$$
=\left\{\begin{array}{cccc}
\underbrace{(\mathrm{SO},\mathrm{SI})}_{\text{saddle}} & \cdots & \underbrace{(\mathrm{SI},\mathrm{SI})}_{\text{sink}} & \underbrace{(\mathrm{SO},\mathrm{SI})}_{\text{saddle}}\\
\underbrace{(\mathrm{DP{:}UP},\mathrm{pF})}_{\text{hyperbolic flow }(+)} & \cdots & \underbrace{(\mathrm{UP{:}DP},\mathrm{pF})}_{\text{hyperbolic-secant flow }(+)} & \underbrace{(\mathrm{DP{:}UP},\mathrm{pF})}_{\text{hyperbolic flow }(+)}\\
\underbrace{(\mathrm{SI},\mathrm{SO})}_{\text{saddle}} & \cdots & \underbrace{(\mathrm{SO},\mathrm{SO})}_{\text{source}} & \underbrace{(\mathrm{SI},\mathrm{SO})}_{\text{saddle}}\\
\vdots & \vdots & \vdots & \vdots\\
\underbrace{(\mathrm{SO},\mathrm{SI})}_{\text{saddle}} & \cdots & \underbrace{(\mathrm{SI},\mathrm{SI})}_{\text{sink}} & \underbrace{(\mathrm{SO},\mathrm{SI})}_{\text{saddle}}\\
\underbrace{(\mathrm{DP{:}UP},\mathrm{pF})}_{\text{hyperbolic flow }(+)} & \cdots & \underbrace{(\mathrm{UP{:}DP},\mathrm{pF})}_{\text{hyperbolic-secant flow }(+)} & \underbrace{(\mathrm{DP{:}UP},\mathrm{pF})}_{\text{hyperbolic flow }(+)}
\end{array}\right\}_{(2n_{11}+2n_{21})\times(2m_{11}+1)}
$$

for $a_{110}>0,\ a_{220}<0$; (3.75)

$$\left\{\begin{array}{cccc} (a_{111}, a_{22(2n_{21})}) & \cdots & (a_{11(2m_{11})}, a_{22(2n_{21})}) & (a_{11(2m_{11}+1)}, a_{22(2n_{21})}) \\ (a_{111}, a_{12(2n_{11})}) & \cdots & (a_{11(2m_{11})}, a_{12(2n_{11})}) & (a_{11(2m_{11}+1)}, a_{12(2n_{11})}) \\ (a_{111}, a_{22(2n_{21}-1)}) & \cdots & (a_{11(2m_{11})}, a_{22(2n_{21}-1)}) & (a_{11(2m_{11}+1)}, a_{22(2n_{21}-1)}) \\ \vdots & \vdots & \vdots & \vdots \\ (a_{111}, a_{221}) & \cdots & (a_{11(2m_{11})}, a_{221}) & (a_{11(2m_{11}+1)}, a_{221}) \\ (a_{111}, a_{121}) & \cdots & (a_{11(2m_{11})}, a_{121}) & (a_{11(2m_{11}+1)}, a_{121}) \end{array}\right\}$$

$$= \left\{\begin{array}{cccc} \underbrace{(\text{SO, SO})}_{\text{source}} & \cdots & \underbrace{(\text{SO, SI})}_{\text{saddle}} & \underbrace{(\text{SI, SI})}_{\text{sink}} \\ \underbrace{(\text{UP:DP, pF})}_{\text{hyperbolic-secant flow (+)}} & \cdots & \underbrace{(\text{DP:UP, pF})}_{\text{hyperbolic flow (+)}} & \underbrace{(\text{UP:DP, pF})}_{\text{hyperbolic-secant flow (+)}} \\ \underbrace{(\text{SO, SO})}_{\text{source}} & \cdots & \underbrace{(\text{SI, SO})}_{\text{saddle}} & \underbrace{(\text{SO, SO})}_{\text{source}} \\ \vdots & \vdots & \vdots & \vdots \\ \underbrace{(\text{SO, SO})}_{\text{source}} & \cdots & \underbrace{(\text{SI, SO})}_{\text{saddle}} & \underbrace{(\text{SO, SO})}_{\text{source}} \\ \underbrace{(\text{UP:DP, nF})}_{\text{hyperbolic-secant flow }(-)} & \cdots & \underbrace{(\text{DP:UP, nF})}_{\text{hyperbolic flow }(-)} & \underbrace{(\text{UP:DP, nF})}_{\text{hyperbolic-secant flow }(-)} \end{array}\right\}_{(2n_{11}+2n_{21})\times(2m_{11}+1)}$$

for $a_{110} < 0, a_{220} < 0.$ (3.76)

Consider $2n_{11} = 2n_{21}$ for illustrations. Phase portraits for the 2-dimensional systems for the hybrid networks of hyperbolic and hyperbolic-secant flows with the simple equilibriums of the saddles, sinks and sources with $x_1^* = a_{111}, a_{112}, \ldots, a_{11(2m_{11}+1)}$ and $x_2^* = a_{221}, a_{121}, \ldots, a_{22(2n_{21})}, a_{12(2n_{11})}$ are presented in Fig. 3.24 (a–d) for $(a_{110} > 0, a_{220} > 0)$, $(a_{110} < 0, a_{220} > 0)$, $(a_{110} > 0, a_{220} < 0)$ and $(a_{110} < 0, a_{220} < 0)$.

$$\left\{\begin{array}{cccc} (a_{111}, a_{12(2n_{11})}) & \cdots & (a_{11(2m_{11})}, a_{12(2n_{11})}) & (a_{11(2m_{11}+1)}, a_{12(2n_{11})}) \\ (a_{111}, a_{22(2n_{21})}) & \cdots & (a_{11(2m_{11})}, a_{22(2n_{21})}) & (a_{11(2m_{11}+1)}, a_{22(2n_{21})}) \\ (a_{111}, a_{12(2n_{11}-1)}) & \cdots & (a_{11(2m_{11})}, a_{12(2n_{11}-1)}) & (a_{11(2m_{11}+1)}, a_{12(2n_{11}-1)}) \\ \vdots & \vdots & \vdots & \vdots \\ (a_{111}, a_{121}) & \cdots & (a_{11(2m_{11})}, a_{121}) & (a_{11(2m_{11}+1)}, a_{121}) \\ (a_{111}, a_{221}) & \cdots & (a_{11(2m_{11})}, a_{221}) & (a_{11(2m_{11}+1)}, a_{221}) \end{array}\right\}$$

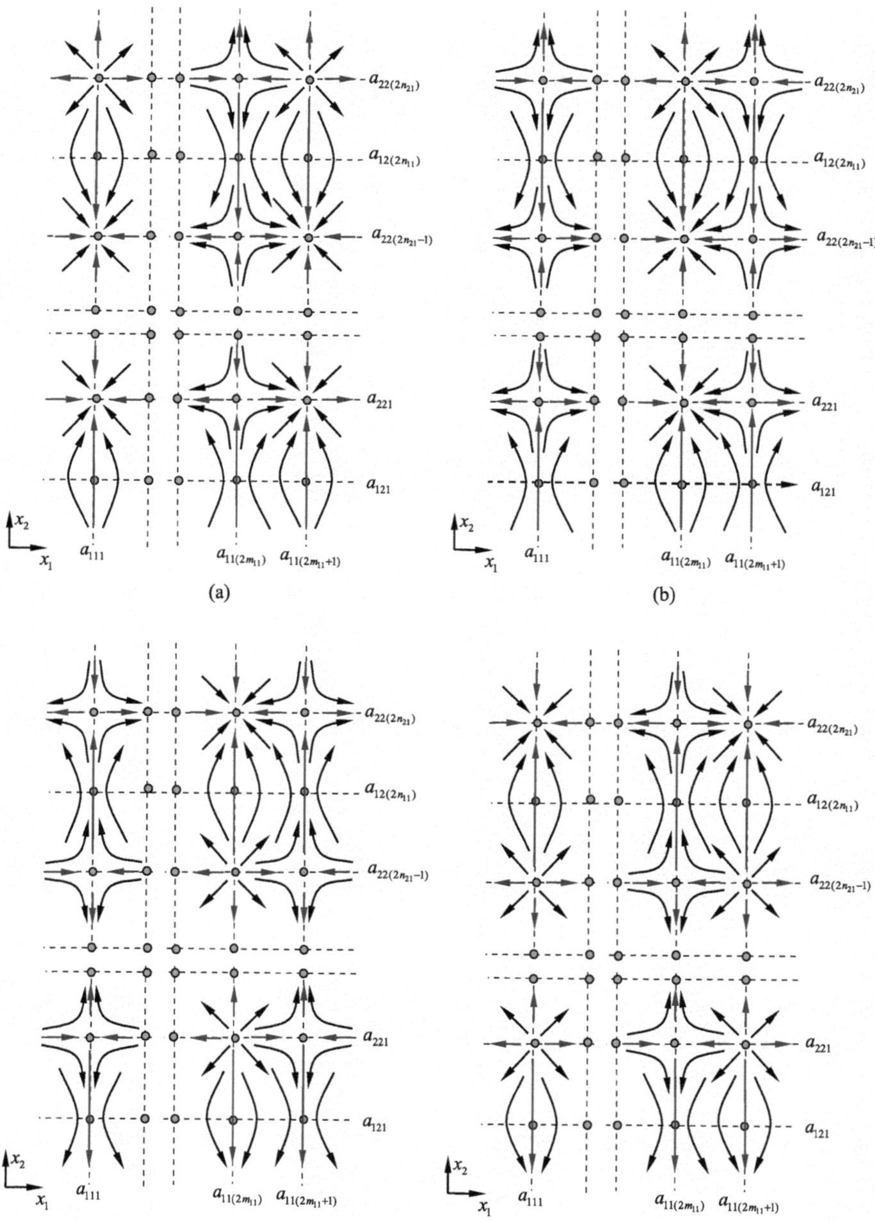

Fig. 3.23 Phase portraits for 2-dimensional systems with $x_1^* = a_{111}, a_{112}, \ldots, a_{11(2m_{11}+1)}$ and $x_2^* = a_{121}, a_{221}, \ldots, a_{12(2n_{11})}, a_{22(2n_{21})}$. The networks of flows with sink, source and saddle: **a** ($a_{110} > 0$, $a_{220} > 0$), **b** ($a_{110} < 0$, $a_{220} > 0$), **c** ($a_{110} > 0$, $a_{220} < 0$), **d** ($a_{110} < 0$, $a_{220} < 0$)

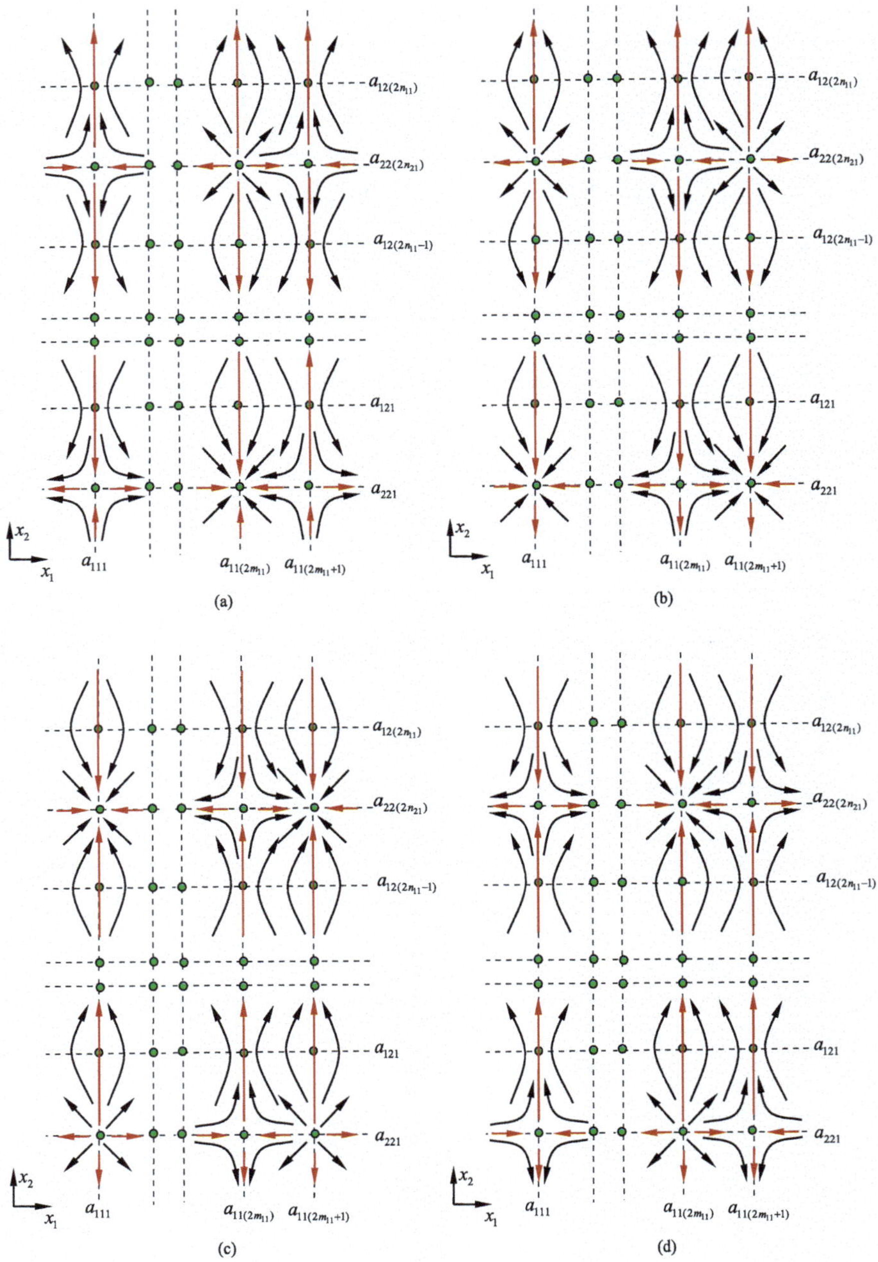

Fig. 3.24 Phase portraits for 2-dimensional systems with $x_1^* = a_{111}, a_{112}, \ldots, a_{11(2m_{11}+1)}$ and $x_2^* = a_{221}, a_{121}, \ldots, a_{22(2n_{21})}, a_{12(2n_{11})}$. The networks of flows with sink, source and saddle: **a** ($a_{110} > 0$, $a_{220} > 0$), **b** ($a_{110} < 0$, $a_{220} > 0$), **c** ($a_{110} > 0$, $a_{220} < 0$), **d** ($a_{110} < 0$, $a_{220} < 0$)

$$
= \left\{ \begin{array}{cccc}
\underbrace{\text{(DP:UP, pF)}}_{\text{hyperbolic flow (+)}} & \cdots & \underbrace{\text{(UP:DP, pF)}}_{\text{hyperbolic-secant flow (+)}} & \underbrace{\text{(DP:UP, pF)}}_{\text{hyperbolic flow (+)}} \\
\underbrace{\text{(SI, SO)}}_{\text{saddle}} & \cdots & \underbrace{\text{(SO, SO)}}_{\text{source}} & \underbrace{\text{(SI, SO)}}_{\text{saddle}} \\
\underbrace{\text{(DP:UP, nF)}}_{\text{hyperbolic flow }(-)} & \cdots & \underbrace{\text{(UP:DP, nF)}}_{\text{hyperbolic-secant flow }(-)} & \underbrace{\text{(DP:UP, nF)}}_{\text{hyperbolic flow }(-)} \\
\vdots & \vdots & \vdots & \vdots \\
\underbrace{\text{(DP:UP, nF)}}_{\text{hyperbolic flow }(-)} & \cdots & \underbrace{\text{(UP:DP, nF)}}_{\text{hyperbolic-secant flow }(-)} & \underbrace{\text{(DP:UP, nF)}}_{\text{hyperbolic flow }(-)} \\
\underbrace{\text{(SO, SI)}}_{\text{saddle}} & \cdots & \underbrace{\text{(SI, SI)}}_{\text{sink}} & \underbrace{\text{(SO, SI)}}_{\text{saddle}}
\end{array} \right\}_{(2n_{11}+2n_{21})\times(2m_{11}+1)}
$$

for $a_{110} > 0, a_{220} > 0$; (3.77)

$$
\left\{ \begin{array}{cccc}
(a_{111}, a_{12(2n_{11})}) & \cdots & (a_{11(2m_{11})}, a_{12(2n_{11})}) & (a_{11(2m_{11}+1)}, a_{12(2n_{11})}) \\
(a_{111}, a_{22(2n_{21})}) & \cdots & (a_{11(2m_{11})}, a_{22(2n_{21})}) & (a_{11(2m_{11}+1)}, a_{22(2n_{21})}) \\
(a_{111}, a_{12(2n_{11}-1)}) & \cdots & (a_{11(2m_{11})}, a_{12(2n_{11}-1)}) & (a_{11(2m_{11}+1)}, a_{12(2n_{11}-1)}) \\
\vdots & \vdots & \vdots & \vdots \\
(a_{111}, a_{121}) & \cdots & (a_{11(2m_{11})}, a_{121}) & (a_{11(2m_{11}+1)}, a_{121}) \\
(a_{111}, a_{221}) & \cdots & (a_{11(2m_{11})}, a_{221}) & (a_{11(2m_{11}+1)}, a_{221})
\end{array} \right\}
$$

$$
= \left\{ \begin{array}{cccc}
\underbrace{\text{(UP:DP, pF)}}_{\text{hyperbolic-secant flow (+)}} & \cdots & \underbrace{\text{(DP:UP, pF)}}_{\text{hyperbolic flow (+)}} & \underbrace{\text{(UP:DP, pF)}}_{\text{hyperbolic-secant flow (+)}} \\
\underbrace{\text{(SO, SO)}}_{\text{source}} & \cdots & \underbrace{\text{(SI, SO)}}_{\text{saddle}} & \underbrace{\text{(SO, SO)}}_{\text{source}} \\
\underbrace{\text{(UP:DP, nF)}}_{\text{hyperbolic-secant flow }(-)} & \cdots & \underbrace{\text{(DP:UP, nF)}}_{\text{hyperbolic flow }(-)} & \underbrace{\text{(UP:DP, nF)}}_{\text{hyperbolic-secant flow }(-)} \\
\vdots & \vdots & \vdots & \vdots \\
\underbrace{\text{(UP:DP, nF)}}_{\text{hyperbolic-secant flow }(-)} & \cdots & \underbrace{\text{(DP:UP, nF)}}_{\text{hyperbolic flow }(-)} & \underbrace{\text{(UP:DP, nF)}}_{\text{hyperbolic-secant flow }(-)} \\
\underbrace{\text{(SI, SI)}}_{\text{sink}} & \cdots & \underbrace{\text{(SO, SI)}}_{\text{saddle}} & \underbrace{\text{(SI, SI)}}_{\text{sink}}
\end{array} \right\}_{(2n_{11}+2n_{21})\times(2m_{11}+1)}
$$

for $a_{110} < 0, a_{220} > 0$; (3.78)

$$
\left\{ \begin{array}{cccc}
(a_{111}, a_{12(2n_{11})}) & \cdots & (a_{11(2m_{11})}, a_{12(2n_{11})}) & (a_{11(2m_{11}+1)}, a_{12(2n_{11})}) \\
(a_{111}, a_{22(2n_{21})}) & \cdots & (a_{11(2m_{11})}, a_{22(2n_{21})}) & (a_{11(2m_{11}+1)}, a_{22(2n_{21})}) \\
(a_{111}, a_{12(2n_{11}-1)}) & \cdots & (a_{11(2m_{11})}, a_{12(2n_{11}-1)}) & (a_{11(2m_{11}+1)}, a_{12(2n_{11}-1)}) \\
\vdots & \vdots & \vdots & \vdots \\
(a_{111}, a_{121}) & \cdots & (a_{11(2m_{11})}, a_{121}) & (a_{11(2m_{11}+1)}, a_{121}) \\
(a_{111}, a_{221}) & \cdots & (a_{11(2m_{11})}, a_{221}) & (a_{11(2m_{11}+1)}, a_{221})
\end{array} \right\}
$$

$$
= \left\{\begin{array}{cccc}
\underbrace{(\text{UP:DP, nF})}_{\text{hyperbolic-secant flow }(-)} & \cdots & \underbrace{(\text{DP:UP, nF})}_{\text{hyperbolic flow }(-)} & \underbrace{(\text{UP:DP, nF})}_{\text{hyperbolic-secant flow }(-)} \\
\underbrace{(\text{SI, SI})}_{\text{sink}} & \cdots & \underbrace{(\text{SO, SI})}_{\text{saddle}} & \underbrace{(\text{SI, SI})}_{\text{sink}} \\
\underbrace{(\text{UP:DP, pF})}_{\text{hyperbolic-secant flow }(+)} & \cdots & \underbrace{(\text{DP:UP, pF})}_{\text{hyperbolic flow }(+)} & \underbrace{(\text{UP:DP, pF})}_{\text{hyperbolic-secant flow }(+)} \\
\vdots & \vdots & \vdots & \vdots \\
\underbrace{(\text{UP:DP, pF})}_{\text{hyperbolic-secant flow }(+)} & \cdots & \underbrace{(\text{DP:UP, pF})}_{\text{hyperbolic flow }(+)} & \underbrace{(\text{UP:DP, pF})}_{\text{hyperbolic-secant flow }(+)} \\
\underbrace{(\text{SO, SO})}_{\text{source}} & \cdots & \underbrace{(\text{SO, SI})}_{\text{saddle}} & \underbrace{(\text{SO, SO})}_{\text{source}}
\end{array}\right\}_{(2n_{11}+2n_{21})\times(2m_{11}+1)}
$$

for $a_{110} > 0, a_{220} < 0$; (3.79)

$$
\left\{\begin{array}{cccc}
(a_{111}, a_{12(2n_{11})}) & \cdots & (a_{11(2m_{11})}, a_{12(2n_{11})}) & (a_{11(2m_{11}+1)}, a_{12(2n_{11})}) \\
(a_{111}, a_{22(2n_{21})}) & \cdots & (a_{11(2m_{11})}, a_{22(2n_{21})}) & (a_{11(2m_{11}+1)}, a_{22(2n_{21})}) \\
(a_{111}, a_{12(2n_{11}-1)}) & \cdots & (a_{11(2m_{11})}, a_{12(2n_{11}-1)}) & (a_{11(2m_{11}+1)}, a_{12(2n_{21}-1)}) \\
\vdots & \vdots & \vdots & \vdots \\
(a_{111}, a_{121}) & \cdots & (a_{11(2m_{11})}, a_{121}) & (a_{11(2m_{11}+1)}, a_{121}) \\
(a_{111}, a_{221}) & \cdots & (a_{11(2m_{11})}, a_{221}) & (a_{11(2m_{11}+1)}, a_{221})
\end{array}\right\}
$$

$$
= \left\{\begin{array}{cccc}
\underbrace{(\text{DP:UP, nF})}_{\text{hyperbolic flow }(-)} & \cdots & \underbrace{(\text{UP:DP, nF})}_{\text{hyperbolic-secant flow }(-)} & \underbrace{(\text{DP:UP, nF})}_{\text{hyperbolic flow }(-)} \\
\underbrace{(\text{SO, SI})}_{\text{saddle}} & \cdots & \underbrace{(\text{SI, SI})}_{\text{sink}} & \underbrace{(\text{SO, SI})}_{\text{saddle}} \\
\underbrace{(\text{DP:UP, pF})}_{\text{hyperbolic flow }(+)} & \cdots & \underbrace{(\text{UP:DP, pF})}_{\text{hyperbolic-secant flow }(+)} & \underbrace{(\text{DP:UP, pF})}_{\text{hyperbolic flow }(+)} \\
\vdots & \vdots & \vdots & \vdots \\
\underbrace{(\text{DP:UP, pF})}_{\text{hyperbolic flow }(+)} & \cdots & \underbrace{(\text{UP:DP, pF})}_{\text{hyperbolic-secant flow }(+)} & \underbrace{(\text{DP:UP, pF})}_{\text{hyperbolic flow }(+)} \\
\underbrace{(\text{SO, SI})}_{\text{saddle}} & \cdots & \underbrace{(\text{SO, SO})}_{\text{source}} & \underbrace{(\text{SI, SO})}_{\text{saddle}}
\end{array}\right\}_{(2n_{11}+2n_{21})\times(2m_{11}+1)}
$$

for $a_{110} < 0, a_{220} < 0$. (3.80)

3.5 [(2m_{11}), (2n_{11} + 1)] and ((2m_{11}), (2n_{21} + 1))-Singular Systems

In this section, there are four global bifurcations of the 1-diemsnional flows through singular hyperbolic-to-hyperbolic-secant flows, and four global bifurcations of equilibriums, through singular saddle-source and saddle-sink bifurcations. The corresponding global bifurcations generate networks of hyperbolic and hyperbolic-secant flows with sources, sinks and saddles. The infinite-equilibriums are the appearing and switching bifurcations for the singular flows and singular equilibriums and for hybrid networks of hyperbolic flows and source, sink and saddles.

3.5.1 [(2m_{11}), (2n_{11} + 1)]-*Flows and* ((2m_{11}), (2n_{21} + 1))-*Equilibriums*

Consider a polynomial dynamical system for $a_{221} \neq a_{121}$ as

$$\begin{aligned} \dot{x}_2 &= a_{220}(x_2 - a_{221})^{2n_{21}+1}, \\ \dot{x}_1 &= a_{110}(x_1 - a_{111})^{2m_{11}}(x_2 - a_{121})^{2n_{11}+1}. \end{aligned} \tag{3.81}$$

The first integral manifold is

$$\begin{aligned} &-\frac{1}{2m_{11}-1}\left[\frac{1}{(x_1-a_{111})^{2m_{11}-1}} - \frac{1}{(x_{10}-a_{111})^{2m_{11}-1}}\right] \\ &= \frac{a_{110}}{a_{220}} \sum_{\alpha,\beta=0}^{\alpha+\beta=2n_{11}+1} \frac{(2n_{11}+1)!}{\alpha!\beta!} \frac{(a_{221}-a_{121})^{\alpha}}{\beta-2n_{21}} \left[(x_2-a_{221})^{\beta-2n_{21}} - (x_2-a_{221})^{\beta-2n_{21}}\right] \\ &\text{for } 2m_{11}-1 \neq 0,\ \beta-2n_{21} \neq 0; \\ &\frac{1}{\beta-2n_{21}}\left[(x_2-a_{221})^{\beta-2n_{21}} - (x_2-a_{221})^{\beta-2n_{21}}\right] = \ln\frac{|x_2-a_{221}|}{|x_{20}-a_{221}|} \\ &\text{for } \beta-2n_{21} = 0. \end{aligned} \tag{3.82}$$

For $(x_1^*, x_2^*) = (a_{111}, a_{121})$ with $\dot{x}_1 = 0$, the [(2m_{11}), (2n_{11} + 1)]-singular flows exist with $\dot{x}_2 = a_{220}(a_{121} - a_{221})^{2n_{21}+1}$. For $(x_1^*, x_2^*) = (a_{111}, a_{121})$, the 1-dimensional flow direction is determined by $\dot{x}_2 \neq 0s$. For $(x_1^*, x_2^*) = (a_{111}, a_{221})$ with $\dot{x}_1 = \dot{x}_2 = 0$, the ((2$m_{11}$), (2$n_{21}$ + 1))- singular equilibriums exist.

(A) For $a_{121} > a_{221}$,

$$\left\{ \begin{array}{l} (a_{111}, a_{121}) \\ (a_{111}, a_{221}) \end{array} \right\} = \left\{ \begin{array}{c} \underbrace{([(2m_{11}), (2n_{11}+1)]\text{UP:UP, pF})}_{[(2m_{11}),(2n_{11}+1)]\text{-hyperbolic-secant-to hyperbolic flow } (+)} \\ \underbrace{((2m_{11})\text{LS}, (2n_{21}+1)\text{SO})}_{[(2m_{11}),(2n_{21}+1)]\text{-lower-saddle-source}} \end{array} \right\},$$

for $a_{110} > 0, a_{220} > 0$;

$$\left\{ \begin{array}{l} (a_{111}, a_{121}) \\ (a_{111}, a_{221}) \end{array} \right\} = \left\{ \begin{array}{c} \underbrace{([(2m_{11}), (2n_{11}+1)]\text{DP:DP, pF})}_{[(2m_{11}),(2n_{11}+1)]\text{-hyperbolic-to-hyperbplic-secant flow } (+)} \\ \underbrace{((2m_{11})\text{US}, (2n_{21}+1)\text{SO})}_{[(2m_{11}),(2n_{21}+1)]\text{-upper-saddle-source}} \end{array} \right\},$$

for $a_{110} < 0, a_{220} > 0$;

$$\left\{ \begin{array}{l} (a_{111}, a_{121}) \\ (a_{111}, a_{221}) \end{array} \right\} = \left\{ \begin{array}{c} \underbrace{([(2m_{11}), (2n_{11}+1)]\text{DP:DP, nF})}_{[(2m_{11}),(2n_{11}+1)]\text{-hyperbolic-to-hyperbolic-secant flow } (-)} \\ \underbrace{((2m_{11})\text{LS}, (2n_{21}+1)\text{SI})}_{[(2m_{11}),(2n_{21}+1)]\text{-lower-saddle sink}} \end{array} \right\},$$

for $a_{110} > 0, a_{220} < 0$;

$$\left\{ \begin{array}{l} (a_{111}, a_{121}) \\ (a_{111}, a_{221}) \end{array} \right\} = \left\{ \begin{array}{c} \underbrace{([(2m_{11}), (2n_{11}+1)]\text{UP:UP, nF})}_{[(2m_{11}),(2n_{11}+1)]\text{-hyperbolic-secant-to-hyperbolic flow } (-)} \\ \underbrace{((2m_{11})\text{US}, (2n_{21}+1)\text{SI})}_{[(2m_{11}),(2n_{21}+1)]\text{-upper-saddle-sink}} \end{array} \right\},$$

for $a_{110} < 0, a_{220} < 0$. (3.83)

(B) For $a_{121} < a_{221}$,

$$\left\{ \begin{array}{l} (a_{111}, a_{221}) \\ (a_{111}, a_{121}) \end{array} \right\} = \left\{ \begin{array}{c} \underbrace{((2m_{11})\text{US}, (2n_{21}+1)\text{SO})}_{[(2m_{11}),(2n_{21}+1)]\text{-upper-saddle-source}} \\ \underbrace{([(2m_{11}), (2n_{11}+1)]\text{DP:DP, nF})}_{[(2m_{11}),(2n_{11}+1)]\text{-hyperbolic-to-hyperbolic-secant flow } (-)} \end{array} \right\},$$

for $a_{110} > 0, a_{220} > 0$;

$$\left\{ \begin{array}{l} (a_{111}, a_{221}) \\ (a_{111}, a_{121}) \end{array} \right\} = \left\{ \begin{array}{c} \underbrace{((2m_{11})\text{LS}, (2n_{21}+1)\text{SO})}_{[(2m_{11}),(2n_{21}+1)]\text{-lower-saddle-source}} \\ \underbrace{([(2m_{11}), (2n_{11}+1)]\text{UP:UP, nF})}_{[(2m_{11}),(2n_{11}+1)]\text{-hyperbolic-secant-to-hyperbolic flow } (-)} \end{array} \right\},$$

for $a_{110} < 0, a_{220} > 0$;

$$\left\{ \begin{array}{l} (a_{111}, a_{221}) \\ (a_{111}, a_{121}) \end{array} \right\} = \left\{ \begin{array}{c} \underbrace{((2m_{11})\text{US},(2n_{21}+1)\text{SI})}_{[(2m_{11}),(2n_{21}+1)]\text{-upper-saddle-sink}} \\ \underbrace{([(2m_{11}), (2n_{11}+1)]\text{UP:UP, pF})}_{[(2m_{11}+1),(2n_{11}+1)]\text{-hyperbolic-secant-to hyperbolic flow } (+)} \end{array} \right\},$$

for $a_{110} > 0, a_{220} < 0$;

$$\left\{\begin{array}{l}(a_{111}, a_{221})\\(a_{111}, a_{121})\end{array}\right\} = \left\{\begin{array}{c}\underbrace{((2m_{11})\text{LS}, (2n_{21}+1)\text{SI})}_{[(2m_{11}),(2n_{21}+1)]\text{-lower-saddle sink}}\\ \underbrace{([(2m_{11}), (2n_{11}+1)]\text{DP:DP, pF})}_{[(2m_{11}),(2n_{11}+1)]\text{-hyperbolic-to-hyperbolic-secant flow }(+)}\end{array}\right\},$$

for $a_{110} < 0, a_{220} < 0$. (3.84)

The $([(2m_{11}), (2n_{11}+1)]$UP:UP, pF) is for a positive singular hyperbolic-secant-to-hyperbolic flow. The $([(2m_{11}), (2n_{11}+1)]$DP:DP, pF) is for a positive singular hyperbolic-to-hyperbolic-secant flow. The $([(2m_{11}), (2n_{11}+1)]$DP:DP, nF) is for a negative singular hyperbolic-to-hyperbolic-secant flow. The $([(2m_{11}), (2n_{11}+1)]$UP:UP, nF) is for a negative singular hyperbolic-secant-to-hyperbolic flow. UP:UP is for a hyperbolic-secant-to-hyperbolic flow, and DP:DP is for a hyperbolic-to-hyperbolic-secant flow. The $((2m_{11})$US, $(2n_{21}+1)$SO] is for a $((2m_{11}), (2n_{21}+1)]$-singular upper-saddle-source. The $((2m_{11})$LS, $(2n_{21}+1)$SO] is for a $((2m_{11}), (2n_{21}+1)]$-lower-saddle-source. The $((2m_{11})$US, $(2n_{21}+1)$SI] is for a $((2m_{11}), (2n_{21}+1)]$-upper-saddle-sink. The $((2m_{11})$LS, $(2n_{21}+1)$SI] is for a $((2m_{11}), (2n_{21}+1)]$-lower-saddle-sink.

The singular hyperbolic-to-hyperbolic-secant flows with singular saddle-sources and saddle-sinks are illustrated in Figs. 3.25 and 3.26. Through such singular flows, the corresponding global bifurcations generate from a hyperbolic (hyperbolic-secant) flow to a network of singular hyperbolic and hyperbolic-secant flows. Through such singular equilibriums, the corresponding global bifurcations generate a network of singular source, sink and saddle equilibriums.

3.5.2 $[(2m_{11}), (2n_{11}+1)]$ and $((2m_{11}), (2n_{21}+1))$-*Infinite-Equilibriums*

Consider a polynomial dynamical system for $a_{221} = a_{121}$ as

$$\begin{aligned}\dot{x}_2 &= a_{220}(x_2 - a_{221})^{2n_{21}+1},\\ \dot{x}_1 &= a_{110}(x_1 - a_{111})^{2m_{11}}(x_2 - a_{121})^{2n_{11}+1}.\end{aligned} \quad (3.85)$$

The first integral manifold is

$$\begin{aligned}&-\frac{1}{2m_{11}-1}\left[\frac{1}{(x_1 - a_{111})^{2m_{11}-1}} - \frac{1}{(x_{10} - a_{111})^{2m_{11}-1}}\right]\\ &= \frac{a_{110}}{a_{220}}\frac{1}{2(n_{11}-n_{21})+1}[(x_2 - a_{121})^{2(n_{11}-n_{21})+1} - (x_2 - a_{121})^{2(n_{11}-n_{21})+1}].\end{aligned} \quad (3.86)$$

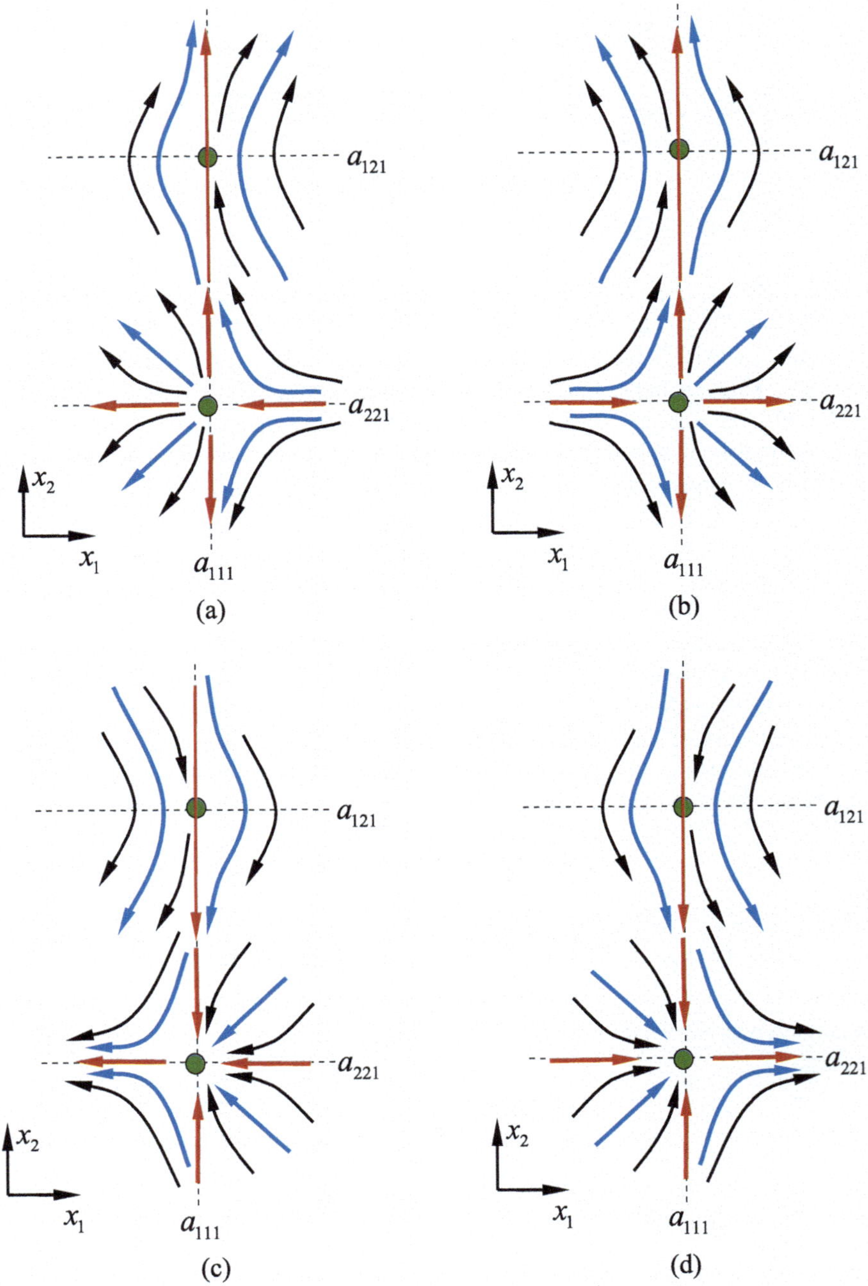

Fig. 3.25 Phase portraits ($a_{121} > a_{221}$) of the ($[(2m_{11}), (2n_{11}+1)]$, xF)-hyperbolic-to-hyperbolic-secant flows at $(x_1^*, x_2^*) = (a_{111}, a_{121})$ and ($(2m_{11}), (2n_{21}+1)$)-saddle-source and saddle-sink at $(x_1^*, x_2^*) = (a_{111}, a_{221})$ for 2-dimensional polynomial systems: **a** ($a_{110} > 0$, $a_{220} > 0$), **b** ($a_{110} < 0$, $a_{220} > 0$), **c** ($a_{110} > 0$, $a_{220} < 0$), **d** ($a_{110} < 0$, $a_{220} < 0$)

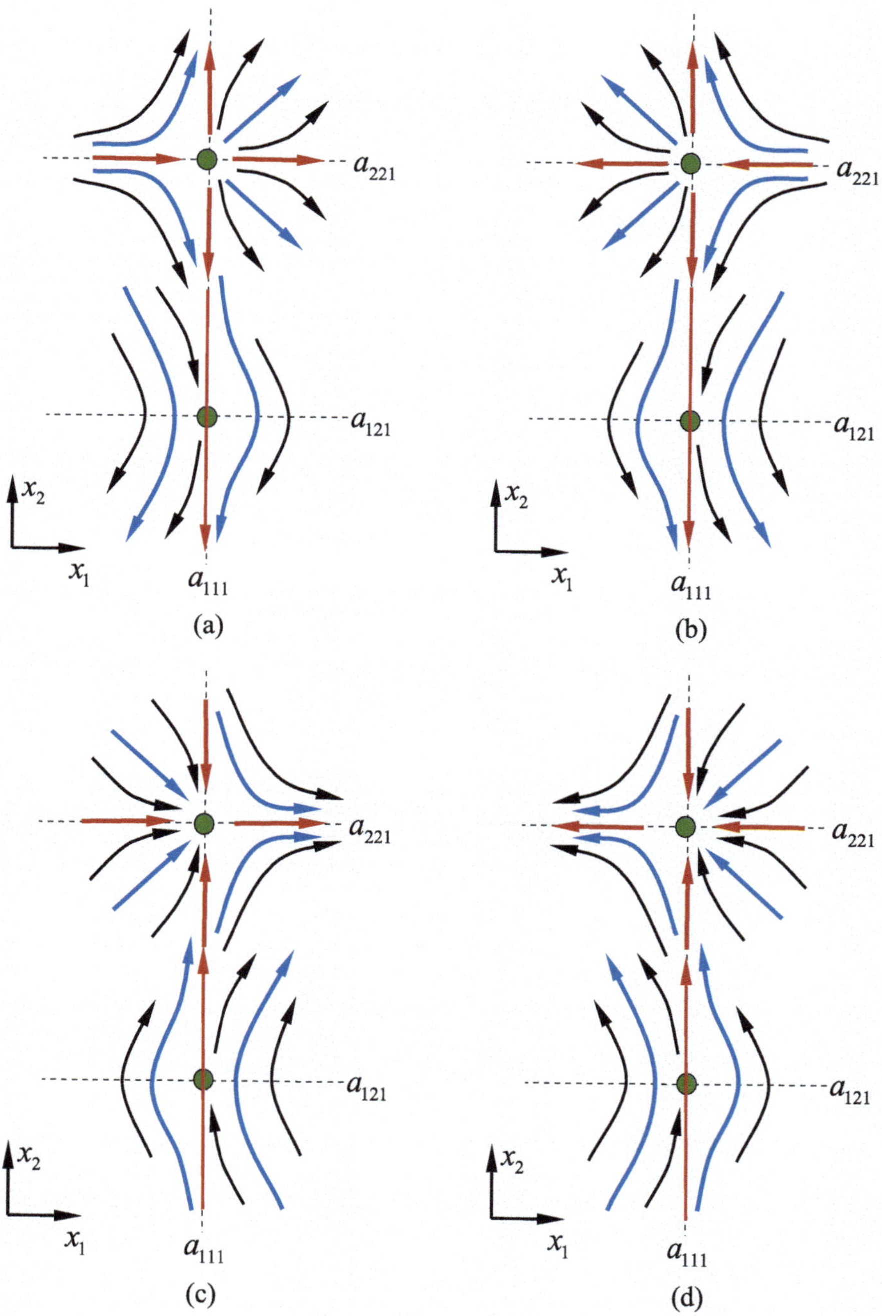

Fig. 3.26 Phase portraits ($a_{221} > a_{121}$) of the ($[(2m_{11}), (2n_{11}+1)]$, xF)-hyperbolic-to-hyperbolic-secant flows at $(x_1^*, x_2^*) = (a_{111}, a_{121})$ and $((2m_{11}), (2n_{21}+1))$-saddle-source and saddle-sink at $(x_1^*, x_2^*) = (a_{111}, a_{221})$ for 2-dimensional polynomial systems: **a** ($a_{110} > 0$, $a_{220} > 0$), **b** ($a_{110} < 0$, $a_{220} > 0$), **c** ($a_{110} > 0$, $a_{220} < 0$), **d** ($a_{110} < 0$, $a_{220} < 0$)

(A) For $2n_{11}+1 \geq 2n_{21}+1$, the corresponding infinite-equilibriums are given.

$$
\begin{aligned}
(a_{111}, a_{221}) &= \underbrace{([2(n_{11}-n_{21})]^{\text{th}}\text{II:II}, [(2m_{11}), (2n_{21}+1)]^{\text{th}}\text{SO})}_{([2(n_{11}-n_{21})],[(2m_{11}+1),(2n_{21}+1)])\text{-increasing inflection-source}}, \\
&\text{for } a_{110}>0, a_{220}>0; \\
(a_{111}, a_{221}) &= \underbrace{([2(n_{11}-n_{21})]^{\text{th}}\text{DI:DI}, [(2m_{11}), (2n_{21}+1)]^{\text{th}}\text{SO})}_{([2(n_{11}-n_{21})],[(2m_{11}),(2n_{21}+1)])\text{-decreasing inflection-source}}, \\
&\text{for } a_{110}<0, a_{220}>0; \\
(a_{111}, a_{221}) &= \underbrace{([2(n_{11}-n_{21})]^{\text{th}}\text{DI:DI}, [(2m_{11}), (2n_{21}+1)]^{\text{th}}\text{SI})}_{([2(n_{11}-n_{21})],[(2m_{11}),(2n_{21}+1)])\text{-decreasing inflection-sink}}, \\
&\text{for } a_{110}>0, a_{220}<0; \\
(a_{111}, a_{221}) &= \underbrace{([2(n_{11}-n_{21})]^{\text{th}}\text{II:II}, [(2m_{11}), (2n_{21}+1)]^{\text{th}}\text{SI})}_{([2(n_{11}-n_{21})],[(2m_{11}),(2n_{21}+1)])\text{-increasing inflection-sink}}, \\
&\text{for } a_{110}<0, a_{220}<0.
\end{aligned}
\tag{3.87}
$$

(B) For $2n_{11}+1 < 2n_{21}+1$, the infinite-equilibriums are given as follows.

$$
\begin{aligned}
(a_{111}, a_{221}) &= \underbrace{([2(n_{21}-n_{11})]^{\text{th}}\text{IC:IC}, [(2m_{11}), (2n_{21}+1)]^{\text{th}}\text{SO})}_{([2(n_{21}-n_{11})],[(2m_{11}+1),(2n_{21}+1)])\text{ -increasing concave-source}}, \\
&\text{for } a_{110}>0, a_{220}>0; \\
(a_{111}, a_{221}) &= \underbrace{([2(n_{21}-n_{11})]^{\text{th}}\text{DC:DC}, [(2m_{11}), (2n_{21}+1)]^{\text{th}}\text{SO})}_{([2(n_{21}-n_{11})],[(2m_{11}+1),(2n_{21}+1)])\text{-decreasing concave-source}}, \\
&\text{for } a_{110}<0, a_{220}>0; \\
(a_{111}, a_{221}) &= \underbrace{([2(n_{21}-n_{11})]^{\text{th}}\text{DC:DC}, [(2m_{11}), (2n_{21}+1)]^{\text{th}}\text{SI})}_{([2(n_{21}-n_{11})],[(2m_{11}+1),(2n_{21}+1)])\text{-decreasing concave-sink}}, \\
&\text{for } a_{110}>0, a_{220}<0; \\
(a_{111}, a_{221}) &= \underbrace{([2(n_{21}-n_{11})]^{\text{th}}\text{IC:IC}, [(2m_{11}), (2n_{21}+1)]^{\text{th}}\text{SI})}_{([2(n_{21}-n_{11})],[(2m_{11}+1),(2n_{21}+1)])\text{-increasing concave-sink}}, \\
&\text{for } a_{110}<0, a_{220}<0.
\end{aligned}
\tag{3.88}
$$

For $x_2^* = a_{121} = a_{221}$, the infinite-equilibrium exists. The infinite-equilibriums are illustrated in Figs. 3.27 and 3.28 for $n_{11} \geq n_{21}$ and $n_{11} < n_{21}$, respectively. For $n_{11} \geq n_{21}$, the singular decreasing-to-increasing and increasing-to-decreasing inflection-source and sink are presented. For $n_{11} < n_{21}$, the singular decreasing-to-increasing and increasing-to-decreasing concave-source and sink are presented.

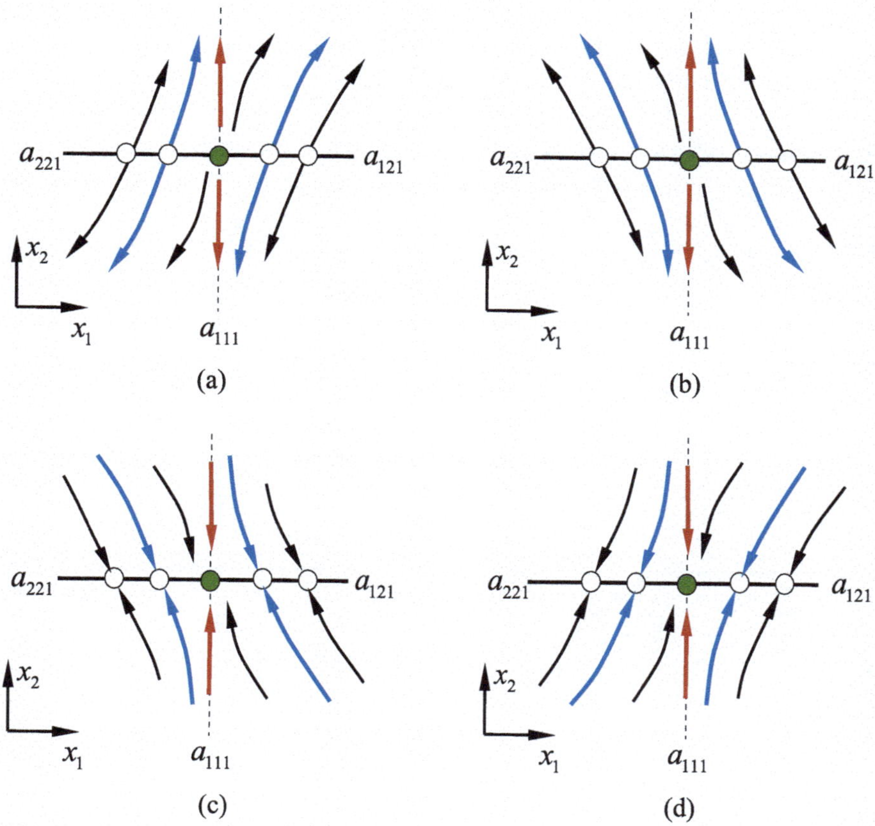

Fig. 3.27 The infinite-equilibriums ($a_{121} = a_{221}, 2n_{11}+1 \geq 2n_{21}+1, m_1 = 2m_{11}$) for 2-dimensional polynomial systems: **a** ($a_{110} > 0, a_{220} > 0$), **b** ($a_{110} < 0, a_{220} > 0$), **c** ($a_{110} > 0, a_{220} < 0$), **d** ($a_{110} < 0, a_{220} < 0$)

3.5.3 $[(2m_{11}), (2n_{11}+1)]$ *and* $((2m_{11}), (2n_{21}+1))$-*Hybrid Networks*

Consider

$$\begin{aligned}\dot{x}_2 &= a_{220}\prod_{s=1}^{2n_{21}+1}(x_2 - a_{22s}),\\ \dot{x}_1 &= a_{110}\prod_{r=1}^{2m_{11}}(x_1 - a_{11r})\prod_{u=1}^{2n_{11}+1}(x_2 - a_{12u}).\end{aligned} \tag{3.89}$$

The first integral manifold is

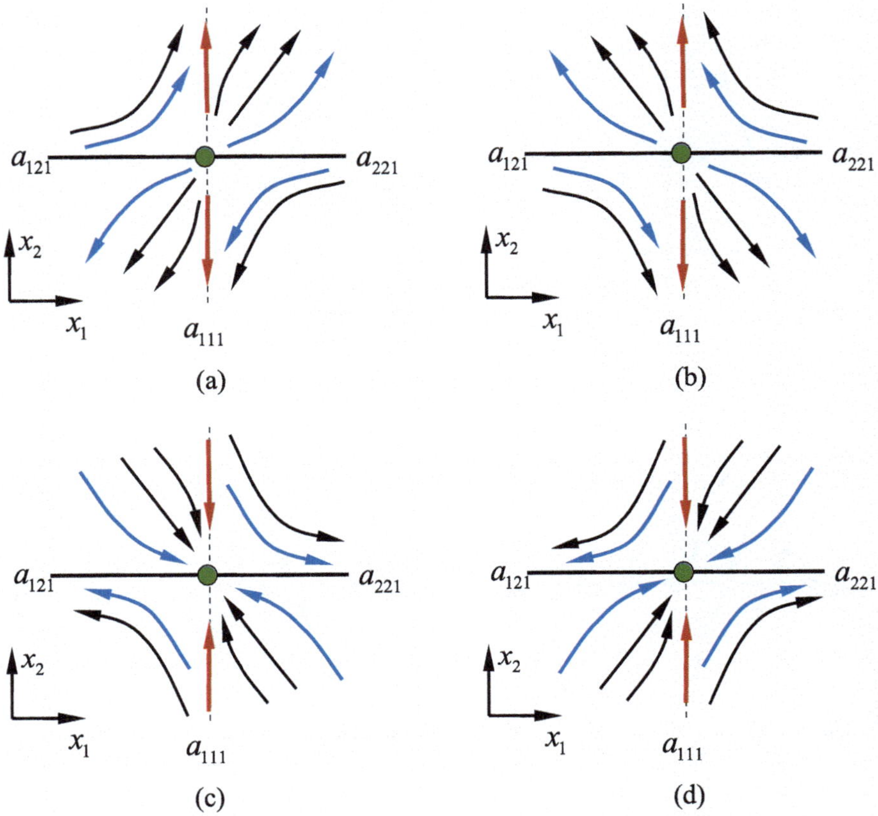

Fig. 3.28 The infinite-equilibriums ($a_{121} = a_{221}, 2n_{11}+1 < 2n_{21}+1, m_1 = 2m_{11}$) for 2-dimensional polynomial systems: **a** ($a_{110} > 0, a_{220} > 0$), **b** ($a_{110} < 0, a_{220} > 0$), **c** ($a_{110} > 0, a_{220} < 0$), **d** ($a_{110} < 0, a_{220} < 0$)

$$\sum_{r_1=1}^{2m_{11}} A_{r_1}\ln\frac{|x_1-a_{11r_1}|}{|x_{10}-a_{11r_1}|} = \frac{a_{110}}{a_{220}}\left\{\sum_{\beta_0=2n_{21}+1}^{2n_{11}+1} b_{12\beta_0} \sum_{\substack{\alpha_1\in[0,\beta_0-(2n_{21}+1)],\\ \beta_1\in[2n_{21},\beta_0-1]}}^{\alpha_1+\beta_1=\beta_0-1} \sum_{\substack{\alpha_2\in[0,\beta_1-(2n_{21}+1)]\\ \beta_2\in[2n_{21}-1,\beta_1-2]}}^{\alpha_2+\beta_2=\beta_1-2} \cdots\right.$$

$$\sum_{\substack{\alpha_{2n_{21}}\in[0,\beta_{2n_{21}-1}-2n_{21}-1]\\ \beta_{2n_{21}}\in[1,\beta_{2n_{21}-1}-2n_{21}]}}^{\alpha_{2n_{21}}+\beta_{2n_{21}}=\beta_{2n_{21}-1}-2n_{21}} \prod_{j=1}^{2n_{21}} \frac{(\beta_{j-1}-j)!}{\alpha_j!\beta_j!}(a_{22s_{j+1}}-a_{22s_j})^{\alpha_j}$$

$$\times\frac{1}{\beta_{2n_{21}}}[(x_2-a_{22s_{2n_{21}+1}})^{\beta_{2n_{21}}}-(x_{20}-a_{22s_{2n_{21}+1}})^{\beta_{2n_{21}}}]$$

$$-\sum_{\beta_0=2n_{21}+1}^{2n_{11}+1} b_{12\beta_0}\sum_{k=2}^{2n_{21}}\left[\sum_{\substack{\alpha_1\in[0,\beta_0-2n_{21}-1),\\ \beta_1\in(2n_{21},\beta_0-1]}}^{\alpha_1+\beta_1=\beta_0-1}\sum_{\substack{\alpha_2\in[0,\beta_1-2n_{21}-1]\\ \beta_2\in(2n_{21}-1,\beta_1-2)}}^{\alpha_2+\beta_2=\beta_1-2}\cdots\right.$$

$$
\begin{aligned}
&\times\Bigg(\sum_{\substack{\alpha_{k-1}\in(\beta_{k-2}-2n_{21},\beta_{k-2}-k+1]\\ \beta_{k-1}\in[0,2n_{21}+1-k)}}^{\alpha_{k-1}+\beta_{k-1}=\beta_{k-2}-k+1}\prod_{j=1}^{k-1}\frac{(\beta_{j-1}-j)!}{\alpha_j!\beta_j!}(a_{22s_{j+1}}-a_{22s_j})^{\alpha_j}\sum_{\rho=k}^{2n_{21}+1}A_{\rho 1}^{(s_k)}\ln\frac{|x_{22}-a_{22s_\rho}|}{|x_{20}-a_{22r_\rho}|}\\
&+\sum_{\substack{\alpha_{k-1}\in(\beta_{k-2}-2n_{21}-1,\beta_{k-2}-2n_{21}]\\ \beta_{k-1}\in[2n_{21}+1-k,2n_{21}-k+2)}}^{\alpha_{k-1}+\beta_{k-1}=\beta_{k-2}-k+1}\prod_{j=1}^{k-1}\frac{(\beta_{j-1}-j)!}{\alpha_j!\beta_j!}(a_{22s_{j+1}}-a_{22s_j})^{\alpha_j}\\
&\times\sum_{\rho=k+1}^{2n_{21}+1}B_{\rho 1}^{(s_k)}\ln\frac{|x_2-a_{22s_\rho}|}{|x_{20}-a_{22s_\rho}|}\Bigg)\Bigg]+b_{j_1j_20}\sum_{\rho=1}^{2n_{21}+1}A_{\rho 1}^{(s_1)}\ln\frac{|x_2-a_{22s_\rho}|}{|x_{20}-a_{22s_\rho}|}\\
&+\sum_{\beta_0=1}^{2n_{21}}b_{j_1j_2\beta_0}\sum_{\rho=2}^{2n_{21}+1}B_{\rho 1}^{(s_1)}\frac{|x_2-a_{22s_\rho}|}{|x_{20}-a_{22s_\rho}|}\Bigg\};
\end{aligned}
\tag{3.90}
$$

where

$$
\begin{aligned}
&A_{r_1}=\frac{1}{\prod_{r_2=1,r_2\neq r_1}^{2m_{11}}(a_{11r_1}-a_{11r_\sigma})};\\
&b_{j_1j_2\beta_0}=\sum_{t_1,l_1=0}^{t_1+l_1=1}\cdots\sum_{t_q,l_q=0}^{t_q+l_q=1}\prod_{k=1}^{2n_{11}+1}(a_{22s_1}-a_{12u_k})^{t_k}\delta_{\beta_0}^{\sum_{r=1}^{2n_1+1}l_r};\\
&\lim_{\beta_{2n_{21}}\to 0}\frac{1}{\beta_{2n_{21}}}[(x_2-a_{22s_{2n_{21}+1}})^{\beta_{2n_{21}}}-(x_{20}-a_{22s_{2n_{21}+1}})^{\beta_{2n_{21}}}]=\ln\frac{|x_2-a_{22s_{2n_{21}+1}}|}{|x_{20}-a_{22s_{2n_{21}+1}}|};
\end{aligned}
\tag{3.91}
$$

$$
\begin{aligned}
&A_{\rho 1}^{(s_k)}=\frac{1}{\prod_{\sigma=k,\sigma\neq\rho}^{2n_{21}+1}(a_{22s_\rho}-a_{22s_\sigma})},\ \text{for }\rho\in\{k,k+1,\ldots,2n_{21}+1\};\\
&B_{\rho 1}^{(s_k)}=\frac{(a_{22s_\rho}-a_{22s_k})^{\beta_{k-1}-k}}{\prod_{\sigma=k+1,\sigma\neq\rho}^{2n_{21}+1}(a_{22s_\rho}-a_{22s_\sigma})},\ \text{for }\rho\in\{k+1,k+2,\ldots,2n_{21}+1\};\\
&k\in\{1,2,\ldots,2n_{21}\}.
\end{aligned}
\tag{3.92}
$$

For all cases, the $(2m_{11})\times(2n_{21}+1)$-simple equilibriums plus $(2m_{11})\times(2n_{11}+1)$-hyperbolic flows are based on the following inflection-source and sink bifurcations for $2n_{11}+1\geq 2n_{21}+1$:

- $[2(n_{11}-n_{21})]^{\text{th}}$II:II, $[(2m_{11}),(2n_{21}+1)]^{\text{th}}$SO)- increasing-inflection source bifurcation for ($a_{110}>0, a_{220}>0$),
- $[2(n_{11}-n_{21})]^{\text{th}}$DI:DI, $[(2m_{11}),(2n_{21}+1)]^{\text{th}}$SO)- decreasing-inflection source bifurcation for ($a_{110}<0, a_{220}>0$),
- $[2(n_{11}-n_{21})]^{\text{th}}$DI:DI, $[(2m_{11}),(2n_{21}+1)]^{\text{th}}$SI)- decreasing-inflection sink bifurcation for ($a_{110}>0, a_{220}<0$),

- $[2(n_{11}-n_{21})]^{\text{th}}$II:II, $[(2m_{11}), (2n_{21}+1)]^{\text{th}}$SI)--increasing-inflection sink bifurcation for ($a_{110} < 0, a_{220} < 0$);

or based on the following concave-source and sink bifurcations for $2n_{11}+1 < 2n_{21}+1$:

- $[2(n_{21}-n_{11})]^{\text{th}}$IC:IC, $[(2m_{11}), (2n_{21}+1)]^{\text{th}}$SO) -increasing-concave source bifurcation for ($a_{110} > 0, a_{220} > 0$),
- $[2(n_{11}-n_{21})]^{\text{th}}$IC:DC, $[(2m_{11}), (2n_{21}+1)]^{\text{th}}$SO) decreasing-concave source bifurcation for ($a_{110} < 0, a_{220} > 0$),
- $[2(n_{11}-n_{21})]^{\text{th}}$DC:DC, $[(2m_{11}), (2n_{21}+1)]^{\text{th}}$SI)-decreasing-concave sink bifurcation for ($a_{110} > 0, a_{220} < 0$),
- $[2(n_{11}-n_{21})]^{\text{th}}$II:II, $[(2m_{11}), (2n_{21}+1)]^{\text{th}}$SI) -increasing-concave sink bifurcation for ($a_{110} < 0, a_{220} < 0$).

Consider $2n_{11}+1 = 2n_{21}+1$ for illustrations, phase portraits for the 2-dimensional systems for the hybrid networks of hyperbolic and hyperbolic-secant flows with the simple equilibriums of the saddles, sinks and sources with $x_1^* = a_{111}, a_{112}, \ldots, a_{11(2m_{11})}$ and $x_2^* = a_{121}, a_{221}, \ldots, a_{12(2n_{11}+1)}, a_{22(2n_{21}+1)}$ are presented in Fig. 3.29 (a–d) for $(a_{110} > 0, a_{220} > 0)$, $(a_{110} < 0, a_{220} > 0)$, $(a_{110} > 0, a_{220} < 0)$ and $(a_{110} < 0, a_{220} < 0)$.

$$
\left\{\begin{array}{cccc}
(a_{111}, a_{22(2n_{21}+1)}) & \cdots & (a_{11(2m_{11}-1)}, a_{22(2n_{21}+1)}) & (a_{11(2m_{11})}, a_{22(2n_{21}+1)}) \\
(a_{111}, a_{12(2n_{11}+1)}) & \cdots & (a_{11(2m_{11}-1)}, a_{12(2n_{11}+1)}) & (a_{11(2m_{11})}, a_{12(2n_{11}+1)}) \\
(a_{111}, a_{22(2n_{21})}) & \cdots & (a_{11(2m_{11}-1)}, a_{22(2n_{21})}) & (a_{11(2m_{11})}, a_{22(2n_{21})}) \\
\vdots & \vdots & \vdots & \vdots \\
(a_{111}, a_{221}) & \cdots & (a_{11(2m_{11}-1)}, a_{221}) & (a_{11(2m_{11})}, a_{221}) \\
(a_{111}, a_{121}) & \cdots & (a_{11(2m_{11}-1)}, a_{121}) & (a_{11(2m_{11})}, a_{121})
\end{array}\right\}
$$

$$
= \left\{\begin{array}{cccc}
\underbrace{(\text{SI, SO})}_{\text{saddle}} & \cdots & \underbrace{(\text{SI, SO})}_{\text{saddle}} & \underbrace{(\text{SO, SO})}_{\text{source}} \\
\underbrace{(\text{DP:UP, nF})}_{\text{hyperbolic flow }(-)} & \cdots & \underbrace{(\text{DP:UP, nF})}_{\text{hyperbolic flow }(-)} & \underbrace{(\text{UP:DP, nF})}_{\text{hyperbolic-secant flow }(-)} \\
\underbrace{(\text{SO, SI})}_{\text{saddle}} & \cdots & \underbrace{(\text{SO, SI})}_{\text{saddle}} & \underbrace{(\text{SI, SI})}_{\text{sink}} \\
\vdots & \vdots & \vdots & \vdots \\
\underbrace{(\text{SI, SO})}_{\text{saddle}} & \cdots & \underbrace{(\text{SI, SO})}_{\text{saddle}} & \underbrace{(\text{SO, SO})}_{\text{source}} \\
\underbrace{(\text{DP:UP, nF})}_{\text{hyperbolic flow }(-)} & \cdots & \underbrace{(\text{DP:UP, nF})}_{\text{hyperbolic flow }(-)} & \underbrace{(\text{UP:DP, nF})}_{\text{hyperbolic-secant flow }(-)}
\end{array}\right\}_{(2n_{11}+2n_{21}+2)\times(2m_{11})}
$$

for $a_{110} > 0, a_{220} > 0$; (3.93)

$$\left\{\begin{array}{cccc}(a_{111},a_{22(2n_{21}+1)}) & \cdots & (a_{11(2m_{11}-1)},a_{22(2n_{21}+1)}) & (a_{11(2m_{11})},a_{22(2n_{21}+1)})\\ (a_{111},a_{12(2n_{11}+1)}) & \cdots & (a_{11(2m_{11}-1)},a_{12(2n_{11}+1)}) & (a_{11(2m_{11})},a_{12(2n_{11}+1)})\\ (a_{111},a_{22(2n_{21})}) & \cdots & (a_{11(2m_{11}-1)},a_{22(2n_{21})}) & (a_{11(2m_{11})},a_{22(2n_{21})})\\ \vdots & \vdots & \vdots & \vdots\\ (a_{111},a_{221}) & \cdots & (a_{11(2m_{11}-1)},a_{221}) & (a_{11(2m_{11})},a_{221})\\ (a_{111},a_{121}) & \cdots & (a_{11(2m_{11}-1)},a_{121}) & (a_{11(2m_{11})},a_{121})\end{array}\right\}$$
$$=\left\{\begin{array}{cccc}\underbrace{(\text{SO, SO})}_{\text{source}} & \cdots & \underbrace{(\text{SO, SO})}_{\text{source}} & \underbrace{(\text{SI, SO})}_{\text{saddle}}\\ \underbrace{(\text{UP:DP, nF})}_{\text{hyperbolic-secant flow }(-)} & \cdots & \underbrace{(\text{UP:DP, nF})}_{\text{hyperbolic-secant flow }(-)} & \underbrace{(\text{DP:UP, nF})}_{\text{hyperbolic flow }(-)}\\ \underbrace{(\text{SI, SI})}_{\text{sink}} & \cdots & \underbrace{(\text{SI, SI})}_{\text{sink}} & \underbrace{(\text{SO, SI})}_{\text{saddle}}\\ \vdots & \vdots & \vdots & \vdots\\ \underbrace{(\text{SO, SO})}_{\text{source}} & \cdots & \underbrace{(\text{SO, SO})}_{\text{source}} & \underbrace{(\text{SI, SO})}_{\text{saddle}}\\ \underbrace{(\text{UP:DP, nF})}_{\text{hyperbolic-secant flow }(-)} & \cdots & \underbrace{(\text{UP:DP, nF})}_{\text{hyperbolic-secant flow }(-)} & \underbrace{(\text{DP:UP, nF})}_{\text{hyperbolic flow }(-)}\end{array}\right\}_{(2n_{11}+2n_{21}+2)\times(2m_{11})}$$

for $a_{110} < 0, a_{220} > 0;$ (3.94)

$$\left\{\begin{array}{cccc}(a_{111},a_{22(2n_{21}+1)}) & \cdots & (a_{11(2m_{11}-1)},a_{22(2n_{21}+1)}) & (a_{11(2m_{11})},a_{22(2n_{21}+1)})\\ (a_{111},a_{12(2n_{11}+1)}) & \cdots & (a_{11(2m_{11}-1)},a_{12(2n_{11}+1)}) & (a_{11(2m_{11})},a_{12(2n_{11}+1)})\\ (a_{111},a_{22(2n_{21})}) & \cdots & (a_{11(2m_{11}-1)},a_{22(2n_{21})}) & (a_{11(2m_{11})},a_{22(2n_{21})})\\ \vdots & \vdots & \vdots & \vdots\\ (a_{111},a_{221}) & \cdots & (a_{11(2m_{11}-1)},a_{221}) & (a_{11(2m_{11})},a_{221})\\ (a_{111},a_{121}) & \cdots & (a_{11(2m_{11}-1)},a_{121}) & (a_{11(2m_{11})},a_{121})\end{array}\right\}$$
$$=\left\{\begin{array}{cccc}\underbrace{(\text{SI, SI})}_{\text{sink}} & \cdots & \underbrace{(\text{SI, SI})}_{\text{sink}} & \underbrace{(\text{SO, SI})}_{\text{saddle}}\\ \underbrace{(\text{UP:DP, pF})}_{\text{hyperbolic-secant flow }(+)} & \cdots & \underbrace{(\text{UP:DP, pF})}_{\text{hyperbolic-secant flow }(+)} & \underbrace{(\text{DP:UP, pF})}_{\text{hyperbolic flow }(+)}\\ \underbrace{(\text{SO, SO})}_{\text{source}} & \cdots & \underbrace{(\text{SO, SO})}_{\text{source}} & \underbrace{(\text{SI, SO})}_{\text{saddle}}\\ \vdots & \vdots & \vdots & \vdots\\ \underbrace{(\text{SI, SI})}_{\text{sink}} & \cdots & \underbrace{(\text{SI, SI})}_{\text{sink}} & \underbrace{(\text{SO, SI})}_{\text{saddle}}\\ \underbrace{(\text{UP:DP, pF})}_{\text{hyperbolic-secant flow }(+)} & \cdots & \underbrace{(\text{UP:DP, pF})}_{\text{hyperbolic-secant flow }(+)} & \underbrace{(\text{DP:UP, pF})}_{\text{hyperbolic flow }(+)}\end{array}\right\}_{(2n_{11}+2n_{21}+2)\times(2m_{11})}$$

for $a_{110} > 0, a_{220} < 0;$ (3.95)

$$\left\{\begin{array}{cccc}(a_{111},a_{22(2n_{21}+1)}) & \cdots & (a_{11(2m_{11}-1)},a_{22(2n_{21}+1)}) & (a_{11(2m_{11})},a_{22(2n_{21}+1)})\\ (a_{111},a_{12(2n_{11}+1)}) & \cdots & (a_{11(2m_{11}-1)},a_{12(2n_{11}+1)}) & (a_{11(2m_{11})},a_{12(2n_{11}+1)})\\ (a_{111},a_{22(2n_{21})}) & \cdots & (a_{11(2m_{11}-1)},a_{22(2n_{21})}) & (a_{11(2m_{11})},a_{22(2n_{21})})\\ \vdots & \vdots & \vdots & \vdots\\ (a_{111},a_{221}) & \cdots & (a_{11(2m_{11}-1)},a_{221}) & (a_{11(2m_{11})},a_{221})\\ (a_{111},a_{121}) & \cdots & (a_{11(2m_{11}-1)},a_{121}) & (a_{11(2m_{11})},a_{121})\end{array}\right\}$$

$$=\left\{\begin{array}{cccc}\underbrace{(\text{SO, SI})}_{\text{saddle}} & \cdots & \underbrace{(\text{SO, SI})}_{\text{saddle}} & \underbrace{(\text{SI, SI})}_{\text{sink}}\\ \underbrace{(\text{DP:UP, pF})}_{\text{hyperbolic flow }(+)} & \cdots & \underbrace{(\text{DP:UP, pF})}_{\text{hyperbolic flow }(+)} & \underbrace{(\text{UP:DP, pF})}_{\text{hyperbolic-secant flow }(+)}\\ \underbrace{(\text{SI, SO})}_{\text{saddle}} & \cdots & \underbrace{(\text{SI, SO})}_{\text{saddle}} & \underbrace{(\text{SO, SO})}_{\text{source}}\\ \vdots & \vdots & \vdots & \vdots\\ \underbrace{(\text{SO, SI})}_{\text{saddle}} & \cdots & \underbrace{(\text{SO, SI})}_{\text{saddle}} & \underbrace{(\text{SI, SI})}_{\text{sink}}\\ \underbrace{(\text{DP:UP, pF})}_{\text{hyperbolic flow }(+)} & \cdots & \underbrace{(\text{DP:UP, pF})}_{\text{hyperbolic flow }(+)} & \underbrace{(\text{UP:DP, pF})}_{\text{hyperbolic-secant flow }(+)}\end{array}\right\}_{(2n_{11}+2n_{21}+2)\times(2m_{11})}$$

$$\text{for } a_{110}<0,\ a_{220}<0. \tag{3.96}$$

Consider $2n_{11}+1=2n_{21}+1$ for illustrations. Phase portraits for the 2-dimensional systems for the hybrid networks of hyperbolic and hyperbolic-secant flows with the simple equilibriums of the saddles, sinks and sources with $x_1^*=a_{111},a_{112},\ldots,a_{11(2m_{11})}$ and $x_2^*=a_{221},a_{121},\ldots,a_{22(2n_{21}+1)},a_{12(2n_{11}+1)}$ are presented in Fig. 3.30 (a–d) for ($a_{110}>0$, $a_{220}>0$), ($a_{110}<0$, $a_{220}>0$), ($a_{110}>0$, $a_{220}<0$) and ($a_{110}<0$, $a_{220}<0$).

$$\left\{\begin{array}{cccc}(a_{111},a_{12(2n_{11}+1)}) & \cdots & (a_{11(2m_{11}-1)},a_{12(2n_{11}+1)}) & (a_{11(2m_{11})},a_{12(2n_{11}+1)})\\ (a_{111},a_{22(2n_{21}+1)}) & \cdots & (a_{11(2m_{11}-1)},a_{22(2n_{21}+1)}) & (a_{11(2m_{11})},a_{22(2n_{21}+1)})\\ (a_{111},a_{12(2n_{11})}) & \cdots & (a_{11(2m_{11}-1)},a_{12(2n_{11})}) & (a_{11(2m_{11})},a_{12(2n_{11})})\\ \vdots & \vdots & \vdots & \vdots\\ (a_{111},a_{121}) & \cdots & (a_{11(2m_{11}-1)},a_{121}) & (a_{11(2m_{11})},a_{121})\\ (a_{111},a_{221}) & \cdots & (a_{11(2m_{11}-1)},a_{221}) & (a_{11(2m_{11})},a_{221})\end{array}\right\}$$

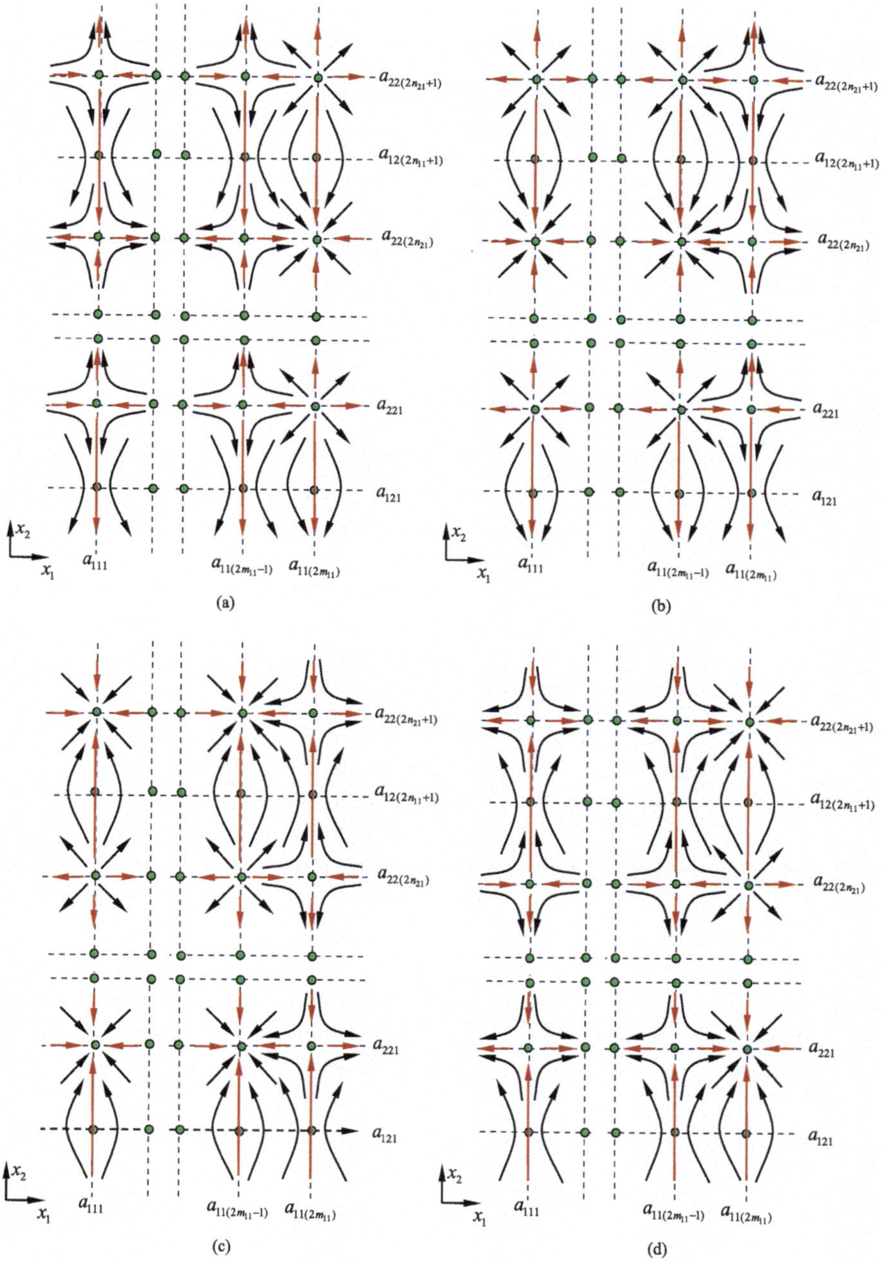

Fig. 3.29 Phase portraits for 2-dimensional systems with $x_1^* = a_{111}, a_{112}, \ldots, a_{11(2m_{11})}$ and $x_2^* = a_{121}, a_{221}, \ldots, a_{12(2n_{11}+1)}, a_{22(2n_{21}+1)}$. The networks of flows with sink, source and saddle: **a** ($a_{110} > 0$, $a_{220} > 0$), **b** ($a_{110} < 0$, $a_{220} > 0$), **c** ($a_{110} > 0$, $a_{220} < 0$), **d** ($a_{110} < 0$, $a_{220} < 0$)

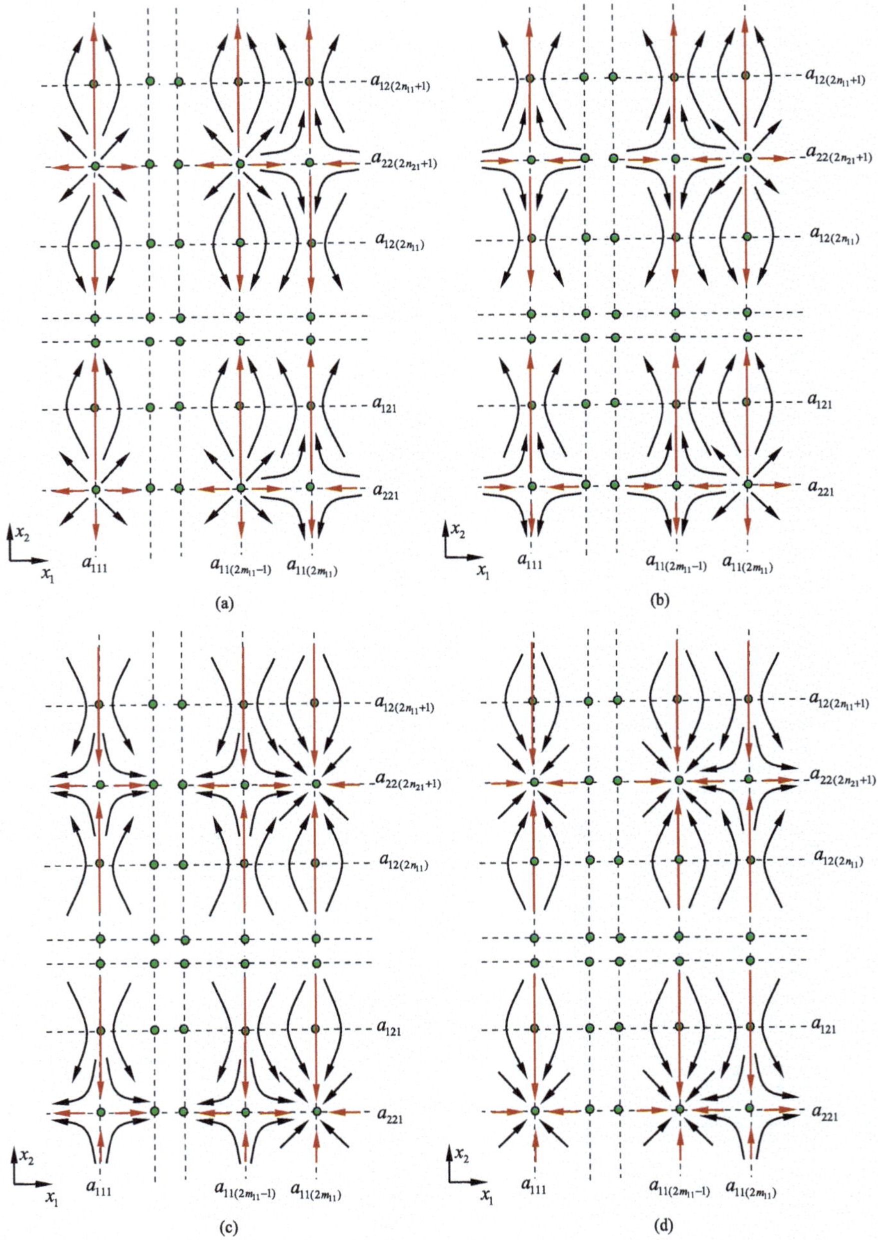

Fig. 3.30 Phase portraits for 2-dimensional systems with $x_1^* = a_{111}, a_{112}, \ldots, a_{11(2m_{11})}$ and . The networks of flows with sink, source and saddle: **a** (a_{110} >0, $a_{220} > 0$), **b** ($a_{110} < 0$, $a_{220} > 0$), **c** ($a_{110} > 0$, $a_{220} < 0$), **d** ($a_{110} < 0$, $a_{220} < 0$)

$$
= \left\{ \begin{array}{cccc}
\underbrace{(\text{UP:DP, pF})}_{\text{hyperbolic-secant flow (+)}} & \cdots & \underbrace{(\text{UP:DP, pF})}_{\text{hyperbolic-secant flow (+)}} & \underbrace{(\text{DP:UP, pF})}_{\text{hyperbolic flow (+)}} \\
\underbrace{(\text{SO, SO})}_{\text{source}} & \cdots & \underbrace{(\text{SO, SO})}_{\text{source}} & \underbrace{(\text{SI, SO})}_{\text{saddle}} \\
\underbrace{(\text{UP:DP, pF})}_{\text{hyperbolic-secant flow }(-)} & \cdots & \underbrace{(\text{UP:DP, pF})}_{\text{hyperbolic-secant flow }(-)} & \underbrace{(\text{DP:UP, nF})}_{\text{hyperbolic flow }(-)} \\
\vdots & \vdots & \vdots & \vdots \\
\underbrace{(\text{UP:DP, pF})}_{\text{hyperbolic-secant flow (+)}} & \cdots & \underbrace{(\text{UP:DP, pF})}_{\text{hyperbolic-secant flow (+)}} & \underbrace{(\text{DP:UP, pF})}_{\text{hyperbolic flow (+)}} \\
\underbrace{(\text{SO, SO})}_{\text{source}} & \cdots & \underbrace{(\text{SO, SO})}_{\text{source}} & \underbrace{(\text{SI, SO})}_{\text{saddle}}
\end{array} \right\}_{(2n_{11}+2n_{21}+2)\times(2m_{11})}
$$

for $a_{110} > 0, a_{220} > 0$; (3.97)

$$
\left\{ \begin{array}{cccc}
(a_{111}, a_{12(2n_{11}+1)}) & \cdots & (a_{11(2m_{11}-1)}, a_{12(2n_{11}+1)}) & (a_{11(2m_{11})}, a_{12(2n_{11}+1)}) \\
(a_{111}, a_{22(2n_{21}+1)}) & \cdots & (a_{11(2m_{11}-1)}, a_{22(2n_{21}+1)}) & (a_{11(2m_{11})}, a_{22(2n_{21}+1)}) \\
(a_{111}, a_{12(2n_{11})}) & \cdots & (a_{11(2m_{11}-1)}, a_{12(2n_{11})}) & (a_{11(2m_{11})}, a_{12(2n_{11})}) \\
\vdots & \vdots & \vdots & \vdots \\
(a_{111}, a_{121}) & \cdots & (a_{11(2m_{11}-1)}, a_{121}) & (a_{11(2m_{11})}, a_{121}) \\
(a_{111}, a_{221}) & \cdots & (a_{11(2m_{11}-1)}, a_{221}) & (a_{11(2m_{11})}, a_{221})
\end{array} \right\}
$$

$$
= \left\{ \begin{array}{cccc}
\underbrace{(\text{DP:UP, pF})}_{\text{hyperbolic flow (+)}} & \cdots & \underbrace{(\text{DP:UP, pF})}_{\text{hyperbolic flow (+)}} & \underbrace{(\text{UP:DP, pF})}_{\text{hyperbolic-secant flow (+)}} \\
\underbrace{(\text{SI, SO})}_{\text{saddle}} & \cdots & \underbrace{(\text{SI, SO})}_{\text{saddle}} & \underbrace{(\text{SO, SO})}_{\text{source}} \\
\underbrace{(\text{DP:UP, nF})}_{\text{hyperbolic flow }(-)} & \cdots & \underbrace{(\text{DP:UP, nF})}_{\text{hyperbolic flow }(-)} & \underbrace{(\text{UP:DP, nF})}_{\text{hyperbolic-secant flow }(-)} \\
\vdots & \vdots & \vdots & \vdots \\
\underbrace{(\text{DP:UP, pF})}_{\text{hyperbolic flow (+)}} & \cdots & \underbrace{(\text{DP:UP, pF})}_{\text{hyperbolic flow (+)}} & \underbrace{(\text{UP:DP, pF})}_{\text{hyperbolic-secant flow (+)}} \\
\underbrace{(\text{SI, SO})}_{\text{saddle}} & \cdots & \underbrace{(\text{SI, SO})}_{\text{saddle}} & \underbrace{(\text{SO, SO})}_{\text{source}}
\end{array} \right\}_{(2n_{11}+2n_{21}+2)\times(2m_{11})}
$$

for $a_{110} < 0, a_{220} > 0$; (3.98)

$$
\left\{\begin{array}{cccc}
(a_{111}, a_{12(2n_{11}+1)}) & \cdots & (a_{11(2m_{11}-1)}, a_{12(2n_{11}+1)}) & (a_{11(2m_{11})}, a_{12(2n_{11}+1)}) \\
(a_{111}, a_{22(2n_{21}+1)}) & \cdots & (a_{11(2m_{11}-1)}, a_{22(2n_{21}+1)}) & (a_{11(2m_{11})}, a_{22(2n_{21}+1)}) \\
(a_{111}, a_{12(2n_{11})}) & \cdots & (a_{11(2m_{11}-1)}, a_{12(2n_{11})}) & (a_{11(2m_{11})}, a_{12(2n_{11})}) \\
\vdots & \vdots & \vdots & \vdots \\
(a_{111}, a_{121}) & \cdots & (a_{11(2m_{11}-1)}, a_{121}) & (a_{11(2m_{11})}, a_{121}) \\
(a_{111}, a_{221}) & \cdots & (a_{11(2m_{11}-1)}, a_{221}) & (a_{11(2m_{11})}, a_{221})
\end{array}\right\}
$$

$$
= \left\{\begin{array}{cccc}
\underbrace{(\text{DP:UP, nF})}_{\text{hyperbolic flow }(-)} & \cdots & \underbrace{(\text{DP:UP, nF})}_{\text{hyperbolic flow }(-)} & \underbrace{(\text{UP:DP, nF})}_{\text{hyperbolic-secant flow }(-)} \\
\underbrace{(\text{SO, SI})}_{\text{saddle}} & \cdots & \underbrace{(\text{SO, SI})}_{\text{saddle}} & \underbrace{(\text{SI, SI})}_{\text{sink}} \\
\underbrace{(\text{DP:UP, pF})}_{\text{hyperbolic flow }(+)} & \cdots & \underbrace{(\text{DP:UP, pF})}_{\text{hyperbolic flow }(+)} & \underbrace{(\text{UP:DP, pF})}_{\text{hyperbolic-secant flow }(+)} \\
\vdots & \vdots & \vdots & \vdots \\
\underbrace{(\text{DP:UP, nF})}_{\text{hyperbolic flow }(-)} & \cdots & \underbrace{(\text{DP:UP, nF})}_{\text{hyperbolic flow }(-)} & \underbrace{(\text{UP:DP, nF})}_{\text{hyperbolic-secant flow }(-)} \\
\underbrace{(\text{SI, SO})}_{\text{saddle}} & \cdots & \underbrace{(\text{SI, SO})}_{\text{saddle}} & \underbrace{(\text{SI, SI})}_{\text{sink}}
\end{array}\right\}_{(2n_{11}+2n_{21}+2)\times(2m_{11})}
$$

for $a_{110} > 0, a_{220} < 0$; (3.99)

$$
\left\{\begin{array}{cccc}
(a_{111}, a_{12(2n_{11}+1)}) & \cdots & (a_{11(2m_{11}-1)}, a_{12(2n_{11}+1)}) & (a_{11(2m_{11})}, a_{12(2n_{11}+1)}) \\
(a_{111}, a_{22(2n_{21}+1)}) & \cdots & (a_{11(2m_{11}-1)}, a_{22(2n_{21}+1)}) & (a_{11(2m_{11})}, a_{22(2n_{21}+1)}) \\
(a_{111}, a_{12(2n_{11})}) & \cdots & (a_{11(2m_{11}-1)}, a_{12(2n_{11})}) & (a_{11(2m_{11})}, a_{12(2n_{11})}) \\
\vdots & \vdots & \vdots & \vdots \\
(a_{111}, a_{121}) & \cdots & (a_{11(2m_{11}-1)}, a_{121}) & (a_{11(2m_{11})}, a_{121}) \\
(a_{111}, a_{221}) & \cdots & (a_{11(2m_{11}-1)}, a_{221}) & (a_{11(2m_{11})}, a_{221})
\end{array}\right\}
$$

$$
= \left\{\begin{array}{cccc}
\underbrace{(\text{UP:DP, nF})}_{\text{hyperbolic-secant flow }(-)} & \cdots & \underbrace{(\text{UP:DP, nF})}_{\text{hyperbolic-secant flow }(-)} & \underbrace{(\text{DP:UP, nF})}_{\text{hyperbolic flow }(-)} \\
\underbrace{(\text{SI, SI})}_{\text{sink}} & \cdots & \underbrace{(\text{SI, SI})}_{\text{sink}} & \underbrace{(\text{SO, SI})}_{\text{saddle}} \\
\underbrace{(\text{UP:DP, pF})}_{\text{hyperbolic-secant flow }(+)} & \cdots & \underbrace{(\text{UP:DP, pF})}_{\text{hyperbolic-secant flow }(+)} & \underbrace{(\text{DP:UP, pF})}_{\text{hyperbolic flow }(+)} \\
\vdots & \vdots & \vdots & \vdots \\
\underbrace{(\text{UP:DP, nF})}_{\text{hyperbolic-secant flow }(-)} & \cdots & \underbrace{(\text{UP:DP, nF})}_{\text{hyperbolic-secant flow }(-)} & \underbrace{(\text{DP:UP, nF})}_{\text{hyperbolic flow }(-)} \\
\underbrace{(\text{SI,SI})}_{\text{sink}} & \cdots & \underbrace{(\text{SI, SI})}_{\text{sink}} & \underbrace{(\text{SO, SI})}_{\text{saddle}}
\end{array}\right\}_{(2n_{11}+2n_{21}+2)\times(2m_{11})}
$$

for $a_{110} < 0, a_{220} < 0$. (3.100)

3.6 $[(2m_{11}),(2n_{11})]$ and $((2m_{11}),(2n_{21}+1))$-Singular Systems

In this section, there are four global bifurcations of the 1-diemsnional flows through *singular inflection* flows, and four global bifurcations of equilibriums, through singular saddle-source and saddle-sink bifurcations. The corresponding global bifurcations generate networks of hyperbolic and hyperbolic-secant flows with sources, sinks and saddles. The infinite-equilibriums are the appearing and switching bifurcations for the singular flows and singular equilibriums and for hybrid networks of hyperbolic flows and source, sink and saddles.

3.6.1 $[(2m_{11}),(2n_{11})]$-*Flows and* $((2m_{11}),(2n_{21}+1))$-*Equilibriums*

Consider a polynomial dynamical system for $a_{221}\neq a_{121}$ as

$$\begin{aligned}\dot{x}_2&=a_{220}(x_2-a_{221})^{2n_{21}+1},\\ \dot{x}_1&=a_{110}(x_1-a_{111})^{2m_{11}}(x_2-a_{121})^{2n_{11}}.\end{aligned}\tag{3.101}$$

The first integral manifold is

$$\begin{aligned}&-\frac{1}{2m_{11}-1}\left[\frac{1}{(x_1-a_{111})^{2m_{11}-1}}-\frac{1}{(x_{10}-a_{111})^{2m_{11}-1}}\right]\\ &=\frac{a_{110}}{a_{220}}\sum_{\alpha,\beta=0}^{\alpha+\beta=2n_{11}}\frac{(2n_{11})!}{\alpha!\beta!}\frac{(a_{221}-a_{121})^{\alpha}}{\beta-2n_{21}}[(x_2-a_{221})^{\beta-2n_{21}}-(x_2-a_{221})^{\beta-2n_{21}}]\\ &\text{for } \beta-2n_{21}\neq 0;\\ &\frac{1}{\beta-2n_{21}}[(x_2-a_{221})^{\beta-2n_{21}}-(x_2-a_{221})^{\beta-2n_{21}}]=\ln\frac{|x_2-a_{221}|}{|x_{20}-a_{221}|}\\ &\text{for } \beta-2n_{21}=0.\end{aligned}\tag{3.102}$$

For $(x_1^*,x_2^*)=(a_{111},a_{121})$ with $\dot{x}_1=0$, the $[(2m_{11}),(2n_{11})]$-singular flows exist with $\dot{x}_2=a_{220}(a_{121}-a_{221})^{2n_{21}+1}$. For $(x_1^*,x_2^*)=(a_{111},a_{121})$, the 1-dimensional flow direction is determined by $\dot{x}_2\neq 0s$. For $(x_1^*,x_2^*)=(a_{111},a_{221})$ with $\dot{x}_1=\dot{x}_2=0$, the $((2m_{11}),(2n_{21}+1))$- singular equilibriums exist.

(A) For $a_{121} > a_{221}$,

$$\left\{\begin{array}{c}(a_{111},a_{121})\\(a_{111},a_{221})\end{array}\right\}=\left\{\begin{array}{c}\underbrace{([(2m_{11}),(2n_{11})]_{\text{II:II}}\text{US, pF})}_{[(2m_{11}),(2n_{11})]\text{-increasing-inflecrtion upper-saddle flow }(+)}\\\underbrace{((2m_{11})\text{US},(2n_{21}+1)\text{SO})}_{[(2m_{11}),(2n_{21}+1)]\text{-upper-saddle-source}}\end{array}\right\},$$

for $a_{110} > 0, a_{220} > 0$;

$$\left\{\begin{array}{c}(a_{111},a_{121})\\(a_{111},a_{221})\end{array}\right\}=\left\{\begin{array}{c}\underbrace{([(2m_{11}),(2n_{11})]_{\text{DI:DI}}\text{LS, pF})}_{[(2m_{11}),(2n_{11})]\text{-decreasing-inflection lower-saddle flow }(+)}\\\underbrace{((2m_{11})\text{LS},(2n_{21}+1)\text{SO})}_{[(2m_{11}),(2n_{21}+1)]\text{-lower-saddle source}}\end{array}\right\},$$

for $a_{110} < 0, a_{220} > 0$;

$$\left\{\begin{array}{c}(a_{111},a_{121})\\(a_{111},a_{221})\end{array}\right\}=\left\{\begin{array}{c}\underbrace{([(2m_{11}),(2n_{11})]_{\text{DI:DI}}\text{US, nF})}_{[(2m_{11}),(2n_{11})]\text{-decreasing-inflection upper-saddle flow }(-)}\\\underbrace{((2m_{11})\text{US},(2n_{21}+1)\text{SI})}_{[(2m_{11}),(2n_{21}+1)]\text{-upper-saddle-sink}}\end{array}\right\},$$

for $a_{110} > 0, a_{220} < 0$;

$$\left\{\begin{array}{c}(a_{111},a_{121})\\(a_{111},a_{221})\end{array}\right\}=\left\{\begin{array}{c}\underbrace{([(2m_{11}),(2n_{11})]_{\text{II:II}}\text{US, nF})}_{[(2m_{11}),(2n_{11})]\text{-increasing-inflection lower-saddle flow }(-)}\\\underbrace{((2m_{11})\text{LS},(2n_{21}+1)\text{SI})}_{[(2m_{11}),(2n_{21}+1)]\text{-lower-saddle-sink}}\end{array}\right\},$$

for $a_{110} < 0, a_{220} < 0$. (3.103)

(B) For $a_{121} < a_{221}$,

$$\left\{\begin{array}{c}(a_{111},a_{221})\\(a_{111},a_{121})\end{array}\right\}=\left\{\begin{array}{c}\underbrace{((2m_{11})\text{US},(2n_{21}+1)\text{SO})}_{[(2m_{11}),(2n_{21}+1)]\text{-upper-saddle-source}}\\\underbrace{([(2m_{11}),(2n_{11})]_{\text{II:II}}\text{US,nF})}_{[(2m_{11}),(2n_{11})]\text{-increasing-inflection upper-saddle flow }(-)}\end{array}\right\},$$

for $a_{110} > 0, a_{220} > 0$;

$$\left\{\begin{array}{c}(a_{111},a_{221})\\(a_{111},a_{121})\end{array}\right\}=\left\{\begin{array}{c}\underbrace{((2m_{11})\text{LS},(2n_{21}+1)\text{SO})}_{[(2m_{11}),(2n_{21}+1)]\text{-lower-saddle-source}}\\\underbrace{([(2m_{11}),(2n_{11})]_{\text{DI:DI}}\text{LS,nF})}_{[(2m_{11}),(2n_{11})]\text{-decreasing-inflection lower-saddle flow }(-)}\end{array}\right\},$$

for $a_{110} < 0, a_{220} > 0$;

$$\begin{Bmatrix} (a_{111}, a_{221}) \\ (a_{111}, a_{121}) \end{Bmatrix} = \begin{Bmatrix} \underbrace{((2m_{11})\text{US},(2n_{21}+1)\text{SI})}_{[(2m_{11}),(2n_{21}+1)]\text{-upper-saddle-sink}} \\ \underbrace{([(2m_{11}),(2n_{11})]_{\text{DI:DI}}\text{US,pF})}_{[(2m_{11}),(2n_{11})]\text{-decreasing-inflection upper-saddle flow }(+)} \end{Bmatrix},$$

for $a_{110} > 0, a_{220} < 0$;

$$\begin{Bmatrix} (a_{111}, a_{221}) \\ (a_{111}, a_{121}) \end{Bmatrix} = \begin{Bmatrix} \underbrace{((2m_{11})\text{LS},(2n_{21}+1)\text{SI})}_{[(2m_{11}),(2n_{21}+1)]\text{-sink}} \\ \underbrace{([(2m_{11}),(2n_{11})]_{\text{II:II}}\text{LS,pF})}_{[(2m_{11}),(2n_{11})]\text{-increasing-inflection lower-saddle flow }(+)} \end{Bmatrix},$$

for $a_{110} < 0, a_{220} < 0$. (3.104)

The $([(2m_{11}),(2n_{11})]_{\text{II:II}}\text{US,pF})$ is for a positive singular increasing inflection-upper-saddle flow. The $([(2m_{11}),(2n_{11})]_{\text{DI:DI}}\text{LS,pF})$ is for a positive singular decreasing-inflection lower-saddle flow. The $([(2m_{11}),(2n_{11})]_{\text{DI:DI}}\text{US,nF})$ is for a negative singular decreasing-inflection upper-saddle flow. The $([(2m_{11}),(2n_{11})]_{\text{II:II}}\text{LS, nF})$ is for a negative singular increasing-inflection lower-saddle flow. The $((2m_{11})\text{US},(2n_{21}+1)\text{SO}]$ is for a $((2m_{11}),(2n_{21}+1)]$-upper-saddle-source. The $((2m_{11})\text{LS},(2n_{21}+1)\text{SO}]$ is for a $((2m_{11}),(2n_{21}+1)]$-lower-saddle-source. The $((2m_{11})\text{US},(2n_{21}+1)\text{SI}]$ is for a $((2m_{11}),(2n_{21}+1)]$-upper-saddle-sink. The $((2m_{11})\text{LS},(2n_{21}+1)\text{SI}]$ is for a $((2m_{11}),(2n_{21}+1)]$-lower-saddle-sink.

The singular inflection-saddle flows with singular source, sink and saddles are illustrated in Figs. 3.31 and 3.32. Through such singular flows, the corresponding global bifurcations generate networks of singular hyperbolic and hyperbolic-secant flows. Through such singular equilibriums, the corresponding global bifurcations generate networks of singular source, sink and saddle equilibriums.

3.6.2 $[(2m_{11}),(2n_{11})]$ and $((2m_{11}),(2n_{21}+1))$-*Infinite-Equilibriums*

Consider a polynomial dynamical system for $a_{221} = a_{121}$ as

$$\begin{aligned} \dot{x}_2 &= a_{220}(x_2 - a_{221})^{2n_{21}+1}, \\ \dot{x}_1 &= a_{110}(x_1 - a_{111})^{2m_{11}}(x_2 - a_{121})^{2n_{11}}. \end{aligned} \tag{3.105}$$

The first integral manifold is

$$-\frac{1}{2m_{11}-1}\left[\frac{1}{(x_1 - a_{111})^{2m_{11}-1}} - \frac{1}{(x_{10} - a_{111})^{2m_{11}-1}}\right]$$

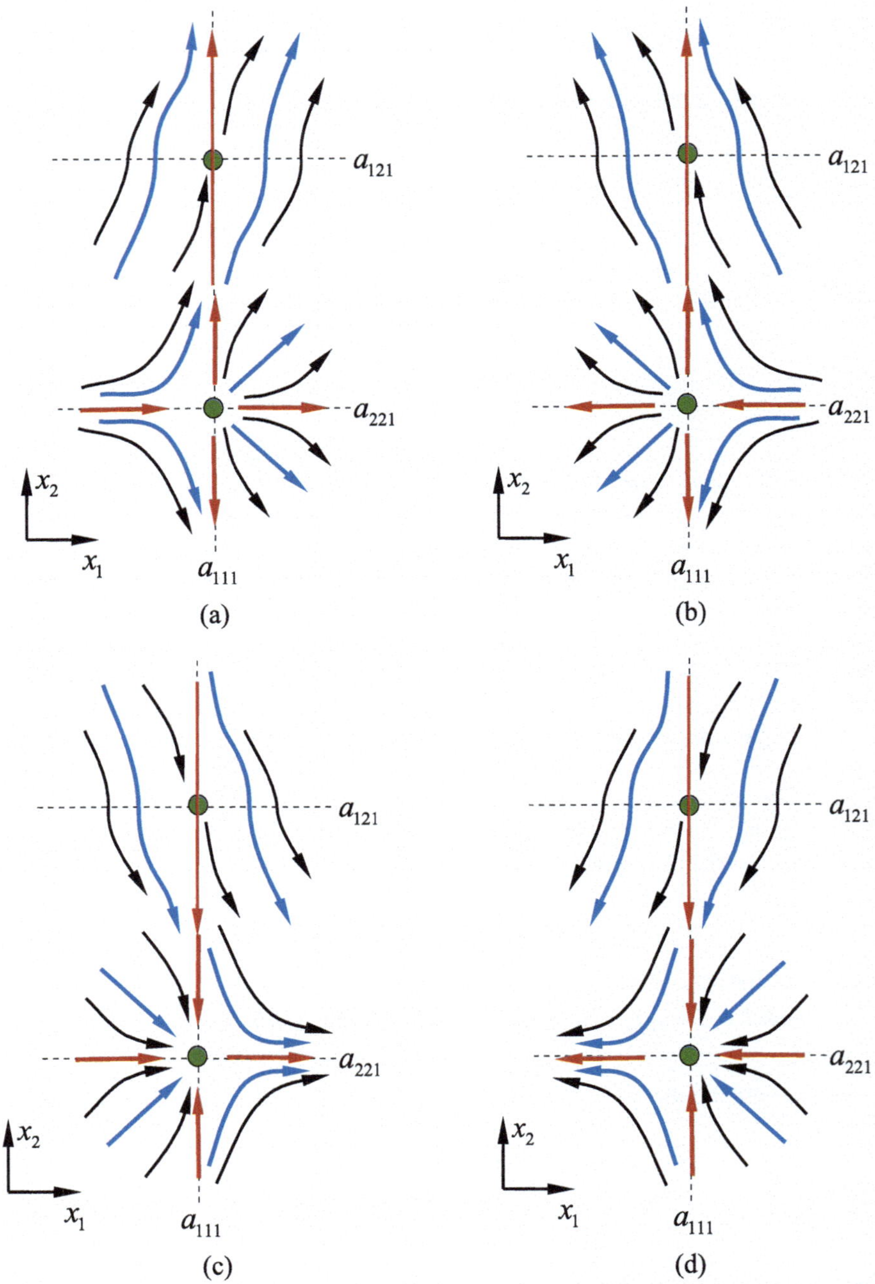

Fig. 3.31 Phase portraits ($a_{121} > a_{221}$) of the ([$(2m_{11})$, $(2n_{11})$], xF)-increasing and decreasing-inflection upper-saddle and lower-saddle flows at $(x_1^*, x_2^*) = (a_{111}, a_{121})$ and ($(2m_{11})$, $(2n_{21}+1)$)-saddle-source and saddle-sink at $(x_1^*, x_2^*) = (a_{111}, a_{221})$ for 2-dimensional polynomial systems: **a** ($a_{110} > 0, a_{220} > 0$), **b** ($a_{110} < 0, a_{220} > 0$), **c** ($a_{110} > 0, a_{220} < 0$), **d** ($a_{110} < 0, a_{220} < 0$)

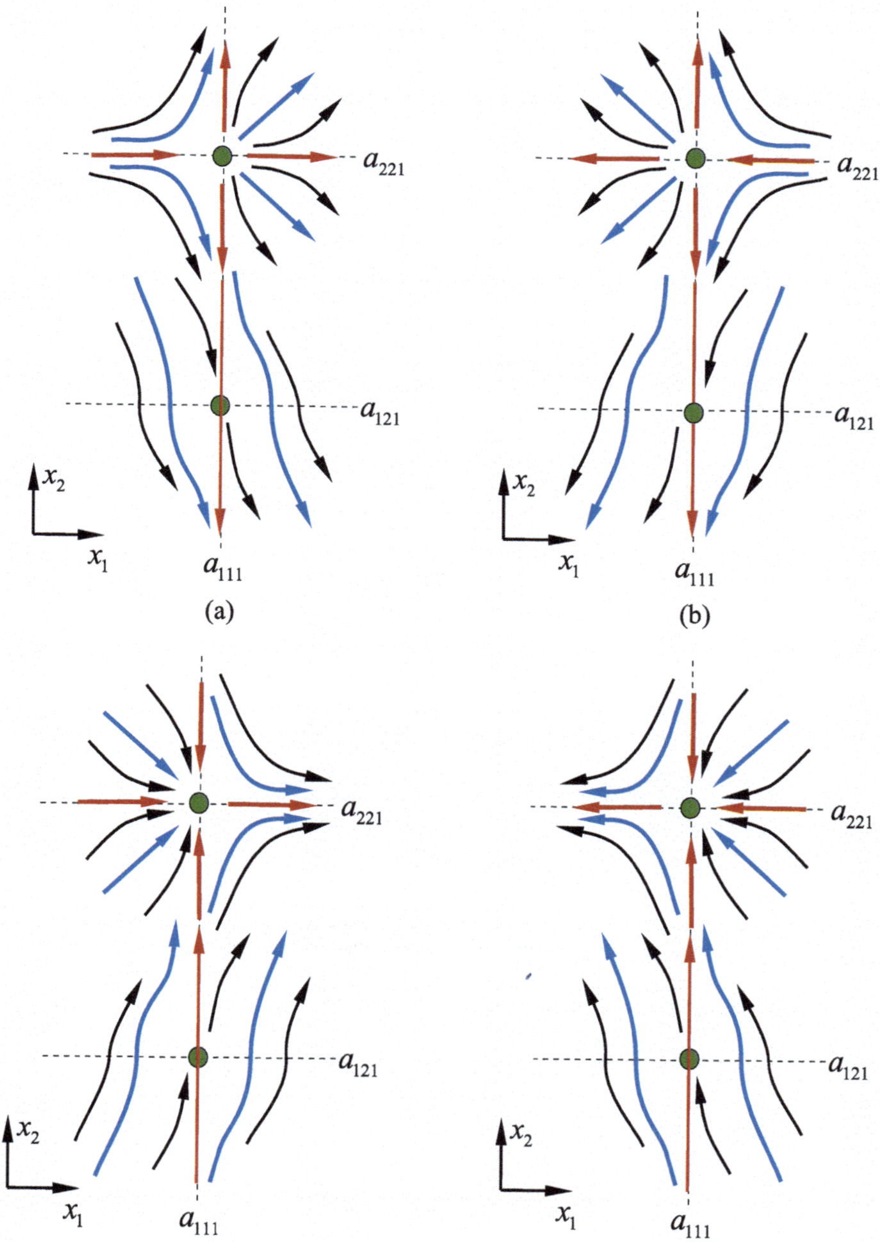

Fig. 3.32 Phase portraits ($a_{221} > a_{121}$) of the ($[(2m_{11}), (2n_{11})]$, xF)-increasing and decreasing inflection-upper-saddle and lower-saddle flows at $(x_1^*, x_2^*) = (a_{111}, a_{121})$ and $((2m_{11}), (2n_{21}+1))$-saddle-source, and saddle-sink at $(x_1^*, x_2^*) = (a_{111}, a_{221})$ for 2-dimensional polynomial systems: **a** ($a_{110} > 0$, $a_{220} > 0$), **b** ($a_{110} < 0$, $a_{220} > 0$), **c** ($a_{110} > 0$, $a_{220} < 0$), **d** ($a_{110} < 0$, $a_{220} < 0$)

$$= \frac{a_{110}}{a_{220}} \frac{1}{2(n_{11}-n_{21})}[(x_2-a_{121})^{2(n_{11}-n_{21})}-(x_2-a_{121})^{2(n_{11}-n_{21})}]$$
for $2(n_{11}-n_{21}) \neq 0$;
$$\frac{1}{2(n_{11}-n_{21})}[(x_2-a_{121})^{2(n_{11}-n_{21})}-(x_2-a_{121})^{2(n_{11}-n_{21})}] = \ln\frac{|x_2-a_{121}|}{|x_{20}-a_{121}|}$$
for $2(n_{11}-n_{21}) = 0$. (3.106)

(A) For $2n_{11} \geq 2n_{21}+1$, the corresponding infinite-equilibriums are given.

$$(a_{111}, a_{221}) = \underbrace{([2(n_{11}-n_{21})-1]^{\text{th}}\text{UP:UP},[(2m_{11}),(2n_{21}+1)]^{\text{th}}\text{SO})}_{([2(n_{11}-n_{21})-1],[(2m_{11}),(2n_{21}+1)])\text{-hyperbolic-secant-to-hyperbolic source}},$$
for $a_{110} > 0, a_{220} > 0$;
$$(a_{111}, a_{221}) = \underbrace{([2(n_{11}-n_{21})-1]^{\text{th}}\text{DP:DP},[(2m_{11}),(2n_{21}+1)]^{\text{th}}\text{SO})}_{([2(n_{11}-n_{21})-1],[(2m_{11}),(2n_{21}+1)])\text{-hyperbolic-to-hyperbolic-secant source}},$$
for $a_{110} < 0, a_{220} > 0$;
$$(a_{111}, a_{221}) = \underbrace{([2(n_{11}-n_{21})-1]^{\text{th}}\text{DP:DP},[(2m_{11}),(2n_{21}+1)]^{\text{th}}\text{SI})}_{([2(n_{11}-n_{21})-1],[(2m_{11}),(2n_{21}+1)])\text{-hyperbolic-to-hyperbolic-secant sink}},$$
for $a_{110} > 0, a_{220} < 0$;
$$(a_{111}, a_{221}) = \underbrace{([2(n_{11}-n_{21})-1]^{\text{th}}\text{UP:UP},[(2m_{11}),(2n_{21}+1)]^{\text{th}}\text{SI})}_{([2(n_{11}-n_{21})-1],[(2m_{11}),(2n_{21}+1)])\text{-hyperbolic-secant-to-hypberbolic sink}},$$
for $a_{110} < 0, a_{220} < 0$. (3.107)

(B) For $2n_{11} < 2n_{21}+1$, the infinite-equilibriums are given as follows.

$$(a_{111}, a_{221}) = \underbrace{([2(n_{21}-n_{11})+1]^{\text{th}}\text{DA:DA},[(2m_{11}),(2n_{21}+1)]^{\text{th}}\text{SO})}_{([2(n_{21}-n_{11})+1],[(2m_{11}),(2n_{21}+1)])\text{-down asymptote-source}},$$
for $a_{110} > 0, a_{220} > 0$;
$$(a_{111}, a_{221}) = \underbrace{([2(n_{21}-n_{11})]^{\text{th}}\text{UA:UA},[(2m_{11}),(2n_{21}+1)]^{\text{th}}\text{SO})}_{([2(n_{21}-n_{11})],[(2m_{11}+1),(2n_{21}+1)])\text{-up asymptote-source}},$$
for $a_{110} < 0, a_{220} > 0$;
$$(a_{111}, a_{221}) = \underbrace{([2(n_{21}-n_{11})+1]^{\text{th}}\text{UA:UA},[(2m_{11}),(2n_{21}+1)]^{\text{th}}\text{SI})}_{([2(n_{21}-n_{11})+1],[(2m_{11}),(2n_{21}+1)])\text{-up asymptote-sink}},$$
for $a_{110} > 0, a_{220} < 0$;
$$(a_{111}, a_{221}) = \underbrace{([2(n_{21}-n_{11})+1]^{\text{th}}\text{DA:DA},[(2m_{11}),(2n_{21}+1)]^{\text{th}}\text{SI})}_{([2(n_{21}-n_{11})+1],[(2m_{11}),(2n_{21}+1)])\text{-down asymptote-sink}},$$
for $a_{110} < 0, a_{220} < 0$. (3.108)

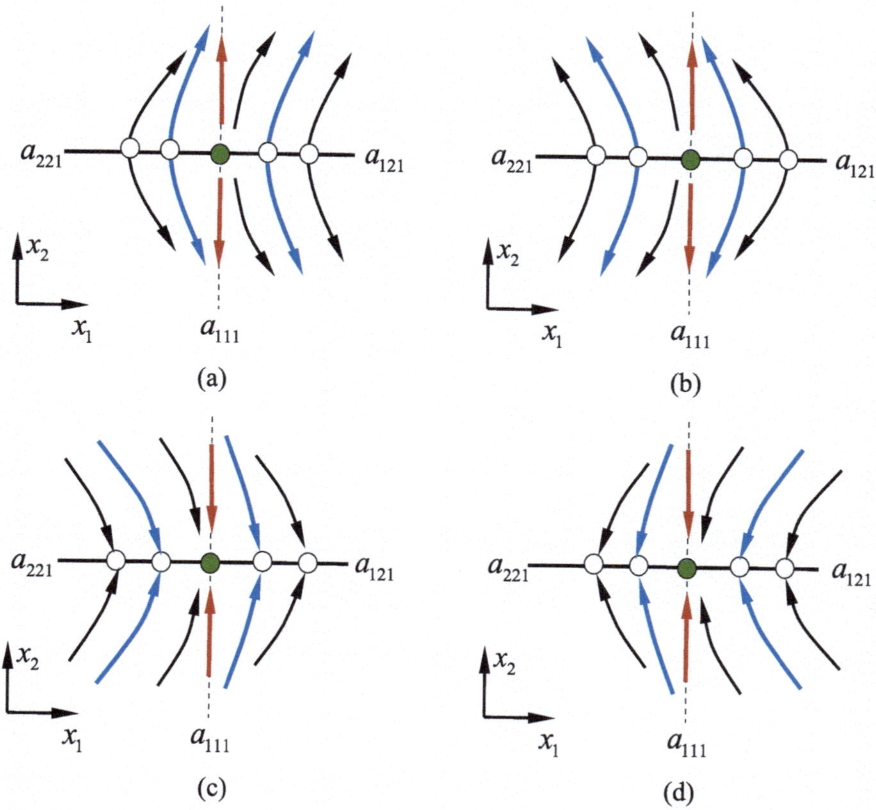

Fig. 3.33 The infinite-equilibriums ($a_{121} = a_{221}$, $2n_{11} \geq 2n_{21}+1$, $m_1 = 2m_{11}$) for 2-dimensional polynomial systems: **a** ($a_{110} > 0$, $a_{220} > 0$), **b** ($a_{110} < 0$, $a_{220} > 0$), **c** ($a_{110} > 0$, $a_{220} < 0$), **d** ($a_{110} < 0$, $a_{220} < 0$)

For $x_2^* = a_{121} = a_{221}$, the infinite-equilibrium exists. The infinite-equilibriums are illustrated in Figs. 3.33 and 3.34 for $2n_{11} \geq 2n_{21}+1$ and $2n_{11} < 2n_{21}+1$, respectively. For $2n_{11} \geq 2n_{21}+1$, the singular increasing and decreasing inflection-source and sink are presented. For $2n_{11} < 2n_{21}+1$, the singular decreasing and increasing asymptote-source and sink are presented.

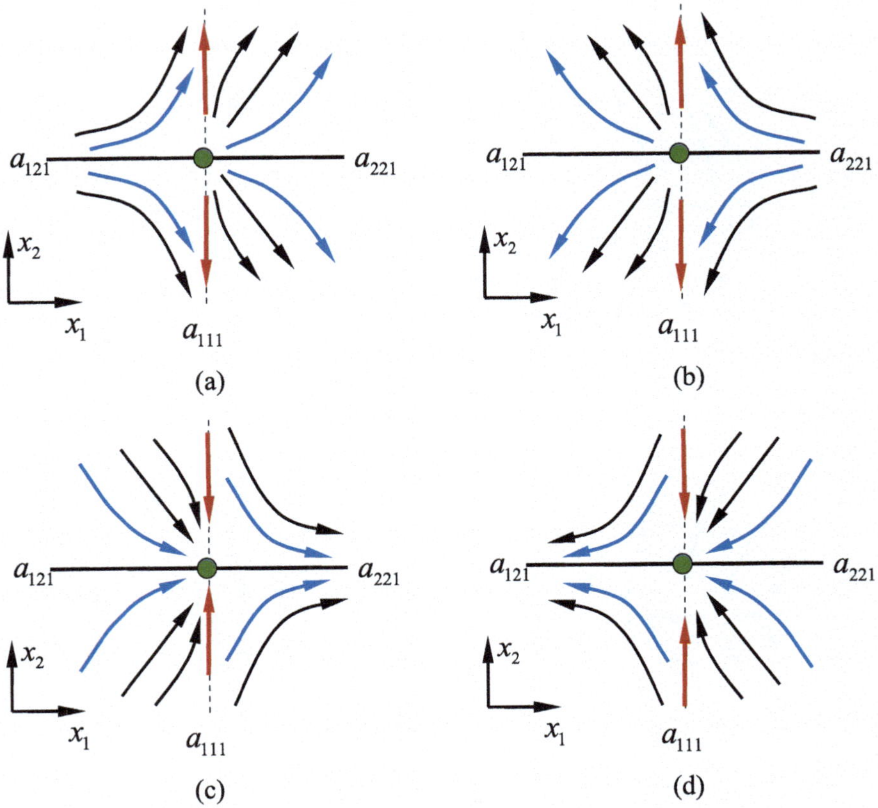

Fig. 3.34 The infinite-equilibriums ($a_{121} = a_{221}$, $2n_{11} < 2n_{21}+1$, $m_1 = 2m_{11}$) for 2-dimensional polynomial systems: **a** ($a_{110} > 0$, $a_{220} > 0$), **b** ($a_{110} < 0$, $a_{220} > 0$), **c** ($a_{110} > 0$, $a_{220} < 0$), **d** ($a_{110} < 0$, $a_{220} < 0$)

3.6.3 $[(2m_{11}), (2n_{11})]$ *and* $((2m_{11}), (2n_{21}+1))$*-Hybrid Networks*

Consider

$$\begin{aligned}\dot{x}_2 &= a_{220}\prod_{s=1}^{2n_{21}+1}(x_2 - a_{22s}),\\ \dot{x}_1 &= a_{110}\prod_{r=1}^{2m_{11}}(x_1 - a_{11r})\prod_{u=1}^{2n_{11}}(x_2 - a_{12u}).\end{aligned} \tag{3.109}$$

The first integral manifold is

$$\sum_{r_1=1}^{2m_{11}} A_{r_1} \ln \frac{|x_1 - a_{11r_1}|}{|x_{10} - a_{11r_1}|} = \frac{a_{110}}{a_{220}} \Bigg\{ \sum_{\beta_0=2n_{21}+1}^{2n_{11}} b_{12\beta_0} \sum_{\substack{\alpha_1+\beta_1=\beta_0-1 \\ \alpha_1\in[0,\beta_0-(2n_{21}+1)],\\ \beta_1\in[2n_{21},\beta_0-1]}} \sum_{\substack{\alpha_2+\beta_2=\beta_1-2 \\ \alpha_2\in[0,\beta_1-(2n_{21}+1)]\\ \beta_2\in[2n_{21}-1,\beta_1-2]}} \cdots$$

$$\sum_{\substack{\alpha_{2n_{21}}+\beta_{2n_{21}}=\beta_{2n_{21}-1}-2n_{21} \\ \alpha_{2n_{21}}\in[0,\beta_{2n_{21}-1}-(2n_{21}+1)]\\ \beta_{2n_{21}}\in[1,\beta_{2n_{21}-1}-(2n_{21}+1)]}} \prod_{j=1}^{2n_{21}} \frac{(\beta_{j-1}-j)!}{\alpha_j!\beta_j!}(a_{22s_{j+1}} - a_{22s_j})^{\alpha_j}$$

$$\times \frac{1}{\beta_{2n_{21}}}[(x_2 - a_{22s_{2n_{21}+1}})^{\beta_{2n_{21}}} - (x_{20} - a_{22s_{2n_{21}+1}})^{\beta_{2n_{21}}}]$$

$$- \sum_{\beta_0=2n_{21}+1}^{2n_{11}} b_{12\beta_0} \sum_{k=2}^{2n_{21}} \Bigg[\sum_{\substack{\alpha_1+\beta_1=\beta_0-1 \\ \alpha_1\in[0,\beta_0-2n_{21}+1),\\ \beta_{2n_{11}}\in(2n_{21},\beta_0-1]}} \sum_{\substack{\alpha_2+\beta_2=\beta_1-2 \\ \alpha_2\in[0,\beta_1-2n_{21}+1]\\ \beta_2\in(2n_{21}-1,\beta_1-2)}} \cdots$$

$$\times \Bigg(\sum_{\substack{\alpha_{k-1}+\beta_{k-1}=\beta_{k-2}-k+1 \\ \alpha_{k-1}\in(\beta_{k-2}-2n_{21},\beta_{k-2}-k+1]\\ \beta_{k-1}\in[0,(2n_{21}+1)-k)}} \prod_{j=1}^{k-1} \frac{(\beta_{j-1}-j)!}{\alpha_j!\beta_j!}(a_{22s_{j+1}} - a_{22s_j})^{\alpha_j} \sum_{\rho=k}^{n_2} A_{\rho 1}^{(s_k)} \ln \frac{|x_{22} - a_{22s_\rho}|}{|x_{20} - a_{22r_\rho}|}$$

$$+ \sum_{\substack{\alpha_{k-1}+\beta_{k-1}=\beta_{k-2}-k+1 \\ \alpha_{k-1}\in(\beta_{k-2}-(2n_{21}+1),\beta_{k-2}-(2n_{21}+1)+1]\\ \beta_{k-1}\in[(2n_{21}+1-k),(2n_{21}+1)-k+1)}} \prod_{j=1}^{k-1} \frac{(\beta_{j-1}-j)!}{\alpha_j!\beta_j!}(a_{22s_{j+1}} - a_{22s_j})^{\alpha_j}$$

$$\times \sum_{\rho=k+1}^{2n_{21}+1} B_{\rho 1}^{(s_k)} \ln \frac{|x_2 - a_{22s_\rho}|}{|x_{20} - a_{22s_\rho}|} \Bigg) \Bigg] + b_{j_1 j_2 0} \sum_{\rho=1}^{2n_{21}+1} A_{\rho 1}^{(s_1)} \ln \frac{|x_2 - a_{22s_\rho}|}{|x_{20} - a_{22s_\rho}|}$$

$$+ \sum_{\beta_0=1}^{2n_{21}} b_{j_1 j_2 \beta_0} \sum_{\rho=2}^{2n_{21}+1} B_{\rho 1}^{(s_1)} \frac{|x_2 - a_{22s_\rho}|}{|x_{20} - a_{22s_\rho}|} \Bigg\}; \tag{3.110}$$

where

$$A_{r_1} = \frac{1}{\prod_{r_2=1, r_2\neq r_1}^{2m_{11}} (a_{11r_1} - a_{11r_\sigma})};$$

$$b_{j_1 j_2 \beta_0} = \sum_{t_1, l_1=0}^{t_1+l_1=1} \cdots \sum_{t_q, l_q=0}^{t_q+l_q=1} \prod_{k=1}^{2n_{11}} (a_{22s_1} - a_{12u_k})^{t_k} \delta_{\beta_0}^{\Sigma_{r=1}^{2n_{11}} l_r};$$

$$\lim_{\beta_{2n_{21}}\to 0} \frac{1}{\beta_{2n_{21}}}[(x_2 - a_{22s_{2n_{21}+1}})^{\beta_{2n_{21}}} - (x_{20} - a_{22s_{2n_{21}+1}})^{\beta_{2n_{21}}}] = \ln\frac{|x_2 - a_{22s_{2n_{21}+1}}|}{|x_{20} - a_{22s_{2n_{21}+1}}|}; \tag{3.111}$$

$$A_{\rho 1}^{(s_k)} = \frac{1}{\prod_{\sigma=k,\sigma\neq\rho}^{2n_{21}+1}(a_{22s_\rho} - a_{22s_\sigma})}, \text{ for } \rho \in \{k, k+1, \ldots, 2n_{21}+1\};$$

$$B_{\rho 1}^{(s_k)} = \frac{(a_{22s_\rho} - a_{22s_k})^{\beta_{k-1}-k}}{\prod_{\sigma=k+1,\sigma\neq\rho}^{2n_{21}+1}(a_{22s_\rho} - a_{22s_\sigma})}, \text{ for } \rho \in \{k+1, k+2, \ldots, 2n_{21}+1\};$$

$$k \in \{1, 2, \ldots, 2n_{21}\}. \tag{3.112}$$

For all cases, the $(2m_{11}) \times (2n_{21}+1)$-simple equilibriums plus $(2m_{11}) \times (2n_{11})$-hyperbolic flows are based on the following up and down parabola-source and sink bifurcations for $2n_{11} \geq 2n_{21}+1$:

- $[2(n_{11}-n_{21})-1]^{\text{th}}$UP:UP, $[(2m_{11}+1), (2n_{11}+1)]^{\text{th}}$SO)-hyperbolic source bifurcation for ($a_{110} > 0, a_{220} > 0$),
- $[2(n_{11}-n_{21})-1]^{\text{th}}$DP:DP, $[(2m_{11}+1), (2n_{11}+1)]^{\text{th}}$SO)-hyperbolic-secant source bifurcation for ($a_{110} < 0, a_{220} > 0$),
- $[2(n_{11}-n_{21})+1]^{\text{th}}$DP:DP, $[(2m_{11}+1), (2n_{11}+1)]^{\text{th}}$SI)-hyperbolic-secant sink bifurcation for ($a_{110} > 0, a_{220} < 0$),
- $[2(n_{11}-n_{21})+1]^{\text{th}}$UP:UP, $[(2m_{11}+1), (2n_{21}+1)]^{\text{th}}$SI)-hyperbolic sink bifurcation for ($a_{110} < 0, a_{220} < 0$);

or based on the following down and up asymptote source and sink bifurcations for $2n_{11} < 2n_{21}+1$:

- $[2(n_{21}-n_{11})-1]^{\text{th}}$DA:DA, $[(2m_{11}+1), (2n_{21}+1)]^{\text{th}}$SO)-up-to-down asymptote-source bifurcation for ($a_{110} > 0, a_{220} > 0$),
- $[2(n_{21}-n_{11})-1]^{\text{th}}$UA:UA, $[(2m_{11}+1), (2n_{21}+1)]^{\text{th}}$SO)-down-to-up asymptote-source bifurcation for ($a_{110} < 0, a_{220} > 0$),
- $[2(n_{21}-n_{11})-1]^{\text{th}}$UA:UA, $[(2m_{11}+1), (2n_{21}+1)]^{\text{th}}$SI)-down-to-up asymptote-sink bifurcation for ($a_{110} > 0, a_{220} < 0$),
- $[2(n_{21}-n_{11})-1]^{\text{th}}$DA:DA, $[(2m_{11}+1), (2n_{21}+1)]^{\text{th}}$SO)-up-to-down asymptote-sink bifurcation for ($a_{110} < 0, a_{220} < 0$).

Consider $2n_{11} = 2n_{21}$ for illustrations, phase portraits for the 2-dimensional systems for the hybrid networks of hyperbolic and hyperbolic-secant flows with the simple equilibriums of the saddles, sinks and sources with $x_1^* = a_{111}, a_{112}, \ldots, a_{11(2m_{11})}$ and $x_2^* = a_{121}, a_{221}, \ldots, a_{12(2n_{11}+1)}, a_{22(2n_{21}+1)}$ are presented in Fig. 3.35(a–d) for $(a_{110} > 0, a_{220} > 0)$, $(a_{110} < 0, a_{220} > 0)$, $(a_{110} > 0, a_{220} < 0)$ and $(a_{110} < 0, a_{220} < 0)$.

$$
\left\{\begin{array}{cccc}
(a_{111}, a_{22(2n_{21}+1)}) & \cdots & (a_{11(2m_{11}-1)}, a_{22(2n_{21}+1)}) & (a_{11(2m_{11})}, a_{22(2n_{21}+1)}) \\
(a_{111}, a_{22(2n_{21})}) & \cdots & (a_{11(2m_{11}-1)}, a_{22(2n_{21})}) & (a_{11(2m_{11})}, a_{22(2n_{21})}) \\
(a_{111}, a_{12(2n_{11})}) & \cdots & (a_{11(2m_{11}-1)}, a_{12(2n_{11}+1)}) & (a_{11(2m_{11})}, a_{12(2n_{11})}) \\
\vdots & \vdots & \vdots & \vdots \\
(a_{111}, a_{221}) & \cdots & (a_{11(2m_{11}-1)}, a_{221}) & (a_{11(2m_{11})}, a_{221}) \\
(a_{111}, a_{121}) & \cdots & (a_{11(2m_{11}-1)}, a_{121}) & (a_{11(2m_{11})}, a_{121})
\end{array}\right\}
$$

$$
= \left\{\begin{array}{cccc}
\underbrace{(\mathrm{SI}, \mathrm{SO})}_{\text{saddle}} & \cdots & \underbrace{(\mathrm{SI}, \mathrm{SO})}_{\text{saddle}} & \underbrace{(\mathrm{SO}, \mathrm{SO})}_{\text{source}} \\
\underbrace{(\mathrm{SI}, \mathrm{SI})}_{\text{sink}} & \cdots & \underbrace{(\mathrm{SI}, \mathrm{SI})}_{\text{sink}} & \underbrace{(\mathrm{SO}, \mathrm{SI})}_{\text{saddle}} \\
\underbrace{(\mathrm{UP{:}DP}, \mathrm{pF})}_{\text{hyperbolic-secant flow }(+)} & \cdots & \underbrace{(\mathrm{UP{:}DP}, \mathrm{pF})}_{\text{hyperbolic-secant flow }(+)} & \underbrace{(\mathrm{DP{:}UP}, \mathrm{pF})}_{\text{hyperbolic flow }(+)} \\
\vdots & \vdots & \vdots & \vdots \\
\underbrace{(\mathrm{SO}, \mathrm{SO})}_{\text{source}} & \cdots & \underbrace{(\mathrm{SO}, \mathrm{SO})}_{\text{source}} & \underbrace{(\mathrm{SI}, \mathrm{SO})}_{\text{saddle}} \\
\underbrace{(\mathrm{UP{:}DP}, \mathrm{nF})}_{\text{hyperbolic-secant flow }(-)} & \cdots & \underbrace{(\mathrm{UP{:}DP}, \mathrm{nF})}_{\text{hyperbolic-secant flow }(-)} & \underbrace{(\mathrm{DP{:}UP}, \mathrm{nF})}_{\text{hyperbolic flow }(-)}
\end{array}\right\}_{(2n_{11}+2n_{21}+1)\times(2m_{11})}
$$

for $a_{110} > 0, a_{220} > 0$; (3.113)

$$
\left\{\begin{array}{cccc}
(a_{111}, a_{22(2n_{21}+1)}) & \cdots & (a_{11(2m_{11}-1)}, a_{22(2n_{21}+1)}) & (a_{11(2m_{11})}, a_{22(2n_{21}+1)}) \\
(a_{111}, a_{22(2n_{21})}) & \cdots & (a_{11(2m_{11}-1)}, a_{22(2n_{21})}) & (a_{11(2m_{11})}, a_{22(2n_{21})}) \\
(a_{111}, a_{12(2n_{11})}) & \cdots & (a_{11(2m_{11}-1)}, a_{12(2n_{11}+1)}) & (a_{11(2m_{11})}, a_{12(2n_{11})}) \\
\vdots & \vdots & \vdots & \vdots \\
(a_{111}, a_{221}) & \cdots & (a_{11(2m_{11}-1)}, a_{221}) & (a_{11(2m_{11})}, a_{221}) \\
(a_{111}, a_{121}) & \cdots & (a_{11(2m_{11}-1)}, a_{121}) & (a_{11(2m_{11})}, a_{121})
\end{array}\right\}
$$

$$
= \left\{\begin{array}{cccc}
\underbrace{(\mathrm{SO}, \mathrm{SO})}_{\text{source}} & \cdots & \underbrace{(\mathrm{SO}, \mathrm{SO})}_{\text{source}} & \underbrace{(\mathrm{SI}, \mathrm{SO})}_{\text{saddle}} \\
\underbrace{(\mathrm{SO}, \mathrm{SI})}_{\text{saddle}} & \cdots & \underbrace{(\mathrm{SO}, \mathrm{SI})}_{\text{saddle}} & \underbrace{(\mathrm{SI}, \mathrm{SI})}_{\text{sink}} \\
\underbrace{(\mathrm{DP{:}UP}, \mathrm{pF})}_{\text{hyperbolic flow }(+)} & \cdots & \underbrace{(\mathrm{DP{:}UP}, \mathrm{pF})}_{\text{hyperbolic flow }(+)} & \underbrace{(\mathrm{UP{:}DP}, \mathrm{pF})}_{\text{hyperbolic-secant flow }(+)} \\
\vdots & \vdots & \vdots & \vdots \\
\underbrace{(\mathrm{SI}, \mathrm{SO})}_{\text{saddle}} & \cdots & \underbrace{(\mathrm{SI}, \mathrm{SO})}_{\text{saddle}} & \underbrace{(\mathrm{SO}, \mathrm{SO})}_{\text{source}} \\
\underbrace{(\mathrm{DP{:}UP}, \mathrm{nF})}_{\text{hyperbolic flow }(-)} & \cdots & \underbrace{(\mathrm{DP{:}UP}, \mathrm{nF})}_{\text{hyperbolic flow }(-)} & \underbrace{(\mathrm{UP{:}DP}, \mathrm{nF})}_{\text{hyperbolic-secant flow }(-)}
\end{array}\right\}_{(2n_{11}+2n_{21}+1)\times(2m_{11})}
$$

for $a_{110} < 0, a_{220} > 0$; (3.114)

$$
\left\{
\begin{array}{cccc}
(a_{111}, a_{22(2n_{21}+1)}) & \cdots & (a_{11(2m_{11}-1)}, a_{22(2n_{21}+1)}) & (a_{11(2m_{11})}, a_{22(2n_{21}+1)}) \\
(a_{111}, a_{22(2n_{21})}) & \cdots & (a_{11(2m_{11}-1)}, a_{22(2n_{21})}) & (a_{11(2m_{11})}, a_{22(2n_{21})}) \\
(a_{111}, a_{12(2n_{11})}) & \cdots & (a_{11(2m_{11}-1)}, a_{12(2n_{11}+1)}) & (a_{11(2m_{11})}, a_{12(2n_{11})}) \\
\vdots & \vdots & \vdots & \vdots \\
(a_{111}, a_{221}) & \cdots & (a_{11(2m_{11}-1)}, a_{221}) & (a_{11(2m_{11})}, a_{221}) \\
(a_{111}, a_{121}) & \cdots & (a_{11(2m_{11}-1)}, a_{121}) & (a_{11(2m_{11})}, a_{121})
\end{array}
\right\}
$$

$$
= \left\{
\begin{array}{cccc}
\underbrace{(\mathrm{SI}, \mathrm{SI})}_{\text{sink}} & \cdots & \underbrace{(\mathrm{SI}, \mathrm{SI})}_{\text{sink}} & \underbrace{(\mathrm{SO}, \mathrm{SI})}_{\text{saddle}} \\
\underbrace{(\mathrm{SI}, \mathrm{SO})}_{\text{saddle}} & \cdots & \underbrace{(\mathrm{SI}, \mathrm{SO})}_{\text{saddle}} & \underbrace{(\mathrm{SO}, \mathrm{SO})}_{\text{source}} \\
\underbrace{(\mathrm{DP{:}UP}, \mathrm{nF})}_{\text{hyperbolic flow }(-)} & \cdots & \underbrace{(\mathrm{DP{:}UP}, \mathrm{nF})}_{\text{hyperbolic flow }(-)} & \underbrace{(\mathrm{UP{:}DP}, \mathrm{nF})}_{\text{hyperbolic-secant flow }(-)} \\
\vdots & \vdots & \vdots & \vdots \\
\underbrace{(\mathrm{SO}, \mathrm{SI})}_{\text{saddle}} & \cdots & \underbrace{(\mathrm{SO}, \mathrm{SI})}_{\text{saddle}} & \underbrace{(\mathrm{SI}, \mathrm{SI})}_{\text{sink}} \\
\underbrace{(\mathrm{DP{:}UP}, \mathrm{nF})}_{\text{hyperbolic flow }(-)} & \cdots & \underbrace{(\mathrm{DP{:}UP}, \mathrm{nF})}_{\text{hyperbolic flow }(-)} & \underbrace{(\mathrm{UP{:}DP}, \mathrm{nF})}_{\text{hyperbolic-secant flow }(-)}
\end{array}
\right\}_{(2n_{11}+2n_{21}+1)\times(2m_{11})}
$$

for $a_{110} > 0, a_{220} < 0$; (3.115)

$$
\left\{
\begin{array}{cccc}
(a_{111}, a_{22(2n_{21}+1)}) & \cdots & (a_{11(2m_{11}-1)}, a_{22(2n_{21}+1)}) & (a_{11(2m_{11})}, a_{22(2n_{21}+1)}) \\
(a_{111}, a_{22(2n_{21})}) & \cdots & (a_{11(2m_{11}-1)}, a_{22(2n_{21})}) & (a_{11(2m_{11})}, a_{22(2n_{21})}) \\
(a_{111}, a_{12(2n_{11})}) & \cdots & (a_{11(2m_{11}-1)}, a_{12(2n_{11}+1)}) & (a_{11(2m_{11})}, a_{12(2n_{11})}) \\
\vdots & \vdots & \vdots & \vdots \\
(a_{111}, a_{221}) & \cdots & (a_{11(2m_{11}-1)}, a_{221}) & (a_{11(2m_{11})}, a_{221}) \\
(a_{111}, a_{121}) & \cdots & (a_{11(2m_{11}-1)}, a_{121}) & (a_{11(2m_{11})}, a_{121})
\end{array}
\right\}
$$

$$
= \left\{
\begin{array}{cccc}
\underbrace{(\mathrm{SO}, \mathrm{SI})}_{\text{saddle}} & \cdots & \underbrace{(\mathrm{SO}, \mathrm{SI})}_{\text{saddle}} & \underbrace{(\mathrm{SI}, \mathrm{SI})}_{\text{sink}} \\
\underbrace{(\mathrm{SO}, \mathrm{SO})}_{\text{source}} & \cdots & \underbrace{(\mathrm{SO}, \mathrm{SO})}_{\text{source}} & \underbrace{(\mathrm{SI}, \mathrm{SO})}_{\text{saddle}} \\
\underbrace{(\mathrm{UP{:}DP}, \mathrm{nF})}_{\text{hyperbolic-secant flow }(-)} & \cdots & \underbrace{(\mathrm{UP{:}DP},\mathrm{nF})}_{\text{hyperbolic-secant flow }(-)} & \underbrace{(\mathrm{DP{:}UP}, \mathrm{nF})}_{\text{hyperbolic flow }(-)} \\
\vdots & \vdots & \vdots & \vdots \\
\underbrace{(\mathrm{SI}, \mathrm{SI})}_{\text{sink}} & \cdots & \underbrace{(\mathrm{SI}, \mathrm{SI})}_{\text{sink}} & \underbrace{(\mathrm{SO}, \mathrm{SI})}_{\text{saddle}} \\
\underbrace{(\mathrm{UP{:}DP}, \mathrm{pF})}_{\text{hyperbolic-secant flow }(+)} & \cdots & \underbrace{(\mathrm{UP{:}DP}, \mathrm{pF})}_{\text{hyperbolic-secant flow }(+)} & \underbrace{(\mathrm{DP{:}UP}, \mathrm{pF})}_{\text{hyperbolic flow }(+)}
\end{array}
\right\}_{(2n_{11}+2n_{21}+1)\times(2m_{11})}
$$

for $a_{110} < 0, a_{220} < 0$. (3.116)

Consider $2n_{11} = 2n_{21}$ for illustrations, phase portraits for the 2-dimensional systems for the hybrid networks of hyperbolic and hyperbolic-secant flows with the simple equilibriums of the saddles, sinks and sources with $x_1^* = a_{111}, a_{112}, \ldots, a_{11(2m_{11})}$ and $x_2^* = a_{221}, a_{121}, \ldots, a_{12(2n_{11})}, a_{22(2n_{21}+1)}$ are presented in Fig. 3.36 (a–d) for ($a_{110} > 0, a_{220} > 0$), ($a_{110} < 0, a_{220} > 0$), ($a_{110} > 0, a_{220} < 0$) and ($a_{110} < 0, a_{220} < 0$).

$$\left\{\begin{array}{cccc}
(a_{111}, a_{22(2n_{21}+1)}) & \cdots & (a_{11(2m_{11}-1)}, a_{22(2n_{21}+1)}) & (a_{11(2m_{11})}, a_{22(2n_{21}+1)}) \\
(a_{111}, a_{12(2n_{11})}) & \cdots & (a_{11(2m_{11}-1)}, a_{12(2n_{11})}) & (a_{11(2m_{11})}, a_{12(2n_{11})}) \\
(a_{111}, a_{22(2n_{21})}) & \cdots & (a_{11(2m_{11}-1)}, a_{22(2n_{21})}) & (a_{11(2m_{11})}, a_{22(2n_{21})}) \\
\vdots & \vdots & \vdots & \vdots \\
(a_{111}, a_{121}) & \cdots & (a_{11(2m_{11}-1)}, a_{121}) & (a_{11(2m_{11})}, a_{121}) \\
(a_{111}, a_{221}) & \cdots & (a_{11(2m_{11}-1)}, a_{221}) & (a_{11(2m_{11})}, a_{221})
\end{array}\right\}$$

$$= \left\{\begin{array}{cccc}
\underbrace{(\text{SI, SO})}_{\text{saddle}} & \cdots & \underbrace{(\text{SI, SO})}_{\text{saddle}} & \underbrace{(\text{SO, SO})}_{\text{source}} \\
\underbrace{(\text{DP:UP, nF})}_{\text{hyperbolic flow }(-)} & \cdots & \underbrace{(\text{DP:UP, nF})}_{\text{hyperbolic flow }(-)} & \underbrace{(\text{UP:DP,nF})}_{\text{hyperbolic-secant flow }(-)} \\
\underbrace{(\text{SO, SI})}_{\text{saddle}} & \cdots & \underbrace{(\text{SO, SI})}_{\text{saddle}} & \underbrace{(\text{SI, SI})}_{\text{sink}} \\
\vdots & \vdots & \vdots & \vdots \\
\underbrace{(\text{DP:UP, pF})}_{\text{hyperbolic flow }(+)} & \cdots & \underbrace{(\text{DP:UP, pF})}_{\text{hyperbolic flow }(+)} & \underbrace{(\text{UP:DP, pF})}_{\text{hyperbolic-secant flow }(+)} \\
\underbrace{(\text{SI, SO})}_{\text{saddle}} & \cdots & \underbrace{(\text{SI, SO})}_{\text{saddle}} & \underbrace{(\text{SO, SO})}_{\text{source}}
\end{array}\right\}_{(2n_{11}+2n_{21}+1)\times(2m_{11})}$$

for $a_{110} > 0, a_{220} > 0$; (3.117)

$$\left\{\begin{array}{cccc}
(a_{111}, a_{22(2n_{21}+1)}) & \cdots & (a_{11(2m_{11}-1)}, a_{22(2n_{21}+1)}) & (a_{11(2m_{11})}, a_{22(2n_{21}+1)}) \\
(a_{111}, a_{12(2n_{11})}) & \cdots & (a_{11(2m_{11}-1)}, a_{12(2n_{11})}) & (a_{11(2m_{11})}, a_{12(2n_{11})}) \\
(a_{111}, a_{22(2n_{21})}) & \cdots & (a_{11(2m_{11}-1)}, a_{22(2n_{21})}) & (a_{11(2m_{11})}, a_{22(2n_{21})}) \\
\vdots & \vdots & \vdots & \vdots \\
(a_{111}, a_{121}) & \cdots & (a_{11(2m_{11}-1)}, a_{121}) & (a_{11(2m_{11})}, a_{121}) \\
(a_{111}, a_{221}) & \cdots & (a_{11(2m_{11}-1)}, a_{221}) & (a_{11(2m_{11})}, a_{221})
\end{array}\right\}$$

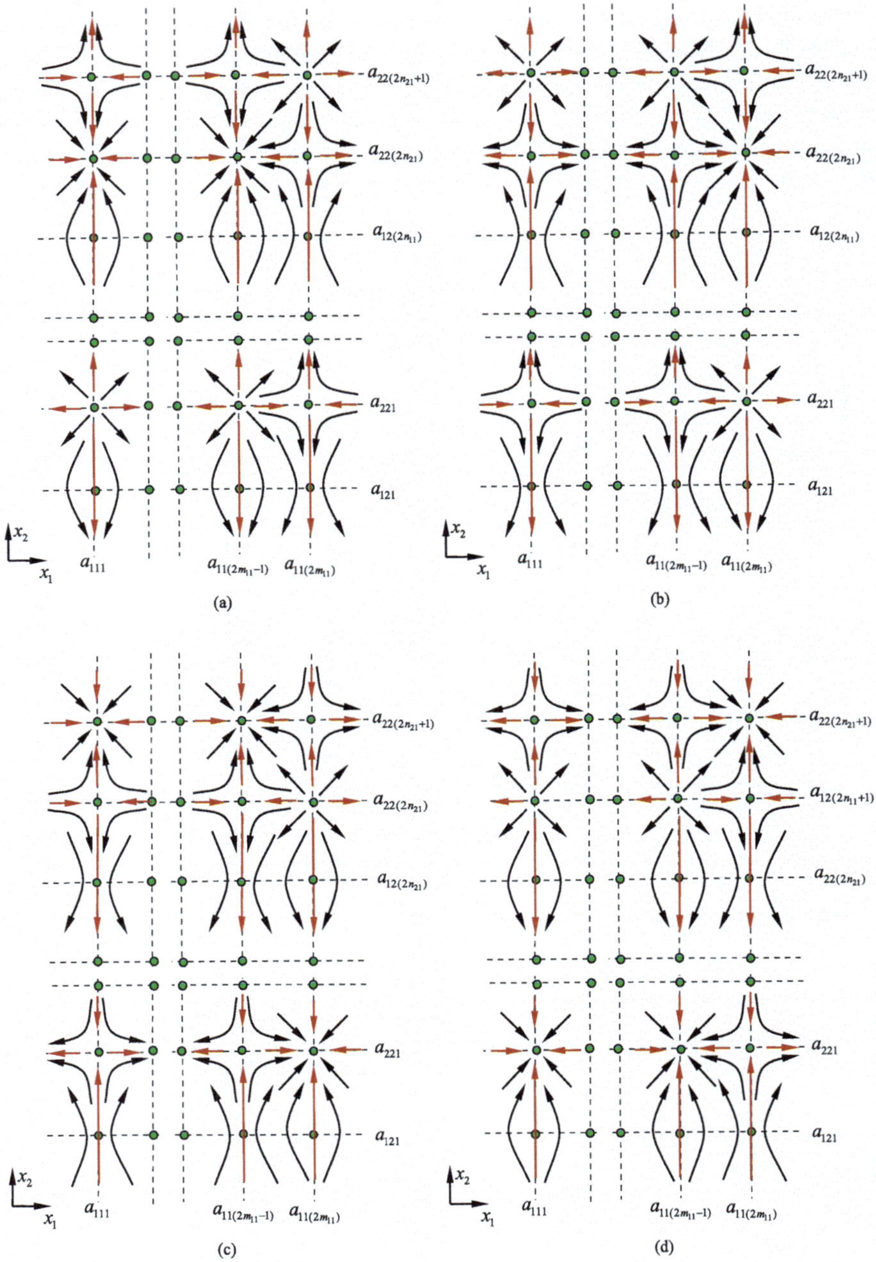

Fig. 3.35 Phase portraits for 2-dimensional systems with $x_1^* = a_{111}, a_{112}, \ldots, a_{11(2m_{11})}$ and $x_2^* = a_{121}, a_{221}, \ldots, a_{12(2n_{11}+1)}, a_{22(2n_{21}+1)}$. The networks of flows with sink, source and saddle: **a** ($a_{110} > 0$, $a_{220} > 0$), **b** ($a_{110} < 0$, $a_{220} > 0$), **c** ($a_{110} > 0$, $a_{220} < 0$), **d** ($a_{110} < 0$, $a_{220} < 0$)

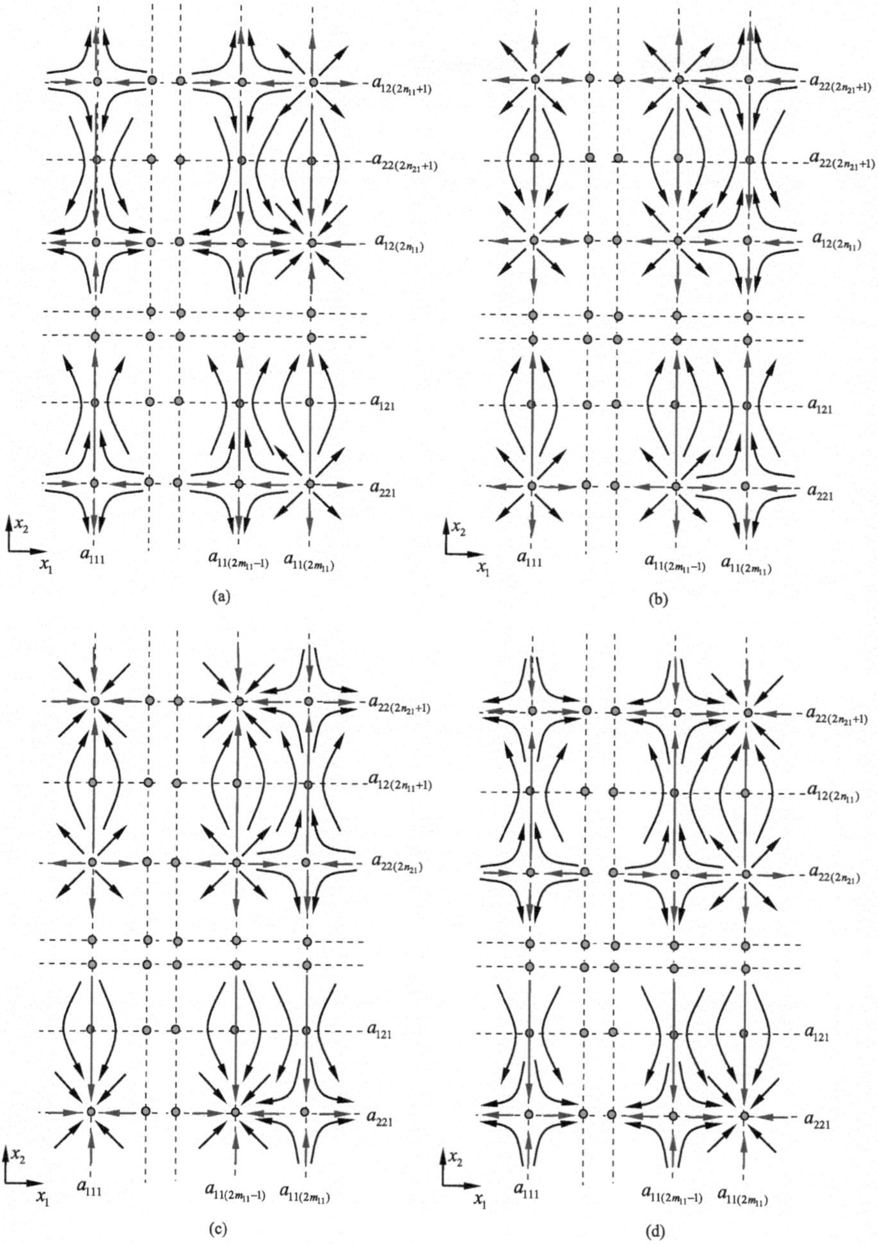

Fig. 3.36 Phase portraits for 2-dimensional systems with $x_1^* = a_{111}, a_{112}, \ldots, a_{11(2m_{11})}$ and . The networks of flows with sink, source and saddle: **a** ($a_{110} > 0$, $a_{220} > 0$), **b** ($a_{110} < 0$, $a_{220} > 0$), **c** ($a_{110} > 0$, $a_{220} < 0$), **d** ($a_{110} < 0$, $a_{220} < 0$)

$$= \left\{\begin{array}{cccc}
\underbrace{(\text{SO, SO})}_{\text{source}} & \cdots & \underbrace{(\text{SO, SO})}_{\text{source}} & \underbrace{(\text{SI, SO})}_{\text{saddle}} \\
\underbrace{(\text{UP:DP, nF})}_{\text{hyperbolic-secant flow }(-)} & \cdots & \underbrace{(\text{UP:DP, nF})}_{\text{hyperbolic-secant flow }(-)} & \underbrace{(\text{DP:UP, nF})}_{\text{hyperbolic flow }(-)} \\
\underbrace{(\text{SI, SI})}_{\text{sink}} & \cdots & \underbrace{(\text{SI, SI})}_{\text{sink}} & \underbrace{(\text{SO, SI})}_{\text{saddle}} \\
\vdots & \vdots & \vdots & \vdots \\
\underbrace{(\text{UP:DP, pF})}_{\text{hyperbolic-secant flow }(+)} & \cdots & \underbrace{(\text{UP:DP, pF})}_{\text{hyperbolic-secant flow }(+)} & \underbrace{(\text{DP:UP, pF})}_{\text{hyperbolic flow }(+)} \\
\underbrace{(\text{SO, SO})}_{\text{source}} & \cdots & \underbrace{(\text{SO, SO})}_{\text{source}} & \underbrace{(\text{SI, SO})}_{\text{saddle}}
\end{array}\right\}_{(2n_{11}+2n_{21}+1)\times(2m_{11})}$$

for $a_{110} < 0, a_{220} > 0;$ (3.118)

$$\left\{\begin{array}{cccc}
(a_{111}, a_{22(2n_{21}+1)}) & \cdots & (a_{11(2m_{11}-1)}, a_{22(2n_{21}+1)}) & (a_{11(2m_{11})}, a_{22(2n_{21}+1)}) \\
(a_{111}, a_{12(2n_{11})}) & \cdots & (a_{11(2m_{11}-1)}, a_{12(2n_{11})}) & (a_{11(2m_{11})}, a_{12(2n_{11})}) \\
(a_{111}, a_{22(2n_{21})}) & \cdots & (a_{11(2m_{11}-1)}, a_{22(2n_{21})}) & (a_{11(2m_{11})}, a_{22(2n_{21})}) \\
\vdots & \vdots & \vdots & \vdots \\
(a_{111}, a_{121}) & \cdots & (a_{11(2m_{11}-1)}, a_{121}) & (a_{11(2m_{11})}, a_{121}) \\
(a_{111}, a_{221}) & \cdots & (a_{11(2m_{11}-1)}, a_{221}) & (a_{11(2m_{11})}, a_{221})
\end{array}\right\}$$

$$= \left\{\begin{array}{cccc}
\underbrace{(\text{SI, SI})}_{\text{sink}} & \cdots & \underbrace{(\text{SI, SI})}_{\text{sink}} & \underbrace{(\text{SO, SI})}_{\text{saddle}} \\
\underbrace{(\text{UP:DP, pF})}_{\text{hyperbolic-secant flow }(+)} & \cdots & \underbrace{(\text{UP:DP, pF})}_{\text{hyperbolic-secant flow }(+)} & \underbrace{(\text{DP:UP, pF})}_{\text{hyperbolic flow }(+)} \\
\underbrace{(\text{SO, SO})}_{\text{source}} & \cdots & \underbrace{(\text{SO, SO})}_{\text{source}} & \underbrace{(\text{SI, SO})}_{\text{saddle}} \\
\vdots & \vdots & \vdots & \vdots \\
\underbrace{(\text{UP:DP, nF})}_{\text{hyperbolic-secant flow }(-)} & \cdots & \underbrace{(\text{UP:DP, nF})}_{\text{hyperbolic-secant flow }(-)} & \underbrace{(\text{DP:UP, nF})}_{\text{hyperbolic flow }(-)} \\
\underbrace{(\text{SI, SI})}_{\text{sink}} & \cdots & \underbrace{(\text{SI, SI})}_{\text{sink}} & \underbrace{(\text{SO, SI})}_{\text{saddle}}
\end{array}\right\}_{(2n_{11}+2n_{21}+1)\times(2m_{11})}$$

for $a_{110} > 0, a_{220} < 0;$ (3.119)

$$\left\{\begin{array}{cccc}
(a_{111}, a_{22(2n_{21}+1)}) & \cdots & (a_{11(2m_{11}-1)}, a_{22(2n_{21}+1)}) & (a_{11(2m_{11})}, a_{22(2n_{21}+1)}) \\
(a_{111}, a_{12(2n_{11})}) & \cdots & (a_{11(2m_{11}-1)}, a_{12(2n_{11})}) & (a_{11(2m_{11})}, a_{12(2n_{11})}) \\
(a_{111}, a_{22(2n_{21})}) & \cdots & (a_{11(2m_{11}-1)}, a_{22(2n_{21})}) & (a_{11(2m_{11})}, a_{22(2n_{21})}) \\
\vdots & \vdots & \vdots & \vdots \\
(a_{111}, a_{121}) & \cdots & (a_{11(2m_{11}-1)}, a_{121}) & (a_{11(2m_{11})}, a_{121}) \\
(a_{111}, a_{221}) & \cdots & (a_{11(2m_{11}-1)}, a_{221}) & (a_{11(2m_{11})}, a_{221})
\end{array}\right\}$$

$$
= \left\{ \begin{array}{cccc}
\underbrace{(\text{SO, SI})}_{\text{saddle}} & \cdots & \underbrace{(\text{SO, SI})}_{\text{saddle}} & \underbrace{(\text{SI, SI})}_{\text{sink}} \\
\underbrace{(\text{DP:UP, pF})}_{\text{hyperbolic flow }(+)} & \cdots & \underbrace{(\text{DP:UP, pF})}_{\text{hyperbolic flow }(+)} & \underbrace{(\text{UP:DP, pF})}_{\text{hyperbolic-secant flow }(+)} \\
\underbrace{(\text{SI, SO})}_{\text{saddle}} & \cdots & \underbrace{(\text{SI, SO})}_{\text{saddle}} & \underbrace{(\text{SO, SO})}_{\text{source}} \\
\vdots & \vdots & \vdots & \vdots \\
\underbrace{(\text{DP:UP, nF})}_{\text{hyperbolic flow }(-)} & \cdots & \underbrace{(\text{DP:UP, nF})}_{\text{hyperbolic flow }(-)} & \underbrace{(\text{UP:DP, nF})}_{\text{hyperbolic-secant flow }(-)} \\
\underbrace{(\text{SO, SI})}_{\text{saddle}} & \cdots & \underbrace{(\text{SO, SI})}_{\text{saddle}} & \underbrace{(\text{SI, SI})}_{\text{sink}}
\end{array} \right\}_{(2n_{11}+2n_{21}+1)\times(2m_{11}+1)}
$$

for $a_{110} < 0, a_{220} < 0.$ (3.120)

3.7 $[(2m_{11}), (2n_{11} + 1)]$ and $((2m_{11}), (2n_{21}))$-Singular Systems

In this section, there are four global bifurcations of the 1-diemsnional flows through singular hyperbolic-to-hyperbolic-secant flows, and four global bifurcations of equilibriums through double-saddle bifurcations. The corresponding global bifurcations generate networks of hyperbolic and hyperbolic-secant flows with source, sink and saddles. The infinite-equilibriums are the appearing and switching bifurcations for the singular flows and singular equilibriums and for hybrid networks of hyperbolic flows and source, sink and saddles.

3.7.1 $[(2m_{11}), (2n_{11} + 1)]$-*Flows and* $((2m_{11}), (2n_{21}))$-*Equilibriums*

Consider a polynomial dynamical system for $a_{221} \neq a_{121}$ as

$$
\begin{aligned}
\dot{x}_2 &= a_{220}(x_2 - a_{221})^{2n_{21}}, \\
\dot{x}_1 &= a_{110}(x_1 - a_{111})^{2m_{11}}(x_2 - a_{121})^{2n_{11}+1}.
\end{aligned} \tag{3.121}
$$

The first integral manifold is

$$
-\frac{1}{2m_{11}-1}\left[\frac{1}{(x_1 - a_{111})^{2m_{11}-1}} - \frac{1}{(x_{10} - a_{111})^{2m_{11}-1}}\right]
$$

$$= \frac{a_{110}}{a_{220}} \sum_{\alpha,\beta=0}^{\alpha+\beta=2n_{11}+1} \frac{(2n_{11}+1)!}{\alpha!\beta!} \frac{(a_{221}-a_{121})^{\alpha}}{\beta-2n_{21}+1}[(x_2-a_{221})^{\beta-2n_{21}+1}$$
$$-(x_2-a_{221})^{\beta-2n_{21}+1}]$$
$$\text{for } \beta-2n_{21}+1\neq 0;$$
$$\frac{1}{\beta-2n_{21}+1}[(x_2-a_{221})^{\beta-2n_{21}+1}-(x_2-a_{221})^{\beta-2n_{21}+1}] = \ln\frac{|x_2-a_{221}|}{|x_{20}-a_{221}|}$$
$$\text{for } \beta-2n_{21}+1=0. \tag{3.122}$$

For $(x_1^*, x_2^*) = (a_{111}, a_{121})$ with $\dot{x}_1 = 0$, the $[(2m_{11}), (2n_{11}+1)]$-singular flows exist with $\dot{x}_2 = a_{220}(a_{121}-a_{221})^{2n_{21}}$. For $(x_1^*, x_2^*) = (a_{111}, a_{121})$, the 1-dimensional flow direction is determined by $\dot{x}_2 \neq 0s$. For $(x_1^*, x_2^*) = (a_{111}, a_{221})$ with $\dot{x}_1 = \dot{x}_2 = 0$, the $((2m_{11}), (2n_{21}))$- singular equilibriums exist.

(A) For $a_{121} > a_{221}$,

$$\left\{\begin{matrix}(a_{111}, a_{121})\\(a_{111}, a_{221})\end{matrix}\right\} = \left\{\begin{matrix}\underbrace{([(2m_{11}), (2n_{11}+1)]\text{UP:UP, pF})}_{[(2m_{11}),(2n_{11}+1)]\text{-hyperbolic-secant-to-hyperbolic flow }(+)}\\\underbrace{((2m_{11})\text{LS}, (2n_{21})\text{US})}_{[(2m_{11}),(2n_{21})]\text{-double-saddle}}\end{matrix}\right\},$$
for $a_{110} > 0, a_{220} > 0$;

$$\left\{\begin{matrix}(a_{111}, a_{121})\\(a_{111}, a_{221})\end{matrix}\right\} = \left\{\begin{matrix}\underbrace{([(2m_{11}), (2n_{11}+1)]\text{DP:DP, pF})}_{[(2m_{11}),(2n_{11}+1)]\text{-hyperbolic-to-hyperbolic-secant flow }(+)}\\\underbrace{((2m_{11})\text{US}, (2n_{21})\text{US})}_{[(2m_{11}),(2n_{21})]\text{-double-saddle}}\end{matrix}\right\},$$
for $a_{110} < 0, a_{220} > 0$;

$$\left\{\begin{matrix}(a_{111}, a_{121})\\(a_{111}, a_{221})\end{matrix}\right\} = \left\{\begin{matrix}\underbrace{([(2m_{11}), (2n_{11}+1)]\text{DP:DP, nF})}_{[(2m_{11}),(2n_{11}+1)]\text{-hyperbolic-to-hyperbolic-secant flow }(-)}\\\underbrace{((2m_{11})\text{LS}, (2n_{21})\text{LS})}_{[(2m_{11}),(2n_{21})]\text{-double-saddle}}\end{matrix}\right\},$$
for $a_{110} > 0, a_{220} < 0$;

$$\left\{\begin{matrix}(a_{111}, a_{121})\\(a_{111}, a_{221})\end{matrix}\right\} = \left\{\begin{matrix}\underbrace{([(2m_{11}), (2n_{11}+1)]\text{UP:UP, nF})}_{[(2m_{11}),(2n_{11}+1)]\text{-hyperbolic-secant-to-hyperbolic flow }(-)}\\\underbrace{((2m_{11})\text{US}, (2n_{21})\text{LS})}_{[(2m_{11}),(2n_{21})]\text{-double-saddle}}\end{matrix}\right\},$$
$$\text{for } a_{110} < 0, a_{220} < 0. \tag{3.123}$$

(B) For $a_{121} < a_{221}$,

$$\left\{\begin{array}{l}(a_{111}, a_{221})\\(a_{111}, a_{121})\end{array}\right\} = \left\{\begin{array}{c}\underbrace{((2m_{11})\text{US}, (2n_{21})\text{US})}_{[(2m_{11}),(2n_{21})]\text{-double-saddle}}\\ \underbrace{([(2m_{11}), (2n_{11}+1)]\text{DP:DP, pF})}_{[(2m_{11}),(2n_{11}+1)]\text{-hyperbolic-to-hyperbolic-secant flow }(+)}\end{array}\right\},$$
for $a_{110} > 0, a_{220} > 0$;

$$\left\{\begin{array}{l}(a_{111}, a_{221})\\(a_{111}, a_{121})\end{array}\right\} = \left\{\begin{array}{c}\underbrace{((2m_{11})\text{LS}, (2n_{21})\text{US})}_{[(2m_{11}),(2n_{21})]\text{-double-saddle}}\\ \underbrace{([(2m_{11}), (2n_{11}+1)]\text{UP:UP, pF})}_{[(2m_{11}),(2n_{11}+1)]\text{-hyperbolic-secant-to-hyperbolic flow }(+)}\end{array}\right\},$$
for $a_{110} < 0, a_{220} > 0$;

$$\left\{\begin{array}{l}(a_{111}, a_{221})\\(a_{111}, a_{121})\end{array}\right\} = \left\{\begin{array}{c}\underbrace{((2m_{11})\text{US}, (2n_{21})\text{LS})}_{[(2m_{11}),(2n_{21})]\text{-double-saddle}}\\ \underbrace{([(2m_{11}), (2n_{11}+1)]\text{UP:UP, pF})}_{[(2m_{11}),(2n_{11}+1)]\text{-hyperbolic-secant-to-hyperbolic flow }(+)}\end{array}\right\},$$
for $a_{110} > 0, a_{220} < 0$;

$$\left\{\begin{array}{l}(a_{111}, a_{221})\\(a_{111}, a_{121})\end{array}\right\} = \left\{\begin{array}{c}\underbrace{((2m_{11})\text{LS}, (2n_{21})\text{LS})}_{[(2m_{11}),(2n_{21})]\text{-double-saddle}}\\ \underbrace{([(2m_{11}), (2n_{11}+1)]\text{DP:DP, pF})}_{[(2m_{11}),(2n_{11}+1)]\text{-hyperbolic-to-hyperbolic-secant flow }(+)}\end{array}\right\},$$
for $a_{110} < 0, a_{220} < 0$. (3.124)

The $([(2m_{11}),(2n_{11}+1)]$UP:UP,pF) is for a positive singular hyperbolic-secant-to-hyperbolic flow. The $([(2m_{11}),(2n_{11}+1)]$DP:DP, pF) is for a positive singular hyperbolic-to-hyperbolic-secant flow. The $([(2m_{11}),(2n_{11}+1)]$DP:DP, nF) is for a negative singular hyperbolic-to-hyperbolic-secant flow. The$([(2m_{11}),(2n_{11}+1)]$UP:UP, nF) is for a negative singular hyperbolic-secant-to-hyperbolic flow. UP:UP is for a hyperbolic-secant-to-hyperbolic flow, and DP:DP is for a hyperbolic-to-hyperbolic-secant flow. The $((2m_{11})$US, $(2n_{21})$US] is for a $((2m_{11}), (2n_{21})]$-double-saddle. The $((2m_{11})$LS, $(2n_{21})$US] is for a $((2m_{11}), (2n_{21})]$- double-saddle. The $((2m_{11})$US, $(2n_{21})$LS] is for a $((2m_{11}), (2n_{21})]$-double-saddle. The $((2m_{11})$LS, $(2n_{21})$LS] is for a $((2m_{11}), (2n_{21})]$-double-saddle.

The singular hyperbolic-to-hyperbolic-secant flows with double-saddles are illustrated in Figs. 3.37 and 3.38. Through such singular flows, the corresponding global bifurcations generate networks of singular hyperbolic and hyperbolic-secant flows. Through such singular equilibriums, the corresponding global bifurcations generate networks of singular source, sink and saddle equilibriums.

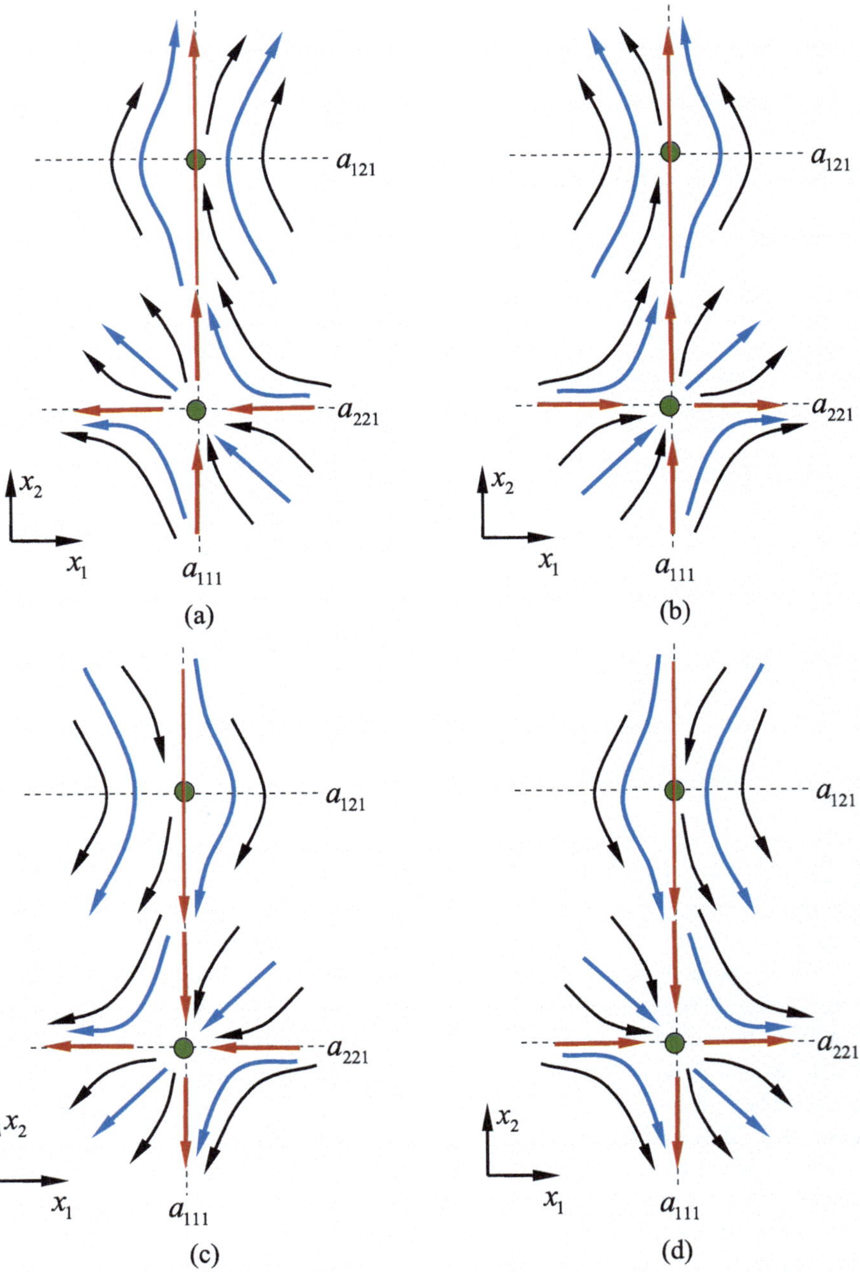

Fig. 3.37 Phase portraits ($a_{121} > a_{221}$) of the ($[(2m_{11}), (2n_{11}+1)]$, xF)-hyperbolic and hyperbolic-secant singular flows at $(x_1^*, x_2^*) = (a_{111}, a_{121})$ and $((2m_{11}), (2n_{21}))$-saddle-source and saddle-sink at $(x_1^*, x_2^*) = (a_{111}, a_{221})$ for 2-dimensional polynomial systems: **a** ($a_{110} > 0$, $a_{220} > 0$), **b** ($a_{110} < 0$, $a_{220} > 0$), **c** ($a_{110} > 0$, $a_{220} < 0$), **d** ($a_{110} < 0$, $a_{220} < 0$)

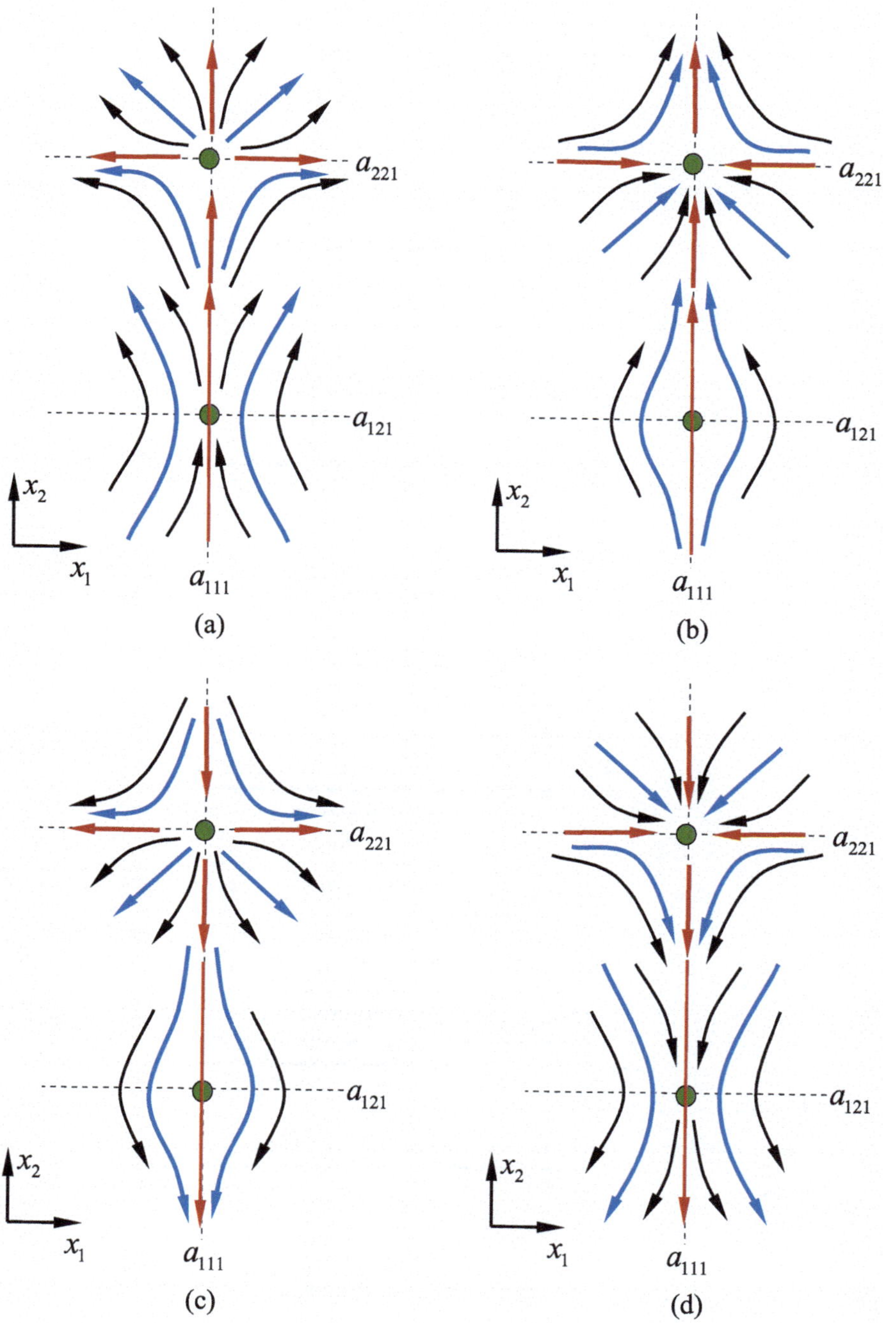

Fig. 3.38 Phase portraits ($a_{221} > a_{121}$) of the ($[(2m_{11}), (2n_{11}+1)]$, xF)-hyperbolic and hyperbolic-secant singular flows at $(x_1^*, x_2^*) = (a_{111}, a_{121})$ and $((2m_{11}), (2n_{21}+1))$-saddle, source and sink at $(x_1^*, x_2^*) = (a_{111}, a_{221})$ for 2-dimensional polynomial systems: **a** ($a_{110} > 0$, $a_{220} > 0$), **b** ($a_{110} < 0$, $a_{220} > 0$), **c** ($a_{110} > 0$, $a_{220} < 0$), **d** ($a_{110} < 0$, $a_{220} < 0$)

3.7.2 $[(2m_{11}), (2n_{11}+1)]$ and $((2m_{11}), (2n_{21}))$-Infinite-Equilibriums

Consider a polynomial dynamical system for $a_{221} = a_{121}$ as

$$\begin{aligned}\dot{x}_2 &= a_{220}(x_2 - a_{221})^{2n_{21}},\\ \dot{x}_1 &= a_{110}(x_1 - a_{111})^{2m_{11}}(x_2 - a_{121})^{2n_{11}+1}.\end{aligned} \tag{3.125}$$

The first integral manifold is

$$\begin{aligned}&-\frac{1}{2m_{11}-1}\left[\frac{1}{(x_1 - a_{111})^{2m_{11}-1}} - \frac{1}{(x_{10} - a_{111})^{2m_{11}-1}}\right]\\ &= \frac{a_{110}}{a_{220}}\frac{1}{2(n_{11}-n_{21})+2}\left[(x_2 - a_{121})^{2(n_{11}-n_{21})+2} - (x_2 - a_{121})^{2(n_{11}-n_{21})+2}\right]\\ &\text{for} 2(n_{11}-n_{21})+2 \neq 0;\\ &\frac{1}{2(n_{11}-n_{21})+2}[(x_2 - a_{221})^{2(n_{11}-n_{21})+2} - (x_2 - a_{221})^{2(n_{11}-n_{21})+2}] = \ln\frac{|x_2 - a_{221}|}{|x_{10} - a_{221}|}\\ &\text{for } 2(n_{11}-n_{21})+2 = 0.\end{aligned} \tag{3.126}$$

(A) For $2n_{11}+1 \geq 2n_{21}$, the corresponding infinite-equilibriums are given.

$$\begin{aligned}(a_{111}, a_{221}) &= \underbrace{([2(n_{11}-n_{21})+1]^{\text{th}}\text{UP:UP}, [(2m_{11}), (2n_{21})]^{\text{th}}\text{US})}_{([2(n_{11}-n_{21})+1],[(2m_{11}),(2n_{21})])\text{-hyperbolic-secant-to-hyperbolic upper-saddle}},\\ &\text{for } a_{110} > 0, a_{220} > 0;\\ (a_{111}, a_{221}) &= \underbrace{([2(n_{11}-n_{21})+1]^{\text{th}}\text{DP:DP}, [(2m_{11}), (2n_{21})]^{\text{th}}\text{US})}_{([2(n_{11}-n_{21})+1],[(2m_{11}),(2n_{21}+1)])\text{-hyperbolic-to-hyperbolic-secant upper-saddle}},\\ &\text{for } a_{110} < 0, a_{220} > 0;\\ (a_{111}, a_{221}) &= \underbrace{([2(n_{11}-n_{21})+1]^{\text{th}}\text{DP:DP}, [(2m_{11}), (2n_{21})]^{\text{th}}\text{LS})}_{([2(n_{11}-n_{21})+1],[(2m_{11}),(2n_{21}+1)])\text{-hyperbolic-to-hyperbolic-secant lower-saddle}},\\ &\text{for } a_{110} > 0, a_{220} < 0;\\ (a_{111}, a_{221}) &= \underbrace{([2(n_{11}-n_{21})+1]^{\text{th}}\text{UP:UP},[(2m_{11}), (2n_{21})]^{\text{th}}\text{LS})}_{([2(n_{11}-n_{21})+1],[(2m_{11}),(2n_{21}+1)])\text{-hyperbolic-secant-to-hyperbolic lower-saddle}},\\ &\text{for } a_{110} < 0, a_{220} < 0.\end{aligned} \tag{3.127}$$

(B) For $2n_{11}+1 < 2n_{21}+1$, the infinite-equilibriums are given as follows.

$$\begin{aligned}(a_{111}, a_{221}) &= \underbrace{([2(n_{21}-n_{11})-1]^{\text{th}}\text{DA:DA}, [(2m_{11}), (2n_{21})]^{\text{th}}\text{US})}_{([2(n_{21}-n_{11})-1],[(2m_{11}),(2n_{21}+1)])\text{-down-asymptote upper-saddle}},\\ &\text{for } a_{110} > 0, a_{220} > 0;\end{aligned}$$

$$
\begin{aligned}
&(a_{111}, a_{221}) = \underbrace{([2(n_{21}-n_{11})-1]^{\text{th}}\text{UA:UA}, [(2m_{11}), (2n_{21})]^{\text{th}}\text{US})}_{([2(n_{21}-n_{11})-1],[(2m_{11}),(2n_{21}+1)])\text{-up asymptote upper-saddle}},\\
&\text{for } a_{110} < 0, a_{220} > 0;\\
&(a_{111}, a_{221}) = \underbrace{([2(n_{21}-n_{11})-1]^{\text{th}}\text{UA:UA}, [(2m_{11}), (2n_{21})]^{\text{th}}\text{LS})}_{([2(n_{21}-n_{11})-1],[(2m_{11}),(2n_{21}+1)])\text{-up asymptote lower-saddle}},\\
&\text{for } a_{110} > 0, a_{220} < 0;\\
&(a_{111}, a_{221}) = \underbrace{([2(n_{21}-n_{11})-1]^{\text{th}}\text{DA:DA}, [(2m_{11}), (2n_{21})]^{\text{th}}\text{LS})}_{([2(n_{21}-n_{11})-1],[(2m_{11}),(2n_{21}+1)])\text{-down asymptote lower-saddle}},\\
&\text{for } a_{110} < 0, a_{220} < 0.
\end{aligned}
\tag{3.128}
$$

For $x_2^* = a_{121} = a_{221}$, the infinite-equilibrium exists. The infinite-equilibriums are illustrated in Figs. 3.39 and 3.40 for $2n_{11} + 1 \geq 2n_{21}$ and $2n_{11} + 1 < 2n_{21}$, respectively. For $2n_{11} + 1 \geq 2n_{21}$, the singular decreasing-to-increasing and increasing-to-decreasing inflection-source and sink are presented. For $2n_{11} + 1 < 2n_{21}$, the singular decreasing-to-increasing and increasing-to-decreasing concave-source and sink are presented.

3.7.3 [(2m_{11}), (2n_{11} + 1)] *and* ((2m_{11}), (2n_{21}))-*Hybrid Networks*

Consider

$$
\begin{aligned}
\dot{x}_2 &= a_{220}\prod_{s=1}^{2n_{21}}(x_2 - a_{22s}),\\
\dot{x}_1 &= a_{110}\prod_{r=1}^{2m_{11}}(x_1 - a_{11r})\prod_{u=1}^{2n_{11}+1}(x_2 - a_{12u}).
\end{aligned}
\tag{3.129}
$$

The first integral manifold is

$$
\begin{aligned}
&\sum_{r_1=1}^{2m_{11}} A_{r_1}\ln\frac{|x_1 - a_{11r_1}|}{|x_{10} - a_{11r_1}|} = \frac{a_{110}}{a_{220}}\Bigg\{\sum_{\beta_0=2n_{21}}^{2n_{11}+1} b_{12\beta_0}\sum_{\substack{\alpha_1\in[0,\beta_0-(2n_{21}+1)],\\ \beta_1\in[2n_{21},\beta_0-1]}}^{\alpha_1+\beta_1=\beta_0-1}\ \sum_{\substack{\alpha_2\in[0,\beta_1-(2n_{21}+1)]\\ \beta_2\in[2n_{21}-1,\beta_1-2]}}^{\alpha_2+\beta_2=\beta_1-2}\cdots\\
&\sum_{\substack{\alpha_{2n_{21}-1}\in[0,\beta_{2n_{21}-2}-2n_{21}]\\ \beta_{2n_{21}-1}\in[1,\beta_{2n_{21}-2}-2n_{21}+1]}}^{\alpha_{2n_{21}-1}+\beta_{2n_{21}-1}=\beta_{2n_{21}-1}-2n_{21}+1}\ \prod_{j=1}^{2n_{21}-1}\frac{(\beta_{j-1}-j)!}{\alpha_j!\beta_j!}(a_{22s_{j+1}} - a_{22s_j})^{\alpha_j}\\
&\times\frac{1}{\beta_{2n_{21}-1}}[(x_2 - a_{22s_{2n_{21}}})^{\beta_{2n_{21}-1}} - (x_{20} - a_{22s_{2n_{21}}})^{\beta_{2n_{21}-1}}]
\end{aligned}
$$

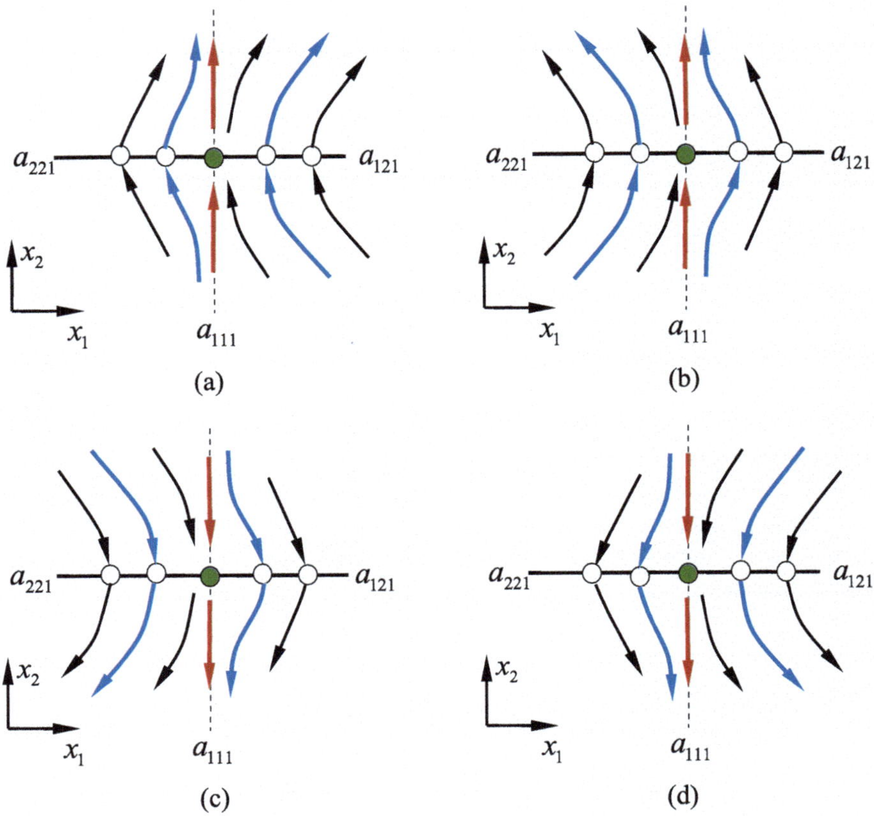

Fig. 3.39 The infinite-equilibriums ($a_{121} = a_{221}, 2n_{11} + 1 \geq 2n_{21}, m_1 = 2m_{11} + 1$) for 2-dimensional polynomial systems: **a** ($a_{110} > 0$, $a_{220} > 0$), **b** ($a_{110} < 0$, $a_{220} > 0$), **c** ($a_{110} > 0$, $a_{220} < 0$), **d** ($a_{110} < 0$, $a_{220} < 0$)

$$
\begin{aligned}
&- \sum_{\beta_0=2n_{21}}^{2n_{11}+1} b_{12\beta_0} \sum_{k=2}^{2n_{21}-1} \Bigg[\sum_{\substack{\alpha_1 \in [0,\beta_0-2n_{21}), \\ \beta_1 \in (2n_{21}-1,\beta_0-1]}}^{\alpha_1+\beta_1=\beta_0-1} \; \sum_{\substack{\alpha_2 \in [0,\beta_1-2n_{21}] \\ \beta_2 \in (2n_{21}-2,\beta_1-2)}}^{\alpha_2+\beta_2=\beta_1-2} \cdots \\
&\times \Bigg(\sum_{\substack{\alpha_{k-1} \in (\beta_{k-2}-2n_{21}+1,\beta_{k-2}-k+1] \\ \beta_{k-1} \in [0,2n_{21}-k)}}^{\alpha_{k-1}+\beta_{k-1}=\beta_{k-2}-k+1} \prod_{j=1}^{k-1} \frac{(\beta_{j-1}-j)!}{\alpha_j!\beta_j!} (a_{22s_{j+1}} - a_{22s_j})^{\alpha_j} \sum_{\rho=k}^{2n_{21}} A_{\rho 1}^{(s_k)} \ln \frac{|x_{22} - a_{22s_\rho}|}{|x_{20} - a_{22r_\rho}|} \\
&+ \sum_{\substack{\alpha_{k-1} \in (\beta_{k-2}-2n_{21},\beta_{k-2}-2n_{21}+1] \\ \beta_{k-1} \in [2n_{21}-k,2n_{21}-k+1)}}^{\alpha_{k-1}+\beta_{k-1}=\beta_{k-2}-k+1} \prod_{j=1}^{k-1} \frac{(\beta_{j-1}-j)!}{\alpha_j!\beta_j!} (a_{22s_{j+1}} - a_{22s_j})^{\alpha_j} \\
&\times \sum_{\rho=k+1}^{2n_{21}} B_{\rho 1}^{(s_k)} \ln \frac{|x_2 - a_{22s_\rho}|}{|x_{20} - a_{22s_\rho}|} \Bigg) \Bigg] + b_{j_1 j_2 0} \sum_{\rho=1}^{2n_{21}} A_{\rho 1}^{(s_1)} \ln \frac{|x_2 - a_{22s_\rho}|}{|x_{20} - a_{22s_\rho}|}
\end{aligned}
$$

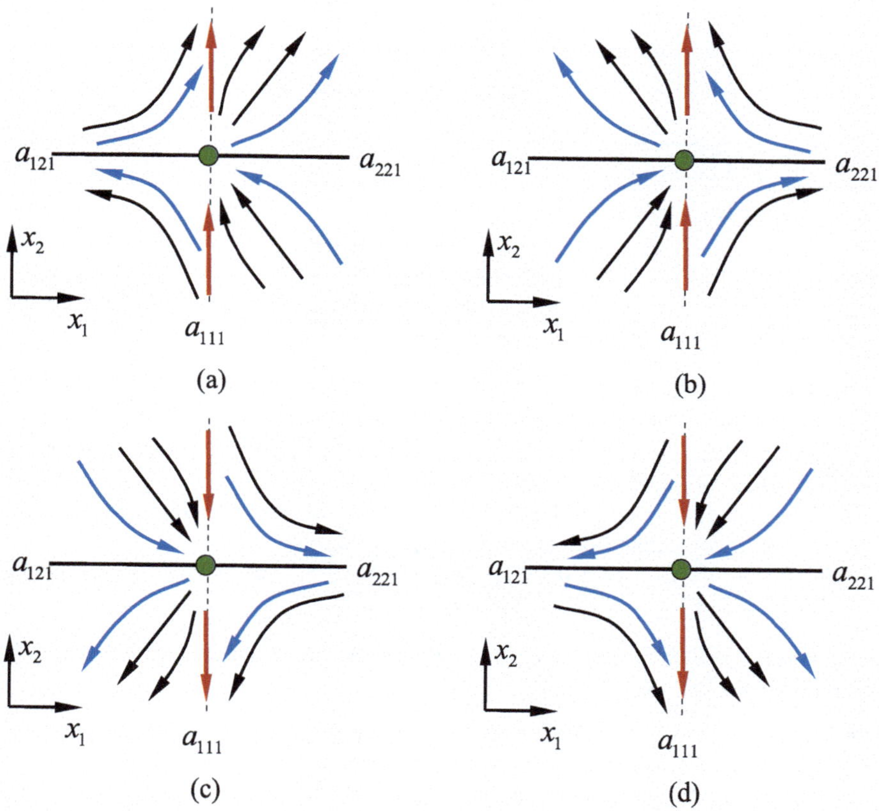

Fig. 3.40 The infinite-equilibriums ($a_{121} = a_{221}, 2n_{11}+1 < 2n_{21}, m_1 = 2m_{11}+1$) for 2-dimensional polynomial systems: **a** ($a_{110} > 0, a_{220} > 0$), **b** ($a_{110} < 0, a_{220} > 0$), **c** ($a_{110} > 0, a_{220} < 0$), **d** ($a_{110} < 0, a_{220} < 0$)

$$+\sum_{\beta_0=1}^{2n_{21}-1} b_{j_1j_2\beta_0}\sum_{\rho=2}^{2n_{21}} B_{\rho 1}^{(s_1)}\frac{|x_2-a_{22s_\rho}|}{|x_{20}-a_{22s_\rho}|}\Bigg\}; \tag{3.130}$$

where

$$A_{r_1}=\frac{1}{\prod_{r_2=1,r_2\neq r_1}^{2m_{11}}(a_{11r_1}-a_{11r_\sigma})};$$

$$b_{j_1j_2\beta_0}=\sum_{t_1,l_1=0}^{t_1+l_1=1}\cdots\sum_{t_q,l_q=0}^{t_q+l_q=1}\prod_{k=1}^{2n_{11}+1}(a_{22s_1}-a_{12u_k})^{t_k}\delta_{\beta_0}^{\Sigma_{r=1}^{2n_{11}+1}l_r};$$

$$\lim_{\beta_{2n_{21}-1}\to 0}\frac{1}{\beta_{2n_{21}-1}}[(x_2-a_{22s_{2n_{21}}})^{\beta_{2n_{21}-1}}-(x_{20}-a_{22s_{2n_{21}}})^{\beta_{2n_{21}-1}}]=\ln\frac{|x_2-a_{22s_{2n_{21}}}|}{|x_{20}-a_{22s_{2n_{21}}}|}; \tag{3.131}$$

$$A_{\rho 1}^{(s_k)} = \frac{1}{\prod_{\sigma=k,\sigma\neq\rho}^{2n_{21}+1}(a_{22s_\rho} - a_{22s_\sigma})}, \text{ for } \rho \in \{k, k+1, \cdots, 2n_{21}\},$$
$$B_{\rho 1}^{(s_k)} = \frac{(a_{22s_\rho} - a_{22s_k})^{\beta_{k-1}-k}}{\prod_{\sigma=k+1,\sigma\neq\rho}^{2n_{21}+1}(a_{22s_\rho} - a_{22s_\sigma})}, \text{ for } \rho \in \{k+1, k+2, \cdots, 2n_{21}\};$$
$$k \in \{1, 2, \cdots, 2n_{21} - 1\}. \tag{3.132}$$

For all cases, the $(2m_{11}) \times (2n_{21})$-simple equilibriums plus $(2m_{11}) \times (2n_{11} + 1)$-hyperbolic flows are based on the following hyperbolic-secant-to-hyperbolic upper-saddle and lower-saddle bifurcations for $2n_{11} + 1 \geq 2n_{21}$:

- $[2(n_{11} - n_{21}) + 1]^{\text{th}}$UP:UP, $[(2m_{11}), (2n_{21})]^{\text{th}}$US) hyperbolic-secant-to-hyperbolic upper-saddle bifurcation for ($a_{110} > 0, a_{220} > 0$),
- $[2(n_{11} - n_{21}) + 1]^{\text{th}}$DP:DP, $[(2m_{11}), (2n_{21})]^{\text{th}}$US) hyperbolic-to-hyperbolic-secant upper-saddle bifurcation for ($a_{110} < 0, a_{220} > 0$),
- $[2(n_{11} - n_{21}) + 1]^{\text{th}}$DP:DP, $[(2m_{11}), (2n_{21})]^{\text{th}}$LS) hyperbolic-to-hyperbolic-secant lower-saddle bifurcation for ($a_{110} > 0, a_{220} < 0$),
- $[2(n_{11} - n_{21}) + 1]^{\text{th}}$UP:UP, $[(2m_{11}), (2n_{21})]^{\text{th}}$LS)-hyperbolic lower-saddle bifurcation for ($a_{110} < 0, a_{220} < 0$);

or based on the following up and down-asymptote upper-saddle and lower-saddle bifurcations for $2n_{11} + 1 < 2n_{21}$:

- $[2(n_{21} - n_{11}) - 1]^{\text{th}}$DA:DA, $[(2m_{11}), (2n_{21})]^{\text{th}}$US)-down-asymptote upper-saddle bifurcation for ($a_{110} > 0, a_{220} > 0$),
- $[2(n_{21} - n_{11}) - 1]^{\text{th}}$UA:UA, $[(2m_{11}), (2n_{21})]^{\text{th}}$US)-up-asymptote upper-saddle bifurcation for ($a_{110} < 0, a_{220} > 0$),
- $[2(n_{21} - n_{11}) - 1]^{\text{th}}$UA:UA, $[(2m_{11}), (2n_{21})]^{\text{th}}$LS)-up-asymptote lower-saddle bifurcation for ($a_{110} > 0, a_{220} < 0$),
- $[2(n_{21} - n_{11}) - 1]^{\text{th}}$DA:DA, $[(2m_{11}), (2n_{21})]^{\text{th}}$LS)-down-asymptote lower-saddle bifurcation for ($a_{110} < 0, a_{220} < 0$).

Consider $2n_{11} + 1 = 2n_{21} + 1$ for illustrations, phase portraits for the 2-dimensional systems for the hybrid networks of hyperbolic and hyperbolic-secant flows with the simple equilibriums of the saddles, sinks and sources with $x_1^* = a_{111}, a_{112}, \ldots, a_{11(2m_{11})}$ and $x_2^* = a_{121}, a_{221}, \ldots, a_{22(2n_{21})}, a_{12(2n_{11}+1)}$ are presented in Fig. 3.41 (a–d) for $(a_{110} > 0, a_{220} > 0)$, $(a_{110} < 0, a_{220} > 0)$, $(a_{110} > 0, a_{220} < 0)$ and $(a_{110} < 0, a_{220} < 0)$.

$$\left\{\begin{array}{cccc}
(a_{111}, a_{12(2n_{11}+1)}) & \cdots & (a_{11(2m_{11}-1)}, a_{12(2n_{11}+1)}) & (a_{11(2m_{11})}, a_{12(2n_{11}+1)}) \\
(a_{111}, a_{22(2n_{21})}) & \cdots & (a_{11(2m_{11}-1)}, a_{22(2n_{21})}) & (a_{11(2m_{11})}, a_{22(2n_{21})}) \\
(a_{111}, a_{12(2n_{11})}) & \cdots & (a_{11(2m_{11}-1)}, a_{12(2n_{11})}) & (a_{11(2m_{11})}, a_{12(2n_{11})}) \\
\vdots & \vdots & \vdots & \vdots \\
(a_{111}, a_{221}) & \cdots & (a_{11(2m_{11}-1)}, a_{221}) & (a_{11(2m_{11})}, a_{221}) \\
(a_{111}, a_{121}) & \cdots & (a_{11(2m_{11}-1)}, a_{121}) & (a_{11(2m_{11})}, a_{121})
\end{array}\right\}$$

$$
= \left\{ \begin{array}{cccc}
\underbrace{(\text{UP:DP, pF})}_{\text{hyperbolic-secant flow }(+)} & \cdots & \underbrace{(\text{UP:DP, pF})}_{\text{hyperbolic-secant flow }(+)} & \underbrace{(\text{DP:UP, pF})}_{\text{hyperbolic flow }(+)} \\
\underbrace{(\text{SO, SO})}_{\text{source}} & \cdots & \underbrace{(\text{SO, SO})}_{\text{source}} & \underbrace{(\text{SI, SO})}_{\text{saddle}} \\
\underbrace{(\text{UP:DP, nF})}_{\text{hyperbolic-secant flow }(-)} & \cdots & \underbrace{(\text{UP:DP, nF})}_{\text{hyperbolic-secant flow }(-)} & \underbrace{(\text{DP:UP, nF})}_{\text{hyperbolic flow }(-)} \\
\vdots & \vdots & \vdots & \vdots \\
\underbrace{(\text{SI, SI})}_{\text{sink}} & \cdots & \underbrace{(\text{SI, SI})}_{\text{sink}} & \underbrace{(\text{SO, SI})}_{\text{source}} \\
\underbrace{(\text{UP:DP, pF})}_{\text{hyperbolic-secant flow }(+)} & \cdots & \underbrace{(\text{UP:DP, pF})}_{\text{hyperbolic-secant flow }(+)} & \underbrace{(\text{DP:UP, pF})}_{\text{hyperbolic flow }(+)}
\end{array} \right\}_{(2n_{11}+2n_{21}+1)\times(2m_{11})}
$$

for $a_{110} > 0, a_{220} > 0$; (3.133)

$$
\left\{ \begin{array}{cccc}
(a_{111}, a_{12(2n_{11}+1)}) & \cdots & (a_{11(2m_{11}-1)}, a_{12(2n_{11}+1)}) & (a_{11(2m_{11})}, a_{12(2n_{11}+1)}) \\
(a_{111}, a_{22(2n_{21})}) & \cdots & (a_{11(2m_{11}-1)}, a_{22(2n_{21})}) & (a_{11(2m_{11})}, a_{22(2n_{21})}) \\
(a_{111}, a_{12(2n_{11})}) & \cdots & (a_{11(2m_{11}-1)}, a_{12(2n_{11})}) & (a_{11(2m_{11})}, a_{12(2n_{11})}) \\
\vdots & \vdots & \vdots & \vdots \\
(a_{111}, a_{221}) & \cdots & (a_{11(2m_{11}-1)}, a_{221}) & (a_{11(2m_{11})}, a_{221}) \\
(a_{111}, a_{121}) & \cdots & (a_{11(2m_{11}-1)}, a_{121}) & (a_{11(2m_{11})}, a_{121})
\end{array} \right\}
$$

$$
= \left\{ \begin{array}{cccc}
\underbrace{(\text{DP:UP, pF})}_{\text{hyperbolic flow }(+)} & \cdots & \underbrace{(\text{DP:UP, pF})}_{\text{hyperbolic flow }(+)} & \underbrace{(\text{UP:DP, pF})}_{\text{hyperbolic-secant flow }(+)} \\
\underbrace{(\text{SI, SO})}_{\text{saddle}} & \cdots & \underbrace{(\text{SI, SO})}_{\text{saddle}} & \underbrace{(\text{SO, SO})}_{\text{source}} \\
\underbrace{(\text{DP:UP, nF})}_{\text{hyperbolic flow }(-)} & \cdots & \underbrace{(\text{DP:UP, nF})}_{\text{hyperbolic flow }(-)} & \underbrace{(\text{UP:DP,nF})}_{\text{hyperbolic-secant flow }(-)} \\
\vdots & \vdots & \vdots & \vdots \\
\underbrace{(\text{SO, SI})}_{\text{saddle}} & \cdots & \underbrace{(\text{SO, SI})}_{\text{saddle}} & \underbrace{(\text{SI, SI})}_{\text{sink}} \\
\underbrace{(\text{DP:UP, pF})}_{\text{hyperbolic flow }(+)} & \cdots & \underbrace{(\text{DP:UP, pF})}_{\text{hyperbolic flow }(+)} & \underbrace{(\text{UP:DP, pF})}_{\text{hyperbolic-secant flow }(+)}
\end{array} \right\}_{(2n_{11}+2n_{21}+1)\times(2m_{11})}
$$

for $a_{110} < 0, a_{220} > 0$; (3.134)

$$\left\{\begin{array}{cccc}(a_{111},a_{12(2n_{11}+1)}) & \cdots & (a_{11(2m_{11}-1)},a_{12(2n_{11}+1)}) & (a_{11(2m_{11})},a_{12(2n_{11}+1)})\\(a_{111},a_{22(2n_{21})}) & \cdots & (a_{11(2m_{11}-1)},a_{22(2n_{21})}) & (a_{11(2m_{11})},a_{22(2n_{21})})\\(a_{111},a_{12(2n_{11})}) & \cdots & (a_{11(2m_{11}-1)},a_{12(2n_{11})}) & (a_{11(2m_{11})},a_{12(2n_{11})})\\\vdots & \vdots & \vdots & \vdots\\(a_{111},a_{221}) & \cdots & (a_{11(2m_{11}-1)},a_{221}) & (a_{11(2m_{11})},a_{221})\\(a_{111},a_{121}) & \cdots & (a_{11(2m_{11}-1)},a_{121}) & (a_{11(2m_{11})},a_{121})\end{array}\right\}$$

$$=\left\{\begin{array}{cccc}\underbrace{(\text{DP:UP, nF})}_{\text{hyperbolic flow }(-)} & \cdots & \underbrace{(\text{DP:UP, nF})}_{\text{hyperbolic flow }(-)} & \underbrace{(\text{UP:DP, nF})}_{\text{hyperbolic-secant flow }(-)}\\\underbrace{(\text{SO, SI})}_{\text{saddle}} & \cdots & \underbrace{(\text{SO, SI})}_{\text{saddle}} & \underbrace{(\text{SI, SI})}_{\text{sink}}\\\underbrace{(\text{DP:UP, pF})}_{\text{hyperbolic flow }(+)} & \cdots & \underbrace{(\text{DP:UP, pF})}_{\text{hyperbolic flow }(+)} & \underbrace{(\text{UP:DP, pF})}_{\text{hyperbolic-secant flow }(+)}\\\vdots & \vdots & \vdots & \vdots\\\underbrace{(\text{SI, SO})}_{\text{saddle}} & \cdots & \underbrace{(\text{SI, SO})}_{\text{saddle}} & \underbrace{(\text{SO, SO})}_{\text{source}}\\\underbrace{(\text{DP:UP, nF})}_{\text{hyperbolic flow }(-)} & \cdots & \underbrace{(\text{DP:UP, nF})}_{\text{hyperbolic flow }(-)} & \underbrace{(\text{UP:DP, nF})}_{\text{hyperbolic-secant flow }(-)}\end{array}\right\}_{(2n_{11}+2n_{21}+1)\times(2m_{11})}$$

for $a_{110} > 0$, $a_{220} < 0$; (3.135)

$$\left\{\begin{array}{cccc}(a_{111},a_{12(2n_{11}+1)}) & \cdots & (a_{11(2m_{11}-1)},a_{12(2n_{11}+1)}) & (a_{11(2m_{11})},a_{12(2n_{11}+1)})\\(a_{111},a_{22(2n_{21})}) & \cdots & (a_{11(2m_{11}-1)},a_{22(2n_{21})}) & (a_{11(2m_{11})},a_{22(2n_{21})})\\(a_{111},a_{12(2n_{11})}) & \cdots & (a_{11(2m_{11}-1)},a_{12(2n_{11})}) & (a_{11(2m_{11})},a_{12(2n_{11})})\\\vdots & \vdots & \vdots & \vdots\\(a_{111},a_{221}) & \cdots & (a_{11(2m_{11}-1)},a_{221}) & (a_{11(2m_{11})},a_{221})\\(a_{111},a_{121}) & \cdots & (a_{11(2m_{11}-1)},a_{121}) & (a_{11(2m_{11})},a_{121})\end{array}\right\}$$

$$=\left\{\begin{array}{cccc}\underbrace{(\text{UP:DP, nF})}_{\text{hyperbolic-secant flow }(-)} & \cdots & \underbrace{(\text{UP:DP, nF})}_{\text{hyperbolic-secant flow }(-)} & \underbrace{(\text{DP:UP, nF})}_{\text{hyperbolic flow }(-)}\\\underbrace{(\text{SI, SI})}_{\text{sink}} & \cdots & \underbrace{(\text{SI,SI})}_{\text{sink}} & \underbrace{(\text{SO, SI})}_{\text{saddle}}\\\underbrace{(\text{UP:DP, pF})}_{\text{hyperbolic-secant flow }(+)} & \cdots & \underbrace{(\text{UP:DP, pF})}_{\text{hyperbolic-secant flow }(+)} & \underbrace{(\text{DP:UP, pF})}_{\text{hyperbolic flow }(+)}\\\vdots & \vdots & \vdots & \vdots\\\underbrace{(\text{SO, SO})}_{\text{source}} & \cdots & \underbrace{(\text{SO, SO})}_{\text{source}} & \underbrace{(\text{SI, SO})}_{\text{saddle}}\\\underbrace{(\text{UP:DP, nF})}_{\text{hyperbolic-secant flow }(-)} & \cdots & \underbrace{(\text{UP:DP, nF})}_{\text{hyperbolic-secant flow }(-)} & \underbrace{(\text{DP:UP, nF})}_{\text{hyperbolic flow }(-)}\end{array}\right\}_{(2n_{11}+2n_{21}+1)\times(2m_{11})}$$

for $a_{110} < 0$, $a_{220} < 0$. (3.136)

Consider $n_{11} = n_{21}$ for illustrations. Phase portraits for the 2-dimensional systems for the hybrid networks of hyperbolic and hyperbolic-secant flows with the simple equilibriums of the saddles, sinks and sources with $x_1^* = a_{111}, a_{112}, \ldots, a_{11(2m_{11})}$ and $x_2^* = a_{221}, a_{121}, \ldots, a_{12(2n_{11})}, a_{12(2n_{11}+1)}$ are presented in Fig. 3.42 (a–d) for ($a_{110} > 0$, $a_{220} > 0$), ($a_{110} < 0$, $a_{220} > 0$), ($a_{110} > 0$, $a_{220} < 0$) and ($a_{110} < 0$, $a_{220} < 0$).

$$
\left\{
\begin{array}{cccc}
(a_{111}, a_{12(2n_{11}+1)}) & \cdots & (a_{11(2m_{11}-1)}, a_{12(2n_{11}+1)}) & (a_{11(2m_{11})}, a_{12(2n_{11}+1)}) \\
(a_{111}, a_{12(2n_{11})}) & \cdots & (a_{11(2m_{11}-1)}, a_{12(2n_{11})}) & (a_{11(2m_{11})}, a_{12(2n_{11})}) \\
(a_{111}, a_{22(2n_{21})}) & \cdots & (a_{11(2m_{11}-1)}, a_{22(2n_{21})}) & (a_{11(2m_{11})}, a_{22(2n_{21})}) \\
\vdots & \vdots & \vdots & \vdots \\
(a_{111}, a_{121}) & \cdots & (a_{11(2m_{11}-1)}, a_{121}) & (a_{11(2m_{11})}, a_{121}) \\
(a_{111}, a_{221}) & \cdots & (a_{11(2m_{11}-1)}, a_{221}) & (a_{11(2m_{11})}, a_{221})
\end{array}
\right\}
$$

$$
= \left\{
\begin{array}{cccc}
\underbrace{(\text{UP:DP, pF})}_{\text{hyperbolic-secant flow }(+)} & \cdots & \underbrace{(\text{UP:DP, pF})}_{\text{hyperbolic-secant flow }(+)} & \underbrace{(\text{DP:UP, pF})}_{\text{hyperbolic flow }(+)} \\
\underbrace{(\text{DP:UP, pF})}_{\text{hyperbolic flow }(+)} & \cdots & \underbrace{(\text{DP:UP, pF})}_{\text{hyperbolic flow }(+)} & \underbrace{(\text{UP:DP, pF})}_{\text{hyperbolic-secant flow }(+)} \\
\underbrace{(\text{SI, SO})}_{\text{saddle}} & \cdots & \underbrace{(\text{SI, SO})}_{\text{saddle}} & \underbrace{(\text{SO, SO})}_{\text{source}} \\
\vdots & \vdots & \vdots & \vdots \\
\underbrace{(\text{UP:DP, nF})}_{\text{hyperbolic-secant flow }(-)} & \cdots & \underbrace{(\text{UP:DP, nF})}_{\text{hyperbolic-secant flow }(-)} & \underbrace{(\text{UP:DP, nF})}_{\text{hyperbolic-secant flow }(-)} \\
\underbrace{(\text{SO, SI})}_{\text{saddle}} & \cdots & \underbrace{(\text{SO, SI})}_{\text{saddle}} & \underbrace{(\text{SI, SI})}_{\text{sink}}
\end{array}
\right\}_{(2n_{11}+2n_{21}+1)\times(2m_{11})}
$$

for $a_{110} > 0, a_{220} > 0$; (3.137)

$$
\left\{
\begin{array}{cccc}
(a_{111}, a_{12(2n_{11}+1)}) & \cdots & (a_{11(2m_{11}-1)}, a_{12(2n_{11}+1)}) & (a_{11(2m_{11})}, a_{12(2n_{11}+1)}) \\
(a_{111}, a_{12(2n_{11})}) & \cdots & (a_{11(2m_{11}-1)}, a_{12(2n_{11})}) & (a_{11(2m_{11})}, a_{12(2n_{11})}) \\
(a_{111}, a_{22(2n_{21})}) & \cdots & (a_{11(2m_{11}-1)}, a_{22(2n_{21})}) & (a_{11(2m_{11})}, a_{22(2n_{21})}) \\
\vdots & \vdots & \vdots & \vdots \\
(a_{111}, a_{121}) & \cdots & (a_{11(2m_{11}-1)}, a_{121}) & (a_{11(2m_{11})}, a_{121}) \\
(a_{111}, a_{221}) & \cdots & (a_{11(2m_{11}-1)}, a_{221}) & (a_{11(2m_{11})}, a_{221})
\end{array}
\right\}
$$

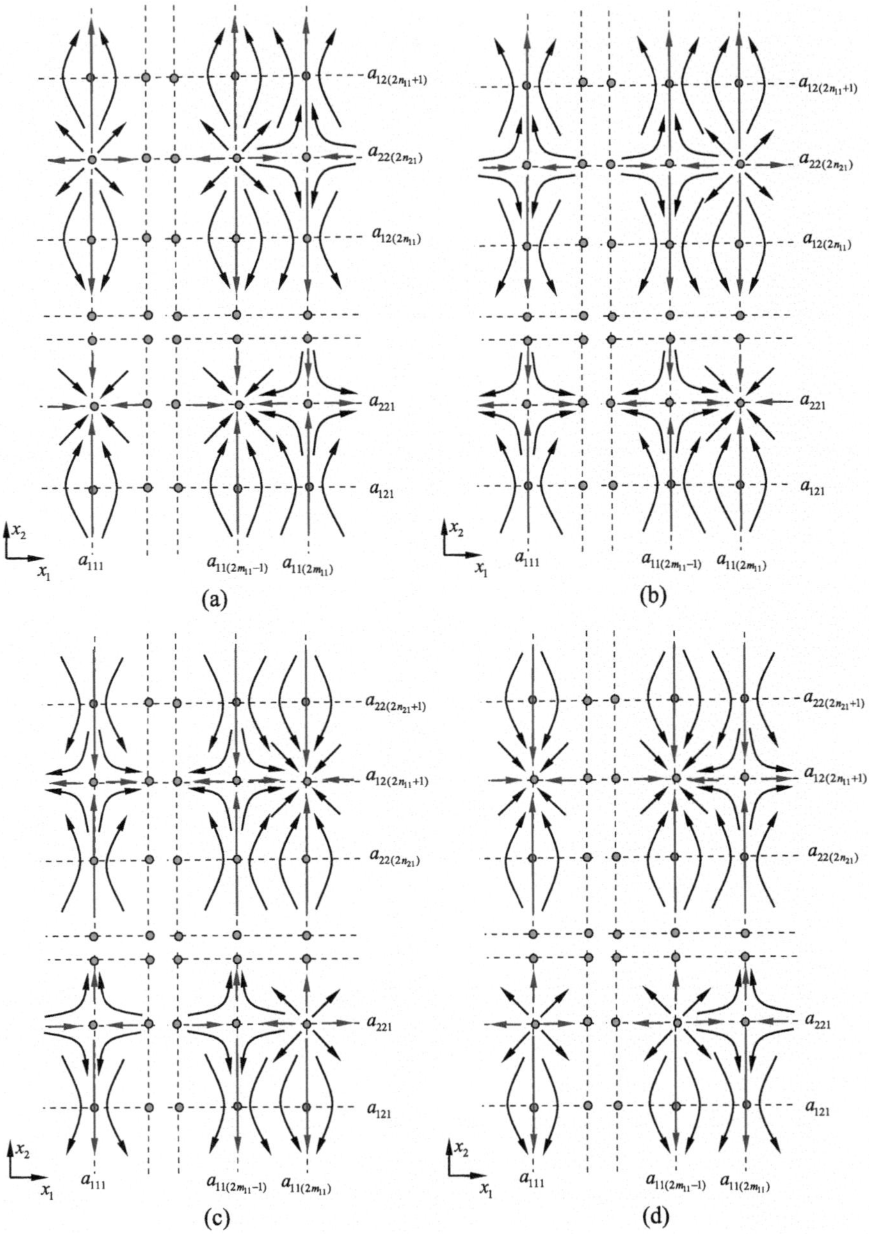

Fig. 3.41 Phase portraits for 2-dimensional systems with $x_1^* = a_{111}, a_{112}, \ldots, a_{11(2m_{11})}$ and $x_2^* = a_{121}, a_{221}, \ldots, a_{22(2n_{21})}, a_{12(2n_{11}+1)}$. The networks of flows with sink, source and saddle: **a** ($a_{110} > 0$, $a_{220} > 0$), **b** ($a_{110} < 0$, $a_{220} > 0$), **c** ($a_{110} > 0$, $a_{220} < 0$), **d** ($a_{110} < 0$, $a_{220} < 0$)

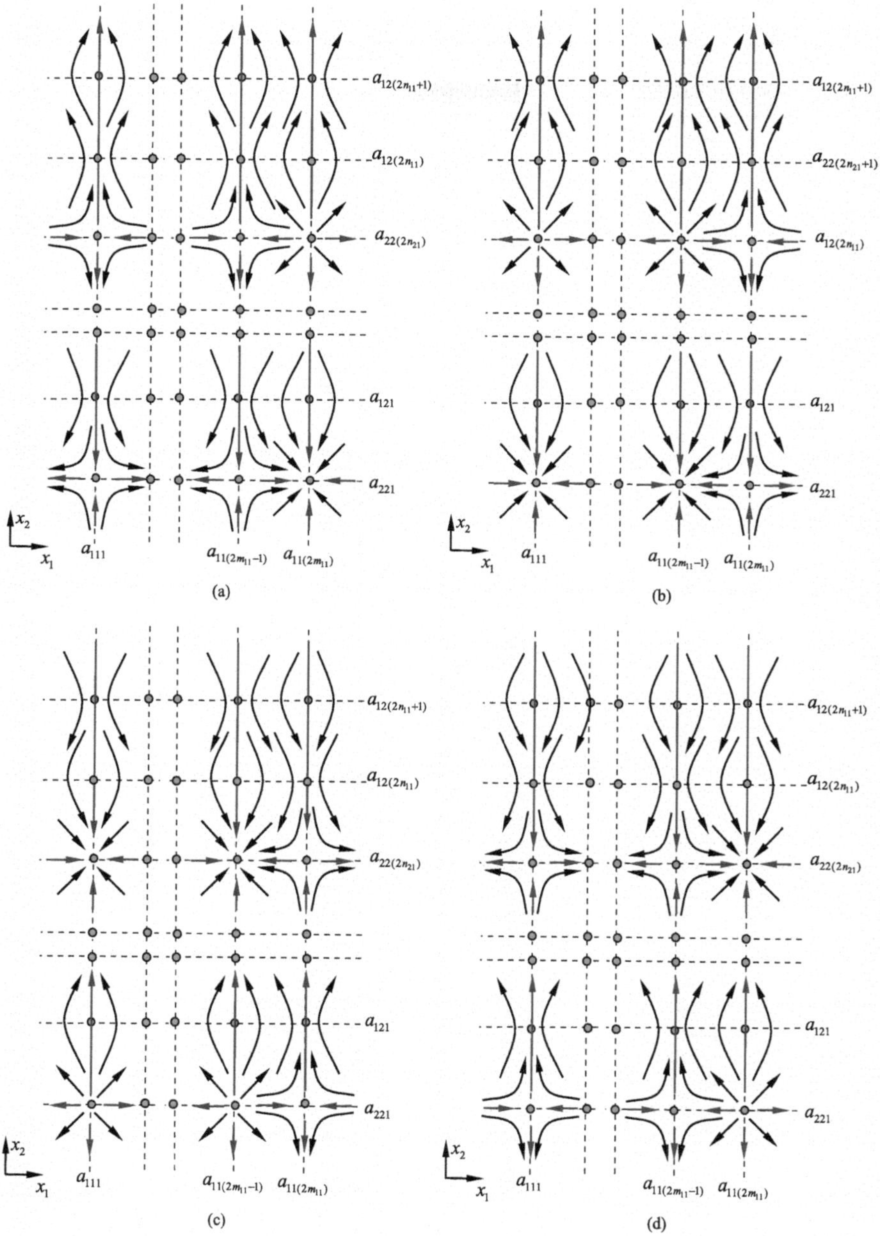

Fig. 3.42 Phase portraits for 2-dimensional systems with $x_1^* = a_{111}, a_{112}, \ldots, a_{11(2m_{11})}$ and $x_2^* = a_{221}, a_{121}, \ldots, a_{12(2n_{11})}, a_{12(2n_{11}+1)}$. The networks of flows with sink, source and saddle: **a** ($a_{110} > 0, a_{220} > 0$), **b** ($a_{110} < 0, a_{220} > 0$), **c** ($a_{110} > 0, a_{220} < 0$), **d** ($a_{110} < 0, a_{220} < 0$)

$$
= \left\{ \begin{array}{cccc}
\underbrace{(\text{DP:UP, pF})}_{\text{hyperbolic flow (+)}} & \cdots & \underbrace{(\text{DP:UP, pF})}_{\text{hyperbolic flow (+)}} & \underbrace{(\text{UP:DP, pF})}_{\text{hyperbolic-secant flow (+)}} \\
\underbrace{(\text{UP:DP, pF})}_{\text{hyperbolic-secant flow (+)}} & \cdots & \underbrace{(\text{UP:DP, pF})}_{\text{hyperbolic-secant flow (+)}} & \underbrace{(\text{DP:UP, pF})}_{\text{hyperbolic flow (+)}} \\
\underbrace{(\text{SO, SO})}_{\text{source}} & \cdots & \underbrace{(\text{SO, SO})}_{\text{source}} & \underbrace{(\text{SI, SO})}_{\text{saddle}} \\
\vdots & \vdots & \vdots & \vdots \\
\underbrace{(\text{UP:DP, nF})}_{\text{hyperbolic-secant flow }(-)} & \cdots & \underbrace{(\text{UP:DP, nF})}_{\text{hyperbolic-secant flow }(-)} & \underbrace{(\text{DP:UP, nF})}_{\text{hyperbolic flow }(-)} \\
\underbrace{(\text{SI, SI})}_{\text{sink}} & \cdots & \underbrace{(\text{SI, SI})}_{\text{sink}} & \underbrace{(\text{SO, SI})}_{\text{saddle}}
\end{array} \right\}_{(2n_{11}+2n_{21}+1)\times(2m_{11})}
$$

for $a_{110} < 0, a_{220} > 0$; (3.138)

$$
\left\{ \begin{array}{cccc}
(a_{111}, a_{12(2n_{11}+1)}) & \cdots & (a_{11(2m_{11}-1)}, a_{12(2n_{11}+1)}) & (a_{11(2m_{11})}, a_{12(2n_{11}+1)}) \\
(a_{111}, a_{12(2n_{11})}) & \cdots & (a_{11(2m_{11}-1)}, a_{12(2n_{11})}) & (a_{11(2m_{11})}, a_{12(2n_{11})}) \\
(a_{111}, a_{22(2n_{21})}) & \cdots & (a_{11(2m_{11}-1)}, a_{22(2n_{21})}) & (a_{11(2m_{11})}, a_{22(2n_{21})}) \\
\vdots & \vdots & \vdots & \vdots \\
(a_{111}, a_{121}) & \cdots & (a_{11(2m_{11}-1)}, a_{121}) & (a_{11(2m_{11})}, a_{121}) \\
(a_{111}, a_{221}) & \cdots & (a_{11(2m_{11}-1)}, a_{221}) & (a_{11(2m_{11})}, a_{221})
\end{array} \right\}
$$

$$
= \left\{ \begin{array}{cccc}
\underbrace{(\text{DP:UP, nF})}_{\text{hyperbolic flow }(-)} & \cdots & \underbrace{(\text{DP:UP, nF})}_{\text{hyperbolic flow }(-)} & \underbrace{(\text{UP:DP,nF})}_{\text{hyperbolic-secant flow }(-)} \\
\underbrace{(\text{UP:DP, nF})}_{\text{hyperbolic-secant flow }(-)} & \cdots & \underbrace{(\text{UP:DP, nF})}_{\text{hyperbolic-secant flow }(-)} & \underbrace{(\text{DP:UP, nF})}_{\text{hyperbolic flow }(-)} \\
\underbrace{(\text{SI, SI})}_{\text{sink}} & \cdots & \underbrace{(\text{SI, SI})}_{\text{sink}} & \underbrace{(\text{SO, SI})}_{\text{saddle}} \\
\vdots & \vdots & \vdots & \vdots \\
\underbrace{(\text{UP:DP, pF})}_{\text{hyperbolic-secant flow (+)}} & \cdots & \underbrace{(\text{UP:DP, pF})}_{\text{hyperbolic-secant flow (+)}} & \underbrace{(\text{DP:UP, pF})}_{\text{hyperbolic flow (+)}} \\
\underbrace{(\text{SO, SO})}_{\text{source}} & \cdots & \underbrace{(\text{SO, SO})}_{\text{source}} & \underbrace{(\text{SI, SO})}_{\text{saddle}}
\end{array} \right\}_{(2n_{11}+2n_{21}+1)\times(2m_{11})}
$$

for $a_{110} > 0, a_{220} < 0$; (3.139)

$$
\left\{\begin{array}{cccc}
(a_{111}, a_{12(2n_{11}+1)}) & \cdots & (a_{11(2m_{11}-1)}, a_{12(2n_{11}+1)}) & (a_{11(2m_{11})}, a_{12(2n_{11}+1)}) \\
(a_{111}, a_{12(2n_{11})}) & \cdots & (a_{11(2m_{11}-1)}, a_{12(2n_{11})}) & (a_{11(2m_{11})}, a_{12(2n_{11})}) \\
(a_{111}, a_{22(2n_{21})}) & \cdots & (a_{11(2m_{11}-1)}, a_{22(2n_{21})}) & (a_{11(2m_{11})}, a_{22(2n_{21})}) \\
\vdots & \vdots & \vdots & \vdots \\
(a_{111}, a_{121}) & \cdots & (a_{11(2m_{11}-1)}, a_{121}) & (a_{11(2m_{11})}, a_{121}) \\
(a_{111}, a_{221}) & \cdots & (a_{11(2m_{11}-1)}, a_{221}) & (a_{11(2m_{11})}, a_{221})
\end{array}\right\}
$$
$$
= \left\{\begin{array}{cccc}
\underbrace{(\text{UP:DP, nF})}_{\text{hyperbolic-secant flow }(-)} & \cdots & \underbrace{(\text{UP:DP, nF})}_{\text{hyperbolic-secant flow }(-)} & \underbrace{(\text{DP:UP, nF})}_{\text{hyperbolic flow }(-)} \\
\underbrace{(\text{DP:UP, nF})}_{\text{hyperbolic flow }(-)} & \cdots & \underbrace{(\text{DP:UP, nF})}_{\text{hyperbolic flow }(-)} & \underbrace{(\text{UP:DP, nF})}_{\text{hyperbolic-secant flow }(-)} \\
\underbrace{(\text{SO, SI})}_{\text{saddle}} & \cdots & \underbrace{(\text{SO, SI})}_{\text{saddle}} & \underbrace{(\text{SI, SI})}_{\text{sink}} \\
\vdots & \vdots & \vdots & \vdots \\
\underbrace{(\text{DP:UP, pF})}_{\text{hyperbolic flow }(+)} & \cdots & \underbrace{(\text{DP:UP, pF})}_{\text{hyperbolic flow }(+)} & \underbrace{(\text{UP:DP, pF})}_{\text{hyperbolic-secant flow }(+)} \\
\underbrace{(\text{SI, SO})}_{\text{saddle}} & \cdots & \underbrace{(\text{SI, SO})}_{\text{saddle}} & \underbrace{(\text{SO, SO})}_{\text{source}}
\end{array}\right\}_{(2n_{11}+2n_{21}+1)\times(2m_{11})}
$$

for $a_{110} < 0$, $a_{220} < 0$. (3.140)

3.8 $[(2m_{11}), (2n_{11})]$ and $((2m_{11}), (2n_{21}))$-Singular Systems

In this section, there are four global bifurcations of the 1-diemsnional flows through singular inflection-saddle flows, and four global bifurcations of equilibriums, through double-saddle bifurcations. The corresponding global bifurcations generate networks of hyperbolic and hyperbolic-secant flows with sources, sinks and saddles. The infinite-equilibriums are the appearing and switching bifurcations for the singular flows and singular equilibriums and for hybrid networks of hyperbolic flows and source, sink and saddles.

3.8.1 $[(2m_{11}), (2n_{11})]$-*Flows* *and* $((2m_{11}), (2n_{21}))$-*Equilibriums*

Consider a polynomial dynamical system for $a_{221} \neq a_{121}$ as

$$
\begin{aligned}
\dot{x}_2 &= a_{220}(x_2 - a_{221})^{2n_{21}}, \\
\dot{x}_1 &= a_{110}(x_1 - a_{111})^{2m_{11}}(x_2 - a_{121})^{2n_{11}}.
\end{aligned} \tag{3.141}
$$

The first integral manifold is

$$-\frac{1}{2m_{11}-1}\left[\frac{1}{(x_1-a_{111})^{2m_{11}-1}}-\frac{1}{(x_{10}-a_{111})^{2m_{11}-1}}\right]$$
$$=\frac{a_{110}}{a_{220}}\sum_{\alpha,\beta=0}^{\alpha+\beta=2n_{11}}\frac{(2n_{11})!}{\alpha!\beta!}\frac{(a_{221}-a_{121})^{\alpha}}{\beta-2n_{21}+1}[(x_2-a_{221})^{\beta-2n_{21}+1}-(x_2-a_{221})^{\beta-2n_{21}+1}]$$
for $\beta-2n_{21}+1\neq 0$;
$$\frac{1}{\beta-2n_{21}+1}[(x_2-a_{221})^{\beta-2n_{21}+1}-(x_2-a_{221})^{\beta-2n_{21}+1}]=ln\frac{|x_2-a_{221}|}{|x_{20}-a_{221}|}$$
for $\beta-2n_{21}+1=0$. (3.142)

For $(x_1^*,x_2^*)=(a_{111},a_{121})$ with $\dot{x}_1=0$, the $[(2m_{11}),(2n_{11})]$-singular flows exist with $\dot{x}_2=a_{220}(a_{121}-a_{221})^{2n_{21}}$. For $(x_1^*,x_2^*)=(a_{111},a_{121})$, the 1-dimensional flow direction is determined by $\dot{x}_2\neq 0s$. For $(x_1^*,x_2^*)=(a_{111},a_{221})$ with $\dot{x}_1=\dot{x}_2=0$, the $((2m_{11}),(2n_{21}))$- singular equilibriums exist.

(A) For $a_{121}<a_{221}$

$$\left\{\begin{matrix}(a_{111},a_{121})\\(a_{111},a_{221})\end{matrix}\right\}=\left\{\begin{matrix}\underbrace{([(2m_{11}),(2n_{11})]_{\text{II:II}}\text{US, pF})}_{[(2m_{11}),(2n_{11})]\text{-increasing inflection upper-saddle flow }(+)}\\\underbrace{((2m_{11})\text{US},(2n_{21})\text{US})}_{[(2m_{11}),(2n_{21})]\text{-double-saddle}}\end{matrix}\right\},$$
for $a_{110}>0,a_{220}>0$;

$$\left\{\begin{matrix}(a_{111},a_{121})\\(a_{111},a_{221})\end{matrix}\right\}=\left\{\begin{matrix}\underbrace{([(2m_{11}),(2n_{11})]_{\text{DI:DI}}\text{US, pF})}_{[(2m_{11}),(2n_{11})]\text{-decreasing inflection lower-saddle flow }(+)}\\\underbrace{((2m_{11})\text{LS},(2n_{21})\text{US})}_{[(2m_{11}),(2n_{21})]\text{-double-saddle}}\end{matrix}\right\},$$
for $a_{110}<0,a_{220}>0$;

$$\left\{\begin{matrix}(a_{111},a_{121})\\(a_{111},a_{221})\end{matrix}\right\}=\left\{\begin{matrix}\underbrace{([(2m_{11}),(2n_{11})]_{\text{DI:DI}}\text{US, nF})}_{[(2m_{11}),(2n_{11})]\text{-decreasing inflection upper-saddle flow }(-)}\\\underbrace{((2m_{11})\text{US},(2n_{21})\text{US})}_{[(2m_{11}),(2n_{21})]\text{-double-saddle}}\end{matrix}\right\},$$
for $a_{110}>0,a_{220}<0$;

$$\left\{\begin{matrix}(a_{111},a_{121})\\(a_{111},a_{221})\end{matrix}\right\}=\left\{\begin{matrix}\underbrace{([(2m_{11}),(2n_{11})]_{\text{II:II}}\text{LS, nF})}_{[(2m_{11}),(2n_{11})]\text{-increasing inflection lower-saddleflow }(-)}\\\underbrace{((2m_{11})\text{LS},(2n_{21})\text{LS})}_{[(2m_{11}),(2n_{21})]\text{-double-saddle}}\end{matrix}\right\},$$
for $a_{110}<0,a_{220}<0$. (3.143)

(B) For $a_{121} < a_{221}$,

$$\left\{\begin{array}{l}(a_{111},a_{221})\\(a_{111},a_{121})\end{array}\right\}=\left\{\begin{array}{c}\underbrace{((2m_{11})\text{US},(2n_{21})\text{US})}_{[(2m_{11}),(2n_{21})]\text{-double-saddle}}\\ \underbrace{([(2m_{11}),(2n_{11})]_{\text{II:II}}\text{US},\text{pF})}_{[(2m_{11}),(2n_{11})]\text{-increasing inflection upper-saddle flow }(+)}\end{array}\right\},$$

for $a_{110} > 0, a_{220} > 0$;

$$\left\{\begin{array}{l}(a_{111},a_{221})\\(a_{111},a_{121})\end{array}\right\}=\left\{\begin{array}{c}\underbrace{((2m_{11})\text{LS},(2n_{21})\text{US})}_{[(2m_{11}),(2n_{21})]\text{-double-saddle}}\\ \underbrace{([(2m_{11}),(2n_{11})]_{\text{DI:DI}}\text{LS},\text{pF})}_{[(2m_{11}),(2n_{11})]\text{-decreasing inflection lower-saddle flow }(+)}\end{array}\right\},$$

for $a_{110} < 0, a_{220} > 0$;

$$\left\{\begin{array}{l}(a_{111},a_{221})\\(a_{111},a_{121})\end{array}\right\}=\left\{\begin{array}{c}\underbrace{((2m_{11})\text{US},(2n_{21})\text{LS})}_{[(2m_{11}),(2n_{21})]\text{-double-saddle}}\\ \underbrace{([(2m_{11}),(2n_{11})]_{\text{DI:DI}}\text{US},\text{nF})}_{[(2m_{11}),(2n_{11})]\text{-decreasing-inflection upper-saddle flow }(-)}\end{array}\right\},$$

for $a_{110} > 0, a_{220} < 0$;

$$\left\{\begin{array}{l}(a_{111},a_{221})\\(a_{111},a_{121})\end{array}\right\}=\left\{\begin{array}{c}\underbrace{((2m_{11})\text{LS},(2n_{21})\text{LS})}_{[(2m_{11}),(2n_{21})]\text{-double-saddle}}\\ \underbrace{([(2m_{11}),(2n_{11})]_{\text{II:II}}\text{LS},\text{nF})}_{[(2m_{11}),(2n_{11})]\text{-increasing-inflection lower-saddle flow }(-)}\end{array}\right\},$$

for $a_{110} < 0, a_{220} < 0$. (3.144)

The $([(2m_{11}),(2n_{11})]_{\text{II:II}}\text{US,pF})$ is for a positive singular increasing-inflection upper-saddle flow. The $([(2m_{11}),(2n_{11})]_{\text{DI:DI}}\text{LS,pF})$ is for a positive decreasing-inflection lower-saddle flow. The $([(2m_{11}),(2n_{11})]_{\text{DI:DI}}\text{US,nF})$ is for a negative decreasing-inflection upper-saddle flow. The $([(2m_{11}),(2n_{11})]_{\text{II:II}}\text{LS,nF})$ is for a negative singular increasing-inflection lower-saddle flow. The $((2m_{11})\text{US},(2n_{21})\text{ US})$ is for a $((2m_{11}),(2n_{21}))$-double-saddle. The $((2m_{11})\text{LS},(2n_{21})\text{US})$ is for a $((2m_{11}),(2n_{21}))$- double-saddle. The $((2m_{11})\text{US},(2n_{21})\text{LS})$ is for a $((2m_{11}),(2n_{21}))$-double-saddle. The $((2m_{11})\text{LS},(2n_{21})\text{LS})$ is for a $((2m_{11}),(2n_{21}))$-double-saddle.

The singular inflection flows with singular saddle-sources and saddle-sinks are illustrated in Figs. 3.43 and 3.44. Through such singular flows, the corresponding global bifurcations generate from an inflection flow to a network of singular hyperbolic and hyperbolic-secant flows. Through such singular equilibriums, the corresponding global bifurcations generate from a flow to a network of singular source, sink and saddle equilibriums.

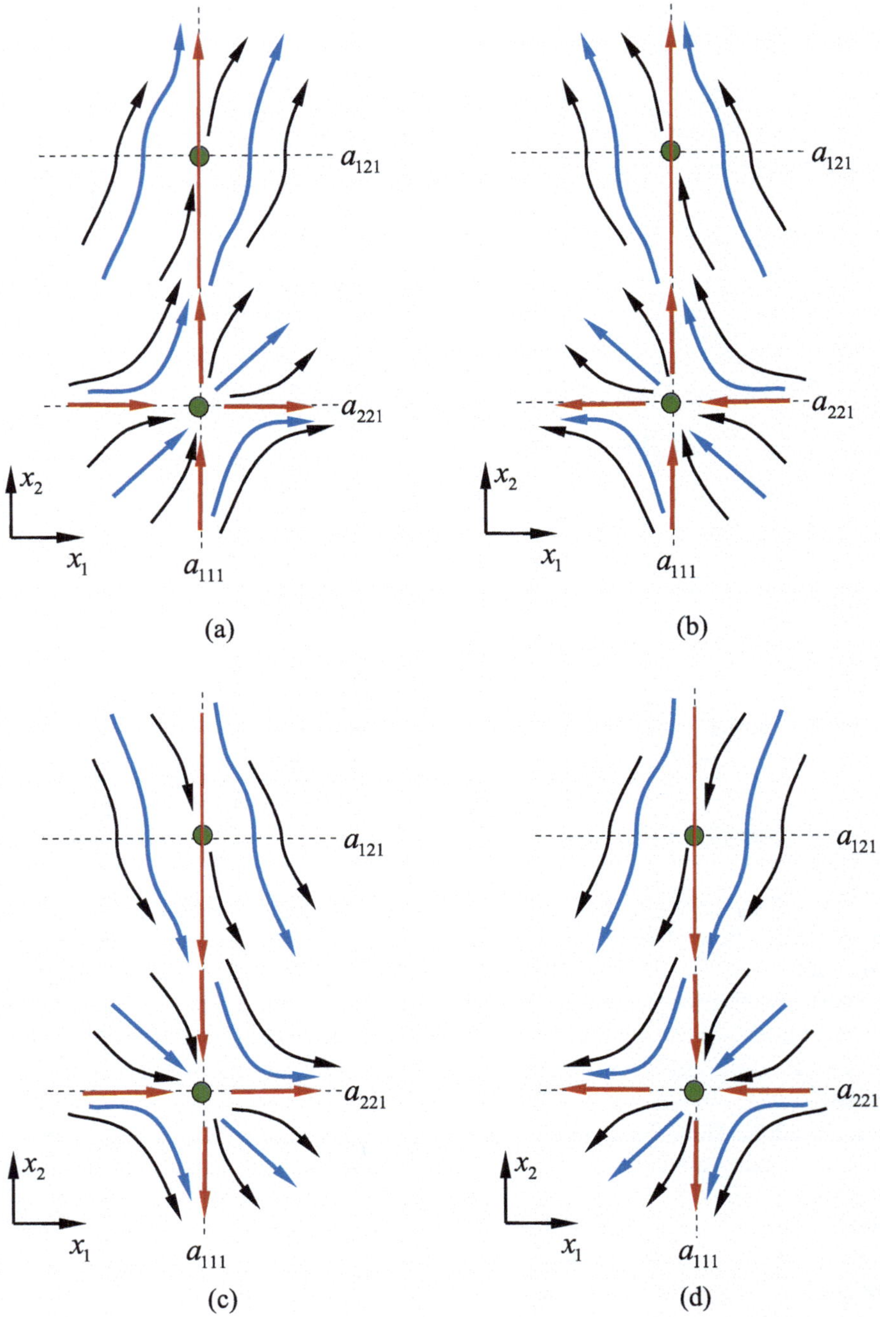

Fig. 3.43 Phase portraits ($a_{121} > a_{221}$) of the $([(2m_{11}), (2n_{11})]$, xF)-inflection-upper-saddle and inflection-lower-saddle flows at $(x_1^*, x_2^*) = (a_{111}, a_{121})$ and $((2m_{11}), (2n_{21}))$-double-saddles at $(x_1^*, x_2^*) = (a_{111}, a_{221})$ for 2-dimensional polynomial systems: **a** ($a_{110} > 0$, $a_{220} > 0$), **b** ($a_{110} < 0$, $a_{220} > 0$), **c** ($a_{110} > 0$, $a_{220} < 0$), **d** ($a_{110} < 0$, $a_{220} < 0$)

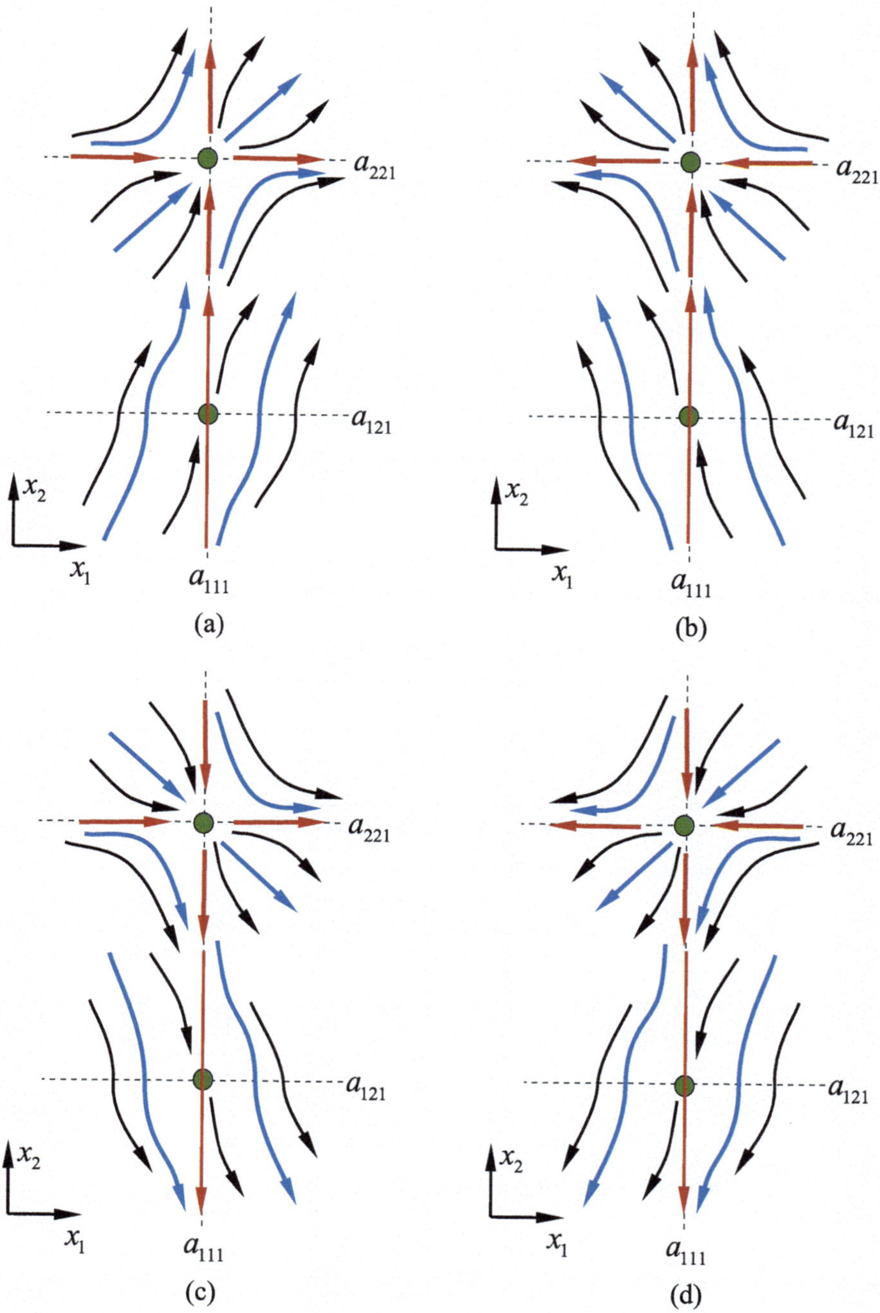

Fig. 3.44 Phase portraits ($a_{221} > a_{121}$) of the ([$(2m_{11})$, $(2n_{11})$], xF)-inflection-upper-saddle and inflection-lower-saddle flows at $(x_1^*, x_2^*) = (a_{111}, a_{121})$ and (($2m_{11}$), ($2n_{21}$))-saddle-sources and saddle-sinks at $(x_1^*, x_2^*) = (a_{111}, a_{221})$ for 2-dimensional polynomial systems: **a** ($a_{110} > 0$, $a_{220} > 0$), **b** ($a_{110} < 0$, $a_{220} > 0$), **c** ($a_{110} > 0$, $a_{220} < 0$), **d** ($a_{110} < 0$, $a_{220} < 0$)

3.8.2 $[(2m_{11}), (2n_{11})]$ and $((2m_{11}), (2n_{21}))$-Infinite-Equilibriums

Consider a polynomial dynamical system for $a_{221} = a_{121}$ as

$$\dot{x}_2 = a_{220}(x_2 - a_{221})^{2n_{21}},$$
$$\dot{x}_1 = a_{110}(x_1 - a_{111})^{2m_{11}}(x_2 - a_{121})^{2n_{11}}. \tag{3.145}$$

The first integral manifold is

$$-\frac{1}{2m_{11}-1}\left[\frac{1}{(x_1 - a_{111})^{2m_{11}-1}} - \frac{1}{(x_{10} - a_{111})^{2m_{11}-1}}\right]$$
$$= \frac{a_{110}}{a_{220}}\frac{1}{2(n_{11}-n_{21})+1}[(x_2 - a_{121})^{2(n_{11}-n_{21})+1} - (x_2 - a_{121})^{2(n_{11}-n_{21})+1}]. \tag{3.146}$$

(A) For $2n_{11} \geq 2n_{21}$, the corresponding infinite-equilibriums are given.

$$(a_{111}, a_{221}) = \underbrace{([2(n_{11}-n_{21})]^{\text{th}}\text{II:II}, [(2m_{11}), (2n_{21})]^{\text{th}}\text{US})}_{([2(n_{11}-n_{21})],[(2m_{11}),(2n_{21})])\text{-increasing inflection-upper-saddle}},$$
for $a_{110} > 0, a_{220} > 0$;
$$(a_{111}, a_{221}) = \underbrace{([2(n_{11}-n_{21})]^{\text{th}}\text{DI:DI}, [(2m_{11}), (2n_{21})]^{\text{th}}\text{US})}_{([2(n_{11}-n_{21})],[(2m_{11}),(2n_{21})])\text{-decreasing inflection-upper-saddle}},$$
for $a_{110} < 0, a_{220} > 0$;
$$(a_{111}, a_{221}) = \underbrace{([2(n_{11}-n_{21})]^{\text{th}}\text{DI:DI}, [(2m_{11}), (2n_{21})]^{\text{th}}\text{LS})}_{([2(n_{11}-n_{21})],[(2m_{11}),(2n_{21})])\text{-decreasing inflection-lower-saddle}},$$
for $a_{110} > 0, a_{220} < 0$;
$$(a_{111}, a_{221}) = \underbrace{([2(n_{11}-n_{21})]^{\text{th}}\text{II:II}, [(2m_{11}), (2n_{21})]^{\text{th}}\text{LS})}_{([2(n_{11}-n_{21})],[(2m_{11}),(2n_{21})])\text{-increasing inflection-lower-saddle}},$$
for $a_{110} < 0, a_{220} < 0$. (3.147)

For $2n_{11} < 2n_{21}$, the infinite-equilibriums are given as follows.

$$(a_{111}, a_{221}) = \underbrace{([2(n_{21}-n_{11})]^{\text{th}}\text{IC:IC}, [(2m_{11}), (2n_{21})]^{\text{th}}\text{US})}_{([2(n_{21}-n_{11})],[(2m_{11}),(2n_{21})])\text{-increasing-concave upper-saddle}},$$
for $a_{110} > 0, a_{220} > 0$;
$$(a_{111}, a_{221}) = \underbrace{([2(n_{21}-n_{11})]^{\text{th}}\text{DC:DC}, [(2m_{11}), (2n_{21})]^{\text{th}}\text{US})}_{([2(n_{21}-n_{11})],[(2m_{11}),(2n_{21})])\text{-decreasing-concave upper-saddle}},$$

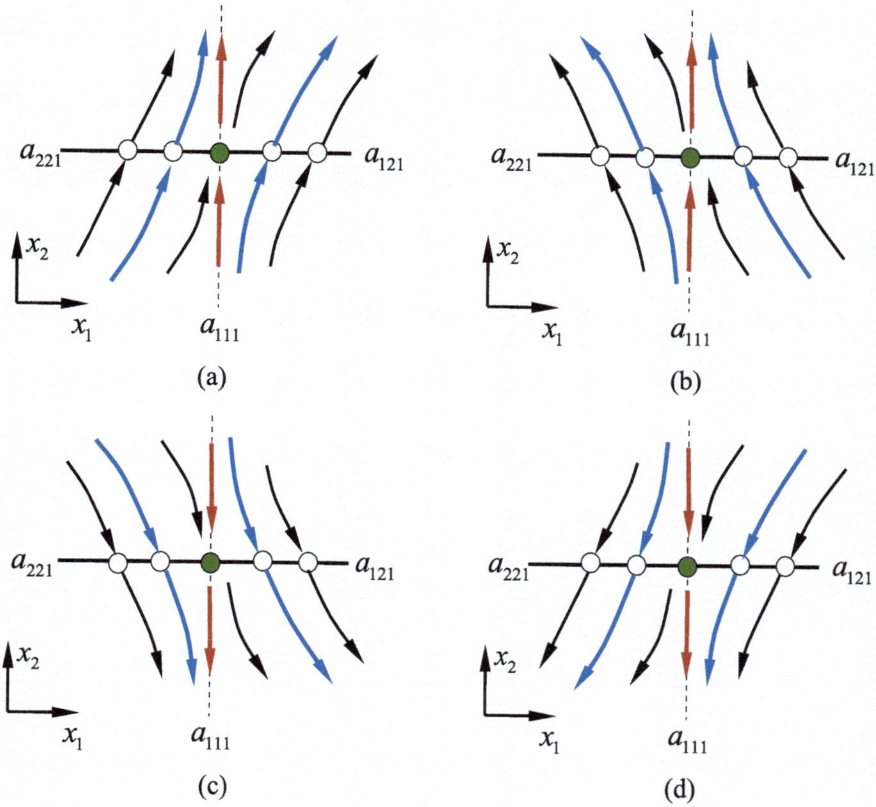

Fig. 3.45 The infinite-equilibriums ($a_{121} = a_{221}, 2n_{11} \geq 2n_{21}, m_1 = 2m_{11}$) for 2-dimensional polynomial systems: **a** ($a_{110} > 0, a_{220} > 0$), **b** ($a_{110} < 0, a_{220} > 0$), **c** ($a_{110} > 0, a_{220} < 0$), **d** ($a_{110} < 0, a_{220} < 0$)

$$\begin{aligned}
&\text{for } a_{110} < 0, a_{220} > 0;\\
&(a_{111}, a_{221}) = \underbrace{([2(n_{21} - n_{11})]^{\text{th}}\text{DC:DC}, [(2m_{11}), (2n_{21})]^{\text{th}}\text{LS})}_{([2(n_{21}-n_{11})],[(2m_{11}),(2n_{21})])\text{-decreasing-concave lower-saddle}},\\
&\text{for } a_{110} > 0, a_{220} < 0;\\
&(a_{111}, a_{221}) = \underbrace{([2(n_{21} - n_{11})]^{\text{th}}\text{IC:IC}, [(2m_{11}), (2n_{21})]^{\text{th}}\text{LS})}_{([2(n_{21}-n_{11})],[(2m_{11}),(2n_{21})])\text{-increasing-concave lower-saddle}},\\
&\text{for } a_{110} < 0, a_{220} < 0.
\end{aligned} \tag{3.148}$$

For $x_2^* = a_{121} = a_{221}$, the infinite-equilibrium exists. The infinite-equilibriums are illustrated in Figs. 3.45 and 3.46 for $n_{11} \geq n_{21}$ and $n_{11} < n_{21}$, respectively. For $n_{11} \geq n_{21}$, the singular decreasing-to-increasing and increasing-to-decreasing inflection-source and sink are presented. For $n_{11} < n_{21}$, the singular decreasing-to-increasing and increasing-to-decreasing concave-source and sink are presented.

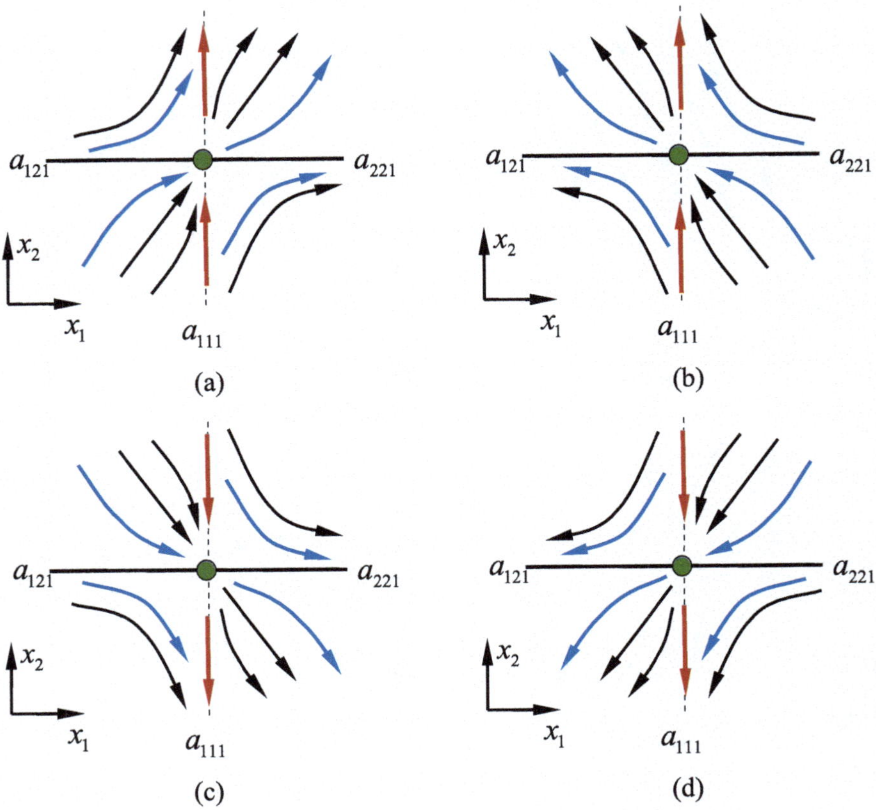

Fig. 3.46 The infinite-equilibriums ($a_{121} = a_{221}, 2n_{11} < 2n_{21}, m_1 = 2m_{11}$) for 2-dimensional polynomial systems: **a** ($a_{110} > 0, a_{220} > 0$), **b** ($a_{110} < 0, a_{220} > 0$), **c** ($a_{110} > 0, a_{220} < 0$), **d** ($a_{110} < 0, a_{220} < 0$)

3.8.3 [(2m₁₁), (2n₁₁)] and ((2m₁₁), (2n₂₁))-Hybrid Networks

Consider

$$\begin{aligned}\dot{x}_2 &= a_{220}\prod_{s=1}^{2n_{21}}(x_2 - a_{22s}),\\ \dot{x}_1 &= a_{110}\prod_{r=1}^{2m_{11}}(x_1 - a_{11r})\prod_{u=1}^{2n_{11}}(x_2 - a_{12u}).\end{aligned} \tag{3.149}$$

The first integral manifold is

$$\sum_{r_1=1}^{2m_{11}} A_{r_1} \ln \frac{|x_1 - a_{11r_1}|}{|x_{10} - a_{11r_1}|} = \frac{a_{110}}{a_{220}} \Bigg\{ \sum_{\beta_0=2n_{21}}^{2n_{11}} b_{12\beta_0} \sum_{\substack{\alpha_1 \in [0,\beta_0-2n_{21}],\\ \beta_1 \in [2n_{21},\beta_0-1]}}^{\alpha_1+\beta_1=\beta_0-1} \sum_{\substack{\alpha_2 \in [0,\beta_1-2n_{21}-1)]\\ \beta_2 \in [2n_{21}-1,\beta_1-2]}}^{\alpha_2+\beta_2=\beta_1-2} \cdots$$

$$\sum_{\substack{\alpha_{2n_{21}-1} \in [0,\beta_{2n_{21}-2}-2n_{21}]\\ \beta_{2n_{21}-1} \in [1,\beta_{2n_{21}-2}-2n_{21}+1)]}}^{\alpha_{2n_{21}-1}+\beta_{2n_{21}-1}=\beta_{2n_{21}-2}-2n_{21}+1} \prod_{j=1}^{2n_{21}-1} \frac{(\beta_{j-1}-j)!}{\alpha_j!\beta_j!} (a_{22s_{j+1}} - a_{22s_j})^{\alpha_j}$$

$$\times \frac{1}{\beta_{2n_{21}-1}} [(x_2 - a_{22s_{2n_{21}}})^{\beta_{2n_{21}-1}} - (x_{20} - a_{22s_{2n_{21}}})^{\beta_{2n_{21}-1}}]$$

$$- \sum_{\beta_0=2n_{21}}^{2n_{11}} b_{12\beta_0} \sum_{k=2}^{2n_{21}-1} \Bigg[\sum_{\substack{\alpha_1 \in [0,\beta_0-2n_{21}),\\ \beta_1 \in (2n_{21}-1,\beta_0-1]}}^{\alpha_1+\beta_1=\beta_0-1} \sum_{\substack{\alpha_2 \in [0,\beta_1-2n_{21}]\\ \beta_2 \in (2n_{21}-2,\beta_1-2)}}^{\alpha_2+\beta_2=\beta_1-2} \cdots$$

$$\times \Bigg(\sum_{\substack{\alpha_{k-1} \in (\beta_{k-2}-2n_{21},\beta_{k-2}-k+1]\\ \beta_{k-1} \in [0,2n_{21}-k)}}^{\alpha_{k-1}+\beta_{k-1}=\beta_{k-2}-k+1} \prod_{j=1}^{k-1} \frac{(\beta_{j-1}-j)!}{\alpha_j!\beta_j!} (a_{22s_{j+1}} - a_{22s_j})^{\alpha_j} \sum_{\rho=k}^{2n_{21}} A_{\rho 1}^{(s_k)} \ln \frac{|x_{22} - a_{22s_\rho}|}{|x_{20} - a_{22r_\rho}|}$$

$$+ \sum_{\substack{\alpha_{k-1} \in (\beta_{k-2}-2n_{21},\beta_{k-2}-(2n_{21}+1]\\ \beta_{k-1} \in [2n_{21}-k,2n_{21}-k+1)}}^{\alpha_{k-1}+\beta_{k-1}=\beta_{k-2}-k+1} \prod_{j=1}^{k-1} \frac{(\beta_{j-1}-j)!}{\alpha_j!\beta_j!} (a_{22s_{j+1}} - a_{22s_j})^{\alpha_j}$$

$$\times \sum_{\rho=k+1}^{2n_{21}} B_{\rho 1}^{(s_k)} \ln \frac{|x_2 - a_{22s_\rho}|}{|x_{20} - a_{22s_\rho}|} \Bigg) \Bigg] + b_{j_1 j_2 0} \sum_{\rho=1}^{2n_{21}} A_{\rho 1}^{(s_1)} \ln \frac{|x_2 - a_{22s_\rho}|}{|x_{20} - a_{22s_\rho}|}$$

$$+ \sum_{\beta_0=1}^{2n_{21}-1} b_{j_1 j_2 \beta_0} \sum_{\rho=2}^{2n_{21}} B_{\rho 1}^{(s_1)} \frac{|x_2 - a_{22s_\rho}|}{|x_{20} - a_{22s_\rho}|} \Bigg\}; \tag{3.150}$$

where

$$A_{r_1} = \frac{1}{\prod_{r_2=1, r_2 \neq r_1}^{2m_{11}} (a_{11r_1} - a_{11r_\sigma})};$$

$$b_{j_1 j_2 \beta_0} = \sum_{t_1,l_1=0}^{t_1+l_1=1} \cdots \sum_{t_q,l_q=0}^{t_q+l_q=1} \prod_{k=1}^{2n_{11}} (a_{22s_1} - a_{12u_k})^{t_k} \delta_{\beta_0}^{\sum_{r=1}^{2n_{21}} l_r};$$

$$\lim_{\beta_{2n_{21}-1} \to 0} \frac{1}{\beta_{2n_{21}-1}} [(x_2 - a_{22s_{2n_{21}}})^{\beta_{2n_{21}-1}} - (x_{20} - a_{22s_{2n_{21}}})^{\beta_{2n_{21}-1}}] = \ln \frac{|x_2 - a_{22s_{2n_{21}}}|}{|x_{20} - a_{22s_{2n_{21}}}|}; \tag{3.151}$$

$$A_{\rho 1}^{(s_k)} = \frac{1}{\prod_{\sigma=k,\sigma\neq\rho}^{2n_{21}} (a_{22s_\rho} - a_{22s_\sigma})}, \text{ for } \rho \in \{k, k+1, \cdots, 2n_{21}\},$$
$$B_{\rho 1}^{(s_k)} = \frac{(a_{22s_\rho} - a_{22s_k})^{\beta_{k-1}-k}}{\prod_{\sigma=k+1,\sigma\neq\rho}^{2n_{21}} (a_{22s_\rho} - a_{22s_\sigma})}, \text{ for} \rho \in \{k+1, k+2, \cdots, 2n_{21}\};$$
$$k \in \{1, 2, \cdots, 2n_{21} - 1\}. \tag{3.152}$$

For all cases, the $(2m_{11}) \times (2n_{21})$-simple equilibriums plus $(2m_{11}) \times (2n_{11})$-hyperbolic flows are based on the following inflection-upper-saddle and lower-saddle bifurcations for $2n_{11} \geq 2n_{21}$:

- $[2(n_{11} - n_{21})]^{\text{th}}$II:II, $[(2m_{11}), (2n_{21})]^{\text{th}}$US)-increasing-inflection upper-saddle bifurcation for ($a_{110} > 0, a_{220} > 0$),
- $[2(n_{11} - n_{21})]^{\text{th}}$DI:DI, $[(2m_{11}), (2n_{21})]^{\text{th}}$US)- decreasing-inflection upper-saddle bifurcation for ($a_{110} < 0, a_{220} > 0$),
- $[2(n_{11} - n_{21})]^{\text{th}}$II:DI, $[(2m_{11}), (2n_{21})]^{\text{th}}$LS)-decreasing-inflection lower-saddle bifurcation for ($a_{110} > 0, a_{220} < 0$),
- $[2(n_{11} - n_{21})]^{\text{th}}$II:II, $[(2m_{11}), (2n_{21})]^{\text{th}}$LS)-increasing-inflection lower-saddle bifurcation for ($a_{110} < 0, a_{220} < 0$);

or based on the following concave-upper-saddle and lower-saddle bifurcations for $2n_{11} + 1 < 2n_{21} + 1$:

- $[2(n_{21} - n_{11})]^{\text{th}}$IC:IC, $[(2m_{11}), (2n_{21})]^{\text{th}}$US)-increasing-concave upper-saddle bifurcation for ($a_{110} > 0, a_{220} > 0$),
- $[2(n_{11} - n_{21})]^{\text{th}}$DC:DC, $[(2m_{11}), (2n_{21})]^{\text{th}}$US)-decreasing-concave upper-saddle bifurcation for ($a_{110} < 0, a_{220} > 0$),
- $[2(n_{11} - n_{21})]^{\text{th}}$DC:DC, $[(2m_{11}), (2n_{21})]^{\text{th}}$LS) -decreasing-concave lower-saddle bifurcation for ($a_{110} > 0, a_{220} < 0$),
- $[2(n_{11} - n_{21})]^{\text{th}}$IC:IC, $[(2m_{11}), (2n_{21})]^{\text{th}}$LS)- increasing-concave lower-saddle bifurcation for ($a_{110} < 0, a_{220} < 0$).

Consider $2n_{11} = 2n_{21}$ for illustrations, phase portraits for the 2-dimensional systems for the hybrid networks of hyperbolic and hyperbolic-secant flows with the simple equilibriums of the saddles, sinks and sources with $x_1^* = a_{111}, a_{112}, \ldots, a_{11(2m_{11}+1)}$ and $x_2^* = a_{121}, a_{221}, \ldots, a_{12(2n_{11})}, a_{22(2n_{21})}$ are presented in Fig. 3.47 (a–d) for ($a_{110} > 0, a_{220} > 0$), ($a_{110} < 0, a_{220} > 0$), ($a_{110} > 0, a_{220} < 0$) and ($a_{110} < 0, a_{220} < 0$).

$$\left\{\begin{array}{cccc} (a_{111}, a_{22(2n_{21})}) & \cdots & (a_{11(2m_{11}-1)}, a_{22(2n_{21})}) & (a_{11(2m_{11})}, a_{22(2n_{21})}) \\ (a_{111}, a_{12(2n_{11})}) & \cdots & (a_{11(2m_{11}-1)}, a_{12(2n_{11})}) & (a_{11(2m_{11})}, a_{12(2n_{11})}) \\ (a_{111}, a_{22(2n_{21}-1)}) & \cdots & (a_{11(2m_{11}-1)}, a_{22(2n_{21}-1)}) & (a_{11(2m_{11})}, a_{22(2n_{21}-1)}) \\ \vdots & \vdots & \vdots & \vdots \\ (a_{111}, a_{221}) & \cdots & (a_{11(2m_{11}-1)}, a_{221}) & (a_{11(2m_{11})}, a_{221}) \\ (a_{111}, a_{121}) & \cdots & (a_{11(2m_{11}-1)}, a_{121}) & (a_{11(2m_{11})}, a_{121}) \end{array}\right\}$$

$$
= \left\{\begin{matrix}
\underbrace{(\text{SI, SO})}_{\text{saddle}} & \cdots & \underbrace{(\text{SI, SO})}_{\text{saddle}} & \underbrace{(\text{SO, SO})}_{\text{source}} \\
\underbrace{(\text{DP:UP, nF})}_{\text{hyperbolic flow }(-)} & \cdots & \underbrace{(\text{DP:UP, nF})}_{\text{hyperbolic flow }(-)} & \underbrace{(\text{UP:DP, nF})}_{\text{hyperbolic-secant flow }(-)} \\
\underbrace{(\text{SO, SI})}_{\text{saddle}} & \cdots & \underbrace{(\text{SO, SI})}_{\text{saddle}} & \underbrace{(\text{SI, SI})}_{\text{sink}} \\
\vdots & \vdots & \vdots & \vdots \\
\underbrace{(\text{SO, SI})}_{\text{saddle}} & \cdots & \underbrace{(\text{SO, SI})}_{\text{saddle}} & \underbrace{(\text{SI, SI})}_{\text{source}} \\
\underbrace{(\text{DP:UP, pF})}_{\text{hyperbolic flow }(+)} & \cdots & \underbrace{(\text{DP:UP, pF})}_{\text{hyperbolic flow }(+)} & \underbrace{(\text{UP:DP, pF})}_{\text{hyperbolic-secant flow }(+)}
\end{matrix}\right\}_{(2n_{11}+2n_{21})\times(2m_{11})}
$$

for $a_{110} > 0, a_{220} > 0$; (3.153)

$$
\left\{\begin{matrix}
(a_{111}, a_{22(2n_{21})}) & \cdots & (a_{11(2m_{11}-1)}, a_{22(2n_{21})}) & (a_{11(2m_{11})}, a_{22(2n_{21})}) \\
(a_{111}, a_{12(2n_{11})}) & \cdots & (a_{11(2m_{11}-1)}, a_{12(2n_{11})}) & (a_{11(2m_{11})}, a_{12(2n_{11})}) \\
(a_{111}, a_{22(2n_{21}-1)}) & \cdots & (a_{11(2m_{11}-1)}, a_{22(2n_{21}-1)}) & (a_{11(2m_{11})}, a_{22(2n_{21}-1)}) \\
\vdots & \vdots & \vdots & \vdots \\
(a_{111}, a_{221}) & \cdots & (a_{11(2m_{11}-1)}, a_{221}) & (a_{11(2m_{11})}, a_{221}) \\
(a_{111}, a_{121}) & \cdots & (a_{11(2m_{11}-1)}, a_{121}) & (a_{11(2m_{11})}, a_{121})
\end{matrix}\right\}
$$

$$
= \left\{\begin{matrix}
\underbrace{(\text{SO, SO})}_{\text{source}} & \cdots & \underbrace{(\text{SO, SO})}_{\text{source}} & \underbrace{(\text{SI, SO})}_{\text{saddle}} \\
\underbrace{(\text{UP:DP, nF})}_{\text{hyperbolic-secant flow }(-)} & \cdots & \underbrace{(\text{UP:DP, nF})}_{\text{hyperbolic-secant flow }(-)} & \underbrace{(\text{DP:UP, nF})}_{\text{hyperbolic flow }(-)} \\
\underbrace{(\text{SI, SI})}_{\text{sink}} & \cdots & \underbrace{(\text{SI, SI})}_{\text{sink}} & \underbrace{(\text{SO, SI})}_{\text{saddle}} \\
\vdots & \vdots & \vdots & \vdots \\
\underbrace{(\text{SI,SI})}_{\text{sink}} & \cdots & \underbrace{(\text{SI, SI})}_{\text{sink}} & \underbrace{(\text{SO, SI})}_{\text{saddle}} \\
\underbrace{(\text{UP:DP, pF})}_{\text{hyperbolic-secant flow }(+)} & \cdots & \underbrace{(\text{UP:DP, pF})}_{\text{hyperbolic-secant flow }(+)} & \underbrace{(\text{DP:UP, pF})}_{\text{hyperbolic flow }(+)}
\end{matrix}\right\}_{(2n_{11}+2n_{21})\times(2m_{11})}
$$

for $a_{110} < 0, a_{220} > 0$; (3.154)

$$
\left\{\begin{matrix}
(a_{111}, a_{22(2n_{21})}) & \cdots & (a_{11(2m_{11}-1)}, a_{22(2n_{21})}) & (a_{11(2m_{11})}, a_{22(2n_{21})}) \\
(a_{111}, a_{12(2n_{11})}) & \cdots & (a_{11(2m_{11}-1)}, a_{12(2n_{11})}) & (a_{11(2m_{11})}, a_{12(2n_{11})}) \\
(a_{111}, a_{22(2n_{21}-1)}) & \cdots & (a_{11(2m_{11}-1)}, a_{22(2n_{21}-1)}) & (a_{11(2m_{11})}, a_{22(2n_{21}-1)}) \\
\vdots & \vdots & \vdots & \vdots \\
(a_{111}, a_{221}) & \cdots & (a_{11(2m_{11}-1)}, a_{221}) & (a_{11(2m_{11})}, a_{221}) \\
(a_{111}, a_{121}) & \cdots & (a_{11(2m_{11}-1)}, a_{121}) & (a_{11(2m_{11})}, a_{121})
\end{matrix}\right\}
$$

$$
= \left\{ \begin{array}{cccc}
\underbrace{(\mathrm{SI}, \mathrm{SI})}_{\text{sink}} & \cdots & \underbrace{(\mathrm{SI}, \mathrm{SI})}_{\text{sink}} & \underbrace{(\mathrm{SO}, \mathrm{SI})}_{\text{saddle}} \\
\underbrace{(\mathrm{UP{:}DP}, \mathrm{pF})}_{\text{hyperbolic-secant flow }(+)} & \cdots & \underbrace{(\mathrm{UP{:}DP}, \mathrm{pF})}_{\text{hyperbolic-secant flow }(+)} & \underbrace{(\mathrm{DP{:}UP}, \mathrm{pF})}_{\text{hyperbolic flow }(+)} \\
\underbrace{(\mathrm{SO}, \mathrm{SO})}_{\text{source}} & \cdots & \underbrace{(\mathrm{SO}, \mathrm{SO})}_{\text{source}} & \underbrace{(\mathrm{SI}, \mathrm{SO})}_{\text{saddle}} \\
\vdots & \vdots & \vdots & \vdots \\
\underbrace{(\mathrm{SI}, \mathrm{SI})}_{\text{sink}} & \cdots & \underbrace{(\mathrm{SI}, \mathrm{SI})}_{\text{sink}} & \underbrace{(\mathrm{SO}, \mathrm{SI})}_{\text{saddle}} \\
\underbrace{(\mathrm{UP{:}DP}, \mathrm{pF})}_{\text{hyperbolic-secant flow }(+)} & \cdots & \underbrace{(\mathrm{UP{:}DP}, \mathrm{pF})}_{\text{hyperbolic-secant flow }(+)} & \underbrace{(\mathrm{DP{:}UP}, \mathrm{pF})}_{\text{hyperbolic flow }(+)}
\end{array} \right\}_{(2n_{11}+2n_{21})\times(2m_{11})}
$$

for $a_{110} > 0, a_{220} < 0$; (3.155)

$$
\left\{ \begin{array}{cccc}
(a_{111}, a_{22(2n_{21})}) & \cdots & (a_{11(2m_{11}-1)}, a_{22(2n_{21})}) & (a_{11(2m_{11})}, a_{22(2n_{21})}) \\
(a_{111}, a_{12(2n_{11})}) & \cdots & (a_{11(2m_{11}-1)}, a_{12(2n_{11})}) & (a_{11(2m_{11})}, a_{12(2n_{11})}) \\
(a_{111}, a_{22(2n_{21}-1)}) & \cdots & (a_{11(2m_{11}-1)}, a_{22(2n_{21}-1)}) & (a_{11(2m_{11})}, a_{22(2n_{21}-1)}) \\
\vdots & \vdots & \vdots & \vdots \\
(a_{111}, a_{221}) & \cdots & (a_{11(2m_{11}-1)}, a_{221}) & (a_{11(2m_{11})}, a_{221}) \\
(a_{111}, a_{121}) & \cdots & (a_{11(2m_{11}-1)}, a_{121}) & (a_{11(2m_{11})}, a_{121})
\end{array} \right\}
$$

$$
= \left\{ \begin{array}{cccc}
\underbrace{(\mathrm{SO}, \mathrm{SI})}_{\text{saddle}} & \cdots & \underbrace{(\mathrm{SO}, \mathrm{SI})}_{\text{saddle}} & \underbrace{(\mathrm{SI}, \mathrm{SI})}_{\text{sink}} \\
\underbrace{(\mathrm{DP{:}UP}, \mathrm{pF})}_{\text{hyperbolic flow }(+)} & \cdots & \underbrace{(\mathrm{DP{:}UP}, \mathrm{pF})}_{\text{hyperbolic flow }(+)} & \underbrace{(\mathrm{UP{:}DP}, \mathrm{pF})}_{\text{hyperbolic-secant flow }(+)} \\
\underbrace{(\mathrm{SI}, \mathrm{SO})}_{\text{saddle}} & \cdots & \underbrace{(\mathrm{SI}, \mathrm{SO})}_{\text{saddle}} & \underbrace{(\mathrm{SO}, \mathrm{SO})}_{\text{source}} \\
\vdots & \vdots & \vdots & \vdots \\
\underbrace{(\mathrm{SI}, \mathrm{SO})}_{\text{saddle}} & \cdots & \underbrace{(\mathrm{SI}, \mathrm{SO})}_{\text{saddle}} & \underbrace{(\mathrm{SO}, \mathrm{SO})}_{\text{source}} \\
\underbrace{(\mathrm{DP{:}UP}, \mathrm{nF})}_{\text{hyperbolic flow }(-)} & \cdots & \underbrace{(\mathrm{DP{:}UP}, \mathrm{nF})}_{\text{hyperbolic flow }(-)} & \underbrace{(\mathrm{UP{:}DP}, \mathrm{nF})}_{\text{hyperbolic-secant flow }(-)}
\end{array} \right\}_{(2n_{11}+2n_{21})\times(2m_{11})}
$$

for $a_{110} < 0, a_{220} < 0$. (3.156)

Consider $2n_{11} = 2n_{21}$ for illustrations. Phase portraits for the 2-dimensional systems for the hybrid networks of hyperbolic and hyperbolic-secant flows with the simple equilibriums of the saddles, sinks and sources with $x_1^* = a_{111}, a_{112}, \ldots, a_{11(2m_{11})}$ and $x_2^* = a_{221}, a_{121}, \ldots, a_{22(2n_{21})}, a_{12(2n_{11})}$ are presented in Fig. 3.48 (a–d) for ($a_{110} > 0, a_{220} > 0$), ($a_{110} < 0, a_{220} > 0$), ($a_{110} > 0, a_{220} < 0$), and ($a_{110} < 0, a_{220} < 0$).

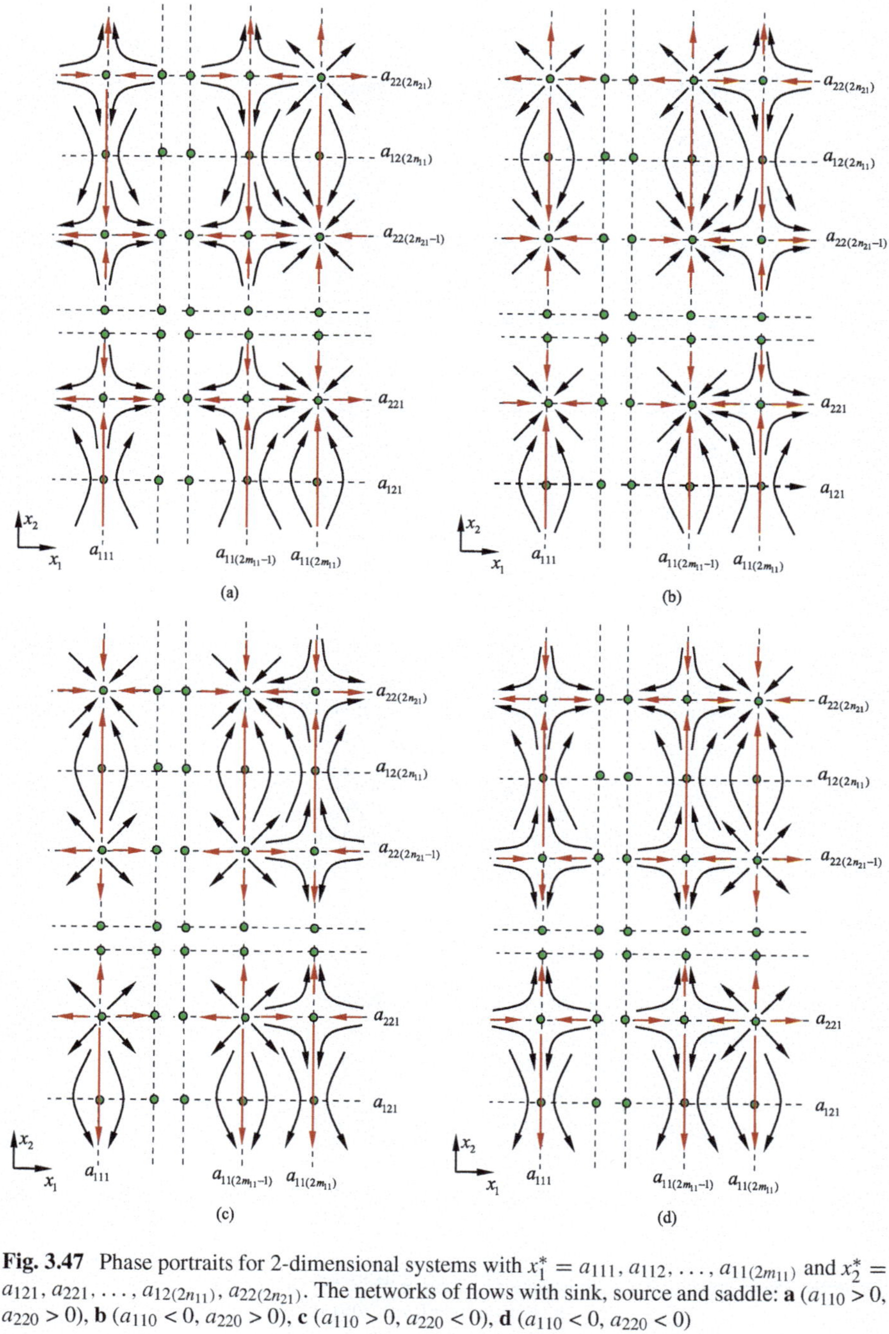

Fig. 3.47 Phase portraits for 2-dimensional systems with $x_1^* = a_{111}, a_{112}, \ldots, a_{11(2m_{11})}$ and $x_2^* = a_{121}, a_{221}, \ldots, a_{12(2n_{11})}, a_{22(2n_{21})}$. The networks of flows with sink, source and saddle: **a** ($a_{110} > 0$, $a_{220} > 0$), **b** ($a_{110} < 0$, $a_{220} > 0$), **c** ($a_{110} > 0$, $a_{220} < 0$), **d** ($a_{110} < 0$, $a_{220} < 0$)

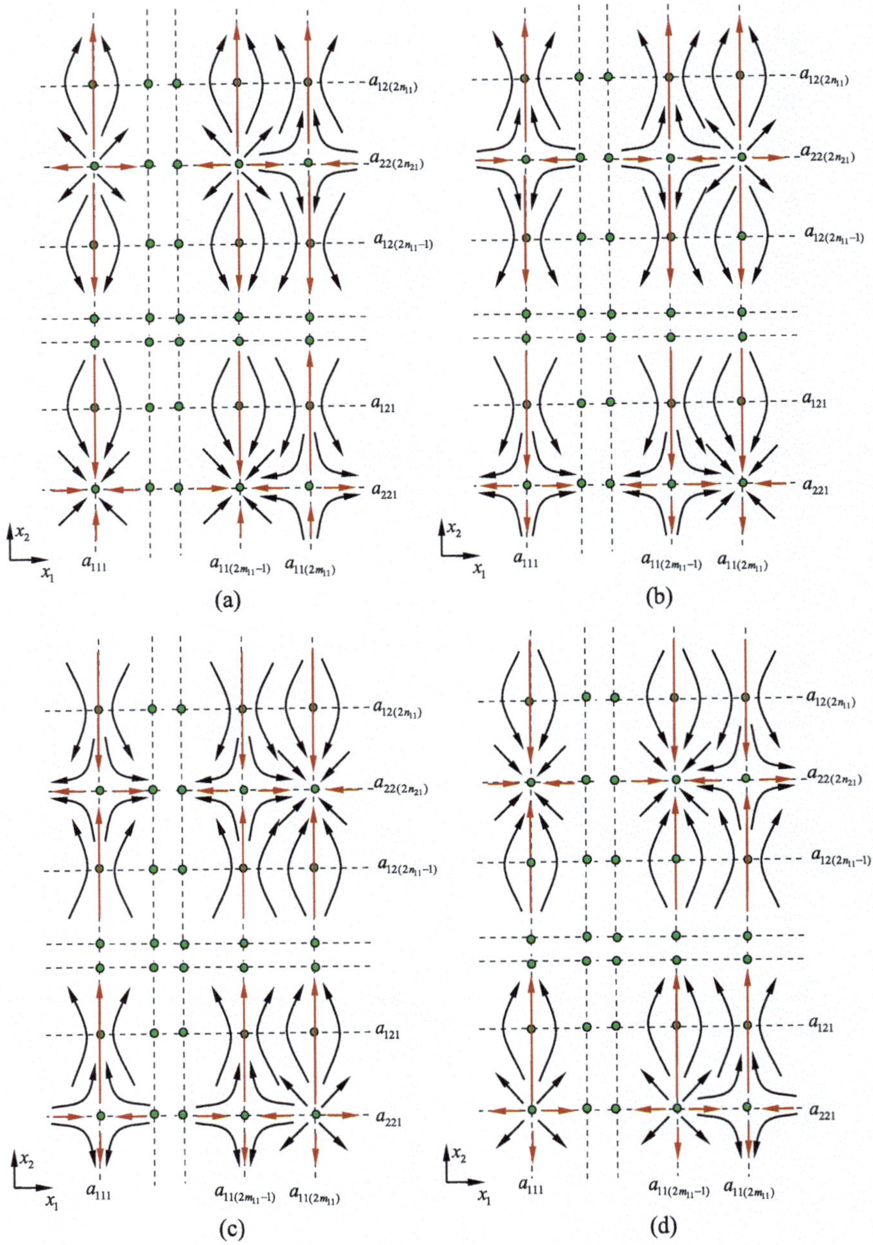

Fig. 3.48 Phase portraits for 2-dimensional systems with $x_1^* = a_{111}, a_{112}, \ldots, a_{11(2m_{11})}$ and $x_2^* = a_{221}, a_{121}, \ldots, a_{22(2n_{21})}, a_{12(2n_{11})}$. The networks of flows with sink, source and saddle: **a** ($a_{110} > 0$, $a_{220} > 0$), **b** ($a_{110} < 0$, $a_{220} > 0$), **c** ($a_{110} > 0$, $a_{220} < 0$), **d** ($a_{110} < 0$, $a_{220} < 0$)

$$\begin{Bmatrix} (a_{111}, a_{12(2n_{11})}) & \cdots & (a_{11(2m_{11}-1)}, a_{12(2n_{11})}) & (a_{11(2m_{11})}, a_{12(2n_{11})}) \\ (a_{111}, a_{22(2n_{21})}) & \cdots & (a_{11(2m_{11}-1)}, a_{22(2n_{21})}) & (a_{11(2m_{11})}, a_{22(2n_{21})}) \\ (a_{111}, a_{12(2n_{11}-1)}) & \cdots & (a_{11(2m_{11}-1)}, a_{12(2n_{11}-1)}) & (a_{11(2m_{11})}, a_{12(2n_{11}-1)}) \\ \vdots & \vdots & \vdots & \vdots \\ (a_{111}, a_{121}) & \cdots & (a_{11(2m_{11}-1)}, a_{121}) & (a_{11(2m_{11})}, a_{121}) \\ (a_{111}, a_{221}) & \cdots & (a_{11(2m_{11}-1)}, a_{221}) & (a_{11(2m_{11})}, a_{221}) \end{Bmatrix}$$

$$= \begin{Bmatrix} \underbrace{(\text{UP:DP, pF})}_{\text{hyperbolic-secant flow }(+)} & \cdots & \underbrace{(\text{UP:DP, pF})}_{\text{hyperbolic-secant flow }(+)} & \underbrace{(\text{DP:UP, pF})}_{\text{hyperbolic flow }(+)} \\ \underbrace{(\text{SO, SO})}_{\text{source}} & \cdots & \underbrace{(\text{SO, SO})}_{\text{source}} & \underbrace{(\text{SI, SO})}_{\text{saddle}} \\ \underbrace{(\text{UP:DP, nF})}_{\text{hyperbolic-secant flow }(-)} & \cdots & \underbrace{(\text{UP:DP, nF})}_{\text{hyperbolic-secant flow }(-)} & \underbrace{(\text{DP:UP, nF})}_{\text{hyperbolic flow }(-)} \\ \vdots & \vdots & \vdots & \vdots \\ \underbrace{(\text{UP:DP, nF})}_{\text{hyperbolic-secant flow }(-)} & \cdots & \underbrace{(\text{UP:DP, nF})}_{\text{hyperbolic-secant flow }(-)} & \underbrace{(\text{DP:UP, nF})}_{\text{hyperbolic flow }(-)} \\ \underbrace{(\text{SI, SI})}_{\text{sink}} & \cdots & \underbrace{(\text{SI, SI})}_{\text{sink}} & \underbrace{(\text{SO, SI})}_{\text{saddle}} \end{Bmatrix}_{(2n_{11}+2n_{21})\times(2m_{11})}$$

for $a_{110} > 0, a_{220} > 0$; (3.157)

$$\begin{Bmatrix} (a_{111}, a_{12(2n_{11})}) & \cdots & (a_{11(2m_{11}-1)}, a_{12(2n_{11})}) & (a_{11(2m_{11})}, a_{12(2n_{11})}) \\ (a_{111}, a_{22(2n_{21})}) & \cdots & (a_{11(2m_{11}-1)}, a_{22(2n_{21})}) & (a_{11(2m_{11})}, a_{22(2n_{21})}) \\ (a_{111}, a_{12(2n_{11}-1)}) & \cdots & (a_{11(2m_{11}-1)}, a_{12(2n_{11}-1)}) & (a_{11(2m_{11})}, a_{12(2n_{11}-1)}) \\ \vdots & \vdots & \vdots & \vdots \\ (a_{111}, a_{121}) & \cdots & (a_{11(2m_{11}-1)}, a_{121}) & (a_{11(2m_{11})}, a_{121}) \\ (a_{111}, a_{221}) & \cdots & (a_{11(2m_{11}-1)}, a_{221}) & (a_{11(2m_{11})}, a_{221}) \end{Bmatrix}$$

$$= \begin{Bmatrix} \underbrace{(\text{DP:UP, pF})}_{\text{hyperbolic flow }(+)} & \cdots & \underbrace{(\text{DP:UP, pF})}_{\text{hyperbolic flow }(+)} & \underbrace{(\text{UP:DP, pF})}_{\text{hyperbolic-secant flow }(+)} \\ \underbrace{(\text{SI, SO})}_{\text{saddle}} & \cdots & \underbrace{(\text{SI, SO})}_{\text{saddle}} & \underbrace{(\text{SO, SO})}_{\text{source}} \\ \underbrace{(\text{DP:UP, nF})}_{\text{hyperbolic flow }(-)} & \cdots & \underbrace{(\text{DP:UP, nF})}_{\text{hyperbolic flow }(-)} & \underbrace{(\text{UP:DP, nF})}_{\text{hyperbolic-secant flow }(-)} \\ \vdots & \vdots & \vdots & \vdots \\ \underbrace{(\text{DP:UP, nF})}_{\text{hyperbolic flow }(-)} & \cdots & \underbrace{(\text{DP:UP, nF})}_{\text{hyperbolic flow }(-)} & \underbrace{(\text{UP:DP,nF})}_{\text{hyperbolic-secant flow }(-)} \\ \underbrace{(\text{SO, SI})}_{\text{saddle}} & \cdots & \underbrace{(\text{SO, SI})}_{\text{saddle}} & \underbrace{(\text{SI, SI})}_{\text{sink}} \end{Bmatrix}_{(2n_{11}+2n_{21})\times(2m_{11})}$$

for $a_{110} < 0, a_{220} > 0$; (3.158)

$$\left\{\begin{array}{cccc}
(a_{111},a_{12(2n_{11})}) & \cdots & (a_{11(2m_{11}-1)},a_{12(2n_{11})}) & (a_{11(2m_{11})},a_{12(2n_{11})})\\
(a_{111},a_{22(2n_{21})}) & \cdots & (a_{11(2m_{11}-1)},a_{22(2n_{21})}) & (a_{11(2m_{11})},a_{22(2n_{21})})\\
(a_{111},a_{12(2n_{11}-1)}) & \cdots & (a_{11(2m_{11}-1)},a_{12(2n_{11}-1)}) & (a_{11(2m_{11})},a_{12(2n_{11}-1)})\\
\vdots & \vdots & \vdots & \vdots\\
(a_{111},a_{121}) & \cdots & (a_{11(2m_{11}-1)},a_{121}) & (a_{11(2m_{11})},a_{121})\\
(a_{111},a_{221}) & \cdots & (a_{11(2m_{11}-1)},a_{221}) & (a_{11(2m_{11})},a_{221})
\end{array}\right\}$$

$$=\left\{\begin{array}{cccc}
\underbrace{(\mathrm{DP{:}UP,\ nF})}_{\text{hyperbolic flow }(-)} & \cdots & \underbrace{(\mathrm{DP{:}UP,\ nF})}_{\text{hyperbolic flow }(-)} & \underbrace{(\mathrm{UP{:}DP,nF})}_{\text{hyperbolic-secant flow }(-)}\\
\underbrace{(\mathrm{SO,\ SI})}_{\text{saddle}} & \cdots & \underbrace{(\mathrm{SO,\ SI})}_{\text{saddle}} & \underbrace{(\mathrm{SI,SI})}_{\text{sink}}\\
\underbrace{(\mathrm{DP{:}UP,\ pF})}_{\text{hyperbolic flow }(+)} & \cdots & \underbrace{(\mathrm{DP{:}UP,\ pF})}_{\text{hyperbolic flow }(+)} & \underbrace{(\mathrm{UP{:}DP,\ pF})}_{\text{hyperbolic-secant flow }(+)}\\
\vdots & \vdots & \vdots & \vdots\\
\underbrace{(\mathrm{DP{:}UP,\ pF})}_{\text{hyperbolic flow }(+)} & \cdots & \underbrace{(\mathrm{DP{:}UP,\ pF})}_{\text{hyperbolic flow }(+)} & \underbrace{(\mathrm{UP{:}DP,\ pF})}_{\text{hyperbolic-secant flow }(+)}\\
\underbrace{(\mathrm{SO,\ SI})}_{\text{saddle}} & \cdots & \underbrace{(\mathrm{SO,\ SI})}_{\text{saddle}} & \underbrace{(\mathrm{SO,\ SO})}_{\text{source}}
\end{array}\right\}_{(2n_{11}+2n_{21})\times(2m_{11})}$$

for $a_{110}>0,\ a_{220}<0;$ (3.159)

$$\left\{\begin{array}{cccc}
(a_{111},a_{12(2n_{11})}) & \cdots & (a_{11(2m_{11}-1)},a_{12(2n_{11})}) & (a_{11(2m_{11})},a_{12(2n_{11})})\\
(a_{111},a_{22(2n_{21})}) & \cdots & (a_{11(2m_{11}-1)},a_{22(2n_{21})}) & (a_{11(2m_{11})},a_{22(2n_{21})})\\
(a_{111},a_{12(2n_{11}-1)}) & \cdots & (a_{11(2m_{11}-1)},a_{12(2n_{11}-1)}) & (a_{11(2m_{11})},a_{12(2n_{11}-1)})\\
\vdots & \vdots & \vdots & \vdots\\
(a_{111},a_{121}) & \cdots & (a_{11(2m_{11}-1)},a_{121}) & (a_{11(2m_{11})},a_{121})\\
(a_{111},a_{221}) & \cdots & (a_{11(2m_{11}-1)},a_{221}) & (a_{11(2m_{11})},a_{221})
\end{array}\right\}$$

$$=\left\{\begin{array}{cccc}
\underbrace{(\mathrm{UP{:}DP,\ nF})}_{\text{hyperbolic-secant flow }(-)} & \cdots & \underbrace{(\mathrm{UP{:}DP,\ nF})}_{\text{hyperbolic-secant flow }(-)} & \underbrace{(\mathrm{DP{:}UP,\ nF})}_{\text{hyperbolic flow }(-)}\\
\underbrace{(\mathrm{SI,\ SI})}_{\text{sink}} & \cdots & \underbrace{(\mathrm{SI,\ SI})}_{\text{sink}} & \underbrace{(\mathrm{SO,\ SI})}_{\text{saddle}}\\
\underbrace{(\mathrm{UP{:}DP,\ pF})}_{\text{hyperbolic-secant flow }(+)} & \cdots & \underbrace{(\mathrm{UP{:}DP,\ pF})}_{\text{hyperbolic-secant flow }(+)} & \underbrace{(\mathrm{DP{:}UP,\ pF})}_{\text{hyperbolic flow }(+)}\\
\vdots & \vdots & \vdots & \vdots\\
\underbrace{(\mathrm{UP{:}DP,\ pF})}_{\text{hyperbolic-secant flow }(+)} & \cdots & \underbrace{(\mathrm{UP{:}DP,\ pF})}_{\text{hyperbolic-secant flow }(+)} & \underbrace{(\mathrm{DP{:}UP,\ pF})}_{\text{hyperbolic flow }(+)}\\
\underbrace{(\mathrm{SO,\ SO})}_{\text{source}} & \cdots & \underbrace{(\mathrm{SO,\ SO})}_{\text{source}} & \underbrace{(\mathrm{SI,\ SO})}_{\text{saddle}}
\end{array}\right\}_{(2n_{11}+2n_{21})\times(2m_{11})}$$

for $a_{110}<0,\ a_{220}<0.$ (3.160)

Index

A. C. J. Luo, *Two-Dimensional Self and Product Polynomial Systems*,
https://doi.org/10.1007/978-981-96-5483-3

D

H

I

L

N

P

S

U

X

Z

The manufacturer's authorised representative in the EU is Springer Nature Customer Service Centre GmbH, Europaplatz 3, 69115 Heidelberg, Germany. If you have any concerns regarding our products, please contact ProductSafety@springernature.com

Printed and bound by CPI Group (UK) Ltd, Croydon, CR0 4YY
29/09/2026
02224601-0008